T0190051

Lecture Notes of the Institute for Computer Sciences, Social Informatics and Telecommunications Engineering

352

More information about this series at http://www.springer.com/series/8197

Honghao Gao · Pingyi Fan ·
Jun Wun · Xue Xiaoping ·
Jun Yu · Yi Wang (Eds.)

Communications and Networking

15th EAI International Conference, ChinaCom 2020
Shanghai, China, November 20–21, 2020
Proceedings

 Springer

Editors
Honghao Gao
Shanghai University
Shanghai, China

Jun Wun
Fudan University
Shanghai, China

Jun Yu
Hangzhou Dianzi University
Hangzhou, China

Pingyi Fan
Tsinghua University
Beijing, China

Xue Xiaoping
Tongji University
Shanghai, China

Yi Wang
Huawei Technologies Co Ltd
Shanghai, China

ISSN 1867-8211   ISSN 1867-822X   (electronic)
Lecture Notes of the Institute for Computer Sciences, Social Informatics
and Telecommunications Engineering
ISBN 978-3-030-67719-0   ISBN 978-3-030-67720-6   (eBook)
https://doi.org/10.1007/978-3-030-67720-6

This Springer imprint is published by the registered company Springer Nature Switzerland AG
The registered company address is: Gewerbestrasse 11, 6330 Cham, Switzerland

Preface

We are delighted to introduce the proceedings of the 15th European Alliance for Innovation (EAI) International Conference on Communications and Networking in China (ChinaCom 2020). This conference brought together researchers, developers and practitioners around the world who are interested in communications and networking from the viewpoint of big data, cloud, sensor network, software-defined network, and so on.

The technical program of ChinaCom 2020 consisted of 54 papers, including 47 full papers and 7 workshop papers in oral presentation sessions in the main conference tracks. The conference sessions were: Session 1 - Transmission Optimization in Edge Computing; Session 2 - Performance and Scheduling Optimization in Edge Computing; Session 3 - Scheduling and Security in 5G; Session 4 - Mobile Edge Network Systems; Session 5 - Communication Routing and Control; Session 6 - Transmission and Load Balancing; Session 7 - Edge Computing and Distributed Machine Learning; Session 8 - Deep Learning. Apart from high-quality technical paper presentations, the technical program also featured two keynote speeches and one technical workshop. The two keynote speeches were delivered by Prof. Caijun Zhong from Zhejiang University and Prof. Feifei Gao from Tsinghua University. The organized workshop was on Data-Intensive Services-Based Applications (DISA). The aim of DISA was to encourage academic researchers and industry practitioners to present and discuss all methods and technologies related to research and practical experience in a broad spectrum of data-intensive services-based applications.

Coordination with the steering chair, Imrich Chlamtac, was essential for the success of the conference. We sincerely appreciate his constant support and guidance. It was also a great pleasure to work with such an excellent organizing committee team for their hard work in organizing and supporting the conference. In particular, the Technical Program Committee, led by our General Chairs and TPC Co-Chairs, Prof. Jun Yu, Prof. Pingyi Fan, Prof. Jun Wu, Dr. Yi Wang, Dr. Honghao Gao, Prof. Shuiguang Deng, Prof. Xiaoping Xue and Dr. Yuyu Yin, who completed the peer-review process of the technical papers and made a high-quality technical program. We are also grateful to the Conference Manager, Viltarė Platzner, for her support and to all the authors who submitted their papers to the ChinaCom 2020 conference and workshops.

We strongly believe that the ChinaCom conference provides a good forum for all researchers, developers and practitioners to discuss all scientific and technical aspects that are relevant to communications and networking. We also expect that future

ChinaCom conferences will be as successful and stimulating, as indicated by the contributions presented in this volume.

Honghao Gao
Pingyi Fan
Jun Wun
Xiaoping Xue
Jun Yu
Yi Wang

Conference Organization

Steering Committee

Chair

Imrich Chlamtac   Bruno Kessler Professor, University of Trento, Italy

Members

Song Guo   The University of Aizu, Japan
Bo Li   The Hong Kong University of Science and Technology
Xiaofei Liao   Huazhong University of Science and Technology
Xinheng Wang   Xi'an Jiaotong-Liverpool University
Honghao Gao   Shanghai University

Organizing Committee

International Advisory Committee

Velimir Srića   University of Zagreb, Croatia
Mauro Pezze   Università di Milano-Bicocca, Italy
Yew-Soon Ong   Nanyang Technological University, Singapore

General Chairs

Ning Gu   Fudan University
Jianwei Yin   Zhejiang University
Xinheng Wang   Xi'an Jiaotong-Liverpool University

TPC Chair and Co-chairs

Honghao Gao   Shanghai University
Yuyu Yin   Hangzhou Dianzi University
Muddesar Iqbal   London South Bank University

Local Chairs

Zhongqin Bi   Shanghai University of Electric Power
Yihai Chen   Shanghai University

Workshops Chairs

Yusheng Xu   Xidian University
Tasos Dagiuklas   London South Bank University
Shahid Mumtaz   Instituto de Telecomunicações

Publicity and Social Media Chairs

Li Kuang	Central South University
Anwer Al-Dulaimi	EXFO Inc
Andrei Tchernykh	CICESE Research Center
Ananda Kumar	Christ College of Engineering and Technology

Publications Chairs

Youhuizi Li	Hangzhou Dianzi University
Azah Kamilah Binti Draman	Universiti Teknikal Malaysia Melaka

Web Chair

Xiaoxian Yang	Shanghai Polytechnic University

Technical Program Committee

CollaborateNet Workshop

Amando P. Singun, Jr.	Higher College of Technology
BalaAnand Muthu	V.R.S. College of Engineering & Technology
Boubakr Nour	Beijing Institute of Technology
Chaker Abdelaziz Kerrache	Huazhong University of Science and Technology
Chen Wang	Huazhong University of Science and Technology
Chi-Hua Chen	Fuzhou University
Fadi Al-Turjman	Near East University
Muhammad Atif Ur Rehman	Hongik University
Rui Cruz	Universidade de Lisboa/INESC-ID
Suresh Limkar	AISSMS Institute of Information Technology

Collaborative Robotics and Autonomous Systems

Craig West	Bristol Robotics Lab
Inmo Jang	The University of Manchester
Keir Groves	The University of Manchester
Ognjen Marjanovic	The University of Manchester
Pengzhi Li	The University of Manchester
Wei Cheah	The University of Manchester

Internet of Things

Chang Yan	Chengdu University of Information Technology
Fuhu Deng	University of Electronic Science and Technology of China
Haixia Peng	University of Waterloo
Jianfei Sun	University of Electronic Science and Technology of China

Kai Zhou	Sichuan University
Mushu Li	University of Waterloo
Ning Zhang	Texas A&M University-Corpus Christi
Qiang Gao	University of Electronic Science and Technology of China
Qixu Wang	Sichuan University
Ruijin Wang	University of Electronic Science and Technology of China
Shengke Zeng	Xihua University
Wang Dachen	Chengdu University of Information Technology
Wei Jiang	Sichuan Changhong Electric Co., Ltd
Wen Wu	University of Waterloo
Wen Zhang	Texas A&M University-Corpus Christi
Wu Xuangou	Anhui University of Technology
Xiaojie Fang	Harbin Institute of Technology
Xuangou Wu	Anhui University of Technology
Yaohua Luo	Chengdu University of Technology
Zhen Qin	University of Electronic Science and Technology of China
Zhou Jie	Xihua University

Main Track

Bin Cao	Zhejiang University of Technology
Ding Xu	Hefei University of Technology
Fan Guisheng	East China University of Science and Technology
Haiyan Wang	Nanjing University of Posts & Telecommunications
Honghao Gao	Shanghai University
Jing Qiu	Guangzhou University
Jiwei Huang	China University of Petroleum
Jun Zeng	Chongqing University
Lizhen Cui	Shandong University
Lizhen Cui	Shandong University
Rong Jiang	Yunnan University of Finance and Economics
Shizhan Chen	Tianjin University
Tong Liu	Shanghai Univerisity
Wei He	Shandong University
Wei Du	University of Science and Technology Beijing
Xiong Luo	University of Science and Technology Beijing
Yu Weng	Minzu University of China
Yucong Duan	Hainan University
Zijian Zhang	University of Auckland, New Zealand

SITN Workshop

| A. S. M. Sanwar Hosen | Jeonbuk National University |
| Aniello Castiglione | Parthenope University of Naples |

Contents

Scheduling and Security in 5G

Mobile Edge Network System

Communication Routing and Control

Transmission and Load Balancing

Edge Computing and Distributed Machine Learning

Deep Learning

Transmission Optimization in Edge Computing

DOA Estimation Based on Intelligent FMCW Radar with Triangle Array Antenna

Xiaoyu Du[1,2] , Guoping Jiang[1(✉)], Chong Han[1],
Chunsong Wang[2], and Yi Zhou[3]

[1] College of Automation and College of Artificial Intelligence,
Nanjing University of Posts and Telecommunications, Nanjing, China
dxy@henu.edu.cn, jianggp@njupt.edu.cn
[2] International Joint Research Laboratory for Cooperative Vehicular Networks
of Henan, Zhengzhou, China
[3] Henan Key Laboratory of Big Data Analysis and Processing,
Henan University, Kaifeng, China

Abstract. In this paper, we propose a target Direction of Arrival (DOA) estimation algorithm through the triangular array antenna to improve the perception accuracy of intelligent vehicular millimeter-wave radar. Firstly, we utilize Total Least Square- Estimation of Signal Parameters to obtain the solution set of target azimuth with the geometrical advantages of the triangular array. Subsequently, the preliminary optimal target azimuth can be derived based on the power spectrum function of the improved Multiple Signal Classification algorithm, which is used to estimate the DOA solution set. Finally, we could estimate the optimal azimuth parameters accurately through searching the preliminary optimal target azimuth. The proposed algorithm avoids a wide range of power spectrum search, and reduces the error of azimuth estimation. Compared with both TLS-ESPRIT and MMUSIC algorithm, the spatial resolution is further improved. The evaluation results indicate that the proposed algorithm reduces the root mean square error and the computational complexity, and meanwhile improves the target azimuth estimation accuracy.

Keywords: Millimeter wave radar · Array antenna · Array signal processing · Spatial resolution · DOA estimation

1 Introduction

The radar should detect the position of the target in front and behind the car and the adjacent lane when the car starts to change lane. Currently, the radar detection methods which applied in car collision prevention mainly include millimeter wave, laser, ultrasonic, infrared, etc., among which millimeter wave radar has the best performance. Frequency modulation Continuous Wave (FMCW) radar is used to measure the distance and speed of the target, it shows the better performance under complex weather conditions, such as fog. At present, it has been used in adaptive cruise, lane change assistance, blind spot detection and forward collision warning system [1, 2].

© ICST Institute for Computer Sciences, Social Informatics and Telecommunications Engineering 2021
Published by Springer Nature Switzerland AG 2021. All Rights Reserved
H. Gao et al. (Eds.): ChinaCom 2020, LNICST 352, pp. 3–16, 2021.
https://doi.org/10.1007/978-3-030-67720-6_1

In some vehicle applications, FMCW radar is required to detect the azimuth of the target. At present, DOA estimation algorithms [3, 4] are divided into two part which include namely traditional methods and subspace methods. Bartlett and capon are part of the method of transmission. The traditional method needs to determine the DOA by searching the peaks in the power spectrum. This method depends on the physical size of the array antenna aperture, which will cause the radar signal resolution to be reduced. Therefore, the high-resolution subspace method is widely used to determine the DOA. Such as the Multiple Signal Classification (MUSIC) and Estimation of Signal Parameters via Rotation Invariance Techniques (ESPRIT), they provide a more accurate signal estimation direction, independent of the physical size of the array antenna aperture.

This paper proposes a low complexity DOA estimation algorithm based on triangular array antenna. The algorithm can be applied to the intelligent car millimeter wave radar which can identify the multi-target in front of the smart car, the complexity of the algorithm is reduced while improving the spatial resolution of the target azimuth, the angle of estimation is also more precise, and overcomes the problem of detecting data storage and large computation. The contributions of this paper are as follows:

1. We proposed a DOA estimation algorithm based on triangle array antenna using the geometric advantage of triangle array and the total least square rotation vector invariance technology, the complexity of the algorithm is reduced, and the azimuth of the target can be determined quickly.
2. Improved ESPRIT algorithm is proposed in this paper, which could increase the accuracy of the algorithm in the case of low SNR.

2   Related Work

Currently, MUSIC and ESPRIT are two classical array signal processing algorithms. Schmidt republished the Multiple Signal Classification (MUSIC) algorithm that he proposed in 1979 in 1986 [5]. The covariance matrix is solved by MUSIC, and then the Eigen-decomposition is carried out to obtain the noise subspace orthogonal to the signal component and the signal subspace corresponding to the signal component. The orthogonality of the two Spaces is used to estimate DOA parameters. Since MUSIC algorithm [6] needs to conduct extensive search of power spectrum, it is more complex and requires higher data storage requirements than ESPRIT algorithm [7]. Roy et al. proposed a rotation-invariant technical algorithm for estimating signal parameters in 1986 [8]. ESPRIT does not need to know the exact information of array manifold guide vector, so it does not need array calibration. Through the rotation invariance of matrix, ESPRIT can overcome the problem of large data storage and calculation.

ESPRIT algorithm estimates DOA parameters differently from MUSIC. ESPRIT algorithm uses the rotation invariance between two uniform subarrays to solve the invariant matrix and estimate the DOA parameters of the target. There are several improved versions of ESPRIT algorithm, such as least-squares ESPRIT (LS-ESPRIT) and overall least-squares ESPRIT (TLS-ESPRIT) [9]. TLS-ESPRIT algorithm is improved on the basis of ESPRIT algorithm. At low SNR TLS-ESPRIT algorithm has better effect, but the angular resolution is lower than the modified MUSIC (MMUSIC) algorithm. MMUSIC algorithm [10, 11] is an improvement of MUSIC algorithm, which reduces the coherence between signals and can estimate the azimuth Angle of coherent and non-coherent signals, improving the angular spatial resolution, but the computing speed is far lower than TLS-ESPRIT.

Due to the high computational load of the 2D-esprit [12] and 2D-music [13] algorithms, the target azimuth cannot be measured in real time in real scenes. The 2D-music algorithm has been improved in literature [14]. Although the complexity is reduced, the computation is still relatively high compared with ESPRIT.

The array antenna includes linear array antenna, circular array antenna, L array antenna, etc., among which the detection range of linear array antenna is front (0°–180°). It's the maximum of the beam gain for the target in the vertical direction of the array antenna. As the beam gain decreases to both sides, the radar detection range will be limited; The detection range of L array antenna is not only the azimuth information of the horizontal plane, but also the height parameters of the target, but it is limited by the detection range like linear array antenna; The circular array antenna detects the surrounding 360° environment, which is not limited by the detection range, but the algorithm is complex.

Considering the position of the FMCW smart car radar which placed on the car, and should detect the target in the forward (0°–180°) range, considering that the parameters of one-dimensional plane can meet the requirements. Based on the line array antenna, this paper adopts the triangular array antenna that overcomes the disadvantage of decreasing beam gain to both sides in the detection range, and enlarges the detection range of radar. Because the MMUSIC algorithm needs a wide range of power spectrum search, it has the disadvantage of large computation and low spatial resolution of TLS-SPRIT algorithm, in order to accurately detect the azimuth between the intelligent car millimeter wave radar and the target, overcome the shortcomings of the TLS-ESPRIT algorithm and MMUSIC algorithm.

3 Triangular Array Signal Model

In this paper, we adopt the method of sending more than receiving. Mm-wave radar uses FM continuous wave radar signal, in which waveform is serrated wave, the cycle is 10 ms, the radar waveform has 24 GHz, and the FM bandwidth is 250 MHz. The triangle array is composed of M-elements evenly, in which the spacing of elements is $d = 1/2*\lambda$ (lambda for wavelength). The noise and signal received by the array elements are independent of each other.

The time-domain emission signal of FMCW MMW radar is expressed as follow:

$$P(t) = A \cos \left\{ 2\pi \left[\left(f_0 - \frac{B}{2} \right) t + \frac{1}{2} \mu t^2 \right] + \varphi_P \right\} \tag{1}$$

And $\mu = B/T$ is the modulation slope, $f_0 = 24$ GHz, $B = 250$ MHz, φ_0 is the initial phase, $T = 10$ ms is the period of the modulation waveform, and A is the amplitude of the signal.

The target moves uniformly with respect to the radar, so R_1 is the relative distance between the target and MMW radar, the speed is V and c is the speed of light, the wavelength is $\lambda = c/f_0$, and the Doppler frequency is $f_d = 2v/\lambda$. The relationship between the distance of the target relative to MMW radar and time is as follows: $R(t) = R_1 - Vt$. The delay formula of the signal received by the radar antenna relative to the transmitted signal is as follow:

$$\tau(t) = \frac{2R(t)}{c} \tag{2}$$

The signal expression reflected back by the radar transmit signal is:

$$Y(t) = A_r \cos\left\{ 2p\left[\left(f_0 - \frac{B}{2}\right)(t - t(t)) + \frac{1}{2}m(t - t(t))^2 \right] + j_Y \right\} \tag{3}$$

Beat signal could be obtained by means of mixing frequency both the radar's reflected and transmit signals:

$$S(t) = P(t) * Y(t) \tag{4}$$

The signal expression on the k-receiving antenna array element is derived by Eq. (5) as:

$$S_k(t) = \frac{1}{2}K_r A^2 \cos\left[2\pi\left(\frac{2BR(t)}{TC} - \frac{2V}{\lambda}\right)t + \varphi_0 \right] \tag{5}$$

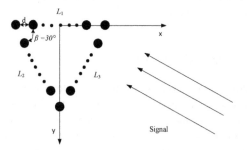

Fig. 1. Triangle array model

Assuming that the DOA estimated by the L_1 array is the direction of the target relative to the antenna, according to the geometric characteristics of the triangle, the Y-axis represents the direction of the smart car's motion, the X-axis represents the L_1 array, and sets the midpoint as the origin, establishes a right-angle coordinate system, and the elements are arranged in a triangular state, as shown in Fig. 1.

For L_1 array antenna, the assumed signal scenario has P $(P < L_1)$ FMCW MMW radar signal sources which are located in the far field, that is to say, the signal is a plane wave before it reaches the array element, and where $\theta_i \in [-\frac{\pi}{2}, \frac{\pi}{2}]$ is the azimuth Angle of reaching L_1 element $(i = 1, 2 \cdots P)$. If the first element of L_1 array is set as the reference element, then the KTH snapshot output signal of L_1 array is:

$$X(k) = AS(k) + N(k) \qquad k = 1, 2 \cdots Q \qquad (6)$$

Where $X(k) = [x_1(k), x_2(k), \ldots, x_{L_1}(k)]^T$ is the output vector of array elements, and A is the array direction matrix, $A = [a(\theta_1) \quad a(\theta_2) \quad \ldots \quad a(\theta_P)]$. The kth snapshot array stream is:

$$a(\theta_k) = \left[1, e^{-\frac{j2\pi d \sin \theta_k}{\lambda}}, \ldots, e^{-\frac{j2\pi d(L_1 - 1)\sin \theta_k}{\lambda}}\right]^T \qquad (7)$$

$S(k) = [s_1(k) \quad s_2(k) \quad \ldots \quad s_P(k)]^T$ denotes the vector of incoming wave signals, and $N(k) = [n_1(k) \quad n_2(k) \quad \ldots \quad n_{L_1}(k)]^T$ denotes the vector of noise signals, where $n_i(k)$ represents the white noise which mean is zero and variance value is σ^2. Similarly, the signal model of L_2 and L_3 array antennas can be established.

The estimated DOA parameters of L_2 and L_3 array antennas need to be converted into the angle of relative radar direction. It is necessary to convert the corresponding DOA into DOA in the relative direction of MMW radar through formula (8). The direction of L_1 antenna is the same as that of radar. $\theta_{Vehicle}$ is the optimal DOA finally estimated by the algorithm in this paper.

$$\theta_{Vehicle} = \begin{cases} \theta_i + \beta, & \text{if } \theta_i \text{ comes from antenna 2} \\ \theta_i - \beta, & \text{if } \theta_i \text{ comes from antenna 3} \\ \theta_i, & \text{if } \theta_i \text{ comes from antenna 1} \end{cases} \qquad (8)$$

The geometric advantage of triangle is used to increase the wide Angle of radar detection target. Since the triangular array antenna is composed of three groups of antennas, the target reflection wave is a vertical antenna, in other word, it coincides with the middle perpendicular line of the antenna, and the beam gain is the strongest. The further away from the mid-perpendicular the smaller the beam gain. Therefore, triangular array antenna is better than linear array antenna when radar detects the same range. The triangle array antenna receives the incoming wave signal reflected by the target, and TEMM algorithm processes three groups of incoming wave signals to find the optimal Angle of the target.

4   DOA Estimation of Triangular Array Antenna Based on TEMM Algorithm

In this paper, the triangular array antenna model is adopted to receive the radar signal reflected by the target, which increases the transverse range of the detected target and improves the beamforming gain of the target. TLS-ESPRIT algorithm and MMUSIC algorithm combined with the triangle array antenna proposed in this paper to estimate DOA, referred to as TEMM (TLS-ESPRIT-MMUSIC) algorithm. Compared with MMUSIC algorithm and 2D-MUSIC algorithm, EMM algorithm reduces the complex spectrum search range, obviously reduces the computational complexity of the algorithm, and solves the storage problem of detection data. Compared with TLS-ESPRIT algorithm and ESPRIT algorithm, the power spectrum function is utilized. The near-optimal solution is searched to find the final optimal solution of the target, which improves the spatial resolution of the target. By using the advantages of the triangular array antenna, the target estimation accuracy with relatively large relative azimuth of the radar can be improved.

4.1   TEMM Algorithm Estimation in Triangular Array

First, TEMM algorithm uses triangle array antenna to obtain incoming signals reflected by three groups of targets, then the rotation invariance between matrices in TLS-ESPRIT algorithm is used to solve the target azimuth. Due to the coherence between the signals, the influence of noise ΔU_X and ΔU_Y is reduced by using the idea of global least squares, and the DOA solution set corresponding to the target is finally solved.

TLS-ESPRIT algorithm assumes that there are two identical matrices, the front (M-2) element of L_1 array is set as X submatrix, and the rear (M-2) element is set as Y submatrix. The formula is shown below:

$$
\begin{aligned}
X(k) &= [x_1(k) \quad x_2(k) \ldots x_{L_1-2}(k)]^T \\
&= [a(\theta_1) \cdots a(\theta_{L_1-2})]S(k) + N_X(k) \\
&= A_X S(k) + N_X(k)
\end{aligned}
\tag{9}
$$

$$
\begin{aligned}
Y(k) &= [x_2(k) \quad x_3(k) \ldots x_{L_1-1}(k)]^T \\
&= [a(\theta_2)e^{j\phi_1} \cdots a(\theta_{L_1-1})e^{j\phi_P}]S(k) + N_Y(k) \\
&= A_X \Phi S(k) + N_Y(k)
\end{aligned}
\tag{10}
$$

Let $A_X = A$, $A_Y = A\Phi$, so there's $A_Y = A_X\Phi$, and $\Phi = diag[e^{-j\frac{w_0 \sin\theta_1}{c}}, \cdots, e^{-j\frac{w_0 \sin\theta_P}{c}}]$. The model of the two submatrices is combined as follows:

$$
X = \begin{bmatrix} X_X \\ X_Y \end{bmatrix} = \begin{bmatrix} A_X \\ A_X\Phi \end{bmatrix} S + N = \overline{A}S + N
\tag{11}
$$

The formula of X's covariance matrix R is shown as follows:

$$R = E[X(k)X^H(k)] = AR_S A^H + R_N \qquad (12)$$

The Eigen decomposition formula of covariance matrix is shown below:

$$R = U_S \Lambda_S U_S^H + U_N \Lambda_N U_N^H \qquad (13)$$

U_S represents the signal subspace of covariance matrix R, and U_N represents the noise subspace of covariance matrix R. At this point, $span\{U_S\} = span(A(\theta))$, and there is a unique non-singular matrix T, which decomposes the signal subspace U_S into a submatrix, so that:

$$U_S = \begin{bmatrix} U_X \\ U_Y \end{bmatrix} = \begin{bmatrix} A_X T \\ A_X \Phi T \end{bmatrix} \qquad (14)$$

Define $U_X \varphi = U_Y$ and substitute it into formula (14) to obtain:

$$\varphi = T^{-1} \Phi T \qquad (15)$$

Where Φ is the eigenvalue of φ, only φ is required, and DOA of the target can be solved through formula (16).

$$\varphi_k = (2\pi d \sin \theta_k)/\lambda \qquad (16)$$

Considering that U_X and U_Y are affected by noise, in order to minimize the 2-Norm of noise ΔU_X and ΔU_Y, U_X and U_Y are corrected, which is the idea of using the global least square method (TLS) and obtained from $U_X \varphi = U_Y$:

$$(U_X + \Delta U_X)\varphi = U_Y + \Delta U_Y \qquad (17)$$

From Eq. (17), it can be concluded that:

$$\begin{cases} \min \|\Delta U\|^2 \\ (U + \Delta U)z = 0 \end{cases} \qquad (18)$$

Among them $z = \begin{bmatrix} 1 \\ \varphi \end{bmatrix}$, $U = [-U_Y \ U_X]$, $\Delta U = [-\Delta U_Y \ \Delta U_X]$. Let $U_S F = 0$ be established, where F is the unitary matrix, making F and U_S orthogonal. According to formula (14), $U_S = [U_X \ U_Y]^T$, it is eigen decomposed:

$$U_S^H U_S = E \Sigma E^H \qquad (19)$$

The eigenvector matrix E is obtained:

$$E = \begin{bmatrix} E_{11} & E_{12} \\ E_{21} & E_{22} \end{bmatrix} \tag{20}$$

$[E_{12} \quad E_{22}]^T$ is the eigenvector matrix with zero eigenvalue, at same time it is the noise subspace. $U_S F = 0$ holds As long as we set $F = [E_{12} \quad E_{22}]^T$. So we can get:

$$ATF_1 + A\Phi TF_2 = 0 \tag{21}$$

Where $F = [F_1 \quad F_2]^T$. $\varphi = -F_1 F_2^{-1}$ and put it into formula (21):

$$T^{-1}\Phi T = \varphi \tag{22}$$

Formula (22) is the same as formula (15), and the azimuth of the target can be obtained. φ_{TLS} can be obtained by formula (18):

$$\varphi_{TLS} = -E_{12} E_{22}^{-1} \tag{23}$$

The direction of signals received by L_1 array antenna can be obtained by TLS-ESPRIT algorithm. Similarly, we can also obtain the azimuth angle of signals received by L_2 and L_3 array antenna respectively. It is obtained what the solution set $(a_i^{L1}, a_i^{L2}, c_i^{L3}), i = 1, 2 \ldots, p$ of three sets of azimuth angles corresponding to the target, and then the initial optimal DOA is obtained from the solution set by TEMM algorithm.

4.2   TEMM Algorithm Obtains Preliminary Optimal DOA

MUSIC algorithm is suitable for high resolution spectrum estimation, but its disadvantage is that it requires eigenvalue decomposition and complex spectral search process. Therefore, compared with TLS-ESPRIT algorithm, its calculation time is long and the calculation speed is slow. MMUSIC algorithm is improved on the basis of MUSIC algorithm and has the same characteristics. Through the power spectrum function of MMUSIC algorithm, the azimuth Angle of the maximum signal power is estimated from three groups of DOA. Because it has an effect on the accuracy of DOA estimation by the geometric position of subarray, the estimation error of Angle is different with different subarray. Through TEMM algorithm to estimate the wave direction of P signals, the DOA solution set Z is obtained as follows:

$$Z_i = (a_i^{L1}, a_i^{L2}, c_i^{L3}) \qquad i = 1, 2, \cdots, P \tag{24}$$

Before the power spectrum function of MMUSIC algorithm estimates the azimuth angles of the three groups, the three groups of DOA sets are respectively sorted in ascending order. The radar waveform reflected has three estimated azimuth angles by each target, and the power spectrum of azimuth Angle of the i-th target can be estimated through formula (25).

$$G(i) = M(Z_i) \tag{25}$$

M is the classical MUSIC power spectrum search function, and its expression is:

$$M(\theta) = \frac{1}{a^H(\theta)E_n E_n^H a(\theta)} \tag{26}$$

By calculating the Angle power spectrum of each target pair, the Angle corresponding to the largest power spectrum is selected as the initial optimal solution of the target. On the basis of the initial optimal solution, TEMM algorithm is used to search near the initial optimal solution, it will calculate the final optimal DOA.

4.3   TEMM Algorithm Obtains the Optimal DOA

It is known from literature [4] that the conjugate reconstruction of the signal matrix $X(k)$ output by the array antenna produces a new matrix $Y(k)$, and the covariance of which is respectively R_X and R_Y. It is obviously that R_X, R_Y and R_Z have the same noise subspace for $R_Z = R_X + R_Y$. We can obtain the target azimuth angle θ_i by substituting the noise subspace into Eq. (26) which is obtained by Eigen decomposition of covariance matrix R_Z. We define $\theta n = 1°$, so the search range is $(\theta_i - \theta_n, \theta_i + \theta_n)$, and the precise target azimuth is obtained by searching near azimuth θ_i through MMUSIC. Due to the different positions of the three groups of triangular array antennas, only the Angle conversion of L_2 and L_3 antennas can be used to realize the true orientation of the target. Through formula (8), the final optimal DOA of the target's relative millimeter wave radar is obtained.

DOA estimation algorithm for MMW radar based on triangular array is as follows:

- 1. The target reflected signal received through the triangular array, the signals $S_1(t)$, $S_2(t)$ and $S_3(t)$ can be obtained from L_1, L_2 and L_3 respectively;
- 2. Calculate submatrices X and Y according to Eq. (9) and (10) and combine submatrices X and Y to estimate covariance matrix R according to Eq. (12);
- 3. Get U_S matrix by Eigen decomposition of R according to Eq. (14);
- 4. The U_S matrix is Eigen decomposed to get matrix E, and E is decomposed;
- 5. φ_{TLS} is obtained from formula (22), and φ_{TLS} is eigen decomposed to obtain the solution of L_1 antenna
- 6. Go to step 2, similarly, we can obtain the solutions of L_2 and L_3 antennas, and DOA solution set $Z_i = (a_i^{L1}, a_i^{L2}, c_i^{L3})$ can be obtained too;
- 7. The three groups of DOA are arranged in ascending order;
- 8. Through formula (25), the maximum power general of the target is solved, and the corresponding angle is set as the initial optimal solution θ_i;
- 9. The final optimal DOA is searched within $(\theta_i - \theta_n, \theta_i + \theta_n)$ range by the power spectrum function of TEMM algorithm.

5 Simulation and Analysis

Simulations are presented to verify the efficacy of the TEMM algorithm which uses geometric features of triangular array antenna to increase the target positioning range (−75°–75°). At the same time, the estimation accuracy of the target azimuth Angle is improved, and the calculation amount of MMUSIC algorithm is reduced under the same accuracy of DOA estimation. In DOA estimation of intelligent vehicle-mounted millimeter-wave radar, signal-to-noise ratio affects the accuracy of the estimated signal DOA, and the sampled fast beats cause errors to the estimation of DOA, and the estimated DOA sizes are different with different accuracy. The Root Mean Square Error (RMSE) is used to evaluate the positioning performance of TEMM algorithm under different SNR, azimuth and fast beats, and the spatial resolution of TEMM algorithm is analyzed with beamforming diagram. MATLAB simulation is conducted to verify the validity of TEMM algorithm in estimation accuracy and resolution.

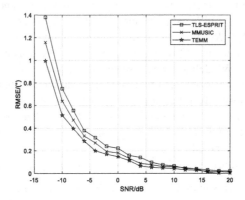

Fig. 2. Performance comparison under different SNR

Case I: The array element of the triangular array was set as M = 26 (L$_1$ = 12, L$_2$ = 7, L$_3$ = 7) and the array element of the linear array antenna was set as 12, the array element spacing was set as half of the wavelength, the target azimuth Angle was set as 50°, 1024 times of fast beats were sampled, and 300 times of Monte Carlo experiments were conducted with the algorithm in this paper and the TLS-ESPRIT algorithm under different SNR. The simulation results were shown in Fig. 2, the three curves respectively showed the change of the root mean square error (RMSE) of DOA with the signal to noise ratio (SNR). With the increase of SNR, the RMS error became smaller and smaller, and the DOA estimation accuracy became higher and higher. When the signal-to-noise ratio was −13 dB, the estimation errors of the three algorithms differ greatly. With the change of SNR from 20 dB–13 dB, it was clear that TEMM algorithm had higher DOA estimation accuracy than TLS-ESPRIT algorithm and MMUSIC algorithm.

Case II: The array element of the triangular array was set as M = 26 (L$_1$ = 12, L$_2$ = 7, L$_3$ = 7) and the array element of the linear array antenna was set as 12, the array element spacing was set as half of the wavelength, the signal-to-noise ratio was set as 8 dB, and 1024 times of fast beats were sampled. Under different target azimuth angles, the algorithm in this paper and the TLS-ESPRIT algorithm were respectively used to conduct 300 times of Monte Carlo experiments. The simulation results were shown in Fig. 3. The two curves respectively showed with the Angle of the target the variation of the RMS error (RMSE) of DOA. As the target azimuth increases, the RMS error became larger and larger, and the DOA estimation accuracy became lower and lower. When the target azimuth Angle was 80°, it was obviously different in the estimation error of the two algorithms. It could be seen from the figure that the DOA estimation accuracy of TEMM algorithm was higher than that of TLS-ESPRIT algorithm with the increase of target azimuth Angle.

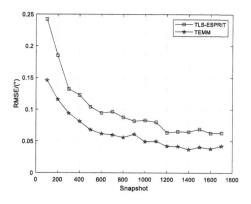

Fig. 3. Comparison of performance at different azimuths

Case III: The array elements of the triangular array were set to M = 26 (L_1 = 12, L_2 = 7, L_3 = 7) and the array element of the linear array antenna was set to 12, the array element spacing was set to half of the wavelength, and the target azimuth angle was set to 50°. The signal-to-noise ratio was 8 dB. Under the different sampling snapshots, 300 Monte Carlo experiments were performed using the algorithm and TLS-ESPRIT algorithm respectively. The simulation results were shown in Fig. 4. It could be seen that the azimuth rms error decreases with the increase of the number of sampling snapshots. When the number of snapshots increases to a certain extent, the root mean square error tends to be stable. In the 100–1700 snapshots, it could be seen that the root mean square error of the TEMM algorithm was small and had a high estimation accuracy.

Case IV: The array elements of the triangular array were set to M = 26 (L_1 = 12, L_2 = 7, L_3 = 7) and the array elements of the linear array antenna were set to 12, the array element spacing was set to half of the wavelength, and the target azimuth angle was set to 60°, 75°, the signal-to-noise ratio was 8 dB, and the sampling snapshot was 1024. Independent experiments were performed using the algorithm and MMUSIC algorithm respectively. The simulation results were shown in Fig. 5.

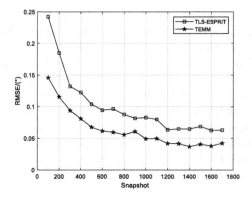

Fig. 4. Effect of different snapshots on root mean square error

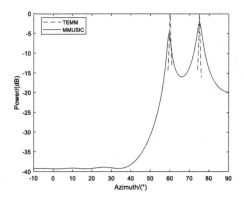

Fig. 5. Spectrum comparison between the algorithm and the MMUSIC algorithm

It could be seen from the Fig. 5 that MMUSIC requires a long time of spectrum search and a large amount of calculation. However, the algorithm in this paper avoids this defect and only needs to search within a short range to get an accurate DOA. Under the same conditions, the algorithm in this paper could more accurately distinguish DOA of adjacent signals, and the spatial resolution was improved. When the Angle reaches 75°, TEMM algorithm had a higher accuracy, and the power spectrum search range was smaller than MMUSIC algorithm.

6 Conclusions

In this paper, theoretical analysis and simulation of TEMM algorithm, MMUSIC algorithm and TLS-ESPRIT algorithm are carried out from the aspects of azimuth, signal-to-noise ratio and snapshot number. From the range of estimated azimuth Angle, it is concluded that the root mean square error of TEMM algorithm is smaller and the accuracy is higher. The TEMM algorithm has a higher accuracy when sampling the same number of snapshots. From the beamforming diagram, the TEMM algorithm has a smaller power spectrum search range than the MMUSIC algorithm, which reduces the amount of computation. However, the target beam estimated by TEMM algorithm is sharper, which is easy to accurately identify the target Angle. In this paper, the improved TEMM algorithm based on TLS-ESPRIT and MMUSIC algorithm can accurately estimate the optimal DOA of the target. Therefore, TEMM algorithm increases the detection range, improves the spatial resolution and reduces the computation, which lays a foundation for the practice of engineering. The mentioned method is mainly aimed at mutually independent signals without considering the influence of coherent signals. However, in practical engineering applications, errors caused by coherent signals cannot be avoided. Therefore, in order to reflect the advantages of TEMM algorithm, coherent signals need to be processed.

Acknowledgment. This work was supported by National Natural Science Foundation of China (61672209, 61701170), Science and technology development plan of Henan Province (192102210282, 202102210327).

References

1. Hyun, E., Jin, Y.S., Lee, J.H.: Moving and stationary target detection scheme using coherent integration and subtraction for automotive FMCW radar systems. In: Radar Conference. IEEE (2017)
2. Hyun, E., Jin, Y.S., Lee, J.H.: Design and development of automotive blind spot detection radar system based on ROI pre-processing scheme. Int. J. Autom. Technol. **18**(1), 165–177 (2017)
3. Cao, R., et al.: A low-complex one-snapshot DOA estimation algorithm with massive ULA. IEEE Commun. Lett. 1–1 (2017)
4. Li, J., Li, D., Jiang, D., et al.: Extended-aperture unitary root MUSIC-based DOA estimation for coprime array. IEEE Commun. Lett. **22**(4), 752–755 (2018)
5. Schmidt, R., Schmidt, R.O.: Multiple emitter location and signal parameters estimation. IEEE Trans. Antennas Propag. **34**(3), 276–280 (1986)
6. Abdalla, M.M., Abuitbel, M.B., Hassan, M.A.: Performance evaluation of direction of arrival estimation using MUSIC and ESPRIT algorithms for mobile communication systems. In: Wireless and Mobile Networking Conference, pp. 1–7. IEEE (2013)
7. Lavate, T.B., Kokate, V.K., Sapkal, A.M.: Performance analysis of MUSIC and ESPRIT DOA estimation algorithms for adaptive array smart antenna in mobile communication. In: Second International Conference on Computer and Network Technology, pp. 308–311 (2010)
8. Roy, R., Kailath, T.: ESPRIT-estimation of signal parameters via rotational invariance techniques. IEEE Trans. Acoustic Speech Sig. Process. **37**(7), 984–995 (1989)
9. Lina, Y.: Study on factors affecting DOA accuracy. Modern Radar **29**(6), 70–73 (2007)
10. Dzielski, J.E., Burkhardt, R.C., Kotanchek, M.E.: Modified MUSIC algorithm for estimating DOA of signals - Comment. Sig. Process. **55**(2), 253–254 (1996)
11. Lou, D., Xie, J., Wang, Y.: Comparison of the statistical characteristics of MUSIC and MMUSIC algorithms. Space Electron. Technol. **4**(3), 31–35 (2007)

12. Zoltowski, M.D., Wong, K.T.: ESPRIT-based 2-D direction finding with a sparse uniform array of electromagnetic vector sensors. IEEE Trans. Sig. Process. **48**(8), 2195–2204 (2000)
13. Belfiori, F., van Rossum, W., Hoogeboom, P.: 2D-MUSIC technique applied to a coherent FMCW MIMO radar. In: IET International Conference on Radar Systems (Radar 2012), Glasgow, UK, pp. 1–6 (2012)
14. Jun, X., Xin, Z., Haitao, H.: An improved two-dimensional music algorithm based on spatial rectangular array. Digit. Manuf. Sci. **69**(02), 5–9 (2018)

Multi-modulation Scheme for RFID-Based Sensor Networks

Zijing Tian[1(✉)] and Zygmunt J. Haas[1,2]

[1] School of Engineering and Computer Science, University of Texas at Dallas, Richardson, TX, USA
zxt140530@utdallas.edu, zhaas@cornell.edu
[2] School of Electrical and Computer Engineering, Cornell University, Ithaca, NY, USA

Abstract. RFID technology is playing an increasingly more important role in the *Internet of Things*, especially in the dense deployment model. In such networks, in addition to communication, nodes may also need to harvest energy from the environment to operate. In particular, we assume that our network model relies on RFID sensor network consisting of Wireless Identification and Sensing Platform (WISP) devices and RFID exciters. In WISP, the sensors harvest ambient energy from the RFID exciters and use this energy for communication back to the exciter. However, as the number of exciters is typically small, sensors further away from an exciter will need longer charging time to be able to transmit the same amount of information than a closer by sensor. Thus, further away sensors limit the overall throughput of the network. In this paper, we propose to use a multi-modulation scheme, which trades off power for transmission duration. More specifically, in this scheme, sensors closer to the exciter use a higher-order modulation, which requires more power than a lower-order modulation assigned to further away sensors, for the same bit error rate of all the sensors' transmissions. This reduces the transmission time of the closer sensors, while also reducing the charging time of the further away sensors, overall increasing the total network throughput. The evaluation results show that the RFID sensor network with our multi-modulation scheme has significantly higher throughput as compared with the traditional single-modulation scheme.

Keywords: RFID systems · IoT · WISP · Energy harvesting · Multi-modulation scheme · Throughput optimization · Sensor network

1 Introduction and Motivation

The *Internet of Things* (IoT) networking paradigm continues to draw a lot of attention, especially with the deployment of novel and exciting applications, many of which requiring dense deployment of sensing/processing/communicating devices ("things"). At the same time, radio frequency identification (RFID) technologies and RFID sensor networks (RSNs), which consist of exciters and RFID nodes, promise to broaden RFID systems by incorporating sensing technologies [1]. Nodes in RSNs are Wireless

H. Gao et al. (Eds.): ChinaCom 2020, LNICST 352, pp. 17–36, 2021.
https://doi.org/10.1007/978-3-030-67720-6_2

Identification and Sensing Platform (WISP)[1] sensor devices which are capable of harvesting their operating energy from RFID transmissions [2]. The problem that we investigate in this paper relates to energy harvesting by sensor nodes in RSN with dense deployment.[2]

In general, an RSN consists of sensor nodes and RFID *exciters*[3] deployed among the nodes. The RFID exciters transmit to the RFID devices by continuous wave (CW) signals, and the sensor nodes convert these signals to DC, charging their on-board energy storage elements (e.g., a super capacitor). When a sensor node accumulates enough energy from an exciter, it will be able to transmit their own signals back to the exciter. In this study, we adopt a combination of RFID tags and sensors [3] functioning as sensor nodes capable of energy harvesting. This kind of sensor nodes could gather information by its sensor, harvest and store energy in the sensor node, and then send the information to the exciter upon receiving permission from the exciter. (The energy harvesting circuit is similar to the circuit in semi-passive tag-based sensor nodes [4]. And although our system model does not employ passive backscattering techniques, other systems models could be considered in future work.

RFID technology is already pervasive in applications such as bar codes [5] and applying the technology to IoT system is becoming widespread in Smart Cities and in Smart Homes projects [6]. RFID technology has also been extensively deployed in various and distinctive fields, such as access control, biomedical implants, identification, tracking, logistics, sensor networks, security, fast payment system, loss prevention and shopping malls [7]. The use of WISP devices as RFID sensor nodes augments the RFID technology with increased computational capabilities [1].

With the growing demand for large size RSNs, the energy harvesting efficiency becomes a constraint that limits the throughput of a sensor network. This paper is aiming at the optimization of RSN communication in dense sensor nodes deployment [8]. In this kind of RSNs, nodes which are deployed further from the exciter experience larger attenuation of the CW signal, and thus will be charged at a slower pace than nodes located closer to the reader. Furthermore, such further-away nodes require more energy to transmit back to the exciter (again because of the signal attenuation with distance) to achieve the same bit error rate (BER) as nodes located close-by to the exciter. Thus, the required charging time is prolonged by the "ineffective energy harvesting" of the far-away nodes, reducing the overall throughput of the whole RSN.

To address this limitation, we propose a multi-modulation scheme, where based on a node's distance from the exciter, different modulation methods are used by the network's nodes. Although RFID system that apply different modulation schemes were previously proposed in [9], however, that work didn't provide any strategy to improve the performance of the RFID system by the use of multi-modulation. In our study, we focus on the RSN design of such multi-modulation-based RFID sensor networks to

---

[1] RSNs and especially the WISP technology is extensively discussed in [1–4].

[2] Imagine that the hardware of a dense IoT device requires changing a battery once in 4 years. In a deployment of 100,000 devices, this means that there is a need to replace 2,500 batteries every year or $\sim$70 batteries every day!

[3] The terms "reader" and "exciter" are interchangeably used in technical literature.

significantly improve the network's performance. Since different modulation methods have different Signal to Noise Ratio (SNR) requirements, we can balance the energy harvesting effectiveness of the different nodes by applying the different modulation schemes according to the sensor nodes' distances to the exciter. In our work, we assume that there is a single channel that is used sequentially by the sensor nodes to transmit. Since one transmission cycle for the RSN consists of the charging time in addition to the transmission times of all the nodes, to maximize the network throughput, we need to reduce the sum of these two components. Because higher order modulation methods result in shorter transmission times (as discussed in Sect. 3), the multi-modulation scheme needs to judiciously trade-off between the transmission time and the required charging time, as to minimize the overall performance. In other words, the multi-modulation scheme allows: (1) saving the overall charging time compared with the case of a single modulation scheme, and (2) saving the overall transmission time compared with the case of a single modulation scheme. In summary, the result shows that our multi-modulation scheme has a significant improvement in throughput of the RSN, as compared with the traditional single-modulation scheme.

The rest of the paper is organized as follows: In Sect. 2, we introduce our multi-modulation scheme, with the theoretical results of the scheme presented in Sect. 3. In Sect. 4, we show how our scheme works in a multi-exciter environment. In Sect. 5, we present numerical results that demonstrate the improvement of our scheme compared with the traditional single-modulation scheme. Section 6 concludes the paper.

2 Multi Modulation Scheme

In the proposed multi-modulation scheme for RFID sensor networks hundreds/thousands of sensor nodes are deployed in the RFID reader coverage area. We assume that nodes in the coverage area follow a uniform distribution, and that the communication requirements (i.e., the amounts of data) that all the nodes need to transmit to the exciter are the same. An RFID exciter (sometimes referred to as a *reader*) emits CW signals to supply energy to sensor nodes in the coverage area. The exciter also serves as a router that is responsible for scheduling data transmissions of the sensor nodes and for receiving messages from nodes. Sensor nodes are designed to harvest energy from the ambient RF signal of the exciter, using known energy harvesting techniques in the RFID systems, as are known to be applied to semi-passive tags [10]. A sensor node can transmit its message only after it has been sufficiently charged and it receives the permission from the exciter. In our model, we assume that sensor nodes cannot harvest energy and transmit data to the exciter at the same time (i.e., all the harvested energy is used to charge the sensor node's energy storage).

In our RSN, all the sensor nodes within the exciter coverage area share one communication channel. As the exciter arranges the schedule of the sensor nodes' transmissions in each cycle, the exciter needs to query each node within its coverage area in each transmission cycle whether the node has data to transmit. The amount of data that each node transmits in a cycle is the same for all the nodes and the symbol rate (discussed in Sect. 3) for all the node-to-exciter transmissions is the same for all the nodes. Thus, the bandwidth requirements of all the nodes are the same too. Since we want to improve the throughput of the RFID sensor network and the throughput of a network has an inverse relation to the cycle time, the cycle time needs to be minimized.

As shown in Fig. 1, there are three major phases in a transmission cycle: Query, Charging, and Transmissions. In the Query phase, the exciter identifies the sensor nodes that have data to transmit. (We note that if the network allows the set of sensor nodes to change over time, this phase could also be used to discover nodes in the coverage area.) Since before any node can transmit it needs to collect sufficient energy, the Query phase includes a short charging time (not shown in the figure) to allow a node to respond to the exciter's query. Based on the nodes' responses, the exciter starts sending CW (in the Charging phase) to sufficiently charge the nodes. The exciter then sequentially sends short polling messages to the next node that is scheduled to transmit (not shown in the figure), so that the nodes can transmit their data. Only nodes who announced to the exciter in the Query phase that they have data to transmit will be polled in this Transmissions phase.

A Cycle

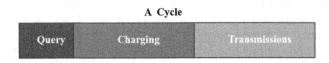

Fig. 1. Phases of a transmission cycle

In general, the charging time period depends on the required charging time of the furthest node, which is determined by the exciter-to-node distance R. To maximize the network throughput, we apply a family of modulation schemes (e.g., PSK) to the nodes in the network, where the further away nodes are assigned such an order of the modulation family (e.g., 8-PSK or 16-PSK) that requires less power to transmit. This way, in each cycle of data transmission, our network will take a shorter time to finish all the transmissions than when a single-modulation scheme is used by all the nodes. In general, the total transmission time of a cycle in this scheme is:

$$t_{total} = t_{query} + \sum t_{transmit} + \text{Max}(t_{charge}), \qquad (1)$$

where t_{query} is the time needed for the Query phase, $\sum t_{transmit}$ is the summation of the transmission times of all the sensor nodes, and $\text{Max}(t_{charge})$ represents the maximum charging time among all the nodes. Thus, the charging time needed for the whole network depends on the charging time of the sensor node that needs the most time to charge, while the transmission time for the whole network depends on the summation of the all the transmission times of the nodes, because we assume a single channel, so that the exciter cannot receive signals from different nodes at the same time.

Since different modulation schemes require different levels of the energy-per-bit to noise-power-spectral-density ratio $(\frac{E_b}{N_0})$ to achieve the same bit error rate (BER), the energy-per-bit, E_b, required for different modulations are also different. If we apply the same modulation scheme to all the sensor nodes, the closer nodes would need to wait for the further nodes to finish charging. However, if we assign to the furthest nodes a modulation scheme that requires less energy to transmit, the charging time for the nodes will decrease; thus, the total cycle time will decrease as well. The motivation behind our scheme is as follows. Since in this network, the symbol rate of the system remains the same for all the nodes, modulation schemes that require less energy to transmit results in a less charging time but longer transmission time (as discussed in Sect. 3). Consequently, assume we have modulation schemes A and B, and nodes applying the scheme A require less energy to transmit per bit than nodes applying the scheme B, while the latter can transmit more bits per unit time. Thus, we can apply scheme A to the further nodes, and apply scheme B to the closer nodes. Since applying modulation scheme B will reduce the transmission time of the closer nodes, the total transmission time will be reduced compared with networks that only apply the modulation scheme A; and because the modulation scheme A reduces the charging time of the whole network, the multi-modulation scheme has a shorter cycle time compared with networks that only apply the modulation scheme B. Hence, the sum of transmission time and charging time (i.e., Eq. (1)) could be optimized under our multimodulation scheme. To summarize, the scheme that we propose in this paper applies different modulation methods to sensor nodes according to their distance to exciter R, as to optimize the throughput of the network.

We first calculate the maximum RFID exciter-to-node range R. There are two main criteria to constrain the maximum range of exciter-to-node communication (further discussed in Sect. 3): (1) the reading range of the RFID exciter, and (2) the required BER of the application. The first criterion determines the furthest possible range of exciter coverage area (R_{cover}), and the second criterion determines the furthest range of exciter-to-node communication for particular modulations (R_{max}) under particular charging time limitation. Although we can extend the charging time to increase R_{max}, R_{max} cannot be larger than R_{cover}.

The references [11–18] outline the main modulation schemes used for RFID UHF schemes: QAM, QPSK, BPSK, ASK, and OOK [4]. We choose PSK, PAM, and QAM [12] modulations with different number of bits per symbol k ($M = 2^k$), because they are the more commonly used schemes in RFID systems.

In the particular example presented in Sect. 5, we apply three different orders of the PSK family of modulations: QPSK, 8-PSK, and 16-PSK to the 100 sensor nodes in the exciter coverage area. These 100 nodes are deployed within the area determined by the QPSK modulation scheme, $R_{max(QPSK)}$. The exciter's coverage area is divided into four areas: (1) the circle with 16-PSK modulation, where the distance of the nodes to the exciter is no larger than $R_{max(16-PSK)}$; (2) the annulus region with 8-PSK modulation, where the distance of the nodes to the exciter is between $R_{max(16-PSK)}$ and $R_{max(8-PSK)}$; (3) the annulus region with QPSK modulation, where the distance of nodes to the exciter is between $R_{max(8-PSK)}$ and $R_{max(QPSK)}$. Figure 2 shows the region partition for our multi-modulation scheme.

---

[4] We do not use the OOK modulation scheme, since with this modulation we cannot apply different number of bits per symbol, k [17].

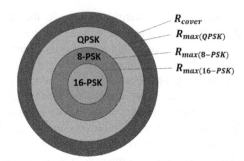

Fig. 2. The region partition and assignments of the modulation schemes

Although we divided the coverage area into the three regions above, the distances of the sensor nodes to the exciter within each ring are not exactly the same. Thus, the length of the charging phase is determined by the node with the largest required charging time; i.e., a node that is the furthest distance from the exciter and is still in its ring area; in other words, a node located on the boundary to the next ring.

3 Derivation of the Multi-modulation Scheme Improvement

In this section, we derive the performance improvement of our multi-modulation scheme. First, we briefly discuss the basic concepts used throughout this section. In this study, the BER required by the application is the main design criterion that limits the node-to-exciter communication range. For a particular modulation scheme, the $\frac{E_b}{N_0}$ has a direct relation to BER. SER represents the symbol error rate, with the energy per symbol $E_s = k \times E_b$ being another parameter in our research, where k represents the number of bits-per-symbol of a particular modulation scheme [19]. We also consider an additive white Gaussian noise (AWGN) in our work [19].

According to the requirement of the modulation methods assignment in Fig. 2, we should first figure out the $\frac{E_b}{N_0}$ of the different modulation schemes based on the same required BER, so that we can determine the partition rings of our scheme. The SER-SNR and BER-SNR formulas of the various modulation schemes are shown in Table 1 [20].

Table 1. SER v. SNR and BER v. SNR for various modulation orders

	K = 1	K = 2	K = 3
SER	$Q\left(\sqrt{2\frac{E_s}{N_0}}\right)$	$2Q\left(\sqrt{2\frac{E_s}{N_0}}\right)$	$2Q\left(\sqrt{2\frac{E_s}{N_0}}\sin\frac{\pi}{M}\right)$
BER	$Q\left(\sqrt{2\frac{E_b}{N_0}}\right)$	$Q\left(\sqrt{2\frac{E_b}{N_0}}\right)$	$\frac{2}{k}Q\left(\sqrt{2k\frac{E_b}{N_0}}\sin\frac{\pi}{M}\right)$

Using the Table 1 and the inverse Q-function, one can derive the required $\frac{E_b}{N_0}$ for a particular value of BER. Then for a choice of M-PSK, we obtain the formulas for $\frac{E_b}{N_0}$ and $\frac{E_s}{N_0}$ in Table 2 [20].

Table 2. $\frac{E_s}{N_0}$ and $\frac{E_b}{N_0}$ as a function of SER for various M-PSK modulations

	BPSK	QPSK	M-PSK(M > 4)
$\frac{E_s}{N_0}$	$(erfc^{-1}(SER))^2$	$2(erfc^{-1}(SER))^2$	$\dfrac{2(erfc^{-1}(SER))^2}{2\left(sin\frac{\pi}{M}\right)^2}$
$\frac{E_b}{N_0}$	$(erfc^{-1}(BER))^2$	$(erfc^{-1}(BER))^2$	$\dfrac{(erfc^{-1}(\frac{k}{2}BER))^2}{k\left(sin\frac{\pi}{M}\right)^2}$

Next, we study the energy transfer in our RFID sensor network to find the exact criterion for R_{max}. The downlink power which shows the exciter-to-node power received at the sensor nodes, P_{node}, is:

$$P_{node} = P_{cw}\left(\frac{\lambda_{cw}}{4\pi R}\right)^2 G_{exciter}G_{node}, \tag{2}$$

where P_{cw} is the power of the CW emitted from the exciter, R is the distance between the exciter and a sensor node, λ_{cw} is the wavelength of the CW signal, $G_{exciter}$ and G_{node} are the power gains of the exciter and node antennas, respectively [21], and all powers are expressed in units of Watts.

For the uplink power, the node-to-exciter power received at the exciter, P_{rd}, is:

$$P_{rd} = P_{back}\left(\frac{\lambda_{cw}}{4\pi R}\right)^2 G_{exciter}G_{node}, \tag{3}$$

where P_{back} is the modulated power emitted by a sensor node. By studying the relationship between P_{back} and P_{node}, we can find the modulation region partition in Fig. 2. If we label the power efficiency of energy harvesting in the nodes as ρ_t, the charging time in each cycle as t_{charge}, and the time duration that a node transmits in each cycle as $t_{transmit}$, the relationship between P_{back} and P_{node} is:

$$P_{back} \cdot t_{transmit} = \rho_t \cdot P_{node} \cdot t_{charge}, \tag{4}$$

where $t_{transmit} = L/R_b$ is the transmission time needed for transmission of L bits (i.e., L is the length of the data needed to be transmitted (in bits)), and R_b is the bit rate. We assume the symbol rate S_b (in units of symbols/sec) is fixed for all the network transmissions. Thus, the data rates R_b of the different modulations with different bits-per-symbol are different, because $R_b = k \times S_b$. The bandwidth B of the PSK/QAM modulated signal is $B = \frac{2R_b}{k}$ [22]. Hence, $B = 2S_b$, and the same bandwidth is used by all the sensor nodes.

In this study, we use the criteria in reference [23] as the value of R_{cover}, which is 30.5 m for semi-passive tags, as we assume that our nodes rely on similar circuitry as semi-passive tags. Next, we concentrate on obtaining R_{max}, which is constrained by the acceptable maximum BER for all the modulations. Equation (5) represents the relationship between P_{rd} and $\frac{E_s}{N_0}$ [24]:

$$\frac{P_{rd}}{N_0} = R_b\left(\frac{E_b}{N_0}\right) = k \times S_b\left(\frac{E_s}{kN_0}\right) = S_b\left(\frac{E_s}{N_0}\right). \tag{5}$$

In our scheme, a sensor node is charged by the exciter as per Eq. (2). After a charging time t_{charge}, the node will have enough power to transmit its data of L bits. Then by Eq. (4), the power sent back from the node is calculated. The back power received in the exciter, P_{rd}, is calculated by Eq. (3). For all the nodes, independent of the modulation scheme, the maximum acceptable BER of the nodes' signals received by the exciter should be the same and is given by the application. Therefore, knowing the required BER, we can find the corresponding required minimum $\frac{E_b}{N_0}$ for PSK modulation with different M by using formulas in Table 1 and Table 2. Then by Eq. (5), we obtain the minimum power P_{rd} needed in the exciter for the different modulations:

$$P_{rd(min)} = N_0 S_b\left(\frac{E_s}{N_0}\right) = N_0 S_b k\left(\frac{E_b}{N_0}\right), \tag{6}$$

where $\frac{E_b}{N_0}$ is calculated by formulas in Table 2. Combining Eq. (2), (3), (4), and (6), we obtain the constraint of the distance R_{max} for the different modulation schemes:

$$\begin{aligned}
N_0 S_b k\left(\frac{E_b}{N_0}\right) &= P_{back}\left(\frac{\lambda_{cw}}{4\pi R_{max}}\right)^2 G_{exciter} G_{node} \\
&= \frac{\rho_t \cdot P_{node} \cdot t_{charge}}{t_{transmit}}\left(\frac{\lambda_{cw}}{4\pi R_{max}}\right)^2 G_{exciter} G_{node} \\
&= \frac{P_{cw} G_{exciter} G_{node} \rho_t t_{charge}}{t_{transmit}}\left(\frac{\lambda_{cw}}{4\pi R_{max}}\right)^4 G_{exciter} G_{node}
\end{aligned}$$

$$\begin{aligned}
R_{max} &= \sqrt[4]{\frac{P_{cw}\rho_t t_{charge}}{N_0 S_{bk}\left(\frac{E_b}{N_0}\right) t_{transmit}}\left(\frac{\lambda_{cw}}{4\pi}\right)^4 (G_{exciter} G_{node})^2} \\[2mm]
&= \sqrt[4]{\frac{P_{cw}\rho_t t_{charge}}{N_0\left(\frac{E_b}{N_0}\right) L}\left(\frac{\lambda_{cw}}{4\pi}\right)^4 (G_{exciter} G_{node})^2}
\end{aligned} \tag{7}$$

Thus R_{max} is determined by the charging time, t_{charge} and by $\frac{E_b}{N_0}$. Since we want the maximum charging time for all the different modulations to be the same, and we assume that all the sensor nodes have the same amount of data L to transmit and the same transmitted symbol rate S_b, we have the relationship between the distances R_1 and R_2 as a function of $\frac{E_{b1}}{N_0}$ and $\frac{E_{b2}}{N_0}$ for two different modulations as:

$$\frac{R_1}{R_2} == \sqrt[4]{\frac{E_{b2}}{N_0} \bigg/ \frac{E_{b1}}{N_0}}. \tag{8}$$

Then the charging time is calculated as:

$$t_{charge} = \frac{N_0 \left(\frac{E_b}{N_0}\right) L}{P_{cw} \rho_t (G_{exciter} G_{node})^2} \left(\frac{4\pi R}{\lambda_{cw}}\right)^4. \tag{9}$$

Assume that we have n nodes distributed throughout the three regions (in Fig. 2) as follows: x nodes using QPSK, y nodes using 8-PSK, and z nodes using 16-PSK, where $n = x + y + z$, and the furthest node is at the distance of R_{max}. Assuming uniform geographic distribution of the nodes in the network, the number of nodes deployed in the different regions is:

$$
\begin{aligned}
x &= n \times \frac{A_{(QPSK)}}{A_{(QPSK)} + A_{(8-PSK)} + A_{(16-PSK)}} = \frac{R^2_{max(QPSK)} - R^2_{max(8-PSK)}}{R^2_{max(QPSK)}}, \\
y &= n \times \frac{A_{(8-PSK)}}{A_{(QPSK)} + A_{(8-PSK)} + A_{(16-PSK)}} = \frac{R^2_{max(8-PSK)} - R^2_{max(16-PSK)}}{R^2_{max(QPSK)}}, \\
z &= n \times \frac{A_{(16-PSK)}}{A_{(QPSK)} + A_{(8-PSK)} + A_{(16-PSK)}} = \frac{R^2_{max(16-PSK)}}{R^2_{max(QPSK)}}.
\end{aligned}
\tag{10}
$$

Based on Eq. (10), we can calculate the duration of the cycles in the multi-modulation and in the single-modulation schemes:

$$
\begin{aligned}
t_{total-mm} &= t_{query} + \sum t_{transmit} + t_{charge(max)} = t_{query} + \frac{L}{k_1 \times S_b} x + \frac{L}{k_2 \times S_b} y \\
&\quad + \frac{L}{k_3 \times S_b} z + t_{charge(Max(QPSK))} \\
t_{total-sm} &= t_{query} + \sum t_{transmit} + t_{charge(max)} = t_{query} + \frac{nL}{k_i \times S_b} \\
&\quad + t_{charge(Max(i-PSK))},
\end{aligned}
\tag{11}
$$

where k_i represents the number of bits-per-symbol in the various modulation schemes (i.e., QPSK, 8-PSK, and 16-PSK), and $t_{charge(Max(QSK))} < t_{charge(Max(8-PSK))} < t_{charge(Max(16-PSK))}$ are calculated by Eq. (9).

Thus, the reduction of the cycle duration time by our scheme compared with single-modulation scheme is:

$$
t_{total-sm} - t_{total-mm} = \frac{nL}{k_i \times S_b} - \frac{L}{k_1 \times S_b}x - \frac{L}{k_2 \times S_b}y - \frac{L}{k_3 \times S_b}z + t_{charge(Max(i-PSK))}
$$
$$
- t_{charge(Max(QPSK))}.
$$

$$(12)$$

We note that when we calculate the improvement of the network throughput, our multi-modulation scheme should be compared with the single-modulation scheme that performs the best among all the three single-modulation schemes which are used in our multi-modulation scheme. Hence, the throughput improvement, B_{multi}, is calculated by Eq. (13):

$$
B_{multi} = \frac{\left(\frac{1}{t_{total(Multi)}} - \frac{1}{MIN\left(t_{total(QPSK)}, t_{total(8-PSK)}, t_{total(16-PSK)} \right)} \right)}{\frac{1}{MIN\left(t_{total(QPSK)}, t_{total(8-PSK)}, t_{total(16-PSK)} \right)}}
$$

$$(13)$$

4 Multi-modulation Scheme with Multiple Exciters

It is likely that in a large deployment of an RSN, there will be more than one exciter present (e.g., multi-exciter network as in [25]). In this section, we assume that the exciters are uniformly deployed in the network (this is a reasonable assumption, given the fact that the sensor nodes are randomly, but uniformly, distributed too). We also assume that the antennas of the sensor nodes are unidirectional, so that the nodes are able to receive power from different directions at the same time. We consider the locations of the exciters at the centers of a hexagonal grid. Then the received power by a node depends on the distances between the node and the exciters (and is calculated by Eq. (2)), and the total received energy of a node is calculated by the sum of their received energy from the all the exciters. Because of the exponential dependence on distance in Eq. (2), in calculating the power collected by a sensor node, we need to consider only the closest several exciters to the node, and we can ignore the power received from the other further away exciters.

Consider an area covered by 7 exciters arranged in a hexagonal grid, as depicted in Fig. 3.

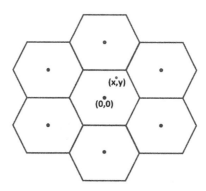

Fig. 3. Structure of a multi-exciter network

Assume that the middle exciter is at the origin, $(0,0)$, while the coordinates of the other six exciters are: $\left(0, \sqrt{3}R\right)$, $\left(\frac{3}{2}R, \frac{\sqrt{3}}{2}R\right)$, $\left(\frac{3}{2}, -\frac{\sqrt{3}}{2}R\right)$, $\left(0, -\sqrt{3R}\right)$, $\left(-\frac{3}{2}R, \frac{\sqrt{3}}{2}R\right)$, and $\left(-\frac{3}{2}R, -\frac{\sqrt{3}}{2}R\right)$. We label the coordinates of an exemplary node in the middle of the hexagonal grid as (x, y). The power received by the exemplary node in the central hexagon is:

$$P_{node}(x,y) = P_{cw} \left(\frac{\lambda_{cw}}{4\pi}\right)^2 G_{exciter} G_{node}$$

$$\times \left(\frac{1}{x^2 + y^2} + \frac{1}{x^2 + \left(y - \sqrt{3R}\right)^2} + \frac{1}{x^2 + \left(y + \sqrt{3R}\right)^2} + \frac{1}{\left(x - \frac{3}{2}R\right)^2 + \left(y - \frac{\sqrt{3}}{2}R\right)^2} + \frac{1}{\left(x - \frac{3}{2}R\right)^2 + \left(y + \frac{\sqrt{3}}{2}R\right)^2} \right.$$

$$\left. + \frac{1}{\left(x + \frac{3}{2}R\right)^2 + \left(y - \frac{\sqrt{3}}{2}R\right)^2} + \frac{1}{\left(x + \frac{3}{2}R\right)^2 + \left(y + \frac{\sqrt{3}}{2}R\right)^2} \right) \tag{14}$$

The level of the received power (displayed only for the central hexagon) as a function of the node's location, (x, y), is shown in Fig. 4.

Next, we study the behavior of Eq. (14) for different locations of the sensor node, (x, y). More specifically, we show that $P_{node}(x, y)$ is a strictly increasing function as we move the location of the node from a boundary of the hexagon to its vertex.

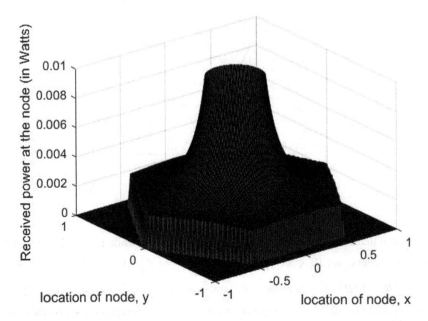

Fig. 4. The received power plot as a function of the sensor node's location in the central hexagon (the plot is truncated at received power of 0.01 W)

Using the polar coordinate system, where $x = r \cos \theta, y = r \cdot \sin \theta$; r is the distance from node to the origin, and θ is the angle of node location relative to the $x-$ axis, Eq. (14) can be rewritten as:

$$
\begin{aligned}
P_{node}(r, \theta) \\
= P_{cw} \left(\frac{\lambda_{cw}}{4\pi} \right)^2 G_{exciter} G_{node} &\times \left(\frac{1}{r^2} + \frac{1}{r^2 + 3R^2 - 2\sqrt{3}rR \sin \theta} + \frac{1}{r^2 + 3R^2 + 2\sqrt{3}rR \sin \theta} \right. \\
&+ \frac{1}{r^2 + 3R^2 - 3rR \cos \theta - \sqrt{3}rR \sin \theta} + \frac{1}{r^2 + 3R^2 - 3rR \cos \theta + \sqrt{3}rR \sin \theta} \\
&\left. + \frac{1}{r^2 + 3R^2 + 3rR \cos \theta - \sqrt{3}rR \sin \theta} + \frac{1}{r^2 + 3R^2 + 3rR \cos \theta + \sqrt{3}rR \sin \theta} \right)
\end{aligned}
\tag{15}
$$

Now, taking a derivative of $P_{node}(r, \theta)$ with respect to r, one can easily verify that $\partial P_{node}(r, \theta) / \partial r < 0$. Hence, the received power gain in any direction is a monotonically increasing function from the boundary to the vertex of a hexagon. Furthermore, as can be demonstrated by Eq. (15) (and is also evident in Fig. 5), there exists r_0 for which the received power at a node is relatively constant for $r_0 \leq r \leq 1$. This indicates that the sensor nodes in the area $r_0 \leq r \leq 1$ could be assigned the same modulation scheme. Combining Eq. (15) with Eqs. (3), (4), and (6), we obtain the region partition formula for any direction θ from the center of a hexagon to its edge:

$$
\frac{\left(\frac{E_{b1}}{N_0}\right)}{\left(\frac{E_{b2}}{N_0}\right)} = \frac{\left(r_2^2\left(\frac{1}{r_1^2} + \frac{1}{r_1^2+3R^2-2\sqrt{3}r_1R\sin\theta} + \frac{1}{r_1^2+3R^2+2\sqrt{3}r_1R\sin\theta} + \frac{1}{r_1^2+3R^2-3r_1R\cos\theta-\sqrt{3}r_1R\sin\theta} + \frac{1}{r_1^2+3R^2-3r_1R\cos\theta+\sqrt{3}r_1R\sin\theta} + \frac{1}{r_1^2+3R^2+3r_1R\cos\theta-\sqrt{3}r_1R\sin\theta} + \frac{1}{r_1^2+3R^2+3r_1R\cos\theta+\sqrt{3}r_1R\sin\theta}\right)\right)}{\left(r_1^2\left(\frac{1}{r_2^2} + \frac{1}{r_2^2+3R^2-2\sqrt{3}r_2R\sin\theta} + \frac{1}{r_2^2+3R^2+2\sqrt{3}r_2R\sin\theta} + \frac{1}{r_2^2+3R^2-3r_2R\cos\theta-\sqrt{3}r_2R\sin\theta} + \frac{1}{r_2^2+3R^2-3r_2R\cos\theta+\sqrt{3}r_2R\sin\theta} + \frac{1}{r_2^2+3R^2+3r_2R\cos\theta-\sqrt{3}r_2R\sin\theta} + \frac{1}{r_2^2+3R^2+3r_2R\cos\theta+\sqrt{3}r_2R\sin\theta}\right)\right)},
$$

$$(16)$$

where r_1 is the furthest exciter-to-node distance for modulation scheme 1, and r_2 is the furthest exciter-to-node distance for modulation scheme 2. For example, using QPSK modulation as modulation scheme 1, $r_1 = r_{\max(QPSK)}$. Let $u_1 = \frac{R}{r_1}$, $u_2 = \frac{R}{r_2}$, Eq. (16) is now changed to:

$$
\frac{\left(\frac{E_{b1}}{N_0}\right)}{\left(\frac{E_{b2}}{N_0}\right)} = \frac{\left(u_1^4\left(1 + \frac{1}{1+3u_1^2-2\sqrt{3}u_1\sin\theta} + \frac{1}{1+3u_1^2+2\sqrt{3}u_1\sin\theta} + \frac{1}{1+3u_1^2-3u_1\cos\theta-\sqrt{3}u_1\sin\theta} + \frac{1}{1+3u_1^2-3u_1\cos\theta+\sqrt{3}u_1\sin\theta} + \frac{1}{1+3u_1^2+3u_1\cos\theta-\sqrt{3}u_1\sin\theta} + \frac{1}{1+3u_1^2+3u_1\cos\theta+\sqrt{3}u_1\sin\theta}\right)\right)}{\left(u_2^4\left(1 + \frac{1}{1+3u_2^2-2\sqrt{3}u_2\sin\theta} + \frac{1}{1+3u_2^2+2\sqrt{3}u_2\sin\theta} + \frac{1}{1+3u_2^2-3u_1\cos\theta-\sqrt{3}u_2\sin\theta} + \frac{1}{1+3u_2^2-3u_2\cos\theta+\sqrt{3}u_2\sin\theta} + \frac{1}{1+3u_2^2+3u_2\cos\theta-\sqrt{3}u_2\sin\theta} + \frac{1}{1+3u_2^2+3u_2\cos\theta+\sqrt{3}u_2\sin\theta}\right)\right)},
$$

$$(17)$$

As represented in Eq. (17), since we use the ratio of r_1 to R, or r_2 to R instead of an actual value of R, the region partition principle is not dependent on the value of R. Thus, in our example of a multi-exciter network, the region partition for the three modulations is shown in Fig. 5. For each direction θ, the relation between $r_{\max(QPSK)}$, $r_{\max(8-PSK)}$, and $r_{\max(16-PSK)}$, should follow Eq. (16) or eq. (17).

In Fig. 5, there are three rings: the black (outer) ring represents the boundary of QPSK; the red (middle) ring represents the boundary for 8-PSK; and the blue (inner) ring represents the boundary for 16-PSK. The area between hexagon boundary (green lines) and the middle ring is the region where QPSK is used by the nodes in this hexagonal area; the area between the middle ring and the inner ring is the region where 8-PSK is used by the nodes in this hexagon; the area within the inner ring is the region where 16-PSK is used by the nodes in this hexagon. The area between the outer ring and hexagon boundary will apply QPSK as well; however, since this area belongs partially to other hexagon regions, some nodes in this area are assigned to the other exciters based on the region partition by the hexagonal grid.

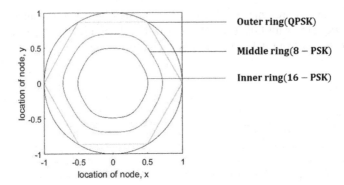

Fig. 5. Partition of the hexagon area by three modulation type (QPSK, 8-PSK, and 16-PSK) in a multi-exciter network

If we have a sensor node using modulation scheme 1 where $u_1 = \frac{R}{r_1}$, we can calculate the charging time for this node in one cycle by Eq. (18) which is extension of Eq. (9) and by Eq. (17) which presents the charging time of sensor nodes in multi-exciter scheme (proof omitted):

$t_{charge-me}$

$$
= \frac{\frac{N_0 \left(\frac{E_{b1}}{N_0}\right) L}{P_{cw} \rho_t (G_{exciter} G_{node})^2} \left(\frac{4\pi}{\lambda_{cw}}\right)^4 \cdot r_1^4}{\left(\begin{array}{c} 1 + \frac{1}{1+3u_1^2-2\sqrt{3}u_1 \sin\theta} + \frac{1}{1+3u_1^2+2\sqrt{3}u_1 \sin\theta} + \frac{1}{1+3u_1^2-3u_1 \cos\theta-\sqrt{3}u_1 \sin\theta} + \\ \frac{1}{1+3u_1^2-3u_1 \cos\theta+\sqrt{3}u_1 \sin\theta} + \frac{1}{1+3u_1^2+3u_1 \cos\theta-\sqrt{3}u_1 \sin\theta} + \frac{1}{1+3u_1^2+3u_1 \cos\theta+\sqrt{3}u_1 \sin\theta} \end{array}\right)}
$$

$$(18)$$

5 Numerical Results

In IEEE 802.15.4 standard [26], the UHF communication frequencies are specified as 868 [MHz], 915 [MHz], and 2.45 [GHz]. Those frequencies could be applied in RFID systems [27]. In our scheme, we also assume a fixed symbol rate of 200 kHz as per [11], which allows at most 32-PSK modulation in the IEEE 802.15.4 standard.

Now we will calculate the numerical result of R_{max} in a single exciter network for QPSK, 8-PSK and 16-PSK, using Tables 1-2 and equations in Sect. 3.

We assume the following parameter values: BER threshold of 10^{-6} [28], $\gamma_{b(QPSK)} = 10.779$ dB, $\gamma_{b(8-PSK)} = 14.205$ dB, $\gamma_{b(16-PSK)} = 18.7$ dB (using Table 2), $P_{cw} = 1$ W, $G_{exciter} = 6$ dBi, $G_{node} = 2$ dBi, $\lambda_{cw} = 0.12245$ m(for frequency of 2.45 GHz), and $N_0 = -140$ dbm/Hz [29, 30]. We further assume that L = 16 kb [31] as the message size to transmit, $S_b = 200$ kHz [11], and $\rho_t = 0.6667$ [24, 32]. We use three levels of PSK modulations: QPSK, 8-PSK, and 16-PSK, so that $t_{transmit} = \frac{L}{k \times S_b}$, $t_{transmit(k=2)} = 0.04$ s, $t_{transmit(k=3)} = 0.00268$ s, and $t_{transmit(k=4)} = 0.02$ s. $R_{cover} = 30.5$ m [23].

The charging time is then calculated by Eq. (9). If, as an example, $R_{\max(QPSK)} = 20$ m, then $t_{charge} = 1.2798$ s..

For $k = 3$, the 8-PSK boundary, which is the middle ring, by eq. (8):

$$R_{\max(8-PSK)} = 20 \times \sqrt[4]{\frac{10^{1.0778}}{10^{1.4205}}} = 16.422 \text{ m}$$

For $k = 4$, the 16-PSK boundary, which is the inner ring,

$$R_{\max(16-PSK)} = 20 \times \sqrt[4]{\frac{10^{1.0778}}{10^{1.87}}} = 12.6774 \text{ m.}$$

Next, we use these $R_{\max}$ values to implement the region partition of our multi-modulation scheme. As a specific example, we again assume that 100 sensor nodes are distributed within the exciter's coverage area following a uniform distribution. Thus, the percentage of nodes with different modulations is proportional to the area of these three regions (as per eq. (10)). For the first region with 16-PSK modulation, the area is $A_{(16-PSK)} = \pi R_{\max(16-PSK)}^2$; for the second region with 8-PSK modulation, the area is $A_{(8-PSK)} = \pi R_{\max(8-PSK)}^2 - \pi R_{\max(16-PSK)}^2$; for the last region with QPSK modulation, the area is $A_{(QPSK)} = \pi R_{\max(QPSK)}^2 - \pi R_{\max(8-PSK)}^2$. The coverage area is $R_{cover} = 30.5$ m, and $R_{\max(QPSK)} = 20$ m, which is the furthest exciter-to-node distance. Thus, using the region partition equations that we obtained above, the 100 nodes are distributed in the range of $0 - 20$ m from the central point which represents the location of the exciter. By Eq. (10), we have 33 nodes using QPSK, 27 nodes using 8-PSK, and 40 nodes using 16-PSK.

Next, we can compare the performance of the proposed multi-modulation scheme (with QPSK, 8-PSK, and 16-PSK in the different regions and with the distribution of sensor nodes following the region partition above) with the three cases of the single-modulation scheme of either QPSK, 8-PSK, or 16-PSK. We assume that at the beginning of the first cycle, all the nodes have no energy stored. The charging time is then calculated by Eq. (9). Assuming $t_{query} = 0.01$ s, the results of the comparison are shown in Table 3.

Table 3. Comparison of cycle times for the three modulation schemes

	Query	Charging time	Transmission Time	Cycle Time
Multi modulation (QPSK, 8-PSK, 16-PSK)	0.01	1.2798	2.8436	4.1334
Single modulation (QPSK)	0.01	1.2798	4	5.2898
Single modulation (8-PSK)	0.01	2.7865	2.68	5.4765
Single modulation (16-PSK)	0.01	7.8445	2	9.8545

Based on the comparison in Table 3, in this example, the proposed multi-modulation scheme results in significantly shorter cycle time relative to all the three single-modulation schemes of either QPSK, 8-PSK, or 16-PSK. If we label the total transmission time in a cycle as t_{total}, the throughput improvement is calculated by Eq. (13). Comparing with the most efficient single-modulation QPSK in Table 3, we gain about 28% in throughput improvement.

In evaluating the performance of the different schemes, we need to take into consideration the maximal distance of a sensor node to the exciter, which will change in different networks, which in the case of M-PSK is $R_{max(QPSK)}$. In the Fig. 6, we show the total cycle time to maximal node-to-exciter distance diagram for multi-modulation and single-modulation schemes of M-PSK.

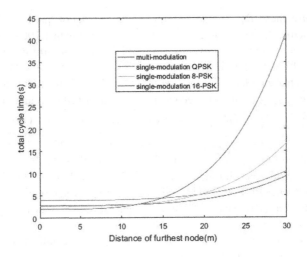

Fig. 6. Comparison of cycle times for different schemes

By Eq. (9) and Eq. (12), the improvement of the cycle time follows equation with only one variable R. As can be seen from Fig. 6, the "best" choice of the scheme depends on the maximal node-to-exciter distance. For instance, for small maximal node-to-exciter distances, the single modulation 16-PSK results in the shortest cycle, while for large maximal node-to-exciter distances, the multi-modulation scheme leads to the shortest cycle time.

To show a more comprehensive performance evaluation of our multi-modulation scheme, we also compare the following three multi-modulation schemes with the PSK multi-modulation scheme of a single-exciter network:

1. Multi-modulation scheme for a <u>multi-exciter</u> network applying M-PSK modulations: In multi-exciter networks, the charging time is calculated by (18), and the region partition follows Fig. 5.

2. Multi-modulation scheme for a single-exciter network applying M-QAM modulations: QAM multi-modulation scheme with M equal to 4, 8, 16.[5]
3. Multi-modulation scheme for a single-exciter network applying M-PAM modulations: PAM multi-modulation scheme with M equal to 2, 4, 8.[6]

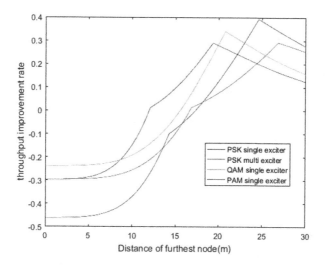

Fig. 7. Comparison of throughput improvement of the different multi-modulation schemes vs. the "best" single-modulation scheme

Next, we evaluate the throughput of all these three multi-modulation schemes, together with the M-PSK scheme with a single-exciter. The throughput improvement is calculated by Eq. (13), following the results in Fig. 6; for any maximal node-to-exciter distance, we should compare the throughput with the "best" (i.e., shortest cycle time) scheme. In Fig. 7, we show the comparison of the throughput improvement or degradation (based on Eq. (13)) for the above three schemes, together with the previously discussed PSK single exciter case, relative to the performance of the multi-modulation schemes. With the exceptions outlined below, most of the parameters remain the same as in Table 3[7]. As can be seen from Fig. 7, in our example, the multi-modulation scheme

---

[5] The function of BER vs $\frac{E_b}{N_0}$ for the QAM and the PAM schemes are evaluated using the Matlab BER tool.

[6] For M-PSK scheme, BPSK and QPSK have the same $\frac{E_b}{N_0}$ at $BER = 10^{-6}$, so we cannot use BPSK, together with QPSK. For M-QAM scheme, 2-QAM cannot be found in the Matlab BER tool, so we use 4-QAM as the lowest order modulation.

[7] Based on Eq. (10), since the $\frac{E_b}{N_0}$ obtained from the Maltab BER tool for the QAM and the PAM schemes are different from PSK, the region partitions for QAM and PAM schemes are different from that of the PSK scheme. As a result, the number of nodes for different regions are changed as well. In addition, since we apply 2-PAM, 4-PAM, and 8-PAM in the PAM scheme, the transmission time for sensor nodes is changed to: 0.08 s, 0.04 s, and 0.0268 s, respectively

results in throughput improvement for networks with the maximal node-to-exciter distance greater than about 17 m. The improvement depends on the maximal node-to-exciter distance, but is most pronounced for larger networks. (The singular points in Fig. 7 are due to the fact that at these points the choice of the "best" scheme changes. For example, in Fig. 6, we observe that the choice of the "best" M-PSK scheme changes at a different maximal exciter-to-exciter distance; those "best" choices will now be used in Eq. (13) for comparison with the multi-modulation scheme. In Fig. 7, the throughput of each curve is calculated at each maximal node-to-exciter distance for the "best" single-modulation scheme of the particular modulation family.)

In conclusion, the throughput improvement by our multi-modulation scheme is significant, as shown in Fig. 7 (e.g., up to nearly 40% improvement for the PAM scheme). However, the improvement depends on the maximal exciter-to-node distance, and for different schemes the throughput improvement curves are different. To maximize the benefits of the multi-modulation scheme, a network designer should choose the best multi-modulation scheme according to the furthest exciter-to-node distance and the throughput improvement curves in Fig. 7.

6 Conclusion and Future Work

In this paper, we propose a multi-modulation scheme to optimize the throughput of RFID Sensor Networks by minimizing the single cycle time of the network. Our scheme is applied to WSIP sensor nodes whose circuits harvest energy for communication and are able to implement different modulation methods. By applying different modulation methods in the nodes according to their distance to the exciter, our multi-modulation scheme minimizes the cycle time by saving the charging time of nodes located further from the exciter or by saving on the transmission time of nodes closer to the exciter, as compared with traditional single-modulation schemes. Our scheme could be applied to both the single-exciter network and multi-exciter network. We demonstrated an improvement of up to 40% of throughput according to our evaluation results. Our proposed scheme is in particular applicable to large scale/ dense RFID Sensor Networks.

References

1. Sample, A., Buettner, M., Greenstein, B.: Revisiting smart dust with RFID sensor networks. In: Proceedings of the 7th ACM Workshop on Hot Topics in Networks (HotNets-VII) (2008)
2. Buettner, M., et al.: RFID sensor networks with the Intel WISP. In: Proceedings of the 6th ACM conference on Embedded network sensor systems (2008)
3. Passive radio frequency tags and sensors for process monitoring in advanced reactors. https://www.energy.gov/sites/prod/files/2019/04/f61/Dirac%20FY2019–1%20Summary-Abstract%20ARD-19-1-16359.pdf. Accessed 28 Oct 2020
4. Kalansuriya, P., Bhattacharyya, R., Sarma, S.: A novel communication method for semi-passive RFID based sensors. In: 2014 IEEE International Conference on Communications (ICC). IEEE (2014)

5. Jia, X., Feng, Q., Fan, T., Lei, Q.: RFID technology and its applications in Internet of Things (IoT). In: 2012 2nd International Conference on Consumer Electronics, Communications and networks (CECNet), pp. 1282–1285. IEEE (2012)
6. Kim, T.H., Ramos, C., Mohammed, S.: Smart city and IoT, pp. 159–162 (2017)
7. Ahson, S.A., Ilyas, M.: RFID handbook: applications, technology, security, and privacy. CRC press, Boca Raton (2017)
8. Currie, I.A., Marina, M.K.: Experimental evaluation of read performance for RFID-based mobile sensor data gathering applications. In: Proceedings of the 7th International Conference on Mobile and Ubiquitous Multimedia, pp. 92–95 (2008)
9. Alhassoun, M., Varner, M. A., Durgin, G.D.: Design and evaluation of a multi-modulation retrodirective RFID tag. In: 2018 IEEE International Conference on RFID (RFID), pp. 1–8. IEEE (2018)
10. Wilas, J., Jirasereeamornkul, K., Kumhom, P.: Power harvester design for semi-passive UHF RFID tag using a tunable impedance transformation. In: 2009 9th International Symposium on Communications and Information Technology, pp. 1441–1445. IEEE (2009)
11. Thomas, S.J., Wheeler, E., Teizer, J., Reynolds, M.S.: Quadrature amplitude modulated backscatter in passive and semipassive UHF RFID systems. IEEE Trans. Microwave Theor. Tech. 60(4), 1175–1182 (2012)
12. Thomas, S.J., Reynolds, M. S.: A 96 Mbit/sec, 15.5 pJ/bit 16-QAM modulator for UHF backscatter communication. In: 2012 IEEE International Conference on RFID (RFID), pp. 185–190. IEEE (2012)
13. d'Errico, R., et al.: An UWB-UHF semi-passive RFID system for localization and tracking applications. In: 2012 IEEE International Conference on RFID-Technologies and Applications (RFID-TA), pp. 18–23. IEEE (2012)
14. Athalye, A., Savic, V., Bolic, M., Djuric, P.M.: Novel semi-passive RFID system for indoor localization. IEEE Sens. J. 13(2), 528–537 (2012)
15. Zhen, B., Mizuno, K., Kobayashi, M., Shimizu, M.: Pulse position modulation for active RFID system. In: 11th International Conference on Parallel and Distributed Systems (ICPADS'05), vol. 2, pp. 58–62. IEEE (2005)
16. Decarli, N., Guidi, F., Conti, A., Dardari, D.: Interference and clock drift effects in UWB RFID systems using backscatter modulation. In: 2012 IEEE International Conference on Ultra-Wideband, pp. 546–550. IEEE (2012)
17. Ramos, A., Lazaro, A., Girbau, D.: Semi-passive time-domain UWB RFID system. IEEE Trans. Microwave Theor. Tech. 61(4), 1700–1708 (2013)
18. Hannan, M.A., Islam, M., Samad, S.A., Hussain, A.: RFID communication using software defined radio technique. In: AIP Conference Proceedings, vol. 1247, no. 1, pp. 41–55. American Institute of Physics (2010)
19. Proloy, M., Hoq, R., Ahmed, S.M.: Comparative analysis of various wireless digital modulation techniques with different channel coding schemes under AWGN channel. Doctoral dissertation, East West University (2016)
20. Meghdadi, V.: BER calculation. Wirel. Commun. (2008)
21. Lee, M.H., Yao, C.Y., Liu, H.C.: Passive tag for multi-carrier RFID systems. In: 2011 IEEE 17th International Conference on Parallel and Distributed Systems, pp. 872–876. IEEE (2011)
22. Wireless network lesson on modulation methods. https://www.usna.edu/ECE/ec312/Lessons/wireless/EC312_Lesson_24_Sup.pdf. Accessed 31 Oct 2019
23. How RFID works on howstuffworks. https://electronics.howstuffworks.com/gadgets/high-tech-gadgets/rfid.htm#pt3. Accessed 13 Aug 2020
24. Modulation, Power and bandwidth introduction. https://www.aticourses.com/sampler/Modulation_Power_Bandwidth.pdf. Accessed 13 Aug 2020

25. Zhou, Z., Gupta, H., Das, S. R., Zhu, X.: Slotted scheduled tag access in multi-reader RFID systems. In: 2007 IEEE International Conference on Network Protocols, pp. 61–70. IEEE (2007)
26. Seetharam, D., Fletcher, R.: Battery-powered RFID. In: 1st ACM Workshop on Convergence of RFID and Wireless Sensor Networks and their Applications (2007)
27. Understanding passive RFID technology. https://rfidworld.ca/understanding-passive-rfid-radio-frequency-identification-technology/1294. Accessed 13 Aug 2020
28. Zhang, Y., Li, X., Amin, M.: Principles and techniques of RFID positioning. In: Rfid Systems, pp. 389 (2010)
29. RFID antenna gain and range. https://www.eetimes.com/document.asp?doc_id=1276310. Accessed 13 Aug 2020
30. Pursula, P., Kiviranta, M., Seppa, H.: UHF RFID reader with reflected power canceller. IEEE Microwave Wirel. Compon. Lett. **19**(1), 48–50 (2008)
31. Thomas, S., Reynolds, M. S.: QAM backscatter for passive UHF RFID tags. In: 2010 IEEE International Conference on RFID (IEEE RFID 2010), pp. 210–214. IEEE (2010)
32. Rembold, B.: Optimum modulation efficiency and sideband backscatter power response of RFID-tags. Frequenz **63**(1), 9 (2009)

Beam-Based Secure Physical Layer Key Generation for mmWave Massive MIMO System

Hao Gao, Yanling Huang, and Danpu Liu[(✉)]

Beijing Laboratory of Advanced Information Networks,
Beijing Key Laboratory of Network System Architecture and Convergence,
Beijing University of Posts and Telecommunications, Beijing
People's Republic of China
dpliu@bupt.edu.cn

Abstract. Massive MIMO system greatly enriches the randomness of the secret keys in the physical layer and increases the rate of key generation. However, it is not practical to obtain full channel state information for key generation in actual communication scenarios due to a large number of additional signaling overhead. In this paper, we proposed a feasible physical layer key generation scheme by using the beam information as a random source. The procedure for key generation is designed based on the current beam management mechanism in 5G NR. Therefore, the secret key is synchronously generated in the process of two-stage beam search between the gNB and the UE before data transmission, and the additional signaling overhead for key generation is little. Furthermore, to cope with the non-uniform distributed characteristics of the beams, we adopt Huffman code in the encoding of the beam index, thereby improving the efficiency of the key generation. Simulation results show that the proposed scheme can achieve mutual information per bit as high as 0.97, which is 2% to 3% better than that of equal length coding. Furthermore, the bit disagreement rate can be less than 1% in a harsh communication environment with a signal-to-noise ratio of -10 dB.

Keywords: Physical layer security · Beam management · MIMO · Secret key generation · Huffman coding

1 Introduction

The Fifth Generation mobile communication (5G) is based on heterogeneous networking and new wireless technologies, providing support for access to massive devices, diverse wireless services, and rapid data traffic growth. Due to the extremely high data transmission rate of 5G mobile communication, people will use it to transmit a large amount of key information, including some private information. If the transmission process is stolen, it is likely to bring hidden dangers to the user security. Therefore, the development of 5G puts forward higher requirements for the performance of wireless communication security such as reliable transmission and privacy protection.

© ICST Institute for Computer Sciences, Social Informatics and Telecommunications Engineering 2021
Published by Springer Nature Switzerland AG 2021. All Rights Reserved
H. Gao et al. (Eds.): ChinaCom 2020, LNICST 352, pp. 37–51, 2021.
https://doi.org/10.1007/978-3-030-67720-6_3

Generating keys based on wireless channel characteristics is a method to ensure information security in the physical layer. During the key generation process, legitimate users can measure dynamic channel parameter information separately, which can effectively avoid the exchange of keys between legitimate users, thereby improving key security. Moreover, the spatial variability and time variability of wireless channels can increase the randomness of secret keys. Owing to low complexity and ease of operation, the key-based physical layer security technology has become a research hotspot for physical layer security.

Many kinds of channel characteristics can be used to generate the secret key, including received signal strength (RSS) [1], channel impulse response (CIR) [2]/ channel state information (CSI) [3], and phases [4]. We note the work in [5], which uses multi-level quantization of MIMO channel measurements to generate secret key bits, as well as [6, 7], which quantize noisy channel measurements at transmitter and receiver to generate secret key bits in conjunction with Slepian-Wolf coding. Other well-known references that study the use of CSI to generate encryption keys are [8–12], which also include information-theoretical analysis of the proposed method.

In recent years, researches on physical layer key generation for large-scale antenna systems have been initially carried out. In [13], for the millimeter-wave(mmWave) Massive MIMO communication system, Long Jiao, Jie Tang et al. proposed to use two new channel characteristics that can reflect the sparsity of the mmWave channel, namely the virtual angle of arrival (AoA) and the angle of departure (AoD) as a random source to generate keys. Besides, the work in [14] proposed a method that uses AoA with a random perturbation angle as a random source for feature extraction. This method has a high key generation rate and key consistency, and to a certain extent can prevent co-eavesdropping. However, the above methods need to work under the condition that the high-dimension full CSI including AoA/AoD is known, and is not compatible with 5G New Radio (NR) protocol where only the equivalent baseband channel with low-dimension is estimated.

In summary, the physical layer secret key generation in the massive MIMO communication system faces the following new challenges due to its unique channel characteristics and the huge number of antennas: 1) The existing key generation algorithms need to obtain full Channel State Information (CSI). High-dimensional channel estimation is challenging as the number of antennas increases. 2) The existing algorithms are not compatible with the two-stage synchronization process in the 5G NR protocol. If designing a key generation mechanism that is independent of the synchronization process, the additional signaling overhead will be introduced.

To address the above issues, we proposed a key generation scheme based on the existing 5G NR protocol framework in this paper. The beam representing the spatial characteristics of the mmWave channel is used to generate secure keys. Choosing the beam instead of AoA/AoD as a random source has the following advantages: 1) The beam information can be obtained by beam sweeping and does not rely on the acquisition of full CSI. 2) The quantized beam index can be directly used as the generated secret keys, thereby avoiding errors caused by the quantization process. Furthermore, the proposed scheme combines the key generation and the two-stage beam search procedure defined in 5G NR protocol, so that the legitimate users simultaneously complete the generation of the physical layer secret key during the existing initial access and beam refinement process, and little extra signaling overhead is required. Finally, considering that the actual beam direction distribution is not uniform, we use Huffman code when encoding the beam index, so that the number of generated secret key bits is as close as possible to the theoretical information entropy, achieving higher coding efficiency than equal length coding. The simulation results show that the mutual information per bit reaches above 0.97 and the key agreement rate reached 99% for the harsh environment with a signal-to-noise ratio (SNR) of -10 dB.

Notation: $\|\mathbf{a}\|_2$ denotes the l_2 norm of vector a. $(\cdot)^H$ means the conjugate transpose of the matrix. $\mathcal{CN}(0, \sigma^2)$ denotes a complex Gaussian distribution with zero mean and variance σ^2. $\cup$ denotes cascade operation.

2 System Model

This paper focuses on a downlink single-user MIMO system. As shown in Fig. 1, Alice and Bob are legitimate users in communication, and Eve is a potential eavesdropper.

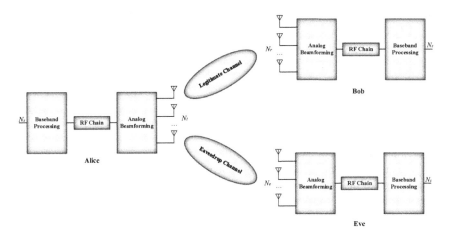

Fig. 1. Key generation system model in massive MIMO.

Alice and Bob can also represent generation node base station (gNB) and user equipment (UE) in the 5G NR protocol. The Alice is mainly composed of a baseband processing unit, N_{RF} Radio Frequency (RF) links, and an analog precoding unit; the Bob or Eve has a similar structure. Assume that N_t antennas deployed on the Alice and N_r antennas deployed on the Bob are connected to one RF link in a fully connected manner, respectively. N_e antennas are deployed on Eve in the same way. The system supports single-stream transmission during communication, which means $N_s = 1$. Generally, the hybrid beamforming system model is as follows:

$$y = \mathbf{w}^H \mathbf{H} \mathbf{f} s + \mathbf{w}^H \mathbf{n} \tag{1}$$

where s is the pilot sequence of symbol, given that $|s|^2 = P_t$, and P_t denotes the transmit power of all antennas. $\mathbf{n} \in \mathbb{C}^{N_r \times N_t}$ is the noise matrix with independent and identically distributed components $\sim \mathcal{CN}(0, \sigma^2)$. Define that $\mathbf{f} \in \mathbb{C}^{N_t \times 1}$ and $\mathbf{w} \in \mathbb{C}^{N_r \times 1}$ are beamforming vectors for analog beamforming (ABF). $\mathbf{H}$ is a wideband geometric mmWave channel that can be modeled as an Extended Saleh-Valenzuela (ESV) channel, and its expression in the time domain is given by

$$\mathbf{h}[d] = \sqrt{\frac{N_t N_r}{\rho}} \sum_{c=1}^{C} \sum_{r_l}^{R_l} \alpha_{r_l} p_{rc}(dT_s - \tau_c - \tau_{r_l}) \boldsymbol{\alpha}_R(\theta_c - \vartheta_{r_l}) \boldsymbol{\alpha}_T^H (\phi_c - \phi_{r_l}) \tag{2}$$

where $\mathbf{h}[d]$ denotes the MIMO channel response when the delay is d. Define that ρ is the path loss, and C denotes the number of clusters. Let the variables $\theta_c, \phi_c \in (0, 2\pi)$ be the physical AoA and AoD respectively. There are R_l paths in each cluster and each path has a relative delay τ_{r_l} and an AOA/AOD offset $\vartheta_{r_l}, \phi_{r_l}$. The variable α_{r_l} denotes the path gain and p_{rc} represents the pulse shaping function corresponding to a sampling interval of T_s at τ second. The vector $\boldsymbol{\alpha}_R \in \mathbb{C}^{N_r \times 1}$ and $\boldsymbol{\alpha}_T \in \mathbb{C}^{N_t \times 1}$ as follows are the antenna array response vectors at Bob and Alice respectively, where uniform linear arrays (ULA) are used.

$$\boldsymbol{a}_R(\theta) = \frac{1}{\sqrt{N_r}} \left[1, e^{j\frac{2\pi}{\lambda}d_s \sin(\theta)}, \ldots, e^{j(N_r-1)\frac{2\pi}{\lambda}d_s \sin(\theta)} \right]^T \tag{3}$$

$$\boldsymbol{a}_T(\emptyset) = \frac{1}{\sqrt{N_t}} \left[1, e^{j\frac{2\pi}{\lambda}d_s \sin(\phi)}, \ldots, e^{j(N_t-1)\frac{2\pi}{\lambda}d_s \sin(\phi)} \right]^T \tag{4}$$

λ denotes the wavelength and d_s denotes the distance between the antenna elements, which is generally taken as half the wavelength.

The design of the beamforming vectors is usually implemented via codebook-based beam sweeping. Each codeword in the codebook corresponds to a beam index and beam direction. The simplest method for legitimate users is to traverse the predetermined beamforming codewords set, and select the best codeword pair that can maximize spectral efficiency to construct beamforming vectors $\mathbf{f}$ and $\mathbf{w}$. The set of beamforming codewords used in this paper is the DFT codebook, where the weighting coefficient of the n-th antenna in the m-th codeword can be expressed as

$$Q_{m,n} = e^{\frac{j2\pi mn}{M}}, m = 0, 1, \cdots, M - 1; n = 0, 1, \cdots, N - 1 \tag{5}$$

where m denotes the number of codewords, and n denotes the number of antennas.

3 Beam-Based Secret Key Generation

In this section, we proposed a key generation scheme for the mmWave massive MIMO system, which will be compatible with the existing NR protocol.

Massive MIMO improves the spectral efficiency of the system while also providing beams as new channel characteristics in spatial dimensions. Unlike the other channel characteristics such as RSS, CIR, and phase, the beam information is discrete and quantified so the error caused by quantization can be avoided in the secret key generation process. Besides, the estimation of the channel matrix will be more challenging as the number of antennas increases, but high-dimensional channel estimation is not required in our scheme because beam information can be obtained through a two-stage synchronization process in 5G NR protocol. Therefore the proposed scheme focuses on beam information and is compatible with the existing beam management framework, devoting to a feasible key generation solution.

3.1 Beam Management in 5G NR

The 5G NR synchronization procedure is based on the two-stage beam management operations to reduce the complexity of beam sweeping:

- **Initial access:** A pair of coarse beams for uplink and downlink communication will be initially established between UE and gNB after coarse beam sweeping. But the beam gain at this time is not high due to its wide width. Therefore, further beam management is required to determine a finer beam pair, thereby obtaining a higher communication rate.
- **Beam refinement:** The best pair of fine beams for communication will be determined in this stage. At first, subdivide the coarse beam determined in the previous stage into several fine beams. Subsequently, UE and gNB perform a fine beam search to ensure accurate beam alignment. Finally, a pair of fine beams having better ABF gain for communication will be determined.

The above two stages constitute beam management. In each stage, beam management is based on four different operations: beam sweeping, beam measurement, beam determination, and beam reporting [15].

The current key generation scheme [13, 14] is independent of the 5G NR beam management framework, which will undoubtedly increase additional signaling overhead. To ensure that the synchronization process and the key generation process are completed at the same time, we redesigned and expanded the existing beam management framework as follows.

3.2 The Beam-Based Secret Keys Generation Procedure

The specific implementation steps can be divided into an initial access stage, a beam refinement stage, and a coding stage, as shown in Fig. 2.

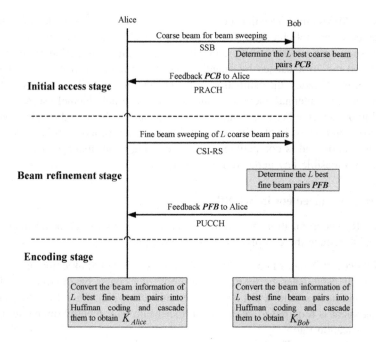

Fig. 2. Beam-based key generation procedure.

- **Initial access stage:** Alice and Bob perform coarse beam sweeping to determine the best coarse beam in this stage. As Bob is not connected to the system, Alice traverses the codewords, using coarse beams to periodically send the reference signal Synchronization Signal Block (SSB), and Bob uses coarse beams to receive and determine the L best coarse beam pairs that maximize the Reference Signal Receiving Power (RSRP), given by

$$\text{RSRP} = \|\mathbf{y}\|_2 \tag{6}$$

Define that $PCB = [TCB_1, RCB_1, TCB_2, RCB_2, \ldots, TCB_L, RCB_L]$ is the L pairs of coarse beam information, where TCB_i and RCB_i are the best transmit and receive coarse beam respectively. One of the best pairs of the coarse beam is used for Bob to access the system, and the remaining $L-1$ pairs of coarse beam index are recorded and used for increasing secret key rate. Subsequently, Bob accesses the system through a physical random access channel (PRACH) and feedbacks PCB to Alice;

- **Beam refinement stage:** Bob can communicate with Alice after the initial access stage. Next, the coarse beam needs to be refined to further align the beam direction to improve communication quality. Alice first traverses the relevant fine beam codewords in TCB_i, and sends the channel state information reference signal (Channel State Information-Reference Signal, CSI-RS) to Bob. Then Bob traverses the relevant fine beam codewords in RCB_i to receive, and selects the pair of beams that maximizes RSRP as the best fine beam pair. After cycling the above steps for the L pairs of coarse beams, Bob gets the L best fine beam pair information $PFB = [TFB_1, RFB_1, TFB_2, RFB_2, \ldots, TFB_L, RFB_L]$, where TFB_i and RFB_i are the best transmit and receive fine beam index respectively. Finally, Bob sends PFB to Alice through the physical uplink control channel (PUCCH);

- **Encoding stage:** Alice and Bob encode the fine beam information PFB into a binary sequence respectively. Due to the different probability of each beam being selected, we will use Huffman coding to convert PFB into binary bits. The specific reasons will be discussed in the next section. Alice and Bob encode each pair of fine beams in PFB, and then concatenate the converted code groups to obtain their respective secret keys as follows:

$$K_{Alice} = \bigcup_{1 \leq i \leq L} \left(THm_i^{Alice} \cup RHm_i^{Alice} \right) \tag{7}$$

$$K_{Bob} = \bigcup_{1 \leq i \leq L} \left(THm_i^{Bob} \cup RHm_i^{Bob} \right) \tag{8}$$

where THm_i and RHm_i are the Huffman code group of the transmitting beam index and the receiving beam index corresponding to the i-th pair of beams.

3.3   Variable Length Coding of Beam Indexes

In [13, 14], the physical AoA and AoD are set to a uniform distribution of $(0, 2\pi)$, and the probability of each beam being selected is equal. However, the codebook-based beam distribution in actual scenarios is not uniform due to the different beam width as shown in Fig. 3. The wider the beam, the higher the probability of being selected. Given that AoA and AoD is uniformly distributed, the probability of being selected for each beam can be expressed by

$$Pb_i = \frac{bw_i}{\sum_{i=1}^{N} bw_i}, \quad i \in 1, 2, \ldots, N \tag{9}$$

where N denotes the number of beams, bw_i is the width of the i-th beam, and Pb_i is the probability of the i-th beam being selected.

According to the above analysis, the actual beam index is a non-uniformly distributed discrete random variable, so it is not appropriate to convert them to binary bits with equal length coding. Huffman coding is a type of variable word length coding that constructs the code group with the shortest average length based on the occurrence probability of characters, which means that the number of codewords is as close as possible to information entropy, thereby improving secret key generation efficiency. From the perspective of information security, even if the beam information has eavesdropped on, Eve cannot get the Huffman code group corresponding to the beam index. In a word, the introduction of Huffman coding can both improve the efficiency of the secret key generation and reduce the risk of eavesdropping.

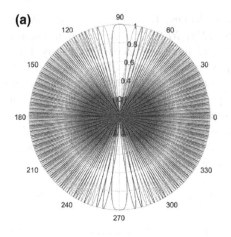

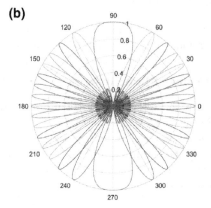

Fig. 3. (a) Beam direction on Alice, $N_t = 128$, (b) Beam direction on Bob, $N_r = 16$

4   Information-Theoretic Analysis

4.1   Mutual Information Analysis

In this section, we discuss the information entropy of random sources under the mmWave channel. Since the selected random source is the beam index that has been quantified to Huffman encoding, the analysis of information entropy can be regarded as the calculation of the information entropy of discrete random variables. Mutual information of beam number can be written as

$$I_k = (\rho_A; \rho_B) \tag{10}$$

where ρ_A and ρ_B are the fine beam indexes at Alice and Bob, respectively, and both of them are discrete random variables. ρ_B can be obtained by two-stage beam sweeping at Bob, ρ_A can be uploaded to Alice by Bob. According to the definition of mutual information, it can be written as

$$I_k = H(\rho_A) + H(\rho_B) - H(\rho_A, \rho_B) \tag{11}$$

As the beam information obtained by Alice depends on the upload process, the joint entropy of ρ_A and ρ_B is

$$H(\rho_A, \rho_B) = H(\rho_B) \tag{12}$$

Combining (11) and (12), mutual information I_k depends on the information entropy of ρ_A. The loss of information entropy between ρ_A and ρ_B mainly results from the transmission errors during the upload process, which can be shown as

$$H(\rho_A) = \sum_{k=0}^{N_{\rho_B}} b\left(k; N_{\rho_B}, ber_{up}\right)[H(\rho_B) - k] \tag{13}$$

where $b\left(k; N_{\rho B}, ber_{up}\right)$ is the binomial distribution, and variable k is the number of error bits. N_{ρ_B} is the length of ρ_B and ber_{up} denotes upload bit error rate. Next we discuss the information entropy of ρ_B in two cases.

Case 1: The distribution of fine beams is uniform, that is, the probability of each fine beam being selected is the same. According to the definition of discrete variable information entropy, we can get

$$H(\rho_B) = log_2 N_B \tag{14}$$

where N_B is the number of values that the random variable ρ_B can take. According to [13],

$$N_B \approx 2^L \cdot N_r \cdot N_t \tag{15}$$

Case 2: The distribution of fine beams is uneven, and that's the case in actual scenarios. The information entropy of ρ_B can be expressed as

$$H(\rho_B) = -\sum_{i=1}^{N_B} p_{Bi} log_2 p_{Bi} \tag{16}$$

where p_{Bi} denotes the possibility of the i-th fine beam combination.

4.2   Security Mutual Information Analysis

In our work, security mutual information can be written as $I_{sk} = (\rho_A, \rho_B | \rho_E)$. Eve has two methods to eavesdrop on secret keys.

One method is that Eve can get the same beam indexes as Bob during the fine beam sweeping process. However, the premise of this method must be that Eve knows the index of candidate beam pairs L and have the same configuration as Bob. Only if the eavesdropping channel must be highly correlated with the legal channel, Eve can get the same scan results as Bob. In the mmWave system, two channels become independent if they are separated by several wavelengths [16]. As is known, the millimeter wavelength is on the scale of the millimeter and the distance between two antenna elements in the Massive MIMO transceiver can be very small (0.5 wavelengths or 5.35 mm) [17]. It is almost impossible to keep Eve within half a wavelength around Bob. So it is very difficult for Eve to eavesdrop on the secret keys through the first method.

The other method for Eve to eavesdrop on the secret keys is to intercept the uploaded information from Bob. However, Bob uploads the beam indexes instead of the secret keys themselves. After receiving the beam indexes, Alice needs to use Huffman encoding to convert them into the final secret key. Even if Eve intercepts the uploaded beam index information, Eve does not know the Huffman code corresponding to each beam number and cannot generate the secret key as the same as a legitimate user. Thus our work guarantees the security of the secret key. We can get $I_{sk} \approx I_k$.

5   Simulation Results

In this section, we will demonstrate the performance of the proposed scheme through simulation. There are three main evaluation indicators: key generation rate, bit disagreement rate, and mutual information per bit. 1) The key generation rate measures the number of bits generated during user access. 2) Bit disagreement rate measures the inconsistent bit rate between legitimate users. Obtaining a lower Bit disagreement rate can effectively reduce the subsequent coordination overhead. 3) Mutual information per bit: the normalized mutual information between Alice and Bob which is the mutual information over the number of generated bits [13]. This metric measures the key generation efficiency.

In our simulation, Alice is equipped with N_t antennas, and Bob is equipped with N_r antennas. Define that $N_t - N_r$ is the antenna configuration and $N_{RF} = 1$. The antenna array structure is ULA. We model a wideband geometric ESV channel, and its AOA and AOD denoted as $\vartheta_{r_l}, \phi_{r_l}$ are evenly distributed in $(0, 2\pi)$. Set that each coarse beam contains 4 fine beams. In order to reduce the impact of beam sweeping on fine beam selection, Bob searches for three more fine beams when performing fine beam search. The 7 fine beams near the center angle of the coarse beam are searched in fine beam search. The number of simulations is set to 500.

Figure 4 shows the fine beam distribution at Alice and Bob under $N_t = 256$, $N_r = 32$, $L = 3$. The probability of each fine beam being selected is equal under ideal conditions, however because of the different width of each fine beam, the wider beam is more likely to be selected. Although AOA and AOD denoted as $\vartheta_{r_l}, \phi_{r_l}$ are evenly distributed in $(0, 2\pi)$, the fine beam distribution in the actual scene is uneven.

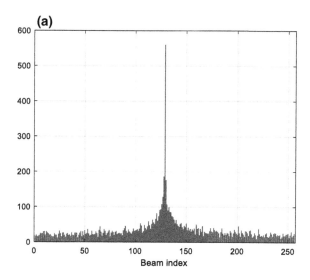

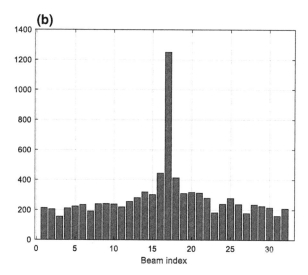

Fig. 4. (a) Fine beam distribution at Alice under $N_t = 256$, $L = 3$, (b) Fine beam distribution at Bob under $N_r = 32$, $L = 3$

Figure 5 shows the comparison of the number of secret key bits in the two antenna configurations. The baseline is the scheme proposed in [13], where full CSI is required and equal length code is used in the encoding of beam indexes. Set that the transmit and receive antenna numbers as $N_t = 128$, $N_r = 16$ and $N_t = 256$, $N_r = 32$ with selected beam pairs number $L = 3$, respectively. It can be seen that more antennas can effectively increase the number of secret key generation. The random variables selected in this paper

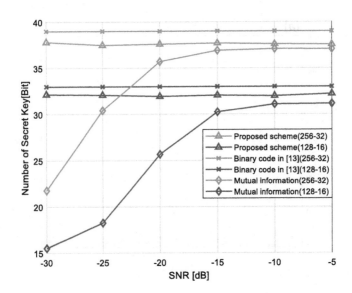

Fig. 5. Secret key bits at different SNRs and antenna configurations

are unequal probability distribution, so the Huffman coding used in our scheme can obtain the minimum code length closer to the mutual information entropy. In other words, we use fewer bits to represent the same amount of information compared with [13].

Figure 6 shows the mutual Information per bit in the two antenna configurations. It can be seen that the mutual information per bit rises with the increase of the signal-to-noise ratio, and the average bit mutual information of the two configurations reaches more than 0.9 at a low SNR = −15 dB; when the SNR is greater than −10 dB, the

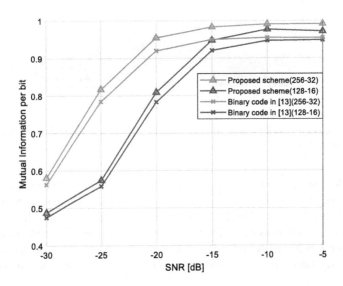

Fig. 6. Mutual Information per bit at different SNRs and antenna configurations

mutual Information per bit tends to be saturated. The increase in the number of antennas is helpful to improve the mutual information per bit. A large number of antennas can obtain a higher beam gain, thereby reducing the bit error rate during the upload process. The encoding method does not affect the mutual information of random variables, therefore the Huffman coding used in this paper can obtain the shortest code length of the secret key, obtaining higher mutual information per bit compared with [13].

Figure 7 shows the bit disagreement rate in the two antenna configurations. It can be seen from the figure that the bit disagreement rate does not perform well at low SNRs. The decrease of bit disagreement rate is obvious with the increase of the SNR. At SNR = −10 dB, the bit disagreement rate drops below 1% under the configuration of 256-32 and 128-16, which shows good robustness of the proposed scheme in a harsh communication environment.

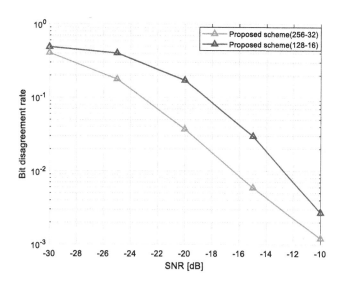

Fig. 7. Bit disagreement rate at different SNRs and antenna number

6   Conclusion

In this paper, we studied the beam-based secure key generation assisted by the two-stage beam sweeping in the massive MIMO system and proposed a key generation scheme compatible with the 5G NR beam management mechanism. In this scheme, we select the beam as the random source to generate the secret key. Bob determines the fine beam information through a two-stage beam sweeping and feedback them to Alice. Considering that the selected random source is unevenly distributed under the ESV channel, we use Huffman variable-length coding to encode the random source to obtain higher key generation efficiency. Even if Eve intercepts the beam index information, Eve cannot generate the secret key as a legitimate user. In the information theory analysis, we discussed the information entropy of our scheme and the two methods for

Eve to obtain the secret keys and put forward reasonable solutions to these two methods. The indicators to measure the effectiveness of the proposed scheme are the key generation rate, the average bit of mutual information, and the key disagreement rate. Simulation results show that Our scheme has higher coding efficiency than equal length coding. When the SNR is −10 dB, the bit disagreement rate is less than 0.01 and the mutual information per bit is greater than 0.97.

Acknowledgement. This work is supported by the Open Project of A Laboratory under Grant No. 2017XXAQ08, the National Natural Science Foundation of China under Grant No.61971069 and 61801051.

References

1. Liu, H., Yang, J., Wang, Y., Chen, Y.J., Koksal, C.E.: Group secret key generation via received signal strength: protocols, achievable rates, and implementation. IEEE Trans. Mob. Comput. **13**(12), 2820–2835 (2014)
2. Mathur, S., Trappe, W., Mandayam, N., Ye, C., Reznik, A.: Radio-telepathy: extracting a secret key from an unauthenticated wireless channel. In: Proceedings of the 14th ACM International Conference on Mobile Computing and Networking, Series MobiCom 2008, New York, NY, USA, pp. 128–139. ACM (2008)
3. Wang, Q., Su, H., Ren, K., Kim, K.: Fast and scalable secret key generation exploiting channel phase randomness in wireless networks. In: 2011 Proceedings IEEE INFOCOM, pp. 1422–1430, April 2011
4. Bakşi, S., Popescu, D.C.: Secret key generation with precoding and role reversal in mimo wireless systems. IEEE Trans. Wirel. Commun. **18**(6), 3104–3112 (2019)
5. Furqan, H.M., Hamamreh, J.M., Arslan, H.: Secret key generation using channel quantization with SVD for reciprocal MIMO channels. In: Proceedings International Symposium Wireless Communication System (ISWCS), pp. 597–602, September 2016
6. Etesami, J., Henkel, W.: LDPC code construction for wireless physical-layer key reconciliation. In: Proceedings 1st IEEE Int. Conference Communication China (ICCC), Beijing, China, pp. 208–213, August 2012
7. Graur, O., Islam, N., Filip, A., Henkel, W.: Quantization aspects in LDPC key reconciliation for physical layer security. In: Proceedings 10th IEEE International ITG Conference System, Communication Coding, Hamburg, Germany, pp. 1–6, February 2015
8. Sun, X., Wu, X., Zhao, C., Jiang, M., Xu, W.: Slepian-wolf coding for reconciliation of physical layer secret keys. In: Proceedings IEEE Wireless Communication Networking Conference, Sydney, Australia, pp. 1–6, April 2010
9. Wallace, J.: Secure physical layer key generation schemes: performance and information theoretic limits. In: Proceedings IEEE International Conference Communication (ICC), Dresden, Germany, pp. 1–5, June 2009
10. Lai, L., Liang, Y., Poor, H.V., Du, W.: Key generation from wireless channels: a review. In: Physical Layer Security Wireless Communication, FL, USA: CRC Press, p. 47C92 (2013)
11. Yaacoub, E.: On secret key generation with massive MIMO antennas using time-frequency-space dimensions. In: 2016 IEEE Middle East Conference on Antennas and Propagation (MECAP), Beirut, pp. 1–4 (2016). https://doi.org/10.1109/mecap.2016.7790086
12. Patwari, N., Croft, J., Jana, S., Kasera, S.K.: High-rate uncorrelated bit extraction for shared secret key generation from channel measurements. IEEE Trans. Mobile Comput. **9**(1), 17–30 (2010)

13. Jiao, L., Tang, J., Zeng, K.: Physical layer key generation using virtual AoA and AoD of mmWave massive MIMO channel. In: IEEE Conference on Communications and Network Security (2018)
14. Jiao, L., Wang, N., Zeng, K.: Secret Beam: robust secret key agreement for mmWave massive MIMO 5G communication. In: 2018 IEEE Global Communications Conference (GLOBECOM), pp. 9–13, December 2018
15. Onggosanusi, E., et al.: Modular and high-resolution channel state information and beam management for 5G new radio. IEEE Commun. Mag. **56**(3), 48–55 (2018)
16. Chen, C., Jensen, M.A.: Secret key establishment using temporally and spatially correlated wireless channel coefficients. IEEE Trans. Mobile Comput. **10**(2), 205–215 (2011)
17. Güvenkaya, E., Hamamreh, J.M., Arslan, H.: On physical-layer concepts and metrics in secure signal transmission. Phys. Commun. **25**, 14–25 (2017)

Weighted Sum Rate Maximization
for NOMA-Based UAV Networks

Zhengqiang Wang[1], Hao Zhang[1], Xiaoyu Wan[1], Zifu Fan[1], Xiaona Yang[2]([✉]),
and Yuanmao Ji[3]

[1] School of Communication and Information Engineering, Chongqing University
of Posts and Telecommunications, Chongqing, People's Republic of China
[2] Huaxin Consulting Company Ltd,
Hangzhou 310014, Zhejiang, People's Republic of China
yangxiaona.hx@chinaccs.cn
[3] Senior Instructor, Ericsson (China) Communication Co., Ltd, Chaoyang, China

Abstract. The unmanned aerial vehicle (UAV) based aerial base station
(BS) has emerged as a feasible solution to the high traffic demands of the
future wireless networks. It is essential that UAV can integrat with non-
orthogonal multiple access (NOMA) to support the massive connection
service. In this paper, we study the problem of weighted sum rate maximization in the downlink for NOMA-based UAV networks. This is a non-
convex optimization problem, which is intractable to be directly solved
using the convex optimization method. To deal with the problem, we propose an efficient iterative algorithm via variable substitution and relaxation methods, then construct the framework of alternating optimization
to solve the problem joint placement and power allocation optimization.
The simulation results show that the proposed algorithm performs better
than other schemes.

Keywords: Unmanned aerial vehicle · Non-orthogonal multiple
access · Power allocation · Placement optimization · Weighted sum rate

1 Introduction

To satisfy the communication service requirements of 5G networks, non-
orthogonal multiple access (NOMA) is proposed to support large-scale access
capability and highly efficient spectrum utilization in [1]. The unmanned aerial
vehicle (UAV) has swift and flexible with good channel conditions in [2]. Therefore, the UAV as a base station (BS) can achieve better communication performance. The communication network of UAV is being studied to meet the
ubiquitous services that are not provided to ground users in crowded or remote
areas. The authors in [3] proposed a coverage plan for multiple UAVs to maximize the average capacity of multiple UAVs, while ensuring that the UAV can

This work was partially supported by the National Natural Science Foundation of P. R.
China(No.61701064), Basic Research and Frontier Exploration Project of Chongqing
(cstc2019jcyj-msxmX0264).

cover the ground users on a large scale. In [4], the author considered multi-UAV wireless network and focused on improving energy efficiency through subchannel assignment and power allocation. Specifically, a NOMA scheme for the power domain of energy-efficient UAV base stations was studied to make better use of limited spectrum resources in [5,6]. Since 5G network will be ultra-dense and heavy-loaded [1], the design of admission control is essential to deal with the infeasibility [7] , and to determine the set of users whose quality-of-service (QoS) requirements can be protected at the same time. The work [8,9] studied how to use NOMA to pair users into each sub-channel, and then maximize the sum rate of the networks. In [10], the scheme for communication with ground users was investigated by trajectory optimization. [11] considered the one-dimensional trajectory optimization of UAVs to maximize energy transmission, where the distribution of users is a linear topology.

So far, for the sake of maximizing the sum rate of users, some works has been done on the placement and power allocation of UAV networks. In [12], jointly optimization of the placement and power allocation of the UAV is a non-convex problem. Firstly, the UAV is deployed at the user's geometric center location, and then the power allocation optimization of UAV is based on Karush-Kuhn-Tucker (KKT) conditions [13]. However,the geometric center locations is suboptimal and the fairness among users is not considered in [12]. Based on this as a motivation, we consider to investigate the weighted sum rate maximization problem for NOMA-based UAV networks. Due to the non-convexity of joint optimization of position and power allocation, it is difficult to solve directly with existing algorithms. We utilize an alternative optimization method to solve the problem of maximizing weighted sum rate. Simulation results show that the proposed algorithm outperform the scheme in In [12].

The remainder of this paper is organized as follows. Section 2 presents the system model of the UAV networks, and formulates the downlink weighted sum rate maximization. Section 3 considers the method of variable substitution and presents the optimal solution to the relaxed problem. In this section, an alternative optimization algorithm is proposed to yield a near-optimal solution to the decoupled problem. Section 4 provides numerical results to validate the effectiveness of our proposed designs. Finally, Section 5 concludes this paper.

2   System Model and Problem Formulation

2.1   System Model

We consider a single antenna UAV in a downlink communication network system. The UAV is deployed at the height of H and sends information to multiple single-antenna users simultaneously. The number of ground users is K. The horizontal position of UAV and ground user i is expressed as (X, Y) and (x_i, y_i), $i = 1, ..., K$, respectively. For simplicity, we assume that the path loss satisfies line-of-sight (LoS) from a single antenna UAV to ground user. UAV sends broadcast information to K users on the ground. Without loss of generality assumption at the same time, the channels are sorted as

$$|h_K|^2 \geq ... \geq |h_k|^2 \geq ... \geq |h_1|^2 \geq 0. \tag{1}$$

$$P_t^{[k]} = P_{total} a_k, a_k \geq 0. \tag{2}$$

where $P_t^{[k]}$ are allocated power for the k-th user, P_{total} is the maximum transmit power of the UAV, and a_k is the allocated power coefficient of the k-th user, h_k represents the channel gain from the UAV to the k-th user. The power allocation coefficient of UAV needs to satisfy the following relationship.

$$\sum_{k=1}^{K} a_k = 1. \tag{3}$$

We assume that the channel model from UAV to the ground user k is given by the following path loss model [12]

$$L^{[k]}_{Los} = 20 \log_{10}(4\pi f_c d_k/c) + \eta_{Los}, k = 1, 2, ..., K. \tag{4}$$

where f_c is the carrier frequency, and d_k is the distance between the UAV and the k-th user, i.e.

$$d_k = \sqrt{H^2 + (X - x_k)^2 + (Y - y_k)^2}. \tag{5}$$

According to the above formula (4) and (5), the received power of the ground k-th user can be obtained as

$$P_r^{[k]} = 10\log_{10}P_t^{[k]} - L_{Los}^{[k]}. \tag{6}$$

The power $P_t^{[k]}$ is transmitted from the UAV to k-th user, and the received power $P_r^{[k]}$ of k-th user can be known. Therefore, the channel power gain expression of the UAV to the k-th user can be given as follows.

$$|h_k|^2 = \frac{\beta}{d_k^2}. \tag{7}$$

$$\beta = (\frac{c}{4\pi f_c})^2 \frac{1}{(\eta_{Los})_{dB}} \tag{8}$$

where β denotes the channels power gain at a reference distance of d=1m, i.e. The UAV use NOMA protocol to broadcast the information to all the ground users. The transmit signal at UAV is given by $s = \sum_{j=1}^{K} s_j$, where s_j is transmission information from UAV to user j such that $E|s_j|^2 = P_t^{[j]} = P_{total} a_j$. The information received by ground k-th user can be given by

$$y_k = h_k s_k + h_k \sum_{j=1, j \neq k}^{K} s_j + n_k. \tag{9}$$

where n_k represents the additive zero-mean Gaussian noise with variance σ^2. According to the successive interference cancellation (SIC) technology of NOMA, the signal-to-interference-plus-noise-ratio (SINR) of the ground k-th user can be obtained as

$$\text{SINR}_k = \frac{P_{total}|h_k|^2 a_k}{P_{total}|h_k|^2 \sum\limits_{j=k+1}^{K} a_j + \sigma^2}. \tag{10}$$

where we have defined $\sum\limits_{j=K+1}^{K} a_j = 0$ to simplify the SINR expression or user K when k equals K. Therefore, the achievable rate for k-th user is given by

$$R_k = \log(1 + \frac{P_{total}|h_k|^2 a_k}{P_{total}|h_k|^2 \sum\limits_{j=k+1}^{K} a_j + \sigma^2}). \tag{11}$$

2.2   Problem Formulation

Given the total power of the UAV, in order to consider the fairness [14] of the information received by the ground users, our goal is to maximize the weighted sum rate of the ground users [15]. We need to optimize the placement and power allocation of UAV jointly [16]. Therefore, the weighted sum rate maximization problem can be written as the following problem.

$$(P1): \underset{a_k, X, Y}{\text{maximize}} \sum_{k=1}^{K} w_k \log(1 + \frac{P_{total}|h_k|^2 a_k}{P_{total}|h_k|^2 \sum\limits_{j=k+1}^{K} a_j + \sigma^2})$$

$$\text{s.t. } C1: \sum_{k=1}^{K} a_k = 1, a_k \geq 0,$$

$$C2: (X - x_{k+1})^2 + (Y - y_{k+1})^2 \\ \leq (X - x_k)^2 + (Y - y_k)^2, \\ k = 1, ..., K-1, \tag{12}$$

$$C3: \log\left(1 + \frac{P_{total}|h_k|^2 a_k}{P_{total}|h_k|^2 \sum\limits_{j=k+1}^{K} a_j + \sigma^2}\right) \geq r_k^{\min}, k = 1, ..., K,$$

$$C4: \min\{x_i\} \leq X \leq \max\{x_i\}, 1 \leq i \leq K,$$

$$C5: \min\{y_j\} \leq Y \leq \max\{y_j\}, 1 \leq j \leq K.$$

where w_k is weighted factor for user k to express the fairness among users for rate allocation. where C1 is the maximum transmit power constraint for the UAV and the non-negative constraint transmit power for the $k-th$ user, C2 is channel gain order for users such that user k can decoding all the user j such that $k > j$, C3 represents the minimum transmission rate requirement from the UAV to the k-th user, C4 and C5 represent the level of the UAV flight regional restrictions. It is observed from the problem formulation that the objective function is not convex with respect to the joint variables of power allocation coefficient a_k, $k = 1, ..., K$ and UAV's coordinate (X, Y). Therefore, problem (P1) is a non-convex optimization problem. Next, we will solve problem (P1) by decoupling the variable of UAV's coordination and user's power allocation with block coordinate descent (BCD) algorithm.

3   Joint Transmit Power and Placement Optimization

First, we introduce auxiliary variables $R_k = \log(1 + \dfrac{P_{total}|h_k|^2 a_k}{P_{total}|h_k|^2 \sum\limits_{j=k+1}^{K} a_j + \sigma^2})$,

$k = 1, \cdots, K$ to problem (P1), the weighted sum rate problem (P1) is equivalent to the following problem.

$$(P2) : \underset{a_k, X, Y, R_k}{\text{maximize}} \sum_{k=1}^{K} w_k R_k$$

$$\text{s.t.} \quad C1 : \sum_{k=1}^{K} a_k = 1, a_k \geq 0,$$

$$C2 : (X - x_{k+1})^2 + (Y - y_{k+1})^2 \leq (X - x_k)^2 + (Y - y_k)^2,$$
$$k = 1, \cdots, K-1,$$

$$C3 : \log \left(1 + \frac{P_{total}|h_k|^2 a_k}{P_{total}|h_k|^2 \sum\limits_{j=k+1}^{K} a_j + \sigma^2} \right) \geq r_k^{\min}, k = 1, ..., K, \qquad (13)$$

$$C4 : \log \left(1 + \frac{P_{total}|h_k|^2 a_k}{P_{total}|h_k|^2 \sum\limits_{j=k+1}^{K} a_j + \sigma^2} \right) \geq R_k, k = 1, ..., K,$$

$$C5 : \min\{x_i\} \leq X \leq \max\{x_i\}, 1 \leq i \leq K,$$
$$C6 : \min\{y_j\} \leq Y \leq \max\{y_j\}, 1 \leq j \leq K.$$

where we have rewritten the equality constraints $R_k = \log(1 + \dfrac{P_{total}|h_k|^2 a_k}{P_{total}|h_k|^2 \sum\limits_{j=k+1}^{K} a_j + \sigma^2}), k = 1, \cdots, K$ by inequality constraints $\log(1 + \dfrac{P_{total}|h_k|^2 a_k}{P_{total}|h_k|^2 \sum\limits_{j=k+1}^{K} a_j + \sigma^2}) \geq R_k, k = 1, \cdots, K$ in C4 for (P2) because the objective function in (P2) is a monotonically increasing function with R_k. Therefore, the optimal solution to (P2) will satisfy the equality constraints $R_k = \log(1 + \dfrac{P_{total}|h_k|^2 a_k}{P_{total}|h_k|^2 \sum\limits_{j=k+1}^{K} a_j + \sigma^2}), k = 1, \cdots, K$. Thus, problem (P2) has the same optimal solution with problem (P1).

Even the object function of (P2) is linear function, problem (P2) is still a non-convex optimization problem due to non-convex constraints C3 and C4. Next, we utilize variable substitution and relaxation method to handle the non-convex constraints. We further decompose problem (P2) into two subproblems by optimizing the placement and power allocation of the UAV separately.

Introduce auxiliary variables $u_k, k = 1, \cdots, K$ to problem (P2), such that u_k satisfies the following equation.

$$\frac{P_{total}|h_k|^2 a_k}{P_{total}|h_k|^2 \sum\limits_{j=k+1}^{K} a_j + \sigma^2} = \mathrm{e}^{u_k}, k = 1, \cdots, K. \tag{14}$$

Substitute (14) into problem (P2), and use the monotonicity of the original problem (P2) with respect to R_k and $R_k = \log(1 + \mathrm{e}^{u_k})$. Then we can relax equation (14) into

$$\frac{P_{total}|h_k|^2 a_k}{P_{total}|h_k|^2 \sum\limits_{j=k+1}^{K} a_j + \sigma^2} \geq \mathrm{e}^{u_k}, k = 1, \cdots, K. \tag{15}$$

As $R_k = \log(1 + \mathrm{e}^{u_k})$ can be viewed as an increasing function of u_k, and problem (P2) can be recast as the following equivalent problem

$$\begin{aligned}
(P3): \ &\underset{a_k,X,Y,R_k,u_k}{\text{maximize}} \ \sum_{k=1}^{K} w_k R_k \\
\text{s.t.} \quad &C1: \sum_{k=1}^{K} a_k = 1, a_k \geq 0, \\
&C2: (X - x_{k+1})^2 + (Y - y_{k+1})^2 \\
&\qquad \leq (X - x_k)^2 + (Y - y_k)^2, \\
&\qquad k = 1, ..., K - 1, \\
&C3: \log(1 + \mathrm{e}^{u_k}) \geq r_k^{\min}, k = 1, ..., K, \\
&C4: \log(1 + \mathrm{e}^{u_k}) \geq R_k, k = 1, ..., K, \\
&C5: \frac{P_{total}|h_k|^2 a_k}{P_{total}|h_k|^2 \sum\limits_{j=k+1}^{K} a_j + \sigma^2} \geq \mathrm{e}^{u_k}, \\
&C6: \min\{x_i\} \leq X \leq \max\{x_i\}, 1 \leq i \leq K, \\
&C7: \min\{y_j\} \leq Y \leq \max\{y_j\}, 1 \leq j \leq K.
\end{aligned} \tag{16}$$

Next, we introduce a new variable substitution s_k, such that $\mathrm{e}^{s_k} = P_{total} a_k, k = 1, ..., K$, is brought into the constraints C1 and C5. Based on the above expressions, problem (P3) is equivalent to

$$(P4): \underset{X,Y,R_k,u_k,s_k}{\text{maximize}} \sum_{k=1}^{K} w_k R_k$$

s.t. $C1 : \sum_{k=1}^{K} e^{s_k} \le P_{total},$

$C2 : (X - x_{k+1})^2 + (Y - y_{k+1})^2$
$\le (X - x_k)^2 + (Y - y_k)^2,$
$k = 1, ..., K - 1,$ (17)

$C3 : \log(1 + e^{u_k}) \ge r_k^{min}, k = 1, ..., K,$

$C4 : \log(1 + e^{u_k}) \ge R_k, k = 1, ..., K,$

$C5 : \dfrac{e^{s_k}|h_k|^2}{|h_k|^2 \sum\limits_{j=k+1}^{K} e^{s_j} + \sigma^2} \ge e^{u_k}, k = 1, ..., K$

$C6 : \min\{x_i\} \le X \le \max\{x_i\}, 1 \le i \le K,$

$C7 : \min\{y_j\} \le Y \le \max\{y_j\}, 1 \le j \le K.$

So far, constraints C4 and C5 are still non-convex. Next, we will convert the non-convex constraints C4-C5 into a form of convex constraints. For the constraint C3, the left-hand side is convex, and the right-hand side is constant. Constraint C3 can be rewritten as

$$1 + e^{u_k} \ge e^{r_k^{min}},$$
$$u_k \ge \log\left(e^{r_k^{min}} - 1\right), k = 1, \cdots, K. \qquad (18)$$

Therefore, C3 are convex constraints because they can be rewritten as $u_k \ge \log\left(e^{r_k^{min}} - 1\right), k = 1, \cdots, K.$ For constrain C4, it can be transformed as the following form by relaxtion.

$$\log(1 + e^{u_k}) \ge \log(e^{u_k}) \ge R_k, k = 1, \cdots, K,$$
$$u_k \ge R_k. \qquad (19)$$

Make the constraint C5 id equivalant to the following mathematical transformation.

$$\dfrac{e^{s_k} \frac{\beta}{H^2 + (X - x_k)^2 + (Y - y_k)^2}}{\frac{\beta}{H^2 + (X - x_k)^2 + (Y - y_k)^2} \sum\limits_{j=k+1}^{K} e^{s_j} + \sigma^2} \ge e^{u_k} \Leftrightarrow$$

$$\dfrac{e^{u_k}\left(\frac{\beta}{H^2 + (X - x_k)^2 + (Y - y_k)^2} \sum\limits_{j=k+1}^{K} e^{s_j} + \sigma^2\right)}{e^{s_k} \frac{\beta}{H^2 + (X - x_k)^2 + (Y - y_k)^2}} \le 1 \Leftrightarrow \qquad (20)$$

Constrain C2 can be transformed into the equivalent linear constraint in (21) by expanding the equations on both sides, and canceling out the quadratic term X^2 and Y^2.

$$(X - x_k)^2 + (Y - y_k)^2 \le (X - x_{k+1})^2 + (Y - y_{k+1})^2,$$
$$2X x_k + 2Y y_k - 2X x_{k-1} - 2Y y_{k-1}$$
$$\le x_k^2 + y_k^2 - x_{k-1}^2 - y_{k-1}^2,$$
$$k = 2, \cdots, K. \qquad (21)$$

According to a series of treatment of the above problems, the problem (P4) is reformulated as the following problem (P5).

$$(P5): \underset{X,Y,R_k,u_k,s_k}{\text{maximize}} \sum_{k=1}^{K} w_k R_k$$

$$\text{s.t.} \quad C1: \sum_{k=1}^{K} e^{s_k} \leq P_{total},$$

$$C2: 2Xx_k + 2Yy_k - 2Xx_{k-1} - 2Yy_{k-1}$$
$$\leq x_k^2 + y_k^2 - x_{k-1}^2 - y_{k-1}^2,$$
$$k = 2, \cdots, K,$$

$$C3: u_k \geq \log\left(e^{r_k^{\min}} - 1\right), k = 1, ..., K,$$

$$C4: u_k \geq R_k, k = 1, ..., K,$$

$$C5: \frac{e^{u_k}\left(\sum_{j=k+1}^{K} e^{s_j} + \sigma^2 \frac{H^2 + (X - x_k)^2 + (Y - y_k)^2}{\beta}\right)}{e^{s_k}} \leq 1, k = 1, ..., K,$$

$$C6: \min\{x_i\} \leq X \leq \max\{x_i\}, 1 \leq i \leq K,$$

$$C7: \min\{y_j\} \leq Y \leq \max\{y_j\}, 1 \leq j \leq K.$$

$$(22)$$

Next, we will give a joint placement and power allocation optimization algorithm to problem (P5).

3.1 Placement Optimization for the UAV

Based on the analysis of the above formula, we first fix the variables u, R, s in (P5) and update the placement coordinates (X, Y) of UAV by problem (P6). The objective function of problem (P6) is indepent of variable (X, Y), but the constraint set is convex. We can use the convex solver (CVX) to find a feasible UAV position, and update the UAV position.

$$(P6): \underset{X,Y}{\text{maximize}} \sum_{k=1}^{K} w_k R_k$$

$$\text{s.t.} \quad C1: 2Xx_k + 2Yy_k - 2Xx_{k-1} - 2Yy_{k-1}$$
$$\leq x_k^2 + y_k^2 - x_{k-1}^2 - y_{k-1}^2,$$
$$k = 2, \cdots, K,$$

$$C2: \log\left(\frac{e^{u_k}\left(\sum_{j=k+1}^{K} e^{s_j} + \sigma^2 \frac{H^2 + (X - x_k)^2 + (Y - y_k)^2}{\beta}\right)}{e^{s_k}}\right) \leq 0,$$

$$C3: \min\{x_i\} \leq X \leq \max\{x_i\}, 1 \leq i \leq K,$$

$$C4: \min\{y_j\} \leq Y \leq \max\{y_j\}, 1 \leq j \leq K.$$

$$(23)$$

Next, after we find the (X, Y), we fix UAV's placement and derive u, s and R in problem (P7). The problem (P7) is a convex optimization problem with respective to u, s, R.

3.2 Transmit Power Optimization for the UAV

The objective function is affine in the following problem (P7), and the left-hand side of the constraint (C4) is convex set with respect to the variables $u_k, R_k, s_k, k = 1, ..., K$ because it satisfies log sum function is convex function.

Therefore, the fixed UAV position coordinates can be solved using the interior point algorithm in convex programing [13] of CVX.

$$(P7) : \underset{R_k, s_k, u_k}{\text{maximize}} \sum_{k=1}^{K} w_k R_k$$

$$\text{s.t.} \quad C1 : \sum_{k=1}^{K} e^{s_k} \leq P_{total},$$

$$C2 : u_k \geq \log \left(e^{r_k^{min}} - 1 \right),$$

$$C3 : u_k \geq R_k,$$

$$C4 : \log\left(\frac{e^{u_k} \left(\sum_{j=k+1}^{K} e^{s_j} + \sigma^2 \frac{H^2 + (X - x_k)^2 + (Y - y_k)^2}{\beta} \right)}{e^{s_k}} \right) \leq 0.$$

(24)

Finally, we constitute an alternate optimization algorithm and obtain solution to problem (P5) as follows.

3.3   Joint Transmit Power and Placement Optimization

The placement and power allocation of the joint optimization UAV is a non-convex problem from the above discussion. It is difficult to find the optimal global solution. Therefore, we decompose the original problem (P1) into two sub-problems (P6) and (P7), and use BCD algorithm to alternately optimize the position and power of the UAV. Then, we can obtain the solution with reasonable accuracy. Based on the above two sub-problems, we propose a joint placement and power allocation optimization (JPPAO) algorithm given in Algorithm 1.

Algorithm 1. Joint placement and power allocation optimization (JPPAO)

Initialization: the UAV's placement (X, Y), the iteration number $i = 0$, and the tolerance of accuracy ε. (X, Y) is the geometric center of the user. Let Π be the set of all possible permutations of K.

Repeat

Let $\pi \in \Pi$ be the decoding order of the users. Let $\pi(i)$, where $i \in \{1, 2, ..., |K|\}$, be its i-th component. SIC decoding order satisfies $h_{\pi(1)} \leq h_{\pi(2)} \leq, ..., \leq h_{\pi(|K|)}$.

   Repeat

   Fix the UAV's placement, find the optimal solution u^*, s^*, R^* at the iteration to problem (P7) by standard convex optimization techniques, and let $u^*, s^*, R^* \rightarrow u^{i+1}, s^{i+1}, R^{i+1}$.

   Fix the UAV's u, s, R, find the optimal solution (X^*, Y^*) to problem (P6) at the i-th iteration, and let $(X^*, Y^*) \rightarrow (X^{i+1}, Y^{i+1})$.

$$\{(u, s, R), (X, Y)\}^i \rightarrow \{(u, s, R), (X, Y)\}^{i+1}, and\ i \rightarrow i + 1.$$

   Until $|R^{i+1} - R^i| \leq \varepsilon$

Until traverse all permutations Π

Find the largest weighted sum rate among all permutations.

Output return the optimal solution of placement (X^*, Y^*) and (u^*, s^*, R^*) to problem (P5).

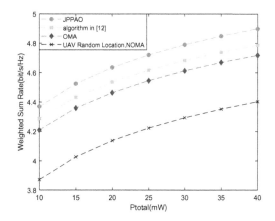

Fig. 1. Weighted sum rate versus maximizing power of UAV

4   Simulation Results and Discussion

In this part, we evaluate the proposed joint placement and power allocation (JPPAO) algorithm performance through simulation results, and users are randomly deployed on the ground. We set K=4, $H = 60$ m, $\sigma^2 = -140$ dBm, $r_k^{min} = 1$ $bit/s/Hz$, $\varepsilon = 10^{-4}$, $\beta = 10^{-3}$, the weight $w_k = 1$, $i = 1,...,K$.

The average weighted sum rate of the vertical axis of Fig. 1 are obtained by taking multiple network topologies. The sum rate of the network is compared for different values of P_{total} with NOMA and orthogonal multiple access (OMA). In order to maximize the weighted sum rate, in this paper, the JPPAO uses an alternative optimization scheme. The algorithm in [12] first obtains the geometric center position of the UAV, and then uses KKT to obtain the power allocation coefficient. The power allocated by the OMA scheme is P_{total}, and OMA scheme is based on the location of the UAV in [12]. It can be seen that the UAV can achieve better performance after optimizing the location of UAV. Compared with other schemes, the proposed scheme can get better performance by optimizing the placement and power allocation of the UAV alternately. When the P_{total} is 20 mw, the performance of JPPAO is 2.6% better than that of algorithm in [12]. Moreover, we can see that the sum rate increases, as the total transmit power of UAV increases. This is because the UAV has more power to allocate to user to maximize the weighted sum rate of system.

From Fig. 2, it can be seen that as the height of UAV increases, the weighted sum rate of the ground users gradually decreases. This is because as the height of the drone increases, the channel gain from UAv to all the user becomes worse and worse.

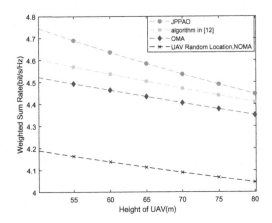

Fig. 2. Weighted sum rate versus height of UAV

5   Conclusion

This paper studies the joint optimization of the UAV's placement and power allocation to maximize the weighted sum rate of the ground users. To solve the non-convex problem, we apply variable substitution and BCD method to the problem, divided into two subproblems. Then, a joint transmit power and placement optimization by convex optimization is developed. In the end, we derive the optimal placement and power allocation of signal UAV for maximizing the weighted sum rate of ground users. The efficiency of the proposed algorithm is vefied by comparison with other benchmark methods.

References

1. Liu, Y., Qin, Z., Cai, Y., Gao, Y., Li, G.Y., Nallanathan, A.: UAV communications based on non-orthogonal multiple access. IEEE Wirel. Commun. **26**(1), 52–57 (2019)
2. Nomikos, N., Michailidis, E.T., Trakadas, P., Vouyioukas, D., Zahariadis, T., Krikidis, I.: FlexNOMA: exploiting buffer-aided relay selection for massive connectivity in the 5G uplink. IEEE Access **7**, 88743–88755 (2019)
3. Chen, R., Li, X., Sun, Y., Li, S., Sun, Z.: Multi-UAV coverage scheme for average capacity maximization. IEEE Commun. Lett. **24**(3), 653–657 (2020)
4. Duan, R., Wang, J., Jiang, C., Yao, H., Ren, Y., Qian, Y.: Resource allocation for Multi-UAV Aided IoT NOMA uplink transmission systems. IEEE Internet Things J. **6**(4), 7025–7037 (2019)
5. Sohail, M.F., Leow, C.Y., Won, S.: Energy-efficient non-orthogonal multiple access for UAV communication system. IEEE Trans. Veh. Technol. **68**(11), 10834–10845 (2019)
6. Nasir, A.A., Tuan, H.D., Duong, T.Q., Poor, H.V.: UAV-enabled communication using NOMA. IEEE Trans. Commun. **67**(7), 5126–5138 (2019)

7. Tang, R., Cheng, J., Cao, Z.: Joint placement design, admission control, and power allocation for NOMA-based UAV systems. IEEE Wirel. Commun. Lett. **9**(3), 385–388 (2020)
8. Nguyen, M.T., Le, L.B.: NOMA user pairing and UAV placement in UAV-based wireless networks. In: 2019 IEEE International Conference on Communications (ICC), Shanghai, China, pp. 1–6 (2019)
9. Masaracchia, A., Da Costa, D.B., Duong, T.Q., Nguyen, M.N., Nguyen, M.T.: A PSO-based approach for user-pairing schemes in NOMA systems: theory and applications. IEEE Access **7**, 90550–90564(2019)
10. Park, J., Lee, H., Eom, S., Lee, I.: UAV-aided wireless powered communication networks: trajectory optimization and resource allocation for minimum throughput maximization. IEEE Access **7**, 134978–134991 (2019)
11. Hu, Y., Yuan, X., Xu, J., Schmeink, A.: Optimal 1D trajectory design for UAV-enabled multiuser wireless power transfer. IEEE Trans. Commun. **67**(8), 5674–5688 (2019)
12. Liu, X., Wang, J., Zhao, N., Chen, Y.: Placement and power allocation for NOMA-UAV networks. IEEE Wirel. Commun. Lett. **8**(3), 965–968 (2019)
13. Body, S., Vandenberghe, L.: Convex Optimization. Cambridge University Press, Cambridge (2004)
14. Dao, V.L., Tran, H., Girs, S., Uhlemann, E.: Reliability and fairness for UAV communication based on non-orthogonal multiple access. In: 2019 IEEE International Conference on Communications Workshops (ICC Workshops), Shanghai, China, pp. 1–6 (2019)
15. Wang, X., Chen, R., Xu, Y., Meng, Q.: Low-complexity power allocation in NOMA systems with imperfect SIC for maximizing weighted sum-rate. IEEE Access **7**, 94238–94253 (2019)
16. Cui, F., Cai, Y., Qin, Z., Zhao, M., Li, G.Y.: Joint trajectory design and power allocation for UAV-enabled non-orthogonal multiple access systems. In: 2018 IEEE Global Communications Conference (GLOBECOM), Abu Dhabi, United Arab Emirates, pp. 1–6 (2018)

Placement Optimization for UAV-Enabled Wireless Power Transfer System

Zhengqiang Wang$^{(\boxtimes)}$, Yang Liu, Hao Zhang, Xiaoyu Wan, and Zifu Fan

School of Communication and Information Engineering, Chongqing University
of Posts and Telecommunications, Chongqing, People's Republic of China
wangzq@cqupt.edu.cn

Abstract. This paper considers an unmanned aerial vehicle (UAV)-enabled wireless power transfer (WPT) system, in which a UAV hovers in a given flying altitude to transfer energy to more than two energy receivers (ERs) on the ground. We consider to maximize the sum energy and weighted sum energy of ERs by UAV placement optimization. As the sum energy and weighted sum energy maximization problem are sum of ratio problems, which are generally NP-hard. It is difficult to give the optimal location of the UAV for those two problems. To tackle those problems, we adopt a novel quadratic transform technique to transfer to an equivalent problem. Based on the equivalent problem, we propose an iterative coordination update algorithm in a closed-form expression, which can converge to the stationary point of the sum energy maximization problem or even the global optimal solution under a sufficient condition of the flight altitude. Simulation results show that the proposed algorithm can achieve nearly the same weighted sum energy for ERs and reduces more than 90% complexity compared to the two-dimensional (2D) exhaustive search method.

Keywords: Unmanned aerial vehicle (UAV) · Wireless power transfer · Placement optimization

1 Introduction

In recent years, with the continuous development of wireless communication, some portable electrical appliances such as laptop computers, mobile devices and music players needing battery charging have had higher and higher constraints on energy consumption. However, the battery life is limited, usually only 3–5 years, which can not meet the needs of people for communication equipment [1]. In addition, in special occasions, such as mining and oil mining, the traditional power transmission has hidden dangers in terms of safety.

This work was partially supported by the National Natural Science Foundation of P.R. China(No.61701064), Basic Research and Frontier Exploration Project of Chongqing(cstc2019jcyj-msxmX0264).

Under the above circumstances, wireless power transmission becomes more and more important and imperative. In the traditional WPT system, energy is usually sent to the energy receiver through a fixed energy beacon (PB) [2]. However, in some practical situations, such energy transfer efficiency is not high. Because the transmission process will be blocked by some obstacles, such as tall buildings and mountains. But against this disadvantage, drones can be effectively circumvented. The high mobility and flexible deployment of UAV can effectively guarantee the line-of-sight link between UAV and ground users, thus achieving reliable and energy-saving energy transmission [3]. Therefore, UAV as a radio frequency energy signal transmitter will be a research hotspot in the future.

We can find that there are many applications of UAV in the WPT system in previous studies. For example, in [4], the authors considered the UAV as an energy source send energy to the ground user, and the ground users exploit the energy received to send information to the UAV. The purpose is to maximize the user's information sum-rate by jointly optimizing the time allocation and position optimization of the UAV. In [5], a parallel wireless information and Power Transmission (SWIPT) technology was proposed to jointly optimize the power allocation and trajectory design of UAV to maximize the minimum energy collected by ground dispersed Internet of Things (loT) devices. However, it is difficult to solve this problem directly, so they focus on finding suboptimal solutions for solving this problem. In addition, WPT was also used in the recently popular UAV mobile edge computing system. [6] considered a kind of edge system, a TDMA working model was proposed, which allows UAV-assisted systems to carry out parallel transmission, downlink energy transfer to charge loT devices, and uplink information transfer. In [7], the authors studied a UAV-supported WPT system, one of which serves two ERs on the ground. The Pareto boundary of the reachable energy region of the two ERs is determined by optimizing the flight path of the UAV under the constraint of maximum velocity. [8] expanded [7] and considered the system model of more than two ERs, where the purpose is to maximize the sum or minimum accepted energy of all ERs by controlling UAV trajectory. Two drones served two ground users was further considered in [9], aim to maximize the uplink shared (minimum) throughput of the two users within a limited UAV mission cycle by jointly optimizing the running trajectories of the two drones and wireless resource allocation. In [10], through the joint optimization of UAV trajectory and wireless resource allocation, the common throughput maximization of wireless powered communication networks (WPCN) supported by UAV was studied.

A sub-problem to be solved for the considered system mode in [8,10] is to find the optimal placement of UAV in a given time so as to maximize the total energy of ground users. For two users case, the authors have given the close-form solution of the location for UAV. They have shown that only one location or two locations are optimal, which depends on the height of the UAV and the distance between the two ground users. For more than two users' case, because the objective function is a non-convex problem, it is difficult to look for the optimal location of the UAV. A two-dimensional exhaustive search was conducted for the user area on the ground. In the case of more than two users, using low complexity algorithm to find the optimal or near optimal UAV position is still a problem to be solved.

In this paper, we tend to maximize the sum energy and weighted sum energy of the ERs by optimizing the placement of UAV in more than two user's case. In general, the problem is NP-hard, we first convert it to an equivalent formulation, which is more easier to handle. On this basis, an iterative coordinate updating algorithm is proposed. The simulation results show that the algorithm has lower complexity and almost the same energy compared with the two-dimensional exhaustive search algorithm.

The rest of this article is organized as follows. Section 2 introduces the WPT system model based on UAV. Section 3, the iterative coordinate updating algorithm for solving sum-energy maximization problems is presented. In section 4, numerical results are given to verify the effectiveness of the proposed algorithm. Finally, the fifth part draws the conclusion.

2   System Model and Problem Formulation

We consider a UAV-enabled WPT system with K users, where a UAV is scheduled to charge K ERs on the ground by transferring wireless energy. We set the power transfer time of the UAV as T. The location of ER $k = 1, \cdots, K$ is fixed on the ground. We assume that UAV flies at a fixed altitude H. And under the condition of transmit power P, we need to find an optimal hovering position to maximize the sum of the energy received by all ERs. We consider the channel model of free-space path loss between the UAV and each ER is same as [8,10]. The channel power gain from the UAV to ER $k \in \{1, \cdots, K\}$ is modeled as $h_k = \beta \frac{1}{d_k^2}$, where $d_k = \sqrt{(x - x_i)^2 + (y - y_i)^2 + H^2}$ is their distance and β denotes the channel power gain at a reference distance of $d = 1$ m.

Firstly, we want to find the best placement of the UAV to maximize the sum energy to the K user for the given time T as follows. Assume the UAVs location is (x, y, H), the k-th ground user's location is (x_k, y_k), $k = 1, \cdots, K$, we want to find the optimal (x, y) to the following problem.

$$\underset{x,y}{\text{maximize}} \quad \sum_{i=1}^{K} \frac{PT\beta}{(x - x_i)^2 + (y - y_i)^2 + H^2} \tag{1}$$

(1) is a non-convex optimization problem. Moreover, (1) can be regarded as a multiple-ratio fractional programming which is generally NP-hard problem [11]. Therefore, global optimization methods such as branch-and-bound or 2D exhaustive search method need to be exploited to find the optimal solution [12,13]. However, the complexity of those methods is too much high. In next section, we will give a simple iteration algorithm based on the equivalent problem of (1) to maximize the sum energy of the senors.

3   Iterative Coordinate Update Algorithm

Because the numerators and denominators of object function are constant and convex function respectively, the constraint set is nonempty convex set. (1) is multiple-ratio concave-convex fractional programming. Since the denominator of object function $PT\beta$ is constant for our optimization problem. Thus, we only need to optimize the following problem.

$$\underset{x,y}{\text{maximize}} \sum_{i=1}^{K} \frac{1}{(x - x_i)^2 + (y - y_i)^2 + H^2} \qquad (2)$$

As problem (2) can be viewed as sum-of-ratios problem, we can use a novel quadratic transform technique to rewritten it as equivalent optimization problem by the lemma in[14]. So, the equivalent problem is as follow.

$$\underset{x,y,\mathbf{z}}{\text{maximize}} \sum_{i=1}^{K} \left(2z_i - z_i^2 \left((x - x_i)^2 + (y - y_i)^2 + H^2\right)\right) \qquad (3)$$
$$\text{subject to} \quad x, y, z_1, \cdots, z_K \in R,$$

where $\mathbf{z}$ is the container for the variables $z_1, \cdots, z_K$. For (3), we can optimize the primal variable x, y and the auxiliary variable $z_1, \cdots, z_K$ iteratively by block coordinate ascent algorithm. For a given $\mathbf{z} = (z_1, \cdots, z_K)$, the object function of (3) is concave function with respective to x and y. Let $w_i = z_i^2 \geq 0 (i = 1, \cdots, K)$, then, (3) is equivalent to the following convex optimization problem for a fixed $(w_1, \cdots, w_K)$.

$$\underset{x,y}{\text{minimize}} \sum_{i=1}^{K} w_i \left((x - x_i)^2 + (y - y_i)^2\right) \qquad (4)$$
$$\text{subject to} \quad x, y \in R,$$

Next, using the first order optimality condition for problem (4), we given the closed-form solution to (4) as follows.

Lemma 1. *For a given $w_i \geq 0, i \in \{1, \ldots, K\}$, the optimal solution to (4) is given by*

$$x = \frac{\sum_{i=1}^{K} x_i w_i}{\sum_{i=1}^{K} w_i}, y = \frac{\sum_{i=1}^{K} y_i w_i}{\sum_{i=1}^{K} w_i}. \qquad (5)$$

Proof. Let $f(x,y) = \sum_{i=1}^{K} w_i \left((x - x_i)^2 + (y - y_i)^2\right)$, we have $\frac{\partial^2 f}{\partial x^2} = 2w_i \geq 0$, $\frac{\partial^2 f}{\partial y^2} = 2w_i \geq 0$, $\frac{\partial^2 f}{\partial x \partial y} = 0$. Therefore, $f(x,y)$ is convex function. Moreover, (4) is a convex optimization problem. The optimal solution $x = \dfrac{\sum\limits_{i=1}^{K} x_i w_i}{\sum\limits_{i=1}^{K} w_i}, y = \dfrac{\sum\limits_{i=1}^{K} y_i w_i}{\sum\limits_{i=1}^{K} w_i}$

to (4) can be obtained by the first order condition $\frac{\partial f}{\partial x} = 0, \frac{\partial f}{\partial y} = 0$. Thus, lemma 1 is held.

For a fixed x, y, because (3) is concave function with respective to $z_1, \cdots, z_n$. Therefore, the optimal solution $z_1, \cdots, z_n$ to (3) is given by

$$z_i = \frac{1}{(x - x_i)^2 + (y - y_i)^2 + H^2}, i = 1, \cdots, K. \tag{6}$$

Using Lemma 1 and (6), we give an iterative coordinate update algorithm as follows.

Because the algorithm proposed in Algorithm 1 can be viewed as the block coordinate descent (BCD) method mentioned in [15] to solve problem (3). Therefore, Algorithm 1 will be convergent due to the object function of (3) obtained by $(x^t, y^t, \mathbf{z}^t)$ is bounded monotonic sequence. Let (x^*, y^*, z^*) be the limit point of the sequence $(x^t, y^t, \mathbf{z}^t), t = 0, 1, \cdots,$. Then, the point (x^*, y^*) will be stationary point of problem (1) by the following theorem.

Algorithm 1. Iterative Coordinate Update Algorithm (ICUA)

Initialization: Let $x^0 = x_{ini}, y^0 = y_{ini}$ such that (x_{ini}, y_{ini}) is located in the convex hull of the region of $(x_i, y_i), i = 1 \cdots, K$;
 compute $z_i^0 = \frac{1}{(x^0 - x_i)^2 + (y^0 - y_i)^2 + H^2}, i = 1, \cdots, K$;
 compute $w_i^0 = (z_i^0)^2, i = 1, \cdots, K$;
 Set $t = 0$;
 repeat
 t=t+1;
 Update (x^t, y^t):
$$x^t = \frac{\sum\limits_{i=1}^{K} x_i w_i^{t-1}}{\sum\limits_{i=1}^{K} w_i^{t-1}}, y^t = \frac{\sum\limits_{i=1}^{K} y_i w_i^{t-1}}{\sum\limits_{i=1}^{K} w_i^{t-1}} \text{ where } w_i^{t-1} = (z_i^{t-1})^2, i = 1, \cdots, K,$$
 Update $\mathbf{z}^t = (z_1^t, \cdots, z_K^t)$:
 $z_i^t = \frac{1}{(x^t - x_i)^2 + (y^t - y_i)^2 + H^2}, i = 1, \cdots, K$;
 until Some stopping criteria is met
 output The coordinate (x, y) of UAV is given by $x = x^t$ and $y = y^t$.

Theorem 1 *Let (x^*, y^*, z^*) be the limit point of the sequence $(x^t, y^t, \mathbf{z}^t), t = 0, 1, \cdots,$ then (x^*, y^*) is the stationary point of problem (1).*

Proof. Let $F(x,y) = \sum_{i=1}^{K} \frac{PT\beta}{(x-x_i)^2+(y-y_i)^2+H^2}$ be the object function of problem (1). The stationary point of problem (1) must satisfied $\frac{\partial F(x,y)}{\partial x} = 0, \frac{\partial F(x,y)}{\partial y} = 0$. Therefore, the stationary point is expressed as follow.

$$
x = \frac{\sum_{i=1}^{K} x_i \left(\frac{1}{(x-x_i)^2+(y-y_i)^2+H^2} \right)^2}{\sum_{i=1}^{K} \left(\frac{1}{(x-x_i)^2+(y-y_i)^2+H^2} \right)^2},
$$

$$
y = \frac{\sum_{i=1}^{K} y_i \left(\frac{1}{(x-x_i)^2+(y-y_i)^2+H^2} \right)^2}{\sum_{i=1}^{K} \left(\frac{1}{(x-x_i)^2+(y-y_i)^2+H^2} \right)^2}.
$$
(7)

From algorithm 1, we have

$$
x^{t+1} = \frac{\sum_{i=1}^{n} x_i w_i^t}{\sum_{i=1}^{K} w_i^t}, y^{t+1} = \frac{\sum_{i=1}^{n} y_i w_i^t}{\sum_{i=1}^{K} w_i^t}
$$
(8)

where $w_i^t = (z_i^t)^2, i = 1, \cdots, K$, and $z_i^t = \frac{1}{(x^t-x_i)^2+(y^t-y_i)^2+H^2}, i = 1, \cdots, K$. Substitute $w_i^t = \left(\frac{1}{(x^t-x_i)^2+(y^t-y_i)^2+H^2} \right)^2$ into (8), we have

$$
x^{t+1} = \frac{\sum_{i=1}^{K} x_i \left(\frac{1}{(x^t-x_i)^2+(y^t-y_i)^2+H^2} \right)^2}{\sum_{i=1}^{K} \left(\frac{1}{(x^t-x_i)^2+(y^t-y_i)^2+H^2} \right)^2},
$$

$$
y^{t+1} = \frac{\sum_{i=1}^{K} y_i \left(\frac{1}{(x^t-x_i)^2+(y^t-y_i)^2+H^2} \right)^2}{\sum_{i=1}^{K} \left(\frac{1}{(x^t-x_i)^2+(y^t-y_i)^2+H^2} \right)^2}.
$$
(9)

From (9), we have

$$
x^* = \lim_{t \to \infty} x^{t+1} = \frac{\sum_{i=1}^{K} x_i \left(\frac{1}{(x^*-x_i)^2+(y^*-y_i)^2+H^2} \right)^2}{\sum_{i=1}^{K} \left(\frac{1}{(x^*-x_i)^2+(y^*-y_i)^2+H^2} \right)^2},
$$
(10)

the last two equalities follow from the continuity of $f_1(x, y) = \dfrac{\sum\limits_{i=1}^{K} x_i \left(\frac{1}{(x - x_i)^2 + (y - y_i)^2 + H^2} \right)^2}{\sum\limits_{i=1}^{K} \left(\frac{1}{(x - x_i)^2 + (y - y_i)^2 + H^2} \right)^2}$. Using the same argument, we have

$$y^* = \frac{\sum\limits_{i=1}^{K} y_i \left(\frac{1}{(x^* - x_i)^2 + (y^* - y_i)^2 + H^2} \right)^2}{\sum\limits_{i=1}^{K} \left(\frac{1}{(x^* - x_i)^2 + (y^* - y_i)^2 + H^2} \right)^2}, \tag{11}$$

From (10) and (11), we can see that the limit point (x^*, y^*) of the sequence (x^t, y^t) must satisfy equation (7). Therefore, Theorem 1 is held.

In proposed algorithm, we have chosen the initial point in the convex hull of the region of (x_i, y_i), where we can choose randomly from (x_i, y_i) for some $i \in 1, \cdots, K$. Indeed, the initial point can even be choosen outside of the the convex hull of the region of (x_i, y_i). After the algorithm runs only one time, the new location of the UAV will be inside the the convex hull of the region of (x_i, y_i). The initial point can be chosen by $x_{ini} = \dfrac{\sum\limits_{i=1}^{n} x_i}{n}, y_{ini} = \dfrac{\sum\limits_{i=1}^{n} y_i}{n}$. However, even for two users' case, the problem has two optimal locations when the hight of the UAV is much less than the distance of the two users. Next, we give a sufficient condition that the problem has a unique optimal solution as follows.

For two ERs case, [8] has proven that there is a unique placement of UAV when the hight of the UAV is larger than a threshold. Next, we give a similar result for more than two users case.

In order to obtain the most essential understanding of the algorithm, we give a sufficient condition among the H and the square of the region of ERs to show that it can converge to a global optimization as follows.

Theorem 2. *The proposed algorithm will be convergent to the global optimal solution when H satisfied the following condition:*

$$H > 2\sqrt{2} \max\{x_{\max}, y_{\max}\} (x_{\max} + y_{\max}) \tag{12}$$

Proof. The main idea is to use the Banach fixed-point theorem [16]. We only need to prove the jacobian matrix of the proposed iteration is less than 1 under some matrix norm or spectral radius of the jacobian matrix is less than 1. From the proposed algorithm and (7), we can see that the iterative algorithm is fixed point of the following function. Consider a function $\mathbf{v} = \mathbf{f}(x, y)$, such that

$$\begin{bmatrix} v_1 \\ v_2 \end{bmatrix} = \begin{bmatrix} f_1(x, y) \\ f_2(x, y) \end{bmatrix} \tag{13}$$

where $f_1(x, y)$ and $f_2(x, y)$ is given by

$$f_1(x, y) = \frac{\sum\limits_{i=1}^{n} x_i \left(\frac{1}{(x - x_i)^2 + (y - y_i)^2 + H^2} \right)^2}{\sum\limits_{i=1}^{n} \left(\frac{1}{(x - x_i)^2 + (y - y_i)^2 + H^2} \right)^2}, \tag{14}$$

$$f_2(x, y) = \frac{\sum\limits_{i=1}^{n} y_i \left(\frac{1}{(x-x_i)^2+(y-y_i)^2+H^2}\right)^2}{\sum\limits_{i=1}^{n} \left(\frac{1}{(x-x_i)^2+(y-y_i)^2+H^2}\right)^2}. \tag{15}$$

Then, from (9), the proposed algorithm can be rewritten as the following iteration: Pick an initial point $(x^0, y^0) \in D$, where $D = [x_{min}, x_{max}] \times [y_{min}, y_{max}]$ and defined for $k = 0, 1, 2, \cdots$,

$$\begin{bmatrix} x^{(k+1)} \\ y^{(k+1)} \end{bmatrix} = \begin{bmatrix} f_1\left(x^{(k)}, y^{(k)}\right) \\ f_2\left(x^{(k)}, y^{(k)}\right) \end{bmatrix} \tag{16}$$

Then, the Jacobian matrix of $\mathbf{f}$ is given by

$$J = \begin{bmatrix} \frac{\partial f_1}{\partial x} & \frac{\partial f_1}{\partial y} \\ \frac{\partial f_2}{\partial x} & \frac{\partial f_2}{\partial y} \end{bmatrix} \tag{17}$$

where

$$\frac{\partial f_1(x,y)}{\partial x} = \frac{-4\sum\limits_{i=1}^{K} x_i(x-x_i)z_i^3(x,y)\sum\limits_{i=1}^{n} z_i^2(x,y) + 4\sum\limits_{i=1}^{n} x_i z_i^2(x,y)\sum\limits_{i=1}^{K} z_i^3(x,y)(x-x_i)}{\left(\sum\limits_{i=1}^{K} z_i^2(x,y)\right)^2}$$

$$\frac{\partial f_1(x,y)}{\partial y} = \frac{-4\sum\limits_{i=1}^{K} x_i(y-y_i)z_i^3(x,y)\sum\limits_{i=1}^{n} z_i^2(x,y) + 4\sum\limits_{i=1}^{n} x_i z_i^2(x,y)\sum\limits_{i=1}^{K} z_i^3(x,y)(y-y_i)}{\left(\sum\limits_{i=1}^{K} z_i^2(x,y)\right)^2}$$

$$\frac{\partial f_2(x,y)}{\partial x} = \frac{-4\sum\limits_{i=1}^{K} y_i(x-x_i)z_i^3(x,y)\sum\limits_{i=1}^{n} z_i^2(x,y) + 4\sum\limits_{i=1}^{n} y_i z_i^2(x,y)\sum\limits_{i=1}^{K} z_i^3(x,y)(x-x_i)}{\left(\sum\limits_{i=1}^{K} z_i^2(x,y)\right)^2}$$

$$\frac{\partial f_2(x,y)}{\partial y} = \frac{-4\sum\limits_{i=1}^{K} y_i(y-y_i)z_i^3(x,y)\sum\limits_{i=1}^{n} z_i^2(x,y) + 4\sum\limits_{i=1}^{n} y_i z_i^2(x,y)\sum\limits_{i=1}^{K} z_i^3(x,y)(y-y_i)}{\left(\sum\limits_{i=1}^{K} z_i^2(x,y)\right)^2}$$

$z_i(x, y) = \frac{1}{(x-x_i)^2+(y-y_i)^2+H^2}$.

We have $z_i(x, y) \in [z_{min}, z_{max}]$ when $(x, y) \in D$, where $z_{min} = \frac{1}{(x_{max}-x_{min})^2+(y_{max}-y_{min})^2+H^2}$, $z_{max} = \frac{1}{H^2}$. Then, the upper bound of $\left|\frac{\partial f_1(x,y)}{\partial x}\right|$ for $(x, y) \in D$ is given by

$$\left|\frac{\partial f_1(x,y)}{\partial x}\right| \leq \frac{4\left|\sum\limits_{i=1}^{K} x_i(x-x_i)z_i^3(x,y)\sum\limits_{i=1}^{n} z_i^2(x,y)\right| + 4\left|\sum\limits_{i=1}^{n} x_i z_i^2(x,y)\sum\limits_{i=1}^{K} z_i^3(x,y)(x-x_i)\right|}{\left(\sum\limits_{i=1}^{K} z_i^2(x,y)\right)^2}$$

$$\leq \frac{4x_{max}^2\left|\sum\limits_{i=1}^{K} z_i^3(x,y)\sum\limits_{i=1}^{n} z_i^2(x,y)\right| + 4x_{max}^2\left|\sum\limits_{i=1}^{n} z_i^2(x,y)\sum\limits_{i=1}^{K} z_i^3(x,y)\right|}{\left(\sum\limits_{i=1}^{K} z_i^2(x,y)\right)^2} \tag{18}$$

$$= \frac{8x_{max}^2\left|\sum\limits_{i=1}^{K} z_i^3(x,y)\sum\limits_{i=1}^{n} z_i^2(x,y)\right|}{\left(\sum\limits_{i=1}^{K} z_i^2(x,y)\right)^2} \leq 8x_{max}^2 z_{max} = \frac{8x_{max}^2}{H^2}$$

Using the same argument, we have

$$\left|\frac{\partial f_1(x,y)}{\partial y}\right| \le \frac{8x_{\max}y_{\max}}{H^2} \tag{19}$$

$$\left|\frac{\partial f_2(x,y)}{\partial x}\right| \le \frac{8x_{\max}y_{\max}}{H^2} \tag{20}$$

$$\left|\frac{\partial f_2(x,y)}{\partial y}\right| \le \frac{8y_{\max}^2}{H^2} \tag{21}$$

Using (18)-(21), we can find an upper bound for $||J||_\infty$.

$$
\begin{aligned}
||J||_\infty &= \max\left(\left|\frac{\partial f_1}{\partial x}\right| + \left|\frac{\partial f_1}{\partial y}\right|, \left|\frac{\partial f_2}{\partial x}\right| + \left|\frac{\partial f_2}{\partial y}\right|\right) \\
&\le \max\left(\frac{8x_{\max}^2}{H^2} + \frac{8x_{\max}y_{\max}}{H^2}, \frac{8x_{\max}y_{\max}}{H^2} + \frac{8y_{\max}^2}{H^2}\right) \\
&\le \frac{8\max\{x_{\max},y_{\max}\}(x_{\max}+y_{\max})}{H^2}
\end{aligned}
\tag{22}
$$

Then, we have $||J||_\infty < 1$ because (12) is held. Therefore, f is a contraction map in D. Therefore, the proposed algorithm will converge to the unique fixed point, which is unique stationary point of the optimization problem (2). As the stationary point is unique, this stationary point is the global optimal for (2) because the minimum value point of (2) can't be inside the region D. This is can be see from that the infimum of (2) is zero as x and y approaches infinity. Thus, Theorem 2 is held.

In the proof of Theorem 2, we have assumed all the coordinate of ERs is located in the first quadrant which means that $x_i \ge 0$ and $y_i \ge 0$ is held. Without loss of generality, this condition can be assumed to be hold because we can always change the coordinate of each ERs by coordinate transformation to satisfy it. Moreover, the sufficient condition given by (12) for H is general not tight. We may choose different matrix norm to find some other lower bound for H.

Next, we extend the proposed algorithm to the weight sum energy maximization problem. The weigh sum energy maximization problem is given as follows.

$$\max_{x,y} \sum_{i=1}^{K} \frac{a_i P T \beta}{(x - x_i)^2 + (y - y_i)^2 + H^2} \tag{23}$$

where $a_i > 0$ is weight factor for user i to emphasizes the different priority for harvesting energy at sensor node i. In real communication system, the weight factor for user i can be choose to balance the harvesting energy between the SNs. For example, for the sensors who has less remain energy, they can choose the higher weight. A simpler weight for each user can be proportion to the percentage of energy it has consumed compare with its battery storage capacity.

Because maximizing the sum of the weighted energies is also a sum-of-ratios problem. We can use the same argument as the sum energy maximization problem to design the iterative coordinate update algorithm. Use Lemma 1 and omit the constant value $PT\beta$ in the numerator in problem (23), then the weight sum energy maximization problem is equivalent to the following problem.

$$\text{maximize}_{x,y,\mathbf{z}} \quad \sum_{i=1}^{K} \left(2z_i\sqrt{a_i} - z_i^2 \left((x-x_i)^2 + (y-y_i)^2 + H^2 \right) \right) \tag{24}$$

$$\text{subject to} \quad x, y, z_1, \cdots, z_K \in R,$$

where $\mathbf{z}$ refers to a collection of variables $z_1, \cdots, z_K$. Using the same method as the sum energy maximization problem, we can extend the ICUA to handle weight sum energy maximization problem by algorithm 2 as follows.

Algorithm 2. ICUA for weight sum energy maximization

Initialization: Let $x^0 = x_{ini}, y^0 = y_{ini}$ such that (x_{ini}, y_{ini}) is located in the convex hull of the region of $(x_i, y_i), i = 1 \cdots, K$;
compute $z_i^0 = \frac{\sqrt{a_i}}{(x^0-x_i)^2+(y^0-y_i)^2+H^2}, i = 1, \cdots, K$;
compute $w_i^0 = \left(z_i^0\right)^2, i = 1, \cdots, K$;
Set $t = 0$;
repeat
t=t+1;
Update $\left(x^t, y^t\right)$:
$$x^t = \frac{\sum\limits_{i=1}^{K} x_i w_i^{t-1}}{\sum\limits_{i=1}^{K} w_i^{t-1}}, y^t = \frac{\sum\limits_{i=1}^{K} y_i w_i^{t-1}}{\sum\limits_{i=1}^{K} w_i^{t-1}} \quad \text{where } w_i^{t-1} = (z_i^{t-1})^2, i = 1, \cdots, K,$$
Update $\mathbf{z}^t = \left(z_1^t, \cdots, z_K^t\right)$:
$z_i^t = \frac{\sqrt{a_i}}{(x^t-x_i)^2+(y^t-y_i)^2+H^2}, i = 1, \cdots, K$;
until Some stopping criteria is met
output The coordinate (x, y) of UAV is given by $x = x^t$ and $y = y^t$.

---

4   Simulation Results

This section presents the simulation results of the proposed algorithm and compares it with the two heuristic algorithm. The main idea of the first heuristic algorithm 2D grid search method is to find an optimal UAV position by searching for M^2 points in an $M \times M$ grid [8,10]. The second heuristic algorithm is searching with four directions under a given step size from the same initial point with proposed algorithm $x_{ini} = \frac{\sum\limits_{i=1}^{n} x_i}{n}, y_{ini} = \frac{\sum\limits_{i=1}^{n} y_i}{n}$ until it reaches the maximum number of iterations. For the sake of brevity, we name the first and second heuristic algorithm as heuristic I and heuristic II.

4.1 Simulation Results Versus the Number of Users

The number of SNs is set from 4 to 60. The weight factor for users is randomly generated from $[0, 1]$. β is set to be 10^{-2}, the height of the UAV H ,the transmission power of UAV P and energy transfer time T are respectively 5 m,40 dBm and 1000 s.The convergence threshold of the proposed algorithm and heuristic II algorithm is set to be 10^{-1}.The random distribution region of the sensor is 30×30 m^2.The simulation results are obtained randomly according to the sensor position. The heuristic I search method have divided the region of the SNs into 30×30 grid-points to find the best location for UAV. The step size for heuristic II is 0.5 m. The maximum iterations for proposed algorithm and heuristic II algorithm is set to be 900, which is the same with number of the searching point as the heuristic I algorithm.

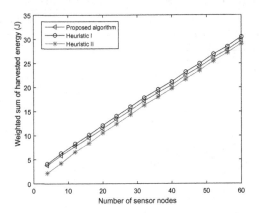

Fig. 1. Weighted sum of energy versus number of SNs

As can be seen from Fig. 1, with the increase in the number of SNs, the energy weighted sum of the three algorithms is increasing. The proposed algorithm has little performance loss compare with the Heuristic I algorithm. In addition, for all SNs quantities, this algorithm is superior to heuristic I algorithm. When the number of SNs is 28, the proposed algorithm can obtained 95.79% and 106.42% weighted sum energy compared with the Heuristic I and II algorithm, respectively.

The normalized CPU time versus number of SNs is given in Fig. 2. And you can see that the number of SNs reaches 32, the proposed algorithm can save about 94% and 34% CPU time compared with the Heuristic I and II algorithm, respectively.

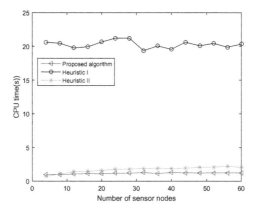

Fig. 2. Normalized CPU time versus number of SNs

4.2   Simulation Results Versus Number of Grids (step Size)

When the grid number (step size) tends to infinity, the heuristic I algorithm can obtain the optimal solution. However, this results in huge CPU computing time. Simulation parameters are as follows:

The number of SNs is set to 30. Other parameters are the same as subsection A. The sensors are randomly distributed in a two-dimensional area of 10×10 m^2,and We can see that in section B, the performance of our proposed algorithm is slightly different from that of heuristic I algorithm. Therefore, the proposed algorithm may not find the optimal solution in some cases. However, we found that as the size of SNs region decreases, our proposed algorithm will find the optimal solution just like the heuristic I algorithm. In brief, as the size of the region for SNs decreases, the sufficient condition for Theorem 2 has more probability to be satisfied. Then, the optimal solution can be sought out by the

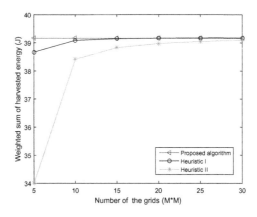

Fig. 3. Weighted sum of energy versus the number of grids

proposed algorithm. The simulation result of 10^3 is obtained randomly according to sensor's position and weighted factor. Next,we will show the influence of grid number on three algorithms.

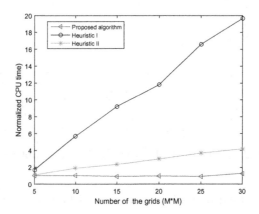

Fig. 4. Normalized CPU time versus number of grids

From above discussion, it can seen that the proposed algorithm can obtain the optimal solution as the heuristic algorithm with much less computation time.

5   Conclusion

In this paper, we studied the placement problem of the UAV that maximizes the sum energy and weighted sum energy of the ERs.Using a novel quadratic transform technique to transfer the non-convex optimization problem to an equivalent problem, we proposed a low-complexity iterative coordinate update algorithm to solve the placement problem. Simulations have shown that the proposed algorithm performs very close to the 2D-exhaustive search algorithm with a significant reduction in run time.

References

1. Niyato, D., Kim, D.I., Maso, M., Han, Z.: Wireless powered communication networks: research directions and technological approaches. IEEE Wirel. Commun. **24**(6), 88–97 (2017)
2. Zhai, C., Li, Y., Qiao, L.: Wireless power transfer based adaptive relaying with energy accumulation. IEEE Trans. Veh. Technol. **69**, 11014–11027 (2020)
3. Chen, J., Ghannam, R., Imran, M., Heidari, H.: Wireless power transfer for 3D printed unmanned aerial vehicle (UAV) systems. In: 2018 IEEE Asia Pacific Conference on Postgraduate Research in Microelectronics and Electronics (PrimeAsia), pp. 72–76 (2018)

4. Jiang, M., Li, Y., Zhang, Q., Qin, J.: Joint position and time allocation optimization of UAV enabled time allocation optimization networks. IEEE Trans. Commun. **67**(5), 3806–3816 (2019)

5. Huang, F., et al.: UAV-assisted SWIPT in internet of things with power splitting: trajectory design and power allocation. IEEE Access **7**, 68260–68270 (2019)

6. Du, Y., Yang, K., Wang, K., Zhang, G., Zhao, Y., Chen, D.: Joint resources and workflow scheduling in UAV-enabled wirelessly-powered MEC for IoT systems. IEEE Trans. Veh. Technol. **68**(10), 10187–10200 (2019)

7. Xu, J., Zeng, Y., Zhang, R.: UAV-enabled wireless power transfer: trajectory design and energy region characterization. In: 2017 IEEE Globecom Workshops (GC Wkshps), pp. 1–7(2017)

8. Xu, J., Zeng, Y., Zhang, R.: UAV-enabled wireless power transfer: trajectory design and energy optimization. IEEE Trans. Wirel. Commun. **17**(8), 5092–5106 (2018)

9. Xie, L., Xu, J.: Cooperative trajectory design and resource allocation for a two-UAV two-user wireless powered communication system. In: 2018 IEEE International Conference on Communication Systems (ICCS), pp. 7–12 (2018)

10. Xie, L., Xu, J., Zhang, R.: Throughput maximization for UAV-enabled wireless powered communication networks. IEEE Internet Things J. **6**(2), 1690–1703 (2019)

11. Freund, R.W., Jarre, F.: Solving the sum-of-ratios problem by an interior-point method. J. Global Optim. **19**(1), 83–102 (2001)

12. Wang, Z., Xiao, W., Wan, X., Fan, Z.: Price-based power control algorithm in cognitive radio networks via branch and bound. IEICE Trans. Inf. Syst. **102**(3), 505–511 (2019)

13. Groppen, V.O.: Experimental verification of efficiency of some composite algorithms. In: 2018 International Conference on Applied Mathematics and Computational Science (ICAMCS.NET), Budapest, Hungary, pp. 43–435 (2018)

14. Shen, K., Yu, W.: Fractional programming for communication systems-part I: power control and beamforming. IEEE Trans. Sig. Process. **66**(10), 2616–2630 (2018)

15. Zhan, C., Zeng, Y., Zhang, R.: Energy-efficient data collection in UAV enabled wireless sensor network. IEEE Wirel. Commun. Lett. **7**(3), 328–331 (2018)

16. Boudaoui, A., Laksaci, N.: Some fixed-point theorems for block operator matrix. In: 2020 2nd International Conference on Mathematics and Information Technology (ICMIT), Adrar, Algeria, pp. 80–85 (2020)

Activate Cost-Effective Mobile Crowd Sensing with Multi-access Edge Computing

Zhengzhe Xiang[1], Shuiguang Deng[2], Yuhang Zheng[2], Dongjing Wang[3], Cheng Zhang[2], Yuanyi Chen[1], and Zengwei Zheng[1(✉)]

[1] Zhejiang University City College, Hangzhou, China
{xiangzz,chenyuanyi,zhengzw}@zucc.edu.cn
[2] Computer College, Zhejiang University, Hangzhou, China
{dengsg,zyhxxds_ludwig,coolzc}@zju.edu.cn
[3] School of Computer Science and Technology, Hangzhou Dianzi University, Hangzhou, China
Dongjing.Wang@hdu.edu.cn

Abstract. Recently, the mobile crowd sensing (MCS) technique is believed to be an important role in multi-source data acquisition tasks. With devices or people with different sensing abilities in the cities, we can easily split and distribute the complex task in an appropriate way so that those devices or people can be stimulated to collect data within different scopes individually, while the results of them can be analyzed and integrated collaboratively to fulfill that complex task. However, in typical centralized architecture, the latency brought by unstable and time-consuming long-distance network transmission limits the development of MCS. The multi-access edge computing (MEC) technique is now regarded as the key tool to solve this problem. By establishing a service provisioning system based at the edge of the network, the latency can be reduced and the analysis or integration can also be conducted in time with the help of corresponding services deployed on nearby edge servers. However, as the edge servers are resource-limited, the sensing abilities vary among devices or people, and the budget of fulfilling a task is determined, we should be more careful in task assignment and service deployment. In this paper, we investigate the relationship between the task quality and the cost in the MEC-based MCS system and propose the analysis framework of it based on two classical cost-performance balancing problems. Besides, we conduct comprehensive experiments to evaluate the performance of our approach. The results show that the proposed approach can easily obtain exact solutions, and the factors that may impact the results are also adequately explored.

Keywords: Multi-access Edge Computing · Mobile crowd sensing · Task assignment · Service deployment · Incentive mechanism

H. Gao et al. (Eds.): ChinaCom 2020, LNICST 352, pp. 78–97, 2021.
https://doi.org/10.1007/978-3-030-67720-6_6

1   Introduction

With the development of mobile computing technology, we are now embracing an era of mobile devices and services [1]. According to the report of GSMA[1], about 5.1 billion people around the world have subscribed to mobile applications, and the number will increase at an average annual growth rate of 1.9% before 2025. As a result, mobile devices and mobile applications are becoming increasingly important and remolded the communication between people and machines. The tremendous increasing number of mobile users and devices has created a huge market that draws the attention of all the world. To make themselves to be the best one among the competitors, mobile application enterprises all want to better understand the preferences of these users and discover their underlying behavior patterns. Therefore, the researchers of these enterprises will always try their best to collect users' behavior records or even interview their target users directly, because they are sure that these structured/unstructured and sequential/non-sequential context data will help them to build a general user portrait model to analyze and predict users' future behaviors. However, as people are rarely willing to provide their data because of their subconscious privacy protection and the worry about the energy consumption of the external computation, it will be hard for the application developers to collect enough high-quality data for their Artificial Intelligence (AI) models legally. To solve this problem, more and more developers turn to the mobile crowd sensing (MCS) technique. Specifically, MCS is a human-oriented technique that leverages the built-in sensors of users' mobile devices as well as the involvement of users to collect data. It does not only care about the effectiveness and accuracy of the data but also focuses on the issues of stimulating users to share their data. With the MCS technique, a reliable publish/subscribe interaction framework is established between users and developers so that high-quality data can be collected with the admissions and willingness of the users if the developers can pay for their cooperation. However, the latency brought by long-distance transmission and traffic congestion of huge data in the network, as well as the energy consumption brought by data pre-processing limits the applications of MCS in typical centralized architecture.

Fortunately, Multi-access Edge Computing (MEC) technique is proposed to solve the aforementioned problems [2–4]. MEC is a novel paradigm that has emerged recently as a reinforcement of mobile cloud computing, to optimize the mobile resource usage and wireless network to provide context-aware services [5,6]. With the help of MEC, computation and transmission between mobile devices and the cloud are partly migrated to edge servers. Therefore, users can easily connect to their nearby edge servers via wireless network [7] and offload their tasks to them. The short-distance connection between users and edge servers can dramatically reduce the latency, and the computation capability of the edge servers are quite qualified to finish those conventional tasks. What's more, with the help of the container platforms in the limelight like **Kubernetes**, it will be easy to manage services (e.g. the data pre-processing services) in the

---

[1] https://www.gsmaintelligence.com/.

MEC environment. However, these advantages cannot be the causes of the carelessness in planning the multi-source data acquisition—if the sensing tasks are not assigned to appropriate users, the data acquisition task may even obtain lower-quality data with much higher cost. More critically, as the edge servers are all resource-constrained [8,9], if the data pre-processing services are not deployed on appropriate edge servers, there would be no enough resources for them to run. Thus, it will be important to design a task assignment scheme as well as a service deployment scheme to balance the quality and cost. The main contributions of this paper are:

1. We investigate the relationship between the task quality and the cost in the MEC-based MCS problem and propose the analysis framework of it based on two classical cost-performance balancing problems.
2. We mathematically model the former problems which aim to balance the task quality and the cost under the constraints of application developers' budget, available resources of edge servers, and the capacities of users as a mixed-integer quadratic programming (MIQP) problem.
3. We conduct a series of experiments to evaluate the results of the solutions and show the improvement compared with other existing baselines. Besides, different configurations of the MEC-based MCS system are investigated to explore the impacts of related factors.

The rest of this paper is organized as follows. In Sect. 2, we use a simple example to show how the task assignment scheme and service development scheme can impact the quality and cost in the MEC-based MCS system. Section 3 presents definitions, concepts and components of the proposed problem. Section 4 describes the approaches we proposed to solve this problem. Section 5 shows the experimental results including the factors that affect our algorithms. Section 6 highlights the related work of the incentive mechanism and task assignment approaches. Finally, Sect. 7 concludes our contribution and outlines future work.

2   Motivation Scenario

In this section, we will briefly introduce the mechanism of how multi-access edge computing techniques can be used to help to optimize the mobile crowd sensing tasks by giving an example of a multi-source data acquisition task in customer portrait construction.

Figure 1 gives an overview of the West Lake business district in Hangzhou city. The wide variety of shops in this business district attract a large number of users every day, and the behavior records of these users in the business district will be a good data source for constructing customer portraits. Therefore, data researchers hope to use mobile crowd sensing technique to collect these data. Specifically, the data that is expected to be collected may include T_1) recent walking distance in the business district T_2) the most frequently visited shops T_3) recent bills in the business district. However, not all stimulated users are willing or able to provide all the data the researchers want, and even these

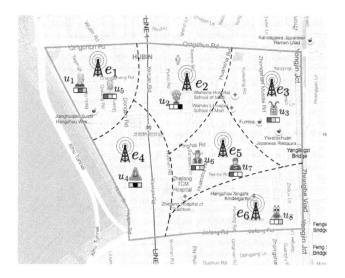

Fig. 1. A customer portrait example

users will have differences in the quality of the provided data. At this time, in order to construct a customer portrait that reflects the "average customer", the data researchers will assume that these customers are sampled from the same "customer space", and assign the three types of data collection tasks to different users, and use the MEC server nearby to pre-process the collected data.

As shown in this figure, edge servers distributed in this business district have their own serving area so that the users can easily connect to the nearest one. There is a rectangle group under every user to describe their willingness for providing different types of data (T_1–T_3), and the user can provide data with better quality if his corresponding rectangles are darker in color. Suppose the users will be equally rewarded if they are assigned with one task. Therefore, we can let user u_6 and u_7 to provide data T_1, let user u_4 to provide data T_2 and let user u_2 to provide data T_3. Meanwhile, the pre-processing module of T_1 can be deployed on e_5, T_2 on e_4 and T_3 on e_2. It is obvious that we can move the pre-processing module for T_3 to e_5 and let u_6 to provide data, but the quality will get worse because user u_6 always pay in cash and the transaction data on his device is not complete.

3   System Model and Problem Description

Although the example in Sect. 2 has given a brief introduction about the scenario, more details like costs and capacities are ignored in it. Therefore, we will give a complete system model in this section and then describe the quality-cost balancing problem (Fig. 2).

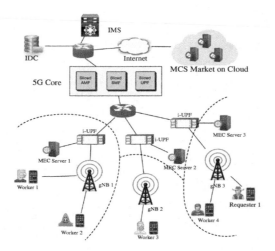

Fig. 2. An illustration of the MEC-based MCS system

3.1 System Entities

In more general MCS systems, the users who submit tasks are called *requesters*, and the users who fulfill the tasks with their mobile devices are called *workers*. Suppose $U = \{u_1, u_2, \ldots, u_M\}$ is the set of mobile users, and then every user can register as a requester or a worker. In this paper, we denote $\mathcal{E} = \{e_1, e_2, \ldots, e_N\}$ as the set of edge servers in the MEC-based MCS system. These edge servers have different serving areas in which the users can connect to them for receiving the sensing tasks. Without loss of generality, the principle of proximity is adopted here so that the users will communicate to their nearest edge server. If we use $\mathcal{L}_{m,j} \in \{0, 1\}$ as an indicator to describe the user coverage of edge servers (namely, whether user u_m is in the serving area of e_j or not), we have

$$\sum_{j=1}^{N} \mathcal{L}_{m,j} = 1, \forall m \in [1, M] \tag{1}$$

3.2 Task Announcement and Deployment

If a user chooses to register as a requester, he/she can then submit a new task T, e.g. a multi-source data acquisition task. When task T is submitted to the MCS market by requester, the MCS market will then decompose task T into several sensing sub-tasks $(T_1, T_2, \ldots, T_K)$. Every sensing sub-task T_k receives the sensing data from mobile devices and conducts a series of analyses to obtain the result that T_k is expected to acquire. For example, the submitted task T may be a multi-source data based machine learning task, and K different kinds of data are needed to be collected from various users or devices. With the help of MEC architecture, the feature pre-processing can be obtained with deployed related services of T_1–T_K on the edge servers in a distributed way. As web services are usually the carriers to fulfill corresponding tasks [10] in practice, we use $S = (s_1, s_2, \ldots, s_K)$ to denote them. These services can be launched as instances on edge servers to deal with the collected data about the sub-tasks. Here we use Q_k to describe how the task T_k is completed in the MEC-based MCS system with s_k, namely the completion quality. In general, there will be a minimum requirement ε_k for the completion quality for given task T_k to ensure that the result of T is satisfying. Because edge servers are resource-constrained, at most $n_j^{\mathcal{E}}$ service instances are allowed to be deployed on edge server e_j. Denote $D_{j,k}$ as the indicator to show whether the instance of s_k is deployed on edge server e_j or not, which is formally defined as:

$$D_{j,k} = \begin{cases} 1, s_k\text{'s instance is deployed on } e_j \\ 0, \text{otherwise} \end{cases}, \tag{2}$$

then the constraints of edge resource can be represented with:

$$\sum_{k=1}^{K} D_{j,k} \le n_j^{\mathcal{E}}, \forall j \in [1, N] \tag{3}$$

3.3 Sub-task Assignment

If a user chooses to register as a worker, he/she then must claim his/her willing tasks via informing the MCS scheduler what kinds of sub-tasks he/she can complete:

$$\mathcal{W}_{m,k} = \begin{cases} 1, u_m \text{ would like to fulfill } T_k \\ 0, \text{otherwise} \end{cases} \tag{4}$$

Besides, the user also needs to tell the MCS scheduler his/her capacity to complete the tasks because the users can never be perpetual-motion machines. For example, the capacity may be the remaining battery of their devices when the sub-tasks are energy-consuming, and it may also be the left physical strength of them if the sub-tasks would exhaust people, like counting the building in a street. Suppose the capacity of user u_m is denoted with n_m^U, and $\mathcal{P}_{m,k}$ denotes how many times user u_m is arranged to complete T_k, then we have:

$$\sum_{k=1}^{K} \mathcal{P}_{m,k} \le n_m^U, \forall m \in [1, M] \tag{5}$$

To evaluate the task completion quality of T and the total cost of the requester, here we use $q_{m,k}$ to denote the task completion quality of sub-task T_k under user u_m because the workers are heterogeneous, and use $c_{m,k}$ to denote the incentive requirement of u_m if he/she would like to complete the sub-task T_k. Then, the task completion quality of T_k can be represented with:

$$Q_k(\mathcal{P}, \mathcal{D}) = \sum_{j=1}^{N} \sum_{m=1}^{M} \mathcal{L}_{m,j} \cdot \mathcal{D}_{j,k} \cdot W_{m,k} \cdot q_{m,k} \cdot \mathcal{P}_{m,k} \tag{6}$$

and when the price of deploying a service instance is ν, the total cost of the requester can be represented with:

$$C_r(\boldsymbol{P}, \boldsymbol{D}) = \sum_{m=1}^{M}\sum_{k=1}^{K} \mathcal{P}_{m,k} \cdot c_{m,k} + \nu \sum_{j=1}^{N}\sum_{k=1}^{K} \mathcal{D}_{j,k} \tag{7}$$

3.4 Problem Definition and Formulation

With the introduction of related concepts, now we can give the definition of this cost-effective service deployment and task assignment for MCS in MEC (CSDATAM$_2$) problem clearly. In this CSDATAM$_2$ problem, the requester would always like to balance the task completion quality and the cost to have his/her submitted task completed, so he/she would like to find the optimal service deployment strategy $\boldsymbol{P}$ and task assignment strategy $\boldsymbol{P}$ to make ends meet. Therefore, we can now formulate the CSDATAM$_2$ problem from the perspective of cost optimization (called CSDATAM$_2$-C):

$$P_C: \quad \min \ C_r(\boldsymbol{P}, \boldsymbol{D}) \tag{8}$$

$$s.t. \quad Q_k(\boldsymbol{P}, \boldsymbol{D}) \geq \varepsilon_k, \forall k \tag{9}$$

$$\sum_{k=1}^{K} \mathcal{D}_{j,k} \leq n_j^{\mathcal{E}}, \forall j \tag{10}$$

$$\sum_{k=1}^{K} \mathcal{P}_{m,k} \leq n_m^{U}, \forall m \tag{11}$$

$$\mathcal{D}_{j,k} \in \{0,1\}, \forall j, \forall k \tag{12}$$

$$\mathcal{P}_{m,k} \in \mathbb{N}, \forall m, \forall k \tag{13}$$

In this problem, every sub-task T_k is required to be completed with a minimum task completion quality ε_k.

Similarly, from the perspective of task completion quality, we can also defined CSDATAM$_2$-Q problem as follows:

$$P_Q: \quad \max \sum_{k=1}^{K} Q_k(\boldsymbol{P}, \boldsymbol{D}) \tag{14}$$

$$s.t. \quad Q_k(\boldsymbol{P}, \boldsymbol{D}) \geq \varepsilon_k, \forall k \tag{15}$$

$$C_r(\boldsymbol{P}, \boldsymbol{D}) \leq C^{\star}, \tag{16}$$

$$\sum_{k=1}^{K} \mathcal{D}_{j,k} \leq n_j^{\mathcal{E}}, \forall j \tag{17}$$

$$\sum_{k=1}^{K} \mathcal{P}_{m,k} \leq n_m^{U}, \forall m \tag{18}$$

$$\mathcal{D}_{j,k} \in \{0,1\}, \forall j, \forall k \tag{19}$$

$$\mathcal{P}_{m,k} \in \mathbb{N}, \forall m, \forall k \tag{20}$$

We can find that in this version, we mainly focus on the total task completion quality while keeping the total cost within an acceptable budget $C^\star$.

4   The Optimal Service Deployment and Task Assignment Strategies

In this section, we reform the representation the problem formulated in Sect. 3.4. First of all, as there are two decision variables $\mathcal{P}$ and $\mathcal{D}$ as well as other constant matrices involved in this problem, here we will vectorize them to obtain the simplified representation. For the decision variables, we denote

$$
\boldsymbol{\mathcal{P}}_k = \begin{bmatrix} \mathcal{P}_{1,k} \\ \mathcal{P}_{2,k} \\ \vdots \\ \mathcal{P}_{M,k} \end{bmatrix}, \boldsymbol{\mathcal{D}}_k = \begin{bmatrix} \mathcal{D}_{1,k} \\ \mathcal{D}_{2,k} \\ \vdots \\ \mathcal{D}_{N,k} \end{bmatrix}
\tag{21}
$$

as the column vectors of the origin matrices, and denote $\boldsymbol{x} = [\boldsymbol{p}\,\boldsymbol{d}]^T$ as the unified decision where $\boldsymbol{p} = [\boldsymbol{\mathcal{P}_1}, \boldsymbol{\mathcal{P}_2}, ..., \boldsymbol{\mathcal{P}_K}]^T$ and $\boldsymbol{d} = [\boldsymbol{\mathcal{D}_1}, \boldsymbol{\mathcal{D}_2}, ..., \boldsymbol{\mathcal{D}_K}]^T$. At the same time, with the constant variables denoted with:

$$
\boldsymbol{\mathcal{W}}_k = \begin{bmatrix} \mathcal{W}_{1,k} \\ \mathcal{W}_{2,k} \\ \vdots \\ \mathcal{W}_{M,k} \end{bmatrix}, \boldsymbol{q}_k = \begin{bmatrix} q_{1,k} \\ q_{2,k} \\ \vdots \\ q_{M,k} \end{bmatrix}, \boldsymbol{c}_k = \begin{bmatrix} c_{1,k} \\ c_{2,k} \\ \vdots \\ c_{M,k} \end{bmatrix}
\tag{22}
$$

and $\boldsymbol{c} = [\boldsymbol{c}_1, \boldsymbol{c}_1, ..., \boldsymbol{c}_K]^T$, $\boldsymbol{n}^{\mathcal{U}} = [n_1^{\mathcal{U}}, n_2^{\mathcal{U}}, ..., n_M^{\mathcal{U}}]^T$, $\boldsymbol{n}^{\mathcal{E}} = [n_1^{\mathcal{E}}, n_2^{\mathcal{E}}, ..., n_N^{\mathcal{E}}]^T$, the task completion quality constraints can be represented as

$$
\sum_{j=1}^{N}\sum_{m=1}^{M} \mathcal{L}_{m,j} \mathcal{W}_{m,k} q_{m,k} \cdot \mathcal{P}_{m,k} \cdot \mathcal{D}_{j,k} = \boldsymbol{\mathcal{P}}_k^T A_k \boldsymbol{\mathcal{D}}_k
$$

$$
= \begin{bmatrix} \boldsymbol{\mathcal{P}}_1 \\ \boldsymbol{\mathcal{P}}_2 \\ \vdots \\ \boldsymbol{\mathcal{P}}_K \\ \boldsymbol{\mathcal{D}}_1 \\ \boldsymbol{\mathcal{D}}_2 \\ \vdots \\ \boldsymbol{\mathcal{D}}_K \end{bmatrix}^T
\begin{bmatrix}
\overset{k\downarrow}{O} & \cdots & \overset{k+K\downarrow}{O} & \cdots & \overset{2K\downarrow}{O} & \cdots & O \\
\vdots & \ddots & \vdots & \ddots & \vdots & \ddots & \vdots \\
\overset{k\rightarrow}{O} & \cdots & O & \cdots & \frac{1}{2}A_k & \cdots & O \\
\vdots & \ddots & \vdots & \ddots & \vdots & \ddots & \vdots \\
\overset{k+K\rightarrow}{O} & \cdots & \frac{1}{2}A_k^T & \cdots & O & \cdots & O \\
\vdots & \ddots & \vdots & \ddots & \vdots & \ddots & \vdots \\
\overset{2K\rightarrow}{O} & \cdots & O & \cdots & O & \cdots & O
\end{bmatrix}
\begin{bmatrix} \boldsymbol{\mathcal{P}}_1 \\ \boldsymbol{\mathcal{P}}_2 \\ \vdots \\ \boldsymbol{\mathcal{P}}_K \\ \boldsymbol{\mathcal{D}}_1 \\ \boldsymbol{\mathcal{D}}_2 \\ \vdots \\ \boldsymbol{\mathcal{D}}_K \end{bmatrix}
\tag{23}
$$

$$
= \boldsymbol{x}^T \widetilde{A}_k \boldsymbol{x}
$$

where A_k is defined as $A_k \triangleq \mathcal{L} \circ W_k \circ q_k$ ($A \circ B$ means the Hadamard product of matrices A and B). Besides these, the capacity constraints can also be transformed with:

$$
\begin{aligned}
\sum_{k=1}^{K} \mathcal{D}_{j,k} \leq n_j^{\mathcal{E}}, \forall j &\Longleftrightarrow \sum_{k=1}^{K} \mathcal{D}_k \leq n^{\mathcal{E}} \\
\sum_{k=1}^{K} \mathcal{P}_{m,k} \leq n_m^{\mathcal{U}}, \forall m &\Longleftrightarrow \sum_{k=1}^{K} \mathcal{P}_k \leq n^{\mathcal{U}}
\end{aligned} \tag{24}
$$

Then we have the simplified form of the CSDATAM$_2$-C:

$$
P_C' : \quad \min \left[\begin{matrix} c \\ \nu \cdot \mathbb{1} \end{matrix} \right]^T x \tag{25}
$$

$$
s.t. \quad x^T \tilde{A}_k x \geq \varepsilon_k, \forall k \tag{26}
$$

$$
\left[\begin{matrix} 1 & O \\ O & 1 \end{matrix} \right] x \leq \left[\begin{matrix} n^{\mathcal{U}} \\ n^{\mathcal{E}} \end{matrix} \right] \tag{27}
$$

$$
\left[\begin{matrix} O & O \\ O & diag(\mathbb{1}) \end{matrix} \right] x \leq \mathbb{1} \tag{28}
$$

$$
x \in \mathbb{N}^{(M+N)\cdot K} \tag{29}
$$

and the simplified form of the CSDATAM$_2$-Q problem:

$$
P_Q' : \quad \max \sum_{k=1}^{K} x^T \tilde{A}_k x \tag{30}
$$

$$
s.t. \quad x^T \tilde{A}_k x \geq \varepsilon_k, \forall k \tag{31}
$$

$$
\left[\begin{matrix} c \\ \nu \cdot \mathbb{1} \end{matrix} \right]^T x \leq C^{\star} \tag{32}
$$

$$
\left[\begin{matrix} 1 & O \\ O & 1 \end{matrix} \right] x \leq \left[\begin{matrix} n^{\mathcal{U}} \\ n^{\mathcal{E}} \end{matrix} \right] \tag{33}
$$

$$
\left[\begin{matrix} O & O \\ O & diag(\mathbb{1}) \end{matrix} \right] x \leq \mathbb{1} \tag{34}
$$

$$
x \in \mathbb{N}^{(M+N)\cdot K} \tag{35}
$$

Obviously, the CSDATAM$_2$M-C and CSDATAM$_2$M-Q problems are quadratic constraint linear programming problems (special cases of the second-order cone programming problem), if ignore the fact that $x \in \mathbb{N}$. Therefore, we can apply the Branch-and-Bound (BnB) framework here to search for the optimal solutions to these problems step by step [11]. And in every step, we will relax the constraint of variable x to $\mathbb{R}$ so that convex solvers like **CPLEX** and **Gurobi** can be applied to solve these classical problems.

5   Experiments and Analysis

Due to the lack of well-adopted platforms and datasets, in this work, we generate our experimental data in a synthetic way. Based on the simulation data, complete experiments are conducted so that the impacts of different factors are explored.

5.1   Baselines and Comparisons

As the problems will be solved with exact solutions after they are translated into the MIQP problem, it is not necessary to compare it with other baselines except the ones that consider the trade-off between accuracy and running time. Thus, we choose some popular and representative mathematical approaches for the programming program problem as baselines. The chosen approaches are:

1. **Genetic algorithm, GA.** Genetic algorithm is one of the famous methods [12] which can be used for this purpose. GA simulates the evolution of populations with operations like selection, crossover, and mutation. It is designed to favor chromosomes with the highest fitness values to produce the next populations (solutions). As a result, the quality of solutions for a problem is gradually improved until the optimal answer is reached.
2. **Tabu search algorithm, TS.** Tabu search algorithm [13] is a meta-heuristic search method employing local search methods used for mathematical optimization, where the local searches take a potential solution to a problem and check its immediate neighbors in the hope of finding an improved solution. Local search methods have a tendency to become stuck in suboptimal regions or on plateaus where many solutions are equally fit.

In Fig. 3, we can find that not all the baseline approaches can find the optimal task assignment and service deployment schemes. Compared to the TS, the GA shows better performance in obtaining the schemes with best quality, but fail to have a good probability to obtain the best cost. However, different from the exact results generated by solving the mixed integer quadratic programming problems, these heuristic algorithms may always sacrifice their accuracy for the time complexity to find the best schemes.

5.2   Impact of System Configurations

To explore how the system configurations will impact the optimal task assignment and service deployment schemes, we conducted a series of parametric experiments in this section.

Impact of User Number: In Fig. 4, we've shown how the average values of total cost and total task completion quality will change with the increasing of user number in problem $CSDATAM_2$-C and $CSDATAM_2$-Q. It can be found that as M increases, the total cost shows a downward trend in $CSDATAM_2$-C while the total task completion quality increase in $CSDATAM_2$-Q. Obviously, both the former decline and the latter rise are mainly due to the potential appropriate users brought by the increase of user number.

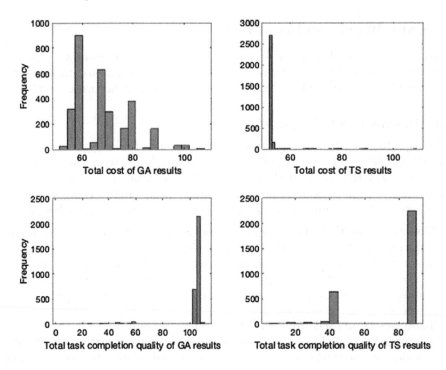

Fig. 3. The comparison with baselines

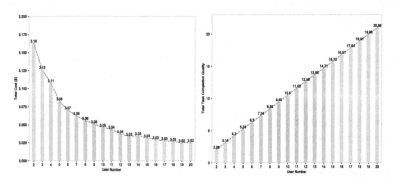

Fig. 4. The impact of user number (left: CSDATAM$_2$-C, right: CSDATAM$_2$-Q)

Impact of Edge Server Number: In Fig. 5, we've shown how the average values of total cost and total task completion quality will change with the increasing of edge server number in problem CSDATAM$_2$-C and CSDATAM$_2$-Q. It can be found that as N increases, the total cost shows a downward trend in CSDATAM$_2$-C while the total task completion quality increases in CSDATAM$_2$-Q. Obviously, both the former decline and the latter rise are mainly due to the potential appropriate edge servers for related services brought by the increase of edge server number. Besides this, we can also find that they become stable

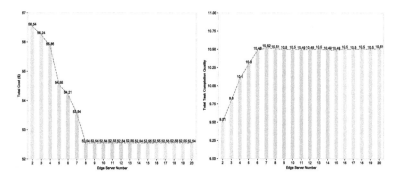

Fig. 5. The impact of edge server number (left: CSDATAM$_2$-C, right: CSDATAM$_2$-Q)

when N is larger than some specific values. This is because we don't need more edge servers if all appropriate users are assigned with matching tasks are related services are deployed near them.

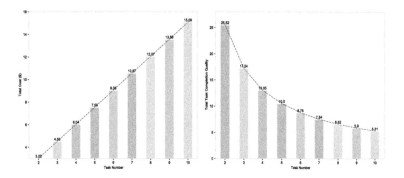

Fig. 6. The impact of sub-task number (left: CSDATAM$_2$-C, right: CSDATAM$_2$-Q)

Impact of Sub-task Number: In Fig. 6, we've shown how the average values of total cost and total task completion quality will change with the increasing of sub-task number in problem CSDATAM$_2$-C and CSDATAM$_2$-Q. It can be found that as K increases, the total cost shows an upward trend in CSDATAM$_2$-C while the total task completion quality decreases in CSDATAM$_2$-Q. The cost rise is due to the increasing number of users to fulfill the task when K increases, so more cost are paid for these additional users. Similarly, the task completion quality decreases because there is no surplus budget to search for better qualities.

Impact of User Willingness: User willingness is ω measured with the percentage of users who would like to be assigned with tasks. In Fig. 7, we've shown how

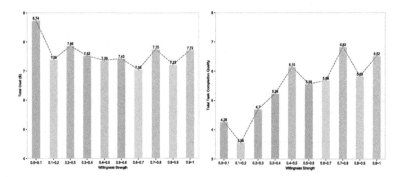

Fig. 7. The impact of user willingness (left: CSDATAM$_2$-C, right: CSDATAM$_2$-Q)

the average values of total cost and total task completion quality will change with the increasing of willing strength ω in problem CSDATAM$_2$-C and CSDATAM$_2$-Q. It can be found that as ω increases, the total cost shows a downward trend in CSDATAM$_2$-C. This decline is mainly due to the external possibilities brought by the increase in ω. However, the task completion quality shows an opposite trend in the result of CSDATAM$_2$-Q, as the external possibilities make it possible for the application developers to select the users with better task completion qualities and lower incentive requirements.

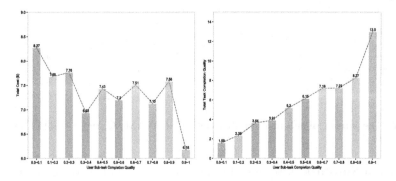

Fig. 8. The impact of users' $\bar{q}$ (left: CSDATAM$_2$-C, right: CSDATAM$_2$-Q)

Impact of User Sub-task Completion Quality: In Fig. 8, we've shown how the average values of total cost and total task completion quality will change with the increasing of users' average task completion quality in problem CSDATAM$_2$-C and CSDATAM$_2$-Q. It can be found that as the user average task completion quality q increases, the total cost shows a downward trend in the result of CSDATAM$_2$-C. This decline is mainly due to the easier satisfaction of the sub-task quality requirements brought by the increase of q. However, the total task completion quality shows an opposite trend in the result of CSDATAM$_2$-Q. This is because the total task completion quality is proportional to the users' task completion qualities.

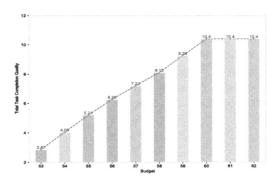

Fig. 9. The impact of cost budget

Impact of Cost Budget: In Fig. 9, we've shown how the average values of total cost will change with the increasing of user average task completion quality in problem CSDATAM$_2$-Q (because it will not effect the results of CSDATAM$_2$-C). It can be found that as the developer's budget on user incentives $C^\star$ increases, the total task completion quality Q shows an upward trend. This increase shows that "the more you pay, the more you get"—the abundant budget make it possible to assign tasks to the users who are competent enough but charge more.

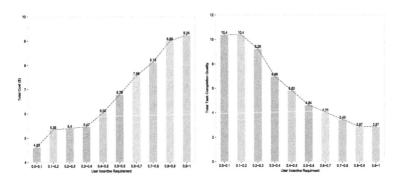

Fig. 10. The impact of users' $\bar{c}$ (left: CSDATAM$_2$-C, right: CSDATAM$_2$-Q)

Impact of User Incentive Requirement: In Fig. 10, we've shown how the average values of total cost and total task completion quality will change with the increasing of user incentives requirement in problem CSDATAM$_2$-C and CSDATAM$_2$-Q. It can be found that as the user incentive requirement c increases, the total cost C_r shows an upward trend in CSDATAM$_2$-C. This is quite easy to understand because the total cost is proportional to the users' incentive requirements. On the other hand, we can also find that the average values of total task completion quality decreases fast with the increasing of c. This is because the limited budget cannot support the pursuit for higher task completion quality—though the optimum can also be found, but the value of it may be smaller than that of a larger space.

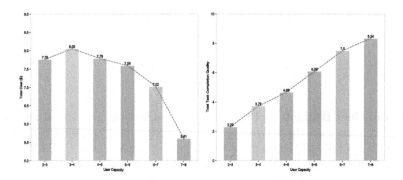

Fig. 11. The impact of user capacity (left: CSDATAM$_2$-C, right: CSDATAM$_2$-Q)

Impact of User Capacity: In Fig. 11, we've shown how the average values of total cost C_r and total task completion quality Q will change with the increasing of user capacity in problem CSDATAM$_2$-C and CSDATAM$_2$-Q. It can be found that as the user capacity n^U increases, the total cost C_r shows a downward trend in CSDATAM$_2$-C. This decline is mainly due to the external possibilities brought by the increase of n^U. This is not difficult to understand because, at first, C_r will not become larger for its optimality. Then if two users have the same task completion quality but different incentive requirement, the sub-task which are originally assigned to the expensive one will be assigned to another. As the reason also works well in the CSDATAM$_2$-Q problem, then we can see that the average values of total task completion quality shows an upward trend.

Impact of Edge Server Capacity: In Fig. 12, we've shown how the average values of total cost and total task completion quality will change with the increasing of edge server capacity in problem CSDATAM$_2$-C and CSDATAM$_2$-Q. It can be found that they do not give exact changing rules with the increasing of the edge server capacity n^U. Actually, this is because here in our model, we've

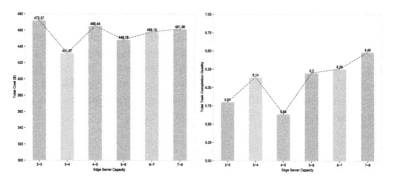

Fig. 12. The impact of edge capacity (left: CSDATAM$_2$-C, right: CSDATAM$_2$-Q)

assumed that the deployment cost on different edge servers are the same so that the objectives will not be effected much when it changes. On the contrary, as the services should be deployed on the edge servers near the users who are assigned with related tasks, the main factor will be the user locations.

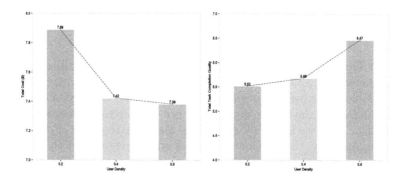

Fig. 13. The impact of user density (left: CSDATAM$_2$-C, right: CSDATAM$_2$-Q)

Impact of User Distribution: To explore the impact of user distribution, here we first use an index called "user density" to measure it. Heuristically, we define it with:

$$\rho_u = 1 - \frac{\min |\{e_{j_i} | \sum_{i=1}^{n} \sum_{m=1}^{M} \mathcal{L}_{m,j_i} \geq 0.8N, e_{j_i} \in \mathcal{E}\}|}{N}. \tag{36}$$

Obviously, the user density will be $1 - 1/10 = 0.9$ if users gather around 1 edge server of 10 edge servers, and will be $1 - 10/10 = 0$ if they are evenly distributed.

In Fig. 13, we've shown how the average values of total cost and total task completion quality will change with the increase of user density in problem

CSDATAM$_2$-C and CSDATAM$_2$-Q. It can be found that as the user density ρ_u increases, the total cost shows a downward trend. This decline is mainly cost in the repeated deployment of services—it will be not necessary for the application developers to deploy the services on several edge servers if the workers of the same sub-tasks are in the same serving area of an edge server. Based on this, we can also see the upward trend of the total task completion quality—as mentioned above, the adequate budget can bring better quality.

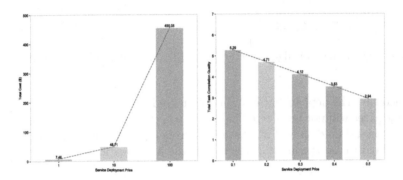

Fig. 14. The impact of deployment price (left: CSDATAM$_2$-C, right: CSDATAM$_2$-Q)

Impact of Deployment Price: In Fig. 14, we've shown how the average values of total cost and total task completion quality will change with the increase of service deployment price ν in problem CSDATAM$_2$-C and CSDATAM$_2$-Q. It can be found that as the service deployment price ν increases, the total cost also increases fast. This rise is mainly due to the fact that the total cost is proportional to ν. Similarly, the developer's budget will strongly limit the total task completion quality with the increasing cost in satisfying the users' requirements, so the decline of it will be obvious.

6   Related Work

In this section, we will review some representative works that are related to our problem to see what efforts are made to solve this kind of problem.

6.1   The Incentive Mechanism in Mobile Crowd Sensing

First, we are to show some research about the incentive mechanism in the MCS environment. Deterding et al. [14] made some of the earliest contributions to the incentive mechanism of group intelligence perception, they introduced economic models into the incentive mechanism and proposed methods of dynamic prices and virtual points to ensure the participation rate and minimize cost.

For users who are rarely selected as participants, there is a possibility of losing interest and withdrawing from the group intelligence perception system. Gao et al. [15] therefore proposed the long-term problem of participation rate, and solved this problem based on the Lyapunov-based auction model. Besides this, Sun et al. introduced the restless multi-armed bandit (MAB) process model and the heterogeneous belief values model [16], they transformed this continuous group intelligence incentive model into a MAB process based on the social status and real-time to solve this long-term incentive problem. The location distribution of the participants affects the quality of the task. The server platform should not only recruit more participants at minimal cost but also consider the location distribution of users. Therefore, Jaimes et al. [17] proposed the GIA algorithm in combination with the Greedy Budgeted Maximum Coverage algorithm to improve the coverage of the area of interest for a given budget.

6.2   The Task Assignment Research

Task assignment is a popular problem in the research about MEC. It is to perform task allocation so that the latency and energy consumption of mobile application execution can be dynamically adjusted to make the task allocation results meet the user requirements. To address the above problem, existing research has proposed several approaches to meet the requirements: Dinh et al. [18] used randomly generated task graphs to allocate some tasks and minimize energy consumption. Kao et al. [19] used integer programming to calculate task allocation and scheduling strategies to save network resources as much as possible. Literature [20] studied to minimize time delay under the constraints of given network resources, and proposed an efficient polynomial-time algorithm to complete task assignment. Sardellitti et al. [21] proposed a task assignment algorithm based on a directed acyclic graph (DAG) model that minimizes the energy consumption of mobile devices under a given time delay constraint. Deng et al. [22] decomposed mobile applications into multiple subtasks, built a DAG model, and modeled the task allocation problem as a nonlinear 0–1 programming problem to achieve the goal of minimizing energy consumption. Kwak et al. [23] designed a dynamic migration algorithm based on Lyapunov's optimization technology to ensure that the system delay is controlled within a given range and to minimize the energy consumption of mobile devices. Jiang and You et al. [24,25] further extended the work of [23] to multi-core mobile devices.

These researches shed light on the fundamental concepts and inspired the thought of related optimizing in mobile crowd sensing systems and multi-access edge computing systems. Based on these works, we try to go further in balancing the performance and cost for the MEC-based MCS system. We will combine the advantages of using incentive mechanisms and the methods of assigning tasks to generate an appropriate strategy that can have a trade-off between performance (quality) and cost.

7   Conclusion and Future Work

In this paper, we first introduce the MEC-based MCS system and describe how the MEC architecture can be used in improving the performance of the mobile crowd sensing system. Then, considering the incentive mechanism and the quality model of complex tasks, we build up a cost-quality analysis framework for the task assignment and service deployment problem in this system. Finally, we solve the problem after translating it into a mixed-integer quadratic programming problem and investigate the factors that may impact the results. In the future, we plan to refine our model and the analysis framework to make it more practical, and we are going to incorporate dynamic factors into this problem to further improve the performance of the proposed approach.

Acknowledgement. This research was partially supported by the National Key Research and Development Program of China (No. 2017YFB1400601), Key Research and Development Project of Natural Science Foundation of China (NO. 61772461, No. 61802343, No. 62072402) and Zhejiang Provincial Natural Science Foundation of China (No. LQ21F020007, No. LQ20F020015, No. LR18F020003).

References

1. Chen, Y., Zhou, M., Zheng, Z., Chen, D.: Time-aware smart object recommendation in social Internet of Things. IEEE Internet Things J. **7**(3), 2014–2027 (2020)
2. Deng, S., Zhao, H., Fang, W., Yin, J., Dustdar, S., Zomaya, A.Y.: Edge intelligence: the confluence of edge computing and artificial intelligence. IEEE Internet Things J. **7**(8), 7457–7469 (2020)
3. Shi, W., Cao, J., Zhang, Q., Li, Y., Xu, L.: Edge computing: vision and challenges. IEEE Internet Things J. **3**(5), 637–646 (2016)
4. Ren, J., Pan, Y., Goscinski, A., Beyah, R.A.: Edge computing for the Internet of Things. IEEE Netw. **32**(1), 6–7 (2018)
5. Wang, D., Xu, D., Yu, D., Xu, G.: Time-aware sequence model for next-item recommendation. Appl. Intell. 1–15 (2020). https://doi.org/10.1007/s10489-020-01820-2
6. Filippini, I., Sciancalepore, V., Devoti, F., Capone, A.: Fast cell discovery in mmwave 5G networks with context information. IEEE Trans. Mob. Comput. **17**(7), 1538–1552 (2017)
7. Fan, Q., Ansari, N.: Application aware workload allocation for edge computing-based IoT. IEEE Internet Things J. **5**(3), 2146–2153 (2018)
8. Chen, Y., Deng, S., Ma, H., Yin, J.: Deploying data-intensive applications with multiple services components on edge. Mob. Netw. Appl. **25**(2), 426–441 (2020). https://doi.org/10.1007/s11036-019-01245-3
9. Zhao, H., Deng, S., Zhang, C., Du, W., He, Q., Yin, J.: A mobility-aware cross-edge computation offloading framework for partitionable applications. In: Bertino, E., Chang, C.K., Chen, P., Damiani, E., Goul, M., Oyama, K. (eds.) 2019 IEEE International Conference on Web Services, ICWS 2019, Milan, Italy, 8–13 July 2019, pp. 193–200. IEEE (2019)
10. Deng, S., Wu, H., Tan, W., Xiang, Z., Wu, Z.: Mobile service selection for composition: an energy consumption perspective. IEEE Trans. Autom. Sci. Eng. **14**(3), 1478–1490 (2015)

11. Gupta, O.K., Ravindran, A.: Branch and bound experiments in convex nonlinear integer programming. Manag. Sci. **31**(12), 1533–1546 (1985)
12. Deng, S., Huang, L., Taheri, J., Yin, J., Zhou, M., Zomaya, A.Y.: Mobility-aware service composition in mobile communities. IEEE Trans. Syst. Man Cybern. Syst. **47**(3), 555–568 (2017)
13. Liu, H., Zhang, J., Zhang, X., Kurniawan, A., Juhana, T., Ai, B.: Tabu-search-based pilot assignment for cell-free massive MIMO systems. IEEE Trans. Veh. Technol. **69**(2), 2286–2290 (2019)
14. Deterding, S., Dixon, D., Khaled, R., Nacke, L.: From game design elements to gamefulness: defining "gamification". In: Proceedings of the 15th International Academic MindTrek Conference: Envisioning Future Media Environments, pp. 9–15 (2011)
15. Gao, L., Hou, F., Huang, J.: Providing long-term participation incentive in participatory sensing. In: 2015 IEEE Conference on Computer Communications (INFO-COM), pp. 2803–2811. IEEE (2015)
16. Sun, J.: An incentive scheme based on heterogeneous belief values for crowdsensing in mobile social networks. In: 2013 IEEE Global Communications Conference (GLOBECOM), pp. 1717–1722. IEEE (2013)
17. Jaimes, L.G., Vergara-Laurens, I., Labrador, M.A.: A location-based incentive mechanism for participatory sensing systems with budget constraints. In: 2012 IEEE International Conference on Pervasive Computing and Communications, pp. 103–108. IEEE (2012)
18. Dinh, T.Q., Tang, J., La, Q.D., Quek, T.Q.: Adaptive computation scaling and task offloading in mobile edge computing. In: 2017 IEEE Wireless Communications and Networking Conference (WCNC), pp. 1–6. IEEE (2017)
19. Kao, Y.-H., Krishnamachari, B., Ra, M.-R., Bai, F.: Hermes: latency optimal task assignment for resource-constrained mobile computing. IEEE Trans. Mob. Comput. **16**(11), 3056–3069 (2017)
20. Zhang, W., Wen, Y., Wu, D.O.: Collaborative task execution in mobile cloud computing under a stochastic wireless channel. IEEE Trans. Wirel. Commun. **14**(1), 81–93 (2014)
21. Sardellitti, S., Scutari, G., Barbarossa, S.: Joint optimization of radio and computational resources for multicell mobile-edge computing. IEEE Trans. Sig. Inf. Process. Netw. **1**(2), 89–103 (2015)
22. Deng, M., Tian, H., Fan, B.: Fine-granularity based application offloading policy in cloud-enhanced small cell networks. In: 2016 IEEE International Conference on Communications Workshops (ICC), pp. 638–643. IEEE (2016)
23. Kwak, J., Kim, Y., Lee, J., Chong, S.: DREAM: dynamic resource and task allocation for energy minimization in mobile cloud systems. IEEE J. Sel. Areas Commun. **33**(12), 2510–2523 (2015)
24. Jiang, Z., Mao, S.: Energy delay tradeoff in cloud offloading for multi-core mobile devices. IEEE Access **3**, 2306–2316 (2015)
25. You, C., Zeng, Y., Zhang, R., Huang, K.: Asynchronous mobile-edge computation offloading: energy-efficient resource management. IEEE Trans. Wirel. Commun. **17**(11), 7590–7605 (2018)

Performance and Scheduling Optimization in Edge Computing

A Survey on Security and Performance Optimization of Blockchain

Dongqing Li[1(✉)], Congfeng Jiang[1], Yin Liu[2], Linlin Tang[2], and Li Yan[2]

[1] College of Computer Science and Technology, Hangzhou Dianzi University,
Hangzhou 310018, China
{ldongqing,cjiang}@hdu.edu.cn
[2] Information and Telecommunications Company,
State Grid Shandong Electric Power Company, Jinan 250000, China
5918874@qq.com, 6335875@qq.com, yanli@sd.sgcc.com.cn

Abstract. This paper investigates the security issues and performance optimization of the blockchain. Security has been a hot topic in blockchain technology. Stealing cryptocurrency and disclosing the privacy of transaction process have exposed the vulnerability of blockchain in different degrees. These vulnerabilities not only caused significant losses to the project team and users, but also raised doubts about the security of the blockchain. As a formalized contract in the code, smart contracts provide a more convenient method than traditional ones, while they increase the risk of blockchain. Moreover, secure transactions should resist external attacks and protect user privacy. In addition, the performance analysis of blockchain has also aroused great interest. Therefore, this paper summarizes the related work of blockchain performance analysis from the following three aspects to promote the further research of blockchain: simulation systems of blockchain, evaluation of blockchain network and optimization of blockchain application.

Keywords: Blockchain · Smart contract · Cryptocurrency · Privacy · Performance analysis

1 Introduction

Blockchain technology is rising, which attracts more people's interest because of its decentralization, tamper proof and anonymity. Blockchain is an encrypted, ordered, block-based and only additional data structure maintained by mutual distrust peers, which have the latest copy of the database. Blockchain perfectly solves the trust crisis in the information system. Blockchain can be divided into public blockchain, private blockchain and consortium blockchain, etc. Public blockchain is that anyone can participate in the system to read data, send confirmable transactions and compete for accounting. Public blockchain is generally considered to be completely decentralized, because no person or organization can control or tamper with data. Private blockchain means that its write permission

H. Gao et al. (Eds.): ChinaCom 2020, LNICST 352, pp. 101–111, 2021.
https://doi.org/10.1007/978-3-030-67720-6_7

is controlled by an organization or institution. Due to the limited participating nodes, private blockchain can often have extremely fast transaction speeds, better privacy protection and lower transaction costs. And Consortium blockchain is jointly managed by several institutions.

There are currently three stages in the development of blockchain technology. The first stage is the era of cryptocurrency such as Bitcoin. The second stage is the main innovation brought by Ethereum, that is, the emergence of smart contracts. Finally, the third stage is the gradual integration of blockchain technology into practical production. Smart contracts reduce other transaction costs related to the contract, but this also brings some problems, such as increased risk. The immutable feature makes it impossible to repair or update once an error occurs. In Ethereum, smart contracts are written using Solidity and then compiled into bytecode, which is executed by the Ethereum virtual machine (EVM). However, multiple vulnerabilities have been found in smart contracts such as solidity, which can make it easy for criminals to steal currency. What's worse, the blockchain is not completely anonymous, because every transaction on the blockchain is public. In other words, anyone can look up all transactions and balances for each public key. Although secret transaction can hide the transaction amount, the computation and communication costs are very high, and it is detrimental to the scalability of cryptocurrency. The traceability of cryptocurrency provides financial crime protection, but it shows that the cryptocurrency lacks confidentiality and anonymity. Therefore, the privacy protection mechanism of blockchain should strike a balance between anonymity and supervision, and achieve transparency and confidentiality at the same time.

In the design and deployment of the blockchain, it is infeasible or impractical to rely on experimentation or trial-and-error to find the best architecture, configuration, or parameters [1]. Fortunately, a simulation-based system allows designers and analysts to easily explore different configurations and their impact on the operation of the blockchain system, which enables researchers to find the optimal system configuration suitable for their goals. As we all know, there is no appropriate measurement standard for the performance of the blockchain [15], but many scholars conducted a lot of research on the performance of blockchain and obtained many useful conclusions. Some work has found some bottlenecks in the performance research of blockchain platform and optimized them. As an emerging technology, blockchain provides open, transparent, traceable, and immutable data protection measures. It guarantees the security and convenience of data usage through a unique encryption and sharing mechanism, and it can also reduce the cost of application and improving efficiency.

The rest of this article is organized as follows. Section 2 outlines the vulnerabilities of smart contracts and transaction security. Section 3 summarizes the simulation, performance evaluation and optimization techniques of the blockchain. Finally, Sect. 4 concludes this paper.

2 Security Issues

As the blockchain becomes more and more popular and valuable, the security issue of blockchain becomes much more prominent. In order to enhance the security of the blockchain, researchers have proposed various methods to help discover and prevent security issues before they cause losses. Blockchain security issues include financial fairness in the presence of fraud and suspension, as well as the cryptographic concepts of confidentiality and authenticity. This section will summarize the current security issues from three aspects: smart contracts, cryptocurrency and privacy security.

2.1 Smart Contract

A smart contract is a program that runs on the blockchain, usually using a high-level programming language called Solidity. Since the contracts stored in the ledger can only be publicly obtained and cannot be modified, the attacker can obtain financial profits from the contract through the vulnerability of the smart contract. The existing work mainly focuses on the research of Ethereum smart contract attacks, because Ethereum is the most famous and common smart contract framework today. Table 1 is a comparison of these jobs. For example, Atzei et al. [4] conducted an investigation on Ethereum smart contracts, and divided them into three classes according to the level of vulnerabilities: Solidity, EVM and Blockchain levels, which were further subdivided into 12 categories of vulnerabilities. Based on the vulnerability classification of Atzei, Cook et al. [10] proposed DappGuard that can protect smart contracts from known attacks and learn new attacks. But DappGuard is just a proof of concept based on development tools. Teether [16] can automatically create exploits by analyzing four key EVM instructions, namely CALL, SELFDESTRUCT, CALLCODE and DELEGATECALL. The first two instructions will cause a direct transfer, and the last two instructions can execute any Ethereum bytecode set by the attacker in the content of the contract. Teether identifies the path leading to the critical instruction and converts the path into a set of constraints. Then, by using the constraints to find the solution, the key transaction that triggered the vulnerability can be inferred. The solc-verify [13] is a source code level verification tool. In addition to the vulnerability analyses of smart contracts on the Ethereum platform, there are also analysis of other blockchain platforms. For example, Wang et al. [30] developed a highly automated formal validator VERISOL for Solidity in the Azure blockchain workbench, which guarantees the security of smart contracts.

2.2 Cryptocurrency

An attacker can exploit the client's insecure settings to attack the blockchain. Specifically, it is to obtain cryptocurrency by executing operations that go against the user's wishes, for example, transferring Bitcoin to the attacker's address. Cheng et al. [8] designed and implemented a system that could capture an

Table 1. Comparison of different methods of protecting smart contracts

Project	Work platform	Support language	Contributions
Atzei et al.	Ethereum	Solidity and EVM bytecode	Analysis of the security of Ethereum smart contracts
DappGuard	Ethereum	Solidity	Propose design solutions for real-time monitoring and protection systems
Teether	Ethereum	EVM bytecode	Automatically exploit contract vulnerabilities
Solc-verify	Ethereum	Boogie	Modular validator for implementing smart contracts
VERISOL	Azure	Solidity	Verify and find errors in smart contracts

attacker's actual malicious behavior in Ethereum, which could be an attack stealing Ether. To avoid the impact of vulnerabilities and protect cryptocurrencies from being stolen, a lot of studies analyzed the vulnerabilities in existing cryptocurrency wallets. Notary [3] is a system composed of both hardware and software architecture, which contains three domains: kernel domain, agent domain and communication domain. Notary's switch design based on separation and reset avoids security loopholes in cryptocurrency hardware wallets, thereby improving application security.

In order to track illegal cryptocurrency activities, DBSCAN clustering technology is applied to the content of fraud websites to find out the types of prepaid and phishing fraud, so as to further understand these cryptocurrency scams. Due to the transparency of the blockchain, it is easy to analyze illegal gains and find further links before activities. Phillips et al. [23] found five different types of fraud on more than 1000 websites.

2.3 Privacy and Security

In the early blockchain systems, such as Bitcoin, Ethereum and Hyperledger, the capital flow and transaction amount of all transactions are publicly disclosed on the blockchain, so as to reach a consensus quickly for each node. Parties can use pseudonym public key to protect their real identity and increase their anonymity; however, these blockchain systems can only provide simple unlinkability. Attackers can link to the user's real identity in the real world through big data analysis, address clustering and network analysis. In fact, the main resistance to the widespread use of decentralized smart contracts and cryptocurrencies is the lack of privacy protection. Therefore, how to achieve a balance between anonymity and minimum regulation is still a challenge for blockchain.

In order to improve the privacy protection of blockchain system, some mixing technology-based solutions have been proposed, but malicious central mixing nodes will leak the privacy of participants and even steal digital assets. For protecting the privacy of participants, Valenta et al. [29] designed Blindcoin, which integrates blind signature into mixcoin. The central mixing node only provides mixing services, but cannot link the input and output of transactions. For preventing the central hybrid node from stealing participants' assets, CoinSwap [20] establishes a connection between the escrow transaction and the corresponding redemption ones through hashing lock, and the Mixcoin [6] uses signature-based accountability technology.

Encryption is the most common solution in privacy protection. A privacy data storage protocol is established based on ring signature on elliptic curve, and the complete anonymity of ring signature is used to ensure the data security and user identity privacy in blockchain applications [18]. However, kumaret et al. [17] emphasized that attackers can still link the transaction address to the real identity when the anonymous size set in the ring signature is small. In order to ensure its unlinkability, a large anonymous set can be selected, but this will significantly increase the storage cost of transactions. RingCT 2.0 [26]is a novel ring signature mechanism designed with an accumulator tool that balances efficiency and privacy issues in Monero.

3   Performance Evaluation and Optimization

For numerous practical reasons, such as multi-parameter and various consensus mechanisms, it's difficult to find the appropriate design and deployment for the blockchain. Simulation of blockchain system is a good choice, which can help designers optimize parameters and evaluate the performance of planned blockchain network by simulating a real operating environment. There is a lot of work to help us better plan and design a scalable, stable and flexible blockchain network. This section will summarize the current blockchain performance issues from three aspects: simulation, performance analysis and performance optimization.

3.1   Simulation

In order to promote the design and research of blockchain networks, many practical simulation tools have been proposed. Table 2 shows the different blockchain goals of the simulation system. For example, BlockSim [1] is a discrete event simulation framework that focuses on modeling methods in the Bitcoin blockchain through the proof-of-work protocol and the longest chain rule. The simulation framework proposed by Chitra et al. [9] is to use agent-based simulation in the

Table 2. Different blockchain goals of the simulation system

Model	Work	Contributions	Condition
BlockSim	Bitcoin	Explore different configurations and their effects on the behavior of blockchain systems through simulation-based models	Simulation
Agent-Based Simulations	Kadena's Chainweb	Evaluate blockchain protocol claims and measure network operations through agent-based adversarial simulation	Simulation
SimBlock	Bitcoin	Simulate a peer-to-peer network of a public blockchain composed of thousands of nodes, of which the parameters of the blockchain and its network can be flexibly configured	Simulation
BlockSIM	Ethereum or Hyperledger	Help blockchain architects better evaluate the performance of planned private blockchain networks and determine the optimal system parameters for their purposes through the run scenario	Simulation
BlockZoom	Blockchain application	Provide a reproducible environment for experimental distributed ledger technology and intelligent contract applications	Real

blockchain algorithm transaction protocol, which is applied to Kadena's Chainweb. This technology can effectively monitor and evaluate the risks in real-time blockchain systems. BlockSIM [22] can accurately model and test the stability time and transaction throughput of the blockchain under a given scenario. SimBlock [5] and BlockZoom [25] are systems for large-scale blockchain networks. Most of the test methods are implemented in a simulator or through a small local blockchain network, but BlockZoom can test the performance of the blockchain under a real scenario.

Table 3. Comparison of different blockchain performance analysis methods

Project	Goal	Scenario	Metrics
Blockchain-Enabled Wireless Internet of Things	Communication and security	Blockchain-enabled wireless IoT system	SINR, TDP transmission success rate, overall communication throughput and security
Scalable Access Management in IoT Using Blockchain	Scalability	Distributed IoT management system based on Ethereum	Throughput, latency and scalability
Blockchain in VANET	The impact of mobility	Vehicle Ad Hoc Network with Blockchain System	The probability of successfully adding a block to the chain, the stability of the rendezvous, and the number of blocks exchanged during the rendezvous
BLOCKBENCH	Data processing capability	Ethereum, Parity and Hyperledger Fabric	Throughput, latency, scalability, fault tolerance and safety indicators
DAGBENCH	Evaluate the performance of the DAG distributed ledger	IOTA, Byteball and Nano	Throughput, latency, scalability, success indicators, resource consumption, transaction data size, transaction fees

SINR: signalto-interference-plus-noise ratio; TDP: transaction data packet

3.2 Performance Analysis

Blockchain has characteristics like invariance and transparency, which means it has the potential to be a data processing platform. Therefore, BLOCKBENCH [11] conducted a comprehensive evaluation of Ethereum, Parity, and Hyperledger Fabric in order to understand the performance of a private blockchain with Turing-complete smart contracts on data processing workloads. The decentralized blockchain uses many encryption technologies to ensure that the data in the ledger cannot be tampered with. It not only provides a way to ensure data security for mutually untrusted nodes of the Internet of Things, but also reduces the high maintenance cost. Therefore, the blockchain can be used in the Internet of Things and vehicle ad hoc networks. However, the existing performance analysis for the blockchain cannot be directly applied to the performance analysis of the Internet of Things system or the vehicle ad-hoc network. Therefore, Sun et al. [27] and Novo et al. [21] focus on the performance of the Internet of Things supporting the blockchain. Similarly, Kim et al. [15] studied the performance of the blockchain system in the vehicle ad-hoc network, because mobility will pose different challenges to the performance of the blockchain.

DAG develops rapidly by virtue of its scalability and resource efficiency. Directed acyclic graphs are block-free, just store transactions as vertices of the graph, so transactions can be processed immediately without waiting for block composition. Most of the current performance analysis methods are for the two representative blockchains, Bitcoin and Ethereum. DAGBENCH [12] is the first framework to comprehensively evaluate DAG distributed ledgers. Table 3 compares these performance analysis methods.

3.3   Performance Optimization

After Sun [27], Javaid [14] and others fully studied, blockchain system models and performance analysis, the performance is further optimized. For example, Sun et al. [27] proposed an algorithm based on performance analysis to optimize the deployment of communication nodes in a blockchain system to achieve maximum transaction throughput. The transaction flow in Hyperledger Fabric follows the execution-sequence-verification model, and many previous performance studies have emphasized that the verification phase is one of its main contributions to latency. Javaid et al. [14] and Thakkar et al. [28] analyzed the delay of Fabric, and then redesigned the verification or submission phase of Fabric.

Although blockchain has many advantages, it also has some obvious shortcomings. To resolve these shortcomings, scholars have put forward corresponding solutions. For example, Setty et al. [24] propose VOLT to improve the insufficient security, slow speed and high cost caused by mining of blockchain. The number of blockchains is constantly increasing, but they are all independent and unconnected. Moreover, the current cross-blockchain operation technology is very limited, which hinders asset transfer and data exchange between different blockchains. This brings challenges to both users and developers. Realizing the interoperability of the blockchain is a great advancement of the blockchain. Dextt [7] implements the interaction between blockchains, in which the deterministic cross-blockchain token transfer protocol is used to achieve the ultimate consistency of cross-blockchain token transfer. It is undeniable that the performance of the blockchain is limited by the global consensus demand. Most solutions assume that the blockchain is accessed synchronously, but Teechain [19] is the first secure payment network that proposes asynchronous blockchain access. Many blockchains are now subject to severe architectural constraints, because of not only the sequential execution of transactions to ensure consistency but also the confidentiality of data. OXII and ParBlockchain [2] are proposed to support distributed applications with conflicting workloads. OXII is a new permissioned blockchain sorting-execution paradigm. It first generates a dependency graph for transactions within a block, allowing parallel execution of non-conflicting transactions, and achieving higher concurrency. ParBlockchain

Table 4. Comparison of different blockchain performance analysis methods

Scenario	Contributions	Optimization design
Blockchain-enabled wireless IoT system [21]	Maximize transaction throughput and determine the best full-featured node deployment of the blockchain system	Design algorithm to find optimal FN density
Hyperledger Fabric [14]	Optimize the delay in the verification phase	Use chaincode caching and parallel database read and write
Hyperledger Fabric [28]	Optimize the delay of verification phase and submission phase	Use MSP cache, block parallel VSCC verification, CouchDB batch read/write during MVCC verification and submission
VOLT [24]	Realize a safe and resource-efficient blockchain network	Building blockchain service providers and Caesar consensus, etc.
Cross-blockchain [7]	Cross-blockchain token transfer	Witness contest, VETO transaction, etc.
Teechain [19]	Implement a payment network that uses asynchronous blockchain access operations and provides dynamic deposits	Dynamic deposits of treasury bonds, payment with asynchronous blockchain access and commission chain
ParBlockchain [2]	Handling workloads with conflicting transactions	Generate dependency graphs for transactions within the block, allowing parallel execution of non-conflicting transactions

is a permissioned blockchain specially designed for the OXII paradigm. Table 4 shows the methods of blockchain performance optimization by different studies.

4 Conclusion

In this survey paper, we analyzed the current research status of blockchain security and performance optimization. Firstly, the classification of blockchain security vulnerabilities and attack prevention technologies are analyzed, then the methods to protect user privacy in blockchain are discussed, and finally, the research on performance evaluation and optimization of blockchain is summarized. With the further development of blockchain, it will have more complete solutions and wider applications.

Acknowledgment. This work is supported by the National Natural Science Foundation of China (No. 61972118).

References

1. Alharby, M., van Moorsel, A.: BlockSim: a simulation framework for blockchain systems. ACM SIGMETRICS Perform. Eval. Rev. **46**(3), 135–138 (2019)
2. Amiri, M.J., Agrawal, D., El Abbadi, A.: ParBlockchain: leveraging transaction parallelism in permissioned blockchain systems. In: 2019 IEEE 39th International Conference on Distributed Computing Systems (ICDCS), pp. 1337–1347. IEEE (2019)
3. Athalye, A., Belay, A., Kaashoek, M.F., Morris, R., Zeldovich, N.: Notary: a device for secure transaction approval. In: Proceedings of the 27th ACM Symposium on Operating Systems Principles, pp. 97–113 (2019)
4. Atzei, N., Bartoletti, M., Cimoli, T.: A survey of attacks on ethereum smart contracts (SoK). In: Maffei, M., Ryan, M. (eds.) POST 2017. LNCS, vol. 10204, pp. 164–186. Springer, Heidelberg (2017). https://doi.org/10.1007/978-3-662-54455-6_8
5. Banno, R., Shudo, K.: Simulating a blockchain network with SimBlock. In: 2019 IEEE International Conference on Blockchain and Cryptocurrency (ICBC), pp. 3–4. IEEE (2019)
6. Bonneau, J., Narayanan, A., Miller, A., Clark, J., Kroll, J.A., Felten, E.W.: Mixcoin: anonymity for bitcoin with accountable mixes. In: Christin, N., Safavi-Naini, R. (eds.) FC 2014. LNCS, vol. 8437, pp. 486–504. Springer, Heidelberg (2014). https://doi.org/10.1007/978-3-662-45472-5_31
7. Borkowski, M., Sigwart, M., Frauenthaler, P., Hukkinen, T., Schulte, S.: DeXTT: deterministic cross-blockchain token transfers. IEEE Access **7**, 111030–111042 (2019)
8. Cheng, Z., et al.: Towards a first step to understand the cryptocurrency stealing attack on ethereum. In: 22nd International Symposium on Research in Attacks, Intrusions and Defenses (RAID 2019), pp. 47–60 (2019)
9. Chitra, T., Quaintance, M., Haber, S., Martino, W.: Agent-based simulations of blockchain protocols illustrated via kadena's chainweb. In: 2019 IEEE European Symposium on Security and Privacy Workshops (EuroS&PW), pp. 386–395. IEEE (2019)
10. Cook, T., Latham, A., Lee, J.H.: DappGuard: active monitoring and defense for solidity smart contracts (2017). Accessed 18 July 2018
11. Dinh, T.T.A., Liu, R., Zhang, M., Chen, G., Ooi, B.C., Wang, J.: Untangling blockchain: a data processing view of blockchain systems. IEEE Trans. Knowl. Data Eng. **30**(7), 1366–1385 (2018)
12. Dong, Z., Zheng, E., Choon, Y., Zomaya, A.Y.: DAGBENCH: a performance evaluation framework for DAG distributed ledgers. In: 2019 IEEE 12th International Conference on Cloud Computing (CLOUD), pp. 264–271. IEEE (2019)
13. Hajdu, Á., Jovanović, D.: SOLC-VERIFY: a modular verifier for solidity smart contracts. In: Chakraborty, S., Navas, J.A. (eds.) VSTTE 2019. LNCS, vol. 12031, pp. 161–179. Springer, Cham (2020). https://doi.org/10.1007/978-3-030-41600-3_11
14. Javaid, H., Hu, C., Brebner, G.: Optimizing validation phase of hyperledger fabric. In: 2019 IEEE 27th International Symposium on Modeling, Analysis, and Simulation of Computer and Telecommunication Systems (MASCOTS), pp. 269–275. IEEE (2019)

15. Kim, S.: Impacts of mobility on performance of blockchain in VANET. IEEE Access **7**, 68646–68655 (2019)
16. Krupp, J., Rossow, C.: teEther: gnawing at ethereum to automatically exploit smart contracts. In: 27th USENIX Security Symposium (USENIX Security 2018), pp. 1317–1333 (2018)
17. Kumar, A., Fischer, C., Tople, S., Saxena, P.: A traceability analysis of Monero's blockchain. In: Foley, S.N., Gollmann, D., Snekkenes, E. (eds.) ESORICS 2017. LNCS, vol. 10493, pp. 153–173. Springer, Cham (2017). https://doi.org/10.1007/978-3-319-66399-9_9
18. Li, X., Mei, Y., Gong, J., Xiang, F., Sun, Z.: A blockchain privacy protection scheme based on ring signature. IEEE Access **8**, 76765–76772 (2020)
19. Lind, J., Naor, O., Eyal, I., Kelbert, F., Sirer, E.G., Pietzuch, P.: Teechain: a secure payment network with asynchronous blockchain access. In: Proceedings of the 27th ACM Symposium on Operating Systems Principles, pp. 63–79 (2019)
20. Maxwell, G.: CoinSwap: transaction graph disjoint trustless trading, October 2013
21. Novo, O.: Scalable access management in IoT using blockchain: a performance evaluation. IEEE Internet Things J. **6**(3), 4694–4701 (2018)
22. Pandey, S., Ojha, G., Shrestha, B.: BlockSim: a practical simulation tool for optimal network design, stability and planning. In: 2019 IEEE International Conference on Blockchain and Cryptocurrency (ICBC), pp. 133–137. IEEE (2019)
23. Phillips, R., Wilder, H.: Tracing cryptocurrency scams: clustering replicated advance-fee and phishing websites. arXiv e-prints (2020)
24. Setty, S., Basu, S., Zhou, L., Stephenson, J., Venkatesan, R.: Enabling secure and resource-efficient blockchain networks with volt. Technical report, MSR-TR-2017-38, Microsoft, August 2017. https://www.microsoft.com/en-us/research/publication/enabling-secure-resource-efficient-blockchain-networks-volt/
25. Shbair, W.M., Steichen, M., François, J., State, R.: BlockZoom: large-scale blockchain testbed. In: 2019 IEEE International Conference on Blockchain and Cryptocurrency (ICBC), pp. 5–6. IEEE (2019)
26. Sun, S.-F., Au, M.H., Liu, J.K., Yuen, T.H.: RingCT 2.0: a compact accumulator-based (linkable ring signature) protocol for blockchain cryptocurrency Monero. In: Foley, S.N., Gollmann, D., Snekkenes, E. (eds.) ESORICS 2017. LNCS, vol. 10493, pp. 456–474. Springer, Cham (2017). https://doi.org/10.1007/978-3-319-66399-9_25
27. Sun, Y., Zhang, L., Feng, G., Yang, B., Cao, B., Imran, M.A.: Blockchain-enabled wireless Internet of Things: performance analysis and optimal communication node deployment. IEEE Internet Things J. **6**(3), 5791–5802 (2019)
28. Thakkar, P., Nathan, S., Viswanathan, B.: Performance benchmarking and optimizing hyperledger fabric blockchain platform. In: 2018 IEEE 26th International Symposium on Modeling, Analysis, and Simulation of Computer and Telecommunication Systems (MASCOTS), pp. 264–276. IEEE (2018)
29. Valenta, L., Rowan, B.: Blindcoin: blinded, accountable mixes for bitcoin. In: Brenner, M., Christin, N., Johnson, B., Rohloff, K. (eds.) FC 2015. LNCS, vol. 8976, pp. 112–126. Springer, Heidelberg (2015). https://doi.org/10.1007/978-3-662-48051-9_9
30. Wang, Y., et al.: Formal specification and verification of smart contracts for azure blockchain. arXiv preprint arXiv:1812.08829 (2018)

Performance Analysis of Blockchain-Based Internet of Vehicles Under the DSRC Architecture

Qilie Liu[1]([✉]), Liang Lin[1], Yun Li[1], and Yongxiang Liu[2]

[1] School of Communication and Information Engineering,
Chongqing University of Posts and Telecommunications, Chongqing, China
{liuql,liyun}@cqupt.edu.cn, linliangsure@foxmail.com
[2] School of Economics and Management, North China Electric Power University,
Beijing, China
yongxiangL@126.com

Abstract. Blockchain technology has shown great potential in the Internet of Vehicles (IoV) in solving the problems of data sharing and information traceability. Understanding the relationship between the IoV communication architecture and the blockchain can facilitate designing dedicated blockchain enabled IoV systems. In this paper, a two-layer wireless blockchain network architecture based on Dedicated Short Range Communication (DSRC) is proposed. Then the M/G/1 queuing model is used to analyze the delivery process of the transaction under the unsaturated condition, and the Markov model is established to analyze the unicast service process. Finally, the verification process of the block in the Tangle consensus network is deduced. The simulation results show that the network load, channel conditions, queuing and backoff service processes in the wireless environment have a significant impact on the delay and throughput of the blockchain network, which further proves that the wireless environment is the main reason for limiting the performance of the Tangle blockchain network.

Keywords: Blockchain · IoV · DSRC · M/G/1

1 Introduction

The rapid development of Internet of Vehicles (IoV) technology has brought a richer driving experience to the drivers. However, in order to improve the user

Q. Liu—The work is supported by the National Natural Science Foundation of China (61671096), Chongqing Natural Fund Key Project (No. cstc2019jcyj-zdxmX0008), Chongqing Science and Technology Innovation Leading Talent Support Program (CSTCCXLJRC201908), Science and Technology Research Program of Chongqing Municipal Education Commission (Grant No. KJZD-K201900605), and State Grid Corporation of China Technology Project (No. 5418-201971184A-0-0-00).

H. Gao et al. (Eds.): ChinaCom 2020, LNICST 352, pp. 112–126, 2021.
https://doi.org/10.1007/978-3-030-67720-6_8

experience, smart car equipment vendors as the service subjects need to collect various data to analyze user requirements, which brings about the problems of data storage, data transmission and privacy leakage. Besides, the traditional IoV architecture relies on centralized data management and needs to face single points of failure and data leakage risks. With the generation of cryptocurrency Bitcoin [1], blockchain technology is proposed as a revolutionary technology to build a distributed and tamper-resistant digital ledger. In the case of removing the third-party institutions, the users who do not trust each other can conduct transactions through the blockchain. The key technologies of blockchain mainly include digital encryption, time stamp server, consensus mechanism, smart contract and incentive mechanism, etc., which can be applied to the fields of cryptocurrency, Internet of Things (IoT), smart city, energy transaction [2,3], etc.

In the IoV, on the one hand, the blockchain consensus mechanism is used to store and verify data, so that the data cannot be tampered with. On the other hand, smart contracts can be used to code complex logical ideas in the IoV, which can improve system execution efficiency.

However, the current application of blockchain in the IoV still faces many challenges. First, extending the blockchain to the IoV need to consider consideration of a suitable consensus mechanism. The traditional consensus mechanism based on Proof of Work (PoW) requires huge computing power, which is not friendly to the IoV nodes with low storage and computing capabilities. In order to solve the problems of limited storage space and computing power, currently, a blockchain consistency mechanism Tangle based on a Directed Acyclic Graph (DAG) algorithm has been widely studied [4]. It supports micro-transactions and has high throughput characteristics, and can be well adapted to the IoT environment. In addition, how to reduce the impact of mobility on network performance is also a challenge when applying blockchain to the IoV.

Nowadays, researchers have done a lot of work on the related issues. In [5], the author analyzed how the blockchain was extended to the IoV and presented a model of the outward transmission of vehicle blockchain data, but the analysis of the blockchain network is defective. In [6], the author established an analysis model of the IoV system based on blockchain. Based on the system model and performance analysis, the author designed an algorithm to determine the optimal full function node deployment for blockchain system under the criterion of maximizing transaction throughput. In [7], the author analyzed the security performance of a wireless blockchain network with malicious interference, and discussed the probability of successful block transmission. In [8], the author used three key indicators to study how mobility affects the performance of the blockchain system in the IoV. The model assumed that the consensus of the blockchain was carried out on moving vehicles. In [9], the author theoretically analyzed the impact of the 802.11 transmission protocol on the performance of the blockchain system, and proposed a random model to analyse the probability of a successful double-spending attack. And in [10], it analyzed why the Tangle blockchain is more suitable for IoT systems than PoW and Proof of Stake (PoS), and discussed the potential problems and challenges of Tangle in the IoT.

Most of the above work consider placing the consensus layer on the fixed IoT nodes, but ignore the mobility of the IoV nodes which will undoubtedly bring about delay fluctuations and reliability problems to the consensus of the blockchain network. Besides, the existing work has not analyzed the performance of the blockchain network on the more mature Dedicated Short Range Communication (DSRC) architecture. Therefore, based on the above work, this article proposes to extend the Tangle blockchain into the IoV. The main contributions are as follows:

- Research the extension of blockchain technology to the IoV under the DSRC architecture, propose a two-layer wireless architecture to avoid the impact of mobility on the blockchain network, and analyzes the blockchain transaction delivery model based on Carrier Sense Multiple Access (CSMA/CA) mechanism.
- We theoretically studied the delay and throughput performance of extending the Tangle blockchain to the IoV. First, we established the M/G/1 queuing model to analyze the delivery process of the transaction, and obtained the average transaction delivery delay and the average number of retransmissions; Then we deduced the block verification process in the Tangle consensus network. Considering the network load, node distribution and channel conditions, the confirmation delay of the transaction from being issued to be verified and the network throughput are obtained.

The rest of this paper is structured as follows. Section 2 introduces prepare knowledge and discusses the system model; Sect. 3 theoretically analyzes the performance of blockchain network in IoV systems, including the wireless transmission model and the blockchain verification process; Sect. 4 presents the evaluation of the proposed model; Finally, Sect. 5 concludes this paper.

2   Preliminaries and System Model

2.1   DAG Based Tangle Blockchain

The Tangle blockchain has been proved to be a very potential distributed network solution. According to reference [4], we know that Tangle uses DAG data structure to store blocks in the network, and each block unit in DAG only contains one transaction. A new transaction entering the Tangle needs to be packaged into block and verifies the two blocks at the end of the DAG. Each block has its own weight and cumulative weight. The cumulative weight is the sum of the own weights of all other blocks that directly and indirectly verify the block. The execution process of the consensus algorithm is the growth process of the cumulative weight, which is related to the load of the network. The block verification delay refers to the time when the cumulative weight reaches the verification threshold.

2.2 System Model

For the IoV network with random distribution and mobility characteristics, if the blockchain consensus layer is placed at the bottom mobile node, this will bring huge challenges to the reliability and delay of the blockchain network, e.g., double-spending attacks are caused by communication delays. In order to reduce the impact of the vehicle mobility on the blockchain network, this paper proposes a two-layer network model with consensus layer moving up. In the model, vehicles as light nodes are only responsible for delivering transactions to the blockchain network, and the fixed Road-Side Unit (RSU) is consensus node, running the blockchain consensus algorithm. In this paper, it assumes that the number of light nodes is subject to $2\beta L_s$ Poisson distribution [11]. β is the unit mileage density of the node in the IoV, the unit is vehicle/m; L_s is the carrier sensing coverage area of the all node, which is determined by the wireless transmission protocol. There are n nodes within the carrier sensing coverage area of the consensus node, which can be expressed as

$$P(n, L_s) = e^{-2\beta L_s} \frac{(2\beta L_s)^n}{n!} \tag{1}$$

Light nodes communicate with neighboring light nodes and consensus nodes through 802.11p wireless transmission protocol. When a light node delivers a transaction to a consensus node, it first needs to compete with other light nodes within its own carrier sensing coverage area, and obtain the right to use the channel through the CSMA/CA mechanism. Only when the nodes meet the two prerequisites that the channel is idle and the backoff counter is 0 can participate in the channel competition. The above process can be modeled as a queuing process, which will be discussed in detail in the next section. The consensus node sends the received blockchain block to other consensus nodes through a wired channel to complete the consensus.

To illustrate the system model of the proposed framework, the detail steps in the Fig. 1 are as follows.

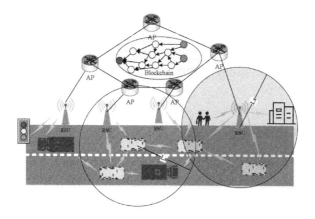

Fig. 1. System model

1) Light nodes classify the data transactions collected by their respective sensors, which can be divided into one-way transactions $id|type|pk|sig$ and two-way transactions $id|type|pk.1|sig.1|pk.2|sig.2$ according to the logic type. The content includes transaction id, transaction business type, transaction information, and signatures of the publisher and recipient. It is assumed that the number of transaction arrivals of each light node follows the Poisson distribution with the parameter λ.

2) The delivery process of blockchain transactions is divided into two stages. The first stage is that light nodes send the transaction to consensus nodes through one-hop unicast. In this process, multiple transactions are first queued in the MAC layer, and monitor the channel to obtain communication resources. Since it is a process of queuing to be served, this paper models it as an M/G/1 queuing system [12]. Therefore, the probability of k transactions reaching the queuing system in time t is

$$P(k,t) = \frac{(\lambda t)^k}{k!} e^{-\lambda t} \tag{2}$$

In this paper, the wireless transmission environment is non-ideal, wireless signals are susceptible to various interference sources, such as buildings and communication distances, and the data packet errors can be caused by channel fading. Therefore, we discusses the channel fading by introducing the packet transmission error probability, and its calculation formula is

$$P_e = 1 - (1 - P_{ber})^{H+L} \tag{3}$$

where P_{ber} is the error probability and $H + L$ is the size of the data packet.

3) The second stage is that the consensus node packs the transaction into the block, then verifies, records, and finally forwards them through the wired communication. According to the network traffic entering the consensus node, which is related to the first stage, the verification process can be divided into low load and high load.

Our analysis framework can be extended to other scenarios with mobility and resource constraints, i.e., the Unmanned Aerial Vehicle (UAV) IoT environment. Note that the focus of this paper is on the uplink transmission from the light node to the consensus node, mainly to analyze the confirmation delay of the transaction being successfully verified. Downlink performance analysis can be done in a similar way.

3   Problem Analysis

To study the performance of the network when extending the blockchain to the IoV, we will focus on the system delay. The transaction confirmation delay T_s

represents the time from when the transaction is issued to being verified, which can be expressed as

$$T_s = T_w + T_v \tag{4}$$

$$T_w = T_q + T_{st} \tag{5}$$

T_w represents the time delay for the transaction to be delivered to the blockchain network. The queuing time T_q is the duration when the transaction reaches the MAC layer cache queue and it enters the CSMA/CA backoff process. The service time T_{st} is the time from entering the CSMA/CA backoff process to successful transmission after queuing, which is determined by wireless environment factors. T_v represents the block verification delay after the transaction enters the blockchain network, that is, the delay when the cumulative weight reaches the verification threshold. Through the Tangle white paper [4], we know that the block verification delay is greatly affected by the transaction load, so it is necessary to study the delivery efficiency and consensus mechanism of transactions in a wireless environment. This section will first analyze the delivery process of transactions in a wireless environment, and then analyze the block verification process.

3.1   Probability Analysis of Transaction Delivery Process

Considering that the delivery of a transaction in a wireless environment is a process of queuing and waiting for service, this paper models it as an M/G/1 queuing model. The transaction generated by the light node of the blockchain first reaches the MAC layer cache queue, and then the light node obtains the right to use the channel through the CSMA/CA back-off mechanism to send the transaction. The back-off process is the service process of the queuing system.

When the light node has a transaction that needs to be uploaded to the blockchain network, it is first necessary to detect whether the wireless channel is idle according to the back-off mechanism. The back-off counter decreases by 1 when it detects that the wireless channel is idle in each slot time σ, otherwise it stays suspended. When the counter reduces to 0, the transaction is sent. If a collision occurs while sending the transaction, the next back-off stage is entered. In this paper, we assume that each back-off stage has the same back-off window and allow the transaction to be retransmitted until the receiver successfully receives it. Because the actual environment is an unsaturated network, the queue may enter the idle period after sending the transaction. Based on the above analysis, we model the backoff behavior of a light node as a Markov process [13].

Since the transaction arrival rate of light nodes obeys the Poisson distribution with the parameter λ, it can be known that the probability that no transaction arrives in the cache queue is

$$P_{idle} = 1 - exp(-\lambda h) \tag{6}$$

Where h is the average slot time length, which will be analyzed later. Within the carrier sensing range of the sending node, the probability that at least one node sends a transaction is

$$P_{tr} = 1 - \sum_{k=0}^{\infty} (1 - \tau)^k e^{-2\beta L_s} \frac{(2\beta L_s)^k}{k!}$$
$$= 1 - e^{-2\beta L_s \tau} \tag{7}$$

where τ represents the probability of a node sending a transaction at a slot time.

Define P_m as the probability that the transaction sent by the light node will fail due to the collision and error at any slot time, i.e., the probability of entering the next back-off stage. It can be expressed as

$$P_m = PP_e + P(1 - P_e) + (1 - P)P_e \tag{8}$$

where $P = 1 - e^{-(2\beta L_s - 1)\tau}$ is the probability that the back-off counter reaches 0 and the channel is detected busy again.

According to the above analysis, we can write the probability expression of a light node sending a transaction in a generic (i.e., randomly chosen) slot time as

$$\tau = \left[\frac{W_0(1 + P_m) + (1 - 2P_{tr})}{2(1 - P_{tr})} + \frac{1 - \rho}{1 - P_{idle}} \right]^{-1} \tag{9}$$

where ρ is the service intensity of the queuing system, which will be analyzed later. Considering (9) is a complicated process, we choose to use an iterative algorithm to solve it. The specific implementation of the algorithm is introduced in the next section.

Therefore, according to (6)–(9), the average length of a slot time can be expressed as

$$h = (1 - P_{tr})\sigma + P_{tr}P_m(T_c[i] + T_e[i]) + P_{tr}(1 - P_m)T_s[i] \tag{10}$$

where $T_c[i]$ and $T_e[i]$ represent the delay of collision and transmission error in unicast, respectively, and the successful transmission delay is $T_s[i]$. These can be expressed as

$$\begin{cases} T_s[i] = T_e[i] = T_{RTS} + T_{SIFS} + T_\delta + T_{CTS} + T_{SIFS} + T_\delta \\ + T_{SIFS} + T(H + L) + T_\delta + T_{ACK} + T_{AIFS[i]} + T_\delta \\ T_c[i] = T_{RTS} + T_{AIFS[i]} + T_\delta \end{cases} \tag{11}$$

where T_δ represents the data propagation delay.

In 802.11p, the Enhanced Distributed Channel Access (EDCA) mechanism is adopted, and Arbitration Inter-frame Spacing ($AIFS$) is used instead of Distributed Inter-frame Spacing ($DIFS$). For different priorities of transactions, the calculation formula of $AIFS$ is as follows

$$AIFS[i] = AIFSN(AC[i]) * \sigma + SIFS \tag{12}$$

Among them, $AIFSN$ is mainly for distinguishing the grade of the transaction, $AC[i]$ represents the type of the transaction, $SIFS$ is the interval between short frames.

The average number of retransmissions in the back-off mechanism is an important indicator for analyzing the transaction delivery efficiency of light nodes. Define P_n as the probability that the transaction is successfully sent after n back-off stages. Since each back-off process is independent of each other, P_n can be expressed as

$$\begin{cases} P_0 = 1 - P_m \\ P_1 = (1 - P_0)P_0 \\ P_2 = (1 - P_0)(1 - P_1)P_0 \\ \cdots \cdots \\ P_n = \prod_{i=0}^{n-1}(1 - P_i)P_0 \end{cases} \tag{13}$$

Finally, the average number of retransmissions caused by collision and error in the process of back-off service can be obtained

$$N_w = \sum_{n=0}^{\infty} n P_n \tag{14}$$

3.2 Daley Analysis of Transaction Delivery Process

On the basis of the probability analysis in the previous section, in order to obtain the average delay of the delivery process in this section, we first need to obtain the service time distribution of the MAC layer. In this paper, we use the Probability Generating Function (PGF) to approximate service time distribution [14]. Denote q_i as the steady state probability that the transaction service time is $i\sigma$. Let $Q(z)$ be the PGF of q_i, which is

$$Q(z) = \sum_{i=0}^{\infty} q_i z^i \tag{15}$$

Due to the simplicity of the symbols in the z transform domain and the one-to-one correspondence between $Q(z)$ and q_i, we discusses how to calculate $Q(z)$ instead of q_i alone, so $Q(z)$ can be expressed as

$$Q(z) = [PH_c(z) + (P_m - P)H_e(z)]^{\lfloor N_w \rfloor} \frac{1}{W_0} \sum_{k=0}^{W_0-1} H_b^k(z) + (1 - P_m) \\ H_s(z) \sum_{i=0}^{\lfloor N_w \rfloor} \{[PH_c(z) + (P_m - P)H_e(z)]^i \frac{1}{W_0} \sum_{k=0}^{W_0-1} H_b^k(z)\} \tag{16}$$

$\lfloor N_w \rfloor$ refers to the rounding operation for the number of retransmissions; $H_c(z) = z^{T_c}$ refers to the time distribution transfer function that causes the transmission failure due to the collision during the transmission process of the transaction; $H_s(z) = z^{T_s}$ refers to the time distribution transfer function that makes the transaction transmitted successfully; $H_e(z) = z^{T_e} = z^{T_s}$ refers to the time distribution transfer function that causes the transmission failure of the transaction due to the channel error code; $H_b(z)$ is the transfer function of the time distribution occupied by the back-off process of the competitive channel. In the scenario conditions of this paper, it can be represented as

$$H_b(z) = (1 - P_{tr})z^\sigma + P_{tr}P_m z^{T_c+T_e} + P_{tr}(1 - P_m)z^{T_s} \tag{17}$$

By differentiating the above (17), we can get the expression and variance formula of the average service time expressed as non-saturated system as follows.

$$\begin{cases} E[T_{st}] = \sum_{i=0} q_i i\sigma = \left.\frac{dQ(z)}{dz}\right|_{z=1} \\ Var[T_{st}] = \left.\frac{d^2 Q(z)}{dz^2}\right|_{z=1} \end{cases} \tag{18}$$

From the Pollaczek-Khintchine (P-K) expression, the average queuing time of the sending node can be obtained in the unsaturated case.

$$E[T_q] = \rho + \frac{\rho^2 + \lambda^2 Q''(z)\big|_{z=1}}{2(1 - \rho)} \tag{19}$$

To derive the average service time distribution, probability τ must be determined. However, the calculation of τ depends on the service parameter $\rho = \frac{\lambda}{\mu}$, where μ is the service rate of the queuing system. Hence, we apply an iterative algorithm to calculate τ and ρ. The iteration steps are shown in Algorithm 1.

Algorithm 1. Time delay double iteration algorithm

1: Initialization: Assume that in the case where the load is saturated with $\rho = 1$, i.e., there are always transaction packets arriving in the queue, the idle state in the Markov model at this time will be removed.
2: **repeat**
3:    Through formulas (6)–(19), the value of $\tau, P_{tr}, h, P_{idle}, P_m, N_w$ can be obtained by the first-level iterative algorithm.
4:    Bring the value iterated in 3 into the PGF function of the service time to obtain the average service time $E[T_{st}]$, the service rate $\mu = \frac{m}{E[T_{st}]}$, and the average queue time $E[T_q]$ of the queue, where m is the maximum number of transactions in a wireless transmission.
5: **until** Update $\rho' = \frac{\lambda}{\mu}$, if $|\rho' - \rho| < \varepsilon$, ε is a predefined minimum error value, output $\rho = \rho'$, and the iteration is completed; otherwise, bring $\rho = \rho'$ into step 2 and repeat the iterative algorithm.

3.3   Daley and Throughput Analysis of Block Verification Process

Through the analysis of the above two subsections, the average number of retransmissions and the average delivery delay of the transaction delivery process have been obtained. Based on this, this section will study the verification process of transactions in the blockchain network, and analyze the verification delay T_{st} and network throughput.

After completing the queuing and back-off process in the wireless environment, the transaction reaches the consensus node firstly, and then is packaged into blocks. Define N_c is the number of all consensus nodes. And then, the consensus node broadcasts blocks to the blockchain network through a wired channel to complete the consensus, and that all blocks in the queue can be broadcast at once. From the above analysis, it can be concluded that the throughput of the wireless environment is the average block arrival rate λ' of consensus nodes in the blockchain. Since the consensus nodes are independent of each other, we assume that the block enters the blockchain network satisfying the Poisson distribution, and the assumption will be more reasonable when the number of consensus nodes increases [15].

In the Tangle network, unlike PoW, blocks will not enter the waiting verification pool, but is directly added to the DAG structure of each consensus node. The cumulative weight is the symbol of block verification. Let $W(t)$ be the expected value of the cumulative weight of the block at time t, and $L(t)$ be the number of tips of the unverified block. The blockchain network update time is D, so the number of tips in the system after a broadcast is

$$L(t) = 2N_c\lambda'D = \begin{cases} 4\beta L_s\lambda N_cD, \rho < 1 \\ \frac{4\beta L_s N_c mD}{E[T_{st}]h}, \rho \geq 1 \end{cases} \tag{20}$$

Since the block arrival rate will affect the block verification delay in the Tangle blockchain network [16]. For this reason, We will discuss the block verification process in saturated and unsaturated wireless environments, respectively. According to the block arrival rate, the cumulative weight verification of the block is divided into low frequency $W_l^{unsat}(t)$ and high frequency $W_h^{unsat}(t)$, as shown in Fig. 2. The verification of low frequency is a linear process with parameter $\lambda_l'\omega$, and the verification of high frequency is first an exponential process, and then a linear process with parameter $\lambda_h'\omega$.

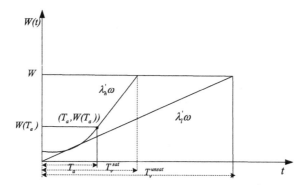

Fig. 2. The cumulative weight growth process of Tangle

Therefore, the cumulative weight under unsaturated wireless environment can be expressed as

$$
\begin{cases}
W_l^{unsat}(t) = \omega \lambda_l' N_c \\
W_h^{unsat}(t) = \begin{cases} 2exp(0.352\frac{t}{D}), t < T_a \\ 2exp(0.352\frac{T_a}{D}) + \omega \lambda_l' N_c(t - T_a), t \geq T_a \end{cases}
\end{cases}
\tag{21}
$$

The critical point of the adaptation period and the linear period is $(T_a, W(T_a))$, where $T_a = \frac{D}{0.352} \ln(2\lambda_l' N_c D)$ represents the end of the adaptation period. If the cumulative weight value of each block reaches W, it is deemed to be successfully verified. Thus the block verification delay under the unsaturated wireless environment can be expressed as

$$
T_v^{unsat} = \begin{cases}
\frac{W}{\omega N_c \lambda_l'}, \lambda_l' N_c D \leq 1 \\
\frac{D}{0.352} \ln(2\lambda_l' N_c D) + \frac{W - W(T_a)}{\omega N_c \lambda_l'}, \lambda_l' N_c D > 1
\end{cases}
\tag{22}
$$

Similarly, in a saturated situation, due to the high block arrival rate, the blockchain network will directly enter the high frequency stage, i.e., the cumulative weight of the block will go through the exponential growth adaptation period and linearity at the speed $\lambda_h' \omega$ growth period. Therefore, the cumulative weight under the saturated wireless environment can be expressed as

$$
W_h^{sat}(t) = 2exp(0.352\frac{T_a}{D}) + \omega \lambda_h' N_c(t - T_a)
\tag{23}
$$

where $T_a = \frac{D}{0.352} ln(2\lambda_h' N_c D)$ represents the end time of the adaptation period at high frequency, and the block verification delay T_v^{sat} under the saturated wireless environment can be obtained.

$$
T_v^{sat} = \frac{D}{0.352} ln(2\lambda_h' N_c D) + \frac{W - W(T_a)}{\omega N_c \lambda_h'}
\tag{24}
$$

For the entire network, the $\rho < 1$ state means that the network is unsaturated, the throughput continues to increase linearly; due to the limitation of channel capacity and delivery efficiency in wireless environment, $\rho \geq 1$ indicates the network reaches saturation state and its throughput tends to be stable.

$$
TPS^{dag} = \begin{cases}
2\beta L_s \lambda N_c, \rho < 1 \\
\frac{2\beta L_s N_c m}{E[T_{st}]}, \rho \geq 1
\end{cases}
\tag{25}
$$

4   Simulation Results

To verify the model proposed in this paper, it is modeled and simulated in MATLAB according to the actual environmental requirements and the 802.11p protocol standard. The system parameters are set as follows: the number of consensus nodes is $N_c = 10$; the communication coverage $L_s = 100\,m$ is controlled by the transmission protocol; the data transmission rate in the wireless channel is 12 Mbps; and the maximum number of transactions for one wireless transmission is set as $m = 32$, the data packet size $E[H + L]$ is 1024. In the blockchain network, the weight of each block is set as $\omega = 3$, and the verification threshold is assumed to be $W = 800$. In this paper, in order to evaluate the performance of the IoV network based on the Tangle blockchain, the influence of parameters, such as the distribution of light nodes, transaction load, and channel attenuation, which are analyzed respectively.

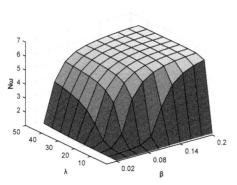

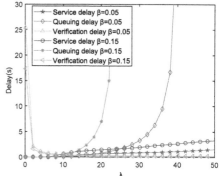

Fig. 3. Average number of retransmissions of transaction delivery

Fig. 4. Transaction queuing delay, service delay, and block verification delay

This article uses a three-dimensional graph to analyze the delivery process of blockchain transactions from light nodes to consensus nodes under the combined effect of the arrival rate λ and node distribution β. As shown in Fig. 3, when λ is fixed, as β increases, the average number of retransmissions increases rapidly. In the same way, after fixing β, the amplitude of N_w also increases with the increase of λ. But it can be clearly seen that β has a greater impact on the average number of retransmissions than λ. This is because λ increases the arrival probability of transactions in the queue and indirectly leads to collision when sending transactions. However, β directly leads to collision by increasing the distribution density of competing nodes. This result is also matched with our previous analysis in (7)–(8). The average number of retransmissions N_w is an important indicator that the wireless environment affects the performance of the blockchain network. As analyzed in (13)–(16), when N_w increases, it means that the average service time T_{st} for transaction delivery becomes larger, and the delivery efficiency decreases, which eventually leads to the block arrival rate of the blockchain network becomes lower in (20).

Figure 4 shows the trend of network queuing delay, service delay and block verification delay with the transaction arrival rate when the light node distribution is 0.01 and 0.15. As shown in the figure, when λ increases to about 25 and 37, the transaction queuing delay T_q increases suddenly, which means that the network is saturated with load. This result is matched with our previous analysis in (19), because the probability of the arrival of transactions in the queue increases with λ, which exceeds the maximum service intensity μ of the queuing system, so new transactions need to be queued to be served. In addition, since λ will indirectly lead to collisions when the transaction is sent, it can also be seen that the service delay T_{st} has a certain increase. Meanwhile, we can also see that $\beta = 0.05$ enters the load saturation stage later than $\beta = 0.15$. Additionally, the figure also shows the trend of block verification delay T_v. According to (14)–(16), when the load is low, the cumulative weight of each block increases slowly, and the block verification delay is large. In contrast, with the increase of λ, the accumulative weight growth rate becomes larger, and thus the block verification delay becomes smaller. Finally, when the network enters the load saturation, the verification delay will stabilize.

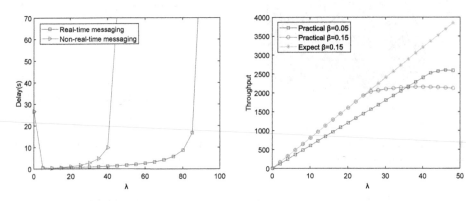

Fig. 5. Confirmation delay of different priority transactions

Fig. 6. Throughput performance

In Fig. 5, we can get the trend of the confirmation delay T_s of different priority transactions. According to (10)–(12), due to the EDCA mechanism of 802.11p, transactions with higher real-time requirements are assigned a higher transmission priority, thereby reducing their service time in the queuing system. From the verification in the figure, the confirmation delay of real-time transactions is lower than that of non-real-time transactions, and it enters the load saturation state later. At the same time, it can be clearly seen in the figure that the confirmation delay of the transaction undergoes a process of falling firstly and then rising, which is caused by the consensus mechanism and the wireless environment, which have been explained in Fig. 4.

In Fig. 6, when the transaction arrival rate is low, the wireless network does not reach the saturated load and the probability of retransmission caused by collision is small, so the delivery efficiency of light node transactions is higher. At this time, the throughput growth rate of $\beta = 0.15$ is higher than that of $\beta = 0.05$. As the transaction arrival rate increases, in the case of $\beta = 0.15$, the practical situation is different from the expect situation in [17], and the throughput performance of the network will eventually stabilize after λ reaches 25. When the maximum value is reached, the throughput will drop to a certain extent due to $E[T_{st}]$, which verifies the analysis in (25). Meanwhile, when the network load is saturated, a larger node distribution density is more likely to collision during transaction delivery, resulting in lower throughput than a blockchain network with a lower node density.

5   Conclusion

In this paper, we studies the performance of the IoV based on the Tangle blockchain, and analyzes how the wireless environment affects the transaction delivery process and consensus process in the Tangle blockchain, which takes into account of the more mature wireless transmission protocol standard DSRC. The simulation results prove: 1) The confirmation delay of the transaction has undergone two stages. In the first stage, the blockchain consensus mechanism leads to a higher confirmation delay, and as the network traffic increases, the confirmation delay decreases. In the second stage, the limitation of the wireless environment causes the confirmation delay to increase; 2) The throughput of the Tangle blockchain network cannot rise infinitely. In fact, as the node distribution density and network load continue to increase, the delivery efficiency of transactions decreases. The throughput of the network eventually stabilizes. The model proposed in this paper shows that it is feasible to extend the Tangle blockchain to the IoV under the DSRC architecture, but its performance is affected by its own consensus mechanism and wireless environmental factors. In the future work, as the IoV standard gradually matures, it will consider studying the blockchain network under the 5G communication architecture. At the same time, the blockchain under the heterogeneous network is also worth studying in the future.

References

1. Nakamoto, S.: Bitcoin: a peer-to-peer electronic cash system. https://bitcoin.org/bitcoin.pdf
2. Lei, A., Cruickshank, H., Cao, Y., et al.: Blockchain-based dynamic key management for heterogeneous intelligent transportation systems. IEEE Internet Things **4**(6), 1832–1843 (2017)
3. Kshetri, N.: Can blockchain strengthen the Internet of Things? IT Prof. **19**(4), 68–72 (2017)
4. Sergio, D.L.: DagCoin: a cryptocurrency without blocks. https://bitslog.com/2015/09/11/dagcoin
5. Jiang, T., Fang, H., Wang, H.: Blockchain-based internet of vehicles: distributed network architecture and performance analysis. IEEE Internet Things J. **6**(3), 4640–4649 (2018)
6. Sun, Y., Zhang, L., Feng, G., et al.: Blockchain-enabled wireless Internet of Things: performance analysis and optimal communication node deployment. IEEE Internet Things J. **6**(3), 5791–5802 (2019)

7. Xu, H., Zhang, L., Liu, Y., et al.: RAFT based wireless blockchain networks in the presence of malicious jamming. IEEE Wirel. Commun. Lett. **9**(6), 817–821 (2020)
8. Kim, S.: Impacts of mobility on performance of blockchain in VANET. IEEE Access **7**, 68646–68655 (2019)
9. Cao, B., Li, M., Zhang, L., et al.: How does CSMA/CA affect the performance and security in wireless blockchain networks. IEEE Trans. Ind. Inform. **16**(6), 4270–4280 (2020)
10. Cao, B., et al.: When Internet of Things meets blockchain: challenges in distributed consensus. IEEE Netw. **33**(6), 133–139 (2019)
11. Yao, Y., Rao, L., Liu, X.: Performance and reliability analysis of IEEE 802.11p safety communication in a highway environment. IEEE Trans. Veh. Technol. **62**(9), 4198–4212 (2013)
12. Yin, X., Ma, X., Trivedi, K.S.: An interacting stochastic models approach for the performance evaluation of DSRC vehicular safety communication. IEEE Trans. Comput. **62**(5), 873–885 (2013)
13. Bianchi, G.: Performance analysis of the IEEE 802.11 distributed coordination function. IEEE J. Sel. Areas Commun. **18**(3), 535–547 (2000)
14. Guo, H., Zhao, H., Shuai, Z.Z.: Platoon architecture VANETs workshop communication process and performance analysis. J. Softw. **30**(4), 1121–1135 (2019)
15. Hao, S., Zhang, H.: Theoretical modeling for performance analysis of IEEE 1901 power-line communication networks in the multi-hop environment. J. Supercomput. **76**(4), 2715–2747 (2019). https://doi.org/10.1007/s11227-019-03065-4
16. Li, Y., et al.: Direct acyclic graph-based ledger for Internet of Things: performance and security analysis. IEEE/ACM Trans. Netw. **28**(4), 1643–1656 (2020)
17. Cao, B., Zhang, Z.H., Feng, D.Q.: Performance analysis and comparison of PoW, PoS and DAG based blockchains. Digit. Commun. Netw. **6**(4), 480–485 (2020)

Cache-Aided Multi-message Private Information Retrieval

Yang Li, Nan Liu$^{(\boxtimes)}$ ⓘ, and Wei Kang

National Mobile Communications Research Laboratory, Southeast University,
Nanjing 210096, China
{younglee,nanliu,wkang}@seu.edu.cn

Abstract. We consider the problem of multi-message private information retrieval (MPIR) from N non-colluding and replicated servers when the user is equipped with a cache that holds an uncoded fraction r from each of the K stored messages in the servers. We assume that the servers are unaware of the cache content. We investigate $D_P^*(r)$, which is the optimal download cost normalized by the message size, as a function of K, N, r, P. For a fixed K, N, we develop an inner bound (converse bound) for the $D_P^*(r)$ curve. The inner bound is a piece-wise linear function in r. For the achievability, we propose specific schemes that exploit the cached as private side information to achieve some corner points. We obtain an outer bound (achievability) for any caching ratio by memory-sharing between these corner points. Thus, the outer bound is also a piece-wise linear function in r. The inner and the outer bounds match for the cases where the number of desired messages P is at least half of the number of the overall stored messages K. Furthermore, the bounds match in two specific regimes for the case $\frac{K}{P} > 2$ and $\frac{K}{P} \in \mathbb{N}$: the very high ratio regime, i.e., $r \geq \frac{1}{N+1}$ and the very low ratio regime, i.e.,
$$r \leq \frac{(N-1)P\alpha_1}{N\left(\sum_{k=2}^{K} \binom{K}{k}\alpha_k - \sum_{k=2}^{K-P}\binom{K-P}{k}\alpha_k\right)+(N-1)P\alpha_1}.$$ Finally, the bounds meet in one specific regime for arbitrarily fixed K, P, N: the very high ratio regime, i.e., $r \geq \frac{1}{N+1}$.

Keywords: Multi-message · Cache · PIR

1 Introduction

With the upgrading of Internet technology and the invention of smart phones, people are using the Internet all the time, leaving more and more traces of each user's preferences. These traces reveal the privacy of users, and when exploited can threaten the safety and well-being of each user. Thus, the protection of user privacy is becoming more and more important.

Along the lines of privacy protection, an important problem called private information retrieval (PIR) has been proposed [2]. It allows each user to retrieve a single message from a database without revealing to the servers that store the database which message the user is interested in. For example, a user interested in a particular stock in the stock market may not want to reveal his/her interest,

H. Gao et al. (Eds.): ChinaCom 2020, LNICST 352, pp. 127–141, 2021.
https://doi.org/10.1007/978-3-030-67720-6_9

worried that this revelation will affect the price of the stock. Or, a user while retrieving some information on a particular disease, do not want to reveal his/her health issues to the outside world.

The private information retrieval (PIR) problem was first considered in [2], where a user wishes to privately download one bit out of a database of K bits. Here, privacy is in the information-theoretic sense, i.e., the queries of the user can not reveal anything about which bit it is interested in. When the user makes queries to one server that stores the database, the only information-theoretic private retrieval way is to download the entire database [2], i.e., all K bits. This incurs a huge download cost and a small download efficiency of $\frac{1}{K}$, as to retrieve 1 bit of desired information, K bits must be downloaded. Hence, it is proposed that the user queries N servers, $N > 1$, that stores the database, without revealing which bit it is interested in to any single server [2]. Reference [3] proposed a private retrieval scheme that can achieve the download efficiency of $\frac{N-1}{N}$, i.e., to download $N - 1$ bits of desired information, N total bits must be downloaded from the N servers.

The recent significant work by Sun and Jafar [4] reformulated the PIR problem from an information-theoretic perspective, and rather than retrieving a bit, they considered retrieving a message from a database of K messages, where each message is arbitrarily long. They found that the maximum efficiency, which they termed *capacity*, of PIR retrieval from N servers is $(1+\frac{1}{N}+\frac{1}{N^2}+\cdots+\frac{1}{N^{K-1}})^{-1}$. It can be seen that the capacity is a monotonically increasing function of the number of servers. The more servers the user retrieves from, the larger the PIR capacity, i.e., by increasing the number of queried servers, the user can effectively reduce the size of the download.

The PIR problem has recently attracted much attention in the information-theoretic community for its capacity characterizations under various constraints and due to limited space, we can not list all of them here. The work that is most related to our paper are references [5] and [1]. Reference [5] considered a cache-aided PIR problem with unknown and uncoded prefetc.hing, i.e., user can prefetc.h Lr number of bits from each message W_k, which is of length L, for all messages, $k = 1, 2, \cdots, K$. Furthermore, the servers are unaware of which bits are stored in the user's cache. The optimal tradeoff between the normalized download cost $\frac{D(r)}{L}$, which is the inverse of capacity, and the caching ratio r was the subject of focus in [5]. The outer and inner bounds are both piece-wise linear curves which consists of K line segments. The inner and the outer bounds meet in general for the cases of very low caching ratio $(r \leq \frac{1}{1+N+N^2+\cdots+N^{K-1}})$ and very high caching ratio $(r \geq \frac{K-2}{(N+1)K+N^2-2N-2})$.

Reference [1] studied the multi-message PIR problem, where the user need to retrieve P messages, $P > 1$, without exposing the identities of the P messages to any one server. Instead of using a single file private information retrieval scheme [4] over and over again for P times, reference [1] showed that the P messages can be retrieved more efficiently by an MDS code.

In this paper, we consider a new PIR scenario, namely cache-aided multi-message PIR (MPIR). In this setup we assume that the user requires multiple messages and at the same time has access to some cached bits, which is a subset of the bits evenly taken from each message. We propose lower and upper bounds on the normalized download cost, which are both piecewise-linear functions of P, K, N, r. We show that the upper and lower bounds match in the following three cases:

1. $P \geq \frac{K}{2}$, i.e., when the number of requested messages is more than half of the total number of messages.
2. $P < \frac{K}{2}$ and $r \geq \frac{1}{N+1}$, i.e., when the caching ratio is sufficiently large.
3. For very low caching ratio and $P < \frac{K}{2}$ and $\frac{K}{P} \in \mathbb{N}$, i.e., when the number of total messages is an integer multiple of the number of retrieved messages and when the caching ratio is sufficiently small.

The rest of the paper is organized as follows. In Sect. 2 we describe the system model. In Sect. 3 our main results for the download cost of the cache-aided MPIR problem is stated. Sections 4 and 5 provide the achievability and converse proofs, respectively. Finally, Sect. 6 concludes the paper.

2   System Model

We consider a PIR problem with K independent messages $W_1, \ldots, W_K$. The length of each message is L bits, i.e.,

$$H(W_1, \ldots, W_K) = H(W_1) + \cdots + H(W_K) = KL. \tag{1}$$

We use the random variable Z to represent the contents of the user's cache. Before sending out retrieval requests to the servers, the user caches Lr bits of each message in advance, where $0 \leq r \leq 1$, and r is called *caching ratio*. The Lr bits are taken randomly and evenly from each L-length message, Therefore,

$$H(Z) = KLr. \tag{2}$$

We use the random variable $\mathbb{H}$ to represent indices of the cached bits, and we assume that all servers know only the cache ratio r and the prefetching rule of the user, but not the value of $\mathbb{H}$.

After the prefetching phase, the user privately generates an index set $\mathcal{P} = \{i_1, \cdots, i_P\} \subseteq \{1, \cdots, K\}$, independent of $\mathbb{H}$, and wants to retrieve the content of the P messages $W_{\mathcal{P}} = (W_{i_1}, W_{i_2}, \cdots, W_{i_P})$ from the servers, while ensuring that the index set of these messages is not known to any single server. Therefore, we have

$$H\left(\mathcal{P}, \mathbb{H}, W_1, \ldots, W_K\right) = H\left(\mathcal{P}\right) + H\left(\mathbb{H}\right) + H(W_1) + \cdots + H(W_K). \tag{3}$$

To retrieve the contents of the message set of interest, i.e., $W_{\mathcal{P}}$, the user sends out request $Q_n^{[\mathcal{P}]}$ to Server n, $n = 1, 2, \cdots, N$. During the retrieval phase, the user does not know anything about the content of any messages in the database. Therefore, the request sent by the user and the messages' contents of the database are independent of each other, i.e.,

$$I\left(W_1, \cdots, W_K; Q_1^{[\mathcal{P}]}, \cdots, Q_N^{[\mathcal{P}]}\right) = 0 \tag{4}$$

Upon receipt of the request $Q_n^{[\mathcal{P}]}$, the n-th server returns an answer string $A_n^{[\mathcal{P}]}$ to the user, determined by the files stored in the server and the request received, hence

$$H(A_n^{[\mathcal{P}]}|Q_n^{[\mathcal{P}]}, W_{1:K}) = 0 \tag{5}$$

User privacy requires that each individual server cannot know anything about the message set $\mathcal{P}$ the user wants to download based on the retrieval request sent by the user. Therefore, for any P messages, the retrieval request and the answer are indistinguishable from any individual server, i.e., $\forall n \in [N], \forall \mathcal{P}_1, \mathcal{P}_2 \subseteq \{1, \cdots, K\}, |\mathcal{P}_1| = |\mathcal{P}_2| = P$,

$$(Q_n^{[\mathcal{P}_1]}, A_n^{[\mathcal{P}_1]}, W_1, \ldots, W_K) \sim (Q_n^{[\mathcal{P}_2]}, A_n^{[\mathcal{P}_2]}, W_1, \ldots, W_K). \tag{6}$$

Here, "$\sim$" means with the same distribution.

After receiving the replies from all the servers, i.e., $A_1^{[\mathcal{P}]}, \ldots, A_N^{[\mathcal{P}]}$, with the help of the cache bits, the user can reliably extract all the required messages $W_\mathcal{P}$. Hence, the following reliability constraint needs to be met:

$$H\left(W_\mathcal{P}|Z, \mathbb{H}, Q_1^{[\mathcal{P}]}, \ldots, Q_N^{[\mathcal{P}]}, A_1^{[\mathcal{P}]}, \ldots, A_N^{[\mathcal{P}]}\right) = o(L), \tag{7}$$

where $o(L)$ denotes a function such that $\frac{o(L)}{L} \to 0$ as $L \to \infty$.

For a fixed N, K, P, and caching ratio r, a pair $(D_P(r), PL)$ is achievable if there exists a PIR scheme for message size L with unknown and uncoded prefetching satisfying the privacy constraint (6) and the reliability constraint (7), where $D_P(r)$ represents the expected number of downloaded bits (over all the queries) from the N servers via the answering strings $A_{1:N}^{[\mathcal{P}]}$, i.e.,

$$D_P(r) = \sum_{n=1}^{N} H\left(A_n^{[\mathcal{P}]}\right). \tag{8}$$

In this work, we aim to characterize the optimal normalized download cost $D^*(r)$ corresponding to every caching ratio $0 \leq r \leq 1$, where

$$D_P^*(r) = \inf\left\{\frac{D_P(r)}{PL} : (D_P(r), PL) \text{ is achievable}\right\}, \tag{9}$$

which is a function of the caching ratio r and the number of interested messages P.

3   Main Results and Discussions

Our first result is the exact characterization of the normalized download cost for the case $P \geq \frac{K}{2}$, i.e., when the user needs to privately retrieve at least half of the messages stored in the servers.

Theorem 1. *For the cache-aided MPIR problem with non-colluding and replicated servers, if the number of desired messages P is at least half of the number of overall stored messages K, i.e., if $P \geq \frac{K}{2}$, then the optimal normalized download cost is given by,*

$$D_P^*(r) = \begin{cases} \left(1 + \frac{K-P}{NP}\right) - \left[1 + \frac{(N+1)(K-P)}{NP}\right]r, & 0 \leq r \leq \frac{1}{1+N} \\ 1 - r, & \frac{1}{1+N} \leq r \leq 1 \end{cases} \tag{10}$$

Remark 1. Our results can be reduced to known results of PIR for replicated servers:

1. When the cache ratio is zero, *i.e.*, r = 0, the above results degenerate into multi-message private information retrieval [1], where the optimal normalized downloads is $1 + \frac{K-P}{NP}$.
2. When the number of required messages is one, *i.e.*, $P = 1$, Theorem 1 becomes single-message cache-aided PIR with unknown and uncoded perfecting [5], where the optimal normalized downloads is shown below.

$$D_1^*(r) = \begin{cases} \left(1 + \frac{K-1}{N}\right) - \left[1 + \frac{(N+1)(K-1)}{N}\right]r, & 0 \leq r \leq \frac{1}{1+N} \\ 1 - r, & \frac{1}{1+N} \leq r \leq 1 \end{cases} \tag{11}$$

Our second main result is that for the case of $P \leq \frac{K}{2}$, we have the following theorem which characterizes the upper and lower bounds on the download cost.

Theorem 2. *For the cache-aided MPIR problem with non-colluding and replicated servers, when $P \leq \frac{K}{2}$, the normalized download cost is lower and upper bounded as,*

$$\underline{D}_P(r) \leq D_P^*(r) \leq \overline{D}_P(r) \tag{12}$$

where the upper bound $\overline{D}_P(r)$ is given by,

$$\overline{D}_P(r) = \max\left\{ \frac{D_2 - D_1}{r_2 - r_1}r + D_1, \frac{D_3 - D_2}{r_3 - r_2}(r - r_3) + D_3, 1 - r \right\} \tag{13}$$

where

$$D_1 = \frac{1 - \left(\frac{1}{N}\right)^{\frac{K}{P}}}{1 - \frac{1}{N}}, \tag{14}$$

$$D_2 = \frac{N\left(\sum_{k=2}^{K} \binom{K}{k}\alpha_k\right)}{N\left(\sum_{k=2}^{K} \binom{K}{k}\alpha_k - \sum_{k=2}^{K-P} \binom{K-P}{k}\alpha_k\right) + (N-1)P\alpha_1}, \tag{15}$$

$$D_3 = \frac{N}{N+1}, \tag{16}$$

$$r_1 = 0, \tag{17}$$

$$r_2 = \frac{(N-1)P\alpha_1}{N\left(\sum_{k=2}^{K} \binom{K}{k}\alpha_k - \sum_{k=2}^{K-P} \binom{K-P}{k}\alpha_k\right) + (N-1)P\alpha_1}, \tag{18}$$

$$r_3 = \frac{1}{N+1}. \tag{19}$$

and $\underline{D}(r)$ *is given by,*

$$\underline{D}_P(r) = \max\left\{ \sum_{i=0}^{\lfloor \frac{K}{P} \rfloor - 1} \frac{1}{N^i} - r \sum_{i=0}^{\lfloor \frac{K}{P} \rfloor - 1} \left(\frac{\lfloor \frac{K}{P} \rfloor - i}{N^i}\right), 1 - r \right\} \tag{20}$$

The achievability proofs for Theorems 1 and 2 are given in Sect. 4, and the converse proofs are given in Sect. 5.

Remark 2. Regarding Theorem 2, we have the following remarks:

1. When $\frac{K}{P} \in \mathbb{N}$, the lower bound of Theorem 2 is reduced to the known result of [5] for $\frac{K}{P}$ messages.
2. For all cases of K and P, we have the result

$$D_P^*(r) \geq 1 - r,$$

for $r \geq \frac{1}{1+N}$. Recall that without the privacy constraint, the download cost is $1 - r$, which means that at $r \geq \frac{1}{N+1}$, requiring privacy does not incur any additional download cost, i.e., there is a smart way to retrieve the desired bits without revealing the privacy of the user. This result was shown to be true for the single-message cache-aided PIR problem [5], and in this paper, we show that this is also true for its multi-message counterpart.

Finally, our third main result is that in the case of $P \leq \frac{K}{2}$ and $\frac{P}{K} \in \mathbb{N}$ *and*

$$r \leq \frac{(N-1)P\alpha_1}{N\left(\sum_{k=2}^{K} \binom{K}{k}\alpha_k - \sum_{k=2}^{K-P} \binom{K-P}{k}\alpha_k\right) + (N-1)P\alpha_1} \tag{21}$$

where $\alpha_k = \frac{1}{N-1} \sum_{i=1}^{P} \binom{P}{i} \alpha_{k+i}$ and the initial conditions of the sequence recursion formula are as follows,

$$\alpha_K = (N-1)^{K-P} \tag{22}$$

$$\alpha_{K-1} = \cdots = \alpha_{K-P+1} = 0 \tag{23}$$

the upper and lower bounds of Theorem 2 match and we therefore have the following corollary.

Corollary 1. *For $P \leq \frac{K}{2}$, $\frac{K}{P} \in \mathbb{N}$ and r satisfying (31), the optimal normalized download cost is given by,*

$$D_P^*(r) = \sum_{i=0}^{\frac{K}{P}-1} \frac{1}{N^i} \left[1 - r \left(\frac{K}{P} - i \right) \right] \tag{24}$$

The proof of the corollary will be provided in the Appendix.

4   Achievability Proofs of Theorems 1 and 2

The achievability proof will be based on the original multi-message PIR [1] achievability scheme, except the corresponding first round of single bits download will now be cached by the user in advance.

4.1   Achievability Proof of Theorem 1

Note that in the case of $P \geq \frac{K}{2}$, we just need to show that when $r = \frac{1}{N+1}$, the retrieval rate matches the upper bound. The rate with the non-vertex can be obtained by memory-sharing at adjacent vertices.

  The scheme requires $L = N + 1$, and is completed in two phases. In the first phase, the user cached a portion of the same size from each message. The user then downloads the new required message bits using the cached private bits. The details of the scheme are as follows.

1. *Index preparation:* Let $[u_i(1), u_i(2), \ldots, u_i(L)]$ denote a random permutation of the L bits of messages $W_i = [w_i(1), \cdots, w_i(L)]^T$ using a random interleaver $\pi_i(.)$ which is known to the user only, i.e.,

$$u_i(m) = w_i(\pi_i(m)), \quad m \in \{1, \cdots, L\} \tag{25}$$

2. *Phase one:* The user caches $(u_1(1), u_2(1), \cdots, u_K(1))$ from external servers which do not collude with the servers in the retrieval phase, i.e., we have cached the first round of the original MPIR [1] scheme.
3. *Phase two:* The user downloads new desired messages mixed with undesired symbols from the cache.
   (a) The user chooses an MDS generator matrix $G \in \mathbb{F}_q^{P \times K}$, where every P columns from G are linearly independent. This implies that the user can decode all the symbols using P linear combinations.

(b) The user picks uniformly and independently at random the permutation matrices $S_1, S_2, \cdots, S_{N-1}$ of size $K \times K$. These matrices shuffle the order of columns of G to be independent of $\mathcal{P}$.

(c) Specially, the user could download P linear combination from first server as follows.

Server1
$u_1(2) + u_2(2) + \cdots + u_P(2) + u_{P+1}(1) + u_{P+2}(1) + \cdots + u_K(1)$
$\lambda_1 u_1(2) + \lambda_2 u_2(2) + \cdots + \lambda_P u_P(2) + \lambda_{P+1} u_{P+1}(1) + \cdots + \lambda_K u_K(1)$
$\vdots$
$\lambda_1^{P-1} u_1(2) + \lambda_2^{P-1} u_2(2) + \cdots + \lambda_P^{P-1} u_P(2) + \lambda_{P+1}^{P-1} u_{P+1}(1) + \cdots + \lambda_K^{P-1} u_K(1)$
$Z = (u_1(1), u_2(1), \ldots, u_K(1))$

where $\lambda_i, \quad i \in \{1, 2, \cdots, K\}$ are non-zero constants that are not equal to each other.

(d) Similarly, the user could download P linear combination from other servers.

4.2 Decodability, Privacy, and Calculation of the Achievable Rate

Next, we verify that the above achievability scheme satisfies cache size constraint, decodability and privacy.

For the cache size constraint: Above scheme cache a bit from each $L = N+1$ bits message, and the cache ratio is $r = \frac{1}{N+1}$, meeting the cache size constraint.

For the reliability: The user can subtract out all the undesired message symbols using the undesired symbols from cached bits. Consequently, the user is left with a $P \times P$ system of equations which is guaranteed to be invertible by the MDS property, hence all symbols that belong to W_P are decodable.

For the privacy: Since the permutation matrices are chosen uniformly and independently from each other, the probability distribution is uniform irrespective to $\mathcal{P}$. Furthermore, the symbols are chosen randomly and uniformly by applying the random interleaver. Hence, the retrieval scheme is private.

To calculate the achievable rate: The user exploits the side information from the cached bits to generate P equations for each side information set. Each set of P equations in turn generates P desired symbols. Hence, the achievable normalized download cost is calculated as,

$$\underline{D}_P(r_3) = \frac{\text{total downloaded equations}}{\text{total number of desired symbols}} \tag{26}$$

$$= \frac{NP}{(N+1)P} \tag{27}$$

$$= \frac{N}{N+1} \tag{28}$$

Therefore, $\underline{D}_P(r_3) = \frac{N}{N+1}$, where $r_3 = \frac{1}{N+1}$. The achievability scheme for non-corner points can be obtained by memory-sharing between the most adjacent interesting caching ratios. Since both $\overline{D}_P(r)$ and $\underline{D}_P(r)$ are piecewise linear function of r, and since $\underline{D}_P(r_3) = \overline{D}_P(r_3)$, $\underline{D}_P(r_1) = \overline{D}_P(r_1)$, and $\underline{D}_P(1) = \overline{D}_P(1)$, we have $\underline{D}_P(r) = \overline{D}_P(r) = D_P^*(r)$ for $r_1 \leq r \leq 1$. Thus we complete the proof of Theorem 1.

4.3 Achievability Proof of Theorem 2

The achievability scheme we propose is the following: consider the scheme proposed in [1] where in round k, the user downloads sums of k terms from different symbols from the database. In our problem, in the prefetching phase, the user caches the bits, which are downloaded in all the stages in round one for every server in the scheme of [1]. In the retrieval phase, the user downloads the same bits as in the other rounds of the scheme proposed in [1]. Obviously, the reliability and the privacy is guaranteed. We only need to verify that when r satisfies (31), the retrieval rate matches the lower bound.

1. *Index preparation:* Let $[u_i(1), u_i(2), \ldots, u_i(L)]$ denote a random permutation of the L bits of messages $W_i = [w_i(1), \cdots, w_i(L)]^T$ using a random interleaver $\pi_i(.)$ which is known to the user only, i.e.,

$$u_i(m) = w_i(\pi_i(m)), \quad m \in \{1, \cdots, L\} \tag{29}$$

2. *Number of stages:* The number of stages needed in each round can be obtained by the following difference equation,

$$y[n] = \frac{1}{N-1} \sum_{i=1}^{P} \binom{P}{i} y[n-i] \tag{30}$$

where the initial conditions of the equation is $y[-P] = (N-1)^{M-P}$, $y[-P+1] = \cdots = y[-1] = 0$. The number of stages in round k is $\alpha_i = y[(K-P)-k]$.
3. *Caching:* The user cache all the stages needed in round one for every server studied by [1].
4. *Retrieval:* The user downloads sums of k terms from different symbols as in the scheme of [1].

4.4 Calculation of the Achievable Rate

First, we verify that the above scheme satisfies cache size constraint. The user cache all the stages needed in round one for every server studied by [1]. In this case, the cache ratio is r_2, which satisfies cache size constraint.

$$r_2 = \frac{(N-1)P\alpha_1}{N \left(\sum_{k=2}^{K} \binom{K}{k}\alpha_k - \sum_{k=2}^{K-P} \binom{K-P}{k}\alpha_k \right) + (N-1)P\alpha_1} \tag{31}$$

In round k, the user downloads the sums of k symbols. The user repeats this round for α_k stages. Each stage contains all the possible combinations of any k symbols. There are $\binom{K}{k}$ such combinations.

$$\overline{D}_P(r_2) = \frac{\text{total downloaded equations}}{\text{total number of desired symbols}} \tag{32}$$

$$= \frac{N\left(\sum_{k=2}^{K}\binom{K}{k}\alpha_k\right)}{N\left(\sum_{k=2}^{K}\binom{K}{k}\alpha_k - \sum_{k=2}^{K-P}\binom{K-P}{k}\alpha_k\right) + (N-1)P\alpha_1} \tag{33}$$

$$= \frac{1}{(1-\frac{1}{N})U}\sum_{i=1}^{P}\gamma_i r_i^{K-P}\left[(N-1)(N^{\frac{K}{P}}-1) - \frac{PK(N-1)}{\gamma_i}\right] \tag{34}$$

where $U = N\left(\sum_{k=2}^{K}\binom{K}{k}\alpha_k - \sum_{k=2}^{K-P}\binom{K-P}{k}\alpha_k\right) + (N-1)P\alpha_1$, $\alpha_k = \sum_{i=1}^{P}\gamma_i r_i^{K-P-k}$, $\boldsymbol{\gamma} = (\gamma_1, \cdots, \gamma_P)^T$ is the solution to the system of linear equations,

$$\begin{bmatrix} r_1^{-P} & r_2^{-P} & \cdots & r_P^{-P} \\ r_1^{-P+1} & r_2^{-P+1} & \cdots & r_P^{-P+1} \\ \vdots & \vdots & \cdots & \vdots \\ r_1^{-1} & r_2^{-1} & \cdots & r_P^{-1} \end{bmatrix} \begin{bmatrix} \gamma_1 \\ \gamma_2 \\ \vdots \\ \gamma_P \end{bmatrix} = \begin{bmatrix} (N-1)^{M-P} \\ 0 \\ \vdots \\ 0 \end{bmatrix} \tag{35}$$

5   Converse Proofs of Theorems 1 and 2

The converse proof will consists of two major steps:

1) the *initialization* step where we connect the normalized download cost $D_P^*(r)$ to $H\left(A_{1:N}^{[\mathcal{P}]}|W_{\mathcal{P}}, Q_{1:N}^{[\mathcal{P}]}, \mathbb{H}, Z\right)$ for some P-size set $\mathcal{P}$ where $\mathcal{P} \subset \{1, 2, \cdots, K\}$, and

2) the *induction* step where we connect $H\left(A_{1:N}^{[\mathcal{P}_i]}|W_{[\mathcal{P}']}, Q_{1:N}^{[\mathcal{P}_i]}, \mathbb{H}, Z\right)$ to $H\left(A_{1:N}^{[\mathcal{P}_0]}|W_{[\mathcal{P}' \cup \mathcal{P}_0]}, Q_{1:N}^{[\mathcal{P}_0]}, \mathbb{H}, Z\right)$ where $\mathcal{P}' = \bigcup_{i=1}^{I}\mathcal{P}_i$, and $\mathcal{P}_i$, $i = 0, 1, \cdots, I$ are all P-size subsets of $\{1, 2, \cdots, K\}$ with no intersections.

The Initialization Step

For any private retrieval scheme with normalized download cost $D_p(r)$, for any P-sized subset of $\{1, 2, \cdots, K\}$, denoted as $\mathcal{P}$, we have

$$D_P(r) = \sum_{n=1}^{N} H\left(A_n^{[\mathcal{P}]}\right) \tag{36}$$

$$\geq H\left(A_{1:N}^{[\mathcal{P}]}|Q_{1:N}^{[\mathcal{P}]}, \mathbb{H}, Z\right) \tag{37}$$

$$= H\left(A^{[\mathcal{P}]}_{1:N}|W_{\mathcal{P}}, Q^{[\mathcal{P}]}_{1:N}, \mathbb{H}, Z\right) - H\left(W_{\mathcal{P}}|A^{[\mathcal{P}]}_{1:N}, Q^{[\mathcal{P}]}_{1:N}, \mathbb{H}, Z\right) \tag{38}$$

$$+ H\left(W_{\mathcal{P}}|Q^{[\mathcal{P}]}_{1:N}, \mathbb{H}, Z\right)$$

$$= H\left(A^{[\mathcal{P}]}_{1:N}|W_{\mathcal{P}}, Q^{[\mathcal{P}]}_{1:N}, \mathbb{H}, Z\right) - o(L) + H\left(W_{\mathcal{P}}|Q^{[\mathcal{P}]}_{1:N}, \mathbb{H}, Z\right) \tag{39}$$

$$= H\left(A^{[\mathcal{P}]}_{1:N}|W_{\mathcal{P}}, Q^{[\mathcal{P}]}_{1:N}, \mathbb{H}, Z\right) - o(L) + H\left(W_{\mathcal{P}}|Z\right) \tag{40}$$

$$= H\left(A^{[\mathcal{P}]}_{1:N}|W_{\mathcal{P}}, Q^{[\mathcal{P}]}_{1:N}, \mathbb{H}, Z\right) - o(L) + PL(1-r) \tag{40}$$

where (36) follows from (8), (37) follows from the chain rule and conditioning reduces entropy, (39) follows from (7), and (40) follows from the uncoded caching scheme and the independence of the messages, the queries and index of the bits prefetched, i.e., (3) and (4).

The Induction Step

Suppose $\mathcal{P}_i$, $i = 0, 1, \cdots, I$ are all P-size subsets of $\{1, 2, \cdots, K\}$ with no intersections. Let $\mathcal{P}' = \bigcup^I_{i=1} \mathcal{P}_i$.

$$H\left(A^{[\mathcal{P}_i]}_{1:N}|W_{[\mathcal{P}']}, Q^{[\mathcal{P}_i]}_{1:N}, \mathbb{H}, Z\right)$$

$$= H\left(A^{[\mathcal{P}_i]}_{1:N}|W_{[\mathcal{P}']}, Q^{[\mathcal{P}_i]}_{1:N}\right) - H\left(\mathbb{H}, Z|W_{[\mathcal{P}']}, Q^{[\mathcal{P}_i]}_{1:N}\right) + H\left(\mathbb{H}, Z|A^{[\mathcal{P}_i]}_{1:N}, W_{[\mathcal{P}']}, Q^{[\mathcal{P}_i]}_{1:N}\right)$$

$$\geq H\left(A^{[\mathcal{P}_i]}_{1:N}|W_{[\mathcal{P}']}, Q^{[\mathcal{P}_i]}_{1:N}\right) - (K - IP)Lr \tag{41}$$

$$\geq \frac{1}{N}\sum^N_{n=1} H\left(A^{[\mathcal{P}_i]}_n|W_{[\mathcal{P}']}, Q^{[\mathcal{P}_i]}_{1:N}\right) - (K - IP)Lr \tag{42}$$

$$= \frac{1}{N}\sum^N_{n=1} H\left(A^{[\mathcal{P}_i]}_n|W_{[\mathcal{P}']}, Q^{[\mathcal{P}_i]}_n\right) - (K - IP)Lr \tag{43}$$

$$= \frac{1}{N}\sum^N_{n=1} H\left(A^{[\mathcal{P}_0]}_n|W_{[\mathcal{P}']}, Q^{[\mathcal{P}_0]}_n\right) - (K - IP)Lr \tag{44}$$

$$\geq \frac{1}{N} H\left(A^{[\mathcal{P}_0]}_{1:N}|W_{[\mathcal{P}']}, Q^{[\mathcal{P}_0]}_{1:N}\right) - (K - IP)Lr \tag{45}$$

$$\geq \frac{1}{N} H\left(A^{[\mathcal{P}_0]}_{1:N}|W_{[\mathcal{P}']}, Q^{[\mathcal{P}_0]}_{1:N}, \mathbb{H}, Z\right) - (K - IP)Lr \tag{46}$$

$$= \frac{1}{N}\left[H\left(A^{[\mathcal{P}_0]}_{1:N}|W_{[\mathcal{P}' \cup \mathcal{P}_0]}, Q^{[\mathcal{P}_0]}_{1:N}, \mathbb{H}, Z\right) + H\left(W_{[\mathcal{P}_0]}|W_{[\mathcal{P}']}, Q^{[\mathcal{P}_0]}_{1:N}, \mathbb{H}, Z\right)\right] \tag{47}$$

$$- \frac{1}{N} H\left(W_{[\mathcal{P}_0]}|A^{[\mathcal{P}_0]}_{1:N}, W_{[\mathcal{P}']}, Q^{[\mathcal{P}_0]}_{1:N}, \mathbb{H}, Z\right) - (K - IP)Lr \tag{48}$$

$$= \frac{1}{N} H\left(A^{[\mathcal{P}_0]}_{1:N}|W_{[\mathcal{P}' \cup \mathcal{P}_0]}, Q^{[\mathcal{P}_0]}_{1:N}, \mathbb{H}, Z\right) + \frac{1}{N} PL(1-r) + \frac{o(L)}{N} - (K - IP)Lr$$

where (41) follows from the uncoded nature of the cache and the nonnegativity of the entropy function, where (43) follows from the fact $A^{[\mathcal{P}]}_n \leftrightarrow (Q^{[\mathcal{P}]}_n, W_{\mathcal{P}}) \leftrightarrow Q^{[\mathcal{P}]}_k (k \neq n)$ is a Markov chain, where (44) follows from privacy constraint (6), where (45) follows from conditioning reduces entropy, where (46) since conditioning reduces entropy, (49) follows from (3)(4)(7).

5.1   Converse Proof for the Case of $P \geq \frac{K}{2}$

Note that in the case of $P \geq \frac{K}{2}$, there can not exists 2 non-intersecting sets $\mathcal{P}_0$ and $\mathcal{P}_1$ that are subsets of $\{1, 2, \cdots, K\}$. Hence, we take the induction step result of (49) for $I = 1$, and perform a union with the complement of set $\mathcal{P}$, denoted as $\bar{\mathcal{P}}$.

$$H\left(A_{1:N}^{[\mathcal{P}]}|W_\mathcal{P}, Q_{1:N}^{[\mathcal{P}]}, \mathbb{H}, Z\right) \tag{49}$$

$$\geq \frac{1}{N}\left[H\left(A_{1:N}^{[\mathcal{P}^*]}|W_{1:K}, Q_{1:N}^{[\mathcal{P}^*]}, \mathbb{H}, Z\right) + H\left(W_{\bar{\mathcal{P}}}|W_\mathcal{P}, Q_{1:N}^{[\mathcal{P}^*]}, \mathbb{H}, Z\right)\right] \tag{50}$$

$$- \frac{1}{N}H\left(W_{\mathcal{P}^*}|A_{1:N}^{[\mathcal{P}^*]}, W_\mathcal{P}, Q_{1:N}^{[\mathcal{P}^*]}, \mathbb{H}, Z\right) - (K-P)Lr$$

$$= \frac{1}{N}(K-P)L(1-r) - (K-P)Lr + \frac{o(L)}{N} \tag{51}$$

where (50) follows the step of induction, where $W_{\bar{\mathcal{P}}}$ corresponds to the complement set of messages of $W_\mathcal{P}$, $\bar{\mathcal{P}} \subseteq \mathcal{P}^*$, $|\mathcal{P}^*| = P$, where (51) follows from (5)(7).

Combining (40) and (51), we conclude the converse proof by dividing by PL and taking the limit as $L \to \infty$, the normalized download cost is lower bounded by maximum value

$$D_P^*(r) \geq \max\{1-r, \left(1 + \frac{K-P}{NP}\right) - \left[1 + \frac{(N+1)(K-P)}{NP}\right]r\} \tag{52}$$

5.2   Converse Proof for the Case of $P < \frac{K}{2}$

Now, we derive the inductive relation for $\frac{K}{P} > 2$. Suppose $\mathcal{P}_i$, $i = 0, 1, \cdots, I$ are all P-size subsets of $\{1, 2, \cdots, K\}$ with no intersections. Let $\mathcal{P}' = \bigcup_{i=1}^{I} \mathcal{P}_i$. Applying induction formula (49) repeatedly.

$$H\left(A_{1:N}^{[\mathcal{P}_1]}|W_{[\mathcal{P}_1]}, Q_{1:N}^{[\mathcal{P}_1]}, \mathbb{H}, Z\right) \tag{53}$$

$$= \frac{1}{N}H\left(A_{1:N}^{[\mathcal{P}_2]}|W_{[\mathcal{P}_1 \cup \mathcal{P}_2]}, Q_{1:N}^{[\mathcal{P}_2]}, \mathbb{H}, Z\right) + \frac{1}{N}PL(1-r) + \frac{o(L)}{N} - (K-P)Lr \tag{54}$$

$$\geq \frac{1}{N}\left[\frac{1}{N}H\left(A_{1:N}^{[\mathcal{P}_3]}|W_{[\mathcal{P}_1 \cup \mathcal{P}_2 \cup \mathcal{P}_3]}, Q_{1:N}^{[\mathcal{P}_3]}, \mathbb{H}, Z\right) + \frac{1}{N}PL(1-r) + \frac{o(L)}{N}\right] \tag{55}$$

$$-\frac{1}{N}(K-2P)Lr + \frac{1}{N}PL(1-r) + \frac{o(L)}{N} - (K-P)Lr \tag{56}$$

$$\geq \cdots \tag{57}$$

$$\geq PL(1-r)\sum_{i=1}^{\lfloor \frac{K}{P} \rfloor -1}\frac{1}{N^i} + L(1-r)\left(K - \left\lfloor\frac{K}{P}\right\rfloor P\right)\frac{1}{N^{\lfloor \frac{K}{P} \rfloor}}$$

$$- Lr\sum_{i=0}^{\lfloor \frac{K}{P} \rfloor -1}\left(\frac{K-(i+1)P}{N^i}\right) + o(L)(\frac{1}{N} + \frac{1}{N^2} + \cdots + \frac{1}{N^{\lfloor \frac{K}{P} \rfloor -1}}) \tag{58}$$

Combining (58) and **the initialization step**, we have

$$D_P(r) + o(L) - PL(1-r) \tag{59}$$

$$\geq PL(1-r)\sum_{i=1}^{\lfloor \frac{K}{P} \rfloor -1}\frac{1}{N^i} + L(1-r)\left(K - \left\lfloor\frac{K}{P}\right\rfloor P\right)\frac{1}{N^{\lfloor \frac{K}{P} \rfloor}}$$

$$- Lr\sum_{i=0}^{\lfloor \frac{K}{P} \rfloor -1}\left(\frac{K-(i+1)P}{N^i}\right) + o(L)(\frac{1}{N} + \frac{1}{N^2} + \cdots + \frac{1}{N^{\lfloor \frac{K}{P} \rfloor -1}}) \tag{60}$$

Consequently, the normalized download cost is lower bounded by

$$D_P(r) \geq \sum_{i=0}^{\lfloor \frac{K}{P} \rfloor -1}\frac{1}{N^i} + (1-r)\left(\frac{K}{P} - \left\lfloor\frac{K}{P}\right\rfloor\right)\frac{1}{N^{\lfloor \frac{K}{P} \rfloor}} - r\sum_{i=0}^{\lfloor \frac{K}{P} \rfloor -1}\left(\frac{\frac{K}{P} - i}{N^i}\right) + o(L) \tag{61}$$

We conclude the converse proof by dividing by PL and taking the limit as $L \to \infty$, the normalized download cost is lower bounded by maximum value

$$D^*(r) \geq \max_{i\in\{2,\cdots,\lfloor \frac{K}{P} \rfloor +1\}}(1-r)\sum_{j=0}^{\lfloor \frac{K}{P} \rfloor +1-i}\frac{1}{N^j} - r\sum_{j=0}^{\lfloor \frac{K}{P} \rfloor -i}\frac{\lfloor \frac{K}{P} \rfloor +1-i-j}{N^j} \tag{62}$$

$$+(1-r)\left(\frac{K}{P} - \left\lfloor\frac{K}{P}\right\rfloor\right)\frac{1}{N^{\lfloor \frac{K}{P} \rfloor}} \tag{63}$$

We note that when $M = \frac{K}{P} \in \mathbb{N}$, the result in (62) reduces to the known result of [5], which is

$$D^*(r) \geq \max_{i\in\{2,\cdots,M+1\}}(1-r)\sum_{j=0}^{M+1-i}\frac{1}{N^j} - r\sum_{j=0}^{M-i}\frac{M+1-i-j}{N^j} \tag{64}$$

6   Conclusion

In this paper, we developed the cache-aided MPIR problem from N non-colluding and replicated servers, when the cache stores uncoded bits that are unknown to the servers. The problem generalizes the cache-aided PIR problem in [5] which retrieves a single message privately and the multi-message PIR scenario with no cache available at the user [1]. We determined inner and outer bounds for the optimal normalized download cost $D_P^*(r)$ as a function of the number of required message P, the total number of messages K, the number of servers N, and the caching ratio r. Both inner and outer bounds are piece-wise linear functions in r (for fixed P, N, K) that consist of two line segments. We determined the exact download cost for this problem when the number of desired messages is at least half of the number of total stored messages. Furthermore, the bounds match in two specific regimes for the case $\frac{K}{P} > 2$: the very high ratio regime,i.e., $r \geq \frac{1}{N+1}$ and the very low ratio regime,i.e., r satisfying (31).

Acknowledgment. This work is partially supported by the National Natural Science Foundation of China under Grants 62071115, 61971135, National Key Research and Development Project under Grant 2019YFE0123600, the Research Fund of National Mobile Communications Research Laboratory, Southeast University (No. 2020A03), and the Six talent peaks project in Jiangsu Province.

A   Proof of Corollary 1

Since both $\overline{D}_P(r)$ and $\underline{D}_P(r)$ are piecewise linear function of r, all we have to do to prove that they are equal at some interval is to prove that the vertices are the same. First, let us note that

$$r_1 = 0 \tag{65}$$

$$r_2 = \frac{(N-1)P\alpha_1}{N\left(\sum_{k=2}^{K}\binom{K}{k}\alpha_k - \sum_{k=2}^{K-P}\binom{K-P}{k}\alpha_k\right) + (N-1)P\alpha_1} \tag{66}$$

Then, we note that

$$\overline{D}_P(r_2) = \frac{N\left(\sum_{k=2}^{K}\binom{K}{k}\alpha_k\right)}{N\left(\sum_{k=2}^{K}\binom{K}{k}\alpha_k - \sum_{k=2}^{K-P}\binom{K-P}{k}\alpha_k\right) + (N-1)P\alpha_1} \tag{67}$$

$$= \frac{1}{(1-\frac{1}{N})U} \sum_{i=1}^{P} \gamma_i r_i^{K-P}\left[(N-1)(N^{\frac{K}{P}}-1) - \frac{PK(N-1)}{\gamma_i}\right] \tag{68}$$

where $U = N\left(\sum_{k=2}^{K}\binom{K}{k}\alpha_k - \sum_{k=2}^{K-P}\binom{K-P}{k}\alpha_k\right) + (N-1)P\alpha_1$, $\alpha_k = \sum_{i=1}^{P}\gamma_i r_i^{K-P-k}$.

Further, we note from (20), by choosing $r = r_2$, we have

$$\underline{D}_P(r_2) \geq \sum_{i=0}^{\frac{K}{P}-1} \frac{1}{N^i}\left[1 - r_2\left(\frac{K}{P} - i\right)\right] \tag{69}$$

$$= \frac{1 - (\frac{1}{N})^{\frac{K}{P}}}{1 - \frac{1}{N}} - \frac{(N-1)P\alpha_1}{U}\left[\frac{\frac{K}{P} - \frac{1}{N}(1 - (\frac{1}{N})^{\frac{K}{P}})}{(1 - \frac{1}{N})^2}\right] \tag{70}$$

$$= \frac{1}{(1 - \frac{1}{N})U}\left[\left(1 - \frac{1}{N^{\frac{K}{P}}}\right)U - (N-1)P\alpha_1\left(\frac{\frac{K}{P} - \frac{1}{N}(1 - (\frac{1}{N})^{\frac{K}{P}})}{(1 - \frac{1}{N})}\right)\right] \tag{71}$$

$$= \frac{1}{(1 - \frac{1}{N})U}\left[(N-1)P\alpha_1\left(\left(1 + \frac{1}{N-1}\right)\left(1 - \frac{1}{N^{\frac{K}{P}}}\right) - \frac{K}{P}\right)\right] \tag{72}$$

$$+ \frac{1}{(1 - \frac{1}{N})U}\left[N\left(1 - \frac{1}{N^{\frac{K}{P}}}\right)\sum_{i=1}^{P}\gamma_i r_i^{K-P}\left(N^{\frac{K}{P}} - N^{\frac{K}{P}-1} - \frac{P}{\gamma_i}\right)\right] \tag{73}$$

$$= \frac{1}{(1 - \frac{1}{N})U}\sum_{i=1}^{P}\gamma_i r_i^{K-P}\left[(N-1)(N^{\frac{K}{P}} - 1) - \frac{PK(N-1)}{\gamma_i}\right] \tag{74}$$

$$= \overline{D}_P(r_2) \tag{75}$$

Thus, since $\underline{D}_P(r_2) \leq \overline{D}_P(r_2)$ by definition, (75) implies $\underline{D}_P(r_2) = \overline{D}_P(r_2)$. We also note that $\underline{D}_P(r_1) = \overline{D}_P(r_1)$. Since both $\overline{D}_P(r)$ and $\underline{D}_P(r)$ are piecewise linear function of r, and since $\underline{D}_P(r_2) = \overline{D}_P(r_2)$ and $\underline{D}_P(r_1) = \overline{D}_P(r_1)$, we have $\underline{D}_P(r) = \overline{D}_P(r) = D_P^*(r)$ for $r_1 \leq r \leq r_2$. Thus we complete the proof of corollary 1.

References

1. Banawan, K., Ulukus, S.: Multi-message private information retrieval: capacity results and near-optimal schemes. IEEE Trans. Inform. Theory **64**(10), 6842–6862 (2018). https://doi.org/10.1109/TIT.2018.2828310
2. Chor, B., Goldreich, O., Kushilevitz, E., Sudan, M.: Private information retrieval. In: Proceedings of IEEE 36th Annual Foundations of Computer Science, pp. 41–50, October 1995. https://doi.org/10.1109/SFCS.1995.492461
3. Shah, N.B., Rashmi, K.V., Ramchandran, K.: One extra bit of download ensures perfectly private information retrieval. In: 2014 IEEE International Symposium on Information Theory, pp. 856–860, June 2014. https://doi.org/10.1109/ISIT.2014.6874954
4. Sun, H., Jafar, S.A.: The capacity of private information retrieval. IEEE Trans. Inform. Theory **63**(7), 4075–4088 (2017). https://doi.org/10.1109/TIT.2017.2689028
5. Wei, Y., Banawan, K.A., Ulukus, S.: Fundamental limits of cache-aided private information retrieval with unknown and uncoded prefetching. CoRR abs/1709.01056 (2017). http://arxiv.org/abs/1709.01056

Evaluation of Dynamic Scheduling for Data Collection in Medical Application Using Firefly Synchronization Algorithm

Norhafizah Muhammad$^{(\boxtimes)}$ and Tiong Hoo Lim

Universiti Teknologi Brunei, Gadong, Brunei
{norhafizah.muhammad,lim.tiong.hoo}@utb.edu.bn

Abstract. This paper proposed a dynamic data transmission scheduling algorithm based on the way a group of firefly communicating with one another. The proposed algorithm will be compared again with a random transmission approach. To evaluate the performance of the algorithms, different numbers of nodes in the networks will be evaluated. The results from the hardware experiments have shown that the firefly algorithm can schedule the transmission of data packet with high delivery rate for a small network. However, as the number of nodes increases, the packet delivery rate decreases. The proposed algorithm can also increase the lifespan of the battery as the nodes will be operating in the sleep mode and will only be awake during the sychronization period for data transmission.

Keywords: Firefly algorithm · Bio-inspired · Sychronization · Transmission scheduling · Health monitoring

1 Introduction

In the vast development of the wireless body sensor networks, one of the applications is the healthcare monitoring system. The monitoring system will be able to sense and alert any abnormality of the patient's health to the personal care, such as the caretaker or emergency center. The sensed data will be collected from the devices that are on or implanted in the patient's body. It is important to collect and transmit the sensed data in synchronous, real-time and continuous manner. Data synchronicity is defined as the ability to collect and transmit sensing data across biosensor networks in an orderly time sequential manner. Time synchronization is important during the transmission of data in time-sensitive applications such as medical health. The data need to be received according to the order of the data are used for diagnostic in an Artificial Intelligence based on the medical application. Djenouri et al. [9] stated that the node should be synchronized to correlate the different reports of information. Zong et al. [4] mentioned the applications of time synchronization can be collaborated, coordinated and localized the position of the nodes. The authors also highlighted that

H. Gao et al. (Eds.): ChinaCom 2020, LNICST 352, pp. 142–160, 2021.
https://doi.org/10.1007/978-3-030-67720-6_10

the sensor nodes are dedicated in cooperating and monitoring the physical or environmental variables, which requires the precise timing among the nodes.

Fixed time synchronization algorithm has been used in the MAC layer to ensure that data can be collected and transmitted reliable at a predetermined interval. In fixed time synchronization, the transmission interval allocated to individual node is equally divided among a set of nodes within a time period. Each node will need to transmit at the assigned interval to avoid packet collisions using time division approaches. However, fixed time synchronization approach is not suitable for medical application as the sensing data needs to be transmitted at any time when a critical event is detected. The default CSMA/CA transmission protocol at the MAC layer is prone to collision when the number of biosensor nodes increases. Hence, there is a need to apply bio-inspired synchronization algorithm at the application layer to ensure that the patient physiological data can be received promptly and reliably.

Bio-inspired algorithms are based on the behaviour of a large biological swarms that is dynamic in nature. These algorithms are shown to tolerate to failure and can dynamically recover from failures. In this paper, a firefly-inspired scheduling algorithm for data transmission is proposed and implemented at the application layer to improve the packet delivery rate while reducing the energy consumption in the nodes through duty cycling. By synchronizing the biosensor to transmit at the same time, it can improve the energy efficiency of each nodes as all the node can off its radio. All the biosensor nodes can wake up at the same time for a short period of time, sample their sensors and transmit immediately to the base station or relay the data along the routing path to the base station without interfering with other nodes [1]. The transmission will be coordinated and interleaved with the awake period. Once all the nodes have transmitted, each node will switch from transmitting mode to sleep/listening mode to reduce energy consumption.

2   Related Works

Bio-inspired synchronicity is a simple model based on the natural phenomenon in the world, for examples colony of ants complete the difficult task, school of fish navigate the sea and swarm of fireflies flashing in a perfect unison [1]. There are related works had been done by using the bio-inspired approach in Wireless Body Sensor Networks. By adapting and implementing a biological inspired systems into a network of nodes, individual node in the network can behave and inherit the adaptive and fault tolerance of the animals or insects. In [2,3], the proposed system can adapt any types of environments, resistance in any internal or external failures, evolve and learn new conditions, and self-organized in a fully distributed and efficient collaboration manners.

Mirollow and Strogatz [7] proposed a mathematical model based on a pulse-coupled oscillator to mimic the behavior of a firefly to synchronize the clock operating in a set of sensor nodes. An individual node will broadcast a pulse periodically to its neighbour. When the neighboring node within its transmission range received the transmitted pulse, it will adjust its internal clock to the time its received the firing pulse or oscillators and respond by sending its own pulse. This process will repeat until all the pulse received are synchronized. In [1] stated that the synchronicity can be used to coordinate the sampling across multiple nodes for the important applications that need a high data rate. However, Zong et al. [4] highlight that the pulse-coupled approach can only achieve synchronization if the internal dynamic oscillators are identical and no coupling delay during the packet exchange among the oscillators. When anode sends out a fire message at time, t_{fire}, the neighboring nodes will not received the message immediately. As a result, those nodes will only respond after $t_{fire} + \alpha$ where α is an unknown propagation delay. An et al. [8] implemented a linear pulsed- coupled oscillator model with linear dynamic with different coupling strength to avoid synchronization error when there is an interruption in the network.

In Werner-Allen et al. [4], the authors developed the Reach-back Firefly Algorithm (RFA) that uses previous transmission information to adjust the future firing phase of the sensor nodes. The authors highlighted that the RFA can overcome the timestamping messages by staggering the messages. Movassaghi et al. [6] found out that the protocol achieved stability and robust to nodes leaving or joining the networks. The system was able to self-adjust at the time of transmissions. However, the authors did not consider the stringent battery and infeasibility memory of the nodes in WBSN [6].

Adhikary et al. [7] designed a Particle Swarm Intelligent Optimzation based on flocking bird to improve the scalability of the system network. The authors showed that by optimizing the transmission power and allocating the bandwidth, it can maximize the system throughput and mitigate the inter WBSN interference. The algorithm was evaluated using MATLAB. The results have shown that the numbers of data loss increase with the transmission rate.

3   Bio-inspired Firefly Scheduling Algorithm for Data Transmission

The firefly inspired synchronization was first introduced by the Mirollo and Strogratz in 1990 [4]. According to Mirollo et al., the M&S model is a synchronization technique based on the pulse-coupled oscillation where the node will only needs to observe the firing events from the neighbor [7]. This model described that the phase node has a function of $f(t)$, which increases to a period time $t = T$ as shown in Fig. 1. When $f(t)$ reaches the threshold, the node will fire and returns to zero.

$$t^{''} = f^{-1}(f(t^{'}) + \epsilon) \tag{1}$$

Figure 1 shows when a node fires, the neighbor's node will respond to the firing event and consequently adjusting its own phase forward. The new phase can be obtained by using Eq. 1, where $t^{''}$ is the new internal time and ϵ is the coupling parameter.

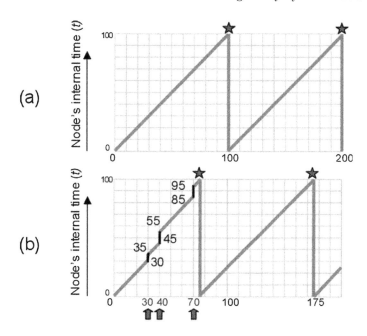

Fig. 1. Adaptive Transmission Scheduling Algorithm (a) Node fires at a time period T. (b) Node responds to neighbors firing [59]

In the proposed algorithm, Eq. 1 is modified to reduce the time of the next firing. This will allow the node to shift its next firing timing closer to its neighbour without interfering. This can be achieved by modifying the epsilon parameter using the Eq. 2.

$$f(x) = f^{-1}(f(t^{'}) - \epsilon) \tag{2}$$

The value of ϵ is determined by 3.

$$\epsilon = (t^{''} - t^{'}) - \alpha \tag{3}$$

where α is a synchronization constant. This will reduce the transmission period for all the nodes in the networks. This mathematical model shows that when a node fired, the neighbors will adjust their position by using the equation (1) above. For example, node B will observes the neighbor's time of firing. Node B will adjust the time by moving forward. This process will proceed for all nodes and this will shorten the time to broadcast the data. Therefore, when all of the nodes broadcasted or fired its information, the nodes will go to sleep until the external reaches the period of time, T.

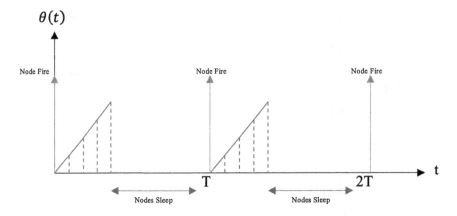

Fig. 2. The Dynamic Data Scheduling Algorithm.

4   Experimental Setup

The setup of the sensor nodes will be deployed similar to the application in the healthcare monitoring systems shown in Fig. 3. The sensor nodes or bio sensors can be connected on or implanted into the human body can be represented as the electrocardiogram (ECG), electroencephalography (EEG), pulse rate, blood pressure, body temperature and so on. The WBAN of the Fig. 3 will be operating in a star topology configuration, where all the sensor nodes are connected to the hub or gateway by using single hop communications.

Synchronization algorithm was installed and evaluated on the TelosB and XM1000 running TinyOS. A TelosB node will be used as a monitoring base station to collect the synchronization statistics and monitor the scheduling of the data transmission. XM1000 mote will be used as the individual sensor nodes that will collect environmental information to be transmitted within a clock cycle, T_{Cycle}. A 32 kHz crystal clock oscillator was used for local time. Each of the sensor nodes will have a unique id.

Different numbers of 5, 10, 15 XM1000 motes were used to evaluate the scalability of the proposed algorithms. The deployment of 5 nodes are shown in Fig. 4. A laptop will be connected to the monitoring node to store and display the statistics collected. The synchronization process will begin when the first node starts to fire. The rest of the sensor nodes receiving the message will adjust its transmission period and transmit its own messages. This process will continue until the experiment ends.

Two experiments are conducted. One experiment will be run without the firefly scheduling algorithm and will be used to compare the stability of the synchronization algorithm and the Packet Delivery Rate.

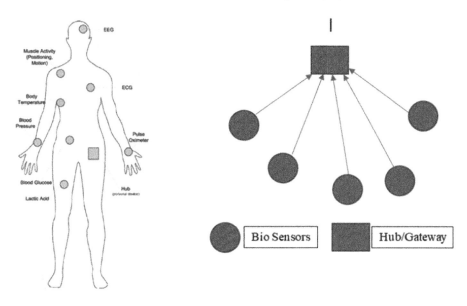

Fig. 3. The Experimental Setup for 5 XM1000 sensing nodes and 1 TelosB monitoring node.

Fig. 4. The hardware used for the experiment with one telosB mote connected to the notebook for data collection

5   Results

In this section, the performance analysis of the dynamic Scheduling for data collection based on the firefly synchronization algorithms applied in Medical Application is compare against normal CSMA/CA random transmission. Two metrics have been used to compare the performance namely the sequencing order of the packet received and the Packet Delivery Rate as the number of nodes increases.

5.1   Sequencing of Packet Delivered for 5 Nodes

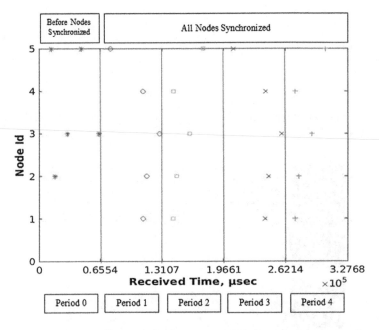

Fig. 5. The order of packet received by the basestation from 5 nodes using the random transmission scheduling algorithm for Synchronity

Figure 5 shows the process of the synchronizing of the nodes. During the synchronization process in period 0, all nodes were collected and identified on how many nodes involved. During this period, all of the sensor nodes will be equally divided and assigned time slot for transmission. By observation from Fig. 5 below shows that in period 1 to period 4, the sensor nodes shows that biosensors were in a synchronized. But there were also one or two nodes were not in synchronized, for node 2 and node 3 of periods 2 and 4 respectively. This means that there were delay in transmitting and broadcasting of the data. In period 4, Node 5 also shows there was delay in transmitting to the basestation. The cause of this delay could be either because of signal interference or traffic during the transmission.

While in Fig. 6 below shows during the synchronization process in period 0, all nodes were collected and identified on how many nodes involved. During this period, all of the sensor nodes will be equally divided and assigned time slot for transmission. By observation it shows that in period 1 to period 4, the sensor nodes showed the data transmitted in synchronized. But there were also one or two nodes were not in synchronized, for node 2 and node 3 of periods 2 and 4 respectively. This means that there were delay in transmitting and broadcasting of the data. In period 4, Node 5 also shows there was delay in transmitting to the basestation. The cause of this delay could be either because of signal interference or traffic during the transmission.

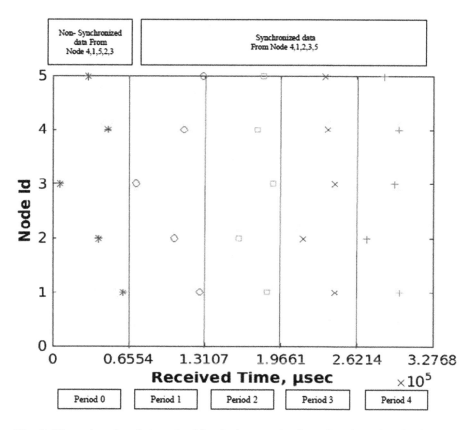

Fig. 6. The order of packet received by the basestation from 5 nodes using the dynamic data transmission scheduling for Synchronicity

5.2 Statistical Test on the Synchronization Period for 5 Nodes

Table 1 and Table 2 showed the results of the average synchronization period for the individual nodes for both experiment settings. These samples were taken for every 100th period of the data received at the basestation. While the *p-value* is the statistical procedure of the *t-test* by taking the average synchronized period of every cycle of each individuals.

Table 1 shows that the synchronization period for every cycles of each individuals were stable. This also shows that the data were propagating between 31999.5 μs to 32319 μs. Whereas for Table 2 showed the average synchronization were inconsistent in every cycle of the individual nodes. This showed that the biosensors were still trying to propagate according to the firefly transmission algorithm.

While the *p-value* from the *t-test* showed that the average synchronized period fof the random time synchronization were consistent and statistically significant at 1.11×10^{-9}, however the dynamic time synchronization showed better results between 1.11×10^{-36} and 9.13×10^{-22}. This also shows that the dynamic time synchronization had higher probability for all the nodes transmitted and received by the basestation. With the consistency of the synchronization period means that the packets delivered and received by the basestation was in synchronized.

Table 1. Average of synchronization period of random time synchronization for 5 nodes.

Node ID	p-value	Average of synchronization period (μsec)				
		Cycle 1	Cycle 2	Cycle 3	Cycle 4	Cycle 5
Node 1	1.14×10^{-9}	31999.5	31998.4	32639.4	32318.2	32319
Node 2	1.14×10^{-9}	31999.5	31998.4	32639.4	32318.2	32319
Node 3	1.14×10^{-9}	31999.5	31998.4	32639.4	32318.2	32319
Node 4	1.14×10^{-9}	31999.5	31998.4	32639.4	32318.2	32319
Node 5	1.14×10^{-9}	31999.5	31998.4	32639.4	32318.2	32319

Table 2. Average of synchronization period of dynamic data transmission scheduling for 5 nodes.

Node ID	p-value	Average of synchronization period (μsec)				
		Cycle 1	Cycle 2	Cycle 3	Cycle 4	Cycle 5
Node 1	1.11×10^{-36}	55915.4	57157.4	56631.5	56036.1	56608.2
Node 2	3.72×10^{-24}	56653.4	49318.9	50668.3	47461.3	52049.4
Node 3	9.13×10^{-22}	51784.2	42139.4	50583.6	50047.2	53968.1
Node 4	1.27×10^{-22}	52591.8	53990.8	53285.6	46202	44065
Node 5	2.87×10^{-23}	47374.6	49972.9	47366.6	52741.6	47858.3

5.3   Sequencing of Packet Delivered for 10 Nodes

Figure 7 and Fig. 8 show the synchronization process of the ten (10) sensor nodes for random and dynamic time synchronization respectively. This shows that in the time synchronization approach will automatically synchronized, once it detected the node transmitting the data to the basestation. Even though the time took for all nodes to be in synchronized was very short, but there will always have a communication overhead which can lead to the node and communication link failure. By observation the communication process, during the initial rounds of the synchronization period only 20% the data transmitted in a synchronized pattern, then in a short period of time one or two sensor nodes rearrange the position. The rearrangement of the sensor nodes caused by the communication overhead, which leads to the delay in transmission of the sensor nodes.

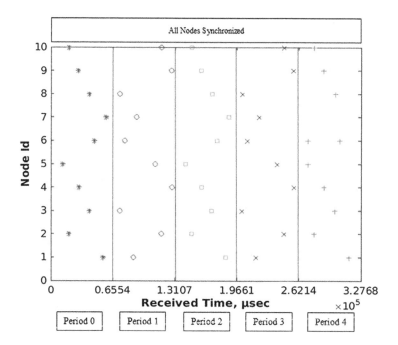

Fig. 7. The order of packet received by the basestation from 10 nodes transmitting at random

Figure 8 shows that the nodes were trying to synchronize the data, hence the unstable received time at the basestation. This basically shows that as the number of nodes increases, the more difficult for the algorithm to synchronize the data for the basestation. During the initial phase, period 0, Node 10 was not detected during the first cycle. While in Period 3, it was observed that Node 4 was not received by the basestation. The reason for Node 4 and Node 10 were not broadcasted, may cause by battery efficiency or unpredictable intermittence connectivity of the sensor node. The unpredictable interference can cause the shift in the propagation time of the transmission. Since Node 4 was not received by the basestation to utilize the bandwidth of Period 3, Node 1 was broadcasted twice. This showed that since the sensor nodes were trying to adjust their phase time in the systems, only 20% of the time the data transmitted in a synchronized pattern. The remaining time, the sensor nodes were still trying to adjust their phase time.

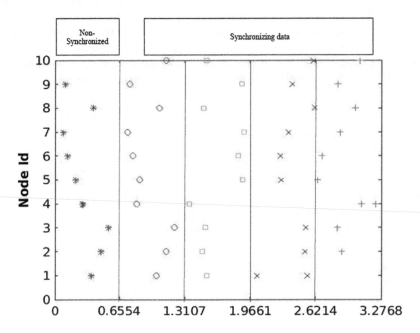

Fig. 8. The order of packet received by the basestation from 10 nodes using dynamic data transmission scheduling algorithm for Synchronicity

5.4   Statistical Test on the Synchronization Period for 10 Nodes

Table 3 shows the results obtained for the average synchronized period for every cycle of each individuals were consistent and statistically significant. This means that the individual nodes transmitted the packets at the same period throughout, this also shows that there was no delay during the transmission of the system.

From Table 4 shows the average synchronization period for every 100th period using dynamic approach. The average of synchronized period shows the average oscillator period was around 34000 to 35000 µs. The data can be seen below that the packet data cannot correctly local clock completely and more time will be needed for the data to converge synchronization. Hence, the inconsistent in the average synchronization period. While the p-$value$ shows that the probability of the sensor nodes to get synchronized was very high, between 6.80×10^{-18} and 4.63×10^{-24}.

Table 3. Average of synchronization period of random time synchronization for 10 Nodes.

Node ID	p-value	Average of synchronization period (μsec)				
		Cycle 1	Cycle 2	Cycle 3	Cycle 4	Cycle 5
Node 1	9.35×10^{-11}	31999	31998.6	31999.2	32318.2	32001.7
Node 2	9.35×10^{-11}	31999	31998.6	31999.2	32318.2	32001.7
Node 3	9.35×10^{-11}	31999	31998.6	31999.2	32318.2	32001.7
Node 4	9.35×10^{-11}	31999	31998.6	31999.2	32318.2	32001.7
Node 5	9.35×10^{-11}	31999	31998.6	31999.2	32318.2	32001.7
Node 6	9.35×10^{-11}	31999	31998.6	31999.2	32318.2	32001.7
Node 7	9.35×10^{-11}	31999	31998.6	31999.2	32318.2	32001.7
Node 8	9.35×10^{-11}	31999	31998.6	31999.2	32318.2	32001.7
Node 9	9.35×10^{-11}	31999	31998.6	31999.2	32318.2	32001.7
Node 10	9.35×10^{-11}	31999	31998.6	31999.2	32318.2	32001.7

Table 4. Average of synchronization period of dynamic data transmission scheduling for 10 Nodes.

Node ID	p-value	Average of synchronization period (μsec)				
		Cycle 1	Cycle 2	Cycle 3	Cycle 4	Cycle 5
Node 1	8.96×10^{-22}	31329.6	26203.5	25801.9	24545.9	27819.6
Node 2	1.12×10^{-22}	35628	35089	34357.3	33266.3	32946.8
Node 3	1.12×10^{-22}	35628	35089	34357.3	33266.3	32946.8
Node 4	1.12×10^{-22}	35628	35089	34357.3	33266.3	32946.8
Node 5	8.89×10^{-22}	31790.9	34183.8	33331.5	34163.6	32969.5
Node 6	7.82×10^{-22}	34719.3	33453.7	33574.7	34102.3	33613.5
Node 7	6.80×10^{-18}	32963.8	32490.8	33815.2	34471.3	38389
Node 8	4.63×10^{-24}	34015.4	34102.5	33598.5	33540.8	34166.5
Node 9	1.55×10^{-21}	34710.1	33250	32960.4	33042.3	32708.3
Node 10	2.39×10^{-19}	35226.8	36789.5	32684.9	31922.1	33407.9

5.5   Sequencing of Packet Delivered for 15 Nodes

In the synchronization process for 15 nodes, the results shown in Fig. 9 and Fig. 10 were for random and dynamic time synchronization respectively. Figure 9 result observed that 45% of the nodes were in synchronized manner. It was also observed that at a certain period the nodes, one of the node was not transmitted and caused the nodes shifted their time of transmission. Once all of the nodes shifted, 20% of the nodes transmitted were synchronized. The reason for the delay in transmission was because there was a delay in transmission of the packets. The reason of the delay can be because of the position of the two sensor nodes, will be the position of sensor nodes either Even though the nodes were in synchronized patterns, the nodes' time shown there were drifted which caused the communication overhead high.

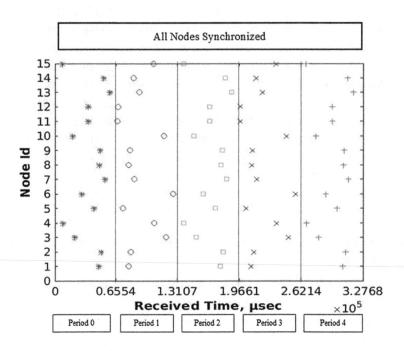

Fig. 9. The order of packet received by the basestation from 15 nodes transmitting at random

Figure 10 shows the synchronization process of the 15 sensor nodes was similar with 10 nodes, where the synchronization process took very long to synchronize the data. It shows that in every period, some of the nodes were transmitted twice in the same cycle. This means that there was a delay in transmitting form one of the biosensors, hence the repeat of biosensor received at the basestation to utilize the bandwidth. The result shows that the sensor nodes in Period 3 and Period 4, the basestation received the data were equally divided within the bandwidth time slot. By observing the results, only 10% of the time the basestation able to receive the data in synchronicity manner. However, it still showed that one or two nodes packets drop or delay in the transmission to the basestation.

5.6   Statistical Test on the Synchronization Period for 15 Nodes

In the next set of the result will be the average synchronization period of the fifteen (15) nodes as shown in Table 5 and Table 6 for random and dynamic time synchronization respectively. Both Tables show that the average synchronization period for each individual nodes were inconsistent, however the random

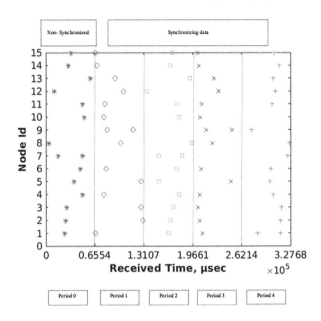

Fig. 10. The order of packet received by the basestation from 5 nodes using the firefly transmission scheduling algorithm for Synchronicity

time synchronization the average synchronization period were between 31997 and 32959.5 μs. This shows that the sensor nodes were propagating along the oscillator clock.

Table 6 below shows that as the number of sensor increases, the average synchronization period for all cycles becomes inconsistent and unstable. The timing of transmission by individual sensor node is different at every cycle which caused the delay by the clock drift. The drift is usually small, but it can be accumulated and compensated to cause message delay or packet loss.

5.7   Evaluation of Packet Delivery Rate

Figure 11 and Fig. 12 below shows the calculated packet delivery ratio (PDR) of the two(2) synchronization algorithms for 5, 10 and 15 sensor nodes. PDR is a ratio of the number of packets received by the basestation to the total number of sensor nodes send to the basestation. The result have shown that the PDR of the random time synchronization was ranges from 98% to 100%, while in the dynamic time synchronization the PDR was between 65% to 100%. This shows that the dynamic time synchronization, the PDR become more unstable as the number of biosensors increases. This also shows that the one or more biosensors will be lost when the system becomes larger.

Table 7 and Table 8 below show the results of the average packet delivery rate for every 100 synchronization rounds. In theory, the value of PDR will be decreased as the number of nodes increased. In the random time synchronization

Table 5. Average of synchronization period of random time synchronization for 15 Nodes.

Node ID	p-value	Average of synchronization period (μsec)		
		Cycle 1	Cycle 2	Cycle 3
Node 1	1.09×10^{-5}	32318.4	31999.8	32318.7
Node 2	7.40×10^{-5}	32637.3	32321	33278.7
Node 3	4.15×10^{-10}	31998.7	32000.1	31997.8
Node 4	1.10×10^{-5}	32318.1	31998.5	31999.6
Node 5	1.15×10^{-9}	31999.3	31996.1	31999.5
Node 6	5.16×10^{-10}	31998.8	32000.9	31998.7
Node 7	1.01×10^{-9}	31997.6	31997	32000.3
Node 8	1.54×10^{-5}	31999.9	31998.3	32000.3
Node 9	7.12×10^{-10}	32000.1	31997.7	32000.4
Node 10	1.10×10^{-5}	32318.2	31998.4	32000.3
Node 11	1.12×10^{-5}	32320.6	31996.1	31999.5
Node 12	4.30×10^{-5}	32318.1	32320.7	32959.5
Node 13	1.53×10^{-9}	31997	31997.8	32001.1
Node 14	2.08×10^{-9}	32001.5	31996.5	31999.7
Node 15	1.50×10^{-9}	31998.1	31997.8	32001.6

Table 6. Average of synchronization period of dynamic data transmission scheduling for 15 Nodes.

Node ID	p-value	Average of synchronization period (μsec)				
		Cycle 1	Cycle 2	Cycle 3	Cycle 4	Cycle 5
Node 1	1.10×10^{-74}	34929.1	32720.9	32689.9	30173.8	33561.9
Node 2	8.44×10^{-82}	38084.9	32870	39859	30647.9	36678.5
Node 3	8.44×10^{-82}	38084.9	32870	39859	30647.9	36678.5
Node 4	8.44×10^{-82}	38084.9	32870	39859	30647.9	36678.5
Node 5	4.57×10^{-86}	29345.3	31797.1	32822.7	33378.1	31611.6
Node 6	1.38×10^{-78}	32981	33050.5	36105.5	35942.5	35517.1
Node 7	1.44×10^{-76}	34558.5	33487	29396.7	32915.3	34461.8
Node 8	7.14×10^{-85}	31981.5	35921.7	32704.8	36109.3	33359.9
Node 9	4.59×10^{-83}	33857.8	35771.4	39013.1	33226.5	31681.6
Node 10	1.96×10^{-78}	29992.7	35813.4	32648.8	37520.6	40158.4
Node 11	2.76×10^{-84}	28539.1	35680.4	32546.6	39652.6	34894.1
Node 12	1.88×10^{-82}	29903.1	35824	32494.9	35365.3	39124.3
Node 13	3.32×10^{-78}	32695.9	39662.8	35504.2	29055.1	33221.2
Node 14	3.32×10^{-78}	32695.9	39662.8	35504.2	29055.1	33221.2
Node 15	3.32×10^{-78}	32695.9	39662.8	35504.2	29055.1	33221.2

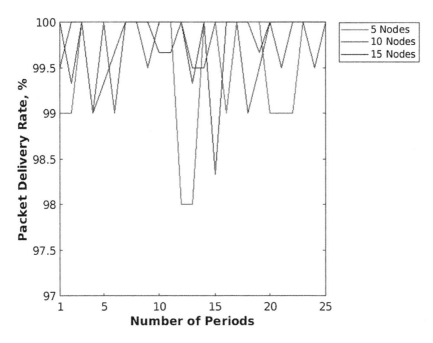

Fig. 11. The Packet Delivery Rate for 5, 10, 15 nodes transmitting at random time synchronization

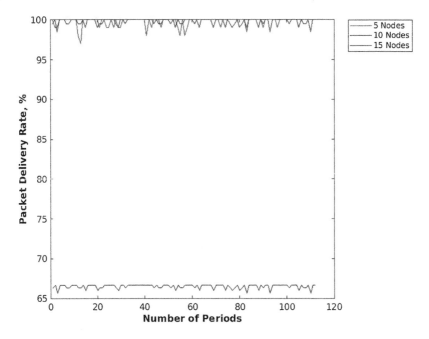

Fig. 12. The Packet Delivery Rate for 5, 10, 15 nodes using the dynamic data transmission scheduling algorithm for Synchronicity

algorithm as shown in Table 7, it shows that when the number of nodes was 5 the average PDR was 99.6% compared with number of nodes of 15 the average was 99.79%. In Period 3, the PDR of 5 nodes shows 99.2%, one of the reasons will be one or more nodes have low energy efficiency, hence loss of data during the transmission. This is caused by retransmitted packet due to collision as the numbers of nodes increases. From the graph also shows that as the number of nodes increases, the numbers of packets decreases during the transmission. This means that some of the packets sent by the sensor did not reach or received by the basestation. This mainly because there were traffic or congestions during the transmissions of the data.

Table 7. Average of Packet Delivery Rate (PDR) for 100 cycles using Random Time Synchronization for 5, 10, and 15 Nodes.

No of Nodes	p-value	Packet deliver rate %					
		Ave	P1	P2	P3	P4	P5
5	1.56×10^{-11}	99.6	99.6	100	99.2	99.6	99.6
10	2.81×10^{-11}	99.74	99.7	99.7	99.8	99.7	99.8
15	1.65×10^{-17}	99.79	99.53	99.87	99.47	99.93	100

Table 8 the result shows that has a better PDR compared to the time synchronization for 5 and 10 sensor nodes between 98% to 100%. However, when the sensor nodes were 15 the PDR of adaptive transmission scheduling was 66.5%. This means that some of the packets sent by the sensor did not reach or received by the basestation. This mainly because there were traffic or congestions during the transmissions of the data. The results show that the allocated time slots for the sensor nodes broadcast were very small for a large scalability. The transmission will be continuously and fast, thus will cause one or more sensor nodes were delay in transmission of the data.

Table 8. Average of Packet Delivery Rate (PDR) for 100 cycles using dynamic data transmission scheduling for 5, 10, and 15 Nodes.

No of Nodes	p-value	Packet deliver rate %					
		Ave	P1	P2	P3	P4	P5
5	1.62×10^{-222}	99.71	100	99	99	100	100
10	1.03×10^{-274}	99.75	99.5	100	98.5	100	100
15	3.93×10^{-267}	66.5	66.33	66.67	65.67	66.67	66.67

6 Discussion

In the random time synchronization, the results shows the packet delivery ratio was in the range of 99.20% to 100%. This PDR results were similar with the PDR in the simulation analysis; where the PDR were ranged between 90% to 100%. This shows that designing a network with random time synchronization can be done using the simulation tools. However, the simulation tools cannot implement the dynamic time synchronizaton scheduling approach.

In comparison of the performance between the time synchronization and the adaptive transmission scheduling approach. It shows that the adaptive transmission scheduling was better than time synchronization, in terms of synchronization pattern and utilizing the time slot of the sensor nodes. This type of synchronization proved that the sensor nodes broadcasted in a synchronized periodic manner and it proved that the behavior followed the mathematical model of Eq. 1. This also proved by using the *t*- *test*, it shows that the adaptive transmission scheduling has higher probability that the nodes able to broadcast the data to the basestation.

In the random time synchronization approach the testbeds show that the synchronization period was better than the adaptive transmission scheduling. This mainly because the random time synchronization nodes will only consider the previous firing nodes and repeat the cycle. While in the adaptive transmission scheduling approach the result shows that nodes were able to transmit the data in a synchronize pattern and utilize the bandwidth. However, this approach will do better in a smaller scale compared to large scale of WBAN. As mentioned in Dressler et al. [5], network with a large scale of wireless communications can exceed the capacity and leads to low reliability due to the packet loss. The loss of packets can be also due to transmission collision, because of the concurrent transmission of the sensor nodes. Hence, the packet delivery rate of the dynamic time transmission synchronization were between 66.5% and 99.71% as shown in Fig. 12 and Table 8.

7 Conclusion

From the results, although the dynamic data transmission synchronization using the firefly algorithm. The results showed that the algorithm can reduce the energy consumption with longer sleeping period, however it is not able to maintain the PDR when the number of nodes increases. In order to improve the PDR, the firefly algorithm needs to be enhanced as a future work. The process of this approach is to have the time of the transmitted data in a decentralized synchronization protocol. The application of this protocol is to have the reliable and accurate distribution of synchronization procedures.

References

1. Ramlall, R.: Wireless Body Area Network Time Synchronization using R Peak Reference Broadcasts (2007)
2. Yildrim, K., Gurcan, O.: Efficient time synchronization in a wireless sensor network by adaptive value tracking. IEEE Trans. Wireless Commun. **13**(7), 3650–3664 (2014)
3. Wu, T.Y., Wang, W.: An improvement mechanism for heartbeat-driven MAC synchronization in wireless body sensor network. Int. J. Commun. Syst. **33**(4), e4224 (2019)
4. Werner-Allen, G., Tewari, G., Welsh, M., Nagpal, R.: Firefly- inspired sensor network synchronicity with realistic radio effects. In: Sensys (2005)
5. Dressler, F., Akan, O.: A survey of bio-inspired approach networking. Comput. Netw. **54**(6), 81–900 (2010)
6. Lee, C.J., Jung, J.Y., Lee, J.R.: Bio-inspired distributed transmission power control considering QoS fairness in wireless body area sensor networks. Sens. J. **17**(10), 2344 (2017)
7. Mirollo, R., Strogatz, S.: Synchronization of pulse-coupled biological oscillator. SIAM J. Appl. Math. **50**(6), 1645–1662 (1990)
8. An, Z., Zhu, H., Zhang, M., Xu, C., Xu, Y., Li, X.: Linear pulse- coupled oscillators model- a new approach for time synchronization in wireless sensor networks. Sci. Res. J. **2**, 108–114 (2010)
9. Adhikary, S., Chattopadhyay, S., Choudhury, S.: A Novel bio-inspired algorithm for increasing throughput in wireless body area network (WBAN) by mitigating inter-wban interference. In: Chaki, R., Cortesi, A., Saeed, K., Chaki, N. (eds.) Advanced Computing and Systems for Security. AISC, vol. 667, pp. 21–37. Springer, Singapore (2018). https://doi.org/10.1007/978-981-10-8183-5_2

Energy Efficient Scheduling and Time-Slot Sharing for Hyper-Dense D2D Networks Using mmWave

Wenson Chang[1]([⊠]) and Bo-Jun Yang[2]

[1] Department of Electrical Engineering, National Cheng-Kung University,
Tainan, Taiwan ROC
wenson@ee.ncku.edu.tw
[2] ASUSTeK Computer Inc., Tainan, Taiwan ROC
ben50106@gmail.com

Abstract. This paper targets on improving the efficiency of the concurrent transmissions for the hyper-dense device-to-device (D2D) networks using the millimeter wave (mmWave) transmission technology. To this end, the increment of the multiple access interference is well controlled by the proposed energy-efficient (EE) power adjustment scheme. Therein, the nonlinear fractional programming technique is firstly applied to transform the nonlinear optimization problem into the linear from. Then, the transmission power of the D2D pairs is formulated as the noncooperative Game. Via the well-known Karush-Kuhn-Tucker condition, the optimal transmission power can then be decided. With the aid of the EE power adjustment, we modify the conventional vertex multi-coloring concurrent transmission scheme to accommodate more D2D pairs. The main concept is to judge the feasibility of the concurrent transmission based the increase of sum data rate rather than the individual rate. The superiority of the proposed scheme is verified by simulations in terms of the EE and data rate.

Keywords: Concurrent transmission · mmwave · D2D · Energy efficiency · Game theory

1 Introduction

Nowadays, the device-to-device (D2D) communication has been widely recognized as a promising technology to enhance the performance of the spectrum efficiency, energy efficiency (EE), network overloading, and transmission delay. However, in the hyper-dense D2D networks, the co-layer interference could be a fatal problem which limits the data rate and wastes energy. Fortunately, the millimeter wave (mmWave) with highly directional transmission characteristic can alleviate the problem of co-layer interference. Moreover, the directionality property can facilitate the concurrent transmissions such that the overall spectrum can be utilized in a more efficient way.

H. Gao et al. (Eds.): ChinaCom 2020, LNICST 352, pp. 161–173, 2021.
https://doi.org/10.1007/978-3-030-67720-6_11

In the literature, several concurrent transmissions schemes have been proposed for the mmWave-based communication systems. In [1], a joint beamwidth selection and scheduling algorithm was proposed to accommodate more users; and consequently the overall network throughput can be boosted. In [2], the concurrent beamforming problem was decomposed into multiple single-link beamforming problems. Via an iterative searching algorithm, the suboptimal beam sets can be obtained such that the system throughput and EE can be improved. In [3], a long-hop transmission path can be divided into multiple short-hop transmissions using the properly selected relay nodes. In this fashion, not only the high rate but also several non-interfering concurrent transmissions can be achieved. In [4], the concurrent transmissions for the backhauling of the multiple small cells was formulated as a mixed integer nonlinear programming (MINLP) problem. By solving the MINLP problem, the scheduling as well as the EE power control algorithm for the concurrent transmissions can be developed.

In addition to the above literatures, the concurrent transmission technique is also an important area for the mmWave-based D2D communications. For example, in [5], the radio accesses and backhauls of the small cells were jointly scheduled to fully exploit the spatial resources. In [6], a transmission path selection method for the multi-hop D2D mmWave communications was designed so that its associated scheduling algorithm can carry out the concurrent transmission using the minimum time slots. Moreover, in [7], the directionality property and the concept of time division multiple access (TDMA) were utilized to carry out the concurrent transmissions. Firstly, the exclusive region (ER) was defined to evaluate the tolerable amount of interference. Based on the ER evaluation, the so-called conflict matrix as well as the graph of vertex coloring (VC) was constructed to describe the ability of concurrent transmission for two arbitrary D2D pairs.

Motivated by [7], we wonder whether the spectrum resources can be shared by more users (i.e. the D2D pairs) by properly adjusting the transmission power such that the additional amount of interference can be tolerable. To this end, different from [7], the sum rate rather than the individual rate is used as the criterion for the concurrent transmission. To be specific, a group of users can be scheduled to share the same time slots if the overall sum rate can be larger than that achieved by using the vertex multi-coloring concurrent transmission (VMCCT) scheme in [7]. Note that in [7], a group of users can share a time slot if each of them can reach higher transmission rate than that using the solely TDMA scheme. Additionally, an EE power control under the sum rate constraint is designed to make the multiple access interference (i.e. the multiple access interference (MAI) caused by sharing the time slot) tolerable so as to accommodate more users during a time slot. Via the simulation results, the superior performance of the proposed concurrent transmission scheme can be verified in terms of the EE and effective user data rate. Although the proposed scheme can only incur a minor improvement of the sum data rate, the EE and effective user data rate can be enhanced by 20.3% and 11.9%, respectively, in one of our considered cases.

The remainder of this paper is organized as follows. Section 2 introduces the system model, including the signal model and time-slotted operation of the

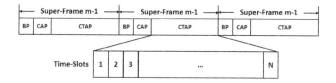

Fig. 1. The super-frame structure for the IEEE 802.15.3c piconet system.

WPAN network. In Sect. 3, the concurrent transmission problem is formulated, and then solved by the proposed EE scheduling and time-slot sharing strategy. Section 4 demonstrates the simulation results, while the concluding remarks as well as some suggestions for future works are given in Sect. 5.

2   System Model

In this paper, the mmWave-based D2D network is developed on top of the IEEE 802.15.3c piconet system [8]; therein, K D2D pairs (denoted by $\mathbf{K} = \{D_i\}\ \forall i = 1, \cdots, K$) are uniformly distributed over the piconet's coverage area of $L \times L\ m^2$. Among the K pairs, one is selected to be the piconet controller (PNC) which arranges all the other pairs to access the network as what follows.

First of all, the D2D network operates according to the time-slotted super-frame (SF) structure. As shown in Fig. 1, an SF can be divided into the beacon period (BP), contention access period (CAP) and channel time allocation period (CTAP). Note that, according the contention result, the PNC can allocate the time slots in the CTAP to the devices who win during the CAP. We name these devices the winning D2D pairs.

In our consider scenario, each device is equipped with an adaptive antenna array such that the beam can be directed toward the receiver. Referring to [9], the cone plus sphere model is applied to generate the radiation beam pattern as

$$G_A(\theta) = \begin{cases} \nu \frac{2\pi}{\theta} & , |\theta| \leq \theta_m \\ (1-\nu)\frac{2\pi}{2\pi-\theta} & , |\theta| > \theta_m \end{cases}, \tag{1}$$

where θ is the incident angle; θ_m is the beamwidth of the mainlobe; ν is the radiation efficiency. Let the i-th D2D pair is the one of interest. Assume that the mainlobes of the transmitting and receiving ends of the i-th D2D pair are aligned with each other. Then, the average receiving power $P_R(i)$ can be written as

$$P_R(i) = k_1 G_T(i) G_R(i) d_i^{-\alpha} P_T(i) , \tag{2}$$

where $k_1 \propto (\lambda/4\pi)^2$ is a constant coefficient dependent on the wavelength; $P_T(i)$ denotes transmitting power; $G_T(i)$ and $G_R(i)$ represent the antenna gain for the transmitting and receiving ends; d_i stands for the distance between the two ends of the i-th D2D pair; α is the path loss exponent. To ease the presentation, let

$g_i = k_1 G_T(i) G_R(i) d_i^{-\alpha}$ and $p_i = P_T(i)$, respectively. Then, the corresponding end-to-end capacity R_i at the n-th time slot can be expressed as

$$R_i(n) = \varphi_i(n) W \log_2 \left[1 + \frac{p_i g_i}{[N_0 W + \Sigma_{i \neq j} I_{i,j}(n)]} \right] , \tag{3}$$

where W and N_0 denote the channel bandwidth and one-sided power spectral density of the additive white Gaussian noise, respectively; $\varphi_i(n) = 1$ if the D_i is scheduled to transmit during the n-th time slot, otherwise $\varphi_i(n) = 0$; $I_{i,j}(n) = \phi_{i,j}(n) p_j g_j$ is the experienced interference from the D_j at the n-th time slot; and $\phi_{i,j}(n) = 1$ means the D_i suffers the MAI from the D_j at the n-th time slot, otherwise $\phi_{i,j}(n) = 0$.

Accordingly, the average data rate of for the i-th D2D pair D_i can be defined as

$$\overline{R}_i = \sum_{n=1}^{N} \frac{R_i(n)}{N} , \tag{4}$$

which leads to the sum data rate as

$$\overline{R} = \sum_{i=1}^{K} \overline{R}_i . \tag{5}$$

Moreover, the utility function of the EE for the i-th pair can be defined as the ratio of the spectrum efficiency (SE) R_i/W over its power consumption P_i as

$$EE = U_i(n, p_i, p_{-i}) = \frac{1}{W} \frac{R_i(n)}{P_i} = \frac{1}{W} \frac{R_i(n)}{\frac{1}{\eta} p_i + 2 p_{cir}} \text{ (bits/J/Hz)} , \tag{6}$$

where p_{-i} stands for the transmission power for all the other D2D pairs; $\eta \in (0,1)$ is the efficiency of the power amplifier; p_{cir} denotes the circuit power and it is assumed to be the same for all the devices. For simplicity, the index of time slot "n" is omitted in the following equations.

3   EE Scheduling and Time-Slot Sharing

3.1   Problem Formulation

Assume that the contention process during the CAP has finished, and the $M \leq N$ winning D2D pairs has been allocated time slots for transmission during the CTAP by using the TDMA scheme, as illustrated in Fig. 1. Let $\mathbf{M}$ denotes the set of these M winning D2D pairs. Then, in the conventional VMCCT algorithm, whether a subset of the D2D pairs (say $\mathbf{m} \subset \mathbf{M}$) can share their $|\mathbf{m}|$ time slots or not (where $|\mathbf{m}|$ measures the volume of the subset $\mathbf{m}$) can be judged according to the following rule

$$|\mathbf{m}| R_i \geq R_{i,o} \ \forall i \in \mathbf{m} , \tag{7}$$

where $R_{i,o}$ is the attainable transmission rate by using the TDMA scheme and it can be written as

$$R_{i,o} = W \log_2 \left(1 + \frac{p_i g_i}{N_0 W} \right) . \tag{8}$$

Moreover, the satisfaction of (7), indicates that the MAI $\Sigma_{i \neq j} I_{i,j}$ in (3) can be tolerable (or even negligible). However, we wonder that whether the MAI can further be controlled by using a certain power control mechanism so that more D2D pairs can share their time slots according to the following criterion

$$\sum_{i \in \mathbf{m'}} R_i \geq \sum_{i \in \mathbf{m}} R_i , \tag{9}$$

where $\mathbf{m} \subset \mathbf{m'}$. To this end, we propose the EE scheduling and time slot sharing scheme as follows.

3.2   Scheduling and Time Sharing

Now, we develop the two-step EE scheduling and time-slot sharing algorithm. For each time slot, the conventional VMCCT algorithm is firstly applied to select the set of users $\mathbf{m}$ in (7) to transmit concurrently. Then, the second step is to decide the additional users $\mathbf{m'}/\mathbf{m}$ to share the time slot with the current users belonging to $\mathbf{m}$, where $\mathbf{A}/\mathbf{B}$ removes the set $\mathbf{B}$ from the set $\mathbf{A}$.

Step 1: VMCCT Algorithm To easy the presentation, let's define some terminologies as follows.

1. D_{ti}: The transmitting end of the pair D_i.
2. D_{ri}: The receiving end of the pair D_i.
3. $D_{ti} \Rightarrow D_{rj}$: The radiation beams of D_{ti} and D_{rj} are aligned.
4. $D_{ti} \nRightarrow D_{rj}$: The radiation beams of D_{ti} and D_{rj} are not aligned.
5. $D_i \Leftrightarrow D_j$: $(D_{ti} \Rightarrow D_{rj})$ or $(D_{tj} \Rightarrow D_{ri})$
6. $D_i \nLeftrightarrow D_j$: $(D_{ti} \nRightarrow D_{rj})$ and $(D_{tj} \nRightarrow D_{ri})$
7. $d(D_{ti}, D_{ri})$: Distance between D_{ti} and D_{ri}.
8. ER_i: The radius of the ER centered at D_{ti}.

Also, according to [9], the ER_i can be defined as

$$ER_i = \left(\frac{k_1 G_T(i) G_R(i) P_T(i)}{N_0 W} \right)^{1/\alpha} . \tag{10}$$

Then, any two arbitrary D2D pairs (say pairs (D_{ti}, D_{ri}) and (D_{tj}, D_{rj})) are allowable to share a time slot if one of the following two conditions (i.e. $\mathcal{C}_1$ and $\mathcal{C}_2$) can be satisfied; otherwise the pairs of $\mathcal{D}_i$ and $\mathcal{D}_j$ conflict with each other.

$$\begin{cases} \mathcal{C}_1 : D_i \nLeftrightarrow D_j \\ \mathcal{C}_2 : (d(D_{ti}, D_{rj}) > ER_i) \, \& \, (d(D_{tj}, D_{ri}) > ER_j) \end{cases} . \tag{11}$$

For the purpose of systematically describing the conflicts between the D2D pairs, the conflict matrix $\mathbf{C} = [\zeta_{i,j}]$ for $i, j \in \mathbf{M}$ can be constructed according to

$$\begin{cases} \zeta_{i,j} = 1 : (D_{ti} \Rightarrow D_{rj}) \& (d(D_{ti}, D_{rj}) \leq ER_i) \\ \zeta_{i,j} = 0 : \text{otherwise}; \end{cases} \quad , \tag{12}$$

where $\zeta_{i,j}$ is the element at the i-th row and j-th column. Take the following conflict matrix as an example:

$$\mathbf{C} = \begin{bmatrix} 0 & 0 & 0 & 0 & 1 & ① \\ 0 & 0 & 0 & 1 & 0 & 0 \\ 0 & 1 & 0 & 1 & 0 & 0 \\ 0 & 1 & 0 & 0 & 0 & 0 \\ ① & 0 & 0 & 0 & 0 & 0 \\ 0 & 0 & 1 & 1 & 0 & 0 \end{bmatrix} . \tag{13}$$

The terms $\zeta_{1,6} = 1$ and $\zeta_{5,1} = 1$ represent the conflict conditions of $(D_{t1} \Rightarrow D_{r6}) \& (d(D_{t1}, D_{r6}) \leq ER_1)$ and $(D_{t5} \Rightarrow D_{r1}) \& (d(D_{t5}, D_{r1}) \leq ER_5)$, respectively. Then, based on the matrix $\mathbf{C}$, the pairs without conflicts can be scheduled to transmit concurrently. For example, the pairs $\mathcal{D}_2$, $\mathcal{D}_5$ and $\mathcal{D}_6$ can share a time slot. In other words, for the subset of $\mathcal{D}_2$, $\mathcal{D}_5$ and $\mathcal{D}_6$, the criterion of (7) can be satisfied.

Moreover, in order to optimize the concurrent transmission and maximize the utilization of the CTAP, the main task of the VMCCT algorithm is to organize all possible subsets (denoted by $\mathbf{M}'_n \subseteq \mathbf{M}$ for $n = 1, \cdots, M'$) of the D2D pairs such that a time slot can be shared among each subset $\mathbf{M}'_n$. Accordingly, each subset can occupy a duration of $N'T_s = \lfloor N/M' \rfloor T_s$, where T_s is the time slot duration. Equivalently, the time slots during the CTAP are reshaped to have M' time slots and each of which is with a duration of $N'T_s$. In addition, to prioritize each D2D pair, the distance weight can be defined as

$$\omega(i) = \frac{\sum_{i \in \mathbf{M}} d(D_{ti}, D_{ri})}{d(D_{ti}, D_{ri})} . \tag{14}$$

Note that the pair with higher weight can be scheduled with higher priority.

Step 2: EE Power Adjustment To facilitate the presentation, the following terminologies are defined.

1. $\mathcal{V}_n$ for $n = 1, \cdots, M'$: the set of the winning D2D pairs which are arranged to transmit during the n-th time slot by using the VMCCT algorithm; also $\mathcal{V}_n(i)$ denotes the i-th member of the set $\mathcal{V}_n$, which can also be denoted by $\mathcal{V}_n = \{\mathcal{V}_n(i)\} \; \forall i \in \mathcal{V}_n$.
2. $\mathcal{W}_n$ for $n = 1, \cdots, M'$: the set of the winning D2D pairs which are not arranged to transmit during the n-th time slot; also $\mathcal{W}_n(i)$ denotes the i-th member of the set $\mathcal{W}_n$, i.e. $\mathcal{W}_n = \{\mathcal{W}_n(i)\} \; \forall i \in \mathcal{W}_n$. Moreover, $\mathcal{W}_n(i) \; \forall i \in \mathcal{W}_n$ are sorted according to the $w(i)$ of (14).
3. $\mathcal{I}_n$ for $n = 1, \cdots, M'$: the set of the D2D pairs among which the MAI exists during the n-th time slot;

4. $\mathcal{V}_n \Leftrightarrow \mathcal{I}_n$: the MAI exists between any particular element of $\mathcal{V}_n$ and any particular element of $\mathcal{I}_n$, i.e. $\mathcal{V}_n(i) \Leftrightarrow \mathcal{I}_n(j) \; \forall i \in \mathcal{V}_n$ and $\forall j \in \mathcal{I}_n$.

Now, we aim to properly select the aforementioned $\mathbf{m'}/\mathbf{m}$ users from the set $\mathcal{W}_n$ so that the sum rate (as list in (9)) can be maximized. To achieve this goal, the EE power adjustment is proposed to alleviate the extra amount of interference caused by accommodating the additional $\mathbf{m'}/\mathbf{m}$ users. Note that accommodating more users may lead to the violation of the two conditions $\mathcal{C}_1$ and $\mathcal{C}_2$ in (11).

According to the $\omega(i)$, let's consider the first candidate D2D pair (i.e. $\mathcal{W}_n(1)$) belonging to $\mathcal{W}_n$. Assume that accommodating the $\mathcal{W}_n(1)$ can cause some extra interference to a subset of the D2D pairs (say $\mathcal{V}'_n \subseteq \mathcal{V}_n$), which leads to $\mathcal{I}_n = \mathcal{W}_n(1) \cup \mathcal{V}'_n$. Then, the optimization problem (denoted by $P1$) of the EE power adjustment can be defined as

$$\max_{p_i} \quad U_i(p_i, p_{-i}) \;, \; \forall \, i \in \mathcal{I}_n \quad (P1)$$

$$\text{(15)}$$

$$s.t. \quad 0 \le p_i \le p_{max}$$
$$R'_{sum} \ge R_{sum} \;,$$

where $R'_{sum} = \sum_{i \in \mathcal{W}_n(1) \cup \mathcal{V}_n} R_i$ and $R_{sum} = \sum_{i \in \mathcal{V}_n} R_i$. To solve this optimization problem, the nonlinear fractional programming (NFP) [10,11] is firstly applied to transform P1 into its linear version (denoted by P2) as

$$\max_{p_i} \quad C_i(p_i, p_{-i}) - q_i^* P_i(p_i) \;, \; \forall \, i \in \mathcal{I}_n(k) \quad (P2)$$

$$\text{(16)}$$

$$s.t. \quad 0 \le p_i \le p_{max}$$
$$R'_{sum} \ge R_{sum} \;,$$

where $q_i^* = U_i(p_i^*, p_{-i}) = \max_{p_i} U_i(p_i, p_{-i})$ and p_i^* is the transmission power which optimizes $U_i(p_i, p_{-i})$ [10]. Also, it leads to

$$\max_{p_i} C_i(p_i, p_{-i}) - q_i^* P_i(p_i) = C_i(p_i^*, p_{-i}) - q_i^* P_i(p_i^*) = 0 \;. \quad \text{(17)}$$

Now, the $P2$ problem can be solved by two phases: (1) determine p_i^* by relaxing $R'_{sum} \ge R_{sum}$; (2) check $R'_{sum} \ge R_{sum}$ to determine whether the D2D pair $\mathcal{W}_n(1)$ can share the time slot with the D2D pairs belonging to $\mathcal{V}_n$.

Phase 1. In fact, the relaxed $P2$ problem (i.e. relaxing $R'_{sum} \ge R_{sum}$) can be modeled by a noncooperative game; and accordingly using the Karush-Kuhn-Tucker condition gives the following Lagrangian function.

$$L_{EE}(p_i, p_{-i}, \mu_i) = C_i(p_i, p_{-i}) - q_i^* P_i(p_i)$$
$$- \mu_i(p_i - p_{max}) \;, \quad \text{(18)}$$

Algorithm 1. EE Scheduling and Time-Slot Sharing

1: Initialize: $\mathcal{V} = \mathcal{I}_n = \mathcal{P} = \mathcal{P}' = \emptyset$
2: Apply the VMCCT algorithm to obtain $\mathcal{V}_n \; \forall n \in M'$
3: **for** $n \in M'$ **do**
4:     $\mathcal{W}_n = \mathbf{M}/\mathcal{V}_n$
5:     **for** $i \in \mathcal{W}_n$ **do**
6:         $\mathcal{I}_n = \mathcal{W}_n(i)$
7:         $\mathcal{P} = \mathcal{V}_n$
8:         **while** $\mathcal{P} \Leftrightarrow \mathcal{I}_n$ **do**
9:             $\mathcal{P}' = \{\mathcal{P}(j)\}$ with $\mathcal{P}(j) \Leftrightarrow \mathcal{I}_n(k) \; \forall j \in \mathcal{P}$ and $\forall k \in \mathcal{I}_n$
10:            $\mathcal{I}_n = \mathcal{I}_n \cup \mathcal{P}'$
11:            $\mathcal{P} = \mathcal{P}/\mathcal{P}'$
12:        **end while**
13:        Apply the EE power adjustment for all the D2D pairs belonging to $\mathcal{I}_n$
14:        **if** $R'_{sum} > R_{sum}$ for $\mathcal{V}_n = \mathcal{V}_n \cup \mathcal{W}_n(i)$ **then**
15:            $\mathcal{V}_n = \mathcal{V}_n \cup \mathcal{W}_n(i)$
16:        **end if**
17:    **end for**
18: **end for**

---

where μ_i is the Lagrange multiplier. Moreover, the equivalent dual problem can be decomposed into the following min max problem as:

$$\min_{\mu_i} \max_{p_i} \; L_{EE}(p_i, p_{-i}, \mu_i) \; . \tag{19}$$

After some derivations, the optimal transmission power p_i^* can be obtained as

$$p_i^* = \left[\frac{\eta \log_2 e}{q_i^* + \eta \mu_i} - \frac{\sum_{i \neq j} I_{i,j} + N_0 W}{g_i} \right]^+ , \; \forall \, i \in \mathcal{I}_n \; , \tag{20}$$

where $j \in \mathcal{I}_n$ and $[x]^+ = \max\{0, x\}$. Note that (20) is indeed the well-known water-filling algorithm and the water level can be decided by the the Lagrange multiplier μ_i. Also, μ_i can be updated by using the gradient method as:

$$\mu_i(\tau + 1) = [\mu_i(\tau) + \epsilon_i(\tau)(p_i^*(\tau) - p_{max})]^+ , \; \forall \, i \in \mathcal{I}_n \; , \tag{21}$$

where τ is the iteration index; ϵ_i is a positive step size. It should be noticed that the power adjusting iteration can be terminated when two conditions are reached: (1) $|p_i(\tau + 1) - p_i(\tau)| \leq \varepsilon$, $\forall i \in \mathcal{I}_n$; or (2) $\tau > \tau_{\max}$, where ε is the convergence threshold.

Phase 2. Check $R'_{sum} \geq R_{sum}$ for the candidate D2D pair $\mathcal{W}_n(1)$. If it can be satisfied, the $\mathcal{W}_n(1)$ can share the time slots with the D2D pairs belonging to $\mathcal{V}_n$, which leads to $\mathcal{V}_n = \mathcal{V}_n \cup \mathcal{W}_n(1)$.

Based on the above description, the Algorithm 1 summarizes the proposed EE scheduling and time-slot sharing scheme.

Table 1. Simulation parameters.

Parameter	Value
System bandwidth (W)	500 MHz
Number of winning D2D Users (M)	$20 \sim 80$
Antenna beamwidth (θ_m)	$30°$, $60°$
Pathloss exponent (α)	4
Constant coefficient k_1 in (2)	-51 dB
PA efficiency η in (6)	35%
Background noise (N_0)	-114 dBm/MHz
Max transmission power ($P_{\max}$)	23 dBm
Constant circuit power (P_{cir})	20 dBm
Radiation frequency (ν)	0.9
Step size ϵ_i in (21)	10^{-4}
Convergence threshold ε for (20)	10^{-5}
Maximum allowable iterations $\tau_{\max}$ for (21)	100
Number of time slot during CTAP (N)	1000

4   Simulation Results

In this section, the proposed scheme and conventional VMCCT algorithm in [7] are compared in terms of the effective D2D data rate of (4), power consumption, EE of (6) and average time utilization for each winning D2D pairs $\mathcal{T}_w$, respectively. To be clear, $\mathcal{T}_w$ can be defined as

$$\mathcal{T}_w = \frac{1}{M} \sum_{i=1}^{M} \frac{1}{N} \sum_{n=1}^{N} \varphi_i(n) . \tag{22}$$

Note that $\varphi_i(n)$ is defined in (3); M' and $\mathbf{M}'_n$ are defined in Sect. 3(B.1). Also, in order to focus on the performance of the concurrent transmission schemes, the impacts of the BP and CAP sessions are omitted. Thus, only the winning D2D pairs are considered to share the time slots during the CTAP. Furthermore, to verify the effectiveness in controlling the incremental interference for the proposed scheme, the impact of the antenna beamwidth is taken into account as well. Note that the VMCCT algorithm used the fixed transmission power, which is taken as the maximum power ($P_{\max}$) for the proposed scheme. The simulations are conducted according to the system descriptions in Sect. 2 and simulation parameters listed in the Table 1, are selected according to [9,10]. All the simulation results are obtained by averaging over 1,000 randomly generated topologies.

Fig. 2 shows the (a) time utilization; and (b) effective D2D data rate with respective to the number of winning D2D pairs M for the antenna beamwidths of $30°$ and $60°$. Thanks to the serious MAI caused by accommodating more D2D

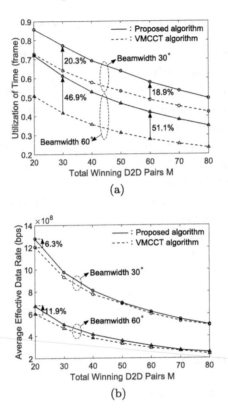

Fig. 2. (a) time utilization; and (b) effective D2D data rate with respective to the number of winning D2D pairs M for the antenna beamwidths of $30°$ and $60°$.

pairs, it is intuitional to expect the lower average data rate of (4) for each D2D pair. Fortunately, the loss of the data rate can be compensated by the higher time utilization. Figure 2 reflects this phenomenon. As shown in Fig. 2(a), the higher time utilization of the proposed scheme does successfully compensate the degraded data rate such that the 11.9% higher effective D2D data rate can be achieved (as illustrated in Fig. 2(b)).

Moreover, it is important to find that the proposed scheme can significantly outperform the VMCCT counterpart in the aspect of power consumption and EE. As shown in Fig. 3(a), the decrement of the transmission power can be 19.2% for the case with $\theta_m = 60°$ and 20 D2D pairs. However, taking the circuit power into consideration, this decrement reduces to 9.6%, as demonstrated in Fig. 3(b). In this case, the corresponding EE can be increased by 20.3%, as shown in Fig. 3(c). It should be noticed that the circuit power is fixed, while the transmission power can be adjusted; and the proposed scheme aims to reduce the transmission power rather than the circuit power.

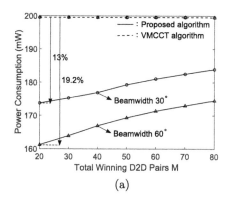

(a)

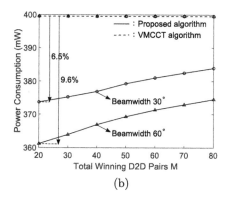

(b)

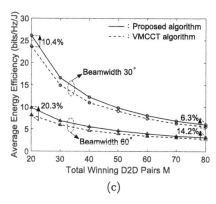

(c)

Fig. 3. (a) Power consumption (without consideration of the circuit power); (b) power consumption (with consideration of the circuit power); and (c) average EE with respective to the number of winning D2D pairs M for the antenna beamwidths of $30°$ and $60°$.

5   Conclusions and Future Works

In this paper, we have proposed the EE scheduling and time-slot sharing algorithm for the hyper-dense D2D networks based on the mmWave transmission technology. In principle, the proposed algorithm consists of two steps. Firstly, the conventional VMCCT algorithm is applied to carry out the concurrent transmission on the premise of no intolerable MAI. Then, in the second step, the proposed EE power adjustment method can effectively alleviate the increment of MAI such that more D2D pairs can be accommodated without causing any loss of the transmission rate. Most importantly, the EE can be significantly improved. In one of our consider cases, the EE can be remarkably raised by 20.3%. Some suggestions for future works include: (1) the fairness issue in the concurrent transmission scheme; (2) some sophisticated MAI elimination schemes to optimize the efficiency of the concurrent transmission; (3) extension of the proposed scheme to the relay-assisted D2D networks.

References

1. Shokri-Ghadikolaei, H., Gkatzikis, L., Fischione, C.: Beam-searching and transmission scheduling in millimeter wave communications. In: IEEE International Conference on Communications (ICC), pp. 1292–1297, June 2015
2. Qiao, J., Shen, X., Mark, J.W., He, Y.: Mac-layer concurrent beamforming protocol for indoor millimeter-wave networks. IEEE Trans. Veh. Technol. **64**(1), 327–338 (2015)
3. Qiao, J., Cai, L.X., Shen, X.S., Mark, J.W.: Enabling multi-hop concurrent transmissions in 60 GHz wireless personal area networks. IEEE Trans. Wire. Commun. **10**(11), 3824–3833 (2011)
4. Y, Niu., et al.: Energy-efficient scheduling for mmWave backhauling of small cells in heterogeneous cellular networks. IEEE Trans. Veh. Technol. **66**(3), 2674–2687 (2017)
5. Niu, Y., Gao, C., Li, Y., Su, L., Jin, D., Vasilakos, A.V.: Exploiting device-to-device communications in joint scheduling of access and backhaul for mmWave small cells. IEEE J. Select. Areas Commun. **33**(10), 2052–2069 (2015)
6. Niu, Y., Su, L., Gao, C., Li, Y., Jin, D., Han, Z.: Exploiting device-to-device communications to enhance spatial reuse for popular content downloading in directional mmWave small cells. IEEE Trans. Veh. Technol. **65**(7), 5538–5550 (2016)
7. ur Rehman, W., Han, J., Yang, C., Ahmed, M., Tao, X.: On scheduling algorithm for device-to-device communication in 60 GHz networks. In: IEEE Wireless Communications and Networking Conference (WCNC), pp. 2474–2479, November 2014
8. Baykas, T., et al.: IEEE 802.15.3c:the first IEEE wireless standard for data rates over 1 Gb/s. IEEE Commun. Mag. **49**(7), 114–121 (2011)
9. Cai, L.X., Cai, L., Shen, X., Mark, J.W.: REX: a randomized exclusive region based scheduling scheme for mmWave WPANs with directional antenna. IEEE Trans. Wire. Commun. **9**(1), 113–121 (2010)

10. Zhou, Z., Ma, G., Xu, C., Chang, Z., Ristaniemi, T.: Energy-efficient resource allocation in cognitive D2D communications: a game-theoretical and matching approach. In: IEEE International Conference on Communications (ICC), pp. 1–6, July 2016
11. Zhou, Z., Ota, K., Dong, M., Xu, C.: Energy-efficient matching for resource allocation in D2D enabled cellular networks. IEEE Trans. Veh. Technol. **66**(6), 5256–5268 (2017)

Adaptive Hybrid MAC Protocol with Novel MOB Backoff Scheme for Massive M2M Communications

Wenson Chang[1](✉) ⓘD and Chun-Wei Huang[2]

[1] Department of Electrical Engineering, National Cheng-Kung University,
Tainan, Taiwan ROC
wenson@ee.ncku.edu.tw
[2] Pegatron Corporation, Tainan, Taiwan ROC
gogoyeah0531@gmail.com

Abstract. This paper targets on solving the high collision problem in the massive machine-to-machine communications. The main idea is to restrict the numbers of allowable contentions to the low-energy devices (LEDs) by using the proposed make-or-break (MOB) backoff scheme such that the unnecessary energy consumption can be reduced. And, consequently, the high-energy devices (HEDs) can have higher probability to attain time slots for data transmissions. However, the restriction mechanism may detain the data forwarding process. To solve this dilemma, the adaptive frame structure is developed to compensate the loss of throughput. The analytical as well as simulation results demonstrate that with a huge amount of machine type devices (MTDs), the proposed scheme can outperform the conventional counterpart in the aspects of the head-of-line delay, energy efficiency and accommodation of the MTDs.

Keywords: Machine-to-Machine (M2M) communications · Media access control (mac) protocol · Collision avoidance · Energy efficiency.

1 Introduction

Nowadays, the Internet of things (IoT) is an evolutionary concept to develop an infrastructure for interconnecting massive objects without humans' interventions. The machine-to-machine (M2M) communications has been regarded as one of the most potential and pivotal innovations to carry out the IoT concept. However, the voluminous data transmissions between massive devices can result in excessive energy consumption and unacceptably high collision probability [1]. To tackle this problem, numerous media access control (MAC) protocols have been proposed in the literature [2–10].

In [2,3], to improve the energy efficiency, the M2M devices were categorized into three types and each type can exclusively occupy its dedicated time slots to forward data packets. In [4], the tree algorithm was applied to proposed a time slot assignment to alleviate the collision problem. In [5], the distributed queueing access technique was developed to avoid the loss of a whole data packet caused by

© ICST Institute for Computer Sciences, Social Informatics and Telecommunications Engineering 2021
Published by Springer Nature Switzerland AG 2021. All Rights Reserved
H. Gao et al. (Eds.): ChinaCom 2020, LNICST 352, pp. 174–185, 2021.
https://doi.org/10.1007/978-3-030-67720-6_12

the collision problem. In [6], the backoff time for the clustered M2M networks was generated according to the volume of the pending data, i.e. the larger the volume, the shorter the backoff time. In [7], the successive-interference-cancellation-based (SIC-based) protocol was proposed to transmit several replicas of the same data packet using variant time slots. In [8], by taking historical records, the access grant time interval (AGTI) was allocated to the clusters which were formed based on the requirements of the quality-of-service (QoS). In [9], an optimized hybrid protocol was proposed to apply the well-known the p-persistent carrier sense multiple access (CSMA) protocol and time division multiple access (TDMA) mechanism to manage the transmissions during contention only period (COP) and transmission only period (TOP), respectively. Moreover, to take the fairness and priority into account, the contending probability for each device can be set and adjusted according to its QoS requirement and its historical accessibility record [10].

In this paper, we propose an adaptive hybrid MAC protocol together with the make-or-break (MOB) backoff scheme by taking the residual energy into account. The main idea is motivated by the biological instinct to reduce the body activities when a creature is weak. Specifically, the unnecessary energy consumption (body activities) caused by collisions is more harmful to the low-energy devices (LEDs, i.e. the weak creatures). The restriction of contending attempts to the LEDs can be an effective method to save energy for lifetime extension, and at the same time effectively increase the successful probability for the high-energy devices (HEDs). In this fashion, the length of the COP can be reduced such that more time slot can be utilized for data transmissions during TOP (more details will be expounded in Sect. 3). Consequently, the detained data forwarding process of an LED can then be expedited whenever it is recharged to become an HED. It should be noticed that using the conventional scheme in [10], a device can repeatedly send the contending request during the COP. Also, the time durations of the COP, TOP and data transmission time for each MTD are fixed rather than adaptively adjusted using the proposed scheme. The simulation results demonstrate the advantage of the proposed scheme in terms of the MTD accommodation, energy efficiency and head-of-line (HOL) delay.

The remainder of this paper is organized as what follows. Section 2 introduces the system model of the massive M2M network, including the MAC protocol and its corresponding time frame structure. In Sect. 3, the problem of high collision probability and its associated energy consumption are firstly defined. Then, we propose the adaptive hybrid MAC protocol together with the MOB backoff scheme. Section 4 demonstrates the simulation and numerical results. Section 5 gives the concluding remarks, including the potential topics for future works.

2   System Model

In this paper, the system model and assumptions are mainly based on those in [9]. Therein, a BS is deployed in the center of the network, and there are K machine-type devices (MTDs) uniformly distributed over the BS's coverage

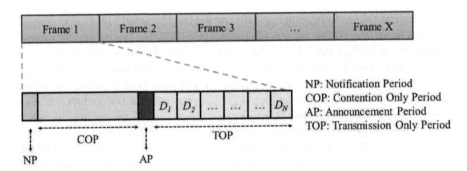

Fig. 1. Time frame structure, where D_n $\forall n = 1, 2, \cdots, N$ are the transmission time slots for the N winning MTDs.

area. Note that a winning MTD is the one who wins the competition during the COP, whereas a losing MTD is the one who fails. Also, the hybrid MAC protocol operates according to the time frame structure shown in Fig. 1. As shown in the figure, a frame consists of four time periods, including the COP, TOP, notification period (NP) and announcement period (AP). Observing the frame structure, one can realize that the term "hybrid" presents the applications of the p-persistent CSMA protocol and TDMA mechanism for the COP and TOP, respectively.

To reach the highest aggregate throughput (denoted by γ), the duration of TOP (T_{TOP}) should be maximized such that it can accommodate the optimal number of winning MTDs (N_{opt}). Based on the structure of the considered time frame, it can be realized that the maximization of T_{TOP} is equivalent to the minimization of T_{COP}. According to the derivations in [9], the T_{COP} can be approximated by

$$T_{COP}(M, N, p) = N \left\{ \frac{1}{Mp} T_{idle} + T_{suc} + T_{col} \left(\frac{1}{Mp(1-p)^{M-1}} - \frac{1}{Mp} - 1 \right) \right\} , \tag{1}$$

where M and N are the numbers of active and winning MTDs; p is the contention probability;

$$T_{suc} = T_{req} + SIFS + T_{ack} + BIFS ; \tag{2}$$

$$T_{col} = T_{req} + BIFS ; \tag{3}$$

$$T_{idle} = T_{req} + BIFS ; \tag{4}$$

$SIFS$ and $BIFS$ are the short interframe spacing and backoff interframe spacing; T_{req} and T_{ack} represent the duration of the request and ACK signals.

Now, the optimization problem can be defined as

$$\max_{M,N,P} \quad \gamma = N R_D T_{tran}$$

$$\text{subject to} \quad T_{cop}(M, N, p) + N T_{tran} \leq T_{frame}$$

$$0 \leq p \leq 1 \tag{5}$$

where T_{tran} and R_D represent a transmission time slot during the TOP and its corresponding data rate of each MTD. Then, T_{COP} is minimized by searching for the optimal contending probability (p_{opt}) and the corresponding number of winning MTDs (N_{opt}). Note that the convexity of T_{COP} with respect to N and p can be proved by showing $\partial^2 T_{COP}/\partial N^2 \geq 0$ and $\partial^2 T_{COP}/\partial p^2 \geq 0$, respectively.

3   Adaptive Hybrid MAC Protocol with MOB Backoff Scheme

3.1   Problem Formulation

Based on the Hybrid MAC protocol (as introduced in Sect. 2), there are two criteria for terminating the COP: (1) T_{COP} expires; (2) the number of winning MTDs reaches the optimal value, i.e. $N = N_{opt}$. However, it is highly possible that T_{COP} expires before $N = N_{opt}$. In this situation, the T_{TOP} reserved for the N_{opt} MTDs can not be fully utilized. Similarly, when the condition of $N = N_{opt}$ is satisfied before the expiry of T_{COP}, the predefined T_{COP} is not fully utilized as well. Furthermore, considering an LED, the stringent collision problem can seriously reduce its lifetime. For example, when the number of active MTDs is large, keen competition can happen during the COP. That means each MTD could possibly win after a lot of request attempts, which can consume extensive energy. Thus, in this paper, we aim to solve the above three problems by proposing an adaptive hybrid MAC protocol and MOB backoff scheme.

3.2   Adaptive Hybrid MAC Protocol with MOB Backoff Scheme

Let $\mathbf{M}$, $\mathbf{M}_L$ and $\mathbf{M}_H$ denote the index sets of all the active MTDs, LEDs and HEDs, respectively. Then, it gives $\mathbf{M} = \mathbf{M}_L \cup \mathbf{M}_H$. Also, let ξ_{th} and ξ_i represent the energy threshold and level of the i-th active MTDs for all $i \in \mathbf{M}$. Thus, one can realize that $i \in \mathbf{M}_L$ if $\xi_i \leq \xi_{th}$; on the contrary, $i \in \mathbf{M}_H$ if $\xi_i > \xi_{th}$. In order to formulate the restriction on the LED's contending privilege, let's define $\eta_{th,i}$ and η_i $\forall i \in \mathbf{M}$ as the number of allowable and accumulated requesting attempts during a COP for the i-th MTD, respectively. Then, it leads to

$$\eta_{th,i} = \begin{cases} \eta_{th} & \forall i \in \mathbf{M}_L \\ \eta_{max} & \forall i \in \mathbf{M}_H \end{cases}, \tag{6}$$

where η_{th} is a predefined limit on the number of allowable contentions for LEDs; η_{max} is the maximal number of allowable contentions for HEDs.

Algorithm 1. The Procedure of Adaptive Hybrid MAC Protocol with MOB Backoff Scheme

1: T: The elapsed time within each frame;
2: **C**: The set includes the MTDs who experience collisions during COP;
3: $\varnothing$: An empty set;
4: Let $T = 0$, $N = 0$ and $M = \rho \times K$;
5: **while** $T < T_{COP}$ or $N < N_{opt}$ **do**
6:      The M MTDs issue their request signals based
7:      on the p-persistent CSMA mechanism with $p = p_{opt}$;
8:      **if** a successful event occurs **then**
9:          $N + +$;
10:         $M - -$;
11:         Remove i from **M**;
12:         $T + = T_{req} + SIFS + T_{ack} + BIFS$;
13:     **else if** an idle event occurs **then**
14:         $T + = T_{req} + BIFS$;
15:     **else if** a collision event occurs **then**
16:         $T + = T_{req} + BIFS$;
17:         Remove i from **M** $\forall i \in \mathbf{C}$ if $\xi_i \leq \xi_{th}$ and $\eta_i =$
18:         $\eta_{th,i}$ are satisfied;
19:         Let $\mathbf{C} = \varnothing$;
20:     **end if**
21: **end while**
22: BS announces the time slot arrangement and the adjusted T_{tran} to all winning MTDs during AP;
23: The N winning MTDs sequentially forward their accumulated data packets during TOP;

It should be emphasized that restricting the contending privilege of the LEDs can extend their lifetime. Moreover, it increases the probability of successful contention since the number of competitors can be decreased by using a proper value of η_{th}. That is to say it can be very helpful to satisfy the second criterion for terminating the COP, i.e. $N = N_{opt}$. If the second criterion can be achieved priori to the first one, i.e. the expiry of T_{COP}, the additional time $\triangle T_{COP} = T_{COP} - T'_{COP}$ can be equally shared by these N_{opt} MTDs, where T_{COP} and T'_{COP} are the predefined and practical time duration of COP, respectively. In this case, $T_{tran} = (\triangle T_{COP} + T_{TOP})/N_{opt}$. However, it is still possible that the first criterion is satisfied earlier. Then, the reserved T_{TOP} can be fully utilized by all the winning MTDs. That means each winning MTD can own $T_{tran} = T_{TOP}/N$ time resources (in this case $N < N_{opt}$). It should be noticed that, by using the proposed scheme, T_{tran} can be adaptively adjusted rather than the fixed as that in (5) such that each winning MTD can forward its accumulated data packets in the more efficient way. Furthermore, it can be realized that the conventional scheme in [9] is a special case of our proposed scheme by setting $\eta_{th} = \eta_{max}$ or the percentage of LEDs to be zero, i.e. $\rho_L = 0$. Algorithm 1 summarizes the proposed Adaptive Hybrid MAC Protocol and MOB Backoff Scheme.

3.3   Semi-Analysis of Head-of-Line Delay

Define p_{sc} as the successful probability for an MTD to win a transmission opportunity during the COP. Then, an arbitrary MTD may spend n time frames to win an opportunity with probability $(1 - p_{sc})^{n-1} p_{sc}$. Thus, the probability mass function (pmf) for an MTD to win at the n-th frame can be written as

$$f(n) = (1 - p_{sc})^{n-1} p_{sc} \quad n \geq 1 \; ; \tag{7}$$

whereas the corresponding cumulative distribution function (cdf) can then be written as

$$F(n) = \sum_{i=1}^{n} (1 - p_{sc})^{n-1} p_{sc} \quad n \geq 1 \; . \tag{8}$$

Owing to the random backoff operations and different allowable contentions for HEDs and LEDs, it is difficult to obtain a closed-form expression for p_{sc}. Thus, we obtain p_{sc} via simulations such that (8) can be utilized to verify the simulation results (as shown in the following Fig. 5).

4   Numerical and Simulation Results

In this section, the impact of the number of allowable requesting attempts (η_{th}) on the proposed scheme is firstly evaluated in terms of the number of winning MTDs. Based on this result, a proper value of η_{th} can be decided. Moreover, to demonstrate the superiority of the proposed scheme over the conventional counterpart in [9], the following performance comparisons are conducted in terms of the energy efficiency, head-of-line delay and number of winning MTDs.

4.1   Simulation Setup

In addition to the system depiction in Sect. 2, the mobility of MTD, differentiated priorities among MTDs and erroneous control messages are not taken into account. Also, to clearly demonstrate the impact of the LEDs, the percentage of LEDs among all MTDs is fixed during the simulations. That means the energy level of an HED will not be significantly reduced to become an LED. Also, an LED will not be charged to become an HED. A winning MTD transmits data packets using a fixed transmission rate $R_D = 250$ kbps. To make the energy consumption model complete, four operation states for each MTD are considered, i.e. the transmission, receiving, idle and sleeping states [5]. As implied by the name, the transmission and receiving states mean the MTD of interest is now transmitting and receiving signals, respectively. Whereas, except issuing request signal, an MTD remains in the idle state if it has not yet won during COP. Also, a winning MTD stays in the idle state if it has not yet finished the data transmissions during TOP. Moreover, it can switch into the sleeping state when it has finished the data transmission task during TOP. Besides, a losing MTD can also switch into the sleeping state when the T_{COP} expires. Note that the time and energy consumption for an MTD to switch between active and inactive modes are ignored here [5]. Table 1 summarizes the parameters in the following simulations.

Table 1. Simulation parameters

Simulation parameters	Values
Number of MTDs (K)	1–1000
Ratio of active MTDs (ρ)	0.5
Percentage of LEDs (ρ_L)	0–100 %
Duration of NP	10.2 μs
Duration of AP	10.2 μs
Duration of time frame (T_{frame})	50 ms
Duration of transmission time slot (T_{tran})	1 ms
$SIFS$	2.5 μs
$BIFS$	7.5 μs
Duration of requesting signal (T_{req})	22.2 μs
Duration of ACK (T_{ack})	7.5 μs
Data transmission rate (R_D)	250 kbps
Transmission power (P_{tx})	100.8 mW
Receiving power (P_{rx})	66.9 mW
Idling power (P_{idle})	525 μW
Sleeping power (P_{sleep})	60 nW
Limit on the allowable contentions (η_{th})	1, 2, 3
Maximum allowable contentions (η_{max}) [11]	16

4.2   Effect of η_{th}

Figure 2 demonstrates the number of winning MTDs N with respect to the total number of MTDs K for various number of allowable requesting attempts η_{th}, where the percentage of LEDs $\rho_L = 50\%$. One can find that, with a small amount of MTDs, a larger η_{th} can contribute to a greater amount of winning MTDs, whereas, when the number of MTDs becomes huge, a smaller η_{th} can accommodate more winning MTDs. For example, at $K = 100$, the numbers of winning MTDs N are 40.7, 36.1 and 26.6 for the cases with $\eta_{th} = 3$, 2 and 1, respectively. However, at $K = 800$, they become 31.4, 31.4 and 36.1. This is because, with a smaller K, the competition is not quite violent. Thus, there is no need to restrict the contending privilege of the LEDs. On the contrary, when the competition becomes stringent, restriction on the LEDs' activity can effectively increase the probability of successful contention. To sum up, one can say that the proposed MOB backoff scheme can be more effective when the number of MTDs is huge. Moreover, in the following simulations, $\eta_{th} = 1$ is adopted.

4.3   Number of Winning MTDs

To demonstrate the effectiveness of our proposed scheme for improving the contention efficiency by limiting the LED's contention attempts, Fig. 3 shows the

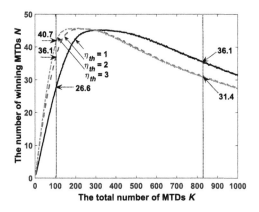

Fig. 2. The number of winning MTDs N with respect to the total number of MTDs K for various numbers of allowable requesting attempts η_{th}, where the percentage of LEDs $\rho_L = 50\%$.

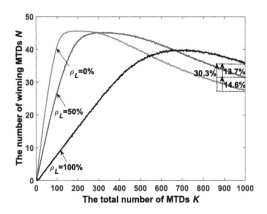

Fig. 3. The number of winning MTDs N with respect to the total number of MTDs K for various percentages of LEDs ρ_L. The curve with $\rho_L = 0\%$ corresponds to the conventional hybrid MAC protocol in [9].

number of winning MTDs N with respect to the total number of MTDs K for various percentages of LEDs ρ_L. Note that the higher the number of winning MTDs, the higher the contention efficiency during COP. Also, the curve with $\rho_L = 0\%$ corresponds to the conventional hybrid MAC protocol in [9]. Apparently, when the traffic is heavy, the proposed MOB backoff scheme with $\rho_L \neq 0\%$ can outperform the conventional counterpart with $\rho_L = 0\%$. Comparing with the conventional scheme at $K = 1000$, the proposed scheme can accommodate 14.6% (from 27.4 to 31.4) and 30.3% (from 27.4 to 35.7) more winning MTDs for the cases with $\rho_L = 50\%$ and 100%, respectively.

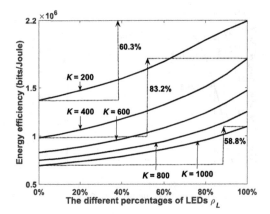

Fig. 4. Performance of average energy efficiency μ with respect to the percentage of LEDs ρ_L for various total numbers of MTDs K.

4.4   Energy Efficiency

To demonstrate the advantage of the proposed scheme in reducing energy consumption and raising the utilization of time resources (i.e. T_{COP} and T_{TOP}), the performance metric of average energy efficiency μ within a frame is defined as

$$\mu = E\left[\frac{\text{BS's total received data bits}}{\text{Total energy consumption of all active MTDs}}\right], \qquad (9)$$

where $E[z]$ is to take the expectation of variable z [12]. Note that the effect of the prolonged lifetime for the LEDs reflects on the reduced energy consumption (as aforementioned which is achieved by eliminating the ineffective contentions). Figure 4 illustrates the performance of average energy efficiency μ with respect to the percentage of LEDs ρ_L for various total numbers of MTDs K. It is obvious that the proposed scheme can effectively improve the average energy efficiency. With $K = 200$ and $K = 1000$, the average energy efficiency for both cases can be improved by an approximately equal amount, i.e. 60.3% (from 1.369×10^6 to 2.194×10^6) and 58.8% (from 6.916×10^5 to 1.098×10^6), respectively. Moreover, this improvement can be 83.2% with $K = 400$. Thus, it can be concluded that the proposed scheme can be more effective in raising the average energy efficiency with an appropriate amount of traffic.

4.5   Head-of-Line Delay

Figure 5 exhibits the (a) cumulative density functions (cdfs), (b) mean and (c) variance of the head-of-line delay with various numbers of MTDs for the conventional and proposed schemes, where $\rho_L = 100\%$ is presumed for the proposed scheme. Firstly, the simulation and numerical results can be supported by each other. Moreover, as shown in the figure, the advantage of the proposed scheme

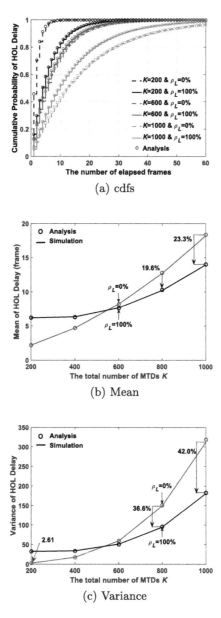

Fig. 5. (a) Cumulative distribution functions (cdfs), (b) mean and (c) variance of the head-of-line delay with various numbers of MTDs for the conventional and proposed schemes, where $\rho_L = 100\%$ is presumed for the proposed scheme.

in the heavy traffic situations is verified again. Phenomenally, as K grows, the curves of proposed scheme (i.e. the ones with $\rho_L \neq 0$) move leftwards relative to the curve of the conventional scheme (i.e. the curve with $\rho_L = 0$). Comparing at

$K = 200$, 90% of the MTDs can transmit their data packets no later than 4 and 13 time frames by using the conventional and the proposed schemes. Conversely, with $K = 1000$, these values become 41 and 31. Furthermore, observing Figs. 5(b) and 5(c), the lower mean and variance values obtained by using the proposed scheme for the cases with more MTDs can also explain this phenomenon. Most importantly, the proposed scheme can be less sensitive to the increase of MTDs, i.e. the lower slope compared with the conventional counterpart.

5   Conclusions and Future Works

In this paper, we have proposed an effective but simple adaptive hybrid MAC protocol to solve the high collision problem in the massive M2M communication systems. Motivated by the biological instinct to restrict the numbers of allowable contentions to the LEDs, not only can the successful contending probability of the HEDs be raised, but also the lifetime of the LEDs can be extended. Moreover, adaptively adjusting the time duration for the COP and TOP can contribute to the higher throughput of data signals. In one of our considered cases, the analytical and simulation results have illustrated the additional 30.3% of MTD accommodation; and the average HOL delay can be reduced by 23.3% as well. Most importantly, the remarkable 83.2% enhancement of the energy efficiency can be attained. It is believed that the proposed scheme can be applicable and helpful to improve the overall system performance of the massive M2M networks. Some suggestions for the possible future works could be: (1) adjusting the contending probability p and the numbers of allowable contentions $\eta_{th,i}$ $\forall i = M$ for the differentiated priorities of MTDs and (2) the complete analysis of the HOL delay, energy efficiency and the number of winning MTDs.

References

1. Rajandekar, A., Sikdar, B.: A survey of MAC layer issues and protocols for machine-to-machine communications. IEEE Internet Things J. **2**(2), 175–186 (2015)
2. Adame, T., Bel, A., Bellalta, B., Barcelo, J., Oliver, M.: IEEE 802.11ah: The WiFi approach for M2M communications. IEEE Wire. Commun. **21**(6), 144–152 (2014)
3. Park, C.W., Hwang, D., Lee, T.-J.: Enhancement of IEEE 802.11ah MAC for M2M communications. IEEE Commun. Lett. **18**(7), 1151–1154 (2014)
4. Vazquez-Gallego, F., Alonso-Zarate, J., Tuset-Peiro, P., Alonso, L.: Energy and delay analysis of contention resolution mechanisms for machine-to-machine networks based on low-power WiFi. In: IEEE International Conference on Communications, pp. 2236–2240, June 2013
5. —, Energy analysis of a contention tree-based access protocol for machine-to-machine networks with idle-to-saturation traffic transitions. In: IEEE International Conference on Communications, pp. 1094–1099, June 2014
6. Park, I., Kim, D., Har, D.: MAC achieving low latency and energy efficiency in hierarchical M2M networks with clustered nodes. IEEE Sens. J. **15**(3), 1657–1661 (2015)

7. Hernandez, A., Vazquez-Gallego, F., Alonso, L., Alonso-Zarate, J.: Performance evaluation of frame slotted-ALOHA with intra-frame and inter-frame successive interference cancellation. In: IEEE Global Communications Conference, pp. 1–6, December 2015
8. Peng, S., et al.: An adaptive massive access management for M2M communications in smart grid. In: IEEE 24th International Symposium on Personal Indoor and Mobile Radio Communications, pp. 3408–3412, September (2013)
9. Liu, Y., Yuen, C., Chen, J., Cao, X.: A scalable hybrid MAC protocol for massive M2M networks. In: IEEE Wireless Communications and Networking Conference, pp. 250–255, April (2013)
10. Liu, Y., Yuen, C., Cao, X., Hassan, N.U., Chen, J.: Design of a scalable hybrid MAC protocol for heterogeneous M2M networks. IEEE Internet Things J. 1(1), 99–111 (2014)
11. Tanenbaum, A. S.: Computer Networks (Fifth Edition). Pearson Education International (2011)
12. Rhee, I., Warrier, A., Aia, M., Min, J., Sichitiu, M.L.: Z-MAC: A hybrid MAC for wireless sensors networks. IEEE/ACM Trans. Netw. 16(3), 511–524 (2008)

Scheduling and Security in 5G

A MIMO Channel Measurement System Based on Delay Lines and Simulations Based on Graph Modeling

Shengnan Xu[✉] [ID]

Tongji University, Shanghai, China
1832953@tongji.edu.cn

Abstract. In this paper, we proposed a novel MIMO channel measurement system architecture for 5G wireless communication based on the delay lines in combination with switches, and we implement propagation graph modeling to simulate the channel measurement procedure. Channel sounders equipped with multiple-element antenna arrays in the transmitter (Tx) and receiver (Rx) usually perform a measurement in two ways: switched channel measurement and parallel channel measurement. The latter usually needs multiple Txs/Rxs which leads a high cost, while the former with a high-speed radio-frequency switch at the transmitter and receiver has a lower cost but it is difficult to realize beamforming due to the Tx/Rx antennas do not transmit/receive signals simultaneously. By adding delay lines to the switched MIMO channel measurement system, the delay in different time slot at every Tx antennas can be compensated so that the multiple Tx antennas (empowered by only one Tx) transmit signals simultaneously. Furthermore, adding phase shifters after delay lines makes it easy to change the phase of each signal, which provides a convenient way for beamforming. The feasibility of the proposed method is preliminarily validated through simulations based on propagation graph modeling, the evaluation of the results is conducted by calculating the channel impulse response (CIR) or power delay profile (PDP) and estimating the direction of arrival (DOA) using Multiple Signal Classification (MUSIC) algorithm.

Keywords: Channel measurement · Delay line · Switch · Propagation graph modeling

1 Introduction

In the development of wireless communication, the evolution of each generation of communication systems is accompanied by an ascent in frequency and bandwidth. The fifth Generation (5G) wireless communication systems will use channels with bandwidth up to 0.5–2 GHz to offer sufficient spectral resources for the users in the frequency range of 0.45–85 GHz [1]. The accurate description of the propagation channel characteristics is the basis of design, analysis and optimization of the 5G wireless communication system. Channel measurement need to be performed accurately. In addition to that, new techniques and scenarios of the 5G wireless communication, such

H. Gao et al. (Eds.): ChinaCom 2020, LNICST 352, pp. 189–199, 2021.
https://doi.org/10.1007/978-3-030-67720-6_13

as massive MIMO, beamforming and millimeter wave communication, require new methodology for channel measurement.

Massive MIMO is a key technique for 5G, which requires a large number of antennas, for example, 32 antennas, 64 antennas, or even more. In this scenario, traditional parallel channel measurement needs a large number of transmitters (TXs). Switched channel measurement is an alternative, but it is inconvenient to implement beamforming, due to it is a time-division-multiplexing (TDM) measurement scheme.

Many schemes for high-frequency channel measurement have been studied. In [2], the authors performed a channel measurement campaign in the frequency domain based on VNA (vector network analyzer), in which the transmitter uses optical fiber for transmission to increase the measurement distance of the VNA-based system. [3] presents a time-domain channel measurement system using a spread spectrum sliding correlator method. The sliding correlator achieves superior multipath time resolution and dynamic range by using pseudo noise sequences operated at slightly different clock speeds at the transmitter and receiver, but it is hard to be extended to multi-band measurement for it is only applicable to fixed frequency bands. It is well known that the multi-antenna channel measurement and the acquisition of spatial angle information are particularly important, and [4] introduces a 28 GHz channel sounder based on automatically rotating horn antennas with time synchronization ability between the transmitter and the receiver. The sounder system in [4] realize a virtual array and the use of high-gain narrow-beam directional antennas compensates the severe attenuation of the millimeter wave channel. [5] presents a novel MIMO channel sounding technique with a fully parallel transceiver architecture that employs a layered scheme of frequency and space-time division multiplexing, and it offers inherent scalability of the number of antennas by a combination of multiple transceiver units and exhibits flexibility for both directional MIMO channel and multi-link MIMO channel measurements. In [6–8], some researches focus on exploring higher efficiency and lower cost MIMO channel measurement systems.

Besides the measurement-based researches, simulation-based methods for channel measurement and modeling are also worth studying. In fact, simulation-based methods become popular in 5G scenarios due to the advantages in efficiency, flexibility and low cost. Graph theory based channel modeling has proved to be an effective approach to simulate the multipath propagation and the diffuse components, which means it is promising to simulate the radio propagation in highly diffuse scattering conditions like propagation at the millimeter wave. It is well known that propagation graph modeling is a good alternative to measurement campaign in environments which are difficult to perform actual measurement, such as the high-speed railway propagation environments. The applicability of graph modeling in high-speed railway propagation environments has been proved in [9]. In [10], the stochastic graph modeling approach is assessed by comparing the statistical characteristics of the channel parameters acquired from graph modeling in an indoor environment with the parameters specified in established WINNER II channel models. In [11], authors proposed a deterministic graph modeling approach by associating the scatterers with realistic environment objects, and by calculating the coefficients of propagation paths based on a proven diffuse scattering theory.

In this paper, we proposed a MIMO channel measurement system with only one Tx/Rx for 5G wireless communication based on delay lines in combination with switches aiming at overcoming the shortcoming that traditional switched channel measurement cannot transmit or receive signals simultaneously which obstructs the realization of beamforming. Then we implement stochastic propagation graph modeling to simulate the channel measurement procedure. In the simulation, delay lines were added to the transmitter to realize transmitting signals simultaneously. The channel impulse response can be calculated analytically with limited computational cost, and we estimate the direction of arrival (DOA) using MUSIC algorithm. The organization of this paper is as follows: Sect. 2 introduces the schematic of the proposed measurement systems and the procedure of simulations for the proposed measurement system. A brief introduction for the graph theory based modeling is also presented. In Sect. 3, the analysis of simulation results is given. Finally, conclusive remarks are addressed in Sect. 4.

2 Simulations for the Proposed Measurement System Based on Graph Modeling

2.1 The Structure of Proposed Measurement System

The structure of proposed MIMO channel measurement system is shown in Fig. 1, the transmitter is equipped with delay lines connecting the switches and phase shifters that help to implement beamforming, hence, the measurement system can transmit signals simultaneously only with one channel sounder and is friendly to beamforming. Antenna arrays at the receiver receive signals parallelly without extra devices as a normal multiple-output.

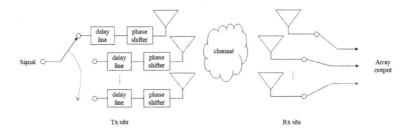

Fig. 1. The structure of proposed MIMO channel measurement system.

2.2 Basic Theory of Propagation Graph Modeling

The propagation graph modeling proposed in [12] calculates the transfer function as:

$$\mathbf{H}(f) = \mathbf{D}(f) + \mathbf{R}(f)(\mathbf{I} - \mathbf{B}(f))^{-1}\mathbf{T}(f) \tag{1}$$

where $\mathbf{D}(f)$ represents the Line-of-Sight (LoS) component of the transfer function and the other terms are the None-Line-of-Sight (NLOS) components of the transfer function. In detail, $\mathbf{D}(f)$ denotes the matrix of the transfer function from the Tx to the Rx directly and $\mathbf{B}(f)$ denotes the bouncing matrix of the transfer function between the scatters, $\mathbf{T}(f)$ and $\mathbf{R}(f)$ denote the matrices of the transfer function from the Tx to the scatters and from the scatters to the Rx, respectively. Detailed derivation of (1) and more information about propagation graph modeling can refer to [12]. The transfer function of a propagation path can be calculated as:

$$A_e(f) = g_e(f)\exp(-j2\pi\tau_e f + j\varphi) \tag{2}$$

where $A_e(f)$ represents the elements of matrices $\mathbf{D}(f)$, $\mathbf{R}(f)$, $\mathbf{B}(f)$ and $\mathbf{T}(f)$ according to different kinds of edges, τ_e denotes the propagation delay between two vertices, and φ is a random phase rotation considered as a random variable uniformly distributed on the interval $[0, 2\pi)$, and $g_e(f)$ is the propagation coefficient depends on different kinds of edges. $g_e(f)$ was originally defined in [13], and a modified deterministic definition of $g_e(f)$ was proposed in [14].

2.3 Procedure of the Simulation

The steps for the simulation of the channel measurement system based on graph modeling in a given propagation environment are generally elaborated as follows:

Step 1: Set the vertex information of the polygonal surfaces according to the digital map of the given environment, and set the positions of the transmitters and receivers as we want;

Step 2: Generate the positions of the scatterers by discretize every object surface into multiple scatterers, record the coordinates of the scatterers, and the normal directions of the surfaces are also recorded and assigned to the corresponding scatterers;

Step 3: Judge the visibility for every pair of two vertices (i.e. a Tx and a scatterer, a scatterer and a Rx) according to their positions and the positions of the other scatterers, which can be implemented by the method that judges whether the edge connecting the two vertices intersects the other scatterers. A propagation graph is now build up, Fig. 2 shows an example of digital map;

Step 4: Calculate the propagation coefficient $g_e(f)$ for each available link and generate the matrices $\mathbf{D}(f)$, $\mathbf{R}(f)$, $\mathbf{B}(f)$ and $\mathbf{T}(f)$ according to (2) for specific frequencies $f = f_{min}, f_{min} + \Delta f, \ldots, f_{max}$;

Step 5: Embed the antenna radiation pattern for the matrices $\mathbf{T}(f)$ and $\mathbf{R}(f)$, here we use omnidirectional antenna radiation pattern;

Step 6: Add time delays to each transmitted signal by adding a delay $(k-1)\tau$ to every element of kth column vector of both $\mathbf{D}(f)$ and $\mathbf{T}(f)$ to simulate the extra delay caused by switches, which means the $A_e(f)$ in both $\mathbf{D}(f)$ and $\mathbf{T}(f)$ change as follow:

$$A_{e,k}(f) = g_{e,k}(f)\exp\left(-j2\pi\big(\tau_{e,k} + (k-1)\tau\big)f + j\varphi\right) \tag{3}$$

where $A_{e,k}(f)$ denotes the elements of kth column vector of $\mathbf{D}(f)$ and $\mathbf{T}(f)$. Then we use $\mathbf{D}'(f)$ and $\mathbf{T}'(f)$ to represent $\mathbf{D}(f)$ and $\mathbf{T}(f)$ after adding time delays, respectively.

Step 7: According to (1), $\mathbf{D}'(f)$, $\mathbf{R}(f)$, $\mathbf{B}(f)$ and $\mathbf{T}'(f)$ are used to calculate the channel transfer function $\mathbf{H}'(f)$ with delays caused by switches, and replace $\mathbf{D}'(f)$ and $\mathbf{T}'(f)$ with $\mathbf{D}(f)$ and $\mathbf{T}(f)$ respectively, we get the channel transfer function $\mathbf{H}(f)$ with delays compensated by delay lines. The CIR is the inverse Fourier transform of the transfer function.

The above procedure shows how to simulate the measurement process based on delay lines and switches. The whole modeling is similar to the original graph modeling procedure [14], however, the processing of calculating CIR is a little different. Finally, we acquire the direction information from CIR adopting MUSIC algorithm.

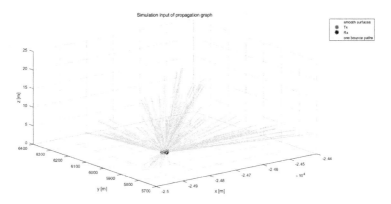

Fig. 2. An example of digital map in simulations.

3   Results of the Simulation

Delay lines are usually tunable in a certain range, but it is still a little difficult to implement the phase shift of a signal by adjusting delay lines, hence we employ the phase shifter as a helper for beamforming after the switch delays compensated by delay lines. Here we consider an N-element linear array composed of isotropic radiating antenna elements, and it is assumed that the signal of the nth element is added with a phase shift $(n - 1)\delta$ by tuning the phase shifter. Additionally, we assume the carrier frequency as $f = 1.75$ GHz, and the spacing of the elements as $d = \lambda/2$ (λ is the wave length). Here, we add different δ to an 8-element linear array ($N = 8$), and the corresponding array factor patterns are exhibited in Fig. 3, 4 and 5. Figure 3 shows the original array factor pattern at $\delta = 0$, while $\delta = \pi/2$ and π respectively in Fig. 4 and 5. Obviously, the three patterns are different from each other. For different δ, adding phase shift $(n - 1)\delta$ on the nth array element after the switch delays compensated by delay lines changes the array factor pattern, which means that beamforming is likely to be realized conveniently. The method of embedding phase shift δ into $\mathbf{D}(f)$ and $\mathbf{T}(f)$ is similar to the step 6 in Sect. 2.3, and we have an expression similar to (3):

$$A_{e,k}(f) = g_{e,k}(f)\exp(-j(2\pi\tau_{e,k}f + (k-1)\delta) + j\varphi) \qquad (4)$$

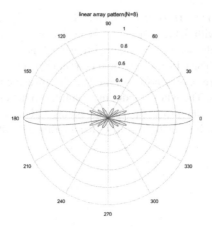

Fig. 3. Array factor patterns with δ = 0.

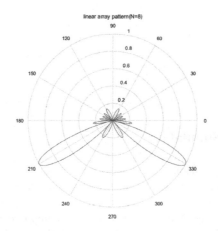

Fig. 4. Array factor patterns with δ = π/2.

Then following the steps in Sect. 2.3, simulations were conducted. The configuration of simulations is described as follows: 3 Tx antennas, 20 Rx antennas, frequency band of 1.74–1.76 GHz with 51 frequency points, the interval of Tx antennas is $\lambda/2$ (λ is the wave length of the carrier), while the interval of Rx antennas is $\lambda/4$, and the relationship of positions of the Tx antennas and Rx antennas is shown in Fig. 6, the switch interval $\tau = 57$ ns (the value of τ can be set according to the actual radio-frequency switch). PDPs are obtained easily from the graph modeling transfer function $H(f)$ or $\mathbf{H}'(f)$, Fig. 7 and 8 show the PDPs acquired from the simulation that has delays caused by radio-frequency switches. Figure 7 is the PDP about the signal component from each Tx antenna to each Rx antenna including 60 curves, and Fig. 8 shows the PDP of the signal component from each Tx antenna to all Rx antennas including 3 curves. It is obvious that main peaks in the PDPs appear respectively at different positions about 50 ns, 100 ns and 150 ns, which means there are fixed delays existing

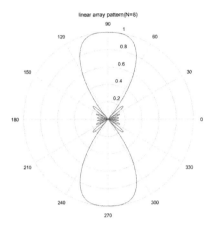

Fig. 5. Array factor patterns with $\delta = \pi$

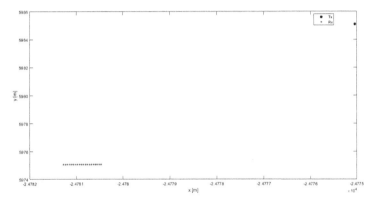

Fig. 6. Positions of Tx antennas and Rx antennas.

in signals transmitted by different Tx antennas. By introducing the delay lines, the fixed time delay of transmitting signals can be compensated. The results of the simulation with delay lines are shown in Fig. 9 and 10, where Fig. 9 describes the PDP of the signal component from each Tx antenna to each Rx antenna and Fig. 10 illustrates the PDP of the signal component from each Tx antenna to all Rx antennas. Both Fig. 9 and 10 display the coincident main peaks at about 50 ns, which reflects the effect brought by delay lines.

For validating the effect of beamforming, phase shift $(n - 1)\delta$ is embedded into the nth Tx antenna under the condition that switch delays are compensated using delay lines, and then MUSIC algorithm is used to estimate the DOA of the simulations. Here we only estimate the azimuth of arrival (AOA) ignoring the elevation of arrival for simplification purpose, and it is reasonable to be based on the premise that the Tx antennas and Rx antennas are almost at the same height. We set $\delta = \pi/2$, and according to the array factor pattern presented in Fig. 4 (the pattern of 3-element linear array we

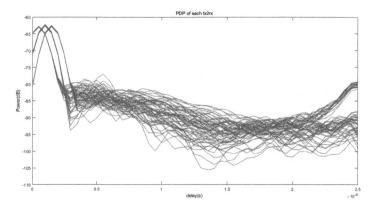

Fig. 7. PDP of the signal component of each Tx to each Rx with delays caused by switches.

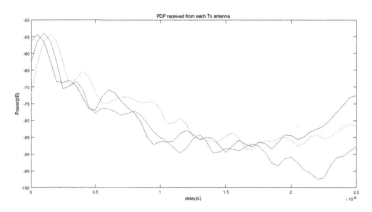

Fig. 8. PDP of the signal component from each Tx antenna with delays caused by switches.

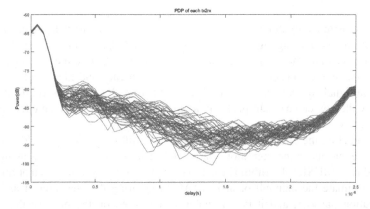

Fig. 9. PDP of the signal component of each Tx to each Rx when delays compensated by delay lines.

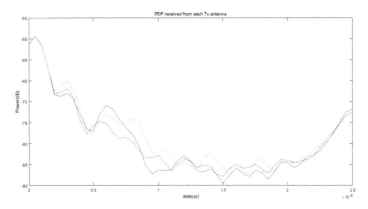

Fig. 10. PDP of the signal component from each Tx antenna when delays compensated by delay lines.

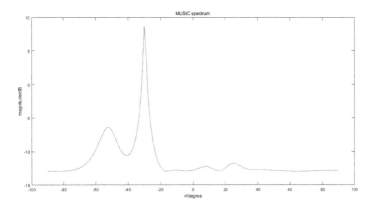

Fig. 11. MUSIC spectrum with delay lines and phase shifter.

adopted is similar) and the positions of Tx and Rx, the AOA is about −30° or −150° theoretically. The estimate results are shown in Fig. 11. We can see that a high peak appears at about −30° in the MUSIC spectrum, which conforms to the theoretical result. Additionally, there is a little peak at about −55°. We consider it reasonable for which the peak is really low and the array pattern with $\delta = \pi/2$ shown in Fig. 4 has a side lobe at −60°. In some way, the simulation results preliminarily prove that the proposed MIMO measurement scheme is applicable.

4   Conclusion

In this paper, we have proposed a MIMO channel measurement system for wireless communication based on delay lines associating with switches. Compared to the traditional switched channel measurement system, the proposed measurement system changes the time-division-multiplexing sounding method adopted by the traditional system. Combining delay lines with switches realizes that multiple Tx antennas transmit signals simultaneously only with one channel sounder in a low-cost way, which also provides a convenient way to beamforming. Simulations for this measurement system based on propagation graph modeling have been implemented, and the results preliminarily show that this MIMO channel measurement system has an applicable potential to channel measurement.

References

1. Ngo, H.Q., Larsson, E.G., Marzetta, T.L.: Energy and spectral efficiency of very large multiuser MIMO systems. IEEE Trans. Commun. **61**(4), 1436–1449 (2013)
2. Naderpour, R., Vehmas, J., Nguyen, S., Järveläinen, J., Haneda, K.: Spatio-temporal channel sounding in a street canyon at 15, 28 and 60 GHz. In: 2016 IEEE 27th Annual International Symposium on Personal, Indoor, and Mobile Radio Communications (PIMRC), Valencia, pp. 1–6 (2016)
3. Ben-Dor, E., Rappaport, T.S., Qiao, Y., Lauffenburger, S.J.: Millimeter-wave 60 GHz outdoor and vehicle AOA propagation measurements using a broadband channel sounder. In: 2011 IEEE Global Telecommunications Conference-GLOBECOM 2011, Houston, TX, USA, pp. 1–6 (2011)
4. Hur, S., Cho, Y., Lee, J., Kang, N.-G., Park, J., Benn, H.: Synchronous channel sounder using horn antenna and indoor measurements on 28 GHz. In: 2014 IEEE International Black Sea Conference on Communications and Networking (BlackSeaCom), Odessa, pp. 83–87 (2014)
5. Kim, M., Takada, J., Konishi, Y.: Novel scalable MIMO channel sounding technique and measurement accuracy evaluation with transceiver impairments. IEEE Trans. Instrum. Measur. **61**(12), 3185–3197 (2012)
6. Papazian, P.B., Gentile, C., Remley, K.A., Senic, J., Golmie, N.: A radio channel sounder for mobile millimeter-wave communications: system implementation and measurement assessment. IEEE Trans. Microwave Theory Tech. **64**(9), 2924–2932 (2016)
7. Salous, S., Feeney, S.M., Raimundo, X., Cheema, A.A.: Wideband MIMO channel sounder for radio measurements in the 60 GHz band. IEEE Trans. Wireless Commun. **15**(4), 2825–2832 (2016)
8. Müller, R., et al.: Simultaneous multi-band channel sounding at mm-Wave frequencies. In: 2016 10th European Conference on Antennas and Propagation (EuCAP), Davos, pp. 1–5 (2016)
9. Tian, L., Yin, X., Zuo, Q., Zhou, J., Zhong, Z., Lu, S.X.: Channel modeling based on random propagation graphs for high speed railway scenarios. In: 2012 IEEE 23rd International Symposium on Personal, Indoor and Mobile Radio Communications-(PIMRC), Sydney, NSW, pp. 1746–1750 (2012)
10. Tian, L., Yin, X., Zhou, X., Zuo, Q.: Spatial cross-correlation modeling for propagation channels in indoor distributed antenna systems. EURASIP J. Wireless Commun. Netw. **2013** (1), 1–11 (2013). https://doi.org/10.1186/1687-1499-2013-183
11. Tian, L., Degli-Esposti, V., Vitucci, E.M., Yin, X., Mani, F., Lu, S.X.: Semi-deterministic modeling of diffuse scattering component based on propagation graph theory. In: 2014 IEEE 25th Annual International Symposium on Personal, Indoor, and Mobile Radio Communication (PIMRC), Washington, DC, pp. 155–160 (2014)

12. Pedersen, T., Fleury, B.H.: Radio channel modelling using stochastic propagation graphs. In: 2007 IEEE International Conference on Communications, Glasgow, pp. 2733–2738 (2007)
13. Pedersen, T., Steinbock, G., Fleury, B.H.: Modeling of reverberant radio channels using propagation graphs. IEEE Trans. Antennas Propag. **60**(12), 5978–5988 (2012)
14. Tian, L., Degli-Esposti, V., Vitucci, E.M., Yin, X.: Semi-deterministic radio channel modeling based on graph theory and ray-tracing. IEEE Trans. Antennas Propag. **64**(6), 2475–2486 (2016)

A DNN-based WiFi-RSSI Indoor Localization Method in IoT

Bing Jia[1], Zhaopeng Zong[1], Baoqi Huang[1(✉)], and Thar Baker[2]

[1] Inner Mongolia University, Hohhot 010021, China
{jiabing,cshbq}@imu.edu.cn, 31809008@mail.imu.edu.cn
[2] SLiverpool John Moores University, Liverpool, UK
t.baker@ljmu.ac.uk

Abstract. Indoor automatic localization technology is very important for the Internet of Things. With the development of wireless technology and the diversification of location service requirements, especially in complex indoor scenarios, users are increasingly demanding location-based services. Traditional Global Positioning System (GPS) location technology is difficult to solve some positioning problems in indoor environments, and WiFi is now available in most indoor environments. Therefore, using WiFi for positioning does not require additional deployment of hardware devices, which is a very cost-effective method. However, WiFi-based indoor positioning requires a large amount of data, so we can use artificial intelligence methods to analyze the data and obtain a positioning model. The traditional indoor positioning methods based on WiFi signals have some problems such as long positioning time and poor accuracy. In order to solve the above problems, this paper proposes an indoor localization method based on Deep Neural Networks (DNN) for WiFi fingerprint. In particular, a DNN-based WiFi-RSSI positioning method is proposed for indoor automatic localization. Besides, in the process of DNN training, a joint training method based on unsupervised learning and supervised learning is adopted and the special loss function is defined. Extensive experiments are carried out in both the UJIIndoor-Loc public database and a real scenario, and a thorough comparison with several existing approaches indicates that the proposed scheme improves the localization accuracy on average.

Keywords: Indoor localization · Deep neural networks · WiFi-RSSI

1 Introduction

Location information is the basis of various Internet of Things (IoT) application systems to achieve service functions. It is an important issue to obtain location information through various location technology. Location issues include

Thanks to the National Natural Science Foundation of China (Grants No. 41761086 and 41871363,), the Natural Science Foundation of Inner Mongolia Autonomous Region of China (Grant No. 2017JQ09), and the Grassland Elite Project of the Inner Mongolia Autonomous Region (Grant No. CYYC5016).

H. Gao et al. (Eds.): ChinaCom 2020, LNICST 352, pp. 200–211, 2021.
https://doi.org/10.1007/978-3-030-67720-6_14

both outdoor localization and indoor localization. Outdoor localization has been widely used, especially Global Positioning System (GPS), which can enable people to travel the world freely. Traditional GPS location technology is difficult to solve some positioning problems in indoor environment due to a lack of line of sight (LoS) transmission channels between satellites and indoor receivers. With the booming development of IoT industry in recent years, the demand for indoor positioning technology has been increased by the various IoT application scenarios have greatly increased. Great efforts have been devoted to developing Indoor Positioning Systems (IPSs) to enable reliable and precise indoor positioning. Nowadays, with the popularity of the Internet, there are many WiFi signals around people. By measuring the received signal strength indicator (RSSI) of the WiFi signal, makes the indoor localization based on RSSI possible. In general, the WiFi-based IPS may estimate the position based on Time of Arrive (TOA) [4], Time Difference of Arrival (TDOA) [3] or Arrival of Angle (AOA) [16]. However, additional specialized equipments must be required to measure the round trip or angle of the WiFi, because the ordinary signal receiving equipment is not accurate enough. In contrast, fingerprint-based methods [1,15] do not require special device and are therefore easier to implement.

The original fingerprint-based positioning system [1] used K Nearest Neighbors (KNN) to find the closest match from the fingerprint database. In order to improve the robustness of the positioning system, a Bayesian-based [6,8,12] filtering method is proposed. Subsequently, the RSSI samples are associated with the fingerprint database by using Support Vector Machine (SVM) [15] and Compressed Sensing (CS) [5,10]. In order to reduce the burden of collecting fingerprint database, Pan et al. [11] proposed transfer learning. Recently, Li et al. [9] proposed a method of shunting short-term memory to solve the problem of location in a dynamic environment. De et al. [13] proposed a multi-hop approach to solve the problem of precise positioning.

However, due to multipath fading and occlusion of objects, the wireless signal has fluctuations, which in turn leads to changes in RSSI, which is a major problem for the positioning accuracy of the fingerprint-based. Even at the same location, the RSSI is different for different time periods. For more precise positioning, we need to collect as much Access Points (AP) information as possible, especially when the environment is particularly large. Therefore, the data of the fingerprint database will also become larger and larger. The challenge of the WiFi-based IPS is how to extract valid features from a large amount of data and find the best match.

The DNN has achieved significant success in solving many similar problems. In the fingerprint method, a machine learning method can be used to extract the main features of the signal. These features are treated as fingerprints and stored in a database. The fingerprint method can not only improve the positioning accuracy, but also reduce the computational complexity. This paper proposes an $8 - layers$ DNN model, which consists of 7 fully connected hidden layers and an output layer to solve indoor positioning problems.

The rest of the paper is organized as follows. Section 2 introduces a general structure of DNN to predict the position, which is a $8 - layers$ DNN model; The WIFI-RSSI positioning model based on DNN is described in Sect. 3. Section 4 gives the WIFI-RSSI database description and the result of experiments. Finally, We conclude this paper and shed light on future works in Sect. 5.

2   The Structure of DNN for Localization

DNN is very effective in the classification and regression of hidden feature extraction, and has been widely used in visual recognition and artificial intelligence in recent years. The popularity of DNN is attributed to the improvement of Graphic Processing Unit (GPU) processing power and the emergence of advanced DNN libraries, which simplify the realization of complex ideas. We can use deep neural networks to model the dependencies between location coordinates and WiFi fingerprints. In this paper, we propose an localization method based on $8 - layers$ DNN to predict the position.

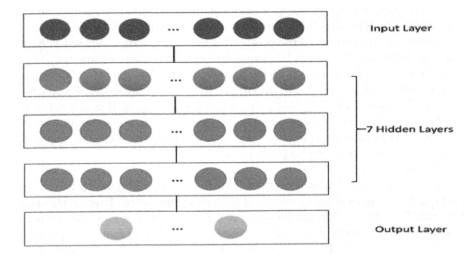

Fig. 1. The structure of the DNN for localization.

We use a fully connected DNN to implement the positioning task with WiFi fingerprint, as shown in Fig. 1. The DNN model has 7 fully connected hidden layers, as well as an input layer and an output layer. Each hidden layer has $k_1, k_2, ..., k_i, ..., k_7$ neurons. The input layer of the DNN is the WiFi signal value $R = (r_1, r_2, ..., r_n)$, and the output layer of the DNN is the predicted coordinates.

3   WiFi-RSSI Localization Model Based on DNN

3.1   The Structure of WiFi-RSSI Localization Model

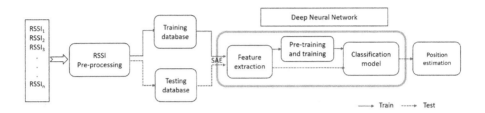

Fig. 2. The structure of WiFi-RSSI localization model.

The localization model, as shown in Fig. 2. The model consists of two parts: the offline-training phase and the online-localization phase. During the offline-training phase, the data of the fingerprint database is used to put into the $8 - layers$ of DNN for training. The training of the DNN is to determine the structure of the deep neural network and the weight of each neuron. First performing feature extraction on the data. Then performing pre-training to initialize network parameters. After a series of fine-turning and dropout, a classification model is finally obtained. In the online-localization phase, we use the testing database to evaluate the performance of the model. First, the WiFi-RSSI data needs preprocessing. Then, inputting the processed data into the trained WiFi-RSSI localization model, and finally outputting the current position coordinates and calculating the Average Localization Errors.

In the localization model, we use Pandas to read the data, Scikit-learn and Numpy to normalize and format the data, and the data flow diagram of Tensor Flow to calculate and update the parameters.

3.2   Encoding and Decoding Based on SAE

To reduce the dimensionality of the WiFi-RSSI data, we used Stacked AutoEncoders(SAE) [2], as shown in Fig. 3. Before encoding, we need to regularize the WiFi-RSSI data. Three hidden layers are used here with dimensions of 256, 128, and 64, and the feature dimensions can be continuously reduced. Finally, a 64-dimensional feature is obtained. This feature is used to perform reverse reconstruction to obtain the reconstructed feature. The reconstructed feature is the same as the original feature. By SAE, we can obtain data that contains many detailed and valuable feature information than the raw data, which is used to train classifiers with specific contexts and is more accurate than training with raw data.

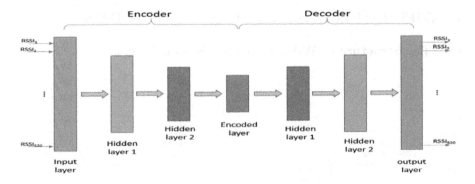

Fig. 3. Stacked AutoEncoder used in DNN.

3.3   The Selection of Activation Function

The activation function introduces non-linearity into the deep neural network so that the deep neural network can have hierarchical non-linear mapping learning ability, which is also an important factor affecting performance. So the activation function is an integral part of the neural network. In the localization model, we use the softmax function [7] as the activation function, the formula is as follows:

$$S_i = \frac{e^i}{\sum_j^n e^j} \tag{1}$$

where, S_i represents the probability that the current output belongs to i, e^i represents the i-th power of e.

3.4   The Definition of Loss Function

We use the loss function based on Back Propagation (BP) to train the DNN weights. In order to obtain better positioning results, it is necessary to continuously reduce the difference between the real coordinates and the predicted coordinates. So we redefine the loss function to calculate the difference between the real coordinates and the output of the DNN model. The formula is as follows:

$$f_{loss} = \sqrt{(x_p - x_t)^2 + (y_p - y_t)^2} \tag{2}$$

where (x_t, y_t) is the real coordinate of reference point sample i, and (x_p, y_p) is the estimated coordinates of reference point sample i. By using the BP algorithm to minimize the value of the loss function f_{loss}, the gradient descent algorithm is used to update the DNN weights until the value of f_{loss} converges.

3.5   Unsupervised Pre-training and Supervised Training

In order to solve the network parameter contribution is small and the update speed is slow, we use unsupervised pre-training. Supervised pre-training uses layer-by-layer greedy training strategy, that is, training the first hidden layer of the network, and then training the second hidden layer, until the last hidden layer is trained. Finally, we use these trained network parameter values as the initial values of the overall network parameters.

The input of the input layer is the n-dimensional preprocessed WiFi-RSSI vector $R_i = (r_1, r_2, ..., r_n)$ and the label L_i. First, the signal value is encoded and decoded through SAE to obtain the representation of the WiFi-RSSI data and reduce the dimensionality of the WiFi-RSSI data. In this way, the results are more accurate than training with raw data. Then, unsupervised pre-training is adopted. After the pre-training is completed, the obtained parameters are used as initial values of the entire network parameters. Finally, supervised training is performed. Using the BP algorithm and the gradient descent algorithm to fine-tuning continuously and obtain the optimal network parameters.

4 Experiments

In this section, extensive experiments have been done to verify the performance of WiFi-RSSI localization model.

4.1 Experimental Data

In order to test and evaluate the performance of the system, a large database containing location tags is required. In this experiment, two databases of self-collected laboratory database and UJIIndoorLoc [14] public database were used for training and testing. The main difference between the two databases is that the self-collected database is a dense database, and the reference points are uniformly distributed; the UJIIndoorLoc public database is a sparse database, the reference points are unevenly distributed, some places are dense, and some places are sparse.

The Self-collected Laboratory Database. The experimental data was collected at the 316 laboratory on the 3rd floor of the School of Computer Science, Inner Mongolia University. The laboratory covers an area of about 78 square meters(the length is 13.5 m, the width is 6 m).

The laboratory data is collected at a sampling interval of 1 m. First, taking the position of the door as the coordinate origin, then dividing the coordinates, from (1,1) to (5,13), a total of 65 points are setting as shown in Fig. 4.

Next, We use mobile phones to collect data at each point. The collecting data from 65 coordinate points as training data; then, at non-integer coordinate points, we randomly select some points to collect data as testing data. Each record of the collected data includes 5 attributes, WiFi name, Media Access Control (MAC) address, RSSI levels, X and Y. Some examples of the collected data are shown in the Table 1.

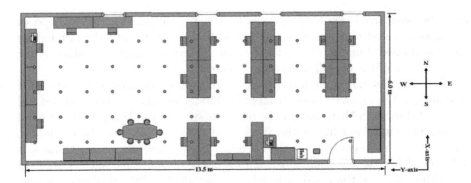

Fig. 4. The coordinate points of 316 laboratory.

Table 1. Some examples of 316 laboratory collected data.

WiFi name	MAC address	RSSI levels	X	Y
IMUDGES Pro	d4:a1:48:a4:87:3c	−83	1	1
network	f4:83:cd:91:16:62	−82	2	1
NETGEAR	00:0f:b5:35:32:a4	−48	3	1

The collected data is processed to establish a fingerprint database. Training database with a total of 10,339 sampled points, each of which includes 84 wireless access points, coordinate points (X, Y), and testing data with 6337 sampled points, the attributes of 316 laboratory database as shown in Table 2.

Table 2. Some examples of 316 laboratory collected data.

Index	Attribute
01–84	RSSI levels
85	X
86	Y

The UJIIndoorLoc Database. The UJIIndoorLoc database covers a surface of 108703 m² including 3 buildings with 4 or 5 floors depending on the building. The number of different places(reference points) appearing in the database is 933. The map of the UJI Riu Sec Campus, as shown in Fig. 5.

The database consists of 21049 sampled points, of which 19938 are used for training and 1111 are used for testing. Each record in the database contains 529 attributes, as shown in the Table 3, the first 520 attributes inform about the RSSI levels of these 520 wireless access points. The remaining 9 attributes contain longitude and latitude of measurement, floor number, building ID, space ID, relative position, user ID, phone ID and timestamp of measurement.

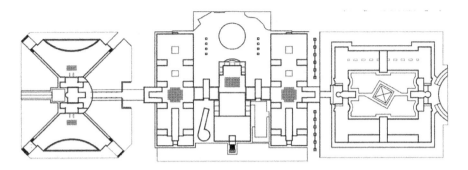

Fig. 5. The map of UJI Riu Sec Campus.

Table 3. The attributes of UJIIndoorLoc database.

Index	Attributes
001–520	RSSI levels
521–523	Real world coordinates of the sample points
524	BuildingID
525	SpaceID
526	Relative position with respect to SpaceID
527	UserID
528	PhoneID
529	Timestamp

To ensure the independence of the database, a validation (or testing) samples are obtained 4 months after the training samples. Therefore, using UJIIndoorLoc database for positioning is a challenge, but the results obtained can be used to estimate the actual performance of the system. For the use of UJIIndoorLoc public database, some methods are used to locate the building and the floor. The method proposed in this paper is used to locate coordinate points.

4.2 Setup

The localization system runs on a PC with the windows 10(64 bit) operating system, which CPU is i7-4690 and memory is 8 GB, the development tool is PyCharm.

This experiment uses Average Localization Errors to measure the performance of the positioning. First, we use the Euclidean distance to calculate the distance between two points, the formula is as follow.

$$d(i, k) = \sqrt{(x_i - x_k)^2 + (y_i - y_k)^2}; \tag{3}$$

where i, k represent the real point and the predicted point, respectively.

Then we sum the distances of all the points and calculate the mean $\bar{d}$ as the average localization errors. The smaller the average localization errors, the better the positioning performance.

$$\bar{d} = \frac{1}{n} \sum_{i=1}^{n} d(i, k) \tag{4}$$

4.3   The Comparison of 316 Laboratory

The experimental results of 316 are shown in the Table 4. The average localization errors of DNN is 2.7 m, and the average localization errors of other methods exceeds 3 m. The positioning effect of DNN is better than other methods. Since the database from the 316 laboratory was collected at a sampling interval of 1 m, the 316 laboratory database has a smaller coordinate range than the UJIIndoorLoc database. Therefore, the average localization errors calculated by these six methods is very close.

Table 4. The comparison of six methods in 316 laboratory.

Method	Average localization errors(m)
DNN	2.7
KNN	3.8
SVM	4.6
DecisionTree	3.4
RandomTree	4.85
ExtraTree	3.65

4.4   The Comparison of UJIIndoorLoc

The UJIIndoorLoc database includes 3 buildings, the building 0 and building 1 with 4 floors and the building 2 with 5 floors. We used the single floor and multiple floors data of three buildings to test the system performance respectively. The main difference between single floor and multiple floors is that single floor only uses the data of a certain floor without considering the influence of the height of the floor. Multiple floors uses the data of multiple floors, and the positioning result will be affected by the height of the floor.

Single Floor. The experimental results of the UJIIndoorLoc database are shown in the Fig. 6. The number 0 to 4 in the figure correspond to different floor of the building. In general, the DNN is relatively stable and has the best positioning performance. The average positioning errors corresponding to the three buildings are 7.53 m, 11.07 m and 10.98 m, respectively. Other methods have large fluctuations and large positioning errors.

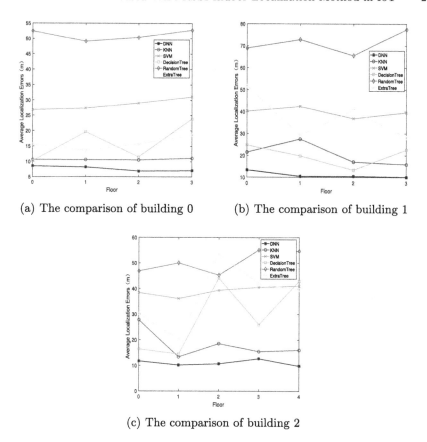

(a) The comparison of building 0   (b) The comparison of building 1

(c) The comparison of building 2

Fig. 6. The comparison of six methods for single floor in UJIIndoorLoc database.

Multiple Floors. The experimental results of the UJIIndoorLoc database are shown in the Fig. 7. The number from 1 to 5 in the figure correspond to the number of floors of the building in the UJIIndoorLoc database. In conclusion, the DNN is stable and has the best positioning performance. The minimum localization errors corresponding to the three buildings are 7.06 m, 10.09 m, and 9.81 meters, respectively. Other methods have large fluctuations and large localization errors.

The positioning effect of DNN is better than other methods. The reason for the high localization errors is include two parts: one is that the position of the UJIIndoorLoc data point is represented by the latitude and longitude of the map, and the value is very large; the second is that when the predicted result is not accurate, it will lead to a large distance between the predicted point and the real point.

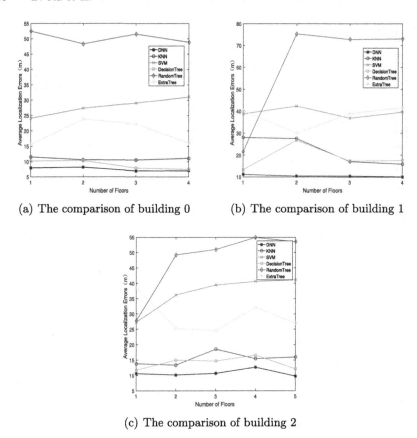

(a) The comparison of building 0   (b) The comparison of building 1

(c) The comparison of building 2

Fig. 7. The comparison of six methods for multiple floors in UJIIndoorLoc database.

5   Conclusion

This paper proposes an DNN-based WiFi-RSSI Indoor Localization Method. The average localization errors on the self-collected database is 2.7 m, which is better than the other five methods. The average localization errors obtained on the UJIIndoorLoc database is 9 m, which is also better than the other methods. Therefore, the DNN-based WiFi-RSSI Localization Method for Indoor Automatic Localization is accurate and adaptable. However, for the positioning of multiple buildings, further researched need to be conducted to improve positioning accuracy.

References

1. Bahl, P., Padmanabhan, V.N.: Radar: an in-building RF-based user location and tracking system. In: Infocom Nineteenth Joint Conference of the IEEE Computer & Communications Societies. IEEE (2000)
2. Bengio, Y., Lamblin, P., Popovici, D., Larochelle, H.: Greedy layer-wise training of deep networks. In: Advances in Neural Information Processing Systems, pp. 153–160 (2007)
3. Cheng, L., Wu, C.D., Zhang, Y.Z.: Indoor robot localization based on wireless sensor networks. IEEE Trans. Consum. Electron. 57(3), 1099–1104 (2011)
4. Ciurana, M., Barceló, F., Cugno, S.: Indoor tracking in WLAN location with TOA measurements. In: Proceedings of the 4th ACM International Workshop on Mobility Management and Wireless Access, pp. 121–125 (2006)
5. Feng, C., Au, W.S.A., Valaee, S., Tan, Z.: Received-signal-strength-based indoor positioning using compressive sensing. IEEE Trans. Mob. Comput. 11(12), 1983–1993 (2012)
6. Fox, D., Hightower, J., Liao, L., Schulz, D., Borriello, G.: Bayesian filtering for location estimation. Pervasive Comput. IEEE 2(3), 24–33 (2003)
7. Gao, B., Pavel, L.: On the properties of the softmax function with application in game theory and reinforcement learning. arXiv preprint arXiv:1704.00805 (2017)
8. Hightower, J., Borriello, G.: Particle filters for location estimation in ubiquitous computing: a case study. In: International Conference on Ubiquitous Computing (2004)
9. Li, Y., Shuai, L., Ge, Y.: A biologically inspired solution to simultaneous localization and consistent mapping in dynamic environments. Neurocomputing 104(104), 170–179 (2013)
10. Nikitaki, S., Tsakalides, P.: Localization in wireless networks via spatial sparsity. In: 2010 Conference Record of the Forty Fourth Asilomar Conference on Signals, Systems and Computers, pp. 236–239. IEEE (2010)
11. Pan, S.J., Zheng, V.W., Yang, Q., Hu, D.H.: Transfer learning for wifi-based indoor localization. In: Association for the Advancement of Artificial Intelligence (AAAI) Workshop, vol. 6. The Association for the Advancement of Artificial Intelligence Palo Alto (2008)
12. Paul, A.S., Wan, E.A.: Wi-fi based indoor localization and tracking using sigma-point kalman filtering methods. In: IEEE/ION Position, Location & Navigation Symposium (2008)
13. Sá, A.O.D., Nedjah, N., Mourelle, L.D.M.: Distributed efficient localization in swarm robotic systems using swarm intelligence algorithms. Neurocomputing 172(C), 322–336 (2016)
14. Torres-Sospedra, J., et al.: Ujiindoorloc: a new multi-building and multi-floor database for WLAN fingerprint-based indoor localization problems. In: 2014 International Conference on Indoor Positioning and Indoor Navigation (IPIN) (2014)
15. Wu, C.L., Fu, L.C., Lian, F.L.: WLAN location determination in e-home via support vector classification. In: IEEE International Conference on Networking (2004)
16. Xu, J., Ma, M., Law, C.L.: Cooperative angle-of-arrival position localization. Measurement 59, 302–313 (2015)

A Downlink Scheduling Algorithm Based on Network Slicing for 5G

Shanwei Wang[✉], Bing Xi, Zhizhong Zhang, and Bingguang Deng

School of Communication and Information Engineering, Chongqing University of Posts and Telecommunications, Chongqing, China
S180101165@stu.cqupt.edu.cn

Abstract. Current cellular mobile network should satisfy the service requirements of the User Equipment (UE) applications through Radio Resource Management (RRM) mechanisms as much as possible. In order to improve the resource utilization rate and Quality of Experience (QoE) for downlink Real-Time (RT) services in 5G system. In this paper, based on the Modified Largest Weighted Delay First (M-LWDF) algorithm, a slicing-oriented resource scheduling algorithm-S-MLWDF is proposed with using 5G network slicing technology. S-MLWDF takes RB groups as the basic units of RA (resource allocation) and takes slices as the allocation object. During the process of in-slice scheduling, on account of the Channel Quality Indication (CQI) obtained from Base Station (BS) feedback and the allocation of RBs over time, the generated weighting factor can guarantee the edge users to get equal scheduling opportunities. Meanwhile, the modified queue delay and HARQ retransmission packets delay can solve the problem of surge in Packet Loss Rate (PLR) near the delay threshold. The simulated results show that the performance of the proposed algorithm is better than the traditional scheduling algorithms. Especially compared with M-LWDF, the fairness and PLR of S-MLWDF are optimized by about 10% and 16.3%, which can better meet the needs of users.

Keywords: 5G · Network-slicing · S-MLWDF algorithm · Resource scheduling · Delay · Resource blocks allocation

1 Introduction

With the global commercial deployment of 5G, the Ministry of Industry and Information Technology (MIIT) issued the occupation permit of radio frequency to China Telecom, China Unicom and China Broadcast Network respectively in February 2020. This means that 5G network has entered the stage of large-scale deployment. In the process of mobile communication system evolving to the fifth generation, it is necessary to meet the great challenges of differentiated service requirements in multiple scenarios. The different requirements such as safety, mobility, transmission delay, instantaneous speed and so on in Enhanced Mobile Broadband (eMBB), Massive Machine Type Communication (mMTC) and Ultra-reliable and Low Latency Communication (uRLLC) should be met as possible. As it happens, network slicing technology can realize the customization capability through effective management of current wireless resources [1, 2].

© ICST Institute for Computer Sciences, Social Informatics and Telecommunications Engineering 2021
Published by Springer Nature Switzerland AG 2021. All Rights Reserved
H. Gao et al. (Eds.): ChinaCom 2020, LNICST 352, pp. 212–225, 2021.
https://doi.org/10.1007/978-3-030-67720-6_15

5G RRM focus on latency-sensitive applications and massive data. As a mechanism to provide Quality of Service (QoS) requirements and improve system throughput in multi-user networks, packet scheduling will meet strict requirements of delay and PLR. In order to meet the QoS demands for RT communication, various packet scheduling algorithms have been used to allocate limited frequency and time sources for all data transfer devices including mobile and wireless networks [3–5]. On the premise of 5G high-speed rate, high-capacity, low latency, high reliability demand, the limited external environment and public resources, the importance of scheduling algorithms to allocate resources reasonably for users is self-evident. Proportional Fair (PF), the classical scheduling algorithm, achieves the trade-off between throughput and user fairness in Non-Real-Time (NRT) services [6]. In Modified Largest Weight Delay First (M-LWDF) algorithm, the delay of packets is the main parameter [7]. However, when the number of users increased, the PLR of queues near delay threshold in buffer increased greatly. The two-stage downlink scheduling algorithm ensured fairness without reducing system throughput [8]. The authors of [9] proposed a downlink channel queue-aware scheduling algorithm based on service curve and overflow probability of buffer, which maximized throughput, provided lower latency and buffer overflow rate. The authors of [10] analyzed the buffer state of user queues which took delay as the weight, and proposed a scheduling scheme based on delay and QoS-aware to minimize the delay of RT traffic. An enhanced delay sensitive algorithm which is used to increase the energy efficiency, network lifetime and throughput was proposed [11]. However, when number of nodes increased, delay also increased. In addition, a new channel-aware scheduling algorithm for improving the cell edge throughput and fairness has been proposed [12], but the problem of PLR growth was neglected. At the same time, scholars at home and abroad have also done research on slice scheduling. A two-layer MAC scheduling framework was proposed in [13] to handle uplink and downlink transmission of network slices with different characteristics in RAN, which can meet the strict latency and reliability requirements in uRLLC. In [14], the authors proposed a strategy which prioritized slices for different service providers. Slices of RT service are scheduled first, and slices of Non-Real-Time (NRT) service are scheduled later. Their studies mainly focused on one aspect of performance Their studies mainly focused on one aspect of performance optimization, however, they did not comprehensively consider the impact of various parameters in 5G different services on RA.

Hence, in this paper, 5G network slicing technology is adopted to divide the wireless resource scheduling process into slice-level scheduling and in-slice resource scheduling. During in-slice scheduling, on the basis of M-LWDF algorithm, we improve the scheduling priority of queues which will approach the delay threshold, consider the current channel quality and RBs allocated to users in the previous period. Simulation results show that the proposed algorithm can keep satisfactory balance between fairness and throughput at the same time of maximizing resource utilization and reduce the PLR effectively.

The rest of the paper is organized as follows. In Sect. 2 the system model of packet scheduling for 5G system is provided, and the resource grid in the time-frequency domain is introduced. The proposed algorithm is discussed in Sect. 3. In Sect. 4 simulation results are shown to compare the proposed algorithm with the existing algorithms in terms of resource utilization, user fairness, system throughput and PLR. Finally, in Sect. 5 we draw our conclusion.

2 System Model

As the smallest unit of resources that can be allocated to a user [15], RBs can form a time-frequency resource grid to represent downlink physical resources. In 5G system, Resource Element (RE) is the basic unit of resource mapping in physical layer. A RE consists of an OFDM symbol in the time domain and a subcarrier in the frequency domain, an RB is composed of OFDM symbols and 12 subcarriers. The length of the RB in time is called the Transmission Time Interval (TTI), all frequency blocks at a given TTI are called a subframe [16]. When the resource blocks take different μ values, it can meet the data transmission rate and throughput rate required by the system for different scenes. The time-frequency wireless resource grid is shown in Fig. 1.

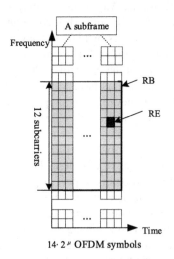

Fig. 1. Wireless resource grid in the time-frequency domain

The MAC scheduler is the controlling entity for multi-users radio RA, which assigns shared physical resources to different users in the cell by prioritizing them. The 5G PHY layer offers a large set of new options for the MAC scheduler, which enable significant improvements for efficiently multiplexing users with highly diverse service requirements [17]. When a service needs to send downlink data, for different service and users, the queue pass through the classifier from the upper layer to different buffer on the base station side. By now the MAC scheduler implements resource scheduling process by accepting CQI reported by UEs and selecting scheduling algorithms (Fig. 2).

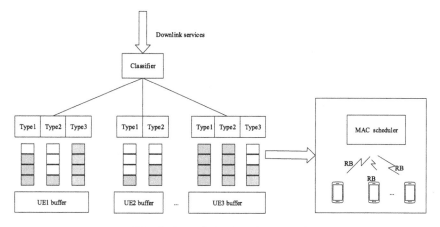

Fig. 2. Downlink scheduling model

During the period of downlink resource scheduling, BS timely adjusts and updates the channel state information of users according to the feedback CQI. Then the Adaptive Modulation and Coding (AMC) module selects the right Modulation and Coding Scheme (MCS) within a given Block Error Rate (BLER), according to the CQI reported, in order to maximize the throughput [18].

3   S-MLWDF Algorithm

Users in different slices in 5G network slicing architecture have different requirements for various services, which makes each user has different priorities in resource scheduling. Therefore, it is necessary to allocate resources adequately according to the differentiated demands of users in slices to ensure the QoE of users in wireless resource scheduling. Based on this, the proposed algorithm includes slice-level scheduling process and in-slice user resource scheduling process.

3.1   Slice-Level Resource Block Group Scheduling

Within a TTI, resources in current time-slot are divided into several RB groups according to the number of candidates, and slices are selected according to the priority order of each slice on each RB group for scheduling until the whole of RB groups are allocated.

We assume that the number of scheduled candidates in current TTI is $I(t)$, the total number of RBs available is R, then the number of an RB groups at time t is $c = \max\left(1, \frac{R}{I(t)}\right)$. Considering that the three scenarios in 5G have different service requirements, we propose the priority metric of slice a in the b-th RB group.

$$m_k = \frac{\sum_{n=1}^{c} \sum_{k=1}^{I_a(t)} (\alpha \cdot r_{m,k}(t) + \beta \cdot W_k(t))}{c} \tag{1}$$

Where m_k is the metric of k-th user, $I_a(t)$ is the number of candidates in slice a, $r_{m,k}(t)$ is the instantaneous transmission rate of the k-th user on the m-th RB. $W_k(t)$ is the delay of the Head of Line (HOL) packet, α, β are transmission rate and weight factor of queue delay respectively, and $\alpha + \beta = 1$. The in-slice scheduling process is as follows:

Algorithm 1 Slice-level Resource Block Group Scheduling

1: **Input:** $T, R, I(t), W_k(t), r_{m,k}(t), \tau_k$;

2: **Output:** m_k

3: **while** $t < T$ **do**

4:    **for** $k = 20, 30, 40, \ldots, 100$ **do**

5:      $c = R / I(t)$;

6:      Calculate the priority of slices m_k on each RB group;

7:      **while** c

8:        Assign RB group to the a-th slice;

9:      **end while**

10: **end for**

11: **end while**

3.2   In-Slice User Scheduling

After the slice-level process, the slices with the highest priority are allocated a set of RBs. At this time, each slice can be regarded as a cell, and the RB group obtained through the slice-level process is used to conduct resource scheduling for users in the corresponding slice. In scheduling process of traditional cells, the M-LWDF is a channel-aware extension of PF and provides bounded packet delivering delay [19]. This algorithm is dedicated for RT services. It can ensure balance between spectrum efficiency, system fairness and QoS. M-LWDF metric can be easily expressed as in Eq. (2).

$$m_k = \arg\max_k \left[\alpha_k W_k(t) \frac{r_k(t)}{\overline{R_k(t)}} \right] \tag{2}$$

Where $r_k(t)$ is the instantaneous transmission rate of user k at time t, the factor $\overline{R_k(t)}$, that represents the past average throughput experienced by the k-th user at time t, and α_k weights metric so that k-th RT user with the most pressing needs in terms of delay threshold and acceptable loss rate, it can be given by

$$\alpha_k = \frac{-\log \sigma_k}{\tau_k} \tag{3}$$

where σ_k is QoS parameter of k-th user, and τ_k is the delay threshold for k-th user.

M-LWDF algorithm is an important guarantee to meet QoS requirements in RT service. The longer a packet waits in the current queue, the higher its priority will be. Nevertheless, packets delay in M-LWDF does not have a significant role. If a user cannot get a scheduling opportunity, a packet will be dropped from the queue of buffer due to deadline expiration. Therefore, on the basis of M-LWDF algorithm, an algorithm proposed in this paper solves the problems of packet loss caused by delay and unfair scheduling opportunity of edge users. An outline of in-slice scheduling as shown in Algorithm 2.

Algorithm 2 In-slice Scheduling

/*Simulation duration T, Number of UEs N_{UE}, Amount of RBs in a-th slice S_{RB} */;

1: **Input:** T, N_{UE}, S_{RB}, $W_k(t)$, τ_k, a;

2: **Output:** m_k

3: **while** $t < T$ **do**

4:    **for** $k = 20, 30, ..., 100$ **do**

5:       Update the priority of k-th UE;

6:       Get the CQI value of UEs feedback;

7:       Calculate $\overline{R_k(t)}$, τ_k, $W_k(t)$ and $r_k(t)$ within t_c;

8:       Calculate $\text{mod\_}W_k(t)$;

9:       Determine HARQ priority factor θ;

10:      Calculate the weighted factor γ and the predicted RBs N_{i+1};

11:      Calculate the priority m_k and allocate RBs;

12:    Delete the allocated RBs from RB group and determine whether the RB group is empty. If so, go to the next TTI; if not, go back to the previous step.

13:    **end for**

14: **end while**

Congestion and queuing are the main causes of packet loss in 5G video service. So, reducing PLR is one of the problems that the proposed algorithm must solve. The packet loss caused by too much queuing delay of users can be replaced by sigmoid function in the original algorithm. The modified queuing delay $\text{mod\_}W_k(t)$ is:

$$\text{mod\_}W_k(t) = \frac{1}{1 + \exp[-\psi(W_k - \tau_k)]} \tag{4}$$

Where ψ is an adjustable threshold factor with the range of $(0, 1)$, which determines the degree of inclination of sigmoid function. The bigger ψ is, the steeper of the function is and the more sensitive the delay is. As can be seen from Fig. 3, with the increase of queue delay, in Eq. (4), users have higher satisfaction with delay. In other words, when the queue delay is close to the maximum delay the user can tolerate (τ_k), it can greatly improve the urgency of the packet, the priority of k-th user, and reduce PLR.

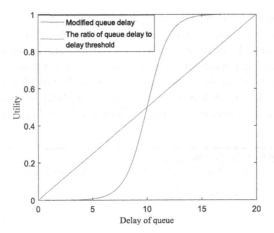

Fig. 3. The utility of W_k/τ_k and $mod\_W_k(t)$

In addition to queuing delay, transmission delay in HARQ process also affect the priority of users. After the terminal receives downlink data, the MAC layer judges whether the data is retransmitted according to the New Data Indicator (NDI) sent by BS. Practically, retransmission packets priority are greater than the new transmission packets in order not to undermine the resources already used in the initial transmission of the packets [20]. Also, the delay of the retransmission packet is larger, so for meeting the demand of 5G low latency, such retransmitted data must be prioritized and scheduled. At this point, a HARQ priority factor θ is introduced to improve the priority of retransmitted data.

$$\theta = \begin{cases} 1, new \\ 2, retransmission \end{cases} \tag{5}$$

The factor θ is used to distinguish whether new or retransmitted packets received due to errors in the previous transmitted packets. In this paper, parameter θ is set as 1 for the new transmission packets and 2 for the retransmitted packets.

Since M-LWDF algorithm continuously schedules a user when current channel quality is great, other users cannot get fair opportunities. Therefore, a weighted factor γ is introduced to influence the fairness of users by combining the current channel quality and the state of RBs allocation in previous time.

$$\gamma_k(t) = \begin{cases} 1 - \frac{S_i}{N'_{i+1}}, CQI_k(t) \geq 10 \\ \exp\left(1 - \frac{S_i}{N'_{i+1}}\right), else \end{cases} \tag{6}$$

where $CQI_k(t)$ represents CQI value feedback from k-th user at the moment, S_i is the number of RBs allocated to k-th user in i-th time-slot, and N'_{i+1} is the number of available RBs predicted in the next slot. If equation $CQI_k(t) \geq 10$ is true, k-th user will be scheduled within several consecutive TTIs. At this point, the RBs allocation in the previous slot should be taken into account to influence the priority of such user. When the channel quality is poor, users at the edge of cell will not be scheduled for a long time. In order to avoid "starvation", an exponential function should be used to appropriately improve the priority of the users on the premise of considering the RA over time. With the increase of RBs allocation, the growth rate of weighting factor becomes slower. In other words, if the current channel quality is poor, the more RBs allocated in previous slot, the less priority of users will be improved. Conversely, the fewer resource blocks allocated, the greater the chance that users will be scheduled. Consequently, we use Bayes' theorem to calculate the probability of users in various states, and then the maximum likelihood estimation method is used to find the possible values of available resource blocks in the current slot. In this paper, the service status of the k-th user is represented by $s_k \cdot s_k = 0$ indicates that the user hasn't been allocated RBs, and $s_k = 1$ indicates that the user has been scheduled successfully. The conditional probability formula for the state of a user is:

$$P(s_k|N_i) = \begin{cases} \left(1 - \frac{1}{N_u}\right)^{N_i}, s_k = 0 \\ 1 - \left(1 - \frac{1}{N_u}\right)^{N_i}, s_k = 1 \end{cases} \tag{7}$$

Where N_u is the number of candidates which will be scheduled, and N_i is the number of available RBs in time t. Imagine there are p_1 users with $s_k = 0$, and p_2 users with $s_k = 1$, the user state probability is given by Eq. (8).

$$P(s|N_i) = P(s_k = 0|N_i)^{p_1} \cdot P(s_k = 1|N_i)^{p_2} \tag{8}$$

When the maximum value N_i of Eq. (8) is obtained, the estimated value N_i' of the number of available RBs is also clear.

$$N_i' = \arg\max \prod_{k=0}^{1} P(s|N_i) \tag{9}$$

The logarithm of both the left and right sides of Eq. (9) can be obtained as follows:

$$\ln N_i' = \arg\max \sum_{k=0}^{1} P(s|N_i) \tag{10}$$

Equation (11) can be obtained by deriving Eq. (10). This function is convex, so when the derivative is 0, N_i is the maximum value obtained by Eq. (10), namely N_i'.

$$\frac{\partial}{\partial N_i}\left(\sum_{k=0}^{1} \ln P(s|N_i)\right) = 0 \tag{11}$$

The number of RBs successfully allocated to users in the i-th slot is denoted as S_i. After the number of available resource blocks are obtained according to the scheduled state of users in i-th slot, the prediction of available RBs in next slot can be calculated by Eq. (12).

$$N'_{i+1} = N'_i - S_i \qquad (12)$$

To sum up, the priority of proposed scheduling algorithm is:

$$m_k = \arg\max_k \left\{ \alpha_k \bmod \_W_k(t) \frac{r_k^\theta(t)}{R_k(t)} \gamma_k(t) \right\} \qquad (13)$$

4 Numerical Results and Analysis

4.1 Simulation Parameters

We consider a wireless system with pedestrian and high moving users in a cell with radius of 500 m [21]. It is also assumed that the number of eMBB, mMTC and uRLLC users in the target cell is the same. Due to the random distribution of BS in a cell and users keep moving, the channel quality of each user is different, so the environment set in this paper is closer to the actual system (Table 1).

Table 1. Simulation parameters

Parameter	Value
Simulation duration	100TTIs
Number of UE	20, 30, 40…
TTI	1 ms
Radius of cell	500 m
Delay threshold	50 ms
Bandwidth	20 MHz
Subcarrier spacing	15 kHz
Number of RBs	100
Delay threshold factor ψ	0.9
Transmission power of UE	0.1 W

4.2 Simulation Results and Analysis

The algorithm proposed in this paper is to calculate the scheduling priority of users in N TTIs. The time complexity of a user's priority within a TTI is $O(n)$, so the time complexity of S-MLWDF algorithm is $O(n^2)$. The same is true for traditional PF and M-LWDF algorithms. Figure 4, Fig. 5, Fig. 6 and Fig. 7 depict the performance of the algorithm for resource utilization, fairness, throughput and PLR with the number of users increase from 20 to 100.

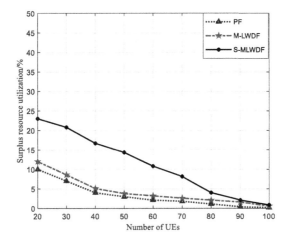

Fig. 4. Surplus resource utilization

Figure 4 presents the utilization of residual resource in the system for these three algorithm strategies. As shown in the simulation, the remaining resource utilization of three algorithms all decreases with the increase of the access users, but there is not much difference between PF and M-LWDF. This is because 5G slicing technology allocates RB groups for slices with the highest priority according to the actual needs in different services, so that the available RBs in each slice can meet the users' needs to the maximum extent. With the increase of users, the resources in the system are severely depleted, so the utilization of surplus resources in these three algorithms tend to be similar. In general, compared with PF and M-LWDF algorithms, S-MLWDF algorithm improves the residual resource utilization by about 8% and 7% on average.

Figure 5 shows the fairness index between users. With the increase of the users, the fairness of PF algorithm and M-LWDF algorithm decreases sharply, while the proposed improved algorithm S-MLWDF performs better than PF and M-LWDF with aggregate percentage is increased by 13% and 10%. Because users are distributed in different locations in cell, the distance from BS is random, and the channel quality is different. When the number of users increases, such users who are at the edge of cell are given priority in the process of in-slice scheduling for take care of them. It can ensure the fairness of scheduling users with poor channel quality or serious queuing delay.

The throughput is presented in Fig. 6. It can be seen that the throughput of three algorithms is proportional to the number of users. Compared with the other two algorithms, S-MLWDF algorithm improves the system throughput by about 33% and 27%. However, as the number of users of the network continue to increase, the RBs available in the system are not sufficient to support all service requirements. By now the network slicing technology can maximize the resource utilization by grouping RBs according to the different demands of access services, so as to make effective use of the empty space resources in the cell. Therefore, slicing-oriented scheduling algorithm is more suitable for the actual scene.

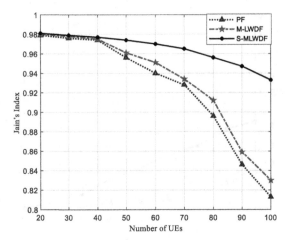

Fig. 5. Fairness

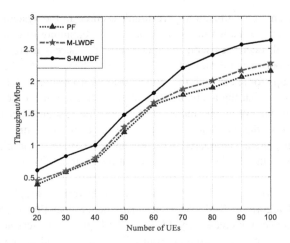

Fig. 6. Throughput

The PLR for all users is shown in Fig. 7. The lower the PLR for each algorithm, the better the performance of the algorithm in terms of PLR. The results show that the PLR of S-MLWDF algorithm is about 22% and 16.3% lower than PF and M-LWDF algorithm, respectively. As more and more users applying for accessing the system, the queuing delay of users increases. When delay approaches the threshold, the corresponding packets will be discarded. Meanwhile, the transmission delay of HARQ retransmission packets will also increase dramatically in PLR. The process of in-slice scheduling in S-MLWDF greatly improves the urgency of the packets and the priority of HARQ retransmission packets when approaching the deadline expiration, so the proposed algorithm can support more users for scheduling, and send more packets in the case of limitation for the number of RBs.

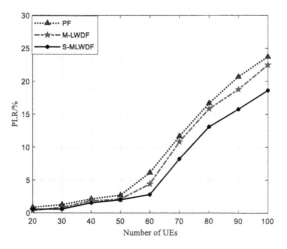

Fig. 7. PLR

5 Conclusion

In this paper, the performance of 5G downlink resource scheduling algorithm is studied. Combing 5G network slicing which is widely used with M-LWDF algorithm, we proposed a novel scheduling strategy named as S-MLWDF algorithm which fully considers the delay, current channel quality status, RBs allocation and the available RBs in next time-slot. At the same time, PF and M-LWDF algorithms are used as comparisons to simulate and analyze for resource utilization, fairness, throughput and PLR. Simulation results show that the proposed algorithm is feasible. Under the environment of multi-users, S-MLWDF algorithm performs better than PF and M-LWDF algorithm in improving fairness between users who are at the edge of cell, reducing PLR caused by delay of queues. Simultaneously, it can satisfy QoE for different users and save a lot of network resources. Besides, the computational complexity is not high, and there are unique advantages for the development of varied services in 5G scenarios. In future work, we need to consider the use of buffer state while scheduling is obligatory to maximize the resource utilization.

Acknowledgement. This work comes from a major project in Chongqing, "R&D and application of 5G road test instruments (No. cstc2019jscx-zdztzxX0002)".

References

1. Gupta, A., Jha, R.K.: A survey of 5G network: architecture and emerging technologies. IEEE Access **3**, 1206–1232 (2015)
2. Foukas, X., Patounas, G., Elmokashfi, A., et al.: Network slicing in 5G: survey and challenges. IEEE Commun. Mag. **55**(5), 94–100 (2017)

3. Comsa, I., Zhang, S.J., Aydin, M.E.: Towards 5G: a reinforcement learning-based scheduling solution for data traffic management. IEEE Trans. Netw. Serv. Manag. **15**(4), 1661–1675 (2018)
4. Wang, C., Shih, K., Li, C.: User location recommendation combined with MLWDF packet scheduling in LTE downlink communication. In: 2018 Asia-Pacific Signal and Information Processing Association Annual Summit and Conference (APSIPA ASC), pp. 450–453. IEEE (2018)
5. Samia, D., Ridha, B.: A new scheduling algorithm for real-time communication in LTE networks. In: 2015 IEEE 29th International Conference on Advanced Information Networking and Applications Workshops, pp. 267–271. IEEE (2015)
6. Shen, L., Wang, T., Wang, S.: Proactive proportional fair: a novel scheduling algorithm based on future channel information in OFDMA systems. In: 2019 IEEE/CIC International Conference on Communications in China (ICCC), pp. 925–930. IEEE (2019)
7. Nasralla, M.M., Rehman, I.U.: QCI and QoS aware downlink packet scheduling algorithms for multi-traffic classes over 4G and beyond wireless networks. In: 2018 International Conference on Innovation and Intelligence for Informatics, Computing, and Technologies (3ICT), pp. 1–7. IEEE (2018)
8. Ferdosian, N., Othman, M., Ali, B.M., et al.: Fair-QoS broker algorithm for overload-state downlink resource scheduling in LTE networks. IEEE Syst. J. **15**(4), 3238–3249 (2018)
9. Rocha, F.G.C., Vieira, F.H.T.: A channel and queue-aware scheduling for the LTE downlink based on service curve and buffer overflow probability. IEEE Wirel. Commun. Lett. **8**(3), 729–732 (2019)
10. Madi, N.K.M., Hanapi, Z.M., Othman, M., Subramaniam, S.K.: Delay-based and QoS-aware packet scheduling for RT and NRT multimedia services in LTE downlink systems. EURASIP J. Wirel. Commun. Netw. **2018**(1), 1–21 (2018). https://doi.org/10.1186/s13638-018-1185-3
11. Padmavathy, C., Jayashree, L.S.: An enhanced delay sensitive data packet scheduling algorithm to maximizing the network lifetime. Wireless Pers. Commun. **94**(4), 2213–2227 (2016). https://doi.org/10.1007/s11277-016-3376-8
12. Ferreira, F.A., Guardieiro, P.R.: A new channel-aware downlink scheduling algorithm for LTE-A and 5G HetNets. In: Chaubey, N., Parikh, S., Amin, K. (eds.) COMS2 2020. CCIS, vol. 1235, pp. 173–183. Springer, Singapore (2020). https://doi.org/10.1007/978-981-15-6648-6_14
13. Ksentini, A., Frangoudis, P.A., Nikaein, N., et al.: Providing low latency guarantees for slicing-ready 5G systems via two-level MAC scheduling. IEEE Network **32**(6), 116–123 (2018)
14. Abdelhamid, A., Krishnamurthy, P., Tilpper, D.: Resource allocation for heterogeneous traffic in LTE virtual networks. In: 2015 16th IEEE International Conference on Mobile Data Management (MDM), vol. 1, pp. 173–178. IEEE (2015)
15. Kayali, M.O., Shmeiss, Z., Safa, H., et al.: Downlink scheduling in LTE: challenges, improvement, and analysis. In: 2017 13th International Wireless Communications and Mobile Computing Conference (IWCMC), pp. 323–328. IEEE (2017)
16. Hadar, I., Raviv, L., Leshem, A.: Scheduling for 5G cellular networks with priority and deadline constraints. In: 2018 IEEE International Conference on the Science of Electrical Engineering in Israel (ICSEE), pp. 1–5. IEEE (2018)
17. Pedersen, K., Pocovi, G., Steiner, J., et al.: Agile 5G scheduler for improved E2E performance and flexibility for different network implementations. IEEE Commun. Mag. **56**(3), 210–217 (2018)

18. Mamane, A., Ghazi, M.E., Barb, G., et al.: 5G heterogeneous networks: an overview on radio resource management scheduling schemes. In: 2019 7th Mediterranean Congress of Telecommunications (CMT), pp. 1–5. IEEE (2019)
19. Capozzi, F., Piro, G., Grieco, L.A., et al.: Downlink packet scheduling in LTE cellular networks: key design issues and a survey. IEEE Commun. Surv. Tutor. **15**(2), 678–700 (2013)
20. Nnamani, C.O., Anioke, C.L., Ani, C.I.: Improved MLWDF scheduler for LTE downlink transmission. Int. J. Electron. **103**(10–12), 1857–1867 (2016)
21. Nguyen, T.T., Ha, V.N., Le, L.B.: Wireless scheduling for heterogeneous services with mixed numerology in 5G wireless networks. IEEE Commun. Lett. **24**(2), 410–413 (2020)

Software Defined Unicast/Multicast Jointed Routing for Real-Time Data Distribution

Shimin Sun[1] , Wentian Huang[1] , Xinchao Zhang[1] ,
and Li Han[2(✉)]

[1] Tiangong University, Tianjin 300387, China
[2] Tianjin University of Technology, Tianjin 300384, China
hanli@tjut.edu.cn

Abstract. The explosive increasing of high bandwidth-consumption traffic puts great pressure on the Internet. Many popular applications springing up work in a manner of one-to-many or many-to-many communication, such as TikTok, Instagram, Tencent conference, and numbers of interactive games. Due to the scalability and applicability problems, existing multicast schemes, e.g. IP multicast, are not widely implemented. Instead, most of those traffic is transmitted through the Internet in unicast, which results in vast redundant traffic in backbone networks. In this paper, we propose a unicast/multicast jointed routing mechanism in software defined networks, SDUM. We devote to achieve unicast data distribution following a dynamic multicast tree, which is managed by centralized control and application plane. Other than OpenFlow protocol, this mechanism doesn't require any specific multicast protocols or software. The network can be a virtualized network with distributed OpenFlow devices interconnected by legacy routers. The evaluation results confirm the efficiency of the proposal in the number of control messages, signaling overhead, occupation of flow table entries, and qualitative comparison.

Keywords: Multicast algorithm · Data manipulation · Group management · Software defined networking

1 Introduction

In recent years, social media applications have made great success; especially those support live webcasting, Facebook, Twitter, Tik-Tok, etc. Online multimedia services are also getting much more pervasive, e.g. video conference, live broadcast, multiplayer gaming, and distributed data caching. Generally, those services distribute data in multicast-liked one-to-many or many-to-many pattern. Although, IP multicast [1] is a desirable approach in a functional viewpoint to avoid parallel delivery of identical packets on the same link. However, in reality, data packets are delivered in unicast in most cases. Large and redundant traffic utilizes significant proportions of network bandwidth resource. Under today's Internet architecture, unicast is still almost the only option for IP data packet transmission. With rapid development of multimedia technology and hardware, high quality streaming traverses the Internet to a large amount of end hosts. It is urgent to design an applicable multicast routing mechanism for those kind of services to reduce the waste of network resource.

© ICST Institute for Computer Sciences, Social Informatics and Telecommunications Engineering 2021
Published by Springer Nature Switzerland AG 2021. All Rights Reserved
H. Gao et al. (Eds.): ChinaCom 2020, LNICST 352, pp. 226–243, 2021.
https://doi.org/10.1007/978-3-030-67720-6_16

The limitation of scalability and rigid network architecture results in the deployment problem of existing multicast mechanisms. IP multicast, as one of the most admirable approach, faces various obstacles, such as global routing, group management, security consideration, QoS guarantee, address allocation, and inter-domain coordination issues [1]. Application Layer Multicast (ALM) [2], also known as overlay multicast, is the alternative approach of IP multicast. It is also not widely deployed due to the requirement of specific software and physical agents.

Software Defined Networking (SDN) [3] is a promising paradigm for elastic network control and resource management. Many internet enterprises, such as Google and Facebook, deployed their Data Center Networks (DCNs) based on SDN architecture. Most of the multicast approaches in SDN still focus on the implementation of IP multicast or group management protocols [4]. Therefore, the problems faced by multicasting in SDN are similar to IP multicast. Since the data is delivered using multicast address, it still requires routers on the multicast tree to support specific protocols, such as IGMP. Besides, a flow table entry of a OpenFlow Switch (OFS) is occupied by one multicast group. However, the storage of flow table, Ternary Content Addressable Memory (TCAM), is capacity constraint and expensive. The number of flow table entries is limited by 10^5 in current stage. For a network with n nodes, the maximum number of multicast group is $O(2^n)$. It is still inadequate to support universal multicast service for ordinary users depending on existing multicast mechanisms.

The motivation of this paper can be unfolded in three aspects. Firstly, instead of urging to design out-of-the-box solutions for IP multicast, we abandon all those basics and devote to realize multicast mechanism depending on pure SDN elements. Secondly, avoiding introducing any specific protocol or mechanism to network entities, we tend to use only OpenFlow protocol and primordial control plane. Finally, we design several modules in application plane to collect, analyze information and make decisions for multicast traffic, which is portable and programmable.

The main contribution of this paper is described as follows.

a. We present a practical multicast scheme for common applications. It is unnecessary for applications to be aware of multicasting. Instead, they can receive data based on unicast sessions.
b. We present a series of modules in application plane. Those modules are responsible for network information collection, resource scheduling, and decision making.
c. We present the detailed operation of packet manipulation and distribution, as well as dynamic group management. Required operations are implemented on OpenFlow-enabled routers without the involvement of end hosts and legacy routers.
d. We simulated the proposal on top of two classic network topologies using Mininet, proved more efficient than other approaches.

The rest of the paper is structured as follows. In Sect. 2, we introduce state-of-the-art literatures about multicast routing algorithms. Section 3 elaborates the proposed architecture, data processing algorithm, and group management mechanisms. Basic Operation of Data Distribution and Tree Algorithm are described in Sect. 4. Section 5 discusses the implementation and analysis, following with conclusion and future work in Sect. 6.

2 Related Work

2.1 Classic Multicast Protocols and Approaches

Various of IP multicast routing protocols exists for diverse application scenarios. Some studies exploit to design branch-ware forwarding technique [4]. DVMRP [5] and MOSPF [6] suit for the network with dense concentration of receivers, while PIM-SM [7] and BIDIR-PIM [8] designed for multicast tree construction in sparse networks. These protocols suffer from several uniformed issues [1], such as group dynamics, network scalability, member security, and multicast capable islands. Mbone project [9] brings multicast closer to reality, which applies IP tunneling to connect each multicast capable islands. However, static IP tunneling stymies the natural growth of multicast service requirement. ALM realizes data dissemination depending on software relay. It is easier to deploy ALM than IP multicast. This is because application owners rather than network providers manage group members and realize multicast functions with unicast relay at specific agent or end hosts. Deploying application in specific servers or end hosts is much more practical than installing network protocols to every network entity. ALM allows dynamic routing relying on network conditions and QoS requirements of applications. For example, Scattercast [10] calculates shortest path tree with minimum latency cost, while Overcast [11] constructs a multicast tree for maximum bandwidth utilization from source to receivers. Some approaches adopt comprehensive consideration of network conditions, such as PeerCast [12] and Vcast [13]. Consistent problems of those solutions are relay agent placement, network discovery, as well as group labeling.

2.2 Software Defined Multicast Approaches

SDN has been verified in practice as a promising network architecture to tackle the problems of IP multicast. SDN does not natively support multicast addressing and multicast routing protocols. Many literatures have been carried out on the efficient delivery of multicast traffic and multicast routing algorithms. However, most of them are still focus on the implementation or amendment of IP multicast in SDN environment. In other words, data distribution is still based on multicast address and suffers from majority of IP multicast problems.

To support IP multicast, MultiFlow [14] presented an approach to enable SDN to parse IGMP messages. OFSs exchange IGMP messages to accomplish group management. Evaluation results showed lower time-consuming for multicast tree creation than DVMRP. PVMC-SDN [15] proposed a virtual platform to interpret MPLS and ARP protocols, in order to forward data along the multicast tree. OFM [16] developed multicast services based on SDN programmable functionalities from clean-slate perspective. The limitation is that OFS need to be aware of joining or leaving message, so that it is able to deliver the message directly to SDN-C. Multicast address was allocated for data distribution through the network, nevertheless lack of detailed process description.

To enhance scalability of IP multicast, Locality-aware Multicast Approach (LAMA) [17] was proposed to reduce the computation for multicast tree creation.

To build a multi-group shared tree, a Rendezvous Point (RP) election algorithm was designed to pick out the proper RP with minimum distance to all sources. However, message exchanging for RP election causes high signaling overhead and challenges RP's flow table size. BAERA [18] presented a branch-aware Steiner tree construction algorithm to minimize the number of branch nodes and edges of the multicast tree. However, authors focus on tree construction algorithm, without presenting data distribution method and flow table capacity of branch nodes. SDM [19] attempted to slice multicast tree to enhance live streaming distribution. It designed SDM domain consisting of pure OFSs to distribute multicast traffic. Similar to IP multicast, a flow table entry was required for each group. Specific messages were elaborated to realize the functions of network entities, such as NL-SDM and Virtual Peer Instance. Outside the domain, packets were transmitted in unicast. A comprehensive survey on SDN based multicast approaches were presented in [20], where most of them still focused on multicast tree construction algorithm as IP multicast.

According to above analysis, SDN based multicast approaches are better than IP multicast and ALM in applicability. However, any introduced message exchanges could increase the load of flow table, which is expensive and volume-limited. For example, mainstream products of OFSs usually have hundreds of thousands of entries using Ternary Content-Addressable Memory (TCAM) [21]. The size of TCAM is considerably scarce comparing with huge amount of traffic flows. This is the great obstacle of SDN based multicast approaches and results in scalability problem.

3   Software Defined Unicast/Multicast Jointed Routing (SDUM)

3.1   Multicast Issues in Current Stage

In current Internet service architecture, one-to-many and many-to-many data distribution schemes are mostly deployed in Client-Server (C/S) or Peer-to-Peer (P2P) approaches. That is, end host fetches data based on unicast session from a single server. Access network of the server bears the maximum flow stress with massive amount of redundant data. Comparing to multicast, the primary advantage of unicast is route aggregation. Route aggregation summarizes route by network address instead of every specific IP address, which can significantly save routing table resource.

To alleviate the traffic load of server-side and backbone networks, multicast is a promising approach to improve the efficiency of data dissemination. The problem is that it is arduous to conduct route aggregation due to the random allocation of multicast addresses and distributed receivers. Every router needs to keep a routing entry for each group, in order to resolve multicast address when processing multicast traffic.

The key idea of this paper is to design agile unicast algorithms to realize data distribution along calculated multicast tree without deploying specific multicast routing protocol. Therefore, route aggregation is executed for those data traverse the network.

3.2   System Architecture

On top of SDN architecture, the system architecture is illustrated in Fig. 1, consisting of data plane, control plane, and application plane. Data plane and control plane follow the basic SDN architecture, while all the designed modules are located at application plane.

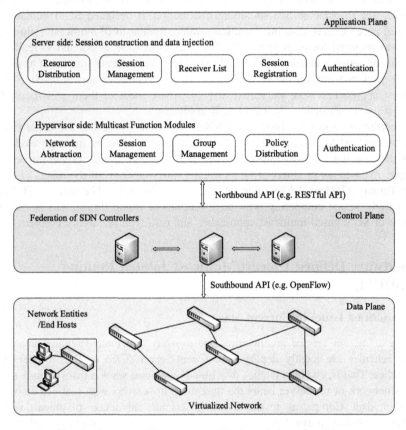

Fig. 1. System architecture to enable formulation of multicast

Data plane is composed of OpenFlow Switches (OFSs) interconnected by legacy routers, which forms a virtualized network. OFSs process data packets following the rules defined in flow table. Besides, they gather link/port status and statistical information of the network, and send to controller using OpenFlow messages through a secure channel.

Control plane is in charge of network management, which is composed of multiple controllers federated to share network information. Controller has the capability to acquire a global overview of network topology and status. Various controllers have already been developed, such as Floodlight, POX, Beacon, and Ryu [22]. They are collaborated by hypervisor to conduct network level traffic engineering. For multi-controller integration, several hypervisor systems were developed, HyperFlow, FlowVisor, Kandoo, Onix, etc. [23].

At application plane, various applications can be developed to realize specific functionalities, such as routing algorithm, load balancing, security concern, network management, and other new features. Application plane interacts with control plane through northbound API (e.g. Restful API). Meanwhile, control plane communicates with data plane through southbound API (e.g. OpenFlow protocol).

We designed two series of modules, server-side modules and hypervisor-side modules. Server-side modules are responsible for data distribution, recording of session information, and authentication. Hypervisor-side modules are responsible for network abstraction, group management, policy distribution, and authentication.

3.3   Functional Description

The roles and responsibilities of the related network entities are described as follows.

a. Receivers. End hosts that send queries and achieve authentication to the server. They receive data in common unicast session.
b. Source. The server that provides streaming service. It performs authentication with applicants as well as its corresponding controller. Besides, it needs to upload group and session information to the controller.
c. Controller. Controller collects network status from OFSs, records session information from Source. It obtains decision from controller-side modules, and issues rules to OFSs.
d. OFSs. Network entities serve as the actors of data forwarding. It also executes data manipulation policy received from Controller.
e. Legacy routers. Current dominant network core entities, which connect OFSs distributed in the network. Legacy routers are ignored in the virtualized network when implementing the proposed mechanisms.

The interactions of entities are illustrated in Fig. 2. For simplicity, we assumed that source (S) only keeps a single multicast group and access to the network through its gateway OFS. The authentication between source and controller is achieved before initiating the data distribution service. Controller issues a multicast address to S as group ID.

Initially, S is ready for service request. When S receives the first request message from host $R1$, it replies with a response message and an authentication message. The authentication could be completed by the common ID/password login identification according to the default design of utilized applications. After the success of authentication, S inserts $R1$'s information to a list, Receiver List (RL). RL list records the session information of each receiver, including source IP address, source port number, destination IP address, destination IP address, destination port number, and an expiration timer. Meanwhile, S sends the requested source to $R1$ based on the unicast session. Since data could be simply transmitted along default route, it is unnecessary to issue rules to OFSs.

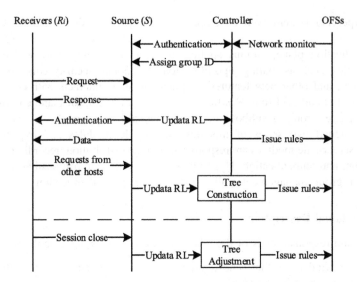

Fig. 2. Brief illustration of functional flowchart

Upon receiving other requests, *S* updates *RL* and keeps update with the controller. The controller make decision for multicast tree construction and data manipulation, which will be described in Sect. 4. Afterwards, it issues the rules to corresponding OFSs to enable data packets transmitted along the constructed multicast tree.

If a session close or expiration timer timeout, a receiver will be marked as departure. *S* and the controller update *RL*. Tree adjustment algorithm calculate the appropriate tree to guarantee data reception and tree optimization.

3.4 Data Structure

The multicast manager (Hypervisor) needs to maintain several tables for each multicast group, which are listed in Table 1 and introduced below.

a. G_i: The multicast group, which is marked by an locally unique multicast address. It stores server-side information: IP address of S (IP_{Si}), port number of S ($PORT_{Si}$), multicast address (IP_{Mi}).

b. RL_i: The table maintains host-side information: IP address of R_i (IP_{Ri}), port number of R_i ($PORT_{Ri}$), and an expiration timer (TTL_{Ri}).

c. $OFSL_i$: The table includes IP address of OFSs that on the multicast tree.

d. $FOFS_i$: The table contains OFSs that need to do specific operation for data delivery along the multicast tree. We call those OFSs, Fork OFS ($FOFS$).

e. $EOFS_i$: The table records OFSs on the multicast tree that is incapable to be a $FOFS$, such as OFSs with insufficient flow table volume.

Table 1. Data structures that deployed in application layer

G_i	$<IP_{Si}, PORT_{Si}, IP_{Mi}>$
RL_i	$<IP_{R1}, PORT_{R1}, TTL_{R1}>, ..., <IP_{Rn}, PORT_{Rn}, TTL_{Rn}>$
$OFSL_i$	$OFS_1, ..., OFS_n$
$FOFS_i$	$<IP_{R1}, OFS_1, Rule_1>, ..., <IP_{Rn}, OFS_n, Rule_n>$
$EOFS_i$	$OFS_1, ..., OFS_n$

4 Operation of Data Distribution and Tree Algorithm

4.1 Data Manipulation Mechanism

To illustrate the basic operations of data dissemination along the multicast tree, we present a brief instance in Fig. 3. Thereby, source is indicated by a triangle icon; OFSs are indicated by circular icon with sequence number inside (from OFS1 to OFS8); and receivers are gray square (R1, R2, and R3).

Source sends packets to the first receiver R1 directly. When R2 joins, according to tree construction algorithm, OFS4 is set to be the operation node for data manipulation. OFS4 duplicates the data packets, and translates the destination IP/Port of duplicated packets from $<IP_{R1}, PORT_{R1}>$ to $<IP_{R2}, PORT_{R2}>$. Then, if R3 joins, OFS2 performs the similar operation, translates duplicated packets from $<IP_{R1}, PORT_{R1}>$ to $<IP_{R3}, PORT_{R3}>$.

Notice that, it is unnecessary to perform data manipulation by a physical fork node or not. It can be any OFS on the tree, especially when the desired fork node capacity overloaded. For example, in Fig. 3, if OFS2 overloaded, OFS1, OFS3, OFS7 or even OFS4 can become the operation node for R3. The hypervisor modular decides which OFS selected as operation nodes, according to tree construction and adjustment algorithm.

4.2 Greedy-Based Multicast Tree Construction Algorithm

In this section, we elaborate multicast tree construction algorithm by host joining group and leaving group. Join group operation is realized by a greedy algorithm to snap a branch to the multicast tree. Leave group operation should take care of the impact of the prune of any branch on other group members. It is possible for the controller to obtain the entire or regional network topology and link status in SDN architecture. Therefore, we can calculate the optimal path and multicast tree in a simulated environment, without injecting any network detection message.

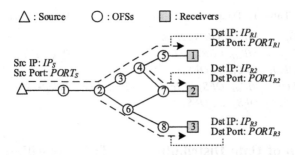

Fig. 3. Basic operations of multicast data distribution

Joining Group Operations

In tree construction stage, multicast tree grows in a greedy way. That is, when new host joins the group, the nearest OFS on the multicast tree has the top priority as the operation node.

As shown in Algorithm 1, we propose a greedy based multicast tree maintenance algorithm. When the first host requests the source, data transmitted to the host along the default route. A crowd of OFSs on the path from the source to the receiver are recorded in table *OFSL*. Then, if other hosts request the source, we need to find a path that can fetch the required data. Simply, it finds the nearest OFS on the path from the source to the first receiver (Eq. (1)). If the OFS is capable to be the operation node, then we assign it as the fork node. Otherwise, delete the OFS from *OFSL* list, and execute Eq. (1) again. If there are more than one OFS that has the same distance to the host. Equation (2) is applied to find an OFS that is closest to the source node.

$$FOFS = Min\big(Dist(\{OFS_i\}, R_j)\big) \tag{1}$$

$$FOFS = Min(Dist(\{FOFS_i\}, S)) \tag{2}$$

After the tree construction, the new receiver is inserted to *RL* list, the selected fork node is added to *FOFS* list. *OFSL* list is updated with new OFSs joining and incapable OFSs deleting. Besides, controller issue rules to the *FOFS* to execute appropriate data manipulation operation.

Algorithm 1: Host Joins a Group
Input: group ID: G, host: R_j, Source: S

---
1.   Add R_j to RL
2.   **if** $j=1$, //the first group member
3.   find all the OFSs on the path from S to R_j
4.   Add above OFSs to $OFSL$
5.   **else**
6.   **for** $\{OFS_i\}$ in $OFSL$, **do**
7. $FOFS_j = Min\left(Dist(\{OFS_i\}, R_j)\right)$
8.   **if** multiple $FOFS_j$ exist
9. $FOFS_j = Min\left(Dist(\{FOFS_i\}, S)\right)$
10.   **if** $FOFS_j$ overloaded
11.   Add $FOFS_j$ to $EOFS$, Delete from $OFSL$
12.   Go to **do**
13.   **else**
14.   Add $FOFS_j$ to $FOFS$ list, add OFSs on the path from $FOFS_j$ to R_j to $OFSL$, issue rules to $FOFS_j$
15.   **endif**
16.   **endif**
17.   **Output:** RL, $OFSL$, $FOFS$, issue rules to $FOFS_j$

Leaving Group Operations

Multicast group member status is dynamic. A member leaves a group when the session closed or network interrupted. If a member is determined as leaving by the controller, it is deleted from RL list. Multicast tree is recalculated according to Algorithm 2. The controller should check whether other receivers acquire data using the branch from $FOFS_j$ to R_j or not.

If no impact on the data transmission to other group members, the branch will be pruned simply. The controller issues deletion rules to $FOFS_j$ to remove related flow entry. Besides, we remove $FOFS_j$ from $FOFS$ table.

If the prune of the branch causes the termination of data dissemination to other members, following operations should be carried out before pruning the branch.

a. According to the greedy algorithm, find an alternative $FOSF$ for the members that will be affected if $FOFS_j$ removed.
b. Issue rules to the selected operation node $FOFS_k$ for R_k.
c. Add those OFSs that connect to R_k to $OFSL$ list.
d. Delete $FOFS_j$ from $FOFS$ list as well as the corresponding flow table entry.

Finally, it outputs updated *RL, OFSL, FOFS* table, and issues rules to $FOFS_k$.

Algorithm 2: Host Leaves a Group	
Input: group ID: *G*, host: R_j, Source: *S*	
1.	Get the set of $\{OFS_i\}$ in *OFSL*
2.	Get *FOFS* of R_j: $FOFS_j$
3.	**for** $\{OFS_i\}$, **do**
4.	Delete OFSs on R_j's branch from *OFSL*
5.	**if** no other branch exists on R_j's branch
6.	Delete $FOFS_j$ from *FOFS*
7.	**else** \\ R_k exists on R_j's branch
8.	$FOFS_k = Min\big(Dist(\{OFS_i\}, R_k)\big)$
9.	**if** multiple $FOFS_k$ exist
10.	$FOFS_k = Min\big(Dist(\{FOFS_i\}, S)\big)$
11.	Issue flow table entry to $FOFS_k$
12.	Add OFSs on R_k's branch to *OFSL*
13.	Delete $FOFS_j$ from *FOFS*
14.	**endif**
15.	**Endif**
16.	Delete flow table entry of R_j from $FOFS_j$
17.	**Output:** *RL, OFSL, FOFS*, issue rules to $FOFS_k$

We can conclude that the application plane achieves the most workload. What the control plane to do is issue rules to corresponding OFSs, while the data plane is responsible for data forwarding. Only the corresponding OFSs are need to execute those received policies. In this way, not all network entities need to support multicast protocol or OpenFlow protocol. AS a contrast, IP multicast requires all routers on the tree to support multicast protocol, while SDN based multicast approaches requires all OFSs on the tree to maintain a forwarding table entry for each group.

5   Methodology, Implementation and Evaluation

Most of multicast approaches in SDN are still explicitly or implicitly to realize IP multicast functions. However, the performance on data delivery efficiency is commonly no better than IP multicast. As a comparison, we chose SDM [19], an SDN based multicast approach; as well as a classical IP multicast routing protocol, PIM-SM [7]. PIM-SM is suitable routing protocol for IP multicast in sparse networks, which is in line with the cases of our proposal.

In the simulation, the default forwarding principle of OFSs was set to be same as the typical routing protocol, Routing Information Protocol (RIP). Default paths from the source to each receiver based on the shortest path. In the experiments, data delivery efficiency was evaluated by counting the number of control messages. Signaling overhead ratios were analyzed to evaluate the overall cost metrics of control packets to enable entire multicast traffic distribution. To evaluate the workload of OFSs in or routers in traditional network, we compared the number of installed flow table rules and routing table entries.

5.1 Simulation Setup

a. **Operating environment:** We implemented SDUM on top of Ryu controller [24] based on OpenFlow protocol version 1.3. The network topology was constructed using Mininet [25] version 2.2.2. Due to function limitation, it is difficult to implement PIM-SM protocol in Mininet. So that, we deployed the same network with exact same link status using NS2 which was able to simulate IP multicast protocols. The testbed was allocated in a desktop PC with Ubuntu Desktop 16.04. S_imulation time was set to 60 s for each scenario.

b. **Network topology:** Two classic network topologies were implemented, DCN topology and realistic WAN topology. A fat-tree DCN topology with three tiers is shown in Fig. 4. There are 14 OFSs and 15 hosts. A realistic WAN topology for Microsoft's data centers is illustrated in Fig. 5, where every OFS connects to two hosts as exhibited in the legend, totally 15 hosts. A single source with a controller were allocated in the network. For each link, bandwidth and delay were set to 1 Gbps and 10 ms respectively in both network. BRITE [26] was used to generate network topologies.

c. **Traffic generation:** The leftmost host was the source to generate constant UDP traffic using iPerf. The multicast address was configured to 224.0.55.55. Data rate injected by the source is constantly 1 Mbps.

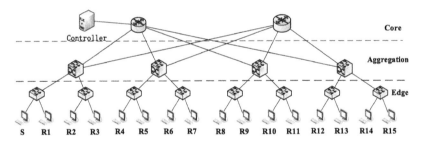

Fig. 4. Typical data center interconnected network topology

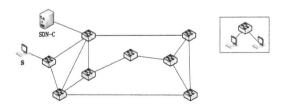

Fig. 5. WAN topology with each switch connected to two hosts

5.2   Evaluation and Analysis

Examined protocols and approaches were based on shortest path tree algorithm. Therefore, they conducted actually the same multicast tree. It is meaningless to compare tree efficiency, data delivery latency or bandwidth consumption with each other. Instead, we evaluated the number of control messages, signaling overhead ratios, and the number of rules and routing table entries installed to enable multicasting. Besides, qualitative comparison of typical multicast approaches is presented.

Control Messaging Overhead

Control messaging overhead was assessed by the number of control messages that injected to the network and its signaling overhead ratio. The number of control messages for the operation of multicast service reveals the load of protocols or approaches imposing on the network. For PIM-SM, control messages are exchanged among routers to maintain multicast tree, such as IGMP messages, Hello messages, and RP election messages. For SDM and SDUM, control messages are mainly OpenFlow messages exchanged between OFSs and controller, including PacketOut messages for rule installation, and PacketIn messages for unknown packet uploading, etc. SDM generates control messages for domain management, virtual peer operation, host joining and leaving, etc. We omitted the Beacon messages between OFSs and controller, because it is inevitable in SDN regardless of whether deploying SDUM or not.

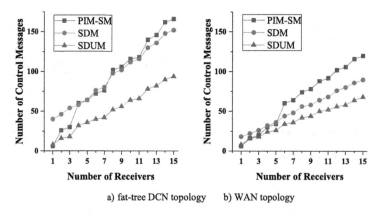

a) fat-tree DCN topology   b) WAN topology

Fig. 6. The number of control messages with incremental receivers

Figure 6 shows the number of control messages variation. It indicates that PIM-SM leads to more reliance on network topology. In Fig. 6(a), the line-symbol graph of PIM-SM in fat-tree DCN topology shows a sharp leap when R4, R8, or R12 joins the tree. This is because the multicast tree topology occurs a significant change. Routers exchanges more control messages to maintain the multicast tree with the growth of network.

The number of control messages of SDM is close to IP multicast of DCN topology, but lower in WAN topology. SDUM generates about 40% reduction of control messages comparing to PIM-SM in both topology, while around 30% reduction comparing with SDM in WAN topology on average.

Signaling Overhead Ratio

Signaling overhead ratios are depicted in Fig. 7. As background traffic, we emulated constant 1Mbps UDP traffic as the streaming. Depending on the basic format of control messages, PIM-SM control messages are around 70 bytes while control messages in SDN are around 200 bytes. As a result, the number of the control message may not reveal the real signaling overhead.

From the evaluation results in Fig. 7, we can see that SDM produces higher signaling overhead than PIM-SM and SDUM. It is on average 47% and 25% higher than SDUM for DCN and WAN topology respectively. The overhead ratio of SDUM is close to PIM-SM due to its lightweight control cost.

OpenFlow Rules Required for Multicast

Flow table utilization is essentially to be measured since the volume of TCAM in OFSs and routing table entries in a router are limited. In the experiments, we counted the number of routing table entries required to realize PIM-SM routing and data delivery, and the flow table entries of OFSs used for multicast packet duplication, address translation, and forwarding. For SDM, the number of rules installed greatly relies on network topology and receiver location. It is necessary to install rules not only on access OFS of each receiver, but also on specific core OFSs. The results in Fig. 8 show that rules increase sharply along with the increase of receivers in both experimental scenarios.

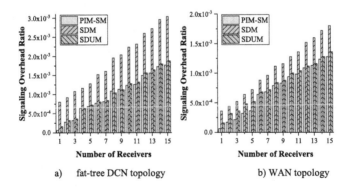

a)   fat-tree DCN topology   b) WAN topology

Fig. 7. Signaling overhead with incremental receivers

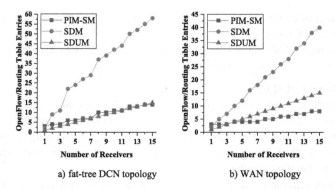

Fig. 8. The number of rules that issued to OFSs with incremental receivers

As a contrast, although PIM-SM generates more control messages as shown in Fig. 8, the occupation of routing table entries is less. This is because IP multicast aims to share multicast routing table with each other, which is one entry per group. Only routers in the multicast tree have the multicast routing table entry. Therefore, the number of routing table entries depend on the quantity of multicast groups deployed in the router. Besides, a RP-Set table exists to store RP connection information with few entries, which we skipped in the experiment.

In SDUM, only *FOFSs* maintain individual entries for each receiver. The number of rules is linearly increasing along with the number of receivers. Other OFSs on the tree only forward packets based on default routing strategies without installing extra rules.

From the experimental results, we can conclude that SDM shows a poor scalability performance with the increasing receivers. Routing table size required by PIM-SM relies on the number of groups. Flow table size required by SDUM highly depends on the number of receivers in a group. Therefore, SDM is suitable for small network with limited receivers; PIM-SM is suitable for sparse network with limited groups; while SDUM is better to be deployed for large networks with a small quantity of receivers in each group.

Qualitative Comparison

It is worthy to analyze the pros and cons of SDUM by making a qualitative comparison with typical protocols and approaches, as listed in Table 2, including PIM-SM, ALM, SDM, and SDUM.

The major difference of them is the deployment requirement. PIM-SM requires all routers to install the multicast protocol, while SDUM has no specific requirement on network entities. ALM needs to deploy specific hardware agents or terminal relay software, which is difficult to be widely deployed with controlled data relay. SDM attempts to build a SDM domain where the designated group ID can be parsed by OFSs, which is implicitly in line with IP multicast.

Those problems lead to deployment difficulty and low scalability to large networks. Distribution management method of IP multicast and ALM results in notoriously difficult to achieve ISP control of multicast traffic. Traffic latency may be high in ALM caused by terminal relay through a longer route. For access control, application layer software development is easier than the design of network layer protocols.

Table 2. Qualitative comparison of typical multicast approaches

	PIM-SM	ALM	SDM	SDUM
Incremental deployment	All routers multicast capable	Relay entities or terminal relay	All OFSs in SDM domain	Only branch nodes are OFS required
Scalability	Low (routers disabled multicasting exist)	High (protocols and entities mature)	Low (inadequate in legacy router coexisting networks	High (deployment with partial SDN)
ISP control	Low (hard to manage distributed routers)	None	High (centralized control of SDM domain)	High (centralized control of FOFSs)
Latency	Low (optimized route)	High (due to terminal relay)	Low (optimized route)	Low (optimized route)
Access control	IGMP-AC and SIGMP required	Application layer authentication	No concern	Application layer source and controller authentication

From experimental results and qualitative evaluation, we distinctly verdict that SDUM performs better than IP multicast (PIM-SM) and SDM in terms of control message overhead, signaling overhead ratio, and flow table entry size. This is an evidence that SDUM is a promising approach for multicast support in evolutional IP networks, conquering the shortage of IP multicast, ALM, and SDM.

6   Conclusion

In this paper, we propose to improve multicast routing by manipulating unicast data transmission on-the-fly to provide a generic multicast service in network layer. All required functionalities are placed on the centralized application plane. With the assistance of the source for information collection and authentication of group members, it offers a transparent service to heterogeneous hosts and network entities. The evaluation results indicate an evident improvement in terms of traffic overhead, flow table occupation, and qualitative comparison. Future work will focus on following studies, including multiple controller coordination mechanism, multicast tree optimization and fast failure recovery mechanism, QoS guarantee, and experiments in random networks with complex topologies.

Acknowledgements. This work is supported by Science and Technology Development Fund of Tianjin Education Commission for Higher Education (No. 2017KJ090).

References

1. Diot, C., Levine, B.N., Lyles, B., Kassem, H., Balensiefen, D.: Deployment issues for the IP multicast service architecture. IEEE Netw. **14**(1), 78–88 (2000)
2. Park, J., Lee, J.M., Kang, S.: Design implementation of overlay multicast protocol for many-to-many multicast services, In: The 9th International Conference on Advanced Communication Technology, pp. 2144–2147, Okamoto, Kobe. IEEE (2007)
3. Nunes, B.A.A., Mendonca, M., Nguyen, X.-N., Obraczka, K., Turletti, T.: A survey of software-defined networking: past, present, future of programmable networks. IEEE Commun. Surv. Tutor. **16**(3), 1617–1634 (2014)
4. Chiang, S., Kuo, J., Shen, S., Yang, D., Chen, W.: Online multicast traffic engineering for software-defined networks, In: Proceedings of IEEE INFOCOM, Honolulu, HI, pp. 414–422 (2018)
5. Waitzman, D., Partridge, C., Deering, S.E.: Distance vector multicast routing protocol, IETF RFC 1075 (Experimental) (1988)
6. Moy, J.: Multicast extensions to OSPF, IETF RFC 1584 (1994)
7. Fenner, B., et al.: Protocol Independent Multicast-Sparse Mode (PIM-SM): Protocol Specification (Revised), IETF RFC 7761 (2016)
8. Speakman, T., Vicisano, L., Handley, M.J., Kouvelas, I.: Bidirectional protocol independent multicast (BIDIR-PIM), IETF RFC 5015 (2007)
9. Almeroth, K.C.: The evolution of multicast: from the MBone to interdomain multicast to Internet2 deployment. IEEE Netw. **14**(1), 10–20 (2000)
10. Yatin, C.: Scattercast: an adaptable broadcast distribution framework. Multimedia Syst. **9**, 104–118 (2003)
11. Jannotti, J., Gfford, D.K., Johnson, K.L., Kaashoek, M.F., O'Toole, J.W.: Overcast: reliable multicasting with an overlay network. In: Proceedings of the 4th Conference on Symposium on Operating System Design & Implementation, San Diego, CA, pp. 197–212. Usenix (2000)
12. Zhang, J., Liu, L., Ramaswamy, L., Pu, C.: PeerCast: churn-resilient end system multicast on heterogeneous overlay networks. J. Netw. Comput. Appl. **31**(4), 821–850 (2008)
13. Hua, K.A., Tran, D.A., Villafane, R.: Overlay multicast for video on demon the internet. In: Proceedings of the 2003 ACM Symposium on Applied Computing, Melbourne, Florida, pp. 935–942. Association for Computing Machinery (2003)
14. Bondan, L., Müller, L.F., Kist, M.: Multiflow: multicast clean-slate with anticipated route calculation on OpenFlow programmable networks. J. Appl. Comput. Res. **2**(2), 68–74 (2012)
15. Liao, S., Wu, C., Hong, X., Zhu, K., Chen, S.: Virtualized platform for multicast services in software defined networks. Chin. J. Electron. **26**(3), 453–459 (2017)
16. Yu, Y., Zhen, Q., Xin, L., Shanzhi, C.: OFM: a novel multicast mechanism based on Openflow. Adv. Inf. Sci. Serv. Sci. **4**(9), 278–286 (2012)
17. Lin, Y.-D., Lai, Y.-C., Teng, H.-Y., Liao, C.-C., Kao, Y.-C.: Scalable multicasting with multiple shared trees in software defined networking. J. Netw. Comput. Appl. **78**, 125–133 (2017)
18. Huang, L.-H., Hung, H.-J., Lin, C.-C., Yang, D.-N.: Scalable bandwidth-efficient multicast for software-defined networks. In: Proceedings of 2014 IEEE GLOBECOM, Austin, TX, USA, pp. 1890–1896. IEEE (2014)
19. Rückert, J., Blendin, J., Hausheer, D.: Software-defined multicast for over-the-top overlay-based live streaming in ISP networks. J. Netw. Syst. Manag. **23**(2), 280–308 (2015)

20. Islam, S., Muslim, N., Atwood, J.W.: A survey on multicasting in software-defined networking. IEEE Commun. Surv. Tutor. **20**(1), 355–387 (2018)
21. Alsaeedi, M., Mohamad, M.M., Al-Roubaiey, A.A.: Toward adaptive scalable OpenFlow-SDN flow control: a survey. IEEE Access **7**, 107346–107379 (2019)
22. Priya, A.V., Radhika, N.: Performance comparison of SDN OpenFlow controllers. Int. J. Comput. Aided Eng. Technol. **11**(4/5), 467–479 (2019)
23. Bannour, F., Souihi, S., Mellouk, A.: Distributed SDN control: survey, taxonomy, challenges. IEEE Commun. Surv. Tutor. **20**(1), 333–354 (2018)
24. Asadollahi, S., Goswami, B., Sameer, M.: Ryu controller's scalability experiment on software defined networks. In: Proceedings of 2018 IEEE International Conference on Current Trends in Advanced Computing (ICCTAC), Bangalore, India. IEEE (2018)
25. Mišić, M.J., Gajin, S.R.: Simulation of software defined networks in Mininet environment. In: Proceedings of 2014 22nd Telecommunications Forum Telfor (TELFOR), Belgrade, Serbia, pp. 1055–1058. IEEE (2014)
26. Medina, A., Lakhina, A., Matta, I., Byers, J.: BRITE: an approach to universal topology generation. In: Proceedings of Proceedings Ninth International Symposium on Modeling, Analysis and Simulation of Computer and Telecommunication Systems, Cincinnati, OH, USA, pp. 346–353. IEEE (2001)

Interference Coordination Using Cell Cluster for 5G Dynamic TDD System

Junping Liu, Zekai Liu, Nan Liu$^{(\boxtimes)}$ (iD), Zhiwen Pan, and Xiaohu You

National Mobile Communications Research Laboratory, Southeast University,
Nanjing 210096, China
{zkliu,nanliu,pzw,xhyu}@seu.edu.cn

Abstract. The deployment of small cells of dynamic time division duplexing (TDD) using orthogonal frequency division multiple access (OFDMA) is one of the key technologies of 5th-generation (5G). The main problem limiting the performance of dynamic TDD is the existence of cross-link interference in these mobile networks. In order to adapt to the new scenario of 5G, on the basis of the traditional clustering scheme, this paper proposes a new clustering criterion considering both the user equipment to user equipment (UE-UE) interference and base station to base station (BS-BS) interference, and realizes dynamic clustering of the cells. The proposed method can simultaneously solve the uplink and downlink interference problems, and improve the system performance. Compared with the traditional clustering algorithm, the scheme proposed in this paper has certain improvement in the throughput of the system, especially in the downlink.

Keywords: Dynamic TDD · Cross-link interference · Cell cluster · Femtocell · Ultra-dense network

1 Introduction

With the rapid development of mobile Internet, the next generation cellular network, the 5th-generation (5G) mobile communication system will emerge in order to cope with the explosive growth of mobile data traffic. One of the key technologies of 5G is the deployment of small cells of dynamic time division duplexing (TDD) using orthogonal frequency division multiple access (OFDMA). The main goal of dynamic TDD is to realize the flexible utilization of the downlink (DL) and uplink (UL) spectrum of small cells, and to meet the needs of asymmetric and dynamically changing upstream and downstream traffic. However, dynamic TDD mode has high and uncontrollable interference, especially the cross-link interference (CLI) between base stations (BSs) and between user equipments (UEs). BSs in UL receiving state will be interfered by other BSs, while UEs in DL receiving state will be interfered by adjacent UEs [1].

Moreover, low-power nodes is introduced in 5G NR system. By placing high-density low-power nodes, the capacity and the coverage rate of the 5G NR system

© ICST Institute for Computer Sciences, Social Informatics and Telecommunications Engineering 2021
Published by Springer Nature Switzerland AG 2021. All Rights Reserved
H. Gao et al. (Eds.): ChinaCom 2020, LNICST 352, pp. 244–255, 2021.
https://doi.org/10.1007/978-3-030-67720-6_17

can be improved greatly. However, dense deployment of low-power nodes can lead to serious inter cell interference (ICI). Inter cell interference coordination technology (ICIC) is the mainstream technology to suppress interference between cells, through collaboration between cells of wireless resource management [2]. As a solution to interference mitigation and traffic adaption (IMTA) problem, cell cluster is widely accepted [3]. With cell cluster, cells with severe mutual interference are grouped in the same cluster, and the same configuration is adopted in the same cluster to reduce interference. In [4], the paper pointed out that the large scale of cell cluster reduced the flexibility of traffic adaptation, resulting in the low DL performance gain of pico cell. Based on the traditional cell clustering, an optimized reconfiguration algorithm was proposed in [5]. [6] proposed an objective function of the cell cluster based on IMTA, combined with the advantages of threshold value and the heuristic algorithm, respectively as the short-term and long-term planning. The scheme is superior to the traditional cell cluster solution, especially in the uplink, and the performance in the downlink are identical. A cluster classification mechanism was proposed in [7] to divide the sub-regions into different groups according to the degree of interference of each base station. The base station could use adaptive adjustment of TDD uplink and downlink configuration to increase the transmission capacity of the system, thus reduce the downlink to uplink (DL-UL) interference and improve the network throughput.

However, the above studies only consider the interference from base station to base station (BS-BS), but ignore the interference from user equipment to user equipment (UE-UE). The main contributions of this paper are as follows. Firstly, in the dynamic TDD network under the 5G scenario, we propose a dynamic clustering algorithm considering the interference between UEs to carry out interference coordination. Then the algorithm we proposed is simulated with the indoor dual stripe-building model, compared with no interference coordination scenario and traditional cell cluster scheme. The results show the proposed scheme can suppress the CLI caused by the dynamic TDD technology in 5G indoor single-layer ultra-dense network (UDN), and improves the system throughput.

The rest of the paper is organized as follows. In Sect. 2, we introduce the interference problem in the dynamic TDD network. In Sect. 3, we introduce the principle of interference coordination by clustering, and elaborate our proposed scheme. In Sect. 4, the proposed scheme is implemented by system-level simulation. Finally, in Sect. 5, we summarize and discuss our work.

2 Interference Problem in Dynamic TDD Network

In dynamic TDD network, there are two kinds of interferences, i.e. co-direction interference and CLI, while there are only co-direction interference in the static TDD mode.

Compared with the static TDD technology, the information transmission direction of the cell may or may not be consistent due to the dynamic change of TDD configuration in time and space. The asynchronous operation of dynamic TDD causes CLI. There are two forms of CLI, UE-UE interference and BS-BS interference.

In this case, the BS on the UL will experience interference from other DL BSs, while the UE on the DL will experience interference from the UL UE of adjacent cell.

An example is given as shown in Fig. 1. Consider two adjacent BSs, each of which has a severely interfered UE at the edge of its coverage area. It is assumed that BS1 and BS2 select frame structure configuration 2 and configuration 1 respectively according to their current traffic loads. In Fig. 1, BS2 is disturbed by DL signal from BS1 at the third sub-frame, resulting in a decrease in signal quality received by BS2. On the other hand, DL signal from BS1 to UE1 is disturbed by UL data transmission from UE2.

Existing researches are mainly aimed at how to eliminate the interference between BSs. Because in a typical deployment scenario, the transmission power of BSs is far higher than the transmission power of UEs. However, the combination of ultra-dense network (UDN) with dynamic TDD allows for the extensive use of ultra-dense cells and low-power nodes. In this case, the UE-UE interference should be paid close attention to, especially when two UEs are located at the nearest edge of two cells, as shown in Fig. 2. In addition, since the location of public facilities is not known, it is unreasonable to just consider the coupling loss between BSs as the clustering criterion. Therefore, the path loss between the closest UEs in the two cells also needs to be considered.

3   Cell Clustering Interference Mitigation

3.1   The Basic Idea of Clustering

The clustering algorithm allocates cells to clusters according to specific rules. Cells are divided into clusters and each cluster can contain one or more cells. The cells in the same cluster adopt the same UL/DL sub-frame configuration to reduce the BS-BS interference and UE-UE interference. However, the transmission direction

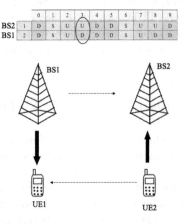

Fig. 1. Dynamic TDD cross-link interference

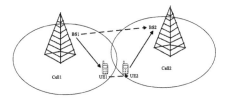

Fig. 2. Serious UE-UE interference scenarios

of cells in different cell clusters can be different. The rules of clustering must take into account the level of interference and asymmetric UL and DL traffic. In different scenarios, there are significant differences in the objectives and criteria of clustering schemes. Figure 3 is an example of an interference coordination scheme using cell cluster, where the femtocells operating at the same frequency. In Fig. 3, the low-power access point network is divided into several clusters, each containing one or more cells. Thus, cells with strong CLI are organized in the same cluster e.g. Cluster1 and Cluster2, and same UL/DL sub-frame configuration is used within a cluster. Therefore, there will be no CLI between cells in the same cluster. And CLI only exists between cells of different clusters with different transmission directions. In 5G ultra-dense network, due to the extensive use and intensive deployment of low-power femtocells, the UE-UE interference must also be considered. Therefore, in the new clustering criterion, the coupling loss between BSs and between UEs should be taken into account at the same time. The goal of the proposed scheme is to improve both the UL and DL throughput of the system.

3.2 New Clustering Scheme

In order to consider both UL and DL performance of the system, we give the formulas for calculating the signal to interference and noise ratio (SINR) of UL and DL respectively. For the j-th UE served by the BS in the i-th cell, SINR in UL is defined as in (1). For the j-th UE connected to the BS in the i-th cell, the DL SINR is defined as in (2).

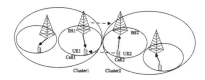

Fig. 3. Dynamic TDD small cluster interference coordination diagram

$$SINR_{i,j}^{UL} = \frac{S_{j,i}^{UL}}{\sum\limits_{k=1,k\neq i}^{K} I_{k,i}^{DL-UL} + \sum\limits_{l=1,l\neq i}^{L} I_{l,i}^{UL-UL} + N_0} \tag{1}$$

$$SINR_{i,j}^{DL} = \frac{S_{i,j}^{DL}}{\sum\limits_{k=1,k\neq i}^{K} I_{k,j}^{DL-DL} + \sum\limits_{l=1,l\neq i}^{L} I_{l,i}^{UL-DL} + N_0} \tag{2}$$

$S_{j,i}^{UL}$ is the power spectral density of the UL signal of the j-th UE received by the i-th BS, $S_{i,j}^{DL}$ is the power spectral density of the DL signal of the i-th BS received by the j-th UE. K is the total number of BSs transmitted in the DL direction, and L is the total number of UEs transmitted in the UL direction. $I_{k,i}^{DL-UL}$ is the power spectral density of the DL interference signal received by the i-th BS from the k-th BS, $I_{l,i}^{UL-UL}$ is the power spectral density of the UL interference signal received by the i-th BS and transmitted by the l-th UE. $I_{k,j}^{DL-DL}$ is the power spectral density of the DL signal transmitted by the k-th BS and received by the equipment of the j-th UE, $I_{l,i}^{UL-DL}$ is the power spectral density of the UL interference signal transmitted by the l-th UE and received by the j-th UE, N_0 is the average power spectral density of AWGN. The BS-BS interference and UE-UE interference can be calculated in (3) (4).

$$I_{k,i}^{DL-UL} = P_k + TAG_k + RAG_i - PL_{k,i}(dB) \tag{3}$$

$$I_{l,j}^{UL-DL} = P_l + TAG_l + RAG_j - PL_{l,k}(dB) \tag{4}$$

P_k and P_l represent the power of the k-th BS and j-th UE respectively, TAG_k and TAG_l represent the transmitting antenna gain of the k-th BS and the l-th UE respectively, RAG_i and RAG_j represent the receiving antenna gain of the i-th BS and j-th UE respectively, $PL_{k,i}$ is the propagation loss from k-th BS to i-th BS. $PL_{l,k}$ is the propagation loss from k-th BS to i-th BS.

$$CL_{k,i}^{BS} = PL_{k,i}^{BS} - TAG_k^{BS} - RAG_i^{BS}(dB) \tag{5}$$

$$CL_{k,i}^{UE} = \frac{\sum\limits_{n_k=1}^{N_k} \sum\limits_{n_i=1}^{N_i} (PL_{n_k,n_i}^{UE} - TAG_{n_k}^{UE} - RAG_{n_i}^{UE})}{N_k \cdot N_i} \tag{6}$$

The coupling loss between the BS of i-th cell and the BS of k-th cell can be calculated in (5). The average coupling loss between UEs in i-th cell and UEs in k-th cell can be calculated in (6). $PL_{k,i}^{BS}$ is the path loss between the BSs in i-th cell and k-th cell. PL_{n_k,n_i}^{UE} represent the path loss between UE n_i in i-th cell and UE n_k in k-th cell. TAG_k^{BS} and $TAG_{n_k}^{UE}$ represent the transmitting antenna gain of the k-th BS and the UE n_k respectively. RAG_i^{BS} and $RAG_{n_i}^{UE}$ represent the receiving antenna gain of the i-th BS and the UE n_i in i-th cell respectively. N_k is the total number of UEs in k-th cell, and N_i is the total number of UEs in i-th cell.

For dynamic cell cluster strategy in dynamic TDD network, we present the definition of different metric in (7). The difference metric (DM) of i-th cell and j-th cells is related to three factors, the coupling loss of BSs between i-th cell and j-th cell, the average path loss of users between i-th cell and j-th cell, and the difference of traffic between i-th cell and j-th cell. The weights of these three factors are δ, β, λ.

$$DM_{i,j} = \delta \frac{CL_{i,j}^{BS}}{\overline{CL^{BS}}} + \beta \frac{CL_{i,j}^{UE}}{\overline{CL^{UE}}} + \lambda \frac{|R_i^{Cell} - R_j^{Cell}|}{\overline{R^{Cdl}}} \qquad (7)$$

$CL_{i,j}^{BS}$ represents the coupling loss between the i-th BS and the j-th BS, $\overline{CL^{BS}}$ represents the mean value of coupling loss between BSs, $CL_{i,j}^{UE}$ represents the average coupling loss between users of the i-th cell and the j-th cell, $\overline{CL^{UE}}$ represents the average coupling loss among users of all cells, $|R_i^{Cell} - R_j^{Cell}|$ is the absolute value of the difference between the ratio of DL and UL of i-th cell and j-th cell, $\overline{R^{Cdl}}$ represents the average ratio of DL and UL in all cells.

The process of cell cluster includes the following steps:

1) DM can be calculated as long as the cell obtains coupling loss and traffic buffer information. A clustering matrix $\mathbf{DM} = [DM_{i,j}]$ can be obtained according to the calculation result of DM. When $i = j$, $DM_{i,j} = 0$.
2) Set the threshold DM_0 for clustering. Firstly, randomly select a cell with SINR lower than the set threshold as the master cell of the cluster, and then check the DM of other cells and the master cell. If the DM between a cell and the master i-th cell is greater than the DM_0, the cell and the master cell are grouped into the same cluster. Based on the above initial cluster, all other cells except the master cell loop the above operation, put the cell with the DM greater than the threshold DM_0 into the initial cluster, to form the first cluster.
3) Then, randomly select a cell whose $SINR$ is lower than the set threshold and is not in the cluster as the master cell of the cluster, and conduct the above operation for all remaining cells until all cells whose $SINR$ is lower than the threshold have been grouped into the clusters.
4) After that, the master cell that is not in the cluster is randomly selected as the master cell of the cluster, and the above operation is carried out for all remaining cells until all cells have been divided into disjoint cell clusters. When the SINR of the UL/DL of any i-th cell are below the threshold or the timer starts again, the cells are regrouped.

After clustering, the UL and DL reconfiguration of cells in the cluster is carried out. Seven frame configurations defined in TDD are shown in Fig. 4. Then, before the start of each configuration cycle, the ratio of DL to UL traffic in each cell is calculated for each cluster, and the average ratio of DL to UL traffic in the cluster is calculated. From the TDD configuration table, select the configuration mode in which the ratio of DL to UL sub-frames is closest to this result, as described in (8).

$$i_x = \underset{i=0,1,2,...,6}{\arg\min} |R_i - R_{Cluster}|. \qquad (8)$$

Config	Subframe number										Up-Link	Down-Link
	0	1	2	3	4	5	6	7	8	9		
0	D	S	U	U	U	D	S	U	U	U	6	2
1	D	S	U	U	D	D	S	U	U	D	4	4
2	D	S	U	D	D	D	S	U	D	D	2	6
3	D	S	U	U	U	D	D	D	D	D	3	6
4	D	S	U	U	D	D	D	D	D	D	2	7
5	D	S	U	D	D	D	D	D	D	D	1	8
6	D	S	U	U	U	D	S	U	U	D	5	3

←————————————————10ms————————————————→

| D | Downlink subframe | | S | Special subframe | | U | Downlink subframe |

Fig. 4. Seven frame configurations for TDD mode.

R_i is the ratio of the DL sub-frames to the upper row subframes of the i-th configuration, R_b is the ratio of the downlink flow to the uplink flow in the actual cache of the cell cluster, R_i is calculated in (9) and $R_{Cluster}$ is calculated in (10). In (10), the ratio of the sum of the DL flow to the sum of the UL flow in the cluster is taken as the ratio of the DL flow to the UL flow of the cluster.

$$R_i = \frac{N_i^{DL} + N_i^S \cdot R_{DwPTS}}{N_i^{UL} + N_i^S \cdot R_{UwPTS}} \tag{9}$$

N_i^{DL} is the number of DL sub-frames in the i-th configuration, N_i^{UL} is the number of UL sub-frames in the i-th configuration, R_{DwPTS} is the ratio of DL pilot time slot (DwPTS) in the special subframe, R_{UwPTS} is the ratio of UL pilot time slot (UwPTS) in a special subframe.

$$R_{Cluster} = \frac{\sum_{n=1}^{N} B_n^{DL}}{\sum_{n=1}^{N} B_n^{UL}} \tag{10}$$

B_n^{DL} and B_n^{UL} represent the DL flow and UL flow of the n-th cell in the cluster respectively.

4   Simulation Results

The first step of simulation is to build the dual strip buildings model. The second step is to decorate femto BSs in combination with UE, and then set the parameters of the simulation according to Table 1. After that, we simulate three schemes in parallel, namely, the original cell without the interference regulation scheme, the IMTA scheme with the traditional clustering method and the scheme we proposed, and finally collected the simulation data and analyzed the results.

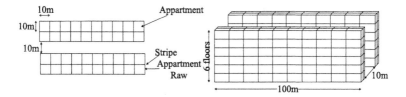

Fig. 5. Dual strip building model

In high-density residential buildings, the deployment of the femtocells is often lack of planning. This is the most complex case, so this article focuses on this case and uses the dual strip model as Fig. 5. Each building has six floors, with two rows of rooms on each floor and 10 rooms in one row. The length, width and height of each room are 10 m, 10 m and 3 m respectively. It is assumed that the femto BS and UE have similar transmitting power. The distribution of femto BS and UE is random. Firstly, the number of i-th cells designated as K, and then K rooms are randomly selected. With each room selected, a femto BS and multiple UEs are randomly scattered.

4.1   System-Level Simulation Model

This paper refers to the path loss models of [8]. In these indoor models, the penetration loss of walls and floors is considered. The penetration loss of signals passing through the walls of buildings as shown in (11).

$$PL = 38.46 + 20 \log_{10} R + 0.7 d_{2d,indoor}$$
$$+ 18.3n((n+2)/(n+1) - 0.46) + p \cdot L_{ow} + q \cdot L_{iw}(dB) \tag{11}$$

R represents the distance between the transmitter and the receiver, $d_{2d,indoor}$ is the distance inside the building. n represents the number of floors between transmitter and receiver. p, q are the number of outer walls and inner walls between the transmitter and receiver respectively, L_{ow} and L_{iw} represents the increased path loss for outer wall and inner wall, and each wall within the building will increase the path loss by 20 dB and 5 dB respectively.

4.2   Metric and Results

The performance metric of this simulation is the throughput of cells. For the convenience of link modelling, the proportional Shannon capacity model proposed in [5] is selected to simulate the influence of different physical layer parameters on link performance. Shannon's theorem can be used to obtain the Shannon capacity as throughput, as shown in (12).

$$U = \begin{cases} 0 & \gamma < \gamma_{min} \\ \alpha \cdot W \cdot \log_2 (1 + \beta \cdot \gamma) & \gamma_{min} \leq \gamma < \gamma_{max} \\ U_{max} & \gamma \leq \gamma_{max} \end{cases} \tag{12}$$

Table 1. Simulation parameters

Parameter	Value
Network deployment	Dual strip model
Carrier frequency	2 GHz
The system bandwidth	10 MHz
Number of single cell users	1
Antenna pattern	Omni antenna
Femto power	23 dBm
Femto noise	13 dB
Antenna gain	0 dBi
Uplink power control	$p_0 = -75\,\text{dBm}, \alpha = 0.8$
UE antenna gain	0 dBi
UE noise	9 dB
UE maximum power	23 dBm
Dynamic configuration cycle	10 ms

Throughput for a given link is denoted by U, and the SINR for such link is expressed by γ. γ_{min} and γ_{max} represent the minimum SINR required for adaptive coding and the SINR when the maximum throughput is reached respectively, U_{max} represents the maximum throughput achieved by adaptive modulation coding, α, β represents the attenuation factors, and W denotes channel bandwidth.

We carried out the simulation for all cases with cells number from 1 to 240 under the carrier frequency of 2 GHz, the values of the UL throughput and DL throughput of all cells in the system with cell number changed were counted. The simulation results are shown in Fig. 6 and Fig. 7.

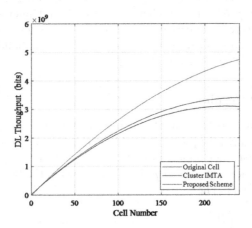

Fig. 6. DL throughput of all cells

Table 2. UL and DL throughput

Scheme	Uplink (Mbits)		Downlink (Mbits)	
	120 cells	240 cells	120 cells	240 cells
Original cell	1739 (baseline)	2854 (baseline)	2437 (baseline)	3091 (baseline)
Cluster IMTA	2029 (16.7%)	3395 (18.9%)	2546 (4.5%)	3405 (10.1%)
Proposed scheme	2039 (17.3%)	3405 (19.3%)	3051 (25.2%)	4740 (53.3%)

Figure 6 shows the change of DL throughput with the number of cells in three schemes. With the increase of the number of cells, the DL throughput of the three simulation schemes increase. However, when the number of cells reached to a certain number, due to serious CLI, the DL throughput of original cells increased slowly or even decreased. The traditional cluster IMTA based on BS-BS coupling loss can slightly improve this situation, and has a certain gain on the DL throughput, but it is unable to solve the interference problem good enough to ensure the throughput continue to increase with the number of cells increasing. The proposed scheme can significantly solve interference problems and greatly improve the system's DL throughput, especially in the case of numerous cells. Figure 7 shows the change of UL throughput with the number of cells in three schemes. Both the traditional cluster IMTA scheme based on BS-BS coupling loss and our scheme can improve the UL throughput of the system largely. Specific data are listed in Table 2. Under the condition of the distribution of 120 cells, original cells without interference coordination have a total UL throughput of 1739 Mbits, traditional cluster IMTA can have a 16.7% gain in UL throughput, and the proposed scheme has a 17.3% gain. For the DL throughput, original cells without interference coordination have a total DL throughput of 2437 Mbits, traditional cluster IMTA can only has about 4.5% gain in DL throughput, while the proposed scheme has a gain of 25.2%. Under the condition of the distribution of 240 cells, original cells without interference have a total UL throughput of 2854 Mbits, cluster IMTA can have

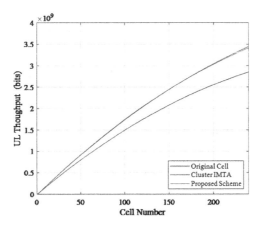

Fig. 7. UL throughput of all cells.

an 18.9% gain in UL throughput, and the proposed scheme has a 19.3% gain. For the DL throughput, original cells without interference coordination have a total DL throughput of 2437 Mbits, and the scheme we proposed have a gain of 53.3%, while cluster IMTA can only has about 10.1% in DL throughput.

5   Conclusion and Discussion

In the 5G UDN scenario, the BS often adopts micro-small BS. In this case, different from the case in LTE, the power of the BS is not significantly different from that of the UE, even in the same order of magnitude. Therefore, the interference between users need to be considered in 5G scenario. Based on the traditional clustering scheme, this paper proposes a new clustering criterion considering both the UE-UE interference and BS-BS interference, and realizes dynamic clustering of the cells. Simulation results show that the proposed method can simultaneously solve the UL and DL interference problems to improve the system performance. Compared with the traditional clustering algorithm, proposed scheme has certain improvement in the SINR and throughput of the system, especially in the DL.

Acknowledgment. This work is partially supported by the National Key Research and Development Project under Grants 2019YFE0123600 and 2020YFB1806800, the National Natural Science Foundation of China under Grants 62071115, and the European Union's Horizon 2020 research and innovation programme under the Marie Sklodowska-Curie grant agreement No 872172 (TESTBED2 project: www.testbed2.org).

References

1. Yun, D.W., Lee, W.C.: LTE-TDD interference analysis in spatial, time and frequency domain. In: 2017 Ninth International Conference on Ubiquitous and Future Networks (ICUFN), pp. 785–787. IEEE (2017)
2. Altabbaa, M.T., Arsan, T., Panayirci, E.: Power control and resource allocation in TDD-OFDM based femtocell networks with interference. In: 2017 IEEE International Black Sea Conference on Communications and Networking (BlackSeaCom), pp. 1–5. IEEE (2017)
3. Zhu, D., Lei, M.: Cluster-based dynamic DL/UL reconfiguration method in centralized ran TDD with trellis exploration algorithm. In: 2013 IEEE Wireless Communications and Networking Conference (WCNC), pp. 3758–3763. IEEE (2013)
4. Sun, F., Zhao, Y., Sun, H.: Centralized cell cluster interference mitigation for dynamic TDD DL/UL configuration with traffic adaptation for HTN networks. In: 2015 IEEE 82nd Vehicular Technology Conference (VTC2015-Fall), pp. 1–5. IEEE (2015)
5. Li, X., Ma, N., Tang, Q., Liu, B.: Buffered DL/UL traffic ratio sensing cell clustering for interference mitigation in LTE TDD system. In: 2018 IEEE Wireless Communications and Networking Conference (WCNC), pp. 1–6. IEEE (2018)
6. Nasreddine, J., Hassan, S.E.H.: Interference mitigation and traffic adaptation using cell clustering for LTE-TDD systems. In: 2016 IEEE International Multidisciplinary Conference on Engineering Technology (IMCET), pp. 155–159. IEEE (2016)

7. Cheng, M.H., Wang, Y.J., Hwang, W.S., Wu, Y.J., Lin, C.H.: Adaptive adjustment of TDD uplink-downlink configuration based on cluster classification in beyond LTE heterogeneous networks. In: 2017 International Conference on Applied System Innovation (ICASI), pp. 1312–1315. IEEE (2017)
8. Jundhare, M.D., Kulkarni, A.: An overview and current development of femtocells in 5G technology. In: 2016 IEEE International Conference on Advances in Electronics, Communication and Computer Technology (ICAECCT), pp. 204–209. IEEE (2016)

CPP-Based Cooperative Defense Against DoS Attacks in Future Non-terrestrial Networks

Zhaori Cong$^{(\boxtimes)}$, Zhilong Zhang, and Danpu Liu

Beijing Laboratory of Advanced Information Network,
Beijing Key Laboratory of Network System Architecture and Convergence,
Beijing University of Posts and Telecommunications,
Beijing 100876, People's Republic of China
694660532@qq.com, {zhangzhilong,dpliu}@bupt.edu.cn

Abstract. In future non-terrestrial networks, satellites with sufficient computing resources are expected to serve as base stations or access points, which changes the structure of traditional satellite networks into a more flexible environment. For such satellite nodes, denial of service (DoS) attacks may become a potential secure threat that should be prevented. Existing researches mainly focus on the defense against DoS attacks on ground nodes and have no consideration of the attacks on future satellites. Moreover, the problem of reducing the access delay when a satellite is under DoS attack has not been addressed. In this paper, we study the DoS attack defense strategy in non-terrestrial networks. By adopting client puzzle protocol (CPP) and load balancing, we propose a cooperative defense strategy where multiple auxiliary nodes are used to help the attacked node to process the intensive attack requests. An access delay minimization problem to optimize the selection of auxiliary nodes, puzzle difficulty as well as traffic offload ratios is then formulated based on queuing theory and solved. Simulation results show that the proposed scheme not only improve the network's anti-attack capability, but also achieves desirable performance in average access delay.

Keywords: Client puzzle · DoS Attack · Non-terrestrial networks · Delay

1 Introduction

Non-terrestrial networks (NTN) has been included in 3GPP's 5G standards due to its wide coverage, strong disaster resistance, and low vulnerability to physical attacks on the ground. The perspective of the future 6G network also includes the NTN architecture, where satellites not only work as relay nodes, but also can be base stations and capable of performing simple calculations. However, non-ground nodes are easy to become targets of DoS attacks due to the limited

H. Gao et al. (Eds.): ChinaCom 2020, LNICST 352, pp. 256–267, 2021.
https://doi.org/10.1007/978-3-030-67720-6_18

computing capability. Therefore, the defense against DoS attacks in the weak authentication process[1] is essential to be concerned.

DoS attack refers to attackers' repeated and intensive requests within a short time, which prevent legal users from being normally served. At present, the main form of DoS attacks is resource-consuming attack. By consuming network bandwidth, computing capacity or other useful resources, attackers will make the systems overwhelmed and paralyzed. Traditional methods of defense against DoS attacks include strengthening the stability of the operating system [1], using firewalls to resist [2,3], bandwidth limitation and QoS guarantee [4], as well as traffic load balancing [5]. These methods are effective in preventing DoS attacks with small traffic and simple structure. However, with the increase of network bandwidth and the use of distributed technologies for DoS attacks, it's difficult for the traditional defense methods to meet the needs. Therefore, the client puzzle protocol (CPP) [6,7] has been proposed to address this issue. In this design, a client has to calculate and obtain the answer of a specified puzzle with a certain degree of difficulty before sending out a request, thus the frequency of requests made by attackers is significantly reduced. There have been some existing studies using CPP to resist DoS attacks on different types of system resources. For example, in [8], the authors propose a light encryption mechanism suitable for the IoT environment. In [9], a lightweight cache reliability problem is addressed. To sum up, the recent research on CPP-based defense mainly focus on designing the function form of the puzzle to make the overall access process more secure.

As in future non-ground networks, satellite nodes may assume the function of authentication. Therefore, some new problems need to be considered. On the one hand, the computing capacity of satellite nodes is limited. The ability to withstand attacks is not as good as ground nodes. On the other hand, the propagation delay of satellite network is large. Weak authentication may increase the access delay of normal users and affect the normal communication. Therefore, security is not the only factor to be considered, the optimization of user's access delay also needs to be addressed.

Motivated by above analysis, we design a weak access authentication mechanism for NTN based on CPP and load balancing. When a DoS attack occurs, multiple nodes around the attacked node are selected to act as auxiliary nodes. The attacked node makes puzzle based on the number of requests it receives, and distributes the solution it receives again to the auxiliary nodes. The auxiliary nodes help verify the correctness of the solution. Thus the nodes anti-attack capability is greatly improved. Furthermore, we establish the delay model based on queuing theory. The model is solved by taking the minimum access delay as the optimization goal, and the optimized selection of auxiliary nodes, puzzle difficulty as well as traffic offload ratios are obtained. Finally, simulation results show that our proposed scheme can effectively reduce user's access delay under DoS attack and improve the ability to resist attacks.

---

[1] Weak authentication is usually performed before strong authentication (i.e., identity authentication).

The rest of this paper is organized as follows. Section 2 describes the proposed CPP-based cooperative weak authentication scheme and formulates an access delay minimization problem. Section 3 provides a solution to this optimization problem. Section 4 simulates the algorithm and verifies the optimization of the algorithm proposed in this paper to the delay. Section 5 summarizes this paper.

2   System Model and Problem Formulation

In future NTN, satellites can serve as base stations or access points. If extra computing resources are deployed on the satellites, they can perform user authorization and other calculations. However, these satellites may be targeted by DoS attackers as well.

As shown in Fig. 1, the NTN includes satellites nodes, ground nodes, normal users and DoS attackers who control a lot of zombie hosts. We assume that each satellite will use CPP against DoS attacks. When the attacked satellite receives a large number of access requests from both attackers and legitimate users, it will collect the resource information of its surrounding satellite nodes. Through these information, the attacked node will determine the difficulty of puzzle and the optimal auxiliary nodes to cooperatively defend the DoS attacks. As a result, when the attacked node receives the puzzle's solutions returned by the attackers and legitimate users, it will distributes these solutions to the auxiliary nodes. After the auxiliary nodes complete the verification of the solutions, the whole weak authentication process ends.

According to above design of cooperative defense among satellite nodes, it is clear that the single satellite nodes ability against DoS attack will be greatly improved with the introduction of multiple auxiliary nodes. However, solving a puzzle will greatly increase the access delay of the legitimate users, especially in the NTN scenario where the propagation delay between the satellite and the ground is very large. In this case, how to improve the legitimate users delay performance through the optimization on the difficulty of the puzzle, as well as the selection of auxiliary nodes and traffic allocation among them becomes an important issue, which will be addressed in the following sections.

2.1   Client Puzzle and Load Balancing Model

Let $\mathcal{N} = \{n_1, n_2, \ldots, n_N\}$ denote the set of all nodes that can assist to balance the traffic load, including satellites and ground stations. When the attacked node n_0 detects an abnormal number of access requests, it selects the auxiliary nodes and determines the traffic offloading ratio for each node. Let k_i denote node selection indicator, which is given by

$$k_i = \begin{cases} 0, \text{node } n_i \text{ is not selected as an auxiliary node} \\ 1, \text{node } n_i \text{ is selected as an auxiliary node} \end{cases} \tag{1}$$

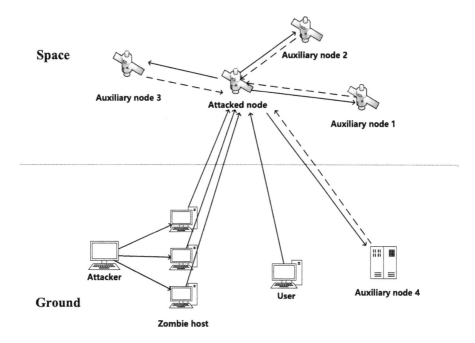

Fig. 1. System model of defense against DoS attack in NTN.

Moreover, let a_i be the offloading ratio of node i and formulate a ratio set $\mathcal{A} = \{a_0, a_1, ..., a_N\}$, which is constrained by

$$\sum_{i=0}^{N} k_i a_i = 1, i \in \{0, 1, 2, ..., N\} \tag{2}$$

Then, node n_0 generates a puzzle with difficulty j and sends it to the users and attackers, where

$$j \in \{0, 1, ..., P\} \tag{3}$$

Both legitimate users and attackers need to consume their own resources to solve the puzzle and send the solutions to node n_0. These solutions are then offloaded to the nodes in the auxiliary network for verification according to the established traffic allocation scheme.

2.2   Delay Model Based on Queuing Theory

Assume that the solutions of users and attackers arrive at node n_0 according to Poisson distribution. Due to the use of CPP for defense, different computing time for solving a puzzle leads to the changing of total average arrival rate, which is determined by difficulty j and denoted as $\lambda_0(j)$. Therefore, the average arrival rate of the traffic offloaded to node i is given by:

$$\lambda_{0i}(j) = a_i \lambda_0(j) \tag{4}$$

Let $f(j)$ denote the computational resource consumed by the puzzle with difficulty j. The resource consumption has an exponential relationship with difficulty, which is given by $f(j) = K_1 2^{K_2 j}$ [10]. Assume that the time for attackers to process the puzzle can be expressed as:

$$t_{pa}(j) = \frac{f(j)}{C_a} \tag{5}$$

where C_a (cycles per second) is the computing capability of attackers.

Similarly, the time for users to process the puzzle can be expressed as:

$$t_{pu}(j) = \frac{f(j)}{C_u} \tag{6}$$

where C_u is the computing capability of a typical user.

For both legitimate users and attackers, calculating the puzzle will result in a lower request arrival rate. But compared with the users' relatively long request interval, the time to calculate the puzzle can be ignored. Therefore, the arrival rate of users' solutions is set to a constant λ_u. On the other hand, the attackers' initial frequency of sending requests is quiet high, so computing puzzle can effectively delay making solutions and reduce the arrival rate of solutions. The larger the difficulty of puzzle it is, the more time attackers takes to compute, and the more the attack frequency decreases. However, the access delay for the legitimate user will also be larger with the increase of puzzle difficulty.

Given that the average time between two requests from the same attacker is approximately the time to process a puzzle $t_{pa}(j)$, the total average arrival rate at a given puzzle difficulty j is:

$$\lambda_0(j) = \lambda_u + \frac{1}{t_{pa}(j)} \tag{7}$$

Assume that the maximal number of solutions verified by satellite node n_i per unit time is μ_i, which is determined by the computing capability of the auxiliary nodes. Further assume that the maximal number of solutions transmitted from n_0 to n_i per unit time is μ_i', which is determined by the bandwidth resources.

The delay for solution verification on the attacked node is different from that on the auxiliary nodes. When a solution is verified on the attacked node, only the processing delay needs to be considered. When a solution is verified on an auxiliary node, the total delay is composed by three parts: the transmission delay between the attacked node and the auxiliary node, the processing delay, and the propagation delay in space transmission. In this paper, we use M/M/1 to model transmission delay and processing delay. Because the above two delays are independent of each other, we use two M/M/1 in series for modeling [11], which is shown in Fig. 2.

Therefore, the average authentication delay of each request is given by:

$$T_i(j) = \begin{cases} \frac{1}{\mu_i - \lambda_{0i}(j)} + \frac{1}{\mu_i' - \lambda_{0i}(j)} + t_{pi}, & \text{on auxiliary nodes} \\ \frac{1}{\mu_i - \lambda_{0i}(j)}, & \text{on attacked node} \end{cases} \tag{8}$$

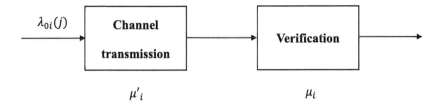

Fig. 2. A queuing model of attacked nodes to auxiliary nodes.

where the first line represents the average authentication delay of requests which are processed on auxiliary nodes. t_{pi} means propagation delay which depends on the distance between node 0 and node i. The second line represents the authentication delay of requests which are processed on the attacked node.

The probability that a typical user's access request is processed at n_i is a_i, so the average access delay of the users in the system can be expressed by:

$$T = t_{pu}(j) + \sum_{i=0}^{N} k_i a_i T_i(j) \tag{9}$$

To minimize the above delay, an optimization problem is formulated according to the Eq. (3) to (9):

$$\min_{j,k_i,a_i} \quad t_{pu}(j) + \sum_{i=0}^{N} k_i a_i T_i(j)$$

$$\text{s.t.} \quad \lambda_{0i}(j) < \mu_i$$
$$\lambda_{0i}(j) < \mu_i' \tag{10}$$
$$j \in \{0, 1, ..., P\}$$
$$\sum_{i=0}^{N} k_i a_i = 1$$

3 Problem Solving

Since problem (10) has 3 variables with interaction effects, it is difficult to obtain the optimal solution directly. To simplify the problem, we first traverse the puzzle difficulty j. For a given difficulty j, the auxiliary node selection scheme and the corresponding traffic allocation scheme can be obtained relatively easily. Therefore, two algorithms are designed for selecting auxiliary node and traffic allocation, respectively. Finally, compare the users' access delays of all difficulties to get the best solution.

3.1 Auxiliary Node Selection

For the construction of the auxiliary node network, it is obvious that the larger the number of auxiliary nodes included, the stronger the network's ability to

---

Algorithm 1. Auxiliary node selection

Input: n_i, μ_i, μ_i', j, $\lambda_0(j)$, t_{pi}
1: $T_i = \frac{1}{\mu_i} + \frac{1}{\mu_i'} + t_{pi}$
2: Sort all nodes in an increasing order according to T_i, and obtain an updated set $\{n_1', n_2', ..., n_N'\}$.
3: **for** $M = 1 : N$ **do**
4:     **if** $\mu_0 + \sum_{m=1}^{M} \min\{\mu_m, \mu_m'\} < \lambda_0(j)$ **then**
5:         Add n_M' to $\mathcal{N}_j'$.
6:     **end if**
7: **end for**
Output: $\mathcal{N}_j'$

---

defend against attacks. Therefore, it is desirable to select as many auxiliary nodes as possible. However, the complexity of the following optimization algorithm increases exponentially with the number of auxiliary nodes. Since the computing resources of the satellite nodes are relatively scarce, it is difficult to support the implementation of algorithms with excessive complexity. In addition, for the actual situation, there is generally a difference in transmission cost between the attacked node and different satellite nodes. Different auxiliary nodes have different processing capabilities. Choosing different auxiliary nodes will also affect the access delay of normal users. Therefore, we have to determine whether a satellite is suitable to serve as an auxiliary node according to the resource information of neighbor satellite nodes and ground nodes.

When the difficulty j is fixed, the arrival rate of access requests $\lambda_0(j)$ is also a fixed value. Through Eq. (8), by setting $\lambda_{0i}(j)$ to 0, we can obtain a lower bound of an auxiliary node's authentication delay, which is given by

$$T_i = \frac{1}{\mu_i} + \frac{1}{\mu_i'} + t_{pi} \tag{11}$$

We sort all satellite nodes in an increasing order according to the above boundary value, and add satellite nodes successively to set $\mathcal{N}_j'$ until Eq. (12) are satisfied.

$$\mu_0 + \sum_{m=1}^{M} \min\{\mu_m, \mu_m'\} > \lambda_0(j) \tag{12}$$

where μ_0 and μ_m represent the average number of requests verified by n_0 and node in $\mathcal{N}_j'$ per unit time. μ_m' represents the average number of requests transmitted between n_0 to n_m' per unit time. M is the number of the finally selected nodes. The auxiliary node selection scheme is summarized in Algorithm 1.

Algorithm 2. Defense algorithm based on CPP and load balancing

Input: n_i, μ_i, μ'_i, $\lambda_0(j)$, t_{pi}, $f(j)$, C_u
1: **for** $j = 0 : P$ **do**
2:     Use algorithm 1 to select the auxiliary nodes $\mathcal{N}'_j = \{n'_1, n'_2, ..., n'_M\}$.
3:     Calculate the minimum of $T_j = t_{pu}(j) + \sum_{m=0}^{M} a'_{mj} T_i(j)$.
4:     Get the traffic allocation scheme $\mathcal{A}'_j = \{a'_{1j}, a'_{2j}, ..., a'_{Mj}\}$.
5: **end for**
6: $T = \min\{T_j\}$
7: Find the corresponding auxiliary nodes $\mathcal{N}'$, traffic allocation schemes $\mathcal{A}'$ and difficult of puzzle j'.
Output: T, $\mathcal{A}'$, j', $\mathcal{N}'$

3.2   Traffic Allocation

After we have determined auxiliary nodes $\mathcal{N}'_j = \{n'_1, n'_2, ..., n'_M\}$, the optimization problem becomes:

$$\min_{j, a'_m} \quad t_{pu}(j) + \sum_{m=0}^{M} a'_m T_i(j)$$

$$\text{s.t.} \quad \lambda_{0m}(j) < \mu_m$$
$$\lambda_{0m}(j) < \mu'_m \qquad (13)$$
$$j \in \{0, 1, ..., P\}$$
$$\sum_{m=0}^{M} a'_m = 1$$

where a'_m represents the traffic allocation scheme for both auxiliary nodes and attacked node, $m \in \{0, 1, 2, ..., M\}$.

With regard to the problem (13), when j is determined, the optimization problem is simplified to a single variable problem. The complexity of the optimization solution is greatly reduced.

The expansion form of the objective function T (taking the solution verified on the auxiliary node as an example) is:

$$T = t_{pu}(j) + \sum_{m=0}^{M} a'_m \left[\frac{1}{\mu_m - \lambda_{om}(j)} + \frac{1}{\mu'_m - \lambda_{om}(j)} + t_{pi}(j) \right] \qquad (14)$$

When j is fixed, the variables $t_{pu}(j)$ and $\lambda_{0m}(j)$ are also constants. T is a function about the multivariate variable a'_m, and the constraints of the independent variable a'_m are:

$$\lambda_{0m}(j) < \mu_i \qquad (15)$$

which can be transformed into a linear constraint.

$$a'_m < \frac{\mu_m}{\lambda_0(j)} \qquad (16)$$

Table 1. Parameter setting

Parameters	Values
μ_m, $m = 0, 1, ..., 6$	70, 75, 200, 82, 66, 65, 10
μ'_m, $m = 0, 1, ..., 6$	null, 120, 200, 150, 150, 60, 50
User's access request arrival rate λ_u	20

The entire optimization problem (13) is simplified to a linear programming problem. It is relatively easy to find the minimum value of the objective function. By comparing the minimum of each difficulty j, we can get the parameters including the difficulty of the puzzle j, auxiliary node $\mathcal{N}'$ and traffic allocation scheme $\mathcal{A}'$.

The whole solving process is summarized in Algorithm 2. First we traversal all the difficulties. For a given difficulty, we can get the node selection scheme through Algorithm 1. Then the whole problem is simplified and the minimum value of the objective function can be found relatively easy. Finally, by comparing the minimum of each difficulty j, we can get the parameters we need.

For NTN, some nodes such as low-orbit satellite nodes are mobile. The period of parameters change caused by movement is much longer than the period of access authentication. Therefore, when the network parameters change, the algorithm must be performed once again.

4   Simulation Results and Analysis

In this section, we verify the effectiveness of the scheme proposed in this paper with two baselines including non-cooperative CPP-based defense and cooperative CPP-based defense with fixed puzzle difficulty.

We use matlab to perform performance comparative simulation of user access delays under different conditions. A total of 66 satellites of the Iridium system and ground nodes are selected. The parameters and motions of the satellites are imported through STK. According to the obtained trajectory data, satellites capable of communicating within a certain period of time are selected to construct a satellite network. Some important parameters are listed in Table 1. The first value of μ'_m is set to null, because it represents the transmission rate of n_0 to n_0. According to the paper [12], the average propagation delay of the satellite node to the ground node of the Iridium system is 15 ms, and the propagation delay between satellites is about 3 ms to 8 ms.

Figure 3 is the comparison result of the cooperative CPP-based defense and the non-cooperative CPP-based defense. It can be seen that under the same attack intensity, the minimum users' access delay of the cooperative CPP-based defense is about 48 ms. However, the non-cooperative CPP-based defense is about 100 ms. In addition, the difficulty corresponding to the minimum delay of the cooperative CPP-based defense is 6, and the non-cooperative CPP-based

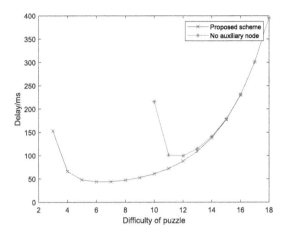

Fig. 3. Comparison of the access delay between the proposed scheme and single node network under the same attack intensity (number of nodes is 6 and number of attacks per second is 10000).

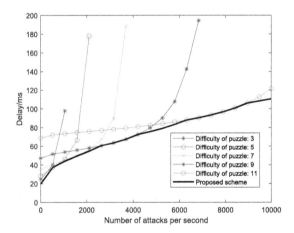

Fig. 4. Comparison of the access delay between the proposed scheme and the fixed puzzle difficulty method under different attack intensities (number of nodes is 6).

defense is 12. Lower difficulty can save the satellite nodes' available computing resources on the basis of delay optimization.

Figure 4 shows the comparison of the access delay between the proposed scheme and fixed puzzle difficulty scheme under the same number of auxiliary nodes. It can be seen from Fig. 4, the users' access delay of the proposed scheme has always been the lowest. In the case of small preset difficulty, the users' access delay will rapidly go up with the increase of attack intensity. Small difficulty may even cause a crash due to insufficient computing resources of the satellites. If the preset difficulty is too large, the satellite nodes' available computing resources are

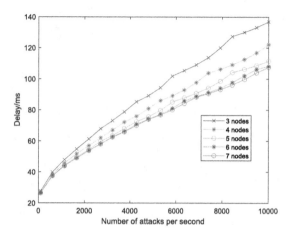

Fig. 5. Comparison of access delays of different numbers of auxiliary nodes using the proposed scheme under different attack intensities.

wasted when there is no attack or the attack intensity is small. Meanwhile, large difficulty will increase the users' access delay. Therefore, it can be judged that dynamically selecting the difficulty of the puzzle can better reduce the access delay and allocate the available computing resources more reasonably.

Figure 5 is a comparison of the users' access delays versus the numbers of auxiliary node. From the simulation result, it is clear that under the same condition, the larger the number of auxiliary nodes, the smaller the users' access delay. This shows that for the auxiliary network, the more nodes, the stronger the attack resistance. However, when the number of auxiliary nodes increases to a certain extent, the reduction in users' access delay becomes small. The reason is that at this time, the available computing resources of the entire network have far exceeded the resource consumption for processing requests. In this case, the main factor affecting the access delay of users is not available computing resources any more, but transmission delay and other factors. In addition, if the nodes with less available resources are used as auxiliary nodes, the access delay of the legitimate user may be increased. Therefore, using the algorithm to select suitable auxiliary nodes can obtain the optimal user's access delay.

5   Conclusion

In this paper, we focus on the potential DoS attacks in future non-terrestrial networks where satellites can perform access authentications. First, the types of DoS attacks and related defense measures is introduced. Then, we propose a model for resisting DoS attacks through the CPP and load balancing technology. After that an algorithm is designed to defend against DoS attacks, including auxiliary node selection algorithm and traffic allocation algorithm to minimize the access delay of legitimate users. Finally, the simulation results verify the effectiveness of our proposed scheme for reducing user's access delay and computation pressure of satellite nodes.

Acknowledgment. This work is supported by the Open Project of A Laboratory under Grant No. 2017XXAQ08 and 2016XXAQ09, the National Natural Science Foundation of China under Grant No. 61971069 and 61801051.

References

1. Chowdhury, N.R., Negi, N., Chakrabortty, A.: A new cyber-secure countermeasure for LTI systems under DoS attacks. In: 2019 27th Mediterranean Conference on Control and Automation (MED), pp. 304–309 (2019)
2. Afianti, F., Wirawan, Suryani, T.: Lightweight and DoS resistant multiuser authentication in wireless sensor networks for smart grid environments. IEEE Access **7**, 67107–67122 (2019)
3. Echevarria, J.J., Garaizar, P., Legarda, J.: An experimental study on the applicability of SYN cookies to networked constrained devices. Softw. Pract. Exper. **48**(3), 740–749 (2018)
4. Abirami, K., Harini, N., Vaidhyesh, P.S., Kumar, P.: Analysis of web workload on QoS to assist capacity. In: Pandian, D., Fernando, X., Baig, Z., Shi, F. (eds.) ISMAC 2018. LNCVB, vol. 30, pp. 573–582. Springer, Cham (2019). https://doi.org/10.1007/978-3-030-00665-5_57
5. Tan, X., Li, H., Wang, L., Xu, Z.: Global orchestration of cooperative defense against DDoS attacks for MEC. In: IEEE Wireless Communications and Networking Conference (WCNC) 2019, pp. 1–6 (2019)
6. Michalas, A., Komninos, N., Prasad, N.R., Oleshchuk, V.A.: New client puzzle approach for DoS resistance in ad hoc networks. In: IEEE International Conference on Information Theory and Information Security 2010, pp. 568–573 (2010)
7. Gupta, D., Saia, J., Young, M.: Proof of work without all the work. In: Proceedings of the 19th International Conference on Distributed Computing and Networking, pp. 1–10 (2018)
8. De Almeida, M.P., de Sousa Júnior, R.T., García Villalba, L.J., Kim, T.-H.: New DoS defense method based on strong designated verifier signatures. Sensors **18**(9), 2813 (2018)
9. Almashaqbeh, G., Kelley, K., Bishop, A., Cappos, J.: CAPnet: a defense against cache accounting attacks on content distribution networks. In: IEEE Conference on Communications and Network Security (CNS) 2019, pp. 250–258 (2019)
10. Aura, T., Nikander, P., Leiwo, J.: DOS-resistant authentication with client puzzles. In: Christianson, B., Malcolm, J.A., Crispo, B., Roe, M. (eds.) Security Protocols 2000. LNCS, vol. 2133, pp. 170–177. Springer, Heidelberg (2001). https://doi.org/10.1007/3-540-44810-1_22
11. Thomopoulos, N.T.: Fundamentals of Queuing Systems: Statistical Methods for Analyzing Queuing Models. Springer, Boston (2012). https://doi.org/10.1007/978-1-4614-3713-0
12. 3GPP: Study on New Radio (NR) to support non-terrestrial networks (Release 15). 3rd Generation Partnership Project (3GPP), Technical Specification Group Radio Access Network (TR) 38.811, version 15.2.0 (2019)

Mobile Edge Network System

Location-Based Multi-site Coordination Beam Tracking for Vehicle mmWave Communications

Xingwen He$^{(\boxtimes)}$, Danpu Liu, and Zhilong Zhang

Beijing Laboratory of Advanced Information Networks, Beijing Key Laboratory
of Network System Architecture and Convergence, Beijing University of Posts
and Telecommunications, Beijing 100876, People's Republic of China
{hexingwen,dpliu,zhangzhilong}@bupt.edu.cn

Abstract. Millimeter wave (mmWave) system with massive multiple input multiple output (mMIMO) meets increasing data traffic requirements. However, fast beam tracking for vehicles with high mobility causes enormous overhead, especially in an ultra-dense network (UDN) with frequently base station (BS) handover. In this paper, we proposed a multi-site coordination beam tracking scheme utilizing the spatial correlation of channel state information (CSI) among different sites to reduce the signaling overhead for beam training and handover. The scenario is a hyper-cellular network (HCN) with one control-BS (CBS) and multiple traffic-BSs (TBSs). The proposed scheme consists of two stages. In the first stage, more accurate position measurement of the moving user equipment (UE) can be achieved by using uniform planar array (UPA), and Extended Kalman Filter (EKF) is exploited in CBS to predict the UE's location in the next slot. In the second stage, the relationship between multi-site and UE's location is used by the CBS to remotely infer the candidate beam between each TBS and the UE, and make a TBS handover decision when necessary. Given that it is the CBS in charge of beam tracking between all the TBSs and the UE centrally, the overhead for beam training and handover are both efficiently reduced. Simulation results based on realistic 3D scenario show that the proposed scheme can achieve 99% of the optimal spectral efficiency with fewer overhead for beam sweeping and handover signaling.

Keywords: Beam tracking · Vehicle network · Multi-site · Extended Kalman filter · Uniform planar array · Handover

1 Introduction

In recent years, the fifth generation (5G) mobile communications developed rapidly. As a typical scenario in 5G, vehicle environment has received widespread attention. Advanced technologies such as high precise road map, real-time position updating and autonomous driving enhance the convenience and safety of driving. However, those applications heavily rely on the vehicle network's support for high capacity and low latency in data sharing.

© ICST Institute for Computer Sciences, Social Informatics and Telecommunications Engineering 2021
Published by Springer Nature Switzerland AG 2021. All Rights Reserved
H. Gao et al. (Eds.): ChinaCom 2020, LNICST 352, pp. 271–287, 2021.
https://doi.org/10.1007/978-3-030-67720-6_19

The mmWave candidates to improve network capacity and increase data rate in 5G, thanks to its huge spectral bandwidth at high frequencies from 30 GHz to 300 GHz. Meanwhile, massive multiple input multiple output (mMIMO) is used to overcome high propagation path loss of mmWave channels [1]. Power is concentrated into one or more beams to communicate with remote users through beamforming. However, the cost of obtaining higher beamforming gain is that the beam gets narrower. Vehicles with high mobility may leave the coverage of one beam easily, resulting in frequent beam misalignment. A traditional approach to address this issue is to periodically transmit training beams to different directions sequentially in time and choose the direction with the largest received signal-to-noise ratio (SNR) [2]. However, the signaling overhead for simple beam sweeping is not acceptable in high mobility scenario, hence more fast and efficient beam tracking methods are needed.

Some prior work of beam tracking considers to utilize the temporal correlation between channel to improve the efficiency of beam alignment. On one hand, authors in [3] define a novel concept called beam coherence time which is an effective measure of beam alignment frequency. On the other hand, Kalman Filter plays an important role in positioning and beam tracking to reduce pilot overhead and required SNR. In [4], a linear Kalman Filter based angle domain channel tracking algorithm was designed to acquire channel state information (CSI) in each coherence time. In [5], authors proposed a Kalman Filter based tracking algorithm which focused on abrupt channel changes (e.g., blockage), slow variations of angles of departure (AoD), and angles of arrival (AoA). [6] proposed an Extended Kalman Filter (EKF)-based algorithm using the position, velocity, and channel coefficient as state variables. Authors in [7] utilized EKF to position prediction and related location-based beamforming for high-speed train (HST) scenario in 5G New Radio (NR) networks. However, most of the above mentioned methods focus on antenna system with uniform linear array (ULA), which resulted in beam alignment errors in real three-dimensional (3D) environment.

Furthermore, in order to meet the rapidly growing demand for mobile data services, ultra-dense network (UDN) reduces the transmission distance between BSs and UE which yields to less path loss and higher traffic density by increasing the number of small BSs [8]. However, there are two major challenge in UDN with high-speed vehicle. 1) A huge number of densely small BSs increase pilot overhead and time for beam alignment. 2) Handover between BSs causes large signaling overhead of core network.

An effective framework in UDN called hyper-cellular network (HCN) is proposed in [9] with separated coverage area of control channel and data channel, where control-BS (CBS) and traffic-BS (TBS) take care of control information and data transmission respectively. In this architecture, CBS performs centralized management and controls TBS selection, which improves system energy efficiency and reduces core network signaling caused by frequent TBS handover in vehicle environment. Because of the common scatters and location-based light-of-sight (LoS) components, the CSIs between geographically separated CBS and TBS emerge nonlinear relevance, which has a great effect on CBS centralized management and scheduling [10]. On this basis, the non-linear correlation between the CSI of TBSs was studied and a neural network-based remote channel interference scheme was proposed to reduce channel acquisition overhead in [11]. Authors in [12] proposed a time-sequence channel learning framework to predict the beam direction of target roadside unit (RSU) in future steps from the CSI of source RSU in the past period of time.

In this paper, we propose a multi-site coordinated beam tracking method in HCN for vehicle environment to reduce the overhead for beam training and TBS handover. The basic idea is to track the beam direction of TBS based on the UE's position information from CBS, and determine TBS handover from serving TBS to target TBS in each slot. The main challenge is that the precision of position prediction based on beam training has a significant effect on the multi-site coordination performance when UE moves with high velocity in practical three-dimensional mMIMO environment. Hence, we adopt an EKF-based position prediction scheme to improve the estimation accuracy. Besides, UPA with flexible 3D beam direction provides more exact position measurement.

The main contributions of this work are summarized as follows.

1) An efficient EKF based positioning and beam tracking method for wideband 5G systems is proposed, and the associated position measurement based on UPA is also developed.
2) Based on the spatial correlation of multi sites, a geometry based beam inference method is applied to reduce the overhead for beam training. Besides, a novel TBS handover scheme is proposed which jointly considered position and reference signal receiving power (RSRP).
3) The performance of the proposed scheme is evaluated based on ray-tracing channel data in real urban scenario, and the simulation result shows that the spectral efficiency for TBS reaches 99% of the upper bound with lower overhead in beam sweeping and TBS handover.

The rest of paper is organized as follows. Section 2 describes the considered system model, including HCN scenario, channel models and transmit and receive signal structure. In Sect. 3, the proposed location-based multi-site coordination beam tracking scheme is developed and presented. Finally, the performance of the proposed approach is evaluated and analyzed in Sect. 4, while the conclusions are drawn in Sect. 5.

Notation: Bold upper and lower case letters represent matrices and vectors, respectively. The transpose and conjugate transpose of a matrix are denoted as $(\cdot)^T$ and $(\cdot)^H$, respectively. Furthermore, $|\cdot|$ and $\|\cdot\|_2$ represent the modulus of a complex number and the 2-norm of a matrix, respectively.

2 System Model

As shown in Fig. 1, we consider an HCN with one CBS and multiple TBSs, and UE moves straightly along the road with high velocity. In order to save overhead for beam training and TBS handover, the CBS predicts the UE's position in future slots based on EKF, and centrally selects candidate beams for each TBS to communicate with the UE based on geometric method. If the link between the serving TBS and the UE satisfies the condition for handover, the CBS controls TBS handover and sends the predicted beam direction to TBS through backhaul link. An efficient TBS handover scheme based on UE's location is also proposed.

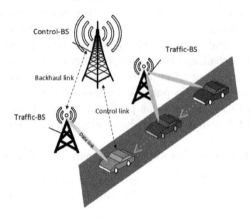

Fig. 1. Schematic of the proposed location-based multi-site coordination beam tracking scheme

Assume that both BSs and the UE are equipped with UPA and single radio frequency (RF) chain. The number of antennas of the BS and UE is $M_{h,BS}$ and $M_{h,UE}$ in each row, $M_{v,BS}$ and $M_{v,UE}$ in each column, respectively. To improve positioning accuracy, a 3D beam steering codebook is adopted to generate the communication beam pairs, which is the Kronecker product of two 2D codebook in vertical and horizontal direction, i.e.

$$\mathbf{F}_{RF} = \mathbf{F}_{RF,v} \otimes \mathbf{F}_{RF,h} \tag{1}$$

where

$$\mathbf{F}_{RF,v}(m,k) = \frac{1}{\sqrt{M_{v,UE}}} \exp\left(j\frac{2\pi}{\lambda} d_v m \, \sin\left(\frac{2\pi k}{2^K}\right)\right) \tag{2}$$

$$\mathbf{F}_{RF,h}(m,k) = \frac{1}{\sqrt{M_{h,UE}}} \exp\left(j\frac{2\pi}{\lambda} d_h m \, \sin\left(\frac{2\pi k}{2^K}\right)\right) \tag{3}$$

where λ is the wavelength of carrier frequency, d_v and d_h denotes the distance between adjacent antennas in vertical and horizontal, respectively, which are equal to the half of wavelength. K denotes quantization resolution of AoA/AoD.

For the uplink, an analog precoder $\mathbf{f}_{RF} \in \mathbb{C}^{M_{v,UE}M_{h,UE} \times 1}$ is applied at UE, and single data stream is precoded and then transmitted on each antenna. As $\mathbf{f}_{RF}$ is implemented with the phase shifter which can only change the phase of transmitted signal, each element of $\mathbf{f}_{RF}$ has a unitary magnitude, i.e. $|\mathbf{f}_{RF}(i,j)| = \frac{1}{\sqrt{M_{v,UE}M_{h,UE}}}$. At the BS, the received signals are firstly processed by an analog combiner $\mathbf{w}_{RF} \in \mathbb{C}^{M_{v,BS}M_{h,BS} \times 1}$, which is generated by the phase shifter similar with $\mathbf{f}_{RF}$ that $|\mathbf{w}_{RF}(i,j)| = \frac{1}{\sqrt{M_{v,BS}M_{h,BS}}}$.

The detected signal at BS can be written as

$$\mathbf{r}_t = \mathbf{w}_{RF}^H \mathbf{H}_t \mathbf{f}_{RF} \mathbf{s}_t + \mathbf{w}_{RF}^H \mathbf{n} \tag{4}$$

where, $\mathbf{s}_t \in \mathbb{C}^{1 \times N_s}$ which consists of N_s sampling points is the uplink orthogonal frequency-division multiplexing (OFDM) signal at time t, and can be written as $\mathbf{s}_t = [\mathbf{s}_t(0), \mathbf{s}_t(1), \ldots, \mathbf{s}_t(N_s)]$, where the n-th sampling point is defined as

$$\mathbf{s}_t(n) = \frac{1}{N} \sum_{k=0}^{N-1} S(k) e^{j\frac{2\pi}{N}kn} \tag{5}$$

where, $S(k)$ is the k-th subcarrier of OFDM symbol, and N denotes the number of subcarriers. $\mathbf{n} \in \mathbb{C}^{M_{v,BS} M_{h,BS} \times N_s}$ denotes the additive white Gaussian noise, each entry of which follows the independent and identically distributed (i.i.d.) complex Gaussian distribution with zero mean and variance σ_n^2.

Furthermore, according to the multi-path time-varying channel model in [13], the uplink channel vector is given by

$$\mathbf{H}_t = \sum_{l=0}^{L-1} \alpha_{l,t} e^{-j\left(\phi_{l,t} + \phi_{f_d,l,t}\right)} \mathbf{a}\left(\theta_{l,t}^{tx}, \varphi_{l,t}^{tx}\right) \mathbf{a}^T\left(\theta_{l,t}^{rx}, \varphi_{l,t}^{rx}\right) \delta\left(t - \tau_{l,t}\right) \tag{6}$$

where, $l = 0$ corresponds LoS path, and other parameters are discussed in detail below. $\alpha_{l,t}$ denotes complex gain based on path loss and shadowing, $\tau_{l,t} = d_{l,t}/c$ denotes corresponding delay for the l-th path whose length is $d_{l,t}$. $\theta_{l,t}^{tx}, \theta_{l,t}^{rx} \in [-\pi, \pi)$ and $\varphi_{l,t}^{tx}, \varphi_{l,t}^{rx} \in [0, \pi)$ denote the azimuth and elevation angle of the l-th path of UE and BS, respectively. $\phi_{l,t} = 2\pi f_c \tau_{l,t}$ denotes the phase caused by carrier frequency f_c, $\phi_{f_d,l,t} = 2\pi f_{d,l,t} t$ denotes Doppler phase shift for the l-th path. The Doppler shift for the l-th path is defined as $f_{d,l,t} = \frac{v \cos \theta_{l,t}^{tx} \cos \varphi_{l,t}^{tx}}{\lambda}$, where v is the velocity of UE. $\mathbf{a}\left(\theta_{l,t}^{tx}, \varphi_{l,t}^{tx}\right)$ and $\mathbf{a}\left(\theta_{l,t}^{rx}, \varphi_{l,t}^{rx}\right)$ denotes the UPA steering vector of the UE and the BS respectively. We assume that the UPA is deployed on y-z plane. By taking the UE side as example, the steering vector can be written as

$$\mathbf{a}\left(\theta_{l,t}^{tx}, \varphi_{l,t}^{tx}\right) = \mathbf{a}_v\left(\theta_{l,t}^{tx}, \varphi_{l,t}^{tx}\right) \otimes \mathbf{a}_h\left(\theta_{l,t}^{tx}, \varphi_{l,t}^{tx}\right) \tag{7}$$

where $\otimes$ denotes the Kronecker product, and the steering vector in vertical and horizontal are written as

$$\mathbf{a}_v\left(\theta_{l,t}^{tx}, \varphi_{l,t}^{tx}\right) = \frac{1}{\sqrt{M_{v,BS}}} \left[1, e^{j\frac{2\pi}{\lambda}d_v \sin \theta_{l,t}^{tx}}, \ldots, e^{j\left(M_{v,BS}-1\right)\frac{2\pi}{\lambda}d_v \sin \theta_{l,t}^{tx}}\right]^T \tag{8}$$

$$a_h\left(\theta_{l,t}^{tx}, \varphi_{l,t}^{tx}\right) = \frac{1}{\sqrt{M_{h,BS}}}\left[1, e^{j\frac{2\pi}{\lambda}d_h\cos\theta_{l,t}^{tx}\sin\varphi_{l,t}^{tx}}, \ldots, e^{j\left(M_{h,BS}-1\right)\frac{2\pi}{\lambda}d_h\cos\theta_{l,t}^{tx}\sin\varphi_{l,t}^{tx}}\right] \tag{9}$$

When data is transmitted from the TBS to the UE, the spectral efficiency can be expressed as

$$R_s = \log_2\left(1 + \frac{P_{TBS}}{\sigma_n^2}\frac{\mathbf{w}_{RF}^H\mathbf{H}\mathbf{f}_{RF}\mathbf{f}_{RF}^H\mathbf{H}^H\mathbf{w}_{RF}}{\mathbf{w}_{RF}^H\mathbf{w}_{RF}}\right) \tag{10}$$

where P_{TBS} is the power of TBS transmitted signal.

3   Location-Based Multi-site Coordination Beam Tracking

In this section, we propose a location-based multi-site coordination beam tracking scheme in HCN for vehicle environment. In spite of mMIMO system with beamforming enhances the spectral efficiency by concentrating signal power on one or more beams, beam misalignment caused by high mobility in vehicle environment leads to frequently beam sweeping, thus increases overhead of beam training. In the proposed framework, we focus on finding the optimal beam of the TBS based on coordinated position prediction by the CBS with low beam training overhead.

In order to introduce the proposed method, we first give some high-level analysis. The optimal beam pattern of TBS is determined by the CSI at TBS, and it can be obtained by the CSI of CBS as well as the spatial correlation of CSI between the CBS and the TBS. However, due to the high complexity of calculating the full-dimensional CSI estimation in mMIMO, we transform the beam inferred problem into angle domain. Note that the correlation of angle domain CSI is mainly decided by the AoA from the UE to the CBS and the TBS. Moreover, the beam direction is strongly connected with the AoA of LoS. Therefore, the optimal beam pattern of each TBS can be inferred by the UE's position. Besides, considering the temporal correlation of channel, we can predict the UE's position in each time step by EKF, so that the positioning accuracy is improved.

Therefore, the proposed method is divided into two stages, i.e. position prediction stage and multi-site coordination stage. In position prediction stage, the CBS estimates the time of arrival (ToA) and AoA of the UE from the received uplink pilot to obtain the UE's position. At the same time, based on the estimated position and the temporal correlation of positions in motion function, the EKF is used to predict the UE's position in the future. In multi-site coordination stage, the CBS determines TBSs' candidate beams for downlink data transmission on the basis of the geometric relationship between the predicted UE's position and multiple TBSs' position. Subsequently, the CBS controls TBS handover based on position and RSRP jointly. As a result, the beam searching complexity and overhead for handover signaling are both reduced.

3.1   EKF-Based Position Prediction

In this part, we propose an EKF-based position prediction scheme. The EKF obtains a more accurate state estimation, according to combining the initial state estimation of the system at the next moment, and the feedback obtained from the measurement. That is why the EKF is an effective method in trajectory prediction problems with nonlinear model function.

In order to reduce measurement error and enhance estimation performance of EKF in the proposed scheme, UPA is equipped on both BSs and UE to provide more accurate spatial position information in 3D mMIMO scenario. In this part, position measurement based on the received signal of UPA is introduced first, then the EKF procedure used for position prediction is presented.

Position Measurement

Consider a UE located at any point P with coordinates (p_x, p_y, p_z) in Space rectangular coordinate system, and the CBS located at the origin. A major concern is that the CBS has difficulty measuring the UE's coordination directly. However, directional beam between the CBS and the UE will provide angle domain information. Consequently, we indicate the UE's location in spherical coordinate (R, θ, φ), where R denotes the distance between the CBS and the UE, θ and φ denotes azimuth and elevation AoA of LoS. The distance is expressed as the product of TOA for LoS path and the speed of light. Therefore, the position measurement is based on TOA and AoA estimation for LoS.

The estimation of ToA and AoA is based on the periodic UL SRS transmitted by the UE, however, the major problem is the multi-path effect. The transmitted signal arrives at receiver side through multi-path with different propagation delays and angles, which causes delay spread and angle spread for received signal. Instead of extracting channel parameters with complex calculation, correlation analysis and beam sweeping are utilized for TOA and AoA estimation.

We assume that the control link between the CBS and the UE is unblocked, the LoS component in the received signal has the maximum power, thus the LoS component in received signal has the strongest correlation with the transmitted signal. Hence, the TOA of LoS can be estimated by cross-correlation between the received signal and the known SRS, and the correlation method takes advantages of processing low SNR signals.

The cross-correlation function observed in the CBS can be written as

$$\mathbf{c}[n] = \sum_{m=0}^{N_s-1} \mathbf{s}^*(n-m)\mathbf{r}(m) \tag{11}$$

where $\mathbf{s}(m)$ is the known SRS transmitted from UE. We neglect the synchronized error between CBS and UE. By finding the sample index where the absolute value of correlation is maximum, the estimation of TOA can be obtained, and is written as $\hat{\tau} = \frac{\hat{n}}{F_s}$, where $\hat{n} = \arg\max_n |\mathbf{c}[n]|$ denotes the estimated TOA in samples and F_s is the sampling frequency.

After the ToA has been estimated, we will execute AoA estimation by beam sweeping. Due to multipath with different AoAs, the received signal consists of LoS and NLoS components, i.e. $\mathbf{y}_{los}$ and $\mathbf{y}_{nlos}$. Beamforming is used to compensate the phase difference of $\mathbf{y}_{los}$ by controlling the phase shifter, so that the amplitude of $\mathbf{y}_{los}$ from different antennas increases after combining. On the contrary, $\mathbf{y}_{nlos}$ will not obtain beamforming gain due to the phase delay of the phase shifter does not match the phase difference of $\mathbf{y}_{nlos}$ on each antenna.

According to the previous analysis, the beam direction with the highest received power can be regarded as the direction of LoS based on the above assumption that LoS exists. The specific process is shown as follows.

Firstly, the CBS chooses the optimal received beam by maximizing received signal power. The corresponding codeword is defined as

$$\hat{\mathbf{w}}_{i_q} = \arg\max_{\mathbf{w}_{i_q}} \left\{ \left(\left\| \mathbf{w}_{i_q}^H y \right\|_2 \right)^2 \right\} \tag{12}$$

where i_q is the candidate beam index. The set of candidate beam index is given as $I = \{i_q | q = 1, \ldots, Q\}$, where Q is the size of candidate beams. In initialization stage, the CBS performs exhaustive beam sweeping to connect with the UE, which means Q equals to the total number of CBS beams. Once the UE's position is predicted, the CBS performs narrow-range beam sweeping around the estimated AOA in last time step, and Q can be set in a small value.

After the best received beam is identified through beam sweeping, we need to determine the azimuth and elevation AOA, i.e. θ and φ, according to the corresponding codeword $\hat{\mathbf{w}}_{i_q}$. We assumed that the number of antenna and the distance between adjacent antennas in horizon and vertical are equal to M and d, respectively. According to the analysis above, $\hat{\mathbf{w}}_{i_q}$ can be expressed as the conjugate transpose of the steering vector, and is written as

$$\hat{\mathbf{w}}_{i_q} = \frac{1}{M} \left[1, \ldots, e^{-j\frac{2\pi}{\lambda}d(m_v \sin\theta + m_h \cos\theta \sin\varphi)}, \ldots, e^{-j(M-1)\frac{2\pi}{\lambda}d(\sin\theta + \cos\theta \sin\varphi)} \right]^T \tag{13}$$

Furthermore, $\hat{\mathbf{w}}_{i_q}$ can be represented as the Kronecker product of $\hat{\mathbf{w}}_{i_q,h}$ and $\hat{\mathbf{w}}_{i_q,v}$, where $\hat{\mathbf{w}}_{i_q,h} \in \mathbb{C}^{M \times 1}$ and $\hat{\mathbf{w}}_{i_q,v} \in \mathbb{C}^{M \times 1}$ are the codeword in horizon and vertical directions, respectively. Therefore, θ and φ can be estimated by the main lobe angle of $\hat{\mathbf{w}}_{i_q,v}$ and $\hat{\mathbf{w}}_{i_q,h}$, i.e. $\hat{\alpha}_{i_q,v}$ and $\hat{\alpha}_{i_q,h}$.

Finally, based on (13), the azimuth and elevation AOA can be estimated by

$$\hat{\theta} = \hat{\alpha}_{i_q,v} \tag{14}$$

$$\hat{\varphi} = \arcsin\left(\frac{\sin \hat{\alpha}_{i_q,h}}{\cos \hat{\alpha}_{i_q,v}} \right) \tag{15}$$

EKF-Based Prediction

In order to reduce the overhead of beam sweeping and improve beam alignment accuracy, the CBS predicts the UE's position based on EKF, which linearized the non-linear measurement model to the currently estimated state [14].

The state model at the kth time step is defined as $\mathbf{x_k} = [p_k^x, p_k^y, v_k^x, v_k^y]^T$, where p_k^x, p_k^y are the x-coordinate and the y-coordinate of the vehicle, and v_k^x, v_k^y are the velocity components in the x-direction and the y-direction, respectively. We initialized the state model as $\mathbf{x_0}$ based on beam sweeping and TOA estimation. Furthermore, the state transition model and measurement model are given as

$$\mathbf{x_k} = f(\mathbf{x_{k-1}}) + \mathbf{w_k} \tag{16}$$

$$\mathbf{z_k} = h(\mathbf{x_k}) + \mathbf{v_k} \tag{17}$$

where $\mathbf{z_k}$ is the measurement vector. Moreover, $f(\cdot)$ and $h(\cdot)$ are the state transition and measurement model function respectively. $\mathbf{w_k} \sim N(\mathbf{0}, \mathbf{Q})$ and $\mathbf{v_k} \sim N(\mathbf{0}, \mathbf{R})$ denote process noise vector and measurement noise vector. $\mathbf{Q} \in \mathbb{R}^{2 \times 2}$ and $\mathbf{R} \in \mathbb{R}^{2 \times 2}$ are the covariance matrices for process noise and measurement noise, respectively.

The assumption that the UE moves straightly with constant velocity results in linear state transition function, which is defined as

$$f(\mathbf{x_{k-1}}) = \mathbf{F}\mathbf{x_{k-1}} = \begin{pmatrix} \mathbf{I}_{2\times 2} & \Delta t \mathbf{I}_{2\times 2} \\ \mathbf{0}_{2\times 2} & \mathbf{I}_{2\times 2} \end{pmatrix} \mathbf{x_{k-1}} \tag{18}$$

where Δt is the time interval between two consecutive states.

Furthermore, the UE's distance and angle is measured by the CBS, which are non-linear with position coordinates. The measurement model function can be expressed as

$$h(\mathbf{x_k}) = \begin{pmatrix} \sqrt{(p_k^x)^2 + (p_k^y)^2} \\ \arctan(p_k^x/p_k^y) \end{pmatrix} \tag{19}$$

The covariance matrix for process noise which is defined based on continuous white noise acceleration model in [15], is given by

$$\mathbf{Q} = \sigma_v^2 \begin{pmatrix} \frac{\Delta t^3}{3}\mathbf{I}_{2\times 2} & \frac{\Delta t^2}{2}\mathbf{I}_{2\times 2} \\ \frac{\Delta t^2}{2}\mathbf{I}_{2\times 2} & \Delta t \mathbf{I}_{2\times 2} \end{pmatrix} \tag{20}$$

where σ_v^2 is the variance of UE's velocity.

In each time step k, the EKF-based estimation method consists of two stages: prediction stage and update stage.

In prediction stage, the priori estimated state vector in current time step $\hat{\mathbf{x}}_{k|k-1}$ is obtained based on transition model function given in (18), i.e. $\hat{\mathbf{x}}_{k|k-1} = f(\hat{\mathbf{x}}_{k-1|k-1})$. Moreover, the priori estimated covariance $\mathbf{P}_{k|k-1}$ can be written as

$$\mathbf{P}_{k|k-1} = \mathbf{F}\mathbf{P}_{k-1|k-1}\mathbf{F}^T + \mathbf{Q} \tag{21}$$

where $\hat{\mathbf{x}}_{k-1|k-1}$ and $\mathbf{P}_{k-1|k-1}$ are the posterior estimated state vector and covariance at the last time step.

In update stage, the priori estimate which is updated based on measurement model function given in (19), is expressed as

$$\hat{\mathbf{x}}_{k|k} = \hat{\mathbf{x}}_{k|k-1} + \mathbf{K}_k(\mathbf{z}_k - h(\hat{\mathbf{x}}_{k|k-1})) \tag{22}$$

$$\mathbf{P}_{k|k} = (\mathbf{I} - \mathbf{K}_k\mathbf{H}_k)\mathbf{P}_{k|k-1} \tag{23}$$

where $\mathbf{K}_k$ is the Kalman gain given by

$$\mathbf{K}_k = \frac{\mathbf{P}_{k|k-1}\mathbf{H}_k^T}{\mathbf{H}_k\mathbf{P}_{k|k-1}\mathbf{H}_k^T + \mathbf{R}_k} \tag{24}$$

where $\mathbf{H}_k = \frac{\partial h}{\partial \mathbf{x}}|\hat{\mathbf{x}}_{k|k-1}$ is the Jacobian matrix of the nonlinear measurement model function $h(\cdot)$ evaluated at $\hat{\mathbf{x}}_{k|k-1}$, and can be derived as

$$\mathbf{H}_k = \begin{bmatrix} \frac{p_k^x}{\sqrt{(p_k^x)^2 + (p_k^y)^2}}, & \frac{p_k^y}{\sqrt{(p_k^x)^2 + (p_k^y)^2}}, 0, 0 \\ \frac{p_k^y}{(p_k^x)^2 + (p_k^y)^2}, & -\frac{p_k^x}{(p_k^x)^2 + (p_k^y)^2}, 0, 0 \end{bmatrix} \tag{25}$$

3.2 Multi-site Coordination Beam Tracking Scheme

We have mentioned that it is quite difficult to acquire the complete CSI for each site individually, which takes up plenty of pilot resource. The location of multiple BSs and the UE can be used as a link for the CSI nonlinear correlation. Therefore, in this part, we consider to use geometric relationship between locations to determine the optimal beam direction between each TBS and the UE. Subsequently, we proposed a novel scheme for TBS handover based on the UE's position to reduce unnecessary switching on the condition of ensuring good performance.

Serving Beam Inference for TBS

As the UE's position is predicted, instead of exhaustive sweeping on the entire beam set, the serving beam direction between each TBS and the UE can be identified through a beam sweeping within a narrow range around the predicted UE's position. The candidate beam set can be obtained by geometric method based on the AOA between the TBS and the UE. Therefore, the size of candidate beam set significantly decreases, thus leads to low beam training overhead for each TBS.

The estimated azimuth and elevation AOA from each TBS to the UE are given by

$$\tilde{\theta}_{TBS_j} = \arctan\left[\frac{p_{z,TBS_j} - \hat{p}_z}{\sqrt{\left(p_{x,TBS_j} - \hat{p}_x\right)^2 + \left(p_{y,TBS_j} - \hat{p}_y\right)^2}}\right] \tag{26}$$

$$\tilde{\varphi}_{TBS_j} = \arctan\left[\frac{p_{y,TBS_j} - \hat{p}_y}{p_{x,TBS_j} - \hat{p}_x}\right] \tag{27}$$

where $\left(\hat{p}_x, \hat{p}_y, \hat{p}_z\right)$ is the UE's predicted position, and $\left(p_{x,TBS_j}, p_{y,TBS_j}, p_{z,TBS_j}\right)$ is the coordinate of the j-th TBS that known by CBS. Since LoS path has the strongest power, the candidate beam pair for each TBS and UE is obtained around $\tilde{\theta}_{TBS_j}$ and $\tilde{\varphi}_{TBS_j}$. Finally, the best beam pair is selected from the candidate set by beam refinement between each TBS and the UE.

TBS Handover Judgement

Besides serving beam prediction, the CBS will judge whether the serving TBS handover occurs when the UE's position changes. According to RSRP measurement for two adjacent TBSs, the serving TBS with stronger signal can be determined. However, RSRP measurement requires beam training between each TBS and the UE, and the overhead is non-negligible. Note that closer communication distance leads to lower path loss, it is reasonable to choose the serving TBS for the UE based on the distance to each TBS. Although closer TBS may experience deeper fading than the remoter TBS, the distance difference makes the RSRP between two TBSs most likely has a small gap, thus the handover leads to limited gain. Consequently, we propose a joint strategy for TBS handover to reduce the overhead. If the distance difference between UE and two TBSs is larger than a certain threshold, the closer TBS is determined as serving TBS directly. Otherwise, RSRP report about each TBS is used as handover condition, the flow chart is introduced in Fig. 2.

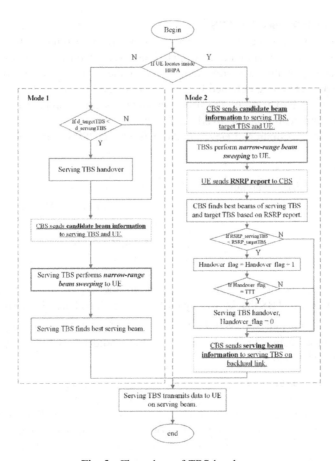

Fig. 2. Flow chart of TBS handover

In the proposed handover method, the serving TBS and target TBS is initialized by CBS before position prediction. We define a high handover probability area (HHPA) whose center at the midpoint of serving TBS and target TBS, and its radius is one-quarter of the distance between two adjacent TBS. The handover condition corresponding to mode 1 and mode 2 depends on whether UE locates inside HHPA, and is introduced as follow.

Mode 1: Firstly, the CBS calculates the distance between each TBS and the estimated UE's position, and chooses the nearest TBS as the serving TBS. Next, the CBS sends control information including candidate beam indexes to the serving TBS and the UE on the backhaul link and PDCCH respectively. Finally, the serving TBS performs beam refinement with the UE on each candidate beam pair and selects the optimal serving beam pair.

Mode 2: The CBS sends candidate beam indexes to each TBS and the UE on the backhaul link and PDCCH respectively. Both the serving TBS and the target TBS perform beam refinement with UE on candidate beams transmitted by CBS. Subsequently, based on the RSRP report of each beam pair from the UE to the CBS on PUCCH, the CBS finds the best beam pair of the serving TBS and the target TBS with UE. The handover process controlled by the CBS starts when the target TBS becomes stronger than the serving TBS, and is defined as a trigger event. After the trigger event, the CBS waits for a certain time duration i.e. time to trigger (TTT). The TBS handover occurs if the RSRP of the target TBS is always stronger than the serving TBS during TTT, and the serving beam index is transmitted to the serving TBS by the CBS on the backhaul link, whereas the beam index at UE side can be determined by UE itself before RSRP report without feedback from the CBS.

4 Simulation

In this section, we evaluate the performance of the proposed location-based multi-site coordination beam tracking method. In order to make our simulation scenario more realistic, the ray-tracing channel parameters for a practical urban environment is generated by Wireless InSite [16]. As shown in Fig. 3, two TBSs located within the coverage of one CBS are equipped with 8 * 8 UPA, while the CBS is equipped with 16 * 16 UPA. The height of the CBS and TBSs are 32 m and 12 m, respectively, and the downlink (DL) transmit power of TBS is 40 dBm. UE is equipped with 4 * 4 UPA, and moves straightly with constant velocity 20 m/s. The spacing of two adjacent route point is 0.3 m, and the number of route point is set to 500. The quantization resolution in 2D steering codebook is set to 8 and 1 for BS and UE, respectively. With regard to the OFDM configurations, the carrier frequency is 28 GHz, and the bandwidth as well as subcarrier spacing are 100 MHz and 120 kHz, respectively. Furthermore, the noise spectral density is set to -174 dBm/Hz.

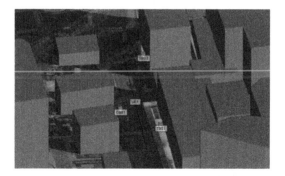

Fig. 3. Simulation scenario

Figure 4 and Fig. 5 show the MSE for estimated distance and AoA against the UL transmit power, respectively. Position is predicted based on UPA in our scheme, while ULA is equipped in [6]. Given that UPA can flexibly adjust beam directions in both horizon and vertical, it leads to more accurate beam alignment and position prediction than ULA in practical 3D mMIMO scenario. It is shown that the MSE of distance and AoA estimation is reduced significantly by using UPA, which are both at the level of 10^{-3} at high transmit power.

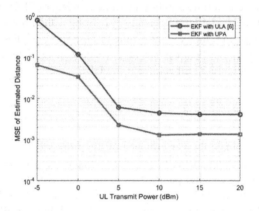

Fig. 4. MSE for distance estimation

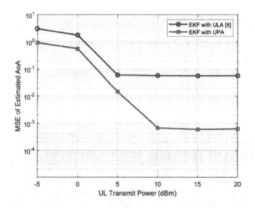

Fig. 5. MSE for AoA estimation

Figure 6 illustrates the average spectral efficiency (SE) in DL from the serving TBS to the UE when the UE moves along the route. The performance of proposed algorithm has 8.8% improvement over the ULA based algorithm in [6] by using the UPA, and achieves about 99% of the upper bound which is defined as the performance with the optimal beamforming vector from UPA codebook by exhaustive beam searching. The result verify that it is reasonable and efficient to track the beam direction for the TBS utilizing the UE's position predicted by the CBS in multi-site scenario, and more accurate position information performs better. In addition, compared with the TBS handover based on RSRP report, the proposed algorithm based on HHPA has almost no performance loss as long as UL transmit power is larger than 10 dBm.

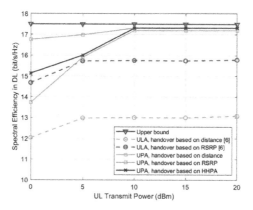

Fig. 6. SE in DL from serving TBS

Figure 7 compares the TBS handover times and beam sweeping complexity for the four schemes. The exhaustive beam searching scheme has the maximum handover times and also the highest beam sweeping complexity which yields unacceptable pilot overhead in each slot. Due to extremely low tracking loss, the complexity of beam training in the proposed multi-site coordination scheme is reduced by about 1000 times because of location-based narrow-range beam sweeping. Furthermore, handover based on HHPA lower the frequent TBS handover, which also promotes the reduction of 1/3 pilot overhead in handover based on RSRP. To sum up, the result shows that the proposed method has a significant advantage in reducing the complexity of beam training and TBS handover times.

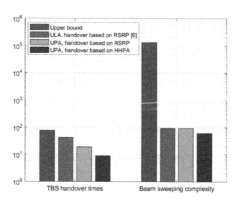

Fig. 7. TBS handover times and beam sweeping complexity, $P_t = 15$ dBm

5   Conclusion

In this paper, we propose a location-based multi-site coordination beam tracking scheme in HCN for vehicle communications. The entire procedure consists of two stages, i.e. position prediction stage and multi-site coordination stage. In the first stage, according to the accurate spatial position measured by UPA, EKF was applied in the CBS to predict the user's position. In the second stage, the optimal beam pattern at each TBS is inferred based on the geometry approach relied on the UE's position, and the TBS handover is determined by position and RSRP report jointly. The numerical results show that the proposed scheme achieves 99% of the maximum spectral efficiency with low beam sweeping complexity and handover overhead.

Acknowledgement. This work is supported by the National Natural Science Foundation of China under Grant No. 61971069, 61801051, and Key R&D Program Projects in Shanxi Province under Grant No. 2019ZDLGY07-10. This work is also supported by Docomo Beijing Lab.

References

1. Shafi, M., et al.: 5G: a tutorial overview of standards, trials, challenges, deployment, and practice. IEEE J. Sel. Areas Commun. **35**(6), 1201–1221 (2017)
2. Wang, J., et al.: Beam codebook based beamforming protocol for multi-Gbps millimeter-wave WPAN systems. IEEE J. Sel. Areas Commun. **27**(8), 1390–1399 (2009)
3. Va, V., Choi, J., Heath, R.W.: The impact of beamwidth on temporal channel variation in vehicular channels and its implications. IEEE Trans. Veh. Technol. **66**(6), 5014–5029 (2017)
4. Shen, Z., Xu, K., Wang, Y., Xie, W.: Angle-domain channel tracking for high speed railway communications with massive ULA. In: 2018 IEEE 18th International Conference on Communication Technology (ICCT), Chongqing, pp. 159–165 (2018)
5. Zhang, C., Guo, D., Fan, P.: Tracking angles of departure and arrival in a mobile millimeter wave channel. In: 2016 IEEE International Conference on Communications (ICC), Kuala Lumpur, pp. 1–6 (2016)
6. Shaham, S., Kokshoorn, M., Ding, M., Lin, Z., Shirvanimoghaddam, M.: Extended Kalman filter beam tracking for millimeter wave vehicular communications. In: 2020 IEEE International Conference on Communications Workshops (ICC Workshops), Dublin, Ireland, pp. 1–6 (2020)
7. Talvitie, J., et al.: Positioning and location-based beamforming for high speed trains in 5G NR networks. In: 2018 IEEE Globecom Workshops (GC Wkshps), Abu Dhabi, United Arab Emirates, pp. 1–7 (2018)
8. Wang, S., Chen, M., Liu, X., Yin, C., Cui, S., Poor, H.V.: A machine learning approach for task and resource allocation in mobile edge computing based networks. IEEE Internet Things J. (2020). https://doi.org/10.1109/jiot.2020.3011286
9. Zhou, S., Zhao, T., Niu, Z., Zhou, S.: Software-defined hyper-cellular architecture for green and elastic wireless access. IEEE Commun. Mag. **54**(1), 12–19 (2016)
10. Jiang, Z., Chen, S., Molisch, A.F., Vannithamby, R., Zhou, S., Niu, Z.: Exploiting wireless channel state information structures beyond linear correlations: a deep learning approach. IEEE Commun. Mag. **57**(3), 28–34 (2019)

11. Chen, S., et al.: Remote channel inference for beamforming in ultra-dense hyper-cellular network. In: GLOBECOM 2017 – 2017 IEEE Global Communications Conference, Singapore, pp. 1–6 (2017)

12. Chen, S., Jiang, Z., Zhou, S., Niu, Z.: Time-sequence channel inference for beam alignment in vehicular networks. In: 2018 IEEE Global Conference on Signal and Information Processing (GlobalSIP), Anaheim, CA, USA, pp. 1199–1203 (2018)

13. Goldsmith, A.: Wireless Communications. Cambridge University Press, Cambridge (2005). https://doi.org/10.1017/CBO9780511841224

14. Talvitie, J., Levanen, T., Koivisto, M., Valkama, M.: Positioning and tracking of high-speed trains with non-linear state model for 5G and beyond systems. In: 2019 16th International Symposium on Wireless Communication Systems (ISWCS), Oulu, Finland, pp. 309–314 (2019)

15. Bar-Shalom, T.K.Y., Li, X.R.: Estimation with Applications to Tracking and Navigation: Theory. Algorithms and Software. Wiley, Hoboken (2002)

16. Remcom, Wireless insite. http://www.remcom.com/wireless-insite

Content-Aware Proactive Caching and Energy-Efficient Design in Clustered Small Cell Networks

Xiang Yu, Huiting Luo$^{(\boxtimes)}$, Long Teng, and Ting Liu

School of Communication and Information Engineering,
Chongqing University of Posts and Telecommunications,
Chongqing 400065, People's Republic of China
luohuiting107@163.com

Abstract. This paper considers clustered small cell networks (SCNs) with combined design of cooperative caching and energy-efficient policy in the Coordinated Multi-Point (CoMP)-enabled cellular network. Small base stations (SBSs) with cache storage are grouped into associative clusters which can communicate with each other. This paper focus on movie on-demand streaming from Internet-based servers and proposed combined caching mode, where every SBS utilizes parts of cache space to cache the most popular contents (MPC), while the remaining is used for cooperatively caching different partitions of the less popular contents (LPC). Instead of the known content popularity, we constructs a content-aware weighted feature matrix (CWFM) in terms of spatiotemporal variation. Based on estimated content popularity and transmission design, we propose a caching scheme that makes a caching decision to maximize the energy efficiency (EE). To tackle this problem, A two-step stepwise optimization method is adopted. First, EE conditioning is optimized with a approach of linear programming and variable recovery. Then, the optimal proportion of cache space for MPC is analyzed by comparing the energy-efficient gain from the MPC with the energy-efficient loss from the discarded contents. Extensive simulation results confirm that our algorithm outperforms state-of-the-art algorithms based on MovieLens data set.

Keywords: Clustered SCNs · Popularity prediction · CWFM · EE · Proactive caching

1 Introduction

In recent years, the emergence of smart mobile devices and multimedia capabilities has resulted in the explosive increase of high data rate applications over wireless networks. According to a recent report from Cisco [1], global mobile data traffic will increase sevenfold between 2017 and 2022 with mobile video traffic accounting for the majority. However, current networks are unable to keep up

© ICST Institute for Computer Sciences, Social Informatics and Telecommunications Engineering 2021
Published by Springer Nature Switzerland AG 2021. All Rights Reserved
H. Gao et al. (Eds.): ChinaCom 2020, LNICST 352, pp. 288–308, 2021.
https://doi.org/10.1007/978-3-030-67720-6_20

with the massive growth of mobile data services in the 5G era. As for wireless networks already investigated in [2], by caching content at wireless edge, edge nodes can deliver the cached content to users directly instead of retrieving it in data center, which can significantly offloading the traffic flowing to the network. The sinking of caching and computing capability can greatly alleviate backhaul traffic and improve the cooperation efficiency of local base stations (BSs). Additionally, it also exhibits strong potential to reduce power consumption and improve energy efficiency (EE) of small cell networks (SCNs) [3].

Proactive caching in SCNs allow BSs to proactively prefetch contents from data center through backhaul links during off-peak hours and deliver the cached content to the user during peak hours. Depending on the availability and placement of the requested content, caching mode can be typically classified into two categories, namely uncoded caching [4] and coded caching [5]. Uncoded caching aims at caching complete content in each BS. Coded caching enables each BS to cooperatively caching different partitions contents. [6] demonstrated that the combination of coded caching and uncoded caching can effectively adapt to poor channels. Most the previous works on proactive caching at the SCN have been developing new methods for accurately predicting content popularity and caching the most popular contents with cooperation. [7] used a generalized Zipf distribution to model content popularity and ignored the influence by spatiotemporal variation. [8] estimated content popularity by using the request statistic of content, namely Least Recently Used (LRU) method. The multi-player multi-armed bandit (MPMAB) learning scheme studied in [9] set tradeoff between exploitation and exploration. The essence is learning in long period learning process to maximize the local content popularity or caching the unrequested content that may be popular. However, on the accuracy of predicting the content popularity, the three methods behaved poorly for ignoring learning the internal relation of user's favor and desired content. [10] utilized feature to preliminarily express content attribute, but it never fully set up the relation between user's favor and content popularity.

In heterogeneous networks, coexistence between SBSs and conventional macro BSs causes additional intercell interference when spectrum resources are shared. Coordinated Multi-Point (CoMP) technique was proposed to limit the intercell interference and cell-edge throughput by allowing geographically separated SBSs to deliver information to users cooperatively [11]. Joint transmission with CoMP in the downlink of heterogeneous cellular networks with randomly located SBSs was studied in [12], where expressions for coverage probability and diversity gain were derived for typical user by using tools from stochastic geometry. Further, recent studies in wireless edge caching with CoMP exhibited new perspectives on the benefits of caching to improve network performance. On cache-level cooperation in CoMP SCNs, the cache space of multiple SBSs in a cluster is utilized as a entity. Parts of cache space is used to cache the most popular contents (MPC) while the remaining selectively cache the less popular contents (LPC) to improve the content diversity. Considering cooperative transmission via caching manager, [13] analyzed the average cache service probability with comprehensive consideration of joint transmission and parallel transmission, and show an optimal inherent tradeoff between transmission diversity and content diversity in cluster-centric network. Nevertheless, none of the existing works provide efficient solutions for the cache utilization policy in cooperative clustered SCNs based on unknown content popularity in prior. By using a more accurate content popularity prediction method and caching cooperatively in the clustered SCNs, the system energy efficiency can be significantly improved.

With the development of the network, SCN with intensive deployment of multiple SBS can serve more users and provide users with higher QoS. Meanwhile, multiple SBSs results in higher energy consumption. Due to the changes in network scenarios, more aspects need to be considered in the design of energy strategy, such as the unknown popularity, collaboration between multiple SBSs, and different caching methods. This paper proposes clustered SCNs with combined design of cooperative caching and energy-efficient policy based on estimated content popularity with spatiotemporal variation in order to maximize EE, which has not been considered in cluster SCNs. The SBSs are grouped into associated clusters, and the SBSs in different cluster can communicate with each other to enhance the performance of cellular network. The overall cache space within a cluster is arranged by central controller so as to either distribute the same MPC in every SBS or cache different partitions of the LPC in different SBSs. In terms of accuracy of predicting content popularity, we use the content features in popularity prediction, which can connect user's favor with content's attributes. Based on estimated popularity, the controllers in each cluster cooperatively assign cache space for the MPC while the remaining cache space for the LPC to achieve largest content diversity. Within a certain cluster, when the requested content is cached in SBSs, depending on whether the content is cached using MPC or LPC strategy, we use two transmission schemes accordingly, namely joint transmission and parallel transmission. When content is cached in MPC, it is delivered by federated transport, and when content is cached in LPC, it will be delivered using parallel transport. We model the average energy efficiency as optimization objective in the clustered SCNs, namely ratio of total transmission rate to total power. For this complex problem in clustered SCNs, we adopt a two-step stepwise optimization method, and quickly search for an inherent tradeoff between cooperation transmission and content diversity with our proposed scheme. We then maximize the average energy efficiency with optimal content placement in clusters for LPC and optimal proportion of cache space for MPC.

This paper is organized as follows. We present the network model and cooperation schemes in Sect. 2. In Sect. 3, we construct a content-aware weighted feature matrix (CWFM) to predict content popularity in terms of spatiotemporal variation. In Sect. 4, we define the average energy efficiency as the main performance metric and give its formulation. Furthermore, we analytically prove the optimization proportion of cache space for MPC in each cluster using our proposed scheme. Simulation results are presented in Sect. 5 and Sect. 6 concludes the paper.

2 System Model and Cooperation Schemes

In this work, we consider cache-enabled clustered SCNs where SBSs in each cluster are distributed according to a two-dimensional homogeneous Poisson Point Process (PPP) [14] and the distribution function is $\Phi_b = \{b_i \in \mathbb{R}^2, \forall i \in \mathbb{N}^+\}$ with intensity λ_b. The set of all SBSs is denoted by $\mathcal{B} = \{1, 2, \cdots, b, \cdots, B\}$. As shown in Fig. 1, geographically adjacent SBSs are grouped into associative clusters where collaboratively deliver contents requested and improve wireless transmission performance. The cache manager (CM) is connected to a data center via a high-speed dedicated link and some clusters are selected to connected CM for efficient operations. The number of connected clusters is determined by parameter ϵ, which is defined as the number of the proportion of the number of selected clusters to the number of all clusters. In order to facilitate the management, there is a cluster controller SBS connected to the other SBSs while the other SBSs are not directly connected. Some controller SBSs are directly connected with some are not. The result of clustering is that all SBSs are clustered into J clusters, and the set of cluster is denoted by $\mathcal{J} = \{1, \cdots, j, \cdots, J\}$, where J is corresponded to the number of small cell. For convenience, the cache device in SBS has the same storage capacity, and the total cache capacity in each cluster is considered as an entity.

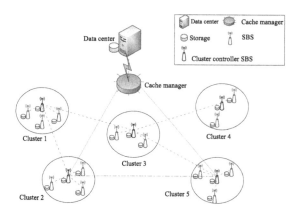

Fig. 1. A cache-enabled clustered SCN.

2.1 Small Cell Clustering

Based on the SCN model as Fig. 1, we consider using a hexagonal grid with inter-cluster center distance $2R_h$, and the area of each cluster is $\mathcal{A} = 2\sqrt{3}R_h^2$ to represent the shape of cell. For a random cluster, the probability mass function that the number of SBSs n inside cluster equal to K follows a Poisson distribution with mean $\lambda_b \mathcal{A}$ is denoted by

$$\mathbb{P}(n = K) = e^{-2\sqrt{3}\lambda_b R_h^2} \frac{(2\sqrt{3}\lambda_b R_h^2)^K}{K!}, \tag{1}$$

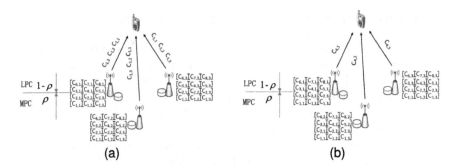

Fig. 2. (a) JT transmission procedure. (b) PT transmission procedure.

there are K SBSs in a certain cluster, the SBS distribution follows a Binomial Point Process. The distance distribution between randomly distributed SBSs and the distance from a SBS to an arbitrary in hexagonal cell have been elaborated in [15], where demonstrated that using a circle with the same area to provide the best approximation of the distribution model, and the radius is $R = R_h \sqrt{\frac{2\sqrt{3}}{\pi}}$. The set of SBSs in j-th cluster is $\mathcal{C}_j = \{b_i \in \Phi_b \bigcap \mathcal{M}(z_j, R)\}$, where $\mathcal{M}(z_j, R)$ denotes the ball centered at z_j with radius R. The j-th cluster with K SBSs is denoted as $\mathcal{C}_j = \{C_{j,1}, \cdots, C_{j,k}, \cdots, C_{j,K}\}$, due to the limitation of cache capacity, each SBS can store up to D contents, then the total available storage capacity of $\mathcal{C}_j$ is u_j, as $KD = u_j$.

The network is operated in time-slotted manner and the time slot is denoted by $\mathcal{T} = \{0, 1, \cdots, t, \cdots, T\}$. The total contents in data center in time slot t are indexed by $\mathcal{F}_t = \{1, 2, \cdots, f, \cdots, F_t\}$, which varies over time for new contents uploaded. In the cluster with K SBSs, each content is divided into K equal-size partitions [16]. At the beginning of each time slot, the CM makes a decision on cooperatively refreshing the cache entities to cache new popular contents. In each cluster, each user makes independent request for contents in each time slot. To be specific, we consider MovieLens, a web-based recommender system based on the rating of movie viewed by users [17] as the dataset. In each time slot, the request statue among SBSs is different.

2.2 Cooperative Transmission

In this work, We assume single antenna at both SBSs and user device. Orthogonal multiple access method is used to tackle simultaneous arrival of content requests. In the SCN, we assume that each content contains S bits, the successful delivery of a content is defined by the event that S bits are successfully delivered using bandwidth W and time T_ϕ. With CoMP, it provides two different transmission modes, namely joint transmission(JT) and parallel transmission(PT), as described in Fig. 2.

Joint Transmission. If the requested content f is in the MPC range, each SBSs in the cluster store entire content. Thus, f is jointly transmitted by K SBSs to user. The purpose is to increase the received signal to interference plus noise ratio (SINR) to enhance the content delivery reliability. It is denoted as JT cooperation scheme with each SBS of a cluster sends S bits to the user using the same bandwidth. The received signals from K SBSs are superimposed and can be regarded as a single stream. The successful content delivery probability (SCDP) [13] is defined as

$$P_K^{\text{JT}} = \mathbb{P}\left[WT_\phi \log_2(1 + \text{SINR}) > S \mid K\right]. \tag{2}$$

$$P_K^{\text{JT}} = \mathbb{P}\left[\text{SINR} > 2^{\frac{R_d}{W}} - 1 \mid K\right], \tag{3}$$

where $R_d = \frac{S}{T_\phi}$ as the expected delivery rate for successful content delivery.

Parallel Transmission. If the requested content f is in the LPC range, each K cooperative SBSs in the cluster stores disjoint partitions. Thus, the different partitions need to be transmitted to the user simultaneously by K streams. It is denoted as PT cooperation scheme. We adopt PT with successive decoding based spectrum sharing case with K SBSs simultaneously send $\frac{S}{K}$ bits to the user by sharing the same W bandwidth. The successive decoding with interference cancellation (SIC) is utilized to decode the signal according to the received signal power order [18]. The SCDP is defined as

$$p_K^{\text{PT}} = \mathbb{P}\left[\bigcap_{i \in \mathcal{C}_j} WT_\phi \log_2(1 + \text{SINR}_i) > \frac{S}{K} \mid K\right], \tag{4}$$

$$p_K^{\text{PT}} = \mathbb{P}\left[\bigcap_{i \in \mathcal{C}_j} \text{SINR}_i > 2^{\frac{R_d}{KW}} - 1 \mid K\right]. \tag{5}$$

where SINR_i is the SINR from the i-th SBS in cluster $\mathcal{C}_j$ of requested content.

Transmission for Sharing Case and Missing Case. In addition to the above two transmission modes, we also consider transmission for sharing and transmission for missing case. When connecting user in $\mathcal{C}_j$ requests content f which cached in $\mathcal{C}_i$, the cluster controller SBS of $\mathcal{C}_i$ retrieve the content f according to the caching mode, and then the cluster controller SBS of $\mathcal{C}_j$ fetches f from the cluster controller SBS of $\mathcal{C}_i$, and shares the decomposed contents to the remaining SBSs within $\mathcal{C}_j$ with PT mode. Meanwhile, if connecting user in $\mathcal{C}_j$ requests content f which is not cached in the local clusters. In this missing case, the cluster controller SBS in $\mathcal{C}_j$ fetches content f from the data center through backhaul links and transmits content f to user with PT mode.

3   CWFM-Based Content Popularity Prediction

Before we design the caching strategy, we need to determine what contents will be cached, which is to predict the popularity. With the edge network has the ability to control, compute and cache, [19] proposed to cache the content purposefully according to the user preferences, the simulation results indicate that a small number of features can also show obvious effects in improving the cache hits. Further, we expand the thought and φ features are extracted to construct a connection between contents and user's favor. The extracted features meet low redundancy between features, and strong correlation between the features and user. The data set MovieLens [17] is utilized and the user's movie scoring process in a timestamp is equivalent to the user's content request process. We use mutual information to measure the correlation and the redundancy, and the selected feature set is defined as $\mathbf{F}^*$.

For the content popularity in different time slot, φ features is constructed into a matrix with M rows and N columns and matrix $\boldsymbol{A}_f$ is used to represent the attribute of content f, where element A_f^{mn} equels 1 if content f has the corresponding feature and 0 otherwise [20]. As the importance of each feature is non-uniform, we determine a weight to each feature and construct a weighted feature matrix (WFM) in terms of spatiotemporal variation. The WFM in $\mathcal{C}_j$ is defined as $\boldsymbol{P}_j(t)$ in time slot t. At the initial time slot, the feature weight is given by the accumulation of the number of historical requests, shown as $p_j^{mn}(0) \triangleq \sum_{f \in \mathcal{F}_0} Q_f^j(\tau) A_f^{mn}$, where $Q_f^j(\tau)$ and $\mathcal{F}_0$ are the request number of content f in $\mathcal{C}_j$ after the training time τ and the set of initial contents respectively. Therefore, the initial WFM can be written as

$$\boldsymbol{P}_j(0) \triangleq \sum_{f \in \mathcal{F}_0} Q_f^j(\tau) \cdot \boldsymbol{A}_f. \tag{6}$$

The popularity of content f in $\mathcal{C}_j$ in time slot t is given as

$$g_{j,f}(t) = (\mathbf{1}^N)^T (\boldsymbol{P}_j(t) \otimes \boldsymbol{A}_f \mathbf{1}^N), f \in \mathcal{F}_t, \tag{7}$$

where $\otimes$ represents the Hadamard multiplication and $\mathbf{1}^N$ is the all one column vector. Thus, it is possible to establish an internal relation between contents in terms of popularity, also a relation between content popularity and user preferences.

It is reasonable to combine the historical feature weight and the content request number in the previous time slot to balance the feature weight in the next time slot. Hence, we learn user's status online and dynamically adjusting weight of feature. If one content is requested, the weight of corresponding feature increases. To measure the increment, we introduce an growth factor $\eta_f^j(t) = \sigma^{d_{j,f}(t)/\sum_f d_{j,f}(t)}$, where $d_{j,f}(t)$ is the requested number of content f in $\mathcal{C}_j$ in time slot t and σ is determined by the specific request status in SCNs, which is no less than 1. It is observed that larger σ makes the prediction of user's favor more inclined to the features in the previous time slot. Different contents in $\mathcal{F}_t$ may have the same features. For simplicity and effectiveness, we adopt statistical method of superimposing the feature weights and averaging them. According to the known $\mathbf{P}_j(t)$ and the request information of $\mathcal{C}_j$ in time slot t, the superimposed CWFM of $\mathcal{C}_j$ at beginning of time slot $t+1$ is

$$\overline{\mathbf{P}_j(t+1)} = \sum_{f\in\mathcal{F}_t} \eta_f^j(t)\, \mathbf{A}_f \otimes \mathbf{P}_j(t). \tag{8}$$

Therefore, the feature matrix of $\mathcal{C}_j$ at the beginning of time slot $t+1$ can be written as

$$\mathbf{P}_j(t+1) = \overline{\mathbf{P}_j(t+1)} \oslash \sum_{f\in\mathcal{F}_t} \mathbf{A}_f, \tag{9}$$

where $\oslash$ represents the Hadamard division. The feature weight of each content can be refreshed with the passage of time slot , and the prediction result of the popularity of content f in $\mathcal{C}_j$ within time slot $t+1$ is

$$g_{j,f}(t+1) = (\mathbf{1}^N)^T (\mathbf{P}_j(t+1) \otimes \mathbf{A}_f \mathbf{1}^N), f \in \mathcal{F}_{t+1}. \tag{10}$$

As a result, by refreshing and iterations, CWFM can achieve the accurate prediction of content popularity, which contains the newly uploaded contents in each time slot.

4   Proposed Energy Efficiency-Based Caching

In clustered SCNs, EE is the most important metric since CoMP can significantly improves SINR and reduces delay, while multiple SBSs cause higher energy consumption. In this paper, we define EE as the ratio of actual delivery rate to energy consumption.

For the requested content f, the delivery policy in time slot t can be expressed as $\mathbf{y}(t) = \{y_{j,f,1}(t), y_{j,f,2}(t), y_{i,j,f}(t), y_{c,j,f}(t), \forall i,j \in \mathcal{J}, i \neq j, \forall f \in \mathcal{F}_t\}$, where $y_{j,f,1}(t)$ represents if content f is fully cached in $\mathcal{C}_j$ and $y_{j,f,2}(t)$ represents if content f is partly cached in $\mathcal{C}_j$. $y_{i,j,f}(t)$ indicates whether $\mathcal{C}_j$ fetches the content f from $\mathcal{C}_i$, $y_{c,j,f}(t)$ indicates whether $\mathcal{C}_j$ fetches the content f from data center. They all take values from $\{0,1\}$. Note that content f is fetched from $\mathcal{C}_i$ only when $\mathcal{C}_i$ caches it, that is

$$y_{i,j,f}(t) \leqslant y_{i,f,1}(t) + y_{i,f,2}(t). \tag{11}$$

Only one case can be used to deliver the requested content f in each time slot, thus

$$y_{j,f,1}(t) + y_{j,f,2}(t) + y_{i,j,f}(t) + y_{c,j,f}(t) \leqslant 1. \tag{12}$$

As the limitation of cache capacity, it needs to be satisfied that

$$\sum_{f\in\mathcal{F}_t} K y_{j,f,1}(t) + y_{j,f,2}(t) \leqslant u_j. \tag{13}$$

Hence, in time slot t, the EE can be formulated as

$$EE(t) = \sum_{f \in \mathcal{F}_t} \sum_{j \in \mathcal{J}} \sum_{i \in \mathcal{J} \setminus j} d_{j,f}(t) \frac{[V_1 y_{j,f,1}(t) + V_2 y_{j,f,2}(t) + V_3 y_{i,j,f}(t) + V_4 y_{c,j,f}(t)]}{[P_1 y_{j,f,1}(t) + P_2 y_{j,f,2}(t) + P_3 y_{i,j,f}(t) + P_4 y_{c,j,f}(t)]},$$

(14)

where V_1, V_2, V_3 and V_4 respectively denote the actual deliver rate of the above four deliver categories and P_1, P_2, P_3, P_4 respectively denote the energy consumption. In clustered SCNs, the energy efficiency-optimized caching strategy is actually maximize the CoMP gain and the content cache gain [21].

4.1   Successful Content Delivery Probability Analysis

As EE is the ratio of actual delivery rate to energy consumption. The actual delivery rate is determined by the actual transmission rate (ATR) and SCDP. Hence, the SCDP in different cases is discussed in this section. The SCDP in JT mode and PT mode have been shown as (3) and (5).

For a typical cluster-center user locates at z_j and requests content with JT, the cooperating SBSs in $\mathcal{C}_j$ transmit the same symbol s with the equal transmit power P_e. Considering the channel loss due to distance, we use a standard distance-dependent power law pathloss attenuation as $d^{-\kappa}$, d is the distance and κ is the pathloss exponent. The channel output at the user is

$$Z = \sum_{b \in \mathcal{C}_j} \sqrt{P_e} d_b^{-\frac{\kappa}{2}} h_b s + \sum_{l \in \mathcal{B} \setminus \mathcal{C}_j} \sqrt{P_e} d_l^{-\frac{\kappa}{2}} h_l s_l + N_o,$$

(15)

where h_b and h_l represent the small-scale Rayleigh fading from the b-th and the l-th SBS to the user respectively, which follows $h_b, h_l \backsim \mathcal{CN}(0,1)$, d_b, d_l respectively represent the distance from the b-th and the l-th SBS. s_l represents the transmitted symbol of the l-th SBS out of the cluster, s is the joint transmitted symbol in $\mathcal{C}_j$, N_o donates the background thermal noise. Considering the interference-limited network and neglecting N_o, the signal to interference ratio (SIR) is

$$\text{SIR}_{JT} = \frac{\left| \sum_{b \in \mathcal{C}_j} h_b d_b^{-\frac{\kappa}{2}} \right|^2}{\sum_{l \in \mathcal{B} \setminus \mathcal{C}_j} |h_l|^2 d_l^{-\kappa}}.$$

(16)

The expected SIR is $\theta_1 = 2^{\frac{R_d}{W}} - 1$, and the SCDP in JT case can be expressed as

$$p_K^{JT}(\theta_1) \simeq \int_0^R \cdots \int_0^R \mathcal{L}_{I \setminus R} \left(\frac{\theta_1}{\sum_{k=1}^K x_k^{-\kappa}} \right) \prod_{k=1}^K \frac{2x_k}{R^2} dx_1 \cdots dx_K,$$

(17)

where $\mathcal{L}_{I \setminus x}$ is the Laplace transform of the interference coming from SBSs located outside of $\mathcal{M}(0, x)$ and has been proven in [13] as

$$\mathcal{L}_{I \setminus x} = \exp \left(-\pi \lambda s^{2/\kappa} \int_{\frac{x^2}{s^{2/\kappa}}}^{\infty} \frac{1}{1 + \omega^{2/\kappa}} \right).$$

(18)

When user locates at z_j and in the case of PT, the cooperating K SBSs in C_j transmit different parts of symbol $[s_1, \cdots, s_k, \cdots, s_K]$ with equal transmit power P_e to the user. The channel output at the user is

$$Z = \sum_{b \in C_j} \sqrt{P_e} d_b^{-\frac{\kappa}{2}} h_b s_b + \sum_{l \in B \backslash C_j} \sqrt{P_e} d_l^{-\frac{\kappa}{2}} h_l s_l + N_o. \tag{19}$$

In order to decode multiple streams simultaneously, we adopt received power ordering to decode different data stream by distance value studied in [22]. The distance vector can be expressed as $\mathbf{d} = [d_1^*, \cdots, d_k^*, \cdots, d_K^*]$, where d_k^* denotes the distance from z_j to the k-th nearest SBS. When decoding the information from the k-th SBS, all signals come from closer $k-1$ SBSs should have been successfully decoded and canceled. Hence, the SIR of the k-th stream is given as

$$\mathrm{SIR}_k \simeq \frac{|h_k|^2 (d_k^*)^{-\kappa}}{\sum_{l \in B \backslash M(0, d_k^*)} |h_l|^2 d_l^{-\kappa}}. \tag{20}$$

Finally, the remaining interference only comes from out-cluster SBSs and the distance is greater than R, thus the SIR of the final decoded stream is

$$\mathrm{SIR}_k \simeq \frac{|h_K|^2 (d_K^*)^{-\kappa}}{\sum_{l \in B \backslash M(0, d_K^*)} |h_l|^2 d_l^{-\kappa}}. \tag{21}$$

The expected SIR is $\theta_2 = 2^{\frac{R_d}{KW}} - 1$, and the SCDP in PT case can be expressed as

$$p_K^{PT}(\theta_2) \simeq \int_{0 < x_k < R} \frac{2K x_K}{R^2} \mathcal{L}_{I \backslash R}(\theta_2 x_K^\kappa) \prod_{k=1}^{K-1} \frac{2k x_k}{R^2} \mathcal{L}_{I \backslash x_k}(\theta_2 x_k^\kappa) dx_1 \cdots dx_K. \tag{22}$$

When user locates at z_j and transmits for sharing case, the cluster controller in C_j fetches f from C_i, $i \in \mathcal{J} \backslash j$, and then transmits with PT mode. Since the inter-cluster data transmission consumes time resources, the maximum transmission time of the wireless side is reduced. We define the delay of fetching requested content f from C_i as $T_{i,j,f}$ and the wireless side maximum transmission time is changed as $\sigma_1 T_\phi$, where $\sigma_1 = 1 - \frac{T_{i,j,f}}{T_\phi}$. Thus, the expected SIR is $\theta_3 = 2^{\frac{R_d}{\sigma_1 W}} - 1$ and the SCDP is $p_K^{PT}(\theta_3)$.

When user locates at z_j and transmits for missing case, the cluster controller in C_j fetches f from data center and then transmits with PT mode. Since the backhaul delay is relatively large, the maximum transmission time on the wireless side is significantly reduced. We define the delay of fetching requested content f from data center as $T_{c,j,f}$ and the wireless side maximum transmission time is changed as $\sigma_2 T_\phi$, where $\sigma_2 = 1 - \frac{T_{c,j,f}}{T_\phi}$. Hence, the expected SIR is $\theta_4 = 2^{\frac{R_d}{\sigma_2 W}} - 1$ and the SCDP is $p_K^{PT}(\theta_4)$.

The actual delivery rate can be defined as the product of SCDP and ATR. For the expected delivery rate R_d, the ATR has a partial probability above

R_d, and a partial probability below as the channel is unstable. Since the total transmission content size remains unchanged, as long as the average transmission rate is greater than R_d, the ATR can be combined by different transmission rate. In addition, we analyze all possible combinations of transmission rates, and the resulting average transmission rate is called the threshold of JT mode or PT mode. If there is a probability that the ATR is greater than R_d, then there must be a minimum transmission rate v^* less than R_d in the different transmission rates combination. When R_d is less than the threshold, we have $v^*=0$, and when R_d is greater than the threshold, in order to meet the condition that the actual delivery rate is greater than or equals to the R_d, $v^*>0$ is necessary, thus the combination of transmission rates interval is $[v^*, \infty]$. The actual delivery rate can be expressed as

$$\sum_{v=v^*}^{\infty} \Delta v \Delta p_K (2^{\frac{v}{KW}} - 1). \tag{23}$$

Wherein, when $v^*=0$, the result of (23) is the threshold value which is greater than or equals to $R_d p_K^{JT}(2^{\frac{v^*}{KW}} - 1)$, the actual delivery rate can be given as

$$\left\{ R_d p_K (2^{\frac{v^*}{KW}} - 1) | R_d p_K (2^{\frac{v^*}{KW}} - 1) \le \sum_{v=v^*}^{\infty} \Delta v \Delta p_K (2^{\frac{v}{Kw}} - 1) \right\}. \tag{24}$$

Hence, the effective delivery rates of above cooperative transmission cases are $R_d p_K^{JT}(\theta_1^*)$, $R_d p_K^{PT}(\theta_2^*)$, $R_d p_K^{PT}(\theta_3^*)$, $R_d p_K^{PT}(\theta_4^*)$, respectively, where $\theta_1^* = 2^{\frac{v_1}{W}}-1$, $\theta_2^* = 2^{\frac{v_2}{KW}}-1$, $\theta_3^* = 2^{\frac{v_3}{\sigma_1 W}}-1$ and $\theta_4^* = 2^{\frac{v_4}{\sigma_2 W}}-1$, v_1, v_2, v_3 and v_4 meet the demand of (24).

4.2   EE-Based Proactive Caching Cooperatively

Energy consumption mainly consists of the wireless side emission energy consumption and the wired side fiber transmission energy consumption. For JT case, it contains only emission energy consumption which is given as KP_e while transmitting unit size content. For PT case, it also contains only emission energy consumption. However, this case has lower transmission rate and the duration of each SBS is similar to the JT case. Thus, the overall energy consumption is also KP_e. The energy consumption of the wired side fiber transmission can be decomposed into the unit size content transmission energy consumption $P_{i,j}$ of the cluster controllers between C_i and C_j and the unit size decomposition part transmission energy consumption P_{km} of the cluster controller to other SBSs. In sharing case, when content f is cached as MPC in C_i, the energy consumption is $(K-1)P_{km}+P_{i,j}+KP_e$, when it is cached as LPC in C_i, the energy consumption is $2(K-1)P_{km} + P_{i,j} + KP_e$. To distinguish the two types, we use $y_{i,j,f,1}(t)$ and $y_{i,j,f,2}(t)$ respectively to indicate whether it occurs in time slot t and we have

$$y_{i,j,f,1}(t) + y_{i,j,f,2}(t) = y_{i,j,f}(t). \tag{25}$$

Similarly, in transmission for missing case, the energy consumption is given as $(K-1)P_{km} + P_{c,j} + KP_e$, where $P_{c,j}$ denotes the energy consumption which data center transmits unit size content to the cluster controller of $\mathcal{C}_j$.

In missing case, the backhaul energy consumption is significantly larger than the other three cases, and the effective delivery rate of the backhaul is significantly smaller. If the local clusters cache the requested content, it should be likely fetched from local clusters rather than the data center. Hence, when the content caching policy is determined, the transmission policy is also determined accordingly, and the total effective delivery rate in the time slot t is formulated as

$$
\begin{aligned}
V_{tot}(t) = &\sum_{f \in \mathcal{F}_t} \sum_{j \in \mathcal{J}} \sum_{i \in \mathcal{J} \backslash j} d_{j,f}(t) \left[p_K^{JT}(\theta_1) R_d y_{j,f,1}(t) + p_K^{PT}(\theta_2) R_d y_{j,f,2}(t) \right] \\
&+ \left[p_K^{PT}(\theta_3) R_d y_{i,j,f,1}(t) + p_K^{PT}(\theta_3) R_d y_{i,j,f,2}(t) + p_K^{PT}(\theta_4) R_d y_{c,j,f}(t) \right].
\end{aligned}
\tag{26}
$$

The total energy consumption is formulated as

$$
\begin{aligned}
P_{tot}(t) = &\sum_{f \in \mathcal{F}_t} \sum_{j \in \mathcal{J}} \sum_{i \in \mathcal{J} \backslash j} d_{j,f}(t) \left\{ KP_e y_{j,f,1}(t) + KP_e y_{j,f,2}(t) + [(K-1)P_{km} + P_{c,j} + KP_e] y_{c,j,f}(t) \right\} \\
&+ \left\{ [(K-1)P_{km} + P_{i,j} + KP_e] y_{i,j,f,1}(t) + [2(K-1)P_{km} + P_{i,j} + KP_e] y_{i,j,f,2}(t) \right\}.
\end{aligned}
\tag{27}
$$

Thus, the maximum EE in time slot t is written as

$$
\max_{\mathbf{y}(t)} \quad \frac{V_{tot}(t)}{P_{tot}(t)}.
\tag{28}
$$

Note that the optimization goal is determined by the 5-dimensional Boolean variable. The problem is actual where to place contents in time slot t, and its cache mode to maximum the EE. However, even if the variables are related, the conditional constraints can not simplify the EE formulation into a function form determined by a single variable.

To optimize the EE, if the requested content gets hit within the cluster, transmission time for wireless is a long period, the effective delivery rate is large, and the energy consumption is small. If the requested content gets hit in the neighboring cluster, the transmission time is reduced, the effective delivery rate becomes smaller, and the energy consumption becomes larger. If the requested content misses, the remaining transmission time is least, the effective delivery rate is smallest, and the energy consumption is largest. Hence, the cache hits become critical. Meanwhile, compared with the LPC, MPC cached fully, although the effective delivery rate increases, the cache space occupied is K times larger than the former. That means one of the MPC is cached, $K-1$ LPC should be discarded. The effective delivery rate for the MPC should be approximately K times than the LPC. Therefore, we should first ensure that the requested content hits as many local clusters as possible, and then gradually analyze and convert the content in LPC range into MPC. Comparing with the non-convex problem which directly solves the optimal value, we adopt a two-step stepwise optimization method. Firstly, all the contents are cached as LPC and transferred by PT mode, which reduces the dimensionality of the variables and the complexity of the computation compared to a direct solution. After that, some of the LPC will be converted to MPC for caching using EE as an indicator. The whole algorithm processing process does not change the cache constraints, but only the variable processing process is solved in two steps, so this method is easy to implement and ensures reliability.

To maximize cache hits, the cache capacity of local cluster is a fixed value, and the optimization objective becomes the maximum effective delivery rate in the PT mode and minimizes the energy consumption. As analyzed above, both in PT mode completely depend on the consumption of fiber transmission and proportional to the time consumption and the energy consumption. Therefore, the optimization objective turns to minimize fiber transmission consumption within time slot t. In this work, we define the transmission consumption is the square root of the product of time consumption and energy consumption. The content caching policy at time slot t is denoted as $\mathbf{y}_r(t) = \{y_{i,j,f,2}(t), \forall i, j \in \mathcal{J}, \forall f \in \mathcal{F}_t\}$. If user associated with $\mathcal{C}_j$ fetches content f from $\mathcal{C}_i$, the consumption is $\sum_{i,j \in \mathcal{J}} \sum_{f \in \mathcal{F}_t} e_{i,j} d_{j,f}(t) y_{i,j,f,2}(t) s_f$, where $e_{i,j}$ represents the consumption that $\mathcal{C}_i$ delivers a unit size content to $\mathcal{C}_j$ and s_f is the size of content f. Note that if $i = j$ represents that user fetches content f from the cluster associated with currently. In addition, if user associated with $\mathcal{C}_j$ fetches content f from data center, the consumption is $\sum_{j \in \mathcal{J}} \sum_{f \in \mathcal{F}_t} e_{c,j} d_{j,f}(t) y_{c,j,f}(t) s_f$, where $e_{c,j}$ represents the consumption that data center delivers a unit size content to $\mathcal{C}_j$. $d_{j,f}(t)$ is the number requests for content f, which is inextricably related with the popularity and can be replaced by $\alpha g_{j,f}(t)$. Thus, our objective is minimizing the consumption as

$$\min_{\mathbf{y}_r(t)} \sum_{i,j \in \mathcal{J}} \sum_{f \in \mathcal{F}_t} \alpha g_{j,f}(t)(e_{i,j} y_{i,j,f,2}(t) + e_{c,j} y_{c,j,f}(t)) s_f. \tag{29}$$

Based on $\sum_{j \in \mathcal{J}} y_{i,j,f,2}(t) + y_{c,j,f}(t) = 1$, we can eliminate $y_{c,j,f}(t)$, the formulation turns to be

$$\max_{\mathbf{y}_r(t)} \sum_{f \in \mathcal{F}_t} \sum_{j \in \mathcal{J}} g_{j,f}(t) s_f \sum_{i \in \mathcal{J}} y_{i,j,f,2}(t)(e_{c,j} - e_{i,j}). \tag{30}$$

As $\mathcal{C}_j$ can retrieve content f from $\mathcal{C}_i$ only when $\mathcal{C}_i$ has cached it before. That is the problem to maximize the objective is actual where to place f, the specific cluster or data center. Hence, $y_{i,j,f,2}(t)$ can be eliminated, the placement policy is given as

$$\max_{\mathbf{y}_r(t)} [y_{1,f,2}(t), \cdots, y_{\mathcal{J},f,2}(t)] \cdot \mathbf{E}_f \cdot [g_{1,f}(t), \cdots, g_{\mathcal{J},f}(t)]. \tag{31}$$

where $\mathbf{E}_f$ is a matrix with a dimension of $J \times J$, and the objective can be written as

$$\max_{\mathbf{y}_r(t)} \mathbf{Y}(t) \cdot \mathbf{E} \cdot \mathbf{G}(t), \tag{32}$$

$\mathbf{Y}(t) = [y_{1,1,2}(t), \cdots, y_{J,1,2}(t); \cdots, y_{j,f,2}(t), \cdots, y_{J,F_t,2}(t)]^T$, $\mathbf{E}$ is a diagonal matrix as $\mathbf{E} = [\mathbf{E}_1, \cdots, \mathbf{E}_f, \cdots, \mathbf{E}_{F_t}]$, $\mathbf{G}(t) = [g_{1,1}(t), \cdots, g_{J,1}(t); \cdots, g_{j,f}(t), \cdots, g_{J,F_t}(t)]$. Thanks to constraints on Boolean variables $y_{j,f,2}(t)$, content size and link consumption, the optimization objective is non-convex. To make it convex, we relax Boolean variable $y_{j,f,2}(t) \in \{0,1\}$ into $[0,1]$ and tackle the problem with linear programming and variable recovery. Therefore, a optimal solution can be achieved within a certain precision range, and then we can obtain a solution denoted as $\mathbf{Y}^*(t)$. Note that $y_{j,f,2}^*(t)$ in $\mathbf{Y}^*(t)$ represents the contribution degree of content f to $\mathcal{C}_j$. Since the dimension of $\mathbf{Y}^*(t)$ is 1, a quick sort algorithm is adopted to map $\mathbf{Y}^*(t)$ into $\mathbf{Y}_v(t) = [y_1(t), \cdots, y_v(t), \cdots, y_{J \cdot F_t}(t)]$ by size of variable value, where $y_{j,f,2}^*(t)$ mapped into $y_v(t)$ means that $y_{j,f,2}^*(t)$ is the v-th largest item in $\mathbf{Y}^*(t)$. The recovery is performed step by step and the recovery policy for v-th variable is

$$y_{j,f,2}(t) = \begin{cases} 1, & s_f \leq u_j - \hat{u}_j, \ y_{j,f,2}(t) \neq 1, \ \forall j \in \mathcal{J}, \\ 0, & \text{otherwise} \end{cases} \tag{33}$$

where $\hat{u}_b$ is the occupied caching space of $\mathcal{C}_j$. Finally, maximum objective value is obtained when CM completes the recovery process, and then substituting the value into (28), we can get the $EE_{ini}(t)$.

4.3 Optimal Design for Maximizing EE

The caching policy can be further improved by discarding some of the LPC and converting the more popular ones into MPC, and using the EE gain of the JT mode to improve the overall EE of the system. The contents cached in $\mathcal{C}_j$ are sorted in descending order of contribution as $[1_1^*, \cdots, F_1^*; \cdots, 1_J^*, \cdots, F_J^*]$. For $\mathcal{C}_j$, f_j^* is converted into the MPC and the increase of effective delivery rate is $d_{j,f_j^*}(t)\left[R_d p_K^{JT}(\theta_1) - R_d p_K^{PT}(\theta_2)\right]$, while energy consumption remains constant. For $\mathcal{C}_i$, the decrease of energy consumption is $\sum_{i \in \mathcal{J} \backslash j} d_{j,f_j^*}(t)(K-1)P_{km}$, while effective delivery rate remains constant. In addition, converting f_j^* to the MPC means that $K-1$ LPC are discarded. For this part of contents, in $\mathcal{C}_j$, the decrease of effective delivery rate is

$$\sum_{l_j^* = F_j^* - K + 2}^{F_j^*} d_{j,l_j^*}(t)\left[R_d p_K^{PT}(\theta_4) - R_d p_K^{PT}(\theta_2)\right], \tag{34}$$

where l_j^* is the discarded less popular content. The increase of energy consumption is

$$\sum_{l_j^* = F_j^* - K + 2}^{F_j^*} d_{j,l_j^*}(t)\left[P_{c,j} + (K-1)P_{km}\right]. \tag{35}$$

In $\mathcal{C}_i$, the decrease of effective delivery rate is

$$\sum_{i \in \mathcal{J} \backslash j} \sum_{l_j^* = F_j^* - K + 2}^{F_j^*} d_{i,l_j^*}(t) \left[Rdp_K^{PT}(\theta_4) - Rdp_K^{PT}(\theta_3) \right], \tag{36}$$

while the increase of energy consumption is

$$\sum_{i \in \mathcal{J} \backslash j} \sum_{l_j^* = F_j^* - K + 2}^{F_j^*} d_{i,l_j^*}(t) \left[P_{c,j} - P_{j,i} - (K-1) P_{km} \right]. \tag{37}$$

Hence, as content f_j^* is converted to the MPC, the change of effective delivery rate is

$$V_{f_j^*}^1(t) = d_{j,f_j^*}(t) \left[Rdp_K^{JT}(\theta_1) - Rdp_K^{PT}(\theta_2) \right] - \sum_{l_j^* = F_j^* - K + 2}^{F_j^*} d_{j,l_j^*}(t) \left[Rdp_K^{PT}(\theta_4) - Rdp_K^{PT}(\theta_2) \right]$$

$$- \sum_{i \in \mathcal{J} \backslash j} \sum_{l_j^* = F_j^* - K + 2}^{F_j^*} d_{i,l_j^*}(t) \left[Rdp_K^{PT}(\theta_4) - Rdp_K^{PT}(\theta_3) \right]. \tag{38}$$

The change of energy consumption is

$$P_{f_j^*}^1(t) = \sum_{l_j^* = F_j^* - K + 2}^{F_j^*} d_{j,l_j^*}(t) \left[P_{c,j} + (K-1) P_{km} \right] - \sum_{i \in \mathcal{J} \backslash j} d_{i,f_j^*}(t)(K-1) P_{km} \tag{39}$$

$$+ \sum_{i \in \mathcal{J} \backslash j} \sum_{l_j^* = F_j^* - K + 2}^{F_j^*} d_{i,l_j^*}(t) \left[P_{c,j} - P_{j,i} - (K-1) P_{km} \right].$$

Hence, the EE can be written as

$$EE_{f_j^*}^1(t) = \frac{V_{f_j^*}^1(t)}{P_{f_j^*}^1(t)}. \tag{40}$$

If content f_j^* is not converted to the MPC and $K-1$ contents as LPC are not discarded, the effective delivery rate is given as

$$V_{f_j^*}^2(t) = d_{j,f_j^*}(t) Rdp_K^{PT}(\theta_2) + \sum_{l_j^* = F_j^* - K + 2}^{F_j^*} d_{j,l_j^*}(t) Rdp_K^{PT}(\theta_2)$$

$$+ \sum_{i \in \mathcal{J} \backslash j} d_{i,f_j^*}(t) Rdp_K^{PT}(\theta_3) + \sum_{i \in \mathcal{J} \backslash j} \sum_{l_j^* = F_j^* - K + 2}^{F_j^*} d_{i,l_j^*}(t) Rdp_K^{PT}(\theta_3). \tag{41}$$

The energy consumption is

$$P_{f_j^*}^2(t) = d_{j,f_j^*}(t)KP_e + \sum_{i \in \mathcal{J} \setminus j} d_{i,f_j^*}(t)\left[KP_e + P_{j,i} + 2(K-1)P_{km}\right]$$

$$+ \sum_{l_j^* = F_j^* - K + 2}^{F_j^*} d_{j,l_j^*}(t)KP_e + \sum_{i \in \mathcal{J} \setminus j} \sum_{l_j^* = F_j^* - K + 2}^{F_j^*} d_{i,l_j^*}(t)\left[KP_e + P_{j,i} + 2(K-1)P_{km}\right].$$

$$(42)$$

The EE can be written as

$$EE_{f_j^*}^2(t) = \frac{V_{f_j^*}^2(t)}{P_{f_j^*}^2(t)}. \tag{43}$$

Whether $\mathcal{C}_j$ converts the content f_j^* to the MPC depends on $EE_{f_j^*}^1(t)$ and $EE_{f_j^*}^2(t)$. If $EE_{f_j^*}^1(t) \geq EE_{f_j^*}^2(t)$, content f_j^* is converted to the MPC and $K-1$ contents with lowest contribution are discarded, and $EE_{ini}(t)$ is refreshed. If $EE_{f_j^*}^1(t) < EE_{f_j^*}^2(t)$, content f_j^* and $EE_{ini}(t)$ remain unchanged. Therefore, at most steps of $J \cdot F$, the entire process of contents conversion can be realized and the maximum value of EE is $EE_{ini}(t)$, and the CM guides the SBSs cooperation of all clusters to maximize the EE of the clustered SCNs.

5   Performance Evaluation

In this section, we compare the performance of proposed energy efficiency caching algorithm based on content-awareness (EECABC) with others: LRU-based caching algorithm with proportion of MPC is 0.2 [8], MPMAB-based caching algorithm [9] and MPC-based caching algorithm with proportion of MPC is 0.1 [13]. In the simulation, we use the MovieLens DataSet [17] which included a total of four full datasets of 100 k, 1 m, 10 m, and 20 m, and in terms of time span as well as data integrity considerations, this paper chooses the latest ml-20m dataset as the simulation data. This dataset has 1000209 ratings of 3952, and assume that the users only request their higher rating movies in requesting process. In addition, considering the integrity of the video information, under the condition of ensuring the reliability of the simulation, we divided the long term timestamp into 20 time slots and 2000 movies were selected as historical contents, and 200 movies were selected as the number of new contents uploaded to the data center in subsequent time slots. We assume that the cache space of each cluster is equal to 9000 M and the unit content size is equal to 300 M. Other parameters in the simulation are showed in Table 1 which refers to the parameter ranges of existing available devices and the parameter settings in [13] [20]. We mainly compare the cache efficiency and system EE, where cache efficiency is defined as the SCDP times the amount of cache hit data divided by the total amount of caching data, which is utilized as a measure of cache hits. The simulation runs 10 times to take the average.

Table 1. Simulation parameters

Parameter	Value
Number of clusters J	7
Proportion of the number of selected cluster controller SBSs (ϵ)	0.3
Density of SBSs (λ_b)	$10^{-4}/\text{m}^2$
Radius of cell (R_h)	100 m
Path fading index (κ)	4
SBS transmit power P_e	1 W
Wired transmission consumption from data center to the selected cluster per content	10 W
Wired transmission consumption between cluster per link per content	[2 W, 2.5 W]
Transmission consumption between cluster controller and the other SBS	0.2 W
Available bandwidth	10 MHz
Transmission delay between interconnected cluster controller SBS per content	5 s
Transmission delay between data center and the connected cluster controller SBS per content	15 s
Average number of contents requested per time slots	20

In order to measure the impact of the number of SBSs on the transmission mode, as shown in Fig. 3, we first compare the relationship between the SCDP and ATR of contents in JT and PT mode when the number of SBSs K is 2, 3 and 4, respectively. In JT mode, the more SBSs in a cluster, the higher SCDP is, and in PT mode is on the contrary. This is because in the JT mode, more SBSs cooperation provide stronger received signal and therefore a higher SIR, the SCDP is increased. In PT mode, the SCDP is defined as the product of the success probabilities of multiple data streams. When the number of parallel streams increase, the reliability of each data stream needs to be ensure. The more SBSs provide a lower SIR, thus the SCDP performance decrease. Obviously, regardless of the value of K in JT or PT mode, as the ATR increases, the SCDP gradually decreases. That is because when the ATR increases, the computational requirements for each SBS and the channel quality requirements increase. In addition, the difference in SCDP between the PT mode and JT mode is quite large, especially when the ATR increase to a large value. In the following simulation, the number of SBSs in a cluster is set to be 3. The threshold of delivery rate in the JT mode is calculated to be 22 Mbps and in the JT mode is 7 Mbps, respectively.

Figure 4 shows the performance comparison among LRU-based, MPMAB-based, MPC-based and proposed EECABC proactive caching algorithm in terms

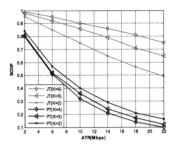

Fig. 3. Comparison among JT and PT in SCDP versus ATR.

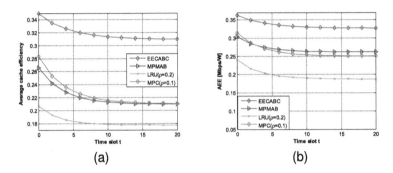

Fig. 4. (a) ACE comparison versus t ($R_d = 10$ Mbps). (b) AEE comparison versus t ($R_d = 10$ Mbps).

of average cache efficiency (ACE) and the average energy efficiency (AEE) as a function of time slot shift with the expected delivery rate R_d is 10 Mbps. As shown in Fig. 4(a), when the time slot changes, the ACE shows a downward trend, and the LRU, MPMAB and MPC cache algorithms decrease obviously. In contrast, the EECABC cache algorithm has a relatively stable rate of decline and eventually tends to be constant. With the passage of time, the new contents are uploaded to the data center gradually. The combination of the historical contents and the new contents improves the difficulty of popularity prediction. Hence, the cache hits is reduced, and the ACE is also reduced. This may prove that the popularity prediction bases on content-awareness has a relatively high prediction accuracy than others and can effectively predict the popularity of new contents. Meanwhile, as shown in Fig. 4(b), the AEE shows a downward trend as well with the time slot changes. The AEE of the EECABC cache algorithm outperforms than the other three algorithms. The combination of the historical contents and the new contents makes the accuracy of content popularity prediction reduce, which has a great impact on the placement of contents. The proposed EECABC cache algorithm achieves a higher prediction accuracy than the others. In addition, it may also prove that in the process of cache algorithm designing, the EECABC algorithm effectively improves the network performance by utilizing the JT mode of the CoMP technology simultaneously.

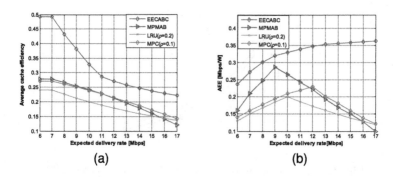

Fig. 5. (a) ACE comparison versus R_d. (b) AEE comparison versus R_d.

Figure 5 shows the performance comparison among LRU-based, MPMAB-based, MPC-based and proposed EECABC proactive caching algorithm in terms of ACE and the AEE as the expected delivery rate R_d is changed. When R_d is in the interval $[6,7]$ which lower than the threshold of 22 Mbps in JT mode and 7 Mbps in PT mode, the SCDP is unchanged, the ACE is also unchanged. When R_d exceeds the threshold of the PT mode, such as in the interval $[7,17]$, the SCDP of the PT mode is rapidly reduced, thus the ACE also decreases. As shown in Fig. 5(a), the ACE of the four cache algorithms show a downward trend when the R_d increases, the proposed EECABC caching algorithm performs better. Meanwhile, when the R_d exceeds 11 Mbps, the downward trend of ACE in proposed EECABC algorithm shows down. This is because our algorithm converts part of LPC into MPC, which can better utilize the advantage of JT mode under high value of R_d.

As AEE is the ratio of actual delivery rate to energy consumption. The actual delivery rate is the product of the SCDP and the expected delivery rate R_d. When R_d increases, the SCDP decreases, thus the AEE fluctuates as the actual delivery rate fluctuates. As shown in Fig. 5(b), the relationship between the AEE and R_d is shown as a downward parabola, and the extreme points are different in different caching algorithms. Comparing the four caching algorithm, our algorithm does not show a rapid decrease after the extreme point, but presents a slow increase and gradually tends to be stable. The reason may be that if the R_d increase gradually, the SCDP of the PT mode is rapidly reduced, due to the relatively high threshold value of JT mode, converting the contents into the MPC can better adapt to the high R_d requirement and we can take advantage of the JT mode. By adaptively adjusting the proportion of the MPC occupied in cache space, our algorithm achieves optimal in EE.

6 Conclusion

In this paper, we use a CWFM-based popularity prediction method with spatiotemporal variation, which can accurately predict the contents popularity, and then we propose the energy-efficient design in the clustered SCNs, and decomposed the EE problem into two sub-problems. Firstly, we performed variable relaxation, LP and variable recovery under the condition of delivery consumption to maximize the cache efficiency of the local cluster. Then, the EECABC iteratively determine whether the contents with higher popularity was converted into MPC, and discarded the corresponding LPC, thereby maximizing the EE of the system within the global cluster. The simulation results showed that the proposed EECABC outperforms the existing strategies in terms of EE.

References

1. (2019). https://www.cisco.com/c/en/us/solutions/collateral/service-provider/visual-networking-index-vni/mobile-white-paper-c11-738429.html
2. Li, Z., Liu, Y., Liu, A., Wang, S., Liu, H.: Minimizing convergecast time and energy consumption in green internet of things. IEEE Trans. Emerg. Topics Comput. **8**(3), 797–813 (2020)
3. Wu, Q., Li, G.Y., Chen, W., Ng, D.W.K., Schober, R.: An overview of sustainable green 5G networks. IEEE Wirel. Commun. **52**, 1897–1903 (2016)
4. Li, X., Wang, X., Leung, V.C.M.: Weighted network traffic offloading in cache-enabled heterogeneous networks. In: IEEE International Conference on Communications (ICC), pp. 1–6, May 2016
5. Piemontese, A., i Amat, A.G.: MDS-coded distributed caching for low delay wireless content delivery. IEEE Trans. Commun. **67**(2), 1600–1612 (2019)
6. Park, S.-H., Simeone, O., Shitz, S.S.: Joint optimization of cloud and edge processing for fog radio access networks. IEEE Trans. Wireless Commun. **15**(11), 7621–7632 (2016)
7. Nguyen, H.T., Tuan, H.D., Duong, T.Q., Poor, H.V., Hwang, W.: Collaborative multicast beamforming for content delivery by cache-enabled ultra dense networks. IEEE Trans. Commun. **67**(5), 3396–3406 (2019)
8. Blasco, P., Güdüz, D.: Learning-based optimization of cache content in a small cell base station. In: Proceedings of IEEE ICC, pp. 1897–1903, June 2014
9. Song, J., Sheng, M., Quek, T.Q.S., Xu, C., Wang, X.: Learning-based content caching and sharing for wireless networks. IEEE Trans. Commun. **65**(10), 4309–4324 (2017)
10. Tanzil, S.S., Hoiles, W., Krishnamurthy, V.: Adaptive scheme for caching YouTube content in a cellular network: a machine learning approach. IEEE Access **5**, 5870–5881 (2017)
11. Chiang, Y., Liao, W., Ji, Y.: RELISH: green multicell clustering in heterogeneous networks with shareable caching. In: IEEE Global Telecommunications Conference (GLOBECOM), pp. 1–7 (2018)
12. Nigam, G., Minero, P., Haenggi, M.: Coordinated multipoint joint transmission in heterogeneous networks. IEEE Trans. Commun. **62**(11), 4134–4146 (2014)
13. Chen, Z., Lee, J., Quek, T.Q., Kountouris, M.: Cooperative caching and transmission design in cluster-centric small cell networks. IEEE Trans. Wireless Commun. **16**(5), 3401–3415 (2017)
14. Zhang, S., He, P., Suto, K., Yang, P., Zhao, L., Shen, X.: Cooperative edge caching in user-centric clustered mobile networks. IEEE Trans. Mobile Comput. **17**(8), 1791–1805 (2018)

15. Zhuang, Y., Luo, Y., Cai, L., Pan, J.: A geometric probability model for capacity analysis and interference estimation in wireless mobile cellular systems. In: IEEE Global Telecommunications Conference (GLOBECOM), December 2011
16. Sourlas, V., Georgatsos, P., Flegkas, P., Tassiulas, L.: Partition-based caching in information-centric networks. In: IEEE Conference on Computer Communications Workshops (INFOCOM WKSHPS), pp. 396–401 (2015)
17. Harper, F.M., Konstan, J.A.: The MovieLens datasets: history and context. IEEE Trans. Commun. **5**(4), 2160–6455 (2015)
18. Sen, S., Santhapuri, N., Choudhury, R.R., Nelakuditi, S.: Successive interference cancellation: carving out MAC layer opportunities. IEEE Trans. Mobile Comput. **12**(2), 346–357 (2013)
19. Müller, S., Atan, O., van der Schaar, M., Klein, A.: Context-aware proactive content caching with service differentiation in wireless networks. IEEE Trans. Wireless Commun. **16**(2), 1024–1036 (2017)
20. Teng, L., Yu, X., Tang, J., Liao, M.: Proactive caching strategy with content-aware weighted feature matrix learning in small cell network. IEEE Commun. Lett. **23**(4), 700–703 (2019)
21. Liu, K., Tao, M.: Exploiting tradeoff between transmission diversity and content diversity in multi-cell edge caching. In: IEEE International Conference on Communications (ICC), pp. 1–6, May 2018
22. Zhang, X., Haenggi, M.: The performance of successive interference cancellation in random wireless networks. IEEE Trans. Inf. Theory **60**(10), 6368–6388 (2014)

Land Cover Classification and Accuracy Evaluation Based on Object-Oriented Spatial Features of GF-2

Xiaomao Chen[1], Jiakun Li[1], and Yuanfa Ji[2($\boxtimes$)]

[1] Guangxi Key Laboratory of Wireless Wideband Communication and Signal Processing, Guilin University of Electronic Technology, Guilin 541004, China
[2] National and Local Joint Engineering Research Center of Satellite Navigation Positioning and Location Service, Guilin 541004, China

Abstract. The urbanization process has changed urban land, which has affected the environmental quality of urban residents. It is very important to obtain urban land cover information. In this paper, Yangshuo, a small country of Guilin City, is used as the research area, and the object-oriented spatial feature extraction module (Feature Extraction, hereinafter referred to as FX) is used to carry out experiments and accuracy evaluation of land cover classification in the research area. Extracting land cover information from the GF-2 remote sensing image, establishing a classification system sample based on the characteristic information of six land cover classification objects such as urban land, waterbody, woodland, farmland, road and other lands, and finally execute Supervising the classification and verify its accuracy. The results show that this method can recognize the land cover accurately and the total accuracy verified is as high as 97.41%.

Keywords: FX · GF-2 · Land cover classification · Accuracy evaluation · Supervision classification

1 Introduction

With the rapid increase of population and the development of urban and rural areas, the depth of land use has shown a trend of rapid expansion compared with the past. Nowadays, grasping and using the land information is the most necessary plan to be implemented by the state and governments, so as to make a more effective judgment and decision on China's economic development plan. Besides, internationally, many countries have made use of remote sensing technology to manage land use and science, and have achieved successful experience.

With the rapid development of remote sensing technology, it has been widely used in many fields. In recent years, remote sensing image has been widely used in land use and land cover classification, which plays a crucial role in land resource management, urban planning, environmental protection, and other applications [1, 2]. With the development of remote sensing technology, Imaging satellites can provide by remote sensing data covered most of the surface of the earth, such as high-resolution remote sensing image has the characteristics of the "three highs" and can realize the day and in

---

The original version of this chapter was revised: the presentation of the authors' name corrected. The correction to this chapter is available at https://doi.org/10.1007/978-3-030-67720-6_55

© ICST Institute for Computer Sciences, Social Informatics and Telecommunications Engineering 2021, corrected publication 2021
Published by Springer Nature Switzerland AG 2021. All Rights Reserved
H. Gao et al. (Eds.): ChinaCom 2020, LNICST 352, pp. 309–323, 2021.
https://doi.org/10.1007/978-3-030-67720-6_21

the aspect of earth observation, all-weather, real-time observation, for the land use and cover classification provides a new opportunity, namely how to make use of high-resolution remote sensing satellites for land use classification.

At present, the most commonly used image classification techniques are pixel-based and object-oriented classification methods. However, due to the lack of spectral information of high-resolution images and the phenomenon of "same thing different spectrum, same spectrum foreign matter", image classification is affected, and the pixel-based classification method cannot be well applied. The object-oriented classification method has obvious advantages in the consideration of spectrum, space, texture and context information, so as to improve the classification accuracy and make the extraction more intuitive for the analysis of the problem. Therefore, this method has attracted wide attention. For example, Meng et al. [3] proposed an object-oriented method to extract urban ecological land cover from the multi-channel images obtained by China's GF-1 satellite. This method can accurately identify the urban land cover with a verification accuracy of 90%. Yu et al. [4] used object-oriented information extraction technology, multi-temporal hj-1a image and other auxiliary data to extract the main land use/land cover types in the research area. Compared with the pixel-based classification method, Yu et al. research results have higher accuracy. Jesus et al. [5] combined object-oriented and pixel-based methods to extract land cover information in the mountains of Mexico. What's more, Wang et al. [6] used object-oriented classification method to test the feasibility and applicability of classifying ecologically sound land, and the overall classification accuracy of Wang et al.'s results was 87.43%. Cahairet et al. [7] used SVM classifier to carry out land cover supervision and classification in Gabes region located in the southeast of Tunisia, with an accuracy of 92.12%. In addition, Shi et al. [8] using satellite remote sensing technology to analyze and study high -resolution remote sensing images are helpful to the effective supervision of urban land use. This paper studies part of Yangshuo county, uses ENVI5.3 platform in object-oriented spatial feature extraction module Example-Based-Feature-Extraction tools to carry out research under Guilin area land cover classification information extraction technology. The advantages of FX include fast, repeatable, accurate, convenient, and accessible, which is suitable for experimental research on object-oriented classification.

The rest of the paper is organized as follows: We describe the data used for research in Sect. 2, and introduce our proposed research area and image preprocessing procedure; In Sect. 3, we introduce the object-oriented classification method; The results obtained when applying the proposed method to selected data sets are reported and discussed in Sect. 4. Finally, we summarize the main conclusions in Sect. 5 and suggest some possibilities for further research.

2   Research Area Overview and Data Preprocessing

2.1   GF-2 Data

The GF-2 satellite is configured with two panchromatic and multispectral CCD camera sensors (PMS) with a resolution of 1 m panchromatic/4 m multispectral. The spatial resolution of the GF-2 satellite can reach 0.8 m, indicating that China's remote sensing satellite has entered the sub-meter "high score era" [9]. GF-2 can provide a combined band area of 45 km, which is reflected in the multispectral image as 6908 × 7300 pixels. The revisit time of GF-2 is 5 days, so it can capture a wide range of detailed information at very short intervals. The GF-2 image has the characteristics of high resolution, wide image coverage, frequent revisit and high image quality, and is an ideal data source for land cover information extraction. GF-2 images at the 2 m and 8 m scales obtained from the sensor PMS1 on March 12, 2019, were used in the study area.

2.2 Study Area Overview

GF-2 has a good performance in information extraction between urban and rural areas. Therefore, this paper selected some urban areas in the GF-2 image of Yangshuo county, Guilin city, Guangxi Zhuang autonomous region as the experimental research area. Yangshuo County is located in the northeast part of Guangxi Zhuang autonomous region and to the south of Guilin city, located at longitude 110° 13'–110° 40' east and latitude 24° 28'–25° 4' north. It is adjacent to Gongcheng county and Pingle in the east, Lipu county in the south, Yongfu county in the west, Lingchuan county in the north and Yanshan district in the north, wild goose mountain area, the county area of 1428.38 sq. km, the experimental study on regional content-rich on the mainland, including urban land, water, wood land, farm land, road and other land use, such as terrain, the terrain has certain representativeness, is advantageous to the object-oriented information extraction experiment research. The study area in this paper is 110°29'–110°30' east longitude and 24°46'–24°47' north latitude. Figure 1 shows the true-color image of the study area, which is composed of red, green and blue wave segments.

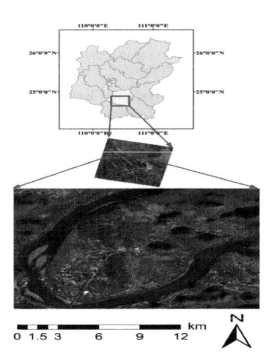

Fig. 1. True color images of the study area

2.3 Preprocessing

Remote sensing image preprocessing is to process remote sensing images and related data through computers, providing a basis for remote sensing information extraction and remote sensing quantitative analysis [10]. Based on the remote sensing image data, this paper carries out the image data pre-processing processes such as orthographic correction, image registration, image fusion and clipping. The basic flow chart of its preprocessing is shown in Fig. 2 below.

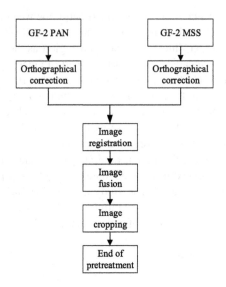

Fig. 2. Flow chart of preprocessing

3   Object-Oriented Land Covers Are Classified

Due to the lack of spectral information of high-resolution images, and the phenomenon of "same thing, different spectrum, same spectrum foreign matter", the result of image classification will be affected, and the pixel-based classification method cannot be well applied. The Object-oriented image classification method is to make full use of high-resolution panchromatic and multispectral data, which has obvious advantages in terms of spectrum, space, texture and context information, so as to improve the classification accuracy and make the extraction of information more intuitive for analysis. Image classification based on samples, namely supervised classification, uses training sample data to identify other unknown objects, including extract texture features, sample definition, classification algorithm selection and output results. The method in this paper is using ENVI5.3 software object-oriented spatial feature extraction module

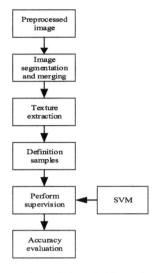

Fig. 3. Flow chart of sample-based object-oriented classification

under Example-Based Feature Extraction tools to carry out the research Guilin area land cover classification. The sample-based object-oriented classification process is shown in Fig. 3.

3.1   Image Segmentation and Merging

FX mainly uses the object-oriented idea to extract the needed information from the remote sensing image, such as urban land, water, wood land, farm land, road, and other lands, etc. It divides the image according to the brightness, texture and color of adjacent pixels. The image segmentation algorithm is according to certain rules to the whole image segmentation research area is a number of small patches of each patch have the characteristics of homogeneous gay, its spectral characteristics, texture characteristics, spatial characteristics, with the same or similarity with - kind of features, image segmentation is the core of segmentation threshold segmentation scale that set up [11]. Due to some threshold values are too low in image segmentation, some features may be misclassified, and a feature may be divided into many parts. Therefore, these problems can be solved by merging, so that better images can be merged. A large number of experiments have been carried out in this paper, and the image merging is obtained on the basis of the segmentation scale of 45. When the merging scale is 85 and the texture kernel is 3, it can be observed that the contours of objects in Fig. 4(a) study area can be well displayed. It can also accurately classify the required objects. Larger areas such as woodland and cultivated land have reached the ideal polygonal range.

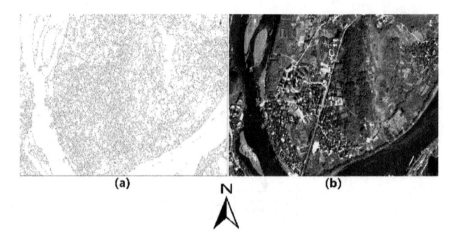

Fig. 4. The best merged effect picture. (a) Segmentation image; (b) Original image

3.2 Extract Texture Features

In order to reduce the possibility of classification errors, we used different strategies to extract land cover. First, water bodies are determined using the normalized difference water index:

$$NDWI = \frac{\rho_{green} - \rho_{nir}}{\rho_{green} + \rho_{nir}} > 0.2 \tag{1}$$

Vegetation through the normalized vegetation index:

$$NDVI = \frac{\rho_{nir} - \rho_{red}}{\rho_{nir} + \rho_{red}} > 0.12 \tag{2}$$

Where ρ_{green}, ρ_{red} and ρ_{nir} are green, red and near infrared band reflectivity; Furthermore, urban land use is identified using spectral information (such as Eq. (3)); Because the spectrum of the road is similar to that of other land, it is difficult to distinguish clearly. In order to reduce this effect, texture features of entropy (such as Eq. (4)) and contrast (such as Eq. (5)) of other land are added to the classification process. Finally, two different classifications are analyzed for accuracy.

$$Max.diff = \frac{i,j \in K_B |c_i(v) - c_j(v)|}{c(v)} \tag{3}$$

where $c_i(v)$ and $c_j(v)$ are the average brightness of the v at the i and j levels, respectively. K_B are all objects. $c(v)$ is the average brightness of the v at the whole level.

$$Ent = \sum_{i=1}^{n} \sum_{j=1}^{n} p(i,j) \times \ln[p(i,j)] \tag{4}$$

where $p(i,j)$ is the probability of pixel value i and j.

$$Q = \sum_{\delta} \delta(i,j)^2 p_\delta(i,j) \tag{5}$$

where, $\delta(i,j) = |i - j|$ | is the sum of the number difference of adjacent pixels and $p_\delta(i,j)$ is the probability of the specified difference δ of adjacent pixels.

3.3   Define the Classification Samples

Taking Yangshuo county in Guilin city as an example, this paper explores the application of object-oriented technology in land cover classification and sets up a classification system sample according to the actual situation of the research area. Through visual interpretation, use ROITOOL to create areas of interest, such as urban land, waterbody, woodland, arable land, road and other lands. In this paper, the land cover classification standard shown in Table 1 will be adopted [12].

Table 1. Sample Standard for Land Cover Classifications

Code	Definition	Meaning
01	Arable land	Refers to the land where crops are grown, including new development, reclamation and consolidation of ripe land; To plant crops (including vegetables); On average, one season of land can be harvested every year. The cultivated land includes a width of less than meters and fixed ditches, channels, roads and bridges. Cultivated land for the temporary cultivation of medicinal herbs, turf, flowers, seedlings, etc.
03	Woodland	Refers to the land on which trees, bamboos, and shrubs grow, including the landscape, the forest land within the residential area, the forest land within the land requisitioned by railways and highways, the berm forest of rivers and ditches, and the land with scattered fruit trees, mulberry trees or other trees
10	Road	Refers to the land that is directly used for the transportation of ground routes. Including civil airports, ports, docks, ground transportation pipelines, and various road land
11	Waterbody	Refers to land areas, ditches, hydraulic structures, etc.
12	Other lands	Refers to other types of land than the above
20	Urban land	Refers to the land used by public institutions such as industrial and mining settlements, independent settlements, industrial and mining sites, places of historic interest, and historic sites other than settlements, including transportation and greening

According to the spectral characteristics of the image, the objects in the image are divided into 6 types of objects shown in Table 1. Then the sample is selected by drawing polygon to select the area of interest of each kind of feature is distinguished by different colors. The interest area is shown in Fig. 5. Red means urban land, lavender means water, cyan means woodland, yellow means wasteland, brown means roads, and light orange means other land.

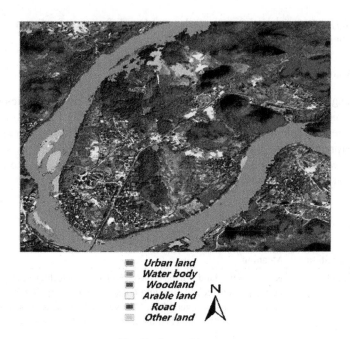

Fig. 5. Area of interest

3.4 Classification of Executive Supervision

According to the classification complexity and accuracy of the research area, the classifier is selected. Different classifiers have different pixel values and can generate regular images according to the parameters of the classification results.

Support vector machine (SVM) classifier: in supervised classification system, SVM is based on statistical learning and SVM is used for classification and regression. SVM classifier separates testing from training. In the training set, each instance has multiple attributes and a target value, SVM the concept of decision surface is adopted. It is used for region classification to maximize the boundaries of classes, and the decision surface is also called the optimal hyperplane. The support vector is defined as a data point close to the decision surface. When creating the training sample set, the key element is the support vector [13]. The solution of SVM depends on the choice of kernel function. The use of different types of kernels in the SVM is linear, polynomial, radial basis function, and Sigmoid function. The linear support vector machine method is used in this paper, as shown in Fig. 6. The preprocessed GF-2 data was used by SVM.

The goal of SVM is to find an optimal hyperplane for classification, which can not only correctly classify each sample, but also make the distance between the closest sample to the hyperplane in each class of samples and the hyperplane as far as possible.

Suppose there are i samples in the training sample set, the feature vector is an n-dimensional vector, and the class label value is +1 or −1, corresponding to positive samples and negative samples, respectively. SVM finds an optimal classification hyperplane for these samples:

$$w^{\mathrm{T}}x + b = 0 \tag{6}$$

where x is the input vector (sample feature vector); w is the weight vector and b is the bias term (scalar). these two sets of parameters are obtained by training.

First, ensure that each sample is correctly classified. For positive samples:

$$w^{\mathrm{T}}x + b \geq 0 \tag{7}$$

For negative samples:

$$w^{\mathrm{T}}x + b < 0 \tag{8}$$

Since the category label of the positive sample is +1 and the category label of the negative sample is −1, it can be uniformly written as the following inequality constraint, $i = 1, 2, \cdots, n$, n is the number of training samples.

$$y_i\left(w^{\mathrm{T}}x_i + b\right) \geq 0 \tag{9}$$

The second requirement is that the distance between the hyperplane and the two types of samples should be as large as possible. According to the distance formula from the midpoint to the plane in analytic geometry, The distance of each sample from the classification hyperplane:

$$d = \frac{\left|w^{\mathrm{T}}x_i + b\right|}{\|w\|} \tag{10}$$

where $\|w\|$ is the L2 norm of the vector. The following constraints may be added to w and b:

$$\min_{x_i}\left|w^{\mathrm{T}}x_i + b\right| = 1 \tag{11}$$

The constraint on the classification hyperplane becomes:

$$y_i\left(w^{\mathrm{T}}x_i + b\right) \geq 1 \tag{12}$$

As a result, the interval between the classification hyperplane and the two types of samples:

$$d(w,b) = \min_{x_i, y_i=-1} d(w,b;x_i) + \min_{x_i, y_i=1} d(w,b;x_i)$$

$$= \frac{2}{\|w\|}$$

(13)

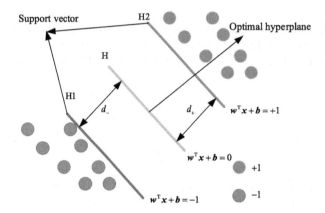

Fig. 6. Linear Support Vector Machine method

A SVM classification method is used to classify images with or without texture features through supervised classification based on sample rules. Because of the complexity and accuracy of classification, a large number of experiments are needed to improve the accuracy of classification. After performing supervised classification, perform post-classification processing such as Majority/Minority Analysis and Clump Analysis on the classification results to eliminate some of the small spots in the classification results and to obtain the final classification results.

3.5 Classification Accuracy Evaluation

In this paper, the confusion matrix in ENVI is used to evaluate the classification accuracy, and the pre-processed image is visually interpreted as the region of interest, so as to serve as a reference for the accuracy evaluation.

The confusion matrix is a standard format for classification accuracy evaluation of remote sensing images. A confusion matrix is a matrix of i rows by i columns, where i represents the number of categories. For the classification confusion matrix of remote sensing image, the basic statistics of the classification accuracy evaluation index include:

(1) Overall classification accuracy: the number of correct classifications divided by the sum of reference numbers;

$$OA = \sum_{i=1}^{k} \frac{x_{ii}}{N} \tag{14}$$

(2) Kappa coefficient: a statistical value of classification accuracy, ranging from 0 to 1. Shows how much better the classification approach is than randomly assigning each pixel to any class;

$$Kappa = \frac{N \sum_{i=1}^{k} x_{ii} - \sum_{i=1}^{k} (x_{hi}x_{li})}{N^2 - \sum_{i=1}^{k} (x_{hi}x_{li})} \tag{15}$$

Where, k is the sum of the total columns of the confusion matrix, that is, the total number of categories; x_{ii} represents the correct classification number of the i category, x_{hi} is the total number of samples in the column of type i, x_{li} is the total number of samples in the row of type i, and N is the total number of pixels participating in the statistics. The relationship between Kappa value and classification accuracy is that the larger the Kappa value, the higher the classification accuracy.

(3) User accuracy: In the same category, the percentage of the correctly classified number x_i to the total number x_{i+} of this category represents the probability that a classified pixel can truly represent this category;

$$UA(c) = \frac{x_i}{x_{i+}} \tag{16}$$

(4) Producer accuracy: It represents the ratio of the correct number x_{ii} of a certain category to the total number of real categories x_{+i} of the category; Reflect the percentage of correctly classified reference data;

$$PA(c) = \frac{x_{ii}}{x_{+i}} \tag{17}$$

(5) Commission error: refers to the proportion of misclassified pixels;

(6) Omission error: refers to the proportion that itself belongs to the real classification of the surface, but is not classified into the corresponding categories by the classifier.

The accuracy of the object-oriented classification method is evaluated by using sample points. Table 2 shows the accuracy evaluation results of the object-oriented classification method of the k-proximity method.

4   Experimental Result

The image was preprocessed by ENVI software V5.3.1 (image data preprocessing process such as orthorectification, image registration, image fusion and cropping). For different urban land cover, the optimal segmentation scale for image segmentation is 45, the merge scale is 85, and the texture kernel is 3. After that, five kinds of urban ecological land cover are classified by ENVI software, and urban land, water body, woodland, cultivated land, road and other land are classified one by one. Because other land and roads are difficult to distinguish, texture information is added to the classification process.

The classification results are shown in Fig. 7 and Fig. 7 (a) and (b) are the results without texture information and with texture information, respectively. They show that the use of texture information turns road pixels into urban land, making the classification result closer to reality. It can be seen from the classification results that some objects of the road are classified as urban land, which makes the display of urban land more complete. The sample points are used to evaluate the accuracy of the object-oriented classification method. Table 2 shows the accuracy evaluation results of the object-oriented classification method with or without texture information in SVM.

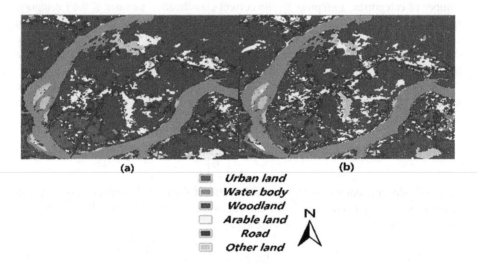

Fig. 7. Classification result map (a): without texture; (b) texture

From the accuracy evaluation results in Table 2, it can be seen that the classification effect of SVM is relatively good, the overall accuracy is above 95%, and the Kappa coefficient is greater than 0.95. Among them, the overall accuracy of the SVM classification results with texture features is the highest, reaching 97.41%, and the Kappa coefficient is 0.9605; the overall accuracy of the untextured SVM classification results is the second, with an overall accuracy of 96.89% and a Kappa coefficient of 0.9539. It is found from the table that the producer accuracy of road and other land with

Table 2. Accuracy evaluation result of classification method with or without texture information of SVM

	Without texture		With texture	
Overall classification accuracy	96.89%		97.41%	
Kappa	0.9539		0.9605	
	User accuracy (%)	Producer accuracy (%)	User accuracy (%)	Producer accuracy (%)
Urban land	93.15	93.31	99.91	97.44
Water body	100	98.28	100	98.29
Woodland	97.36	99.89	94.87	99.40
Arable land	88.55	99.61	94.31	99.84
Road	82.91	84.99	86.32	99.44
Other lands	91.89	79.42	100	88.33

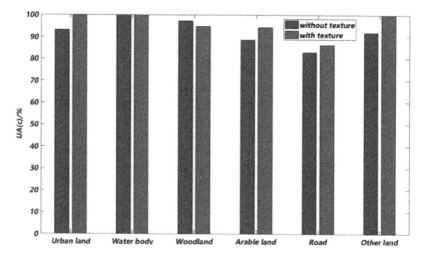

Fig. 8. User accuracy comparison with and without texture

texture information is 99.44% and 88.33% respectively, while the producer accuracy without texture information is only 84.99% and 79.42% respectively.

From the two Fig. 8 and Fig. 9, it can be clearly found that the user accuracy and producer accuracy of all kinds of objects with texture information have good performance. Using the texture information in the classification rules, the accuracy of roads and other land use is greatly improved. From this point of view, the object-oriented classification method with texture information in this paper can improve the accuracy of urban areas, which is a promising method and should be extended to other urban land cover.

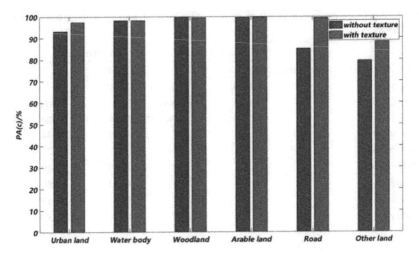

Fig. 9. Producer accuracy comparison with or without texture

5   Conclusion

Based on the object-oriented method to extract spatial characteristics of land cover classification, GF-2 in the image is utilized to extract the urban land cover information, on the basis of image segmentation, the object using cover features information, Though the establishment of the classification system to classify urban land, water, woodland, arableland, road, and other land six land cover and evaluate it accuracy. The results show that the classification effect with texture feature information is better, and the classification accuracy is as high as 97.41%. But this object-oriented method still has some errors, which may be caused by the setting of the segmentation threshold, the availability of supervised classification and the complexity of land. Therefore, the next work will further refine the classification system to improve the accuracy and increase the segmentation and recognition of other types of data.

Acknowledgment. The authors thanks national key R & D program funding (2018YFB05-05103), Major Special Plan of Guilin Science and Technology Plan Project in 2019 (20190219-1), Innovation Project of GUET Graduate Education(2020YCXS026, 2020YCXS028) and Guangxi Key Laboratory of Wireless Wideband Communication and Signal Processing 2019 Director Fund Project (GXKL06190111) for providing the necessary support and funds. The authors also thank the satellite navigation team for all the validation data provided during the experiment.

References

1. Jia, T., Luo, Y., Chen, J., Dong, W.: Present Situation and Trend of Remote Sensing Land Use/Cover Classification_Extraction 15 (2018). https://doi.org/10.1109/geoinformatics. 2018.8557159
2. Clerk Maxwell, J.: A Treatise on Electricity and Magnetism, 3rd ed., vol. 2, Clarendon, Oxford, pp. 68–73 (1892)
3. Meng, J., et al.: Urban ecological land extraction from Chinese Gaofen-1 data using object-oriented classification techniques. In: IEEE International Geoscience and Remote Sensing Symposium (IGARSS), Milan, 2015, pp. 3076–3079 (2015). http://doi.org/10.1109/IGARSS.2015.7326466
4. Yu, H., Wang, C., Ren, C.: Object-oriented information extraction using HJ-1 remote sensing: the case on Changbai Mountain, Northeast China. In: 2012 2nd International Conference on Remote Sensing, Environment and Transportation Engineering, Nanjing, 2012, pp. 1–4 (2012). https://doi.org/10.1109/rsete.2012.6260646
5. Jesus, A., Arie, C., Joost, F.: Optimizing land cover classification accuracy for change detection, a combined pixel-based and object-based approach in a mountainous area in Mexico. Appl. Geography 34, 29–37 (2012)
6. Wang, J., Zhang, X., Du, Y., Jia, X., Lin, Y.: Object-Oriented Classification for Ecologically Sound Land Based on High-Resolution Images, pp. 7476–7479 (2018). https://doi.org/10.1109/igarss.2018.8518608
7. Chairet, R., Ben Salem, Y., Aoun, M.: Features extraction and land cover classification using Sentinel 2 data. In: 2019 19th International Conference on Sciences and Techniques of Automatic Control and Computer Engineering (STA), Sousse, Tunisia, pp. 497–500 (2019). https://doi.org/10.1109/sta.2019.8717307
8. Huien, S., Wentao, F., Huang, J.: Building segmentation in mountainous environment based on improved watershed algorithm. In Proceedings of the 3rd International Conference on Video and Image Processing (ICVIP 2019). USA: Association for Computing Machinery (2019)
9. She, Y.: Automatic extraction of rocky desertification information based on GF-2 spectral characteristics. Central south University of Forestry and Technology (2017)
10. Zhang, D., Zhang, L., Jiang, Y.G.: Extraction method based on ENVI in the mining collapse area of Ezhou. Land and Natural Resources Research, 02, 37–38 (2013)
11. Ning, Z.: Rice planting information extraction and dynamic monitoring in Shenyang city based on Landsat8 remote sensing image. Shenyang Jian Zhu University (2018)
12. Yi, Z.: Land use information extraction based on the SPOT5 panchromatic image. China University of Geosciences (Beijing) (2010)
13. Jin, H.: Application research of object-oriented remote sensing image classification method in land use information extraction. Chengdu University of Technology (2010)

A Signaling Monitor Scheme of RRC Protocol in 5G Road Tester

Bingying Zhang$^{(\boxtimes)}$, Fang Cheng, and Bingguang Deng

School of Communication and Information Engineering, Chongqing University of Posts and Telecommunications, Chongqing 400065, China
s180131221@stu.cqupt.edu.cn

Abstract. Faced with the growth of massive data in the 5G network and the low query efficiency in the existing signaling synthesis algorithm, the traditional LTE/LTE-A signaling monitor scheme has been unable to satisfy with the new 5G network architecture. The Radio Resource Control (RRC) protocol is the core of the control plane, which manages and controls the wireless resources of the network. Based on this, a signaling monitor scheme of RRC protocol suitable for 5G road tester was proposed and the specific functions of its main submodule were introduced in detail. Additionally, this paper discussed an improved dynamic hash signaling synthesis algorithm based on AVL tree under a new network architecture. The tree structure is used to reduce the query time of traditional algorithm in the hash table, so as to quickly deal with hash collisions and improve the real time on the signaling synthesis. At present, the proposed scheme was applied to the real network testing of air interface data in the 5G road tester. The experimental results show that the improved algorithm can efficiently solve the issue of low efficiency of Call Detail Recording (CDR) synthesis and the average query time can be reduced by 49.3% and 33.1% compared with the two traditional algorithms. The signaling synthesis scheme described above achieves the expected effect and signaling messages in RRC protocol can be accurately decoded and monitored in real time.

Keywords: 5G road tester · Signaling synthesis · Signaling monitoring · Hash collisions · AVL tree

1 Introduction

In order to adapt to the growth of massive data and accelerate the development of various new service application scenarios, the fifth generation mobile communication network (5G) has arised [1]. More open and flexible 5G network will meet the requirements of mobile network development. 5G will support a diverse range of requirements such as the handling of low latency (in the order of 1 ms), massive Machine-Type Communication (mMTC) and extreme Mobile Broadband (xMBB) services. At present, international organizations such as 3GPP and ITU are actively promoting the standardization process of 5G technology. 5G is at a critical stage of the evolution from technology research to technology verification [2].

© ICST Institute for Computer Sciences, Social Informatics and Telecommunications Engineering 2021
Published by Springer Nature Switzerland AG 2021. All Rights Reserved
H. Gao et al. (Eds.): ChinaCom 2020, LNICST 352, pp. 324–336, 2021.
https://doi.org/10.1007/978-3-030-67720-6_22

Faced with the higher services test requirements and more complex road test data than 4G network, the traditional Long Term Evolution-Advanced (LTE-A) signaling monitoring analyzer is difficult to adapt to the 5G network structure and handle massive mobile signaling data. 5G road tester with independent intellectual property rights will meet more high-performance test requirements of 5G network in the high rate and low delay scenario and promote the rapid development of 5G test industry. Signaling monitor is one of the core technologies of 5G road tester. It mainly completes the whole process from signaling processing to CDR synthesis. By monitoring the CDR signaling messages, it can provide an analysis basis for the various services of the 5G network.

In the recent years, many scholars have researched on signaling in different interfaces or protocols of mobile communication network. In [3], it used secondary re-hashing to complete CDR synthesis by studying the signaling monitor of layer 3 for Uu interface in LTE network. This method can avoid the clustering of hash collision elements in a certain area, but increased the search time. In [4], the authors proposed an effective monitoring scheme based on the study of signaling for S6a interface in LTE network, which used hash table instead of binary tree to store CDR key-value. In [5], NAS protocol monitoring in LTE-A network are researched; In [6], a multi-protocol association technology was proposed for Uu interface to monitor signaling association and synthesis in LTE-A network. Both of them completed the CDR synthesis by using separate chaining. This method can effectively reduce the clustering of hash elements when hash collisions occurred, but the disadvantage is that it will waste a lot of time to frequently visit memory because of using sequential search.

Therefore, this paper will propose a new scheme based on the traditional LTE-A signaling monitor, which is suitable for RRC protocol in 5G network to realize real-time monitoring and data analysis. Furthermore, in order to solve problems such as long search time and low index efficiency in the traditional signaling synthesis algorithm, an improved dynamic hash signaling synthesis algorithm based on AVL tree is proposed. It can efficiently and quickly process signaling messages under the massive user data in 5G network, so as to accurately and fully complete signaling monitor. Analysis results show that the improved signaling synthesis algorithm applied to the signaling monitor module in the 5G road tester can significantly improve the CDR synthesis efficiency and reduce memory consumption.

2 Overview of RRC Protocol

The RRC protocol which is the most complex one in the air interface is responsible for the management and control of the radio resources of 5G network and it's the control center of the whole assess layer [7]. The functions of RRC protocol mainly include radio bears control and mobility management, system message broadcast, paging, RRC connection management [8]. In order to meet the requirements of mMTC service for low power consumption and large connection, a new RRC state (RRC connected inactive) has been proposed in 5G systems. The high-level characteristics of RRC Inactive are similar to LTE's Light Connected [9]. This new state enables the UE to maintain a connection to the core network at all times and save the UE context so that it can quickly return to RRC Connected state. Figure 1 is the RRC state model for the 5G radio access network.

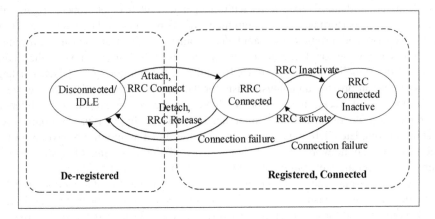

Fig. 1. RRC state model for the 5G radio access network.

3 System Module Design

The signaling monitoring module in 5G road tester combines the traditional signaling monitor technology in the LTE-A with data acquisition technology to complete the whole process from data acquisition to signaling analysis. Considering the characteristics of signaling in the RRC protocol and the application requirements, module is divided into four parts: data preprocessing, protocol decoding, CDR synthesis and statistics module. Adopting the thought of module design greatly improved independence of modules and subsequent extensibility. The design of signaling monitor module in 5G road tester is shown in Fig. 2.

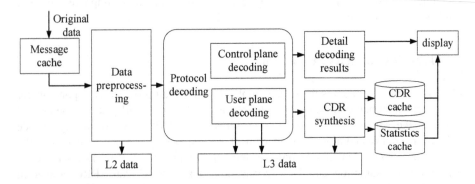

Fig. 2. Signaling monitor module framework in 5G road tester.

3.1   Data Preprocessing Module

The function of the module is mainly to receive the original signaling data from the data acquisition module and convert them into data structures such as binary code stream that can be recognized by the system, including three units of data elimination, protocol recognition and storage. Data elimination is to eliminate the underlying protocol data irrelevant to subsequent RRC protocol decoding and synthesis in the original signaling data. The control plane data mainly includes RRC and NAS protocol, and system information block (SIB), master information block (MIB), signaling radio bears such as SRB0, SRB1 and SRB2. User plane data mainly includes PDCP, SDAP and IP, HTTP protocols. The data type of messages can be judged by identifying the channel type contained in the data header. Then the message is respectively stored at the end of the queue for control plane data and user plane data. The channel types and data types in the data header are shown in Table 1.

Table 1. The relationship of data types and channel types.

Data types	Channel types
Control plane data	BCCH
	PCCH
	CCCH
	DCCH
	MCCH
User plane data	DTCH

3.2   Protocol Decoding Module

RRC protocol messages are transmitted through Signaling Radio Bear (SRB) and it's divided into six processes: paging, RRC connection establishment, secure activation, RRC connection configuration, RRC connection re-establishment and release. The RRC decoding module uses the ASN.1 compiler to generate the decoding function of each signaling message in RRC protocol, and then completes the simple decoding and detailed decoding. The Protocol Data Unit (PDU) of the RRC protocol is expressed in an abstract language and multiple types of information elements are defined nested within the message. The decoding module puts the data structure of the decoded results on the stack, which can be called by the synthesis module, statistics module. When a valid RRC message is received, the corresponding data type of this message is obtained by judging the logical channel type (BCCH、PCCH、CCCH、DCCH、MCCH), transmission channel type (BCH、PCH、DL-SCH) and transmission direction (uplink、downlink) of RRC message, the decoding function of the data type is called to decode the message. The specific decoding process is shown in Fig. 3.

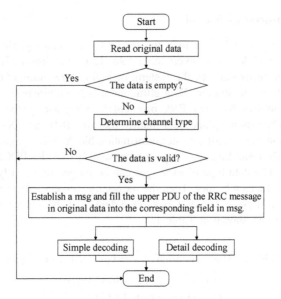

Fig. 3. Decoding process.

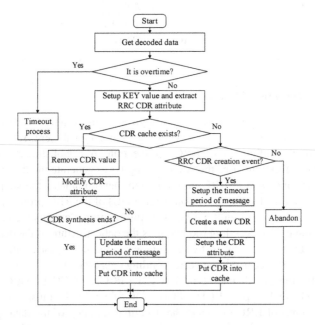

Fig. 4. RRC protocol CDR synthesis process.

3.3 CDR Synthesis Module

CDR synthesis module is the core module in the RRC protocol monitor scheme. It contains four steps: create, modify, store and delete. The key field information needs to be extracted from the protocol decoding module before signaling processes are analyzed. The information includes protocol type, CDR creation event, CDR completion event, associated key value. The specific implementation process is shown in Fig. 4. Each message from protocol decoding module needs to complete a timeout check. If it's overtime, the node associated with the key value in this message will be deleted and exit the synthesis process. Otherwise, extract the key value and CDR attribute. The message type is determined by the key value. Then, judge whether the CDR structure cache corresponding to the key value exists. If it exists, the message will be placed at the end of the corresponding CDR structure queue and modify its CDR attribute. The end of CDR synthesis is determined by judging whether the message type belongs to a CDR completion event in RRC signaling. If not, update the timeout period of message and put CDR of cache. Otherwise, the CDR synthesis process is ending. If the CDR cache isn't exist, a new CDR is created and its property value is placed in the CDR structure cache by identifying whether the message belongs to a CDR creation event in RRC signaling. Repeat the process until all the subsequent messages are associated and synthesized into the created CDR records.

3.4 Statistics Module

Statistics module is responsible for collecting related protocol processes and signaling in air interface. The messages are classified according to service types, procedure types and specific message types, which form a set to calculate the number of protocol messages and some statistical indicators. It is convenient for users to monitor and check abnormal data and print error information in real time. Because this module is used to synthesize the related information for the CDR synthesis, the trigger of the statistic function is merged into the interface function in the CDR synthesis module.

4 Algorithm Design

4.1 Hash Signaling Synthesis Algorithm

There is a huge amount of signaling data in 5G network at any moment. In order to increase efficiency of CDR synthesis, received signaling messages need to establish a hash index by certain rules and realize the one-to-one mapping between the Key value and the hash table. Firstly, the algorithm selects the Key from the RRC protocol association identifier as the independent variable, and calculate value which must be the same and unique by hash function. Then the Key value is mapped to the hash table to visit the CDR synthesis record when the query begins.

When the number of queried Key is much larger than the size of the hash table, hash collisions are inevitable. Traditional hash signaling synthesis algorithm (HSSA) used to resolve hash collisions include Open Addressing (Quadratic Probing, Linear Probing, and Rehashing/Double Hashing) and Separate Chaining [10, 11]. Open Addressing is simple to insert and search data. But the disadvantage is that the hash table will be full and memory will be accumulated if hash collision occurs. By contrast, Separate Chaining can avoid these problems. It's easier to change the linked list when inserting or deleting some elements. But it will waste some time to visit memory when the Key value is queried in the hash table because it uses Sequential storage structure.

4.2   Improved AVL-HSSA Algorithm

In order to solve problems such as long search time and low index efficiency in traditional hash signaling synthesis algorithm, this paper proposed an improved hash signaling synthesis algorithm based on AVL tree (AVL-HSSA). The tree structure is used to reduce the time of querying CDR cache in the hash table, so as to quickly deal with hash collisions and improve the timeliness of signaling synthesis.

The improved algorithm need to construct hash function and select the Key value before building hash table. Because there is a large amount of irregular signaling data from different users and protocols before signaling synthesis, this paper uses remainder of division method to construct hash function, which can quickly and uniformly distribute the data and realize the mutual mapping between Key and value [12]. Specifically speaking, the same user is identified by the source IP, the source port number, the destination IP address, and the destination port number. And we select IMSI, C-RNTI and CellID as Key value to realize CDR synthesis in the same signaling process from the same user. KEY_i is the Key of a signaling message. The Key value is divided by the size of hash table–N, then the result is $H_i(KEY)$, also known as Hash address. Its specific expression is as follows:

$$KEY = KEY_i.IMSI + KEY_i.C-RNTI + KEY_i.CellID \qquad (1)$$

$$H_i(KEY) = KEY \bmod N \qquad (2)$$

$$Hashed\_Addr = H_i(KEY) \qquad (3)$$

The HSSA algorithm stores the root node of the tree in a hash table without a hash collision. Once collisions are occurred, it will use Hashed-AVL structure to store collision node. Each node in the tree contains a pointer for the left and right subtrees, a Key value for comparing node size, and a balance factor. The Key value of each node in the left subtree is smaller than the value of the root node, and the Key value of each node in the right subtree is larger than the value of the root node. The height difference between left and right subtrees isn't more than one [13]. The hash table structure when the AVL-HSSA algorithm handles collisions is shown in Fig. 5.

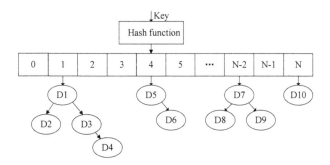

Fig. 5. Hash table structure of AVL-HSSA.

The specific search steps of the algorithm are as follows:

Step 1: extract the signaling association identification *KEY*ᵢ from the protocol decoder before CDR synthesis, and find the corresponding hash address by hash function.

Step 2: judge whether hash collision occurs. If not, execute step3; otherwise, execute step4.

Step 3: insert the Key as root node into the hash table if the hash address hasn't any Key cache.

Step 4: insert the Key with the same hash address into the corresponding AVL tree and compare it with the Key value in the root node. If the queried Key is larger than the root node, it is compared with the nodes of the right subtree until the queried Key find its CDR cache and modify the CDR property [14, 15].

Algorithm: Searching of key in AVL-HSSA
1: **procedure** SEARCH(key)
2: *hash = hashValue(key)*
3: *key_node = searchForHash*
4: **if** key_node is NULL **then**
5:     create a new root node in AVL and set data
6: **else**
7:     *Node = searchfordata*
8:     **while** the key value is large than Node
9:             *Node = Node.rchild*
10:    **while** the key value is less than Node
11:            *Node = Node.lchild*
12:    return Node

4.3   Analysis of Algorithms

Usually, we use average search length (ASL) to evaluate the performance of the hash signaling synthesis algorithm in the process of searching signaling. Both Chain-HSSA and AVL-HSSA are analyzed in Table 2 when the Key value is successfully queried.

Table 2. Comparison of signaling synthesis algorithm

Algorithm	KEY search method	Average search length	Average time complexity
Chain-HSSA	sequential search	$(n+1)/2$	$O(n)$
AVL-HSSA	AVL tree search	$\log_2(n+1)-1$	$O(\log n)$

Chain-HSSA algorithm uses a sequential search method to compare the queried Key with each Key in the single linked list in turn until the same Key value is found. Its average search length is (n + 1)/2 and its average time complexity is O(n). It will waste a lot of time to compare CDR cache if there are a lot of number in linked list, so the efficiency of query is low. AVL-HSSA proposed in this paper breaks through the original storage structure, so that the AVL search tree can be used in the hash table. Using AVL tree to query the Key value may improve the efficiency of signaling synthesis and reduce the query time. Its average search length is $\log_2(n+1)-1$ and its average time complexity is O(logn).

5   Analysis of Experimental Results

In order to evaluate the performance of the proposed AVL-HSSA algorithm in signaling synthesis, we respectively compared Open-HSSA and Chain-HSSA in the traditional hash signaling synthesis algorithm with AVL-HSSA to finish the test. The experimental environment of this paper is: Windows 10 operating system, Visual Studio 2017 compiler running environment and the hardware configuration is 64-bit operating system, the processor is Inter(R) Core(TM) i5-8500 CPU 3.00 GHZ. The C-RNTI allocated by the gNB to the UE is used as the Key value in the signaling synthesis. The test data source of this experiments is obtained by extracting the dynamic identifier C-RNTI from the decoding module.

5.1   Search Time Analysis

The load factor α is a marker to measure how full a hash table is. When α is 1, we respectively select 1000, 2000, 3000, 4000, 5000, 6000, 7000, 8000 data from the test data with LP-HSSA, Chain-HSSA and AVL-HSSA to test. The test results are shown in Fig. 6.

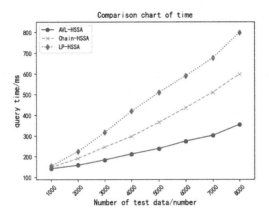

Fig. 6. Comparison of search time ($\alpha = 1$).

When there are the same number of data, Open-HSSA consumes the most search time and owns the least efficiency. By contrast, Chain-HSSA consumes less search time than Open-HSSA, but it is still higher than AVL-HSSA algorithm. As the number of data increases, the comparison of search time required by the three algorithms becomes more obvious. In the above test results, the average search time of the improved AVL-HSSA is reduced by 49.3% and 33.1% compared with Open-HSSA and Chain-HSSA.

We select 8000 test data from the data source and use three algorithms under different loading factors to finish the test. As shown in Fig. 7, when α is 1/10, the probability of hash collision is small because the length of hash table is longer than the number of queried Key. For Open-HSSA, the method can directly locate into the hash table by hash function when the queried Key exists in the hash table, so its search time of 269.7 ms is shortest. Using AVL-HSSA to locate into hash table is the same as Open-HSSA, the difference of them is that AVL-HSSA use the AVL tree to search data, so the search speed of the algorithm is slightly slower than that of Open-HSSA when α is 1/10, and the search time is 276.4 ms. By contrast, Chain-HSSA uses sequential search to visit linked list. Therefore, it needs to consume more time to visit memory.

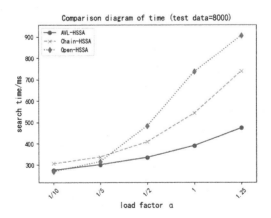

Fig. 7. Comparison of search time (test data = 8000).

When α is 1/2, the search time of Open-HSSA is obviously higher than the other two algorithms and its efficiency is lowest. By contrast, the AVL-HSSA algorithm has the shortest search time. And the greater the loading factor becomes, the more obvious the advantages of this method is. Its stability and adaptability are much higher than the other two hash signaling synthesis algorithms.

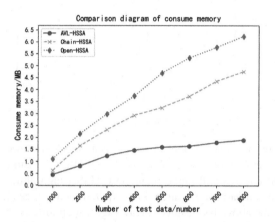

Fig. 8. Comparison of consume memory.

5.2 Memory Consumption Analysis

The comparison diagram of consume memory for AVL-HSSA algorithm and LP-HSSA, Chain-HSSA algorithm in the process of signaling synthesis are shown in Fig. 8. It can be intuitively seen from the figure that AVL-HSSA algorithm has an advantage in memory consumption, which occupies the lowest memory. The LP-HSSA algorithm allocates memory for all table items when it starts to create a hash table, and it will consume the most memory because of memory accumulated when the number of data source is more in the signaling synthesis. Chain-HSSA algorithm dynamically applies for memory when inserting into a hash table, and the memory occupied will increase with the more number of test data.

5.3 Analysis of Signaling Monitoring Scheme Test Results

The signaling decoding results show the protocol data information with the form of tree structure in Fig. 9. According to the definition of RRC BCCH-DL-DCH message in the 3GPP standard, each field in the decoding tree matches the definition of the message. The field value of mcc-mnc-digit is 4, and the bitmask is 0100.

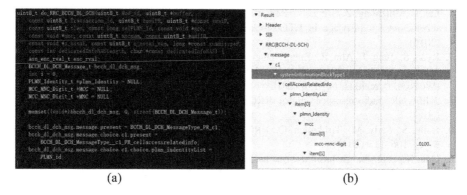

(a)   (b)

Fig. 9. RRC BCCH-DL-DCH message decoding. (a) The function declarations. (b) Detail decoding result

6 Conclusion

By studying the RRC protocol in a new 5G network architecture, this paper proposed a signaling monitor scheme of RRC protocol based on Hashed-AVL and discussed the specific function of its main submodule. The implementation process of the protocol decoding module and the CDR synthesis module was introduced in detail. Focused on the low query efficiency and high average traversal time complexity in the traditional signaling synthesis algorithm, an improved dynamic hash signaling synthesis algorithm under a new network architecture is used to reduce the query time of traditional algorithm in the hash table, so as to quickly deal with hash collisions and improve the real time on signaling synthesis. It shows that the system is effective and feasible. The signaling synthesis scheme described above achieves the expected effect and signaling messages in RRC protocol can be accurately decoded and monitored in real time.

Acknowledgment. It is supported by the Science and Technology Major Project in Chongqing (R&D and application of 5G road test instruments: No.cstc2019jscx-zdztzxX0002).

References

1. Hucheng, W., Hui, X., Zhimi, C.: Current research and development trend of 5G network technologies. Telecommun. Sci. **9**, 149–155 (2015)
2. Navarro-Ortiz, J., Romero-Diaz, P., Sendra, S., et al.: A survey on 5G usage scenarios and traffic models. IEEE Commun. Surv. Tutorials, **22**(2), 905–929 (2020). Secondquarter
3. Jiang, Y.: Design and implementation of LTE Uu Interface Layer 3 Protocol Monitor System. Beijing University of Posts and Telecommunications (2014)
4. Yingying, H., Bing, X., Zhizhong, Z.: Research and implementation on synthetic scheme of S6a interface in the LTE network monitoring system. Appl. Mech. Mater. **3670**, 237–240 (2015)
5. Pei, P., Longhan, C., Zhizhong, Z.: Research of NAS protocol monitor scheme of air Interface in LTE-A network monitoring instrument. Video Eng. **39**(21), 44–48 + 60 (2015)

6. Lei, L., Zhizhong, Z., Bing, X.: Research and implementation of multi-protocol association scheme on Uu interface in LTE-Advanced network. Telecommun. Sci. **32**(6), 167–176 (2016)
7. Ryoo, S., Jung, J., Ahn, R.: Energy efficiency enhancement with RRC connection control for 5G new RAT. In: 2018 IEEE Wireless Communications and Networking Conference (WCNC), Barcelona, pp. 1–6 (2018)
8. 3GPP, NR., radio resource control (RRC) protocol specification (release 15).: 3GPP 38.331 (2019)
9. Hailu, S., Saily, M., Tirkkonen, O.: RRC state handling for 5G. IEEE Commun. Magazine **57**(1), 106–113 (2019)
10. Gou, X., et al.: Single hash: use one hash function to build faster hash based data structures. In: 2018 IEEE International Conference on Big Data and Smart Computing (BigComp), Shanghai, pp. 278–285 (2018)
11. Atighehchi, K., Rolland, R.: Optimization of tree modes for parallel hash functions: a case study. IEEE Trans. Comput. **66**(9), 1585–1598 (2017)
12. Agrawal, A., Bhyravarapu, S., Venkata Krishna Chaitanya, N.: Matrix hashing with two level of collision resolution. In: 2018 8th International Conference on Cloud Computing, Data Science & Engineering (Confluence), Noida, pp. 14–15 (2018)
13. Chinnaiyan, R., Kumar, A.: Construction of estimated level based balanced binary search tree. In: 2017 International conference of Electronics, Communication and Aerospace Technology (ICECA), Coimbatore, pp. 344–348 (2017)
14. Dhar, S., Pandey, K., Premalatha, M., Suganya, G.: A tree based approach to improve traditional collision avoidance mechanisms of hashing. In: 2017 International Conference on Inventive Computing and Informatics (ICICI), Coimbatore, pp. 339–342 (2017)
15. Cao, Y., et al.: Binary hashing for approximate nearest neighbor search on big data: a survey. IEEE Access **6**, 2039–2054 (2018)

Snoop Through Traffic Counters to Detect Black Holes in Segment Routing Networks

Marco Polverini$^{(\boxtimes)}$, Antonio Cianfrani, and Marco Listanti

University of Roma "Sapienza", Via Eudossiana 18, 00184 Roma, Italy
{marco.polverini,antonio.cianfrani,marco.listanti}@uniroma1.it

Abstract. The new Segment Routing paradigm provides network operator the possibility of highly increasing network performance exploiting advanced Traffic Engineering features and novel network programability functions. Anyway, as any new solutions, SRv6 has a side effect: the introduction of unknown service disruption events. In this work we focus on packet lost events due to the incorrect computation of the Maximum Transmission Unit (MTU) value of an end-to-end path in an SRv6 network. This event, referred to as *MTU dependent SR Black Hole*, cannot be detected by known monitoring solutions based on active probing: the reason is that in SRv6 probe packets and user data can experience different network behaviors. In this work we propose a passive monitoring solution able to exploit the SRv6 Traffic Counters to detect links where packets are lost due to MTU issues. The performance evaluation shows that the algorithm proposed is able to identify the link affected by the blackhole with a precision equal to 100%; moreover, the flow causing the blackhole cannot be detected with the same precision, but it is possible to identify a restricted set of flows, referred to as suspected flows, containing the target one.

Keywords: Segment routing · Network failures · Black holes · Interface counters

1 Introduction

The introduction of a new technology in the networking area leads to new or optimized functions with respect to existing one, and it translates in improved performance for users. As a side effect, it also represents a possible vector for the creation of new flaws and bugs. These kind of effects can be considered as technology dependent failures, which can determine a service disruption. Due to their nature, technology dependent failures are usually transparent to existing monitoring solutions. In literature, they have been identified with the term *network black holes*. As an example of technology dependent black hole, in [9] the service disruption due to LSP failure in MPLS network is reported.

Recently, Segment Routing Architecture [8] has been introduced as a practical realization of the source routing paradigm. SR allows for a better usage

Published by Springer Nature Switzerland AG 2021. All Rights Reserved
H. Gao et al. (Eds.): ChinaCom 2020, LNICST 352, pp. 337–350, 2021.
https://doi.org/10.1007/978-3-030-67720-6_23

of the bandwidth, increasing the Traffic Engineering capabilities of traditional networks, completely reusing existing data plane functionalities. Currently, two implementations of SR, named SR-MPLS and SRv6 are available [19], using MPLS and IPv6 data planes, respectively. This work is focused on SRv6.

SR comes with a set of tools for Operation and Maintenance (OAM) [11], which allow to troubleshoot the network and to identify possible anomalies (such as failures and misconfigurations) in the current configuration. These tools extend the existing ones, by providing further capabilities and enhancement to the performance. As an example, in [1] SR is exploited to realize a link failure detection infrastructure, named SCMon, able to check the status of a layer 3 link realized as a bundle of layer 2 connections.

One of the main innovation introduced by SR is that the flow states are moved from the routers to the packets headers. This enables SR networks to highly scale, at the cost of increasing the packet length (i.e., overhead); in fact, the length of a packet can increase during the journey from the source up to the destination nodes, due to the application of specific SR information (Segment List, TLV, etc.). Depending on the complexity of the flow states, the packet size can eventually become greater than the minimum MTU, causing drop phenomena, which in turns determine QoS degradation and, in some cases, service disruption. Unfortunately, existing OAM tools are not able to detect this type of failure. We refer to it as *MTU dependent SR Black Hole*.

As we will explained in Sect. 3, classical methods fail in detecting this type of failure since they rely on an active approach, i.e., they are probe based techniques. Probe messages are sent along an end to end path to assess the Path MTU, or to detect a possible black hole in the Path MTU Detection (PMTUD) procedure. Due to the fact that SR forwarding is based on the concept of policy routing, it is not guaranteed that the sent probes follow the same end to end path of the flow hit by the *MTU dependent SR Black Hole*, or that they are processed in the exact same way. On the contrary, in this paper we present a novel algorithm, called *Segment Routing Black Holes Hound* ($SRBH^2$), that uses a passive approach to diagnose the current network status and possibly detect *MTU dependent SR Black Holes*. In particular, $SRBH^2$ exploits the information gathered by the SR Traffic Counters [6], a new OAM tool available in all SR capable nodes. Since SR Traffic Counters collect statistics directly on the data traffic, if a traffic flow falls into a black hole, then this event is somehow coded into the traffic counters. Then $SRBH^2$ represents a method to search and extract this information from the Traffic Counters.

The proposed solution is a dedicated solution to identify *MTU dependent SR Black Holes* and, in a real scenario, will run in parallel with different solutions able to detect different packet losses causes, such as physical failures and traffic congestion. It is important to remark that $SRBH^2$ is based on data traffic information, i.e. SRv6 traffic counters, already available at the OAM module, with no need of generating additional traffic, such as packet probes and traffic mirroring.

To summarize, the main contributions of this work are:

- the identification of a new type of black hole, related to the deployment of SRv6 technology;
- the characterization of classical OAM tools weakness in detecting such types of failures;
- the definition of a low complexity algorithm exploiting SR Taffic Counters to detect *MTU dependent SR Black Holes*.

The rest of the paper is organized as follows: In Sect. 2 we provide a background on SRv6, in Sect. 3 we explain in detail the *MTU dependent SR Black Hole* and show an example of such a failure, in Sect. 4 it is presented the $SRBH^2$ algorithm able to detect them. Sect. 5 provide a performance evaluation, and finally Sect. 6 concludes the paper.

2 Segment Routing Background

Segment Routing (SR) [8] is a novel network paradigm based on source routing, i.e., the source node decides the path that each packet has to go through. The end to end paths are encoded as an ordered list of instructions, also referred to as segments. Thus the full list of segments is named Segment List (SL). Each segment can represent any topological (e.g., send the packet to a given node) or service based (e.g., duplicate the packet) instruction. Segments are expressed as labels, named Segment IDentifiers or SIDs. If the underlay data plane is based on IPv6, a SID has a is represented by an IPv6 address.

In SRv6, SR related information are included in a new defined extension header, named Segment Routing Header (SRH) [7], which is added to each incoming packet in two different ways. The first option is the *insert mode*, i.e., the SRH is included directly on the original IPv6 packet header. The second method is the *encap mode*, i.e., the original packet is encapsulated in an outer IPv6 packet, having the SRH. SL is a special field of the SRH, and is represented by a list of IPv6 addresses, having the meaning of SIDs.

The SRv6 packet forwarding works as follows. When the border router receives a packet, it has to steer it over a specific SL. This last is chosen according to a given *SR Policy*. Specifically, each node of the SR domain maintains an *SR Policy* table, where all the policies available at that node are stored. After a packet has been processed according to a matched *SR Policy*, its most external IPv6 header is extended by the inclusion of an SRH. This last containing the SL and a pointer named *segment left* (sl), which indicates what is the active segment (*actsgm*). It represents the current instruction to be applied on the packet. In particular, the *actsgm* is copied also in the destination address field of the outer IPv6 header, while the SID of the node that has performed the encapsulation is used as source address. As a consequence, the path followed by the packet from a SID to the next one is the path computed by IGP routing protocols (usually the shortest one). Transit routers forward incoming packet by inspecting the IPv6 destination address of the outer header. Once the node having the same SID

of the *actsgm* is reached, the so called *MyLocalSID* table is inspected, in order
to verify what function has to be applied. A common function is the END one,
which implies that the *actsgm* has to be updated, thus the *sl* is decremented
and the new *actsgm* is copied into the IPv6 destination field. Finally, before the
packet leaves the SR domain, the SRH has to be removed.

3   Introducing the *MTU Dependent SR Black Hole*

In this section we provide a discussion about the main causes of the *MTU dependent SR Black Hole*. Then, we show an example of such a black hole, and we
provide a discussion about the weakness of existing monitoring tools when an
MTU dependent SR Black Hole occurs. Finally, guidelines for the definition of
an algorithm able to identify *MTU dependent SR Black Hole* are identified.

3.1   MTU Related Issues in IPv6

We recall that SRv6 reuses the IPv6 dataplane, by defining a new extension
header called SRH. As known, IPv6 [4] discourages the use of fragmentation,
limiting this operation to be performed only at the source node. Specifically,
the source node performs the Path MTU Discovery technique [15] (PMTUD) to
acquire the knowledge of the maximum packet size that do not cause violations of
the MTU constraint. This procedure is based on the following steps: i) the source
node starts emitting packets having a length equal to the MTU of the output
link (the one used by the source to reach the next hop); ii) each router along the
path checks whether the size of the received packet meets the requirement on the
maximum allowed size, with respect to the MTU on the link toward the next
hop; if MTU constraints are satisfied, then the packet is normally forwarded,
otherwise, an *ICMP Packet Too Big* (PTB) message is sent back to the source
node, which notify the maximum allowed size. The source node is now able to
fragment the packet taking into account the MTU constraint.

One of the main weakness of this mechanism is the fact that it assumes that
all intermediate routers participate to the discovery process. PMTUD failures
are a well studied problem in IPv6 networks. Some studies [3,13] have shown
how different failure modes are possible. Major causes are due to routers that do
not send PTB messages, either due to configuration issues or because of MTU
mismatch. In other cases, firewalls might discard fragmented packets, creating
a service disruption. In particular [3] has pointed out that a common policy
adopted by service provider is represented by TCP Maximum Segment Size
(MSS) clamping. This technique forces servers to advertise a MSS smaller than
the maximum allowed one, so that to force the clients to generate small packets,
eventually avoiding problems related to MTU constraint violation. Clearly, UDP
flows cannot exploit MSS clamping, increasing the chances of the creation of a
black hole.

To overcome these limitations, the so called Packetization Layer Path MTU
Discovery [14] (PLPMTUD) is used in conjunction with the regular PMTUD.

PLPMTUD procedure is the same one of PMTUD, but it is performed at the transport layer; in this way no ICMP messages are required. Clearly this mechanism can be performed only between end systems.

In the context of an SRv6 network, the problems related to the packet fragmentation in IPv6 are further enhanced by the following aspects: i) SRv6 policies require the encapsulation of packets, ii) the size of the packets can change during the journey from source to destination, and iii) due to network programming, end to end paths can change frequently and in an unplanned manner. In the next we detail how, each of these SR features, can lead to the creation of an *MTU dependent SR Black Hole*. Considering the first point, when the encap mode is used to apply the SRH on packets, the source address of the external IPv6 header is set equal to the address of the head end node. This last, as specified in [7], is in charge of performing the PMTUD to discover maximum packet size that can be used to reach the end point of the SR policy. The main consequence of this fact is that PLPMTUD cannot be used, since the MTU discovery is not carried out between end systems, forcing the procedure to rely on the correct delivery of ICMP PTB messages.

Due to the fact that the PMTUD could fail, the following recommendation is specified in [7]: *"deploy a greater MTU value within the SR Domain than at the ingress edges"*. Even though this last expedient can often prevent from critical situations in classical IPv6 networks, it is less likely to work in SRv6 environments. In fact, as previously stressed, the size of the packets can change during the journey from source to destination. There are different factors that can cause an increase of the overall packet size, such as the application of a further level of SR encapsulation and the insertion of TLVs. As a consequence, even though a packet meets the MTU constraint when injected in the SR domain, it can eventually grow up to reach a size that overcomes the allowed threshold. To better explain this problem, let us refer with the term *margin* (M) as the difference between the minimum MTU among the links internal to the SR domain and the one on the links at the edge. If during its journey inside the SR domain, the length of a packet grows more than M, then the MTU constraint is violated and an *MTU dependent SR Black Hole* might happens.

A possible countermeasure to this problem could be to limit the number of extra bytes added to each packet, by properly configuring the SR policy to be applied on each traffic flow: knowing the end to end path followed by a packet, it is possible to configure the SR processing applied on it, so that to limit the maximum amount of extra information to add on it. Unfortunately, the set of functions applied on a traffic flow is not static, due to the network programming paradigm, which can make end to end paths change in an unpredictable manner. As an example, in [5] it is described a procedure to provide zero-loss Virtual Machine migration. This process is based on the definition of a special SID (named *forward to local IF present*), which forces a router to decide the next node to visit based on the verification of a condition. Another example of unpredictable path change is represented by the Topology Independent Loop Free Alternate (TI-LFA), which consists in a backup scheme to provide protection against link failures: when a link fails, the head end node react by re-routing the traffic flows over a backup path. This re-routing is performed by means of an SR encapsulation. Then, it is not possible to plan in advance the set of functions applied on a traffic flow and consequently limit the amount of extra information added to each packet.

3.2 An Example of MTU Dependent SR Black Hole

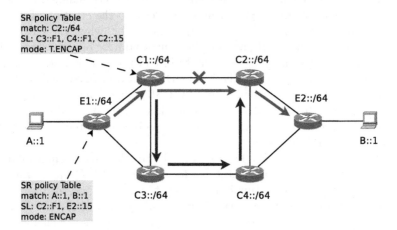

Fig. 1. Example of creation of an *MTU dependent SR Black Hole*. (Color figure online)

In Fig. 1 we present a simple example showing how all the aforementioned events can interact to generate an *MTU dependent SR Black Hole*. Let us assume that, in the scenario reported in Fig. 1 the margin is $M = 100$ Bytes, and that all the links internal to the SR domain have the same MTU. The traffic flow originated by the host A and directed to the host B is steered through an SR policy installed at node $E1$, whose end to end path is shown in red. This SR policy increases the size of each packet of 64 Bytes, thus no fragmentation is needed. In normal conditions, the considered traffic flow is steered over the red path and delivered to the host B. Now, let us assume that the link between nodes $C1$ and $C2$ fails. A backup scheme is implemented at node $C1$ by means of an SR policy, that encapsulates packets that were supposed to be routed over the failed link, with an SRH having a SL of depth 3. The SR policy at node $C1$ adds 80 Bytes of overhead to the incoming packets. Note that the application of this policy is unplanned, since it is used only when a link failure happen. Globally, each packet is increased in size (with respect to the length before being injected in the SR domain) of 144 Bytes, which is a quantity greater than the margin. Consequently the MTU constraint is violated. If ICMP PTM messages are not generated, the head end node of the outer policy ($C1$ in the example) is unaware of this situation and do not fragments the packets. Thus an *MTU dependent SR Black Hole* is created.

3.3   Weakness of Classical Monitoring Tools in Detecting MTU Dependent SR Black Holes

In this subsection we describe known monitoring tools used to identify black holes when the PMTUD procedure fails.

One of the most reliable tools to discover PMTUD failures is Scamper [12]. Scamper in a two steps procedure to determine either the largest MTU that can be used on a end to end path, and to discover (in case of a failure) what is the router that is not participating to the PMTUD. Both the phases of Scamper are based on the enforcement of probes along the end to end path to check. These probes are represented by a set of UDP segments destined to an unused port, when performing the first step, and a set of ICMP Echo messages destined to intermediate routers, in the second phase.

Netalyzr is presented in [10], it is a network measurement and debugging tool to monitor the Internet. The architecture is provided with a set of applets pre-installed, and one of these aims at determining the path MTU toward a destination server. This search is based on a process that require to send a set of UDP probes to the target destination.

Ripe Atlas [18] is a worldwide monitoring infrastructure based on the use hardware probes placed in the so called vantage point. In [2] Ripe Atlas has been used to discover path MTU black holes in the Internet, with the specific focus of assessing the main causes and the most affected data plane protocol. The obtained results show that black holes due to failure in the PMTUD procedure affect both the IPv4 and the IPv6 data planes. Specifically, Ripe Atlas was able to detect the main causes and the location of these failures, such as PTB messages and fragmented packets filtered by firewalls.

Unfortunately, none of the aforementioned tools is able to detect *MTU dependent SR Black Holes*. The reason is that all tools are based on an active approach, i.e. probe packets must be sent through an end-to-end path to be tested. The use of probes is inefficient when dealing with SR for two reasons: i) the presence of routing policies makes possible that the SR path followed by user data could be different than the one followed by the probes, even if the source and the destination of the path are the same; ii) even if the same end to end path is used for data and probes, the network programming feature of SRv6 could results to different SR policies for user data flows and probes, which turns in different SRH applied on probes (and different MTU constraints).

To overcome this limitation, in the next we introduce $SRBH^2$, a passive monitoring system able to detect *MTU dependent SR Black Holes* simply relying on data plane information gathered by the network devices.

4   *Segment Routing Black Holes Hound* Algorithm

The idea behind *Segment Routing Black Holes Hound* ($SRBH^2$) is to use a passive approach: instead of actively probe destination addresses, it searches possible black holes by analyzing passive measurements collected at the data plane. Specifically, the proposed algorithm is based on the Segment Routing

Traffic Counters [6], a new OAM feature available in SR capable routers. SR
Traffic Counters, described in [17], allow to collect statistics on the processed
SR traffic on SR capable nodes. Among the different types of available counters,
we consider only the Base Pcounters ones. A Base Pcounter is able to account
the amount of traffic crossing a given node and having a specific SID as active
segment. As an example, referring to the scenario reported in Fig. 1, the Base
Pcounter at node $C1$ related to the active segment $C2$ is accounting all the
packets sent from host A to host B. With the notation $y_B(i, a)$ we refer to
the value assumed by the Base Pcounter instantiated at node i and accounting
packets having a as active segment.

$SRBH^2$ is a two steps algorithm which is able to detect *MTU dependent SR
Black Hole* based on the analysis of the SR Traffic Counters. In the first step, the
link where the *MTU dependent SR Black Hole* occurs is detected; in the second
step, the affected flow (flows) is (are) found. $SRBH^2$ exploits the relationships
existing among different counters to detect a possible *MTU dependent SR Black
Hole*.

To better explain the algorithm, let's start describing the effect of an *MTU
dependent SR Black Hole* on SR Traffic Counters. Let $\mathcal{G}(\mathcal{N}, \mathcal{L})$ be the direct
graph modeling the network, where $\mathcal{N}$ is the set of nodes, and $\mathcal{L}$ is the set of links.
The link connecting nodes i and j is referred to as $l_{i,j}$. A traffic counter $y_L(l_{i,j})$
(classical link count) is associated to each link and measures the aggregated
amount of data sent through it. In the considered scenario, when the node i
receives a packet having the SID a as active segment, the shortest path between
nodes i and a is used to forward it. This path is represented to as $P_{i,a}$. A real
number ranging between 0 and 1 is used to specify the percentage of the total
traffic received at node i and destined at node a that is forwarded over the link
$l_{i,j}$. This quantity is referred to as $P_{i,a}(l_{i,j})$, namely routing split coefficient,
and can be easily determined by knowing the IGP paths used in the underlay
IPv6 network. Using the presented notation, considering the interface counter
associated to the link $l_{i,j}$, the following condition holds:

$$y_L(l_{i,j}) = \sum_{a \in \mathcal{N}} P_{i,a}(l_{i,j}) y_B(i, a) \tag{1}$$

Equation (1) shows that the traffic sent over the link $l_{i,j}$ is equal to the
summation of the traffic received by the node i and that it is forwarded over
the target link. Specifically, each Base Pcounter $y_B(i, a)$ accounts the traffic
received by node i having a as active segment. This quantity is then multiplied
for the routing coefficient $P_{i,a}(l_{i,j})$. In normal condition, i.e., in absence of *MTU
dependent SR Black Hole*, Eq. 1 holds, since all the received traffic is regularly
forwarded over the output link. On the contrary, if the MTU constraint over the
link $l_{i,j}$ is violated, then a fraction of the received packets is dropped, leading to
a discrepancy in the two sides of Eq. 1.

After the detection of the link affected by the *MTU dependent SR Black
Hole*, the next step consists in determining the traffic flow involved. It is clear
that the involved flow is one of the flows routed over the target link, referred to

as suspected flows. SR Traffic Counters can be exploited to reduce the size of the set of suspected flows. In particular, referring to as $\mathcal{F}(l_{i,j})$ to the set of flows crossing link $l_{i,j}$, this can be further divided into disjoint subsets, one for each possible active segment. Then, in order to check whether the affected flow is in the subset related to the active segment a, the validity of the following condition has to be verified:

$$y_B(i,a) = \sum_{n \in \mathcal{N}_i} P_{n,a}(l_{n,i}) y_B(n,a) \tag{2}$$

where i is the ID of the node where the target link is entering on, and $\mathcal{N}_i$ represents the set of neighbors of the node i. Equation 2 shows the relationship arising between Base Pcounters instantiated in neighboring nodes. Specifically, considering the active segment a, the Base Pcounter at node i accounts all the packets forwarded by its neighbors, properly weighed by the routing split coefficients. In presence of an *MTU dependent SR Black Hole* on the link between the node i and one of its neighbors, there will be a mismatch on the value of the SR counters due to the fact that some of the traffic received at the previous hop are dropped. In this way it is possible to detect the SID used as active segment by the target flow; the set of suspected flows can be now reduced to the set of flows having the previous SID as active segment.

To summarize, $SRBH^2$ works as follows: i) for each link, Eq. (1) is checked to detect *MTU dependent SR Black Holes*; ii) for each link violating Eq. (1), the SID used by the flow affected by the *MTU dependent SR Black Hole* is identified using Eq. (2). The final outcome of the procedure will be a set of suspected flows. In the performance evaluation, the algorithm will be characterized by its ability in detecting the link affected by a *MTU dependent SR Black Hole*, and by the size of the set of suspected flows obtained.

Before concluding, it is important to remark that, as far as Eqs. 1 and 2 are concerned, the presented approach is valid under the hypothesis that the only source of packet loss is due to the presence of an *MTU dependent SR Black Hole*. Anyway, the applicability of $SRBH^2$ can be easily extended also to cases where multiple causes of packet loss (such as congestion and other types of failures) exist. This can be done by considering different available statistics (all types of SR Traffic Counters, and packet drop on router interfaces), and using dedicated monitoring system (e.g. the approach proposed in [1]) to detect different causes of packet loss causes, such as physical failures, running in parallel to $SRBH^2$. The analysis of such an architecture is currently under investigation.

5   Performance Evaluation

In this section we provide a preliminary performance evaluation of $SRBH^2$ algorithm when used to detect *MTU dependent SR Black Holes* (simply blackholes in the following). The experiments are conducted considering the France Telecom (38 nodes and 144 links) and the Brain (161 nodes and 166 links) networks [16] as the underlay IPv6. For these networks, real traffic matrices are available. Shortest paths are calculated using the Dijkstra algorithm and assuming all the links having the same cost. Segment lists are enforced at the ingress node on the incoming traffic flows according to the egress node and a specific traffic class. A single traffic class is considered, i.e., the best effort one. Best effort traffic is routed over IGP paths, consequently, the resulting SL is composed by a single SID indicating the egress node.

The performance evaluation is carried out in a simulated environment, using Matlab. A blackhole is simulated by dropping all the packet of a target flow when transiting over a target link. For each target flow a set of simulation is run. In each simulation the location of the blackhole is chosen among the links belonging to the path followed by the target flow. In order to get more realistic results, the first link of the path (i.e., the one leaving from the source node) is not considered as a possible location for the black hole, since the source node knows the value of the MTU over this link, then can fragment the packets accordingly. Finally, it is assumed that the only cause of packet loss is due to the presence of a black hole.

In the next, it is presented a set of experiments aiming at evaluating the performance of the $SRBH^2$ algorithm. The performance are evaluated in terms of *precision* and *recall*. The *precision* represents the fraction of correctly detected blackhole with respect to the total number of black holes (both true and false) computed by $SRBH^2$. The *recall* express the percentage of blackholes that have been correctly detected.

These parameters are used to determine the ability of $SRBH^2$ in assessing the link (first analysis) where the black hole occurs, and the flow (second analysis) causing it. Regarding the assessment of the link where the failure is happening, we have found that $SRBH^2$ always gets *precision* and *recall* equal to 100%. It means that the proposed algorithm is always able to isolate the link affected by the blackhole from the rest of the devices, highly reducing the portion of the network to be checked. This is a valuable result, since it highlights that blackholes can be identified and located with no errors using the proposed passive approach.

The second analysis regards the target flow: it has to be found among the ones routed over the link affected by the black hole. To give an idea of the impact of Eq. 2 on the performance of $SRBH^2$, we have reported in Fig. 2 the *precision* obtained by reporting as output the full set of flows traversing the affected link (referred to as *precision 2*) and in case the suspects are limited only to the subset of flows which violate the condition of Eq. 2 (referred to as *precision 1*). As it can be seen from the results reported in Fig. 2, the *recall* achieved by the proposed approach is always equal to 100%, which means that the affected flow is always among the suspected ones (our algorithm has no false negatives). This finding is encouraging to support the validity of $SRBH^2$ as a method to detect blackholes. Concerning the *precision*, from the analysis reported in Fig. 2 emerges that: i) it is affected by the size of the network, ii) $SRBH^2$ has a poor *precision* in assessing the ID of the flow affected by the black hole, iii) in some cases it reaches 100%, and iv) in general there are some failures that are more difficult to analyze. Furthermore it is clear the improvement on the *precision*

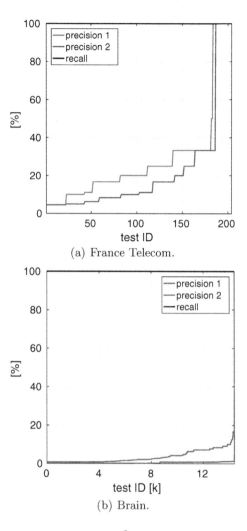

(a) France Telecom.

(b) Brain.

Fig. 2. Precision and Recall of $SRBH^2$ in the detection of the affected flow.

due to the use of Eq. 2 to reduce the size of the set of suspected flows. With reference to point one, the relation between the size of the network and the obtained *precision* is due to the number of traffic flows. When this last grows, the average number of flows routed over a link increases, making harder to detect the target flow by means of Eq. 2.

In order to better investigate the results, in Fig. 3 the CDF of the percentage of suspected flows detected by $SRBH^2$ is reported. In the France Telecom network (Fig. 3(a)), it is possible to see that in the 80% of the considered failures, the percentage of suspected flows is less than 6%. Considering that the total number of flows in this case is equal to 154, then the set of suspected flows (which contain the true positive) has a size of 22 flows at maximum for the 80%

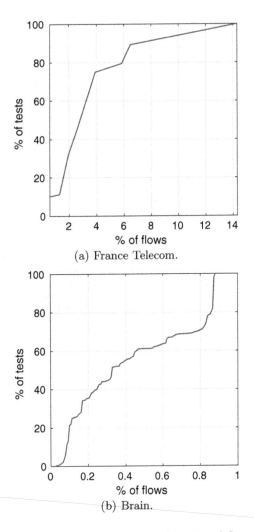

(a) France Telecom.

(b) Brain.

Fig. 3. CDF of the percentage of suspected flows.

of cases. Similarly, in case of the Brain network (Fig. 3(b)), the percentage of suspected flows never exceeds the 1% of the overall number of flows. Since in this case there are more than $14K$ traffic relations, then the number of false positives is equal to 140. Thus, even though the *precision* is poor, actually $SRBH^2$ is greatly reducing the size of the suspected flows. To conclude, the performance evaluation has shown that $SRBH^2$ algorithm is a suitable approach (due to the high recall) for the detection of *MTU dependent SR Black Holes*, even if further work is needed to improve the precision.

6 Conclusion

In this paper we demonstrated that a new type of failure related to MTU can be generated in a SRv6 network. This failure, referred to as *MTU dependent SR Black Hole*, cannot be detect by known monitoring tools since are based on active probing. We proposed $SRBH^2$, a passive measurement based algorithm able to detect links affected by blackholes only looking at SRv6 Traffic Counters collected by network SR capable routers. The main outcome of performance evaluation is that our solution is always able to detect the link affected by the blackhole; it is still not possible to univocally detect the flow causing the blackhole, but a significant reduction of the number of suspected flows is obtained.

As a future steps we aim at improving the precision and enable the use of $SRBH^2$ when multiple sources of packet loss are present. This imply to consider a wider set of network statistics, such as different SR Traffic Counters and interface counters, and to run in parallel additional monitoring tools.

References

1. Aubry, F., Lebrun, D., Vissicchio, S., Khong, M.T., Deville, Y., Bonaventure, O.: Scmon: leveraging segment routing to improve network monitoring. In: IEEE INFOCOM 2016-The 35th Annual IEEE International Conference on Computer Communications. IEEE (2016)
2. de Boer, M., Bosma, J.: Discovering path MTU black holes in the Internet using ripe atlas (2012)
3. Custura, A., Fairhurst, G., Learmonth, I.: Exploring usable path MTU in the internet. In: 2018 Network Traffic Measurement and Analysis Conference (TMA), pp. 1–8. IEEE (2018)
4. Deering, D.S.E., Hinden, B.: Internet protocol, version 6 (IPv6) specification. RFC 8200, July 2017. https://doi.org/10.17487/RFC8200
5. Desmouceaux, Y., Townsley, M., Clausen, T.H.: Zero-loss virtual machine migration with IPv6 segment routing. In: 2018 14th International Conference on Network and Service Management (CNSM), pp. 420–425. IEEE (2018)
6. Filsfils, C., Ali, Z., Horneffer, M., Voyer, D., Durrani, M., Raszuk, R.: Segment routing traffic accounting counters. Internet-Draft draft-filsfils-spring-sr-traffic-counters-00, Internet Engineering Task Force, June 2018. Work in Progress
7. Filsfils, C., Dukes, D., Previdi, S., Leddy, J., Matsushima, S., Voyer, D.: IPv6 segment routing header (SRH). Internet-Draft draft-ietf-6man-segment-routing-header-21, Internet Engineering Task Force, June 2019. Work in Progress
8. Filsfils, C., Previdi, S., Ginsberg, L., Decraene, B., Litkowski, S., Shakir, R.: Segment routing architecture. RFC 8402, July 2018. https://doi.org/10.17487/RFC8402
9. Kompella, R.R., Yates, J., Greenberg, A., Snoeren, A.C.: Detection and localization of network black holes. In: IEEE INFOCOM 2007–26th IEEE International Conference on Computer Communications, pp. 2180–2188. IEEE (2007)
10. Kreibich, C., Weaver, N., Nechaev, B., Paxson, V.: Netalyzr: illuminating the edge network. In: Proceedings of the 10th ACM SIGCOMM Conference on Internet Measurement, pp. 246–259. ACM (2010)

11. Kumar, N., Pignataro, C., Swallow, G., Akiya, N., Kini, S., Chen, M.: Label switched path (LSP) ping/traceroute for segment routing (SR) IGP-prefix and IGP-adjacency segment identifiers (SIDs) with MPLS data planes. RFC 8287, December 2017. https://doi.org/10.17487/RFC8287

12. Luckie, M., Cho, K., Owens, B.: Inferring and debugging path MTU discovery failures. In: Proceedings of the 5th ACM SIGCOMM conference on Internet Measurement. USENIX Association (2005)

13. Luckie, M., Stasiewicz, B.: Measuring path MTU discovery behaviour. In: Proceedings of the 10th ACM SIGCOMM Conference on Internet Measurement, pp. 102–108. ACM (2010)

14. Mathis, M., Heffner, J.: Packetization layer path MTU discovery. RFC 4821, March 2007. https://doi.org/10.17487/RFC4821

15. McCann, J., Deering, S.E., Mogul, J., Hinden, B.: Path MTU discovery for IP version 6. RFC 8201, July 2017. https://doi.org/10.17487/RFC8201

16. Orlowski, S., Pióro, M., Tomaszewski, A., Wessäly, R.: SNDlib 1.0-survivable network design library. In: Proceedings of the 3rd International Network Optimization Conference (INOC 2007), Spa, Belgium, April 2007

17. Polverini, M., Cianfrani, A., Listanti, M.: Interface counters in segment routing v6: a powerful instrument for traffic matrix assessment. In: 2018 9th International Conference on the Network of the Future (NOF), pp. 76–82, November 2018. https://doi.org/10.1109/NOF.2018.8597768

18. Staff, R.: Ripe atlas: a global internet measurement network. Internet Protocol J. 18(3), 1–31 (2015)

19. Ventre, P.L., et al.: Segment routing: a comprehensive survey of research activities, standardization efforts and implementation results. CoRR abs/1904.03471 (2019). http://arxiv.org/abs/1904.03471

Network Select in 5G Heterogeneous Environment by M-F-U Hybrid Algorithm

Haodong Liu[✉], Fang Cheng, and Bingguang Deng

School of Communication and Information Engineering, Chongqing University
of Posts and Telecommunications, Chongqing, China
1423072611@qq.com

Abstract. Heterogeneous network convergence, as the current development
trend of wireless communication network systems, has attracted the attention
and research of many experts. In order to solve the problem of incomplete
handover decision parameters and single decision algorithm in 5G heterogeneous network handover system, an M-F-U hybrid algorithm based on the
multiple attribute decision making (MADM), fuzzy logic, and utility function is
proposed. First, the decision parameters are divided into two parts, which are
calculated by the MADM and fuzzy logic methods, the results obtained as the
input of the utility function, secondly, the risk attitude coefficient is introduced
into the utility function to describe the user's tolerance for switching risk, then,
Then calculate the value of the comprehensive utility function, and finally,
choose the optimal network scheme according to the comprehensive utility
value. The simulation results show that compared with the traditional algorithm,
the M-F-U algorithm can improve the handover accuracy, reduce the number of
handovers, and complete the switching decision in a short time.

Keywords: 5G heterogeneous network · MADM · Fuzzy logic · Utility
function · The optimal network scheme

1 Introduction

There has been an exponential growth in mobile data usage over the last 15 years (over
400 million times) that is expected to go up nearly 6-fold between 2017–2022 reaching
77 Exabyte per month by 2022 [1]. In addition to providing high data rates, it is equally
important to provide reliable handover (HO) mechanisms as this directly impacts on the
perceived quality of experience (QoE) for the end-user [2]. During the communication
process, mobile terminals inevitably cross the cell boundaries of networks of the same
structure or networks of different structures, which is more frequent in the network
environment of multi-network convergence. As one of the key technologies of communication networks, handover technology has a great influence on improving the
effectiveness and reliability of the entire system, and plays an important role in modern
communication systems. The whole handover process is divided into three stages,
namely handover discovery, handover decision and handover execution. The handover
decision phase is the most important stage to solve the handover problem. Therefore,
how to improve the handover decision algorithm, optimize the handover execution

© ICST Institute for Computer Sciences, Social Informatics and Telecommunications Engineering 2021
Published by Springer Nature Switzerland AG 2021. All Rights Reserved
H. Gao et al. (Eds.): ChinaCom 2020, LNICST 352, pp. 351–367, 2021.
https://doi.org/10.1007/978-3-030-67720-6_24

mechanism, provide the best experience for users at the lowest price, and achieve the purpose of ensuring communication quality and service requirements has important practical significance [3].

Based on the development in recent decades, domestic and foreign scholars have never stopped the research on network switching algorithms. Based on mobile behavior, literature [4] categorizes frequent handover-experience users as either fast-moving or ping-pong users. Fast-moving users are then handed over to the macro layer, and ping-pong users are managed by adjustment of handover parameters. The method that leverages device-level caching along with the capabilities of dual-mode base stations to minimize handover failures has proposed in [5]. Literature [6] combines the Analytical Hierarchy Process (AHP) technique to obtain the weight of the handover metrics and the Grey Relational Analysis (GRA) method to rank the available cells for the best handover target. References [7–9], including the above references, increase the algorithm complexity to a certain extent.

With the increase of decision parameters, different users and the types of services required by different users have different requirements for decision parameters. On the other hand, the computational complexity of their switching algorithms also greatly increases. Fuzzy algorithm can comprehensively consider many parameters, so it has been widely used in various fields. However, if a single decision parameter or a single processing method is still used to process the decision parameters [10, 11], once the decision parameters are increased, the number of fuzzy rules will increase by a geometric multiple.

Therefore, this paper proposes an M-F-U algorithm based on multi-attribute and sub-module, the algorithm divides eight parameters into several modules, each module uses different processing methods according to its characteristics. The algorithm introduces the S-type utility function into the decision-making process, use the S-type utility function to express user preferences, assist the decision-making system to make more reasonable decisions, reduce decision risks, and make decisions more scientific and effective. according to the principle of utility maximization, it provides decision-makers with the choice of the highest utility, which solves the problems of long handover time, high algorithm complexity and poor handover performance in the traditional network.

The rest of this paper is outlined as follows. In the next section, Heterogeneous network model will be discussed. the proposed M-F-U algorithm are provided in Sect. 3. Section 4 gives the simulation results and analysis. Finally, Sect. 5 concludes this paper.

2 System Model

This paper considers a common heterogeneous network model. As shown in Fig. 1, the network covers WLAN, WiMAX, 5G and LTE from inside to outside in the same area, serving as network access points for users. Users are randomly distributed in a heterogeneous network and follow a certain speed. Move in random directions.

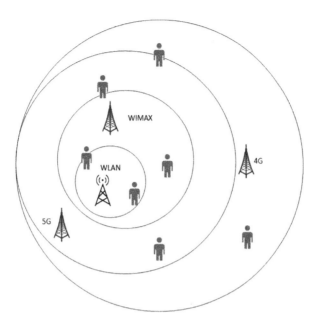

Fig. 1. Heterogeneous network model

Based on the above model, the received signal strength (RSS) of the system can be expressed as:

$$RSS_i = S_i - S_{loss} - \delta \tag{1}$$

Where S_i indicates the transmit power, δ is the shadow fading, S_{loss} is the path loss, and can be given by:

$$S_{loss} = S + 10 * n * \log(d) \tag{2}$$

Where S is the constant path loss, n stands for path loss index, and the distance between the user and the network center is d. When the user moves from the edge of the network to the other end, the moving speed is V, and the coverage radius of the network is R. According to [12], the formula for the probability of handover failure is

$$p_f = \frac{2}{\pi} \left(\sin^{-1}\left(\frac{V\tau_i}{2R}\right) - \sin^{-1}(\frac{VT_1}{2R}) \right) (0 \le T_1 \le \tau_i) \tag{3}$$

T_1 is the time threshold. When the residence time of the user entering the network is calculated to be greater than T_1, the handover is started, and when the user's movement time in the cell is less than the required handover delay time, considered handover failure.

When the user's movement time in the cell is less than the sum of the handover time when entering and leaving the cell, an unnecessary handover occurs. For the unnecessary handover probability, the time threshold T_2 is introduced, the probability of non-essential switching is as follows:

$$P_f = \frac{2}{\pi} \left(\sin^{-1} \left(\frac{V(\tau_i + \tau_0)}{2R} \right) - \sin^{-1}(\frac{VT_2}{2R}) \right) (0 \leq T_2 \leq \tau_i) \qquad (4)$$

Where τ_o indicates the delay in switching from the current network to other networks, the handoff delay from the other network to current network is τ_i.

3   M-F-U Algorithm

3.1   Handover Decision Process

The network selection algorithm of M-F-U is shown in Fig. 2. In the selection of parameters, the network conditions and user selection are comprehensively considered here, and some parameters are selected as switching decision indicators. According to the characteristics of the parameters, MADM and fuzzy logic inference methods are used respectively, and the final network selection depends on the value of the utility function. Before network selection, it is necessary to make a preliminary screening based on the mobile terminal speed and network load. The purpose is to eliminate the network that does not meet the standard and reduce the unnecessary switching, and shorten the calculation time of algorithm.

Screening of mobile speed, Compare the current speed of the user with the maximum movement speed. The calculation of the maximum moving speed can refer to [14]. If it is greater than the maximum mobile speed supported by the network, then remove the network from the list of candidate networks.

Network load screening, the network load reflects the usage of users in a network. If the number of network users exceeds the maximum available number of the network, this network is saturated and need to select the candidate network.

Figure 2 illustrates the specific handover decision process.

This paper abandons the method of single processing decision parameters, and divides the handover decision model into Network QoS (NQ) module, Spend Engine (SE) module, Benefit Engine (BE) module and utility function module.

The NQ model includes the calculation of received signal strength (RSS), transmission rate, delay, and packet loss. Based on the highly sensitive characteristics of the Technique for Order Preference by Similarity to an Ideal Solution (TOPSIS) method for data, this paper takes the TOPSIS method as the basis of judgment in NQ model. The BE and SE models include fuzzy inferences about velocity, power consumption, coverage and battery life. For SE and BE model, most of the indicators contained in it can be expressed by degree quantifier, which are processed by fuzzy reasoning. Finally, the S-type utility function is introduced to express user preferences, and the ideal network is selected according to the utility function.

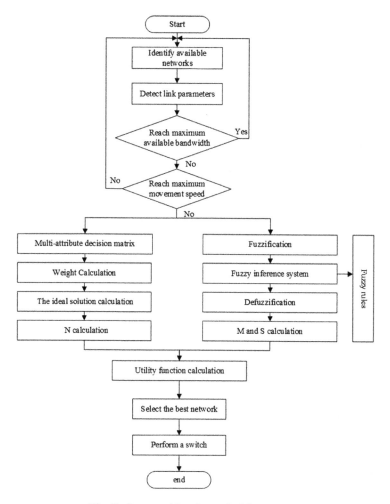

Fig. 2. Improved handover decision process

3.2 NQ Module

Consider the HetNets where the user is located with m available network alternatives, and n network parameters are selected. The candidate network set can be expressed as $S = \{S_1, S_2, \ldots, S_m\}$, $G = \{G_1, G_2, \ldots, G_n\}$ is the decision parameter set, The NQ module can be formulated as a multi-attribute decision matrix A as follows

$$A = \begin{bmatrix} d_{11} & d_{12} & \cdots & d_{1j} & d_{1n} \\ d_{21} & d_{22} & \cdots & d_{2j} & d_{2n} \\ \vdots & \vdots & \ddots & \vdots & \vdots \\ d_{i1} & d_{i2} & \cdots & d_{ij} & d_{in} \\ d_{m1} & d_{m2} & \cdots & d_{mj} & d_{mn} \end{bmatrix} \qquad (5)$$

Where d_{ij} indicates the value of the decision parameter j in the network candidate network i.

Due to the difference in measurement methods between different parameters, normalization is essential. For benefit parameter, the normalization formula is as follows:

$$E = \frac{a_i - a_{min}}{a_{max} - a_{min}} \tag{6}$$

For spend parameter, the normalized formula is

$$E = \frac{a_{max} - a_i}{a_{max} - a_{min}} \tag{7}$$

The value range of the attribute obtained by this normalization method is range from 0 to 1, which can also reflect the performance of the attribute value to a certain extent.

Then, the standardized multi-attribute decision-making matrix B can be expressed as:

$$B = \begin{bmatrix} b_{11} & b_{12} & \cdots & b_{1j} & b_{1n} \\ b_{21} & b_{22} & \cdots & b_{2j} & b_{2n} \\ \vdots & \vdots & \cdots & \vdots & \vdots \\ b_{i1} & b_{i2} & \cdots & b_{ij} & b_{in} \\ b_{m1} & b_{m2} & \cdots & b_{mj} & b_{mn} \end{bmatrix} \tag{8}$$

The value from matrix B is the Normalized value.

Determine the weight values of different decision parameters. According to the preference of QoS performance under different service types, the weight relations can be expressed as:

$$W_i = \frac{\left(\prod_{j=1}^{n} b_{ij}\right)^{\frac{1}{n}}}{\sum_{i=1}^{n} (\prod_{j=1}^{n} b_{ij})^{\frac{1}{n}}} \tag{9}$$

Thus, the set of network parameter is given by $W = \{w_1, w_2 \ldots, w_N\}$.

After obtaining the weight of each handover decision parameter, the consistency test should be carried out on the decision matrix C. As shown in Table 1, the consistency ratio (CR) is calculated regarding the value of the random consistency index (RI). If $CR < 0.1$, the inconsistency of the matrix is within the allowable range, and the weight W derived from the matrix is available.

Table 1. RI values

N	1	2	3	4	5	6	7	8	9
RI	0	0	0.58	0.90	1.12	1.24	1.32	1.41	1.45

The standard decision matrix $C = (C_{ij})_{mn}$ with weights is expressed as:

$$C = \begin{bmatrix} \omega_1 b_{11} & \omega_2 b_{12} & \cdots & \omega_i b_{1j} & \omega_m b_{1n} \\ \omega_1 b_{21} & \omega_2 b_{22} & \cdots & \omega_i b_{2j} & \omega_m b_{2n} \\ \vdots & \vdots & \cdots & \vdots & \vdots \\ \omega_1 b_{i1} & \omega_2 b_{i2} & \cdots & \omega_i b_{ij} & \omega_m b_{in} \\ \omega_1 b_{m1} & \omega_2 b_{m2} & \cdots & \omega_i b_{mj} & \omega_m b_{mn} \end{bmatrix} \tag{10}$$

In the standard weighted decision matrix $C_{ij} = \omega_j b_{ij}$, the positive ideal solution C^+ and negative ideal solution C^- of matrix C can be calculated by the following formula:

$$C_j^+ = \max\{\omega_j b_{ij} | i = 1,2,\cdots,m\} = b_j^+ \omega_j \tag{11}$$

$$C_j^- = \min\{\omega_j b_{ij} | i = 1,2,\cdots,m\} = b_j^- \omega_j \tag{12}$$

Where b_j^+ in it is the optimal value of column j in the matrix B, b_j^- is the worst value of column of the matrix B, they can be expressed as:

$$b_j^+ = \max\{b_{ij} | i = 1,2,\cdots,m\} \tag{13}$$

$$b_j^- = \min\{b_{ij} | i = 1,2,\cdots,m\} \tag{14}$$

Referring to (13) and (14), the final positive ideal solution and negative ideal solution can be calculated according (11) (12) as follows:

$$C^+ = \left(b_1^+ \omega_1, b_2^+ \omega_2, \cdots, b_n^+ \omega_n\right) \tag{15}$$

$$C^- = \left(b_1^- \omega_1, b_2^- \omega_2, \cdots, b_n^- \omega_n\right) \tag{16}$$

Then, the distance of each value in the standard weighted decision matrix C to the positive ideal solution and the negative ideal solution, which can be expressed by the sum of squares of errors:

$$d^+ = \sum_{j=1}^{n} \left(C_{ij} - C_j^+\right)^2 = \sum_{j=1}^{n} \omega_j^2 \left(b_{ij} - b_j^+\right)^2, i = 1,2,\cdots,m \tag{17}$$

$$d^- = \sum_{j=1}^{n} \left(C_{ij} - C_j^-\right)^2 = \sum_{j=1}^{n} \omega_j^2 \left(b_{ij} - b_j^-\right)^2, i = 1,2,\cdots,m \tag{18}$$

According to (17) and (18), Calculate the relative proximity to the ideal solution. the final output $N_{(value)}$ is calculated as follows:

$$N = \frac{d_i^-}{d_i^+ + d_i^-}, (0 \leq N \leq 1)$$ (19)

As NQ module output, $N_{(value)}$ determines the network performance.

3.3 BE and SE Modules

For SE modules, two input variables (Velocity and Power consumption) are assigned as its inputs and $S_{(value)}$ as its output using the FIS Editor. which is shown in Fig. 3. In Fig. 3, the $V_{(value)}$ is defined within the range of 0 to 50 km/h, where the range of $P_{(value)}$ is defined from 0 to 3700 W and $M_{(value)}$ is from 0 to 1. Three fuzzy membership functions (L, M, and H) with trapezoidal shapes are used to indicate each of the input variables. While, in the case of output variable, three fuzzy membership functions (L, M, and H) with triangular shape. Since, there are two variables or two fuzzy sets at the input of SE modules, and each of them has three FMFs, need $3^2 = 9$ fuzzy rules to specify the behavior of the mentioned fuzzy engine.

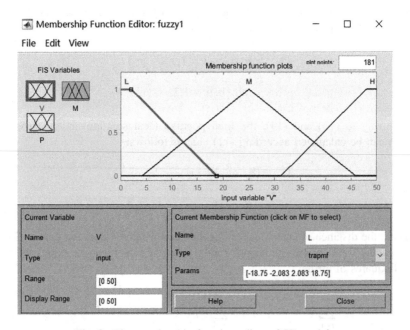

Fig. 3. The membership function editor of SE module

The applied rules in this fuzzy engine are listed in Table 2, where indicate the changes of $S_{(value)}$ or output fuzzy membership functions with different values of input fuzzy membership functions.

Table 2. SE module fuzzy rules

Number	Velocity	Power consumption	Output
1	Low	Low	High
2	Low	Medium	Medium
⋮	⋮	⋮	⋮
9	High	High	Low

Then, performing the defuzzification process, convert the aggregated fuzzified data back into crisp value by applying centroid method, which can be given by:

$$S = \frac{\int \tilde{S}E(x) \cdot x dx}{\int \tilde{S}E(x) dx} \tag{20}$$

Where x is a continuously changing quantity in the value range of the fuzzy set, represents the membership function of fuzzy sets, the $S_{(value)}$ is the final output value of the SE module.

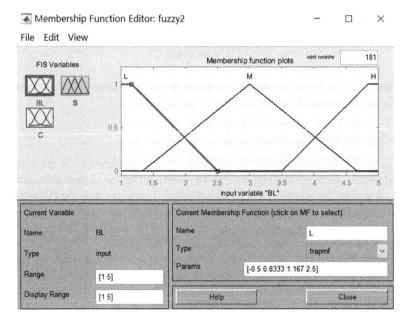

Fig. 4. The membership function editor of BE module

Similarly, two input variables (Battery Life and Coverage) are assigned as its inputs and $B_{(value)}$ as its output using the FIS Editor. which is shown in Fig. 4. In Fig. 4, the $BL_{(value)}$ is defined within the range of 1 to 5 day, where the range of $C_{(value)}$ is defined from 0 to 2000 m and $B_{(value)}$ is from 0 to 1. Three fuzzy membership functions (L, M, and H) with trapezoidal shapes are used to indicate each of the input variables. While, in the case of output variable, three fuzzy membership functions (L, M, and H) with triangular shape. Since, there are two variables or two fuzzy sets at the input of BE modules, and each of them has three FMFs, it needs $3^2 = 9$ fuzzy rules to specify the behavior of the mentioned fuzzy engine.

The applied rules in this fuzzy engine are listed in Table 3, where indicate the changes of $B_{(value)}$ or output fuzzy membership functions with different values of input fuzzy membership functions.

Table 3. BE module fuzzy rules

Number	Battery life	Coverage	Output
1	Low	Low	Low
2	Low	Medium	Low
⋮	⋮	⋮	⋮
9	High	High	High

Then, performing the defuzzification process, convert the aggregated fuzzified data back into crisp value by applying centroid method, which can be given by

$$B = \frac{\int \tilde{BE}(y) \cdot y dy}{\int \tilde{BE}(y) dy} \tag{21}$$

Where y is a continuously changing quantity in the value range of the fuzzy set. The $B_{(value)}$ is the final output value of the BE module.

3.4   S-Type Utility Function Module

Since the utility function allows user to express their preference for each standard involved in the decision-making process by determining the degree of satisfaction, which In line with the problem of network selection. This paper uses the utility function as the criterion for evaluating degree of satisfaction. Three input variables (M(value), S (value), and B(value)) are assigned as the inputs of utility function. The multi-attribute decision steps under the S-type utility function are as follows:

Consider $\theta = (\theta_1, \theta_2, \cdots, \theta_l)$ is the network's risk state set, such as handover failed, unnecessary handover, etc. P_t express as the probability of risk occurs which satisfies $0 \leq P_t \leq 1$, the evaluation information of the status θ_t of each attribute in the network is $\left(\mu_{ij}^t, \sigma_{ij}^t\right)$, a fuzzy number. Thus, the risk state matrix can be express as:

$$D_1 = \left(\mu_{ij}^1, \sigma_{ij}^1\right)_{m \times n}, D_2 = \left(\mu_{ij}^2, \sigma_{ij}^2\right)_{m \times n}, \cdots, D_l = \left(\mu_{ij}^l, \sigma_{ij}^l\right)_{m \times n} \tag{22}$$

Step 1: Change the risk state decision matrix into scoring function matrix, $\left(\mu_{ij}^t, \sigma_{ij}^t\right)$ can be convert to real number S_{ij}^t as follow:

$$S_{ij} = \frac{\exp(\mu(x) - \sigma(x))}{\pi(x) + 1} \tag{23}$$

Where $\mu(x)$, $\sigma(x)$, and $\pi(x) = 1 - \mu(x) - \sigma(x)$ expressed as membership, non-membership, and hesitation for fuzzy sets of evaluation information, respectively. The scoring function is given by:

$$S_1 = \left(s_{ij}^1\right)_{m \times n}, S_2 = \left(s_{ij}^2\right)_{m \times n}, \cdots, S_l = \left(s_{ij}^l\right)_{m \times n} \tag{24}$$

In the process of switching decisions, the decision system has corresponding expected values for each attribute. This expected value can be used as a reference point for system decision-making, and the reference point reflects user preferences well. When the attribute value is higher than the reference point of the decision system for this attribute, it will be regarded as a loss to the user, and the higher the reference point, the greater the loss. Conversely, when the attribute value is lower than the reference point of the decision system for the attribute, the user will consider it to be profitable, and the lower the reference point, the greater the value of the income.

Step 2: To calculate the value matrix, this article takes the mean as a reference point and calculates the utility value of each attribute in the network to obtain the value matrix:

$$U = \left(u_{ij}\right)_{m \times n} = \sum_{t=1}^{l} \omega(P_t) u\left(s_{ij}^t\right) \tag{25}$$

Where $u\left(s_{ij}^t\right)$ indicates the value function which reflect the subjective utility value of the decision system, the formula is expressed as:

$$u\left(s_{ij}^t\right) = \begin{cases} \left(s_{ij}^t - \bar{s}_j^t\right)^\alpha & s_{ij}^t \geq \bar{s}_j^t \\ -\theta\left(\bar{s}_j^t - s_{ij}^t\right)^\beta & s_{ij}^t \leq \bar{s}_j^t \end{cases} \tag{26}$$

Where α, β indicate the risk attitude coefficient which satisfies $0 < \alpha, \beta < 1$, as the increase of α and β indicate that the degree of risk that users can take, and θ show the loss avoidance coefficient. Research shows that the decision results are basically consistent with the empirical data when $\alpha = \beta = 0.88, \theta = 2.25$ [13].

Where $\omega(p)$ is a monotonically increasing probability weight function which reflects the overestimation or underestimation of risk events as follow:

$$\omega(P) = \begin{cases} \frac{P^\chi}{(P^\chi + (1-P^\chi)^\chi)^{1/\chi}} & s_{ij}^t \geq \bar{s}_j^t \\ \frac{P^\delta}{\left(P^\delta + (1-P^\delta)^\delta\right)^{1/\delta}} & s_{ij}^t \geq \bar{s}_j^t \end{cases} \tag{27}$$

Where P is the objective probability of risk occurrence, the coefficient of risk-return attitude and the coefficient of risk-loss attitude can be express as χ and δ, which satisfies $0 < \chi, \delta < 1$, and usually set to $\chi = 0.61, \delta = 0.72$.

Step 3: Calculate the attribute weights $G = \{M, S, B\}$ with square root method.

Step 4: Calculate the comprehensive utility value. The greater the comprehensive utility value, the better the network scheme. The final formula given by:

$$u_i = 10 \sum_{j=1}^{n} G_j \cdot \omega_j \cdot \mu_{ij}, i = 1, 2, \cdots, m \tag{28}$$

If the utility value of the candidate network is almost the same as the utility value of the original network, it indicates that the user has the same preference for the two networks, and in this case, the handover is not performed. If the value of the original network's comprehensive utility function is the largest of all available networks, the user prefers to maintain the original network service without being forced by external factors.

4 Simulation and Analysis

In this section, we evaluate the performance of the proposed heterogeneous network handover algorithm in a heterogeneous network environment with multi-network convergence by MATLAB simulation platform. The simulation scenario of the heterogeneous network consists of 5G, LTE, WiMAX, and WLAN, the cell radius of 5G base stations (BS) is set to 500 m, the cell radius of LTE is 1000 m, WIMAX is 300 m, and WLAN is 75 m (Table 4).

Table 4. Network-side handover parameters

Access network	RSS (dBm)	Transmission rate (Mbit/s)	Delay (ms)	Drop rate (%)
5G	$-95 \sim -20$	200	20	0.015
LTE	$-95 \sim -20$	50	100	0.04
WiMAX	$-95 \sim -20$	55	100	0.04
WLAN	$-95 \sim -20$	1000	25	0.025

The 5G network base station is taken as the origin coordinate, and the rest available networks are randomly distributed with reference to 5G base stations. By consulting the existing network parameter standards in the literature, the simulation parameters are set as Table 5.

According to (6) and (7), normalizing RSS, transmission rate, delay, and drop rate. The final result is as follows:

Table 5. Normalized network side handover parameters

Access network	RSS (dBm)	Transmission rate (Mbit/s)	Delay (ms)	Drop rate (%)
5G	0.853	0.728	0.679	0.15
LTE	0.744	0.384	1	0.4
WiMAX	0.702	0.421	1	0.4
WLAN	0.54	1	0.623	0.25

Then, the output of NQ module is calculated according to the formula from (8) to (19), and the final $N_{(\text{value})}$ is

$$N = \{0.6834, 0.5514, 0.5411, 0.6003\}$$

Where the $N_{(\text{values})}$ are stored in the order of 5G, LTE, WiMAX, and WLAN.

The parameters in the SE and BE modules undergo the process of fuzzification, fuzzy reasoning, and defuzzification, $S_{(\text{values})}$, $B_{(\text{values})}$ are given by

$$S = \{0.4967, 0.1343, 0.3574, 0.5260\}$$

$$B = \{0.4610, 0.6194, 0.5053, 0.4769\}$$

Then, $N_{(\text{values})}$, $S_{(\text{values})}$, $B_{(\text{values})}$ are assigned as the input of the comprehensive utility function, the output of the utility function is the evaluated result, and the final obtained utility function values are sorted as follows:

$$u_i = \{0.0335, 0.013, 0.0149, 0.0205\}$$

From the simulation results, under the comprehensive consideration of network quality and user preferences, network users' preference for 5G is higher than that of other traditional networks, followed by WLAN, which is also the development trend of future communication networks.

Considering that different users have different preferences for network attributes, the difference of α, β and θ have an impact on the optimal network scheme, this section conducts a sensitivity analysis on the value parameter (α, β or θ) that reflects the user's risk attitude. As the values of the parameters are different, the change trend of the comprehensive utility value is shown in Fig. 5.

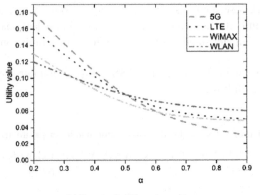

(a) Trend of utility value with α

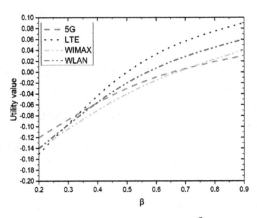

(b) Trend of utility value with β

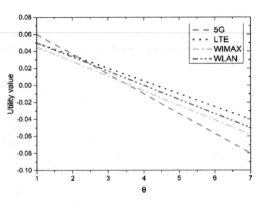

(c) Trend of utility value with θ

Fig. 5. Trend of utility value

As shown in Fig. 5(a), when α is increased to 0.5, the 5G network is no longer the optimal network scheme. As the user's risk tolerance increases, the utility value curve of the income area is flatter, indicating that the user is more biased risk. As shown in Fig. 5(b), when β is increased to 0.36, the LTE network becomes the best choice. For the increased risk tolerance of the loss area, the smoother the utility curve of the loss area, indicating that users also prefer risk. As shown in Fig. 5 (c), as the value of θ becomes larger, users are more sensitive to losses, indicating that users will be more conservative when facing risks. Therefore, as the value of and decreases and the value increases, users are more inclined to avoid risks.

In order to further evaluate the performance of the algorithm proposed in this paper, the M-F-U algorithm is compared with the traditional RSS-based handover algorithm and fuzzy algorithm. Consider that the network users are randomly distributed in the coverage area of networks. The performance comparison is shown in the following figure:

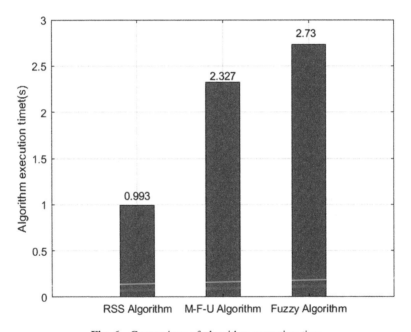

Fig. 6. Comparison of algorithm execution time

Figure 6 is a comparison of the execution time of algorithms based on the RSS algorithm, M-F-U algorithm, and fuzzy algorithm. It can be seen that due to the single handover decision parameter, the RSS-based handover decision algorithm has the shortest execution time, while under the condition that the same number of decision parameters, the M-F-U algorithm has a performance advantage of about 0.4 s in execution time compared with the single fuzzy logic algorithm.

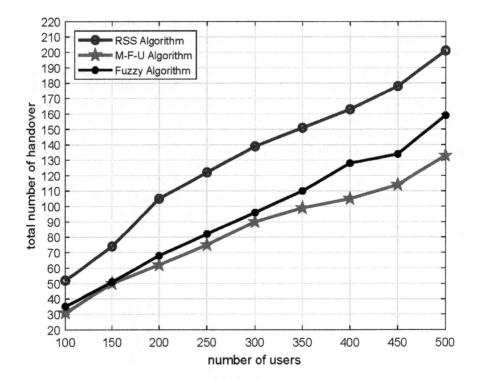

Fig. 7. Total handover times

In Fig. 7, When the number of users is different, It can be seen that the difference between the handover time of the MFU algorithm and the fuzzy algorithm is small when the number of users is about 300. Infer that in the case of the same decision model and the same decision parameters, processing complexity is almost equivalent. As the number of users increases in a certain range, the calculation of a single fuzzy algorithm will increase exponentially. The results show that the M-F-U algorithm can effectively reduce the number of handovers and improve the accuracy of handover.

5   Conclusions

In this paper, a hybrid algorithm based on traditional multi-attribute decision algorithm is proposed. The algorithm balances the weight of the network and the user. In the evaluation of the network, the value of the S-type utility function is introduced into the decision-making process. The S-type utility function is used to express user prefer-ences, which assists the decision-making system to make more reasonable decisions and reduce decision risk, so decision-making can be more scientific and effective. Compared with the traditional handover algorithm, the M-F-U algorithm takes a more comprehensive consideration of the parameters, effectively reducing the total number of handover and shortening the execution time of the algorithm.

Acknowledgment. This work comes from a major project in Chongqing, "R&D and application of 5G road test instruments (No. cstc2019jscx-zdztzxX0002)".

References

1. Cisco: Cisco Visual Networking Index: Global Mobile Data Traffic Forecast Update 2017–2022. Cisco. White Paper, San Jose, CA, USA (2019)
2. Tayyab, M., Gelabert, X., Jantti, R.: A survey on handover management: from LTE to NR. IEEE Access **7**, 118907–118930 (2019)
3. IMT-2020 (5G) Program: White paper on 5G Concept (2015)
4. Hasan, M.M., Kwon, S., Oh, S.: Frequent-handover mitigation in ultra-dense heterogeneous networks. IEEE Trans. Veh. Technol. **68**, 1035–1040 (2019)
5. Semiari, O., Saad, W., Bennis, M., Maham, B.: Caching meets millimeter wave communications for enhanced mobility management in 5G networks. IEEE Trans. Wireless Commun. **17**, 779–793 (2018)
6. Alhabo, M., Zhang, L.: GRA-based handover for dense small cells heterogeneous networks. IET Commun. **13**(13), 1928–1935 (2019)
7. Alhabo, M., Zhang, L.: Unnecessary handover minimization in two-tier heterogeneous networks. In: 13th Wireless On-demand Network systems and Services Conference, Jackson Hole, Wyoming, pp. 160–164. IEEE (2017)
8. Barmpounakis, S., Kaloxylos, A., Spapis, P., et al.: Context-aware, user-driven, network-controlled RAT selection for 5G networks. Comput. Netw. **113**(11), 124–147 (2017)
9. Calabuig, D., Barmpounakis, S., Gimenez, S., et al.: Resource and mobility management in the network layer of 5G cellular ultra-dense networks. IEEE Commun. Mag. **55**(6), 162–169 (2017)
10. Thumthawatworn, T., Tillapart, P., Santiprabhob, P.: Adaptive multi-fuzzy engines for handover decision in heterogeneous wireless networks. Wireless Pers. Commun. **93**(4), 1005–1026 (2017). https://doi.org/10.1007/s11277-017-3963-3
11. Zineb, A.B., Ayadi, M., Tabbane, S.: Fuzzy MADM based vertical handover algorithm for enhancing network performances. In: 23rd International Conference on Software, Telecommunications and Computer Networks (SoftCOM), SplitSupetar, Croatia, pp. 153–159. IEEE (2015)
12. Almosbahi, R., Elalem, M.: Optimization of coverage and handover for heterogeneous networks. Int. J. Adv. Res. **3**(3), 213–219 (2019)
13. Zhang: Research on Cell Handover Algorithm under UDN. Chongqing University of Posts and Telecommunications (2017)
14. Tversky, K.A.: Prospect theory: an analysis of decision under risk. Econometrica **47**(2), 263–291 (1979)

Communication Routing and Control

Channel Estimation Algorithm Based on Demodulation Reference Signal in 5G

Bingguang Deng, Xiaofang Min$^{(\boxtimes)}$, Siyi Yu, and Qianqian Ye

School of Communication and Information Engineering, Chongqing University of Posts and Telecommunications, Chongqing, China
S180101183@stu.cqupt.edu.cn

Abstract. In order to track 5G downlink shared channel in real time and meanwhile to reduce computational complexity, a linear minimum mean square error algorithm based on demodulation reference signal adaptive parameter estimation is proposed. Firstly, the SNR nonlinear centralized optimization problem is transformed into a multivariable linear programming problem due to the restriction of non-uniform energy distribution in time-domain channel. Secondly, considering the uncertainty of multipath delay channel, the combination of negative exponential distribution model and generalized correlation algorithm is taken advantage of so that the original problem is turned into a specific parameter optimization problem. At the same time, according to the obtained delay parameters and SNR, the most appropriate interpolation coefficient is selected for the LMMSE channel estimation by combining with the sliding window, which avoids the matrix inversion process, realizes the real-time matching of parameters, and reduces the computational complexity. The simulation results show that the proposed algorithm has better system performance compared with the classical channel estimation algorithm.

Keywords: 5G · Linear minimum mean square error · SNR · Time delay · Channel estimation

1 Introduction

With the function of effectively improving spectrum utilization and realizing multi-user access, Orthogonal Frequency Division Multiplexing (OFDM) is frequently applied in 5G system. The quality of channel estimation determines whether OFDM technology can be implemented efficiently. In a consequence, it is very important for communication system to obtain effective data information by appropriate methods. The data processing of the channel estimation module of Physical Downlink Shared Channel (PDSCH) has a direct effect on the performance of the whole 5G system, so it is of great importance to find a method easy to implement and with better system performance. In literature [1], an algorithm based on Fast Fourier Transformation (FFT) operation, circular shift operation and fast Toplitz matrix vector multiplication to obtain the autocorrelation matrix has been put forward. In literature [2], it is pro-posed that the method of estimating the noise variance in the time domain through the leading symbol has strong robustness, yet multiple iterations increase the complexity. In

H. Gao et al. (Eds.): ChinaCom 2020, LNICST 352, pp. 371–380, 2021.
https://doi.org/10.1007/978-3-030-67720-6_25

literature [3], a linear minimum mean square error channel estimation algorithm based on compressed sensing has been raise, which improves the frequency band utilization. These techniques may increase the computational complexity, so there are difficulties to apply them in practice. Literature [4] proposes two-dimensional linear minimum mean square error channel estimation method for OFDM systems, which can reduce the computational complexity, but this algorithm is only suitable for slow fading channels. In literature [5], the channel delay parameters are estimated, and then the channel estimation is carried out by using the lookup table, under the condition of high SNR, this algorithm suffers from flat effect, which may degrade system performance. In literature [6], a new low rank minimum mean square error method is pro-posed to simplify the filtering matrix and diagonal matrix, which has good system performance, but involves the acquisition of channel statistics.

Aiming at the difficulty and high complexity of channel prior information acquisition, an adaptive parameter channel estimation method based on Demodulation Reference Signal (DRMS) is proposed in this work. According to the characteristics of energy and power delay profile distribution in 5G time-domain channel, the methods of impulse point decision and difference approximation are adopted to estimate the delay parameters and Signal to Noise Ratio (SNR) in real time respectively, and the channel estimation is carried out in combination with the sliding window. In the light of the simulation results, this method achieves a good compromise between system performance and computational complexity compared with the classical channel estimation algorithm.

2 System Module

Estimation of DRMS in 5G downlink Shared channel is divided into two parts. At the pilot, the algorithm of Least Square (LS) or Linear Minimum Mean Square Error (LMMSE) is adopted; at the data, interpolation algorithm is adopted [7, 8]. Including:

- Channel estimation at pilot adopts LS algorithm;
- LMMSE interpolation is used for data locations in the frequency domain;
- Linear interpolation is adopted at the location of time-domain data. After the above steps, channel responses of all reference signals are finally obtained.

5G system has made a new design for reference signal. The function of the most important Cell Reference Signal (CRS) in LTE system will be carried by DRMS, Channel State Information Reference Signal (CSI-RS), Phase Tracking Reference Signal (PTRS) et al. In particular, DRMS is used for channel estimation of PDSCH in 5G systems, which uses LS algorithm to complete the estimation of the initial channel frequency response.

Based on the existing Minimum Mean Square Error (MMSE) algorithm, the instantaneous power of each frame is replaced by the average power of each sub-channel, so LMMSE denotes $\hat{H}_{LMMSE} = R_{HH} \left(R_{HH} + \frac{\beta}{SNR} I \right)^{-1} \hat{H}_{LS}$ as the channel responses, where $W = R_{HH} \left(R_{HH} + \frac{\beta}{SNR} I \right)^{-1}$ represents the interpolation matrix, R_{HH}

represents the correlation matrix, I is an unit matrix, β is a constant, which depends on what kind of modulation is used. When the modulation is QPSK, $\beta = 1$; 16QAM, $\beta = 17/9$; 64QAM, $\beta = 2.6854$; 256QAM, $\beta = 3.4371$.

It can be seen from the expression that the emphasis lies in the calculation of correlation matrix and SNR, as well as the inverse operation of the matrix [9, 10]. The former constitutes the interpolation filter coefficient, which is related to the system performance, while the computational complexity of LMMSE algorithm is mainly reflected in the process of matrix inversion.

This paper mainly studies the frequency domain interpolation algorithm based on LMMSE to apply in practical engineering. Considering its practicability from different perspectives, there is a design scheme.

3   Proposed Algorithm

The expression of LMMSE channel estimation correlation matrix is as follows:

$$R_{HH} = E\left[HH^H\right] = \left[r_{x,y}\right] \tag{1}$$

$$r_{x,y} = \frac{1 - \exp\left\{-\tau_m\left[j\frac{2\pi}{N}(x-y) + \frac{1}{\tau_r}\right]\right\}}{\tau_r\left[1 - \exp\left(-\frac{\tau_m}{\tau_r}\right)\right]\left(\frac{1}{\tau_r + j\frac{2\pi}{N}(x-y)}\right)} \tag{2}$$

The maximum diameter delay τ_m and root mean square delay τ_r determine $r_{x,y}$ [11, 12]. The traditional method uses a channel model with uniform delay distribution and negative exponential power delay profile, and assumes that the root mean square delay is a fixed constant.

3.1   Proposed Algorithm Design

The flow chart of 5G downlink Shared channel DRMS channel estimation algorithm are shown in Fig. 1. Specific steps of the algorithm are given as:

- Obtain DRMS reference signals and then use LS algorithm for DRMS estimation.
- Conduct the LMMSE frequency domain interpolation at the data points according to the estimated channel value obtained in the previous step.
- According to the characteristics of energy distribution in the time-domain of the channel, the SNR is calculated by tracking the distribution of the channel in real time, and the delay parameters are obtained by using correlation method.
- Calculate the frequency domain LMMSE Wiener filtering coefficient through the obtained delay parameters and SNR.
- In the frequency domain, the sliding window and LMMSE algorithm are combined to estimate the data signals.
- Using linear interpolation to estimate the data signals in the time domain.

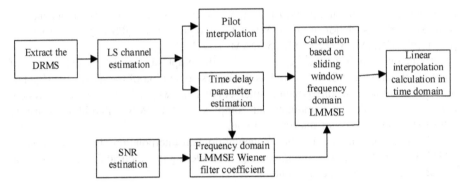

Fig. 1. Adaptive LMMSE channel estimation

3.2   Proposed Adaptive Delay Parameter Estimation

Real-time estimation of delay parameters can realize adaptive adjustment of the number of pilots at the sending end, and the channel estimation performance can be improved effectively. In order to solve the channel mismatch problem caused by delay parameters, this paper randomly assigns the delay value and estimates parameters in the channel model with negative exponential distribution of power delay profile. The delay power is calculated as $P(\tau) = Ce^{-\tau/\tau_r}$, C represents the maximum value of multipath power spectrum. The relation between delay power and multipath delay can convert $r_{x,y}$ into

$$r_{x,y} = \frac{1 - \theta \, \exp\left(\tau_m 2\pi j \frac{x-y}{N}\right)}{(1-\theta)\left[1 - \frac{-\tau_m}{\ln\theta} 2\pi j \frac{x-y}{N}\right]} \tag{3}$$

The correlation matrix problem is transformed into the maximum path delay problem, which improves the accuracy and reliability of parameters.

In order to obtain the instantaneous correlation value at pilot frequency of OFDM symbol, Correlation operation is carried out for frequency domain channel response at pilot. The instantaneous correlation value is presented as

$$\hat{R}_H(\Delta) = E_i\left[\hat{H}(m)\hat{H}^*(m+\Delta)\right] \tag{4}$$

Considering the characteristic of concentrated energy distribution of time-domain channel, the power delay spectrum is obtained by Inverse Discrete Fourier Transform (IDFT). This can formally be expressed as

$$\hat{r}_h(l) = IDFT\{\hat{R}_H^n(\Delta)\} = \sum_{\Delta=0}^{N-1} \hat{R}_H^n(\Delta)e^{-j2\pi\Delta/N} \tag{5}$$

Where $l = 0, 1, \ldots, N - 1$. Since the power delay spectrum is approximately symmetric with $N/2$ as the center, the channel impulse points are determined by the following decision

$$\hat{r}_h(l) > (\hat{r}_h(l - 1) + \hat{r}_h(l + 1))/2 \tag{6}$$

The position parameters of the peak element of the impulse point are obtained, $l \times T_s$ is the delay time of each path, and T_s is the tap interval.

The maximum path delay spread value of the channel is a relative delay value, which is actually the value of the power attenuation udB of the strongest multipath signal. So define $E = \hat{r}_h(0) \cdot 10^{-u/10}$ to determine the number of taps that is set as $G = \underset{l \in [1,2,\ldots,N/2]}{\arg\max} \{\hat{r}_h(l) > E\}$, where u is the threshold value of power attenuation corresponding to the maximum diameter delay. After the impulse point decision and threshold setting, the maximum diameter delay extension value τ_m is obtained by multiplying the tap interval and the tap number. The process is shown in Fig. 2.

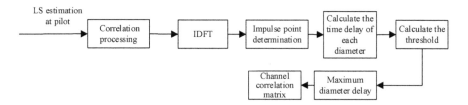

Fig. 2. The process of maximum diameter delay estimation

3.3 Proposed SNR Estimation

The SNR is estimated according to the characteristic that channel energy sparsely distributes on M channel paths, and always tends to be distributed on K paths of what the size we don't know. Therefore, this paper proposes an adaptive SNR estimation method to track the channel energy distribution in real time. This method can determine not only the size of K, but also its corresponding specific location. It is also suitable for the channel with decentralized energy distribution.

Receiver signal in time domain is $r(i) = y(i) + w(i)$, where $r(i) = r_I(i) + jr_Q(i)$ represents the received signal. By splitting the real part and the imaginary part, the received signal power can take the form

$$P = \sum_{i=0}^{i=M-1} P(i) = \sum_{i=0}^{i=M-1} \left[|real(r(i))|^2 + |imag(r(i))|^2 \right] \tag{7}$$

The noise on each path in the channel is different, but its noise power always approaches one value. Assuming that the noise power of each path is σ^2, there are M paths. So $\bar{P} = \frac{P}{M} > \sigma^2$ is the average power, where P represents the total power of received signal. According to the above analysis, in order to determine the exact location of the signal, the algorithm gives the decision condition $P(i) - \bar{P} > 0$, where $P(i)$ represents the received signal power of path i, which must be the value of 0, 1, 2, ..., K.

Exclude the determined signal path, calculate the average power of the remaining $M - i$ paths, and repeat the first two steps until it meets $|P(i) - \bar{P}| < \delta$ (δ is a positive number approaching 0). After determination, the remaining path, which does not meet the above conditions, is the path of noise, so as to determine the K paths of signal distribution.

Given that the noise power of each path is basically the same, the average noise power of the last $M - K$ paths is calculated as

$$\sigma^2 = \frac{\sum_{i=K}^{M-1} P(i)}{M - K} \tag{8}$$

Therefore, the SNR is estimated with

$$SNR = \frac{P - N\sigma^2}{N\sigma^2} = \frac{P}{N\sigma^2} - 1 \tag{9}$$

The process of adaptive SNR estimation is shown in the Fig. 3.

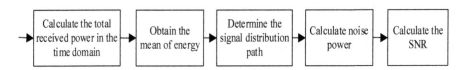

Fig. 3. The process of adaptive SNR estimation

In practical engineering application, to save resources as much as possible, the consideration of computational complexity is the key. Suppose that there are N pilots in OFDM symbol, and the initial subspace dimension of the estimated signal is assumed to be r through adaptive estimation. Each OFDM symbol channel estimation achieves a moving average instead of statistical autocorrelation values. The next time estimation is an orthogonal iteration, the value is the average of the previous estimation. The total computation is $O(N(r)^2)$, but the traditional LMMSE estimation computation is $O(N^3)$. This method reduces the dimensions of the correlation matrix inversion, and the number of calculations for plural matrix multiplication and addition.

4 Simulation Results and Discussion

4.1 Simulation Parameter

In order to verify the validity and stability of the proposed estimation algorithm, MATLAB is used to simulate it. Simulation parameters are shown in Table 1, and the channel multipath number is generated randomly between timely delays.

Table 1. Simulation parameters

Parameter	Value
System bandwidth	20 M Hz
Modulation method	64QAM
Channel model	Negative exponential distribution
Subcarrier number	512
Subcarrier interval	7.815 kHz
Pilot model	Comb
FFT	2048
Sampling frequency	30.76 MHz
Cyclic prefix	Normal CP

Figure 4 shows the simulation of power delay spectrum under different SNR. It can be seen from the figure that the change of SNR has little influence on the delay spectrum, and only when the SNR is very small (0–10 dB) may generate noise impulse points leading to misjudgment. It is shown that the multipath time delay estimation algorithm presented in this paper is not affected by channel noise and has good stability.

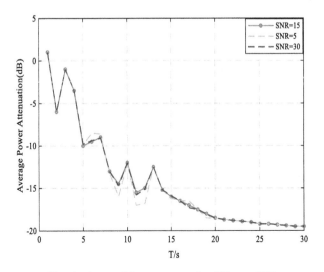

Fig. 4. Power delay spectrum for different SNR

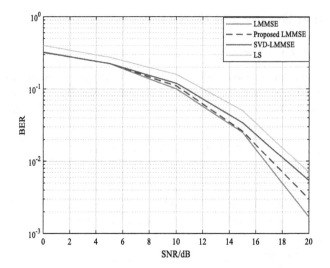

Fig. 5. Comparison of BER for different algorithms

Figure 5 depicts Bit Error Rates (BER) generated by different channel estimation algorithms. Compared to Singular Value Decomposition (SVD) LMMSE interpolation algorithm, the improved LMMSE algorithm has a smaller BER. The BER performance of LMMSE algorithm is the most ideal. Although the performance of the improved LMMSE interpolation algorithm is not completely consistent with the ideal channel, its system performance is relatively better than that of other interpolation algorithms. The simulation results show that when the SNR is relatively small, the improved LMMSE algorithm has lower BER and better performance. However, BER decreases faster when SNR is greater than 15 dB. This is because the algorithm makes full use of the correlation between symbols to track the signal information in real time. And it estimates the dimension of signal subspace within a small error range to reduce the estimation error.

Figure 6 shows the Mean Square Error (MSE) performance of different algorithms. With the increase of SNR, the MSE of the four algorithms decreases. Under the same SNR, the MSE of LS interpolation algorithm is the largest. The improved LMMSE algorithm is superior to SVD-LMMSE algorithm. Moreover, with the increase of SNR, the improved LMMSE algorithm is very close to the characteristic curve of the ideal LMMSE algorithm. The reason is that the improved algorithm takes full account of the characteristics of uneven delay distribution and concentrated energy distribution in the channel to realize the adaptive estimation of parameters, and make the actual estimation as close as possible to the ideal channel.

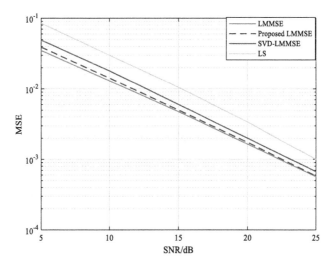

Fig. 6. Comparison of MSE for different algorithms

5 Conclusion

Based on the characteristics of the 5G downlink shared channel, the time delay parameter and SNR estimation of the LMMSE channel estimation algorithm are optimized under the condition of not requiring the channel statistics information, and better system performance is obtained. A corresponding model is established for the proposed time delay estimation optimization method. Moreover, the methods of frequency domain correlation method and impulse point decision are combined to simplify the multivariate solution to a single variable by using the negative exponential distribution channel model. For SNR optimization, considering the characteristics of channel energy distribution, channel distribution decision conditions are set to track the channel information in real time, and fixed variable estimation is converted into adaptive estimation, so as to improve the information accuracy. Ultimately, the slid-ing window LMMSE algorithm is made use of to obtain the channel response of the whole channel. This paper also attempts to compare the proposed method with other different methods of channel estimation. Simulation results demonstrate that com-pared with the traditional LMMSE algorithm, the improved algorithm reduces the computational complexity and effectively reduces the waste of hardware resources.

Acknowledgment. This work is supported by National Key Research and Development Project (No. 2018YFB2100200).

References

1. Khlifi, A., Bouallegue, R.: A very low complexity LMMSE channel estimation technique for OFDM systems. In: Vehicular Technology Conference (VTC Spring), pp. 1–5, May 2015

2. He, P., Li, Z., Wang, X.: A low-complexity SNR estimation algorithm and channel estimation method for OFDM systems. In: International Conference on Information Science and Technology, pp. 698–701, April 2014
3. Ge, L., Zhang, Y., Chen, G., Tong, J.: Compression-based LMMSE channel estimation with adaptive sparsity for massive MIMO in 5G systems. IEEE Syst. J. **13**(4), 3847–3857 (2019)
4. Choi, Y.J., Bae, J.H., Lee, J.W.: Low-complexity 2D LMMSE channel estimation for OFDM systems. In: International on Vehicular Technology (VTC2015-Fall), pp. 1–5, September 2015
5. Duan, H.G., Tian, M., Li, Y.L.: DMRS based channel estimation algorithm in LTE-A system. Commun. Technol. **49**(11), 1424–1428 (2016)
6. Tang, R., Zhou, X., Wang, C.: A novel low rank LMMSE channel estimation method in OFDM systems. In: International Conference on Communication Technology (ICCT), pp. 249–253, October 2017
7. Savaux, V., Bader, F., Louët, Y.: A joint MMSE channel and noise variance estimation for OFDM/OQAM modulation. IEEE Trans. Commun. **63**(11), 4254–4266 (2015)
8. Liu, W., Li, X.: An improved LMMSE channel estimation algorithm of LTE system. In: International Conference on Computational and Information Sciences (ICCIS), pp. 231–234, August 2012
9. Hei, S.S., Wang, X.Q.: Block sliding window channel estimation algorithm based on LMMSE. Comput. Eng. Appl. **53**(16), 79–83 (2017)
10. Kalakech, A., Berbineau, M., Dayoub, I., Simon, E.P.: Time-domain LMMSE channel estimator based on sliding window for OFDM systems in high-mobility situations. IEEE Trans. Veh. Technol. **64**(12), 5728–5740 (2015)
11. Senol, H.: Joint channel estimation and symbol detection for OFDM systems in rapidly time-varying sparse multipath channels. Wireless Pers. Commun. **82**(3), 1161–1178 (2015)
12. Sheng, Z., Tuan, H.D., Nguyen, H.H., Fang, Y.: Pilot optimization for estimation of high-mobility OFDM channel. IEEE Trans. Veh. Technol. **66**(10), 8795–8806 (2017)

Constrained Multipath Routing Algorithm Based on Satellite Network

Pan Liu[(⊠)] and Tao Zhang

School of Electronic and Information Engineering, Beihang University,
Beijing, China
541094973@qq.com

Abstract. Due to dynamic changes in network topology and constant changes in links between satellites, the routing paths calculated by the traditional shortest multipath routing algorithm are not updated in time, resulting in that the problematic paths are still transmitting data, so a large number of problems such as packet loss and service failure exist in the network. This paper designs a multipath routing algorithm for the satellite network topology which changes frequently, constrained multipath routing algorithm (CMRA). By calculating all feasible path of the entire network topology, CMRA will choose the multiple paths which satisfy the constraint conditions. After that, it will choose the lowest cost value of these paths. Repeatedly performing these operations, select multiple high-quality paths which satisfy the constraint conditions for traffic. Simulation results show that compared with the traditional shortest multipath routing algorithm and single path routing algorithm, the proposed routing algorithm is better in packet loss rate and average time delay.

Keywords: Satellite network · Constraint condition · CMRA

1 Introduction

Satellite communication systems, which cover a wide range of areas and are not affected by geographical and climatic conditions, have received widespread attention [1]. Satellite networks are characterized by frequent changes of network topology caused by the high-speed movement of satellite nodes, high density and short bursts of data received in places with large population, and unequal distribution of traffic, resulting in a large number of packet loss and service failure. The so-called multi-path routing refers to finding multiple paths from the source node to the destination node in the network through certain constraint rules [2]. The traditional shortest multipath routing algorithm selects multiple paths with the smallest path length after relaxing all the edges. It does not take the problem of route consumption, time delay and path failure caused by the change of satellite link into account [3]. Literature [4] proposes a perceived more energy path load balancing routing algorithm which uses energy perception to choose node which meets the condition as a routing node, and establishes more efficient way of connecting the source node and destination node. Through the analysis of the hop count of path and the buffer occupancy of nodes, it chooses the high quality path to transmit traffic. Literature [5] proposes a LEO load balancing routing

H. Gao et al. (Eds.): ChinaCom 2020, LNICST 352, pp. 381–391, 2021.
https://doi.org/10.1007/978-3-030-67720-6_26

algorithm based on ant colony. It makes routing decisions by collecting physical layer information and uses multi-objective optimization model to realize load balancing, which performs well in balancing load and improving message transmission rate. Literature [6] proposed a tunable accuracy multiple constraints routing algorithm, TAMCRA. Although the algorithm can solve the multi constraint routing problem, the performance and cost of the algorithm depend on the number of shortest paths, K. If K is large, the algorithm can provide good performance, but at the cost of sacrificing too much computing cost. The routing algorithms above can do well in traffic balance, but they do not take the comprehensive quality of the path, the routing path failure problems caused by the changes of satellite links and the failure of satellite nodes into account.

In this paper, constrained multipath routing algorithm (CMRA) is proposed. After calculating all the possible paths in the network, CMRA filters the feasible paths by multiple constraints, and chooses the path with the lowest cost which meets the set constraints. Then, the network topology is simplified. Repeat the previous operation many times, and more paths which meet the set constraints will be selected. Finally, the traffic is carried by these paths according to a certain traffic allocation ratio. The aim of simplifying the network topology is that all selected paths do not interfere with each other, which means that if there are some problems with one node in the path, other paths can carry the traffic normally. Simplifying the network topology improves the reliability of multiple paths and reduces the possibility that multiple paths cannot transmit data due to node failure. Experimental results show that CMRA is superior to the shortest multipath routing algorithm in terms of packet loss and average time delay.

2   Constrained Multipath Routing Algorithm

2.1   General Introduction

The routing algorithm proposed in this paper is a multi-path routing algorithm based on Bellman_Ford algorithm combined with constraint mechanism, named constrained multipath routing algorithm (CMRA). CMRA mainly includes three aspects: the determination of multiple constraints, the network topology simplification and the allocation of traffic.

After relaxing the network topology, CMRA selects all the routing paths that meet the conditions according to the multiple constraints, and selects the path with the lowest cost value from the paths. The network topology simplification operation is performed on the network topology according to the selected routing path so that the first path and the second path do not interfere with each other. The second routing path that meets the constraints is selected after the operation of simplifying network topology until there is no path that meets the multiple constraints. At last, the traffic is carried according to a certain traffic allocation ratio.

The general process is shown in the figure below (Fig. 1):

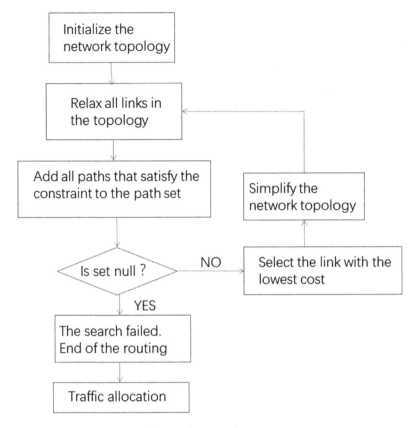

Fig. 1. The general process.

Its concrete realization is as follows:

(1) Relax the initialized network topology;
(2) According to Bellman_Ford routing algorithm, all feasible paths are calculated, and all paths that satisfy multiple constraints are found. If the path does not exist, the search fails and the route ends. if there is, the next step is carried out;
(3) Find the path with the lowest cost value among all the paths selected in the previous step and set it as the first path;
(4) The simplified network topology G_1 is obtained by simplifying operation of network topology G;
(5) Relax the network topology G_1;
(6) According to Bellman_Ford routing algorithm calculates all feasible paths in G_1 and finds out all paths that satisfy multiple constraints. If there are no paths, the search fails and the route ends; if there is, the next step is carried out;
(7) Find the path with the lowest cost value among all the paths selected in the previous step and set it as the second path;
(8) Go back to (4) to get the third path, the fourth path until the end of the route.

2.2   The Initialization of Network Topology and Relaxation Operation

Firstly, the whole network topology is initialized, that is to update all node information including location information and link connection information.

Relaxation operation is if there is another path between two points, and the distance between the two points is less than the current distance between the two points, then the shortest path length between the two points is updated to this smaller distance. If there is a vertex that can not be relaxed, it is said that the vertex has converged. As shown in Fig. 2, the distance from a to b is 5, but the distance from a to b via c is 3 which is smaller than 5 for $d_{ac} + d_{cb} = 1 + 2 = 3 < d_{ab} = 5$. Therefore, the distance between a and b is updated to 3, and all points have converged.

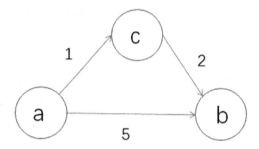

Fig. 2. A network topology for showing relaxation operation.

2.3   The Determination of Multiple Constraints

Multiple constraint condition is a routing mechanism which can select feasible paths that satisfy various constraints according to the different requirements of QoS measurement parameters in network state information [6]. The commonly used QoS metrics parameter include three types: additive metrics parameter, such as typical cost, hops, time delay, jitter and forwarding hops; multiplicative metrics parameter, such as loss rate and reliability; concave metrics parameter, such as bandwidth.

For a path P in the network, if $w(e)$ is an additive metric parameter for each link on path P, then the value of the QoS metric parameter of path P is the cumulative value of the QoS metric parameters of each link on the path P, namely:

$$w(P) = \sum w(e) \tag{1}$$

If $w(e)$ is the multiplicative metric parameter for each link on path P, then the value of the QoS metric parameter of path P is the product value of the QoS metric parameters of each link on the path P, namely:

$$w(P) = \prod w(e) \tag{2}$$

If $w(e)$ is the concave metric parameter for each link on path P, then the value of the QoS metric parameter of path P is the minimum value of the QoS metric parameters of each link on the path P, namely:

$$w(P) = \min w(e) \tag{3}$$

There are many classical multi constraint routing algorithms such as heuristic multi constraint optimal path routing algorithm, named H_ MCOP algorithm. H_ MCOP algorithm is to ensure to find the path that satisfies multiple constraints and costs the least [7]. Its comprehensive measurement parameters are as follows:

$$g_\lambda(p) = \left(\frac{w_1(p)}{c_1}\right)^\lambda + \left(\frac{w_2(p)}{c_2}\right)^\lambda + \cdots + \left(\frac{w_k(p)}{c_k}\right)^\lambda \tag{4}$$

Where $w_k(p)$ is the k'th metric parameter of path P and C_k is the k'th constraint parameter of path P. As λ tends to be positive infinity from 1, the success rate of the algorithm to find the path that satisfies multiple constraints and costs the least is also increasing [8].

G(V, E) is assumed to be the basic model of satellite network topology, where $V = \{v_1, v_2, \cdots, v_N\}$ represents the set of N switching nodes and $E = \{e_1, e_2, \cdots, e_N\}$ represents N links in the satellite network topology, and constraint parameters are added for each link. In this paper, constraint parameters include the reliability in the multiplicative measure parameters, which are represented by d(e), and the hops and time delay in the additive measure parameters, which are represented by $w_k(e), k = 1, 2, \cdots, K$. Each parameter is:

(1) Reliability: conditions of actual communication links;
(2) The number of hops: the number of the output ports of routing node through which the packet will pass from the source node to the destination node;
(3) Time delay: the time delay is the total delay of data packet transmission, which is the sum of propagation delay, sending delay, queuing delay and waiting delay. The propagation delay is the physical distance between the source node and the destination node divided by the speed of light. The transmission delay is the processing time of the link layer. The queuing delay is the sum of all the data transmission delays waiting to be forwarded before the data is to be transmitted. The waiting delay is the time waiting for the link to be connected to forward the data.

If the source node is S and the destination node is T in the satellite network topology, the parameter is $C(C_1, C_2, \cdots, C_K)$ for K additional constraints. Let P1 and P2 be the paths from source node S to destination node T, (i, j) be the subpath on path P1, and (a, b) be the subpath on path P2. Then the cost of P1 is:

$$\text{cost}(P1) = \sum_{(i,j) \in P1} \cos t(i,j) \tag{5}$$

So, using the cost of P1, we can know that its constraint condition is:

$$w_k(P1) = \sum_{(i,j)\in P1} w_k(i,j) \le C_k, k = 1, 2, \cdots, K \tag{6}$$

In the same way, the constraint condition of P2 is obtained as follows:

$$w_k(P2) = \sum_{(i,j)\in P2} w_k(a,b) \le C_k, k = 1, 2, \cdots, K \tag{7}$$

Finally, the paths are then filtered again based on the overall reliability metrics of P1 and P2.

The general process is shown in the Fig. 3.

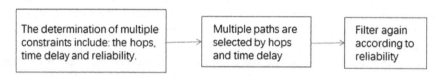

Fig. 3. The process of the determination of multiple constraints.

2.4   The Network Topology Simplification

Due to the change of satellite network topology, in order to reduce the impact of satellite link changes and node failure resulting that routing path can not normally transmit data, this paper designs a simplification of network topology. Network topology simplification is to delete some links or nodes according to a routing path selected, so that there is no common link between the routing path calculated by the network topology before simplifying and the routing path calculated by the simplified network topology. This also realizes that when the link changes or the node fails on a certain routing path, it will not cause multiple paths to be unable to transmit data normally.

The main ideas of network topology simplification are as follows:

(1) Find out the subnet of source node and destination node;
(2) After calculating the first path satisfying the constraint conditions, the sub network of each node in the path except the source node and the destination node is found;
(3) Judge whether the sub network is the same as the sub network of the source node or the sub network of the destination node. If yes, delete the node and the link associated with the node; if not, delete all the nodes in the sub network and the link associated with each node in the sub network.

For example, if the network topology is as follows:

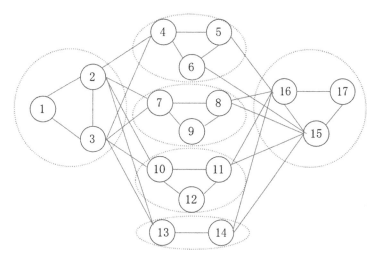

Fig. 4. A network topology.

The source node is node 1 and the destination node is node 17. The subnet of each node is shown in the Fig. 4. The first path is assumed to be: 1->2->4->5->16->17. As shown in the Fig. 4, the subnet of node 2 coincides with that of source node 1. Then delete node 2 and link 1-2, link 2-3, link 2-4, link 2-7, link 2-10 and link 2-13. The subnet of node 4 and node 5 is different from that of source node 1 and destination node 17. Therefore, delete all nodes of the subnet, including node 4, node 5, node 6, and their associated links, including link 4-5, link 4-6, link 5-6, link 4-2, link 4-3, link 5-16, and link 6-15. For node 16, its subnet is the same as that of destination node 17. So delete node 16 and the links associated with node 16, including links 5-16, link 8-16, link 11-16, link 14-16, link 16-17, and link 16-15.

Therefore, the new network topology after simplifying the above network topology is shown in the Fig. 5:

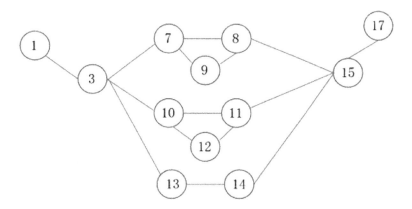

Fig. 5. The simplified network topology.

The process of the network topology simplification is (Fig. 6):

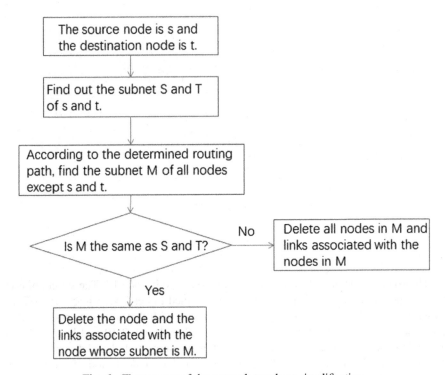

Fig. 6. The process of the network topology simplification.

2.5   The Allocation of Traffic

In order to avoid congestion and achieve traffic balance, it is necessary to allocate traffic according to the information of path when transmitting data through multi-path routing. Firstly, two definitions are given.

(1) Link bandwidth utilization: For $\forall (i,j) \in E$, the link bandwidth utilization of (i, j) is:

$$U_{ij} = \left[1 - (R_{ij} - b)/C_{ij}\right] \times 100 \qquad (8)$$

(2) Path bandwidth utilization: For $\forall (i,j) \in P1$, the path bandwidth utilization of P1 is:

$$U(P_1) = \max_{(i,j) \in P_1} U_{ij} \qquad (9)$$

R_{ij} is the residual bandwidth of link, b is the number of bandwidth requested by the call and C_{ij} is the link capacity.

Link bandwidth utilization reflects the bandwidth usage of a link, and usually also reflects the load of the link. The path bandwidth utilization is represented here by the bandwidth utilization of the most heavily loaded link on the path [9].

In this paper, in order to better balance the traffic, the path is weighted according to the path bandwidth utilization, and the weight is called the path equalization weight, named W. A smaller W indicates a lighter load on the path, while a bigger W indicates a heavier load on the link. Finally, carry on the traffic allocation according to W.

3   The Simulation Test

3.1   The Simulation Scene

The simulation in this paper is based on the SNSim simulation platform independently developed by the laboratory. The TELEDESIC satellite constellation is the satellite network scenario used in this paper. The satellites in the TELEDESIC constellation orbit the Earth at a nominal altitude of 1,375 km and have 12 circular orbits with near-polar dip angles and adjacent ascending points. Each orbit plane contains at least 24 active satellites with uniform spacing. That is to say, there are 288 satellites in the whole satellite constellation with an orbit inclination of 84.7°. Set up three domestic earth stations, one foreign station. They are Beijing, Sanya, Kashi, and Greenbell in the United States. In this paper, three domestic ground stations receive services and send them to Greenbell ground stations. The size of each packet is 1000 bytes.

3.2   Simulation Data Analysis

In this paper, we simulate the constrained multipath routing algorithm in the above satellite network scenarios, and compare with RIP routing algorithm and classical shortest multipath routing algorithm in packet loss rate and average time delay.

The simulation data of constrained multipath routing algorithm (CMRA), shortest multipath routing algorithm (SMRA) and single path RIP routing algorithm are as follows:

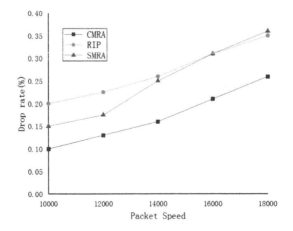

Fig. 7. Comparison of packet loss rate of different algorithms.

From the simulation data in Fig. 7, it can be seen that with the increase of packet speed, the packet loss rate of CMRA, RIP and SMRA increases, but the performance of multipath routing algorithm under constraints is better. This shows that with the increase of the number of packets received by nodes, CMRA can reasonably consider the path of data transmission according to the network link conditions, which greatly reduces the packet loss rate (Fig. 8).

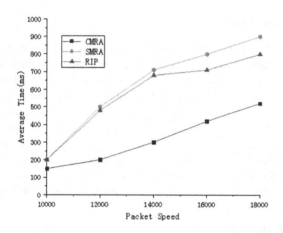

Fig. 8. Comparison of average time delay of different algorithms.

It can be seen from the simulation data, with the increase of package speed, the average time delay of CMRA, RIP and SMRA increases, but on the whole, CMRA greatly reduces the average time delay. This shows that CMRA can choose the idle path when a certain path's load is too heavy.

In general, with the increase of packet speed, the performance of CMRA in packet loss rate and average time delay is better than that of traditional multipath routing algorithm. The constrained multipath routing algorithm fully considers the specific situation of the link, and selects a better path for traffic transmission to alleviate the pressure of congested path. Constrained multipath routing algorithm can greatly reduce the impact of node failure or link changes which does not reflect in the routing path in time. With the increase of packet speed, the effect is more obvious.

4 Conclusion

In order to solve the problems of packet loss and average time delay caused by the dynamic changes of satellite network topology, this paper proposes a multipath routing algorithm under the constraint which can effectively solve the link frequency change, satellite node failure and the overall traffic balance problem in satellite network. The algorithm selects the high-quality path to transmit traffic through the constraint conditions, and simplifies the network topology.

This paper completes the simulation of CMRA, RIP and SMRA, and compares the three routing algorithms in packet loss rate and average time delay. From the simulation data, we can see that CMRA greatly reduces the packet loss rate and average time delay, and can better achieve the overall traffic balance of satellite network.

Acknowledgment. This work was supported by the National Key R&D Program of China (No. 2016YFB1200100), the National Nature Science Foundation of China (No. 91638301).

References

1. Xu, D., Meng, F., Xie, J., Zhang, X.: Analysis of CO frequency interference between terrestrial internet of things and satellite Internet of things. Telev. Technol. **42**(09), 42–46+51 (2018)
2. Zheng, H., Chen, S., Liu, L.: Analysis and improvement of multipath routing protocol based on DSR. J. Shandong Univ. (Eng. Ed.) (01), 115–118 (2007)
3. Zuo, X., Wang, H.: Research on an improved rip routing algorithm. Comput. Program. Skills Maint. (04), 156–157+166 (2019)
4. Wang, X., Su, S., Lin, Q., Wang, L.: Energy aware multipath load balancing routing algorithm. J. Air Force Eng. Univ. (Nat. Sci. Ed.) **18**(03), 85–91 (2017)
5. Wang, H., Zhang, Q., Xin, X., et al.: Cross-layer design and ant-colony optimization based routing algorithm for low earth orbit satellite networks. China Commun. **10**(10), 37–46 (2013)
6. De Neve, H., Van Mieghem, P.: TAMCRA: a tunable accuracy multiple constraints routing algorithm. Comput. Commun. **23**(7), 667–679 (2000)
7. Xu, C., Zhou, J.: Constrained routing algorithm in MPLS traffic engineering. Comput. Knowl. Technol. (30), 598–600 (2008)
8. Chen, L., Zhou, J., Jiang, H., Yan, P.: Path selection algorithm for QoS multi constrained optimization. Comput. Appl. (04), 900–902 (2005)
9. Li, M.: simulation of routing algorithm based on link traffic assignment weights. Intelligent automation professional committee of Chinese society of automation. In: Proceedings of 2007 China Intelligent Automation Conference. Intelligent automation professional committee of Chinese society of automation: Intelligent Automation Professional Committee of Chinese society of automation, pp. 612–616 (2007)

Container Performance Prediction: Challenges and Solutions

Jiwei Wang[1]([⊠]), Yuegang Li[1], Congfeng Jiang[1], Chao Ma[2], Linlin Tang[2], and Shuangshuang Guo[2]

[1] School of Computer Science and Technology, Hangzhou Dianzi University, Hangzhou 310018, China
{wangjiwei,lyg,cjiang}@hdu.edu.cn
[2] Information & Telecommunications Company, State Grid Shandong Electric Power Company, Jinan 250000, China
machao@sd.sgcc.com.cn, 6335875@qq.com, 416054882@qq.com

Abstract. With popularity of cloud computing services, more and more tasks and services are deployed on large-scale clusters. As an emerging technology in cloud computing field, containers make virtualization extremely lightweight. However, lack of prediction causes scheduling decisions lag behind the dynamics of clouds. Thus, how to carry out performance prediction before container scaling has become an urgent problem to be resolved. Here we emphasized the necessity of container performance prediction and summarized the current research progress and effort of container performance modeling. Finally, we compared pros and cons of numerical analysis and machine learning in terms of practice.

Keywords: Virtualization · Container · Containerized cloud · Performance prediction · Machine learning

1 Introduction

With rise of social network and popularity of IoT (Internet of Things) devices, amounts of data generated around the world is increasing exponentially. However, most of the existing cloud systems are unable to satisfy such capacity of demands effectively. Compared to direct way of running services on separate physical machines, virtualization alleviates the suffering of inflexible problems and make the whole system more cost-efficient. However, full virtualization introduces non-negligible overhead when running services, as VMs (virtual machines) simulate not only the whole operating system, but also hardware architecture and network environments. As result of this, full virtualization is not suitable for compute-intensive and request-intensive scenarios.

Recently, with demand of resource effectiveness, a light-weight virtualization technology called container is actively promoting the development of the next generation of cloud services [17]. Different from full virtualization technologies

H. Gao et al. (Eds.): ChinaCom 2020, LNICST 352, pp. 392–402, 2021.
https://doi.org/10.1007/978-3-030-67720-6_27

(represented by VMs), several containers share the same kernel of the operating system, by which much performance loss resulted from hardware virtualization can be eliminated drastically. In early practice of containers, scheduling strategies are simply transferred from existing works on VMs. However, with containers widely deployed in production practice, many evidences have proved that scheduling containers is a more open question compares to VMs. As container provides flexibility to cloud environment, it brings new challenges like low-isolation, scheduling complexity and compatibility with existing systems.

Above challenges are promising to be tackled by sufficient study of characteristics of containers and novel performance prediction methods.

The remainder of this paper is organized as follows. Section 2 illustrates challenges existing in each modeling methods, while Sect. 3 contrasts numerical analysis with machine learning methods in detail and Sect. 4 concludes the paper and comes up with new potential optimization direction of predictions.

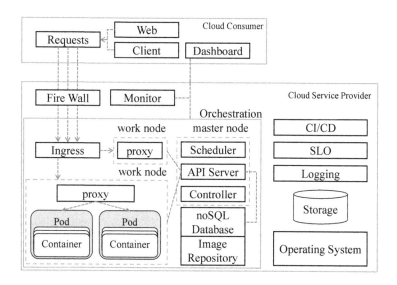

Fig. 1. Paradigm of containerized cloud.

2   Challenges in Performance Modeling

As paradiam of containerized cloud illustrated in Fig. 1, orchestration plays the core role inside a containerized cloud. In essence, container-orchestration systems is the manager of container life-circle, and their strategies determine the QoS (Quality of Service) and resource efficiency of the long-run services. In this section, we studied existing container modeling solutions. To our knowledge, container performance modeling methods can be divided into two main parts: traffic modeling methods that study outer factors and resource modeling methods that capture inner characteristics from containers to whole containerized cloud. Their benefits and challenges are concluded in Table 1.

Table 1. Modeling Solutions for containerized cloud

Category	Solutions	Benefits	Challenges
Incoming	Requests;	User behaviors characterized;	Lack of difference between tasks;
	Task structure;	Inner relationship of tasks captured;	Dependencies are hard to be obtained;
	Scientific computing;	Accurate processing time prediction;	Scenarios are specified and highly limited;
Runtime	Physical constraints identification;	Clarified boundaries;	Heterogeneous problems;
	Dynamic resource consumption;	Fine-grained model;	Absence of external factors;

2.1 Incoming Modeling

Request. Service calls, usually raised by requests event-driven, are strongly related to workload. However, every single container has limited request process capacity. The study of requests characteristics provided lots of inspiration for how to rule the performance of containers based on the real-time requests situation [15].

Task Structure/Topology. As dependencies existing between tasks, order of tasks processing is always strictly defined in advance. Bao et al. [1] modeled every application as DAGs (Directed Acyclic Graphs), in which services inside of applications are defined as verticals and volume of data delivered between each service is defined as weight. Yang et al. [27] transformed DAG-formed tasks to queue models, thus high complexity is avoided. However, dependencies aren't always explicitly declared, so such modeling methods' application scenario is highly restricted.

Scientific Computing. Nowadays, researchers in physics, chemistry, and biology start to upload their datasets to the scientific computing cloud for faster processing speed and lower monetary cost. Baughman et al. [3] found that algorithm configuration, datasets and the load level almost totally determined the performance of applications. However, in most of the general cloud systems, incoming data is organized as streams. So, modeling for datasets is limited to a narrow scenario.

2.2   Resource Modeling

As different services reveal diverse resource consumption patterns [7], one of the most tricky challenges is how to properly place containers with different resource patterns so that resource utilization and communication latency will be well balanced.

To better explain above issues, we assume that there are three services, which are named as A1S1, A1S2 and A1S3 separately and construct one single application together. Two manners of placement strategy are showed on Fig. 2 and distinguished by gray dashed line. The first represents the anti-affinity manner. Services are distributed to different containers, which will maximize the resource balancing (services with different patterns are co-located on the same machine) and disaster recovery (a few of failed nodes no longer matters). However, the extreme anti-affinity manner introduces significant communication overhead. On the right of the dashed line is the affinity manner, which represents that all services of one application are deployed into one single node. Affinity manner minimizes the communication overhead, but it worsens the resource efficiency, as some physical resources will be idle or saturate (memory and I/O in the example respectively). To achieve a better trade-off between resource efficiency and QoS violation, sufficient resource modeling study should be carried out.

Physical Constraints Identification. Profiling physical resource is challenging in practice, as it's difficult to decide which level of abstraction can best satisfy demand. At the same time, as legacy servers still deployed inside the clusters, heterogeneous problems make it tougher to handle resource modeling. In order to deal with the relationship between resource demand and resource availability, physical resource boundaries should be firstly clarified [12]. Moreover, heterogeneous resource like GPU (Graphic Processing Unit) hasn't been well supported by virtualization. To accommodate this, reasonable methods of allocating GPU resources were proposed to avoid contentions [6, 26].

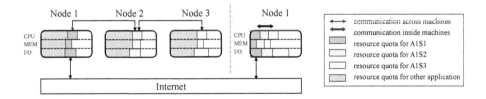

Fig. 2. Different manners of placement

Dynamic Resource Consumption. Dynamic features of consumption capture the status of the resource (e.g. CPU, memory, disk I/O, network bandwidth, etc.) allocated to containers or left for idle. Rahman et al. [22] modeled end-to-end latency for micro-service by utilizing collected CPI (Clock per Instruction) of the VMs which host containerized micro-services. CPI describes the contention on physical resources caused by different containers or applications. However, most resource modeling methods make predictions without attention to incoming traffics. Lack of concern about external factors will limit the range of observation and thus hard to make accurate predictions in advance.

Table 2. Related Works in Performance Modeling and Prediction

Reference	CPU	MEM	NET	TOP	INC	LAT	BEH	ETC	Predicted objects	Constraints**	Base algorithm	Evaluation***
Meng et al. [18]	•								Container	\	ARIMA	Simulated
Zhao et al. [28]	•								Pod	\	ARIMA	Simulated
Kalim et al. [15]	•		•						Distributed systems	\	Prophet	Simulated
Callara et al. [5]					•				Application	Sampling rate	Probabilistic	Real-world
Bauer et al. [2]	•	•	•		•				Micro-service	\	Hybrid decomposition	Simulated
Hu et al. [12]	•	•	•		•				Container	Multi-resource	MCFP***	Simulated
Bao et al. [1]	•	•			•		•	•	Micro-service	Budget	Regression	Simulated
Baughman et al. [3]	•	•			•				Scientific workflow	\	Regression	Real-world
Thinakaran et al. [26]						•			Container	Resource	Regression	Simulated
Rahman et al. [22]	•					•			Micro-service	SLO	Regression/LSTM	Simulated
Jindal et al. [14]	•			•	•			•	Micro-service	SLO	Regression	Simulated
Scheuner et al. [23]	•	•	•	•				•	VM	\	Regression	Real-world
Podolskiy et al. [21]	•			•	•				Container	SLO	Regression	Real-world
Yang et al. [27]			•			•	•		Micro-service	Consumer number	LSTM	Real-world
Tang et al. [24]	•	•	•						Container	\	LSTM	Real-world

* "MEM" stands for "memory", "NET" for "network", "TOP" for "throughput", "INC" for "Incoming traffic/ request", "LAT" for "latency", "BEH" for "behavior", "ETC" for "other metrics".
** "\" means "not defined or quantified explicitly in the paper".
*** "MCFP" stands for "minimum-cost flow problem".

3 Solution Comparation: Numerical Analysis or Machine Learning?

Some studies supposed that workloads of cloud environment are changing irregularly, so its performance is unable to be predicted [4]. Under such assumptions, most works make scheduling decision base on performance-aware mechanism. However, their methods inevitably lag behind current demands of containers. To pre-empt the delays in scheduling and further improve the resource-efficiency, performance prediction is necessary.

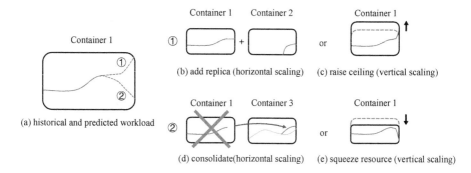

Fig. 3. Why prediction?

Benefits of container performance prediction can be illustrated in Fig. 3. In (a), historical workload is represented in solid line, while predicted workload in dashed line. Let's think about two predicted workload trend: increase and decrease. As light-weight nature of containers, horizontal and vertical scaling decision can be done in advance to avoid significant performance fluctuation. If coming spike can be foreseed by schedulers, replica of a container can be added to balance the additional workload, as shown in (b). Although action of activating new container replicas is transparent to cloud users, activating itself is still a CPU-intensive work. Therefore, as illustrated by in (c), raising resource ceilings for containers is an easier way to meet the surging demand. Above actions aim to maintain high-available services, in contrast, if workload will go down in the future, fractional idle resources can be re-collected for higher resource efficiency. In horizontal way demonstrated in (d), containers with low workload can be consolidated into one. Also, shown in (e), resource can be squeezed for other containers by vertical scaling way.

In this section, we will take an insight of state-of-art solutions for container performance prediction. As two main methods of implements, numerical analysis and machine learning work with totally different manner and have their own advantages and drawbacks (summarized in Table 3).

Table 3. Predicting solutions for containerized cloud

Category*	Solution	Advantages	Disadvantages
NA	ARIMA	Low complexity; Fast process time;	Sensitive to outliers; Lack of internal dynamics;
	Prophet	Support for seasonality & holiday effect;	Lack of internal dynamics;
ML	Regression	Training rapidly; Interpretable model;	Sensitive to outliers;
	LSTM	Well support for multi-variables and multi-objectives problems;	Large training datasets are necessary; Fail in some corner cases;

* "NA" stands for Numerical Analysis, while "ML" represents for Machine Learning.

3.1   Numerical Analysis

On behalf of a wide range of methods to search for mathematical laws underlying raw data, numerical analysis plays an important role in the design of the prediction mechanism. Although many researchers blamed traditional numerical analysis methods for their inflexibility and suboptimal solutions, short calculation time and problem-specific design make numerical analysis a competitive candidate for solving performance prediction in containerized cloud.

ARIMA (Autoregressive Integrated Moving Average)-Based Methods. ARIMA is commonly adopted in stationary time series scenario. Recently, efforts in adopting ARIMA in cloud environment harvested the good results by forecasting coming workload spikes [18,28]. However, as can be seen from Table 2, above ARIMA-based methods only took CPU as the indicator. To our knowledge, it is hard to capture internal dynamics of the system by observing a few features. Also, data source fed in ARIMA-based methods should be regularly spaced. That means, high-quality monitoring and elaborately preprocess are necessary. Although time series analysis methods' fatal flaws hampered their future development, their low complexity makes it a light-weight way to predict rapidly without too much computing/energy consumption.

Prophet. To avoid drawbacks (e.g. observed data should be spaced equally and outliers should be removed in advance) existing in ARIMA, Facebook released a temporal sequence prediction framework called Prophet [9]. Here, we briefly explain the mechanism of Prophet.

$$y(t) = g(t) + s(t) + h(t) + \epsilon_t \tag{1}$$

$g(t)$ represents the trend model, $s(t)$ stands for seasonality (e.g. fluctuate daily, weekly or yearly), $h(t)$ is the holiday effect (e.g. festival or big sale), and ϵ_t is a normally distributed error item. Prophet generates future time series based on the yearly, weekly, and daily seasonality of the historical data, so it outperformed most of other similar frameworks in scale and flexibility [25]. By integrating Prophet framework as traffic forecast methods, Kalim et al. [15] established a model to obtain the throughput rate between each pair of components. However, Prophet framework is still lack of concerning of internal dynamics, and makes prediction from the surface.

As indicated above, the goal of numerical analysis is not to solve problems in far-ranging scenarios. In contrast, they are designed to fit in a highly specific area. However, elaborate adjustments of numerical analysis methods will gain certain accuracy in prediction and thus release positive effects on system performance.

Table 4. Works in prediction optimization

Reference	Target	FS*	AD*	EL*	Evaluation	Baseline
Liu et al. [16]	Adaptability	×**	●	×	Real-world	Basic version model (err)
Iqbal et al. [13]	Adaptability	×	●	●	Real-world	Liu et al. [16] (err)
Grohmann et al. [10]	Manual overhead	●	●	●	Simulated	Manually labelling (err)
Eismann et al. [8]	Manual overhead	●	●	×	Simulated	Basic version model (speed)
Rahman et al. [22]	Manual overhead	●	\	×	Simulated	SLO target

* "FS" stands for feature selection, while "AD" for adaptive design, "EL" for ensemble learning.
** "●" stands for fully support, while "×" for not support, "\" for not clearly claimed.

3.2 Machine Learning

Cloud services iterate rapidly, approaches without adaptability will soon become obsolete. Every update of the numerical analysis costs much domain expert efforts. To avoid such defect, machine learning methods, which don't need too much prior knowledge, are becoming increasingly popular [11, 20].

Regression. Many works utilize regression models to explore quantified relationship between monitored metrics and container service performance [1, 3, 14, 21–23, 26]. With benefits of simplicity, training of most regression can be done rapidly and models are interpretable enough and easy to be generalized to more areas. However, regression methods are sensitive to outliers. As load of containers is highly dynamic, high latency or crash down of communication between containers may cause logging system to collect certain ratio of abnormal data. Regression model is not good at handling such exceptions, so preprocess of outliers is necessary if regression is applied in the containerized scenario.

LSTM (Long Short-Term Memory). Traditional time series methods can't deal with multi-variables and multi-objectives problems well. To tackle this problem, many researchers start to employ LSTM into containerized cloud. LSTM works well in the field of time series and thus suitable in prediction. Tang et al. [24] captured features between metrics by bidirectional LSTM, which outperformed basic LSTM in the ability of context-awareness. In the scenario of Web service and database, the accuracy of prediction improved over 50% compared to traditional ARIMA and LSTM-based methods. Miao et al. [19] utilized LSTM to make traffic prediction with the concern of heterogeneous characteristics of edge nodes. Although LSTM offers an excellent way to solve the multi-variables and multi-objectives problems, as back-propagation process will fail when the layers are too deep, LSTM can't deal with periods exhibited in days or longer. Moreover, effective LSTM models are based on large training sets, and even highly manually adjusted machine learning methods may fail in some corner cases. Therefore, before we apply a new prediction solution, sufficient test and comparison between numerical analysis and machine learning methods should be done in several scenarios. Also, as listed in Table 4, manually tuning can be replaced by automatic methods gradually. Automatic methods not only save time and efforts, but also make predicting models adaptive enough to new environment even after certain modification in configurations.

4   Conclusion

In this paper, we clarified challenges in container performance modeling and emphasized a lot in the state-of-art achievements in terms of profiling and predicting containers. We illustrated pros and cons in solutions proposed by researchers and provided suggestions according to different scenarios. For future work, practical experiments will be carried out to compare each solution quantitatively. Moreover, based on sufficient insights, we plan to propose our own design of container orchestration platform.

Acknowledgments. This work is supported by the National Natural Science Foundation of China (No. 61972118).

References

1. Bao, L., Wu, C., Bu, X., Ren, N., Shen, M.: Performance modeling and workflow scheduling of microservice-based applications in clouds. IEEE Trans. Parallel Distrib. Syst. **30**(9), 2114–2129 (2019)
2. Bauer, A., Lesch, V., Versluis, L., Ilyushkin, A., Herbst, N., Kounev, S.: Chamulteon: coordinated auto-scaling of micro-services. In: 2019 IEEE 39th International Conference on Distributed Computing Systems (ICDCS), pp. 2015–2025. IEEE (2019)
3. Baughman, M., Chard, R., Ward, L.T., Pitt, J., Chard, K., Foster, I.T.: Profiling and predicting application performance on the cloud. In: UCC, pp. 21–30 (2018)
4. Boza, E.F., Abad, C.L., Narayanan, S.P., Balasubramanian, B., Jang, M.: A case for performance-aware deployment of containers. In: Proceedings of the 5th International Workshop on Container Technologies and Container Clouds, pp. 25–30 (2019)
5. Callara, M., Wira, P.: A probabilistic learning approach for predicting application launches in cloud computing architectures. In: 2019 IEEE/SICE International Symposium on System Integration (SII), pp. 584–589. IEEE (2019)
6. Chen, Q., Oh, J., Kim, S., Kim, Y.: Design of an adaptive GPU sharing and scheduling scheme in container-based cluster. Cluster Comput. **23**, 1–13 (2019)
7. Chen, W.Y., Ye, K.J., Lu, C.Z., Zhou, D.D., Xu, C.Z.: Interference analysis of co-located container workloads: a perspective from hardware performance counters. J. Comput. Sci. Technol. **35**, 412–417 (2020)

8. Eismann, S., Grohmann, J., Walter, J., Von Kistowski, J., Kounev, S.: Integrating statistical response time models in architectural performance models. In: 2019 IEEE International Conference on Software Architecture (ICSA), pp. 71–80. IEEE (2019)
9. Facebook Open Source: Prophet: Automatic forecasting procedure. https://github. com/facebook/prophet. Accessed 7 Oct 2020
10. Grohmann, J., Eismann, S., Elflein, S., Kistowski, J.V., Kounev, S., Mazkatli, M.: Detecting parametric dependencies for performance models using feature selection techniques. In: 2019 IEEE 27th International Symposium on Modeling, Analysis, and Simulation of Computer and Telecommunication Systems (MASCOTS), pp. 309–322. IEEE (2019)
11. He, H., Hu, J., Da Silva, D.: Enhancing datacenter resource management through temporal logic constraints. In: 2017 IEEE International Parallel and Distributed Processing Symposium (IPDPS), pp. 133–142. IEEE (2017)
12. Hu, Y., Zhou, H., de Laat, C., Zhao, Z.: Concurrent container scheduling on heterogeneous clusters with multi-resource constraints. Future Gener. Comput. Syst. **102**, 562–573 (2020)
13. Iqbal, W., Berral, J.L., Erradi, A., Carrera, D., et al.: Adaptive prediction models for data center resources utilization estimation. IEEE Trans. Netw. Serv. Manag. **16**(4), 1681–1693 (2019)
14. Jindal, A., Podolskiy, V., Gerndt, M.: Performance modeling for cloud microservice applications. In: Proceedings of the 2019 ACM/SPEC International Conference on Performance Engineering, pp. 25–32 (2019)
15. Kalim, F., et al.: Caladrius: a performance modelling service for distributed stream processing systems. In: 2019 IEEE 35th International Conference on Data Engineering (ICDE), pp. 1886–1897. IEEE (2019)
16. Liu, C., Liu, C., Shang, Y., Chen, S., Cheng, B., Chen, J.: An adaptive prediction approach based on workload pattern discrimination in the cloud. J. Netw. Comput. Appl. **80**, 35–44 (2017)
17. Maenhaut, P.J., Volckaert, B., Ongenae, V., De Turck, F.: Resource management in a containerized cloud: status and challenges. J. Netw. Syst. Manag. **28**(2), 197–246 (2020)
18. Meng, Y., Rao, R., Zhang, X., Hong, P.: CRUPA: a container resource utilization prediction algorithm for auto-scaling based on time series analysis. In: 2016 International Conference on Progress in Informatics and Computing (PIC), pp. 468–472. IEEE (2016)
19. Miao, Y., Wu, G., Li, M., Ghoneim, A., Al-Rakhami, M., Hossain, M.S.: Intelligent task prediction and computation offloading based on mobile-edge cloud computing. Future Gener. Comput. Syst. **102**, 925–931 (2020)
20. Orhean, A.I., Pop, F., Raicu, I.: New scheduling approach using reinforcement learning for heterogeneous distributed systems. J. Parallel Distrib. Comput. **117**, 292–302 (2018)
21. Podolskiy, V., Mayo, M., Koey, A., Gerndt, M., Patros, P.: Maintaining SLOs of cloud-native applications via self-adaptive resource sharing. In: 2019 IEEE 13th International Conference on Self-Adaptive and Self-Organizing Systems (SASO), pp. 72–81. IEEE (2019)
22. Rahman, J., Lama, P.: Predicting the end-to-end tail latency of containerized microservices in the cloud. In: 2019 IEEE International Conference on Cloud Engineering (IC2E), pp. 200–210. IEEE (2019)

23. Scheuner, J., Leitner, P.: Estimating cloud application performance based on micro-benchmark profiling. In: 2018 IEEE 11th International Conference on Cloud Computing (CLOUD), pp. 90–97. IEEE (2018)
24. Tang, X., Liu, Q., Dong, Y., Han, J., Zhang, Z.: Fisher: an efficient container load prediction model with deep neural network in clouds. In: 2018 IEEE International Conference on Parallel & Distributed Processing with Applications, Ubiquitous Computing & Communications, Big Data & Cloud Computing, Social Computing & Networking, Sustainable Computing & Communications (ISPA/IUCC/BDCloud/SocialCom/SustainCom), pp. 199–206. IEEE (2018)
25. Taylor, S.J., Letham, B.: Forecasting at scale. Am. Stat. **72**(1), 37–45 (2018)
26. Thinakaran, P., Gunasekaran, J.R., Sharma, B., Kandemir, M.T., Das, C.R.: Kube-knots: resource harvesting through dynamic container orchestration in GPU-based datacenters. In: 2019 IEEE International Conference on Cluster Computing (CLUSTER), pp. 1–13. IEEE (2019)
27. Yang, Z., Nguyen, P., Jin, H., Nahrstedt, K.: MIRAS: model-based reinforcement learning for microservice resource allocation over scientific workflows. In: 2019 IEEE 39th International Conference on Distributed Computing Systems (ICDCS), pp. 122–132. IEEE (2019)
28. Zhao, A., Huang, Q., Huang, Y., Zou, L., Chen, Z., Song, J.: Research on resource prediction model based on kubernetes container auto-scaling technology. In: IOP Conference Series: Materials Science and Engineering, vol. 569, p. 052092. IOP Publishing (2019)

A Random Access Control Scheme for a NOMA-Enabled LoRa Network

Wei Wu[1,2], Wennai Wang[1,2(✉)], Jihai Yang[1], and Bin Wang[1,2]

[1] Nanjing University of Posts and Telecommunications, Nanjing, China
`wangwn@njupt.edu.com`
[2] Key Laboratory of Broadband Wireless Communication and Sensor Network
Technology, Ministry of Education, Nanjing, China

Abstract. LoRa is one of the most prominent Low-Power Wide-Area
Network (LPWAN) technologies, to accommodate pervasive Internet-of-
Things (IoT) connectivities. However, its service capacity and scalability
are limited due to the scarce channel resources and the Aloha-like random access mechanism specified by LoRaWAN. We propose a NOMA-
enabled LoRa gateway, which permits multiple end-devices to transmit
their data at the same time over a shared channel. The whole random
access process are provided in detail, including collision resolution and
transmission scheduling based on a Distributed Queuing (DQ) method.
In addition to that, Spreading Factor (SF) allocation in the transmission
scheduling phase is also considered and an optimal problem is formulated to achieve maximum data transmission rate. In order to solve the
problem efficiently, an SF allocation algorithm is developed based on
the matching theory. Numerical results show that our proposed scheme
significantly enhances the sum achievable user rate when the number of
users increases.

Keywords: LoRa · Distributed queueing · Non-orthogonal multiple
access · Spreading factor allocation · Optimal solution

1 Introduction

Internet of Things (IoT) is a world wide network with pervasive interconnected
smart devices [1,2]. It requires advanced technologies, like low-power wide-area
Networks (LPWANs), to enable energy-efficient long-range communications.
Among LPWAN technologies, LoRa is widely adopted [3].

LoRa is based on chirp spread spectrum (CSS) modulation and spreading
factor (SF) is one of the most important transmission parameters. Six SFs (7–
12) are available, among which smaller SFs provide higher data rates but reduced
ranges, and vise versa. On top of LoRa Physical (PHY) layer, LoRaWAN defines
the system architecture and upper layers [5]. The architecture is a star-of-stars
topology where gateways (GWs) transparently relay the messages from end-
devices (EDs) to network server (NS). The medium access control (MAC) layer
protocol is based on Aloha.

© ICST Institute for Computer Sciences, Social Informatics and Telecommunications Engineering 2021
Published by Springer Nature Switzerland AG 2021. All Rights Reserved
H. Gao et al. (Eds.): ChinaCom 2020, LNICST 352, pp. 403–420, 2021.
https://doi.org/10.1007/978-3-030-67720-6_28

The service capacity and scalability of LoRa networks depend on available channel resources and random access mechanism. On one hand, the SF-domain provides another channel resource domain for LoRa, apart from the time- and frequency-domain, since the orthogonality of SFs were demonstrated theoretically [4]. Multiple users can access to the shared time-frequency channel simultaneously using different SFs. However, the number of usable SFs is limited and in practice their complete orthogonality can not be ensured [6]. On the other hand, the throughput performance of Aloha-like MAC protocol is notorious under high traffic load [3], which is directly responsible for LoRa's scalability issue.

With short supply of channel resources, non-orthogonal multiple access (NOMA) has potential to enhance spectral efficiency and support massive connectivity by multiplexing users in power- or code-domain [1]. For the power-domain NOMA, superposition coding (SC) and successive interference cancellation (SIC) are employed to code and decode multiple users' signals. We focus on the power-domain NOMA, which is simply called NOMA later in this paper. In the view of the random access mechanisms, distributed queuing (DQ) is an alternative to Aloha due to its stable and scalable performance in densely loaded networks [9,10].

With that in mind, we propose to leverage the advantages of NOMA and DQ technologies in LoRa networks, in order to address the scalability issue and improve the service capacity. In our NOMA-enabled LoRa network, the gateway coordinates users' contention resolution and data transmission based on a DQ method, and it is NOMA-enabled to permits multiple users to overlap their signals on a shared channel. We firstly provide the whole random access process in detail, including the collision resolution and transmission scheduling. Then, we consider the SF assignment problem for the transmission scheduling phase and formulate it as a maximization problem of achievable rate, under the constraints of inter-SF and successive interferences. For the mathematical intractability of the optimization problem, we propose an SF allocation algorithm based on matching theory.

The remainder of this paper is organized as follows. Section 2 reviews the related works. Section 3 describes the DQ-based access process in details. The SF allocation problem and allocation algorithm are given out in Sect. 4. Numerical results are shown in Sect. 5. Finally, conclusions are made in Sect. 6.

2   Related Works

NOMA has been applied to LPWANs for improving system performance [1,2,7, 8]. Authors in [1] model the uplink NOMA in LPWAN and demonstrate that by employing their proposed resource allocation algorithm NOMA can significantly enhance LPWAN. Authors in [2,7,8] leverage NOMA in the context of NB-IoT. [2] proposed a user clustering scheme for a power-domain NOMA aided NB-IoT system, while [7] focuses on the subcarrier and power allocation to maximize the connection density of NB-IoT systems with NOMA. [8] developed an access algorithm that features cluster-based reusable preamble allocation, to enhance the access efficiency for NB-IoT networks.

DQ protocol is first introduced by Xu and Campbell for cable TV distribution [9]. This basic protocol mechanism has been extended to LPWANs with large number of devices [11–13]. [11] proposed a highly-efficient low-power MAC protocol (LPDQ) for wireless data collection scenarios, while [12] and [13] designed DQ-based protocols for crowdsourcing LPWAN and LoRa, respectively. All these works show high throughput regardless of the number of connected devices, due to DQ's stable and scalable performance for densely loaded networks.

Besides, various SF allocation algorithms have been proposed for different objectives, namely, avoiding near-far problems [15], improving network performance [16], enhancing energy efficiency [17], and throughput fairness [18,19]. Among them, the schemes in [17–19] are based on matching theory.

3   Random Access Scheme

3.1   DQ Basics

DQ protocol is based on tree-split algorithm, in which contention resolution and data transmission are separated into two phases: At first, active devices (i.e., the devices who have data to transmit) contend the shared channel by sending short access request (AR). Then, successful devices are allowed to transmit their data in order, whereas collided devices wait for subsequent contention resolution. To inform the devices about the state of each contention, a central coordinator is usually required to broadcast feedback information. Therefore, the frame structure is divided into three parts: (i) m minislots for collision resolution, (ii) one data slot (DS) for uplink collision-free data transmission, and (iii) one feedback slot for downlink feedback information.

Besides, there are two logical queues to manage the collided and successful devices, collision resolution queue (CRQ) and data transmission queue (DTQ). The rules are as follows. The devices that collide in the same CS are organized into one group (called collided group hereafter). All the collided devices are split into different groups and queued into CRQ. Whereas, each successful device enters into one position of the DTQ, waiting for collision-free data transmission. More details about DQ can be found in [9].

3.2   Random Access Process

We consider a NOMA-enabled LoRaWAN system with one GW and N EDs, as depicted in Fig. 1. The GW is located at the center of the cell whose radius is R km, and the EDs randomly distributed in it. One SF can be reused by two or more EDs multiplexed in the power-domain. The sets of EDs and available SFs are denoted by $\mathcal{N} = \{1, 2, ..., N\}$ and $\mathcal{S} = \{7, 8, ..., 12\}$, respectively.

The MAC layer access scheme in our NOMA-enabled LoRa network is developed based on DQ mechanism. Note that the rules of DTQ in this paper is a bit different from the native DQ, since multiple users' data streams can be transmitted in the same DS by combining the power- and SF-domain resources.

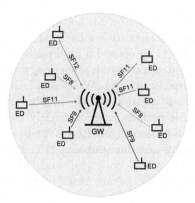

Fig. 1. A NOMA-enabled LoRa network, where multiple EDs use the same SF channel in power-domain and multiple SFs are used simultaneously.

The modified rules are as follows. Among the successful devices in one minislot, some of them are allowed to transmit in the same DS and they are organized into one group (called successful group later). Each successful group enters into one position of DTQ. For that, the feedback information should also indicate, apart from the state of each CS, the allocated SF for each successful ED and whether it is permitted to transmit in the next frame.

Figure 2 shows an example to illustrate the whole random access process. In this example, there are ten EDs and one GW, and three available SFs (SF7–9). Besides, we make two assumptions: (i) Six contention slots (CSs) are set in each contention window (CW). Since the speed of collision resolution is faster than data transmission in native DQ when three or more CSs are used [10], it is reasonable to set more CSs when multiple users' data can be transmitted in one DS. (ii) Each SF can be assigned to at most two EDs. More number of users' signals superimposed will increase the complexity of receiver.

- **Synchronization:** a beacon period starts after the GW broadcasts a beacon (BCN) signal. Ten active devices (ED1–10) are synchronized and switch to Class B mode.
- **Superframe 1:** all the devices contend in the CW by sending ARs. ED3, 5, 6, and 9 succeed in the second, forth, fifth and sixth CS respectively, while others collide. After that, the GW broadcasts feedback (FB) in the FS. Note that there is no data transmission in this frame because DTQ is empty at that time. All the devices receive the FB.
  Two collided groups ({1, 4, 7, 8} and {2, 10}) enter into the first two positions of CRQ, while two successful groups ({3, 5, 6} and {9}) enter into the first two positions of DTQ. Note that although ED9 succeeds in this CW, it queues in the second position since it is not allowed to transmit in the next frame.
- **Superframe 2:** the first group in the CRQ (ED1, 4, 7 and 8) contend in the CW and they both succeed. The first group in the DTQ (ED3, 5 and 6) transmit their data using assigned SFs in the DS. GW broadcasts FB in the FS.

ED1, 4, 7 and 8 are allowed to transmit in the next superframe and enter into the first position of DTQ.

- **Superframe 3:** ED2 and 10 contend in the CW. Subsequently, ED1, 4, 7, 8 and 9 transmit data using different SFs. GW still broadcasts FB in the FS. ED2 and 10 enter into the first position of DTQ.
- **Superframe 4:** the CW is empty since the CRQ is empty at that time. Then, ED2 and 10 transmit data in the DS, and the GW broadcasts FB in the FS.

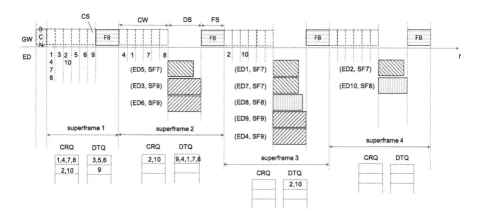

Fig. 2. An example to illustrate the random access process in the NOMA-enabled LoRa network.

4   Spreading Factor Allocation

4.1   Problem Formulation

For each ED n assigned to SF s, it would be impaired by (one of) two kinds of interferences: (a) inter-SF interference, caused by the EDs using SF j ($j \neq s$); (b) co-SF interference, resulting from the EDs using SF s. Note that when the NOMA-enabled GW performs SIC, the strongest signal is decoded at first from the composite received signal and then it will be reconstructed and subtracted from the received signal, after that, the next strongest and so on. Therefore, among the co-SF interferences, only the EDs with lower received power than ED n will affect ED n's reception, which is called successive interference. So that we consider inter-SF (θ_s) and successive interferences in this paper, and the inter-SF and SIC (μ) capture thresholds are listed in Table 1 [19,22].

The channel gain between ED n ($n \in \mathcal{N}$) and the GW, considering both path-loss and small-scale fading, can be expressed as,

$$h_n = \frac{|g_n|^2 \, \mathrm{A}(f_c)}{d_n^\alpha}, \tag{1}$$

Table 1. Inter-SF and SIC capture thresholds

SF	7	8	9	10	11	12
Inter-SF threshold θ_s (dB)	−7.5	−9	−13.5	−15	−18	−22.5
SIC threshold μ (dB)	6					

where g_n, f_c, d_n, and α are the small-scale channel fading gain, carrier frequency, distance between ED n and the GW, and path-loss exponent, respectively. $A(f_c) = (f_c^2 \times 10^{-2.8})^{-1}$, is the deterministic path-loss term [19].

Then, the signal-to-interference-plus-noise-ratio (SINR) can be expressed as,

$$\text{SINR}_{ns} = \frac{p_n \gamma_n}{I_{\text{iSF}} + I_{\text{SI}} + 1}, \tag{2}$$

where p_n is the transmit power of ED n and $\gamma_n = h_n/\sigma_c^2$, where σ_c^2 is the power of additive white gaussian noise. I_{iSF} and I_{SI} are the power of inter-SF and successive interferences, which can be expressed as

$$I_{\text{iSF}} = \sum_{j \in \mathcal{S}\setminus\{s\}} \sum_{i \in \mathcal{N}\setminus\{n\}} x_{ij} p_i \gamma_i, \tag{3}$$

$$I_{\text{SI}} = \sum_{i=n+1}^{|\mathcal{U}_s|} x_{is} p_i \gamma_i, \tag{4}$$

where x_{ij} is an indicator variable for SF assignment, 1 when ED i assigned to SF j and 0 otherwise. $\mathcal{U}_s$ is the set of EDs assigned to SF s, and they are sorted according to the descending order of received power. Considering Rayleigh fading channels, γ_n can be modeled as an exponential random variable with mean $\bar{\gamma}_n = A(f_c)/(d_n^\alpha \sigma_c^2)$ [19].

It is worth to mention that Eq. (2) is a general SINR expression. In some cases, the I_{iSF} or I_{SI} term not exists. For the detailed expression, there are three cases when $N > 1$ (let $|\mathcal{S}_A| > 1$ denotes the set of used SFs):

- **Case 1:** $|\mathcal{S}_A| > 1$ and $|\mathcal{U}_s| = 1$. ED n is only impaired by inter-SF interference, i.e., $\text{SINR}_{ns}^{(\text{case1})} = p_n \gamma_n/(I_{\text{iSF}} + 1)$. Therefore, its transmission can be successfully decoded if it satisfies the inter-SF capture condition and signal reception condition.
- **Case 2:** $|\mathcal{S}_A| = 1$ and $|\mathcal{U}_s| > 1$. ED n is only impaired by successive interference, i.e., $\text{SINR}_{ns}^{(\text{case2})} = p_n \gamma_n/(I_{\text{SI}} + 1)$. The transmission can be successfully decoded if ED n satisfies the SIC and signal reception conditions.
- **Case 3:** $|\mathcal{S}_A| > 1$ and $|\mathcal{U}_s| > 1$. ED n is subject to both inter-SF and successive interferences, i.e., $\text{SINR}_{ns}^{(\text{case3})} = p_n \gamma_n/(I_{\text{iSF}} + I_{\text{SI}} + 1)$. The transmission can be successfully decoded when the inter-SF, SIC, and signal reception conditions are all satisfied.

According to [20], the relationship of these three conditions is "SIC capture threshold> SIC threshold> reception condition".

As the access process shown in Fig. 2, SF allocation should be performed every superframe, during which the path-loss term can be assumed to be fixed. Therefore, we consider achievable short-term average rate when formulate the SF allocation problem. The achievable short-term average rate (simply achievable rate hereafter) is defined as the multiplication of successful probability and data bitrate [20], i.e.,

$$\mathrm{AR}_{ns} = R_s \times P_{ns}, \tag{5}$$

where P_{ns} is the probability of successful reception and $R_s = (s \cdot \mathrm{CR} \cdot \mathrm{BW})/2^s$ is the data bitrate, where BW is the bandwidth and CR is the coding rate. In LoRa, there are three typical bandwidths (125, 250, and 500 kHz) and four coding rates (4/5, 4/6, 4/7, and 4/8) [14].

In the first case, the successful probability is,

$$P_{ns}^{(\mathrm{case1})} = P\left(\mathrm{SINR}_{ns}^{(\mathrm{case1})} \geq \theta_s\right), \tag{6}$$

According to [19], Eq. (6) can be written as,

$$P_{ns}^{(\mathrm{case1})} = e^{-\frac{\theta_s \sigma_c^2 d_n^2}{A(f_c)p_n}} \prod_{j \in \mathcal{S} \setminus \{s\}} \prod_{i \in \mathcal{N} \setminus \{n\}} \frac{1}{\theta_s x_{ij} \frac{p_i}{p_n} \left(\frac{d_n}{d_i}\right)^\alpha + 1}. \tag{7}$$

In the second case, the successful probability is changed to,

$$P_{ns}^{(\mathrm{case2})} = P\left(\mathrm{SINR}_{ns}^{(\mathrm{case2})} \geq \mu\right), \tag{8}$$

We derive $P_{ns}^{(\mathrm{case2})}$ with similar calculations as in [19] (see Appendix 1),

$$P_{ns}^{(\mathrm{case2})} = e^{-\frac{\mu \sigma_c^2 d_n^2}{A(f_c)p_n}} \prod_{i=n+1}^{|\mathcal{U}_s|} \frac{1}{\mu x_{is} \frac{p_i}{p_n} \left(\frac{d_n}{d_i}\right)^\alpha + 1}. \tag{9}$$

In the last case, the successful probability is,

$$P_{ns}^{(\mathrm{case3})} = P\left(\mathrm{SINR}_{ns}^{(\mathrm{case3})} \geq \mu\right), \tag{10}$$

which can be written as (see Appendix 2),

$$P_{ns}^{(\mathrm{case3})} = e^{-\frac{\mu \sigma_c^2 d_n^2}{A(f_c)p_n}} \prod_{j \in \mathcal{S} \setminus \{s\}} \prod_{i \in \mathcal{N} \setminus \{n\}} \frac{1}{\mu x_{ij} \frac{p_i}{p_n} \left(\frac{d_n}{d_i}\right)^\alpha + 1}$$
$$\times \prod_{i=n+1}^{|\mathcal{U}_s|} \frac{1}{\mu x_{is} \frac{p_i}{p_n} \left(\frac{d_n}{d_i}\right)^\alpha + 1}. \tag{11}$$

Given the above analysis, the general expression of successful probability can be written as,

$$P_{ns} = \mathrm{I}\left(|\mathcal{S}_A| > 1, |\mathcal{U}_s| = 1\right) \times P_{ns}^{(\text{case1})}$$
$$+ \mathrm{I}\left(|\mathcal{S}_A| = 1, |\mathcal{U}_s| > 1\right) \times P_{ns}^{(\text{case2})} \qquad (12)$$
$$+ \mathrm{I}\left(|\mathcal{S}_A| > 1, |\mathcal{U}_s| > 1\right) \times P_{ns}^{(\text{case3})}$$

where $\mathrm{I}(*)$ is an indicator function, 1 when the condition $(*)$ is verified and 0 otherwise.

Finally, the SF allocation optimization problem is formulated as follows,

$$(\text{P}) \max \sum_{s \in \mathcal{S}} \sum_{n \in \mathcal{N}} x_{ns} \mathrm{AR}_{ns} \qquad (13a)$$

$$\text{s.t.} \quad \text{C1}: \sum_{s \in \mathcal{S}} x_{ns} \leq 1, \quad \forall n \in \mathcal{N} \qquad (13b)$$

$$\text{C2}: \sum_{n \in \mathcal{N}} x_{ns} \leq \lambda_s, \quad \forall s \in \mathcal{S} \qquad (13c)$$

$$\text{C3}: x_{ns} = \{0, 1\}, \quad \forall n \in \mathcal{N}, \forall s \in \mathcal{S} \qquad (13d)$$

where λ_s is the maximal number of EDs that can be allocated to SF s. Our objective function is the maximization of sum achievable rate of all served EDs (i.e., $\sum_{s \in \mathcal{S}} x_{ns} \neq 0, \forall n \in \mathcal{N}$). Constraints C1 ensures that an ED n is assigned to at most one SF, while C2 makes that the maximal number of EDs sharing SF s is λ_s. Constraint C3 defines the binary SF allocation variable x_{ns}.

4.2   Proposed SF Allocation Algorithm

For simplicity, we assume that $p_n = p_0$ ($\forall n \in \mathcal{N}$), then the optimization problem (P) is an integer programming problem with a non-linear objective function, which is difficult to obtain its optimal solution. Therefore, we propose an SF allocation algorithm based on matching theory [21], which is a promising tool for resource allocation in wireless networks. According to the matching theory, our allocation problem (P) is classified as a many-to-one matching problem with conventional externalities and peer effects. The basic concepts of the matching theory that have been used in our algorithm are listed in Table 2.

The steps of our proposed SF allocation algorithm are as follows.

Initialization: the preference lists of EDs and SFs are initialized at first. The preference list of ED n is,

$$\mathcal{PL}_n = \{s \in \mathcal{S}, \text{s.t. } d_n \leq l_s\}, \qquad (14)$$

where l_s is the distance threshold of SF s, and $\mathcal{PL}_n$ is arranged according to the ascending order of l_s. We assume that all EDs are in the coverage area of the largest SF, hence $\mathcal{PL}_n \neq \emptyset$ ($\forall n \in \mathcal{N}$).
The preference list of SF s is,

$$\mathcal{PL}_s = \{n \in \mathcal{N}, \text{s.t. } d_n \leq l_s\}. \qquad (15)$$

Table 2. Basic concepts used in our algorithm

Concept name	Definition
Matching pair	A couple (n, s) assigned to each other
Quota of a player	The maximum number of players with which it can be matched. Each ED has a quota of 1 and each SF s has a quota of λ_s
Utility of an ED n	The achievable rate of ED n, i.e., $f_n = R_s \times P_{ns}$
Utility of an SF s	The sum achievable rate of the EDs assigned to SF s, i.e., $f_s = \sum_{n \in \mathcal{U}_s} R_s \times P_{ns}$
Preference relation	For ED n, it prefers SF s_1 over SF s_2, if the utility of n is higher when it is matched to s_1 than when it is matched to s_2. For an SF, the criterion is same
Blocking pair	A matching pair (n, s) which makes new utility $\sum_{s \in \mathcal{S}} f'_s$ higher than the current utility $\sum_{s \in \mathcal{S}} f_s$. In this case, n will leave its current match to make pair with s
Two-sided exchange stable matching	A matching solution where there is no blocking pair

$\mathcal{PL}_s$ is arranged in the ascending order of d_n. Unmatched devices are added to $\mathcal{L}_u$.

Initial Matching: for each ED n in the unmatched list $\mathcal{L}_u$, it is removed from $\mathcal{L}_u$ whenever it starts requesting the first SF in $\mathcal{PL}_n$. If the quota of the kth $(1 \leq k \leq |\mathcal{PL}_u|)$ SF allows it to accept the request, ED n is matched to this SF and stops requesting, otherwise it moves to request the $(k+1)$th SF. If ED n is not accepted by any SF in $\mathcal{PL}_n$, it is added into the set $\mathcal{L}_{ru}$. The process is repeated until $\mathcal{L}_u$ becomes empty.

SF Matching Refinement: for each matching pair (n, s), the algorithm calculates $\sum_{s \in \mathcal{S}} fs$. Firstly, if there exists an SF k whose quota allows it to accept n, the algorithm makes new pair (n, k) and calculates the new utility $\sum_{s \in \mathcal{S}} fs'$. If $\sum_{s \in \mathcal{S}} fs' > \sum_{s \in \mathcal{S}} fs$, ED n leaves SF s to be matched with SF k. Then, the algorithm makes a swap between every matched pair (m, k) and (n, s). If (n, k) or (m, s) is a blocking pair, the swap is validated. This swapping step is repeated until reaching a two-sided stable matching.

Algorithm 1. Initial Matching

Initialization: $\mathcal{L}_u \leftarrow \mathcal{N}, \mathcal{L}_{ru} \leftarrow \emptyset$.

1: **while** $\mathcal{L}_u \neq \emptyset$ **do**
2: **for** $i \in \mathcal{L}_u$ **do**
3:   $\mathcal{L}_u \leftarrow \mathcal{L}_u \backslash \{i\}$;
4:   $c = 1$;
5:   **while** $\sum_{j \in \mathcal{S}} x_{ij} < 1$ and $c \leq |\mathcal{PL}_i|$ **do**
6:    $j \leftarrow$ the cth SF in $\mathcal{PL}_i$;
7:    **if** $|\mathcal{U}_j| < \lambda_j$ **then**
8:     $x_{ij} = 1$;
9:     $\mathcal{U}_j \leftarrow \mathcal{U}_j \cup \{i\}$;
10:   **else**
11:    $c = c + 1$;
12:   **if** $\sum_{j \in \mathcal{S}} x_{ij} = 0$ **then**
13:    $\mathcal{L}_{ru} \leftarrow \mathcal{L}_{ru} \cup \{i\}$;

User Matching Refinement (when $\mathcal{L}_{ru} \neq \emptyset$): Firstly, for each ED i in $\mathcal{L}_{ru}$ and an SF j ($j \in \mathcal{PL}_i$) whose quota allows it to accept ED i, the algorithm makes a new pair (i, j). If (i, j) makes the new utility $\sum_{s \in \mathcal{S}} fs'$ higher, the new pair is validated. Then, for each SF l ($l \in \mathcal{PL}_i$) whose quota not allows it to accept ED i, the algorithm makes a swap between every (k, l) ($k \in \mathcal{U}_l$) and $(i, \emptyset)$. If (i, l) is a blocking pair, the swap is validated. This swapping step is repeated until reaching a two-sided stable matching.

Complexity: The running time of our proposed SF allocation algorithm is upper bounded by $\mathcal{O}\left(\lambda_{\max}^2 S^2 + (1 + \lambda_{\max})NS\right)$, where $\lambda_{\max} = \max_{s \in \mathcal{S}} \lambda_s$.

Proof. (i) Initial matching complexity: in the worst case, every ED has the same preference list including all the SFs and it has to search until the last one in its preference list. Therefore, the complexity of the initial matching is upper bounded by $\mathcal{O}(NS)$.

(ii) SF matching refinement complexity: for each SF s, the algorithm considers at most λ_s assigned EDs and examines $\sum_{j \in \mathcal{S} \backslash \{s\}} \lambda_j$ swap operations for each ED. Then the number of swap operations that are examined in one iteration is upper bounded by $\mathcal{O}\left(\lambda_s \sum_{j \in \mathcal{S} \backslash \{s\}} \lambda_j\right)$. The complexity of the SF matching refinement is upper bounded by $\mathcal{O}\left(\sum_{s \in \mathcal{S}}\left(\lambda_s \sum_{j \in \mathcal{S} \backslash \{s\}} \lambda_j\right)\right)$. Let $\lambda_{\max} = \max_{s \in \mathcal{S}} \lambda_s$, it can be represented as $\mathcal{O}\left(\lambda_{\max}^2 S(S - 1)\right)$.

(iii) User matching refinement complexity: in the worst case, each ED n in the set $\mathcal{L}_{ru}$ has the same preference list containing all the SFs, and at most λ_s swap operations are examined for each SF s in ED n's preference list. Hence, in each

Algorithm 2. SF Matching Refinement

```
1: change←ture;
2: while change==ture do
3:     change←false;
4:     for j ∈ S do
5:         for i ∈ U_j do
6:             for l ∈ S\{j} do
7:                 if U_l == ∅ then
8:                     Swap (i, j) and (∅, l);
9:                     if (i, l) is blocking pair then
10:                        Validate the swap;
11:                        change←true;
12:                 else
13:                     for k ∈ U_l do
14:                         Swap (i, j) and (k, l);
15:                         if (i, j) or (k, l) is blocking pair then
16:                             Validate the swap;
17:                             change←true;
```

interation the number of swap operations is upper bounded by $\mathcal{O}(\lambda_{\max} S)$. Considering at most N EDs in $\mathcal{L}_{ru}$, the complexity of the user matching refinement is upper bounded by $\mathcal{O}(\lambda_{\max} N S)$.

Therefore, the computational complexity of the SF allocation algorithm is upper bounded by $\mathcal{O}\left(\lambda_{\max}^2 S^2 + (1 + \lambda_{\max}) N S\right)$.

Table 3. Simulation parameters

Carrier frequency (f_c)	868 MHz
Bandwidth (BW)	125 kHz
ED transmit power (p_0)	14 dBm
Cell radius (R)	1 km
Path loss exponent (α)	4
Available SFs	7–12
Quota of each SF s (λ_s)	2

5   Numerical Results

We consider two baseline schemes for performance comparison with our proposed SF allocation scheme.

- **Random SF allocation:** for each ED n, it randomly requests a usable SF (i.e., the SFs in its preference list), and the SF accepts the request if its quota allows. The process stops untill ED n has been matched with an SF, or all the usable SFs has been requested.

Algorithm 3. User Matching Refinement

```
1: change←ture;
2: while change==ture do
3:     change←false;
4:     for i ∈ 𝓛_ru do
5:         for j ∈ 𝓟𝓛_i do
6:             Calculate ∑_{s∈𝓢} f_s;
7:             if |𝓤_j| < λ_j then
8:                 Make the new pair (i, j);
9:                 Calculate the new utility ∑_{s∈𝓢} f'_s;
10:                if ∑_{s∈𝓢} f'_s ≥ ∑_{s∈𝓢} f_s then
11:                    Validate the new pair;
12:                    change←true;
13:            else
14:                for k ∈ 𝓤_j do
15:                    Swap (i, ∅) and (k, j);
16:                    if (i, j) is blocking pair then
17:                        Validate the new pair;
18:                        change←true;
```

– **Distance SF allocation:** for each ED n, it requests the usable SFs in order, from the smallest to the largest SF. The process stops untill ED n has been matched with an SF, or all the usable SFs has been requested.

Simulation parameters are set under European regulations, shown in Table 3. Note that we consider a lossy urban environment with path loss exponent 4.

The performance comparison of our proposed scheme and the baseline schemes are shown in Fig. 3, where the sum achievable rate and the number of served EDs are plotted, against the number of EDs. We can observe from Fig. 3(a) that our proposed algorithm outperforms the baseline algorithms regardless of the number of devices. The reason is that our algorithm considers the effects of interferences (both inter-SF and successive interferences) and tries to suppress them by careful SF assignment. Note that all the curves become flat when $N \geq 16$, which can be explained by Fig. 3(b). The number of served devices under the random and distance algorithms reach the maximum served devices (12) when $N \geq 16$, whereas that under the proposed algorithm keeps around 8. The results are rational since the proposed scheme exhausts itself suppressing the interferences rather than serving users. The relationship between the number of EDs and the served EDs would give instructions to the optimal number of CSs setting in each CW.

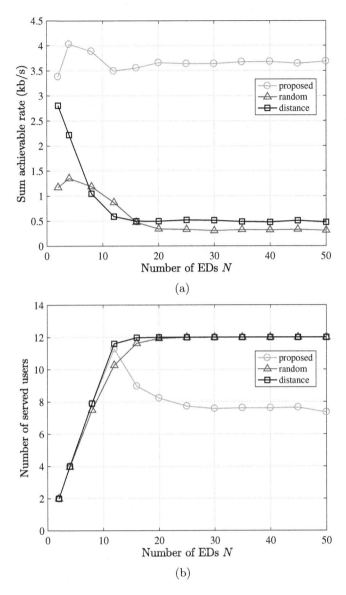

Fig. 3. Performance comparison between the proposed SF allocation scheme and baseline schemes. (a) Sum achievable rate. (b) Number of served users.

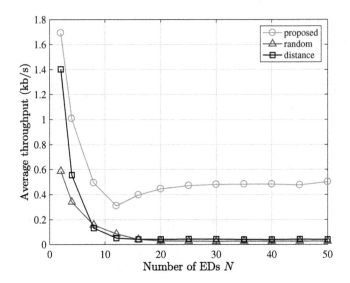

Fig. 4. Average throughput.

When $4 < N < 16$, the sum achievable rate under the distance algorithm is slightly higher than the random one. The reason is that the distance algorithm tries to allocate the smallest available SF for each ED, which makes (i) smaller gap between the used SFs, and (ii) a higher probability that one SF is assigned to more than one device. Therefore, interference rises in the distance algorithm case. However, the sum achievable rate under the distance algorithm becomes slightly higher than the random one, when the number of served devices increases. The reason is that devices suffer serious interferences in both baseline algorithms when the served devices increases, but distance algorithm allocate the smallest available SF to each device, making the data rate higher.

Figure 4 shows the average throughput. The average throughput of the proposed scheme first descends with the increasing number of EDs when $N \leq 12$, then it rises slightly and goes flat. The average throughput of the baseline schemes drops fast when $N \leq 12$, after that it keeps around 0.05 kb/s. Our proposed algorithm is superior than the baseline ones. In addition, in terms of the throughput, our proposal is also better than that in [18]. As shown in Fig. 3 of [18], the average throughput of their proposal decreases against the number of nodes and is lower than 0.2 kb/s when the number of nodes is greater than 15. Whereas, our proposal makes the average throughput always higher than 0.3 kb/s.

6 Conclusion

To accommodate massive connectivity in IoT applications, we propose a novel LoRa network combined with NOMA and DQ technologies, in which a NOMA-enabled gateway allows multiple devices to transmit data simultaneously over the shared channel. The whole random access process of EDs is divided into collision resolution and transmission scheduling phases, which is coordinated by the GW based on the DQ mechanism. The details of access process are provided. In addition, an SF allocation algorithm for the transmission scheduling phase is proposed. We first formulate an SF allocation problem under the constraints of inter-SF and successive interferences, then develop the allocation algorithm based on the optimization problem and matching theory. Numerical results show that our SF allocation scheme outperforms the random and distance based schemes.

A Appendix 1

We now prove Eq. (9). Under the assumption of Rayleigh fading, γ_n can be modeled as exponential random variable with mean $\bar{\gamma}_n = A(f_c)/(d_n^\alpha \sigma_c^2)$, then Eq. (8) can be developed as

$$P_{ns}^{(\text{case2})} = P\left(\frac{\gamma_n p_n}{\sum\limits_{i=n+1}^{|\mathcal{U}_s|} x_{is}\gamma_i p_i + 1} \geq \mu \middle| \gamma_{n+1} \cdots \gamma_{|\mathcal{U}_s|}\right) \times P\left(\gamma_{n+1} \cdots \gamma_{|\mathcal{U}_s|}\right)$$

$$= \int_{\gamma_{n+1}} \cdots \int_{\gamma_{|\mathcal{U}_s|}} P\left(\gamma_n \geq \frac{\mu}{p_n}\left(\sum\limits_{i=n+1}^{|\mathcal{U}_s|} x_{is}\gamma_i + 1\right)\right)$$
$$\times P\left(\gamma_{n+1} \cdots \gamma_{|\mathcal{U}_s|}\right) d\gamma_{n+1} \cdots d\gamma_{|\mathcal{U}_s|}. \tag{16}$$

Let $Z = P\left(\gamma_n \geq \frac{\mu}{p_n}\left(\sum\limits_{i=n+1}^{|\mathcal{U}_s|} x_{is}\gamma_i + 1\right)\right)$, then

$$Z = \int\limits_a^\infty p(\gamma_n)\, d\gamma_n = \int\limits_a^\infty \frac{1}{\bar{\gamma}_n} e^{-\frac{\gamma_n}{\bar{\gamma}_n}}\, d\gamma_n$$

$$= e^{-\frac{\mu\left(\sum\limits_{i=n+1}^{|\mathcal{U}_s|} x_{is}\gamma_i p_i + 1\right)}{\bar{\gamma}_n p_n}}, \tag{17}$$

where $a = \frac{\mu}{p_n}\left(\sum\limits_{i=n+1}^{|\mathcal{U}_s|} x_{is}\gamma_i p_i + 1\right)$. Substituting Z in Eq. (16), we obtain that

$$
\begin{aligned}
P_{ns}^{(case2)} &= \int_{\gamma_{n+1}} \cdots \int_{\gamma_{|\mathcal{U}_s|}} e^{-\frac{\mu\left(\sum\limits_{i=n+1}^{|\mathcal{U}_s|} x_{is}\gamma_i p_i + 1\right)}{\bar{\gamma}_n p_n}} P\left(\gamma_{n+1}\cdots\gamma_{|\mathcal{U}_s|}\right) \\
&\quad \times d\gamma_{n+1}\cdots d\gamma_{|\mathcal{U}_s|} \\
&= e^{-\frac{\mu}{\bar{\gamma}_n p_n}} \prod_{i=n+1}^{\mathcal{U}_s} \int_{\gamma_i} e^{-\frac{\mu x_{is}\gamma_i p_i}{\bar{\gamma}_n p_n}} p(\gamma_i)\, d\gamma_i \\
&= e^{-\frac{\mu}{\bar{\gamma}_n p_n}} \prod_{i=n+1}^{\mathcal{U}_s} \frac{1}{\bar{\gamma}_i} \int_{\gamma_i} e^{-\frac{\mu x_{is}\bar{\gamma}_i p_i + \bar{\gamma}_n p_n}{\bar{\gamma}_n \bar{\gamma}_i p_n}}\gamma_i\, d\gamma_i \\
&= e^{-\frac{\mu}{\bar{\gamma}_n p_n}} \prod_{i=n+1}^{\mathcal{U}_s} \frac{1}{\bar{\gamma}_i} \cdot \frac{\bar{\gamma}_n \bar{\gamma}_i p_n}{\mu x_{is}\bar{\gamma}_i p_i + \bar{\gamma}_n p_n} \\
&\quad \times \int_{\gamma_i} \frac{\mu x_{is}\bar{\gamma}_i p_i + \bar{\gamma}_n p_n}{\bar{\gamma}_n \bar{\gamma}_i p_n} e^{-\frac{\mu x_{is}\bar{\gamma}_i p_i + \bar{\gamma}_n p_n}{\bar{\gamma}_n \bar{\gamma}_i p_n}\gamma_i}\, d\gamma_i \\
&= e^{-\frac{\mu}{\bar{\gamma}_n p_n}} \prod_{i=n+1}^{\mathcal{U}_s} \frac{\bar{\gamma}_n p_n}{\mu x_{is}\bar{\gamma}_i p_i + \bar{\gamma}_n p_n} \\
&= e^{-\frac{\mu \sigma_c^2 d_n^2}{A(f_c)p_n}} \prod_{i=n+1}^{\mathcal{U}_s} \frac{1}{\mu x_{is}\frac{p_i}{p_n}\left(\frac{d_n}{d_i}\right)^{\alpha} + 1}.
\end{aligned}
\tag{18}
$$

B   Appendix 2

We derivate Eq. (11) with similar calculations in Appendix 1.

$$
\begin{aligned}
P_{ns}^{(case3)} &= P\left(\frac{\gamma_n p_n}{\sum\limits_{j\in\mathcal{S}\setminus\{s\}}\sum\limits_{i\in\mathcal{N}\setminus\{n\}} x_{ij}\gamma_i p_i + \sum\limits_{i=n+1}^{|\mathcal{U}_s|} x_{is}\gamma_i p_i + 1} \geq \mu \,\middle|\, \gamma_1\cdots\gamma_{n-1}\gamma_{n+1}\cdots\gamma_N\right) \\
&\quad \times P\left(\gamma_1\cdots\gamma_{n-1}\gamma_{n+1}\cdots\gamma_N\right) \\
&= \int_{\gamma_1}\cdots\int_{\gamma_N} P\left(\gamma_n \geq \frac{\mu}{p_n}\left(\sum\limits_{j\in\mathcal{S}\setminus\{s\}}\sum\limits_{i\in\mathcal{N}\setminus\{n\}} x_{ij}\gamma_i p_i + \sum\limits_{i=n+1}^{|\mathcal{U}_s|} x_{is}\gamma_i p_i + 1\right)\right) \\
&\quad \times P\left(\gamma_1\cdots\gamma_N\right)\, d\gamma_1\cdots d\gamma_N.
\end{aligned}
\tag{19}
$$

Let $Z' = P\left(\gamma_n \geq \frac{\mu}{p_n}\left(\sum_{j\in\mathcal{S}\setminus\{s\}}\sum_{i\in\mathcal{N}\setminus\{n\}} x_{ij}\gamma_i + \sum_{i=n+1}^{|\mathcal{U}_s|} x_{is}\gamma_i + 1\right)\right)$,

$$Z' = \int_{a'}^{\infty} \frac{1}{\bar{\gamma}_n} e^{-\frac{\gamma_n}{\bar{\gamma}_n}}\, d\gamma_n \qquad (20)$$

$$= e^{-\frac{\mu\left(\sum_{j\in\mathcal{S}\setminus\{s\}}\sum_{i\in\mathcal{N}\setminus\{n\}} x_{ij}\gamma_i p_i + \sum_{i=n+1}^{|\mathcal{U}_s|} x_{is}\gamma_i p_i + 1\right)}{\bar{\gamma}_n p_n}},$$

where $a' = \frac{\mu}{p_n}\left(\sum_{j\in\mathcal{S}\setminus\{s\}}\sum_{i\in\mathcal{N}\setminus\{n\}} x_{ij}\gamma_i p_i + \sum_{i=n+1}^{|\mathcal{U}_s|} x_{is}\gamma_i p_i + 1\right)$. Substituting Z' in Eq. (19), we obtain

$$
\begin{aligned}
P_{ns}^{(case3)} &= \int_{\gamma_1}\cdots\int_{\gamma_N} e^{-\frac{\mu\left(\sum_{j\in\mathcal{S}\setminus\{s\}}\sum_{i\in\mathcal{N}\setminus\{n\}} x_{ij}\gamma_i p_i + \sum_{i=n+1}^{|\mathcal{U}_s|} x_{is}\gamma_i p_i + 1\right)}{\bar{\gamma}_n p_n}} \\
&\quad \times P\left(\gamma_1\cdots\gamma_N\right) d\gamma_1\cdots d\gamma_N \\
&= e^{-\frac{\mu}{\bar{\gamma}_n p_n}}\prod_{j\in\mathcal{S}\setminus\{s\}}\prod_{i\in\mathcal{N}\setminus\{n\}}\int_{\gamma_i} e^{-\frac{\mu x_{ij}\gamma_i p_i}{\bar{\gamma}_n p_n}} p(\gamma_i)\, d\gamma_i \\
&\quad \times \prod_{i=n+1}^{\mathcal{U}_s}\int_{\gamma_i} e^{-\frac{\mu x_{is}\gamma_i p_i}{\bar{\gamma}_n p_n}} p(\gamma_i)\, d\gamma_i \\
&= e^{-\frac{\mu}{\bar{\gamma}_n p_n}}\prod_{j\in\mathcal{S}\setminus\{s\}}\prod_{i\in\mathcal{N}\setminus\{n\}}\frac{\bar{\gamma}_n p_n}{\mu x_{is}\bar{\gamma}_i p_i + \bar{\gamma}_n p_n} \qquad (21) \\
&\quad \times \prod_{i=n+1}^{\mathcal{U}_s}\frac{\bar{\gamma}_n p_n}{\mu x_{is}\bar{\gamma}_i p_i + \bar{\gamma}_n p_n} \\
&= e^{-\frac{\mu\sigma_c^2 d_n^2}{A(f_c)p_n}}\prod_{j\in\mathcal{S}\setminus\{s\}}\prod_{i\in\mathcal{N}\setminus\{n\}}\frac{1}{\mu x_{is}\frac{p_i}{p_n}\left(\frac{d_n}{d_i}\right)^\alpha + 1} \\
&\quad \times \prod_{i=n+1}^{\mathcal{U}_s}\frac{1}{\mu x_{is}\frac{p_i}{p_n}\left(\frac{d_n}{d_i}\right)^\alpha + 1}
\end{aligned}
$$

References

1. Li, K., Benkhelifa, F., Mccann, J.: Resource allocation for non-orthogonal multiple access (NOMA) enabled LPWA networks. arXiv: 1908.09336 (2019)
2. Shahini, A., Ansari, N.: NOMA aided narrowband IoT for machine type communications with user clustering. IEEE IoT J. **6**(4), 7183–7191 (2019)
3. Adelantado, F., Vilajosana, X., Tuset-Peiró, P., et al.: Understanding the limits of LoRaWAN. IEEE Commun. Mag. **55**(9), 34–40 (2017)
4. Vangelista, L.: Frequency shift chirp modulation: the LoRa modulation. IEEE Signal Process. Lett. **24**(12), 1818–1821 (2017)
5. LoRa Alliance. V1.0.3. LoRaWANTM Specification (2018)
6. Corce, D., Gucciardo, M., Tinnirello, I., et al.: Impact of spreading factor imperfect orthogonality in LoRa communications. In: Piva, A., Tinnirello, I., Morosi, S. (eds.) Digital Communication. Towards a Smart and Secure Future Internet. CCIS, vol. 766, pp. 165–179. Springer, Cham (2017). https://doi.org/10.1007/978-3-319-67639-5

7. Mostafa, A.E., Zhou, Y., Wong, V.W.S.: Connection density maximization of narrowband IoT systems with NOMA. IEEE Trans. Wireless Commun. **18**(10), 4708–4722 (2019)
8. Wu, F., et al.: An enhanced random access algorithm based on the clustering-reuse preamble allocation in NB-IoT system. IEEE Access. **7**, 183847–183859 (2019)
9. Xu, W., Campbell, G.: A near perfect stable random access protocol for a broadcast channel. In: Proceedings of Discovering New World Communication (SUPERCOMM/ICC), pp. 370–374. IEEE Press, Chicago (1992)
10. Zhang, X., Campbell, G.: Performance analysis of distributed queueing random access protocol–DQRAP. DQRAP Research Group report 93-1 (1994)
11. Tuset-Peiro, P., Vazquez-Gallego, F., Alonso-Zarate, J., et al.: LPDQ: a self-scheduled TDMA MAC protocol for one-hop dynamic low-power wireless networks. Pervasive Mobile Comput. **20**, 84–99 (2015)
12. Zhang, K., Marchiori, A.: Crowdsourcing low-power wide-area IoT networks. In: Proceedings of the IEEE International Conference on Pervasive Computing and Communications (PerCom), pp. 41–49. IEEE Press, Kona (2017)
13. Wu, W., Li, Y., Zhang, Y., et al.: Distributed queueing based random access protocol for LoRa networks. IEEE IoT J. **7**(1), 763–772 (2020)
14. SX1272/3/6/7/8 LoRa Modem Design Guide AN1200.13 Revision 1 (2013)
15. Reynders, B., Meert, W., Pollin, S.: Power and spreading factor control in low power wide area networks. In: 2017 IEEE International Conference on Communications (ICC), pp. 1–6. IEEE Press, Paris (2017)
16. El-Aasser, M., Elshabrawy, T., Ashour, M.: Joint spreading factor and coding rate assignment in LoRaWAN networks. In: 2018 IEEE Global Conference on Internet of Things (GCIoT), pp. 1–7. IEEE Press, Alexandria (2018)
17. Su, B., Qin, Z., Ni, Q.: Energy efficient resource allocation for uplink LoRa networks. In: 2018 IEEE Global Communications Conference (GLOBECOM), pp. 1–7. IEEE Press, Abu Dhabi (2018)
18. Amichi, L., Kaneko, M., Rachkidy, N. E., et al.: Spreading factor allocation strategy for LoRa networks under imperfect orthogonality. In: 2019 IEEE International Conference on Communications (ICC), pp. 1–7. IEEE Press, Shanghai (2019)
19. Amichi, L., Kaneko, M., Fukuda, E.H., et al.: Joint allocation strategies of power and spreading factors with imperfect orthogonality in LoRa networks. IEEE Trans. Commun. **68**(6), 3750–3765 (2020)
20. Waret, A., Kaneko, M., Guitton, A., et al.: LoRa throughput analysis with imperfect spreading factor orthogonality. IEEE Wireless Commun. Lett. **8**, 408–411 (2019)
21. Gu, Y., Saad, W., Bennis, M., et al.: Matching theory for future wireless networks: fundamentals and applications. IEEE Commun. Mag. **53**, 52–59 (2015)
22. Laporte-Fauret, B., Temim, M.A.B., Ferre, G., et al.: An enhanced LoRa-like receiver for the simultaneous reception of two interfering signals. In: 2019 IEEE 30th Annual International Symposium on Personal, Indoor and Mobile Radio Communications (PIMRC), pp. 1–6, Istanbul (2019)

Fast Power Spectrum Estimation with Sparse Learning for Wideband Spectrum Sensing

Shuai Liu[(✉)], Wen Xiao, Yao Zhang, Jing He, and Jixin Wu

Xi'an Jiaotong University, Xi'an 710049, China
`sh_liu@mail.xjtu.edu.cn`

Abstract. The Compressed Sensing technology in wideband spectrum sensing (WSS) has greatly improved the utilization of spectrum resources. Based on this, we combining sparse learning and fast power spectrum estimation to achieve WSS in this paper. Sparsity adaptive matching pursuit (SAMP) algorithm is exploited to obtain the sparse sample representation for WSS. Then the limitations of power spectrum estimation in WSS are considered. To ease the limitations, the computational tasks are decomposed by multiple fast Fourier transforms. Theoretical performance analysis is made to further explain the proposed method. By improving the process of sample collection and power spectrum estimation, the proposed method can effectively achieve the pur-pose of fastly and exactly sensing. The final simulation results are utilized to verify the applicability of the proposed method and its advantages over other methods.

Keywords: Wideband spectrum sensing · Power spectrum estimation · Sparse learning

1 Introduction

In recent years, 5G technology has become more and more important in many new applications and use cases, the emergence of 5G technology puts forward higher requirements for the number of wireless device connections, which makes us have to pay attention to the effective utilization of spectrum resources. As we all know, its development is restricted by the scarcity of spectrum resources, so there is an urgent need to resolve the resource shortage by effectively utilizing the spectrum resource. WSS enables cognitive radio to be applied to a wider frequency band to meet the development of modern wireless communication. Specifically, WSS is a process which scans a wide frequency band consisting of several sub-bands of narrow frequency band, so as to find the spectral cavity in the whole bandwidth and improve the spectrum utilization. Now WSS is generally known and has become a field of interest for many researchers [1, 2]. As an important research method in WSS, CS can reduce the sampling burden and realize efficient and fast wideband spectrum sensing.

At present, there are two kinds of WSS technologies. One is the method based on Nyquist sampling [3, 4]. Pei et al. proposed the sequential scanning method in [3] which identifies the sensing order and sensing-access strategies to achieve maximum energy efficiency. In [4], filter banks are introduced as a tool in CR systems, which can achieve for spectrum sensing by choosing the proper one from various filter banks. Its

H. Gao et al. (Eds.): ChinaCom 2020, LNICST 352, pp. 421–433, 2021.
https://doi.org/10.1007/978-3-030-67720-6_29

performance is evaluated theoretically and through numerical examples. But the methods based on Nyquist sampling usually suffer from either long sensing latency or enormous hardware complexity.

Another one based on sub-Nyquist sampling is proposed [5, 6]. Among these, more and more researchers achieve the WSS by exploring the compressed sensing theory [7–12] to sense a wide frequency band via a low sampling rate. Utilizing the sparse or statistical characteristics, the wideband spectrum signals can be reconstructed at a lower sampling rate than the Nyquist rate. In [10], Zhao et al. schedules the sequential compressed spectrum sensing by jointly exploring compressed sensing and sequential periodic detection techniques to achieve more accurate and timely wideband sensing. Using the heterogeneous wideband spectrum, the latent block-like structure is exploited to construct efficient compressive spectrum sensing models for obtaining the well heterogeneous wideband spectrum in [11]. Fast matching pursuit, a fast and accurate greedy recovery algorithm, is proposed for compressed wideband spectrum sensing in [12].

However, these compressed wideband spectrum sensing approaches also suffer several limitations, including noise, the unavoidably high computational complexity of reconstruction, and the low signal-to-noise ratio.

To solve these problems, reconstructing the power spectrum of the wideband spectrum signal is proposed to replace the original signal, by using sub-Nyquist [13–15]. The feature-based method is proposed for the primary user's spectrum identification by exploring the interference immunity with a reduced amount of data in [14]. The approach in [15] exploited the statistical structure of random processes to obtain the effective signal compression and showed an alternative perspective for sparsity-agnostic inference.

In this paper, we reconstruct the power spectrum of the wideband signal through the sub-Nyquist sampling, which is achieved by exploring the sparsity of the wideband spectrum signal. Firstly, the proposed method uses the sparsity adaptive matching pursuit algorithm (SAMP) to collect signals sparsely and reserve the latent information contained in the original signal as much as possible [16, 17]. Secondly, the fast Fourier transform (FFT) is exploited to obtain a fast power spectrum estimation [18]. In this way, compared to the existing method, the proposed method can improve the computational efficiency of power spectrum reconstruction and can be effectively applied to the field of compressed wideband spectrum sensing.

2   Algorithm Design

2.1   The Proposed Method

The details of the proposed method are shown in Fig. 1. In this part, we will introduce the sample collection by using the sparse sampling method and give a detailed derivation process of how to realize the power spectrum estimation of signals with FFT, includes using FFT to decompose computing tasks to reduce computational complexity.

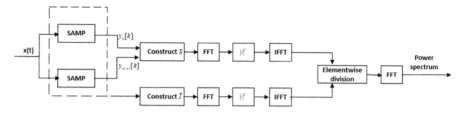

Fig. 1. The proposed method.

First of all, some samples of l moment are collected in the m th channel by using SAMP sparse sampling, it is given by

$$y_m(l) = x(lNT) \tag{1}$$

Then, since the collected data $\{y_m[l]\}_{m=1,l=0}^{M,L-1}$ is a partial subset of sub-Nyquist sampled data $\{x[n]\}_{n=0}^{LN-1}$. In order to establish the relationship between the Nyquist sampling and the sub-Nyquist sampling data, an indicator sequence $\{I[n]\}_{n=0}^{LN-1}$ and a data sequence $\{h[n]\}_{n=0}^{LN-1}$ are defined as follows

$$h[n] = \begin{cases} y_m(l), & n = lN \\ 0, & otherwise \end{cases} \tag{2}$$

$$I[n] = \begin{cases} 1, & n = lN \\ 0, & otherwise \end{cases} \tag{3}$$

In power spectrum estimation, the widely-used unbiased estimation of the signal $x[n]$ is given as

$$r_x[k] \approx \frac{1}{|Q_k|} \sum_{n \in Q_k} (x[n]x^*[n-k]) \tag{4}$$

where $Q_k = \{n | 0 \le n - k \le LN - 1, 0 \le n \le LN - 1\}$, Only data $\{h[n]\}$ can be obtained in the experiment, so we get a new unbiased estimation which is computed by

$$r_x[k] \approx \frac{1}{Q_k} \sum_{n \in \hat{Q}_k} (x[n]x^*[n-k])$$
$$= \frac{1}{Q_k} \sum_{n \in Q_k} (h[n]h^*[n-k]) \tag{5}$$

where

$$Q_k \triangleq |\hat{\mathbf{Q}}_k| \triangleq \{n|I[n]I[n-k]=1\} \tag{6}$$

once $\{r_x[k]\}$ is calculated, the power spectrum can be calculated by DFT.

In order to improve the spectrum resolution, L should be as large as possible within the allowable range, but in this way, it will significantly increase the computational complexity. Therefore, FFT is introduced in the solution process to solve this problem. we divide (5) into two parts and introduced FFT for calculation respectively.

The first part can be defined as

$$r_h[k] = \sum_{n=0}^{LN-1} (h[n]h^*[n-k]) \tag{7}$$

where $n < 0$ or $n \ge LN$, let $h[n] = 0$. At this point, we define $\{\hat{h}[n]\}_{n=-LN+1}^{LN-1}$ to be the reverse for $\{\bar{h}[n]\}_{n=-LN+1}^{LN-1}$ and define $\{\bar{h}[n]\}_{n=-LN+1}^{LN-1}$ as follows

$$\bar{h}[n] = \begin{cases} h[n], & LN-1 \ge n \ge 0 \\ 0, & -LN+1 \le n < 0 \end{cases} \tag{8}$$

Then $r_h[k]$ can be written as

$$r_h[k] = \begin{cases} \sum_{n=-LN+1+k}^{LN-1} (\bar{h}[n]\hat{h}^*[k-n]), k \ge 0 \\ \sum_{n=-LN+1}^{LN-1+k} (\bar{h}[n]\hat{h}^*[k-n]), k < 0 \end{cases}$$

$$\overset{(a)}{=} \sum_{n=-LN-1}^{LN-1} (\bar{h}[n]\hat{h}_P^*[k-n]) \tag{9}$$

$$= (\bar{h}*\hat{h}^*)[k]$$

where $P \triangleq 2NL - 1$. As the periodic sum of $\hat{h}[n], \hat{h}_P[n]$ can be shown as

$$\hat{h}_P[n] \triangleq \sum_{k=-\infty}^{+\infty} \hat{h}[n-kP] \tag{10}$$

The symbol $*$ of (9) represents cyclic convolution, and (a) represents: When $-LN+1 \le n < 0, \bar{h}[n] = 0$ and when $-2LN+1 \le n < -LN$, $\hat{h}_P[n] = \hat{h}[n-NL] = 0$. From this We can get as follows

$$r_h \triangleq [r_h[-LN+1]...r_h[LN-1]]^T,$$
$$\bar{h} \triangleq [\bar{h}[-LN+1]...\bar{h}[LN-1]]^T, \tag{11}$$
$$\hat{h} \triangleq [\hat{h}[-LN+1]...\hat{h}[LN-1]]^T.$$

By referring to the theorem of circular convolution, it can be obtained: $F_{2NL-1}r_h = (F_{2NL-1}\bar{h}) \circ (F_{2NL-1}\hat{h})$. F_{2NL-1} represents the discrete Fourier Transform (DFT) at $(2NL - 1)$-point. $\circ$ represents the product of corresponding elements in two vectors. Since $\{\hat{h}[n]\}$ represents the time reversal for $\{\bar{h}[n]\}$, consider the property of complex-conjugate and time-reversal for DFT, the sequence $\{\hat{h}^*[n]\}$ is the DFT complex conjugate of $\{\bar{h}[n]\}$. So we can get

$$(F_{2NL-1}\bar{h}) \circ (F_{2NL-1}\hat{h}) = \left| F_{2NL-1}\bar{h} \right|^2 \tag{12}$$

Where $|\bullet|^2$ represents the square of the complex vector module. Finally, we can get the final form of r_h shown as

$$r_h = F_{2NL-1}^{-1} \left| F_{2NL-1}\bar{h} \right|^2 \tag{13}$$

According to (6), the second part is defined as:

$$Q_k = \sum_{n \in \mathbf{Q}_k} I[n]I[n-k] \tag{14}$$

At this point, we calculate Q_k by introducing FFT, firstly, we define

$$q \triangleq [Q_{-LN+1} \cdots Q_{LN-1}]^T, \\ \bar{I} \triangleq [\bar{I}[-LN+1] \ldots \bar{I}[LN-1]]^T. \tag{15}$$

where

$$\bar{I}[n] = \begin{cases} I[n], & LN-1 \geq n \geq 0 \\ 0, & -LN+1 \leq n < 0 \end{cases} \tag{16}$$

Note that the definition of the second part is similar to the first part, according to the computational process of (7), we can get $q = F_{2NL-1}^{-1} \left| F_{2NL-1}^{-1}\bar{I} \right|^2$.

After we compute $\{Q_k\}$ and $\{r_h[k]\}$, we get the result of (5): $r_x[k] = r_h[k]/Q_k$. The final power spectrum estimation can be obtained by performing DFT on this result.

2.2 Theoretical Performance Analysis

We found that the proposed method only involves FFT, IFFT and some simple multiplication operations. It can be easily calculated through floating-point operations in total. Obviously, compared with existing methods, the proposed method is dominant in the calculation and can compute efficiently in parallel. In this part, we will calculate the mean square error (MSE) for the proposed method, and make a theoretical analysis of it.

$\hat{s}$ represents the estimated power spectrum, and s represents the actual power spectrum. MSE can be solved as

$$E[\|\hat{s} - s\|_2^2] \overset{(a)}{=} \|E[r_x]\|_2^2 + 1^T D(r_x) - 2E[\hat{s}]^H s + \|s\|_2^2 \qquad (17)$$
$$= \|E[\hat{s}] - s\|_2^2 + 1^T D(r_x)$$

In (17), r_x represents the autocorrelation vector estimation, F represents the DFT matrix, and (a) represents: $E[\|r_x\|_2^2] = \|E[r_x]\|_2^2 + 1^T D(r_x)$, 1 is a vector where all terms are 1, $D(r_x)$ is the variance of each element for r_x. Therefore, the mean square error of the estimator can be given by

$$MSE = 1^T D(r_x) \qquad (18)$$

we simplify the main calculation task to the variance of the r_x element of the vector, therefore, we obtain

$$D(r_x[k])$$
$$= E\left[\left|\frac{1}{Q_k^2} \sum_{n \in Q_k, m \in Q_k} (x[n]x^*[n-k]x[m]x^*[m-k])\right|^2\right] - |E[r_x[k]]|^2 \qquad (19)$$

we assume a special case where the signal $x[t]$ is a white signal with zero mean and variance σ^2. When $k \neq 0$, there is $D(r_x[k]) = \frac{1}{Q_k}\sigma^4$; Where $k = 0$, $D(r_x[k]) = \frac{\sigma^4}{Q_0^2}$. Therefore, for this special case, MSE can be simplified as

$$MSE = 1^T D(r_x[k]) = \sigma^4 \sum_k 1/Q_k \qquad (20)$$

It is easy to conclude that the increase in the number of data samples will reduce MSE, while the more sparse learning sampling data, the larger $\{Q_k\}$ can be obtained, thus estimating accurately.

3   Simulation Experiments

In this part, this paper provides experimental results to prove the accuracy and effectiveness of the proposed method.

3.1   The Setting of Experimental Parameters

During the experiment, this paper uses the band-pass FIR filter to filter zero-unit variance White Gaussian Noise and produces 5 wide-sense stationary signals in the frequency range of [0, 1] GHz. The purpose of this experiment is to identify the frequency position of the signal propagating over the band [0, 1] GHz. For the system parameters, we set the signal component bandwidth as 10 MHz, and 2 GHz for the

Nyquist sampling rate, the carrier frequencies are set as 100, 210, 360, 450, and 670 MHz respectively. SAMP algorithm is adopted to collect the sub-Nyquist samples in this proposed method, so we set the number of sampling channels as $M = 8$, the sampling rate as 100 MHz in each channel, and we collect the samples with a duration of 1 ms for each channel in this experiment. The downsampling factor is set as $N = fnyq/(100MHz) = 20$. The spectral resolution is set to 60 kHz, corresponding to the 33,335-length power spectrum. Therefore, there is $2NL \geq 33335$, and $L \geq 32000/2N \approx 667$ data samples need to be collected for each sampling channel. Finally, we consider the selection of window function, which can reduce the spectrum energy leakage. We use a hanning window to the autocorrelation sequence in this proposed method, in order to achieve a more accurate power spectrum estimation.

3.2   The Analysis of Experimental Results

First, we design an experiment with the results shown in Fig. 2. The signal-to-noise ratio (SNR) is set as −5 dB in each experiment. Zero mean Gaussian noise is used to corrode the original signal to generate noise signals. Here, SNR is given as

$$\text{SNR} = 10 \log_{10} \frac{\sum_{n=1}^{N_t} |x[n]|^2}{N_t \sigma^2} \tag{21}$$

It can be seen from Fig. 2 that the proposed method can recover the true power spectrum accurately. In addition, the results show that for the sub-Nyquist sampling data within 10 ms, the recovery accuracy of the proposed method is slightly higher than the condition of 1 ms samples. The results show that the accuracy of power spectrum estimation with sparse learning can be improved by increasing the sampling time.

Figure 3 shows the ROC curve of the method proposed in this paper with different SNRs, this advantage is particularly suitable for WSS, the receiver must work in a low SNR state to obtain sensitivity in the practical sense, it can be calculated as

$$S = 10 \log(kT_{syst}) + 10 \log(B) + NF_{RX} + SNR \tag{22}$$

In (22), B represents the bandwidth of the signal, NF_{RX} means the noise figure of the receiver, and its value is 6 dB. Therefore, we can calculate the receiver sensitivity is $S = -174\text{dBm/Hz} + 10 \log(1\text{GHz}) + 6\text{dB} + SNR = -78\text{dBm}$. In order to achieve the receiver sensitivity of −80 dbm, in this case, we take the SNR as $-80 - (-78) = -2$ dB. The sampling rate of ADC is set as 100 MHz, 125 MHz, and 200 MHz respectively. The corresponding downsampling factor n are 20, 16, and 10, respectively. As we can see, the result of Fig. 4 is based on the condition that SNR is set as −15 dB. Figure 4 shows that the performance of the method is gradually improved with the increase of the sampling rate.

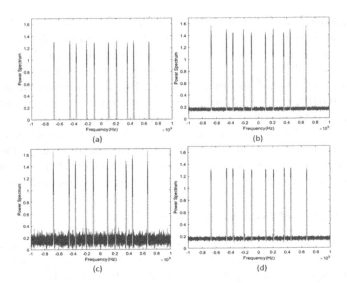

Fig. 2. The results of power spectrum reconstruction with different samples and method: (a) noiseless Nyquist samples collected within 1 ms; (b) noisy Nyquist samples collected within 1 ms;(c) 1 ms noise sub-Nyquist samples via the proposed method; (d) 10 ms sub-Nyquist samples via the proposed method.

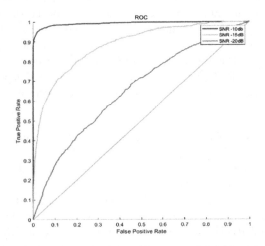

Fig. 3. True positive rate and false positive rate in different SNRs.

From the running time of the experiment shown in Fig. 5, it can be seen that the proposed method significantly improves the computational efficiency compared with the frequency-domain method. We find that there is the same number of channels and sampling rate for MWC and SAMP algorithms, so the comparison is reasonable. For these two methods, though collecting data samples at an interval of 1 ms, this experiment considers that the frequency resolution is set as 62.5 kHz. The operating

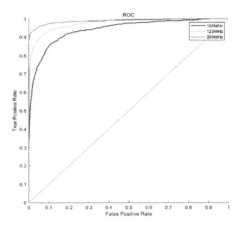

Fig. 4. True positive rate and false positive rate in different compression ratios.

conditions of the experiment are as follows: a 2.60 GHz Intel i7CPU and 16.0 GB of RAM laptop using MATLAB R2020a. The average running time and ROC curves are showed in Fig. 5, which can be seen that our method attains convergence faster than the frequency- domain method. In addition, the time required by the proposed method is about half of that of the frequency-domain method, therefore, we can conclude, our method effectively reduces the computational complexity, the practical system based on FPGA can be better implemented [19].

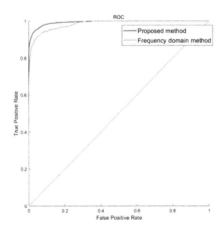

Fig. 5. The ROC of the frequency-domain method and our method, the running times are 0.36 s and 0.18, respectively.

We also set a experiment based on the assumption that the time-domain method uses sparse learning [20], to compare the proposed method with the traditional time-domain method [21], its results are shown in Fig. 6. The values of some system parameters are the same as those described above. Specially, we set the SNR as −12 dB, and the frequency resolution as 1 MHz. Data samples continuously collected within 0.1 ms were used in this experiment. Figure 6 shows the ROC curve and the average running time of this experiment. As we can see from the experimental results in Fig. 6, the proposed method has better performance than the traditional time-domain method.

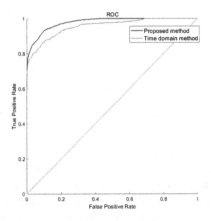

Fig. 6. The ROC of the proposed method and the time-domain method, the running times are 0.03 s and 0.13 s, respectively.

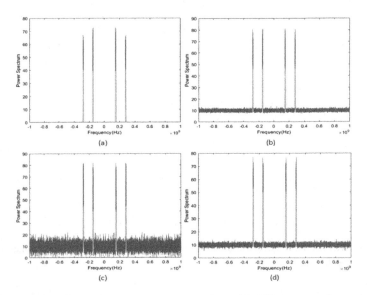

Fig. 7. The results of power spectrum reconstruction with different samples and method: (a) 1 ms noiseless Nyquist samples; (b)1 ms noisy Nyquist samples; (c) 1 ms noise sub-Nyquist samples, under the proposed method; (d) 10 ms sub-Nyquist samples, under the proposed method.

The proposed method is based on the assumption that multi-band signals are generalized stationary, and the generalized stationarity is a widely used assumption for spectrum sensing, also for compressed power spectrum estimation [22]. However, the actual signal may be either cyclostationary or non-stationary. The statistical properties of non-stationary random signals, such as mean and variance, vary with time. In other words, it is characterized by slow time-varying statistics. So they can be regarded as generalized stationery signals for a short time in these circumstances, and the proposed method is also suitable for cyclostationary communication signals. In order to prove these results, we simulate two cyclostationary communication signals, QAM16 signal and BPSK signal whose frequency range is [0, 1] GHz, The carrier frequencies are set as 280 MHz and 150 MHz respectively, the SNR is set as −10 dB, and the sparse learning sampling architecture is set as mentioned earlier in this section. From Fig. 7, we can conclude that the proposed method is suitable for cyclostationary signals and can produce accurate power spectrum estimation for it.

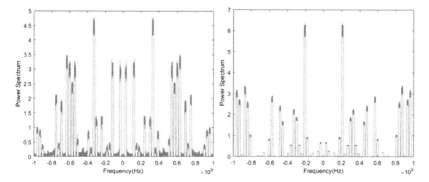

Fig. 8. Left: Reconstruct the power spectrum using 10 ms noiseless Nyquist samples under the proposed method; Right: Reconstruct the power spectrum using noiseless Nyquist samples collected within 1 ms.

In order to illustrate that the method proposed in this paper does not impose a sparse constraint on the monitored spectrum when reconstructing the power spectrum [23], we carried out this experiment in a multiband signal consisting of 32 narrowband components in a frequency range of [0, 1] GHz and set the bandwidth of narrowband signal as 20 MHz. The spectrum occupancy is computed as $(32 \times 20$ MHz$)/$ $(1$ GHz$) = 64\%$ from the parameter settings described earlier, it shows that the monitored power spectrum is not sparse. Two comparative experiments are shown in Fig. 8. In the results of the reconstructed power spectrum showed in Fig. 8, we set the NMSE as 0.0014 and 0.0055 respectively. As we can see, this method also provides a precise power spectrum estimation even if the monitored spectrum is not sparse.

4 Conclusion

In this paper, we introduce the sparse sampling method and FFT into the fast power spectrum estimation. Aiming at the high computational complexity of traditional methods and unable to meet the requirements of the real-time spectrum sensing system, a fast compressed power spectrum estimation method is proposed. By collecting the sub-Nyquist samples with sparse sampling, we improve the sampling rate in WSS. Next, the method reconstructs the power spectrum estimation model, specially, when L is large, it will provide better spectral resolution, but increase the computational complexity in this model. To address the problem, FFT is introduced into the computational process, the computational complexity is reduced by using three FFT. Experimental data analysis shows that the proposed method can meet the requirements of a real-time spectrum sensing system with low computational complexity and can be effectively implemented in the actual system, the final experimental results listed above also prove the effectiveness and computational efficiency of our method.

Acknowledgement. This research was funded by the National Natural Science Foundation of China (No. 61703328), the China Postdoctoral Science Foundation funded project (No. 2018M631165), the Fundamental Research Funds for the Central Universities (No. XJJ2018254).

References

1. Sun, H., Nallanathan, A., Wang, C.X., Chen, Y.: Wideband spectrum sensing for cognitive radio networks: a survey. IEEE Wirel. Commun. Mag. **20**(2), 74–81 (2013)
2. Ali, A., Hamouda, W.: Advances on spectrum sensing for cognitive radio networks: theory and applications. IEEE Commun. Surv. Tutorials **19**(2), 1277–1304 (2017)
3. Pei, Y., Liang, Y.C., The, K.C., Li, K.H.: Energy-efficient design of sequential channel sensing in cognitive radio networks: optimal sensing strategy, power allocation, and sensing order. IEEE J. Sel. Areas Commun. **29**, 1648–1659 (2011)
4. Farhang-Boroujeny, B.: Filter bank spectrum sensing for cognitive radios. IEEE Trans. Signal Process. **56**, 1801–1811 (2008)
5. Zhao, Y., Chen, Y., Zheng, Y.: Wideband power spectrum estimation based on sub-nyquist sampling in cognitive radio networks. IEEE Access **7**, 115339–115347 (2019)
6. Shaban, M., Perkins, D., Bayoumi, M.: Application of compressed sensing in wideband cognitive radios when sparsity is unknown. In: 15th IEEE Wireless and Microwave Technology Conference (2014)
7. Donoho, D.L.: Compressed sensing. IEEE Trans. Inf. Theory **52**(4), 1289–1306 (2006)
8. Laska, J., Wen, Z., Yin, W., Baraniuk, R.: Trust, but verify: fast and accurate signal recovery from 1-bit compressive measurements. IEEE Trans. Signal Process. **59**(11), 5289–5301 (2011)
9. Khalfi, B., Hamdaoui, B., Guizani, M., Zorba, N.: Efficient spectrum availability information recovery for wideband DSA networks: a weighted compressive sampling approach. IEEE Trans. Wirel. Commun. **17**, 2162–2172 (2018)
10. Li, Z., Xu, W., Zhang, X., Lin, J.: A survey on one-bit compressed sensing: Theory and applications. Front. Comput. Sci. **12**(2), 217–230 (2018)
11. Peng, X., Bin, L., Xiaodong, H., Quan, Z.: 1-bit compressive sensing with an improved algorithm based on fixed-point continuation. Sig. Process. **154**, 168–173 (2019)

12. Zhang, X., Du, H., Qiu, B., Chen, S.: Fast sparsity adaptive multipath matching pursuit for compressed sensing problems. J. Electron. Imaging **26**(3), 033007 (2017)
13. Cohen, D., Eldar, Y.C.: Sub-Nyquist sampling for power spectrum sensing in cognitive radios: a unified approach. IEEE Trans. Signal Process. **62**, 3897–3910 (2014)
14. Lagunas, E., Nájar, M.: Spectral feature detection with sub-Nyquist sampling for wideband spectrum sensing. IEEE Trans. Wirel. Commun. **14**, 3978–3990 (2015)
15. Romero, D., Ariananda, D.D., Tian, Z., Leus, G.: Compressive covariance sensing: structure-based compressive sensing beyond sparsity. IEEE Sig. Process. Mag. **33**, 78–93 (2016)
16. Liquan, Z., Ke, M., Yanfei, J.: Improved generalized sparsity adaptive mathching pursuit algortithm based on compressive sensing. J. Electr. Comput. Eng. (2020)
17. Khobahi, S., Soltanalian, M.: Signal recovery from 1-bit quantized noisy samples via adaptive thresholding. In: 52nd Asilomar Conference on Signals, Systems, and Computers (2018)
18. Kumar, A., Saha, S.: FFT-based multiband spectrum sensing in SIMO in-band full-duplex cognitive radio networks. Radio Sci. **55**(6) (2020)
19. Wang, P., You, F., He, S., Zhao, C.: A double screening orthogonal-matching-pursuit algorithm for compressed sensing receiver with high column correlation sensing matrix. IEICE Electron. Express **16**(18) (2019)
20. Eldafrawy, M., Boutros, A., Yazdanshenas, S., Betz, V.: FPGA logic block architectures for efficient deep learning inference. ACM Trans. Reconfigurable Technol. Syst. **13**(3), 1–34 (2020)
21. Manohar, C., Halaki, A., Gurugopinath, S.: Cooperative spectrum sensing based on flatness in spectral domain under noise variance uncertainty
22. Beck, E., Bockelmann, C., Dekorsy, A.: Compressed edge spectrum sensing: extensions and practical considerations. at-Automatisierungstechnik **67**(1), 51–59 (2019)
23. Wang, Y., Pandharipande, A., Polo, Y.L., Leus, G.: Distributed compressive wide-band spectrum sensing. In: Information Theory and Applications Workshop, pp. 178–183. IEEE (2009)

QoS-Guaranteed AP Selection Algorithm
in Dense IEEE 802.11 WLANs

Zhihui Weng, Zhibin Xie$^{(\boxtimes)}$, and Haoran Qin

Jiangsu University of Science and Technology, Zhenjiang, China
xiezhibin@just.edu.cn

Abstract. IEEE 802.11 wireless local area network (WLAN) is a popular connectivity method because of its convenient deployment, low cost and flexibility. Due to the limited coverage of single access point (AP), multiple APs are often arranged in current WLANs to meet the coverage requirement. In such dense WLANs, the actual situation is that every wireless station (WS) has different quality of service (QoS) requirements for the actual acquired throughput. Thus, this paper proposed a QoS-Guaranteed AP selection algorithm to increase the overall QoS of WSs for the actual acquired throughput. The proposed algorithm relies on a centralized framework and considers the diverse QoS requirements. According to these concrete requirements, the proposed algorithm can distribute WSs to APs fairly and achieve the near-optimal access of AP based on the game theory. Finally, the numerical simulation results verify that the proposed algorithm can effectively improve the overall QoS of WSs.

Keywords: AP selection algorithm · QoS-Guaranteed · Dense IEEE 802.11 WLANs · Game theory

1 Introduction

IEEE 802.11 wireless local area networks (WLANs) have been widely deployed as wireless infrastructures, which can provide data access services in home, corporate and public environments.

Due to the limited coverage of a single access point (AP), multiple APs are often arranged in WLAN to meet the coverage requirement. In such dense WLANs, every wireless station (WS) has different quality of service (QoS) requirements for the actual acquired throughput according to its own situation. The QoS of a WS is strongly influenced by which AP it associates with. Thus, for each WS, how to select a suitable AP to obtain satisfactory service is very important.

The conventional AP selection strategy for WS is to associate with the AP with the highest received signal strength indication (RSSI), which has been proved to be not effective in dense WLANs. Because location distribution of WSs is usually quite unevenly around APs [1], and this simple strategy might lead to bad performance and load imbalance [2].

H. Gao et al. (Eds.): ChinaCom 2020, LNICST 352, pp. 434–443, 2021.
https://doi.org/10.1007/978-3-030-67720-6_30

Existing AP selection can be classified as distributed or centralized strategies. In the case of distributed strategies, the WSs gather some metrics from APs and then choose the best AP based on these metrics. The authors in [3] presented a distributed strategy based on game theory to maximize the overall network throughput. The authors in [4] presented a classification of works and introduced a distributed strategy, which addressed quality of experience (QoE) enhancement.

In the case of centralized strategies, the decision on the selection of the best AP is performed by a controller. The controller has an overall view of the managed network. For centralized strategies, the authors in [5, 6] used SDN-based platforms to implement centralized approaches to address AP selection for WSs. The authors in [7] proposed a cloud-based access node selection approach using a potential game.

However, the traditional works did not consider the different QoS requirements of WSs and cannot distribute WSs to APs fairly. In order to increase the overall QoS of WSs instead of individual QoS of WS, a QoS-Guaranteed AP selection algorithm is proposed for dense WLANs in this paper. In the proposed algorithm, the different QoS requirements of WSs and the distance between WSs and APs are considered, and a new performance metric are proposed to evaluate the QoS of WS. In addition, based on the fair evaluation, the centralized framework and two-tier approximation game theory are adopted to manage the network to increase the overall QoS. The numerical simulation results show the effectiveness of the proposed algorithm.

2   System Model

2.1   Network Model

As shown in Fig. 1, this paper considers a centralized dense IEEE 802.11 WLAN. The network has one access point controller (AC), M APs and N ($N > M$) WSs. The AC is used to manage all APs and has an overall view of the managed network. Let $\mathcal{M} = \{1, 2,..., M\}$ and $\mathcal{N} = \{1, 2,..., N\}$ denote the sets of APs and WSs, respectively. Assume that neighboring APs operate at different (non-overlapping) frequency channels, so that there is no interference among them.

The network transmission protocol adopts IEEE 802.11b. IEEE 802.11b is a multi-rate protocol and the supported transmission rates are 11 Mbps, 5.5 Mbps, 2 Mbps and 1 Mbps. The transmission rates are determined by channel conditions. The MAC layer protocol of IEEE 802.11b adopts distributed coordination function (DCF). Each WS i ($i \in \mathcal{N}$) has a QoS requirement for the actual acquired throughput r_i and thus it has to select a suitable AP to meet its r_i. Assume that each WS is in saturation mode, which means it always transmits data to AP.

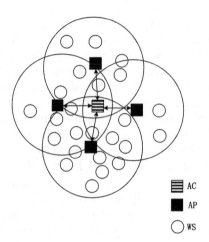

Fig. 1. A centralized dense IEEE 802.11 WLAN

2.2 Communication Model

The link capacity $c_{i,j}$ between WS i and AP j can be expressed as:

$$c_{i,j} = B \log_2(1 + \frac{p_i h_i}{\sigma}),$$ (1)

where B represents the bandwidth of the selected channel, p_i represents the transmission power of WS i, h_i represents the channel gain between WS i and AP j, and σ represents the thermal noise power.

Once $c_{i,j}$ is determined, the bit rate $b_{i,j}$ between WS i and AP j can be obtained. The $b_{i,j}$ is calculated by mapping $c_{i,j}$ to the closest but below the bit rate level provided by the orthogonal frequency division multiple access (OFDMA) modulation scheme supported by IEEE 802.11b.

When multiple WSs associate with the same AP and transmit data to the AP through the DCF provided by the MAC protocol of the IEEE multi-rate protocol at the same time, the performance anomaly occurs. And the high-rate WS will be dragged down by the low-rate WS, so that the actual acquired throughput of the high-rate WS will be greatly reduced.

The Fig. 2 illustrates a performance anomaly scenario. There are two IEEE 802.11b APs and five WSs. WS_4 has two choices, AP_1 or AP_2. The Table 1 shows that if WS_4 selects AP_2, the achievable throughput of WS_5 is dragged down by WS_4, from 7.73 Mbps to 0.81 Mbps. It can also be seen that if WS_4 selects AP_1, it can have a better throughput 1.96 Mbps, although AP_1 is more crowded than AP_2. AP_1 is also a better choice from the perspective of total throughput 15.57 Mbps. The achievable throughputs in Table 1 are calculated based on the analysis of [8].

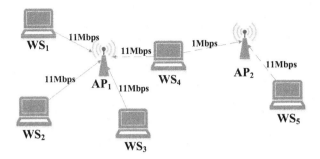

Fig. 2. Scenario illustrating performance anomaly

Table 1. Achievable throughputs based on the choice of WS$_4$

Choice of WS$_4$	Achievable throughput (Mbps)					Total (Mbps)
	WS$_1$	WS$_2$	WS$_3$	WS$_4$	WS$_5$	
AP$_1$	1.96	1.96	1.96	1.96	7.73	15.57
AP$_2$	2.66	2.66	2.66	0.81	0.81	9.60

3   Problem Formulation

In IEEE 802.11b environment, $X_{i,\,j}$ represents the obtained throughput after WS i selecting AP j and can be expressed as:

$$X_{i,j} = U_{i,j} \times \frac{s_d}{b_{i,j}T_{i,j}} \times b_{i,j}, \tag{2}$$

where $U_{i,\,j}$ is the fraction of time that WS i is able to access the medium of AP j, s_d is the length (in bits) of the data frame and its value is 12272, $T_{i,\,j}$ is the overall transmission time (counting protocol overhead, transmission time, and the time spent in contention procedure) for a single frame sent by WS i and $b_{i,\,j}$ is the bit rate between WS i and AP j.

Let $\mathbb{W}_j$ denote all the WS sets (WS $i \in \mathbb{W}_j$) of AP j. K represents the number of WS in $\mathbb{W}_j$. The $T_{i,j}$ can be expressed as:

$$T_{i,j}(K) = t_{tr} + t_{ov} + t_{cont}(K), \tag{3}$$

where t_{tr} represents the frame transmission time, t_{ov} represents constant overhead and t_{cont} represents the time spent in the competition.

The t_{tr} can be expressed as:

$$t_{tr} = \frac{s_d}{b_{i,j}}, \tag{4}$$

The t_{ov} varies according to the bit rate used by WS. If it transmits at 1, 2, 5.5 or 11 Mbps, then the $t_{ov} = 541, 305, 271, 262$ μs respectively. (these parameters are for 802.11b).

The t_{cont} can be written as:

$$t_{cont}(K) \simeq SLOT \times \frac{1 + P_c(K)}{2K} \times \frac{CW_{min}}{2}, \tag{5}$$

where the value of $SLOT$ is 20 μs, the congestion window CW varies between $CW_{min} = 31$ and $CW_{max} = 1023$. $P_c(K)$ is the proportion of collisions of each data packet successfully acknowledged at the MAC layer $(0 \le P_c(K) < 1)$. The $P_c(K)$ can be expressed as:

$$P_c(K) = 1 - (1 - \frac{1}{CW_{min}})^{K-1}, \tag{6}$$

The $U_{i,j}$ can be expressed as:

$$U_{i,j} = \frac{T_{i,j}}{\left(\sum_{WS \ k \in W_j} T_{k,j}\right) + P_c(K) \times t_{jam} \times K}, \tag{7}$$

where t_{jam} represents the average time spent in a collision.

The duration of the collision depends on the type of WS collision (slow or fast) involved. Thus, the t_{jam} can be obtained by considering all possible WS pairs between at different rates, that is as follows:

$$t_{jam} = P_1 T_{1,j} + P_2 T_{2,j} + P_3 T_{3,j} + P_4 T_{4,j}, \tag{8}$$

where $T_{1,j}, T_{2,j}, T_{3,j}$, and $T_{4,j}$ represent the overall transmission time of a single frame for WS with transmission rates of 1, 2, 5.5 and 11 Mbps, respectively. P_1, P_2, P_3 and P_4 represent the probability of having a packet sent at the slow transmission rates of 1, 2, 5.5, and 11 Mbps, respectively.

P_1, P_2, P_3 and P_4 can be computed as the ratio between the number of WS pairs that contain the slow host and the total number of pairs that can be formed in the set of all WSs. They can be calculated as below formulas:

$$P_1 = \frac{K_1(K_1 - 1) + 2K_1(K_2 + K_3 + K_4)}{K(K - 1)}, \tag{9}$$

$$P_2 = \frac{K_2(K_2 - 1) + 2K_2(K_3 + K_4)}{K(K - 1)}, \tag{10}$$

$$P_3 = \frac{K_3(K_3 - 1) + 2K_3 K_4}{K(K - 1)}, \tag{11}$$

$$P_4 = \frac{K_4(K_4 - 1)}{K(K - 1)}, \tag{12}$$

where K_1 is the number of WS with a bit rate of 1 Mbps in W_j, K_2 is the number of WS with a bit rate of 2 Mbps, K_3 is the number of WS with a bit rate of 5.5 Mbps, K_4 is the number of WS with a bit rate of 11 Mbps.

Each WS i has a QoS requirement for the actual acquired throughput r_i. Define $Q_{i,j}$ as the QoS of WS i to the selected AP j for the actual acquired throughput, and can be expressed as:

$$Q_{i,j} = \begin{cases} \frac{X_{i,j}}{r_i}, & \frac{X_{i,j}}{r_i} < 1 \\ 1, & \frac{X_{i,j}}{r_i} \geq 1 \end{cases}, \tag{13}$$

Each WS i needs to select the appropriate AP in the WLAN for data transmission. Let $\lambda_{i,j} \in \{0,1\}$ denote the selection decision between the WS i and the AP j, i.e. $\lambda_{i,j} = 1$ means that the WS i decides to associate with AP j, otherwise $\lambda_{i,j} = 0$.

Based on this, the problem of maximizing the overall QoS for the actual acquired throughput can be described as follows:

$$\max_{\lambda_{i,j}} \quad \sum_{i=1}^{N} \sum_{j=1}^{M} Q_{i,j} \lambda_{i,j}, \tag{14}$$

4 QoS-Guaranteed AP Selection Algorithm

Game theory has been widely used as an efficient method to address the decision-making problems among multiple game players [9]. In order to maximize the overall QoS for the actual acquired throughput, we adopt a two-tier approximation algorithm called QoS-Guaranteed AP selection algorithm (QASA). In the proposed algorithm, we regard WSs as the rational game players. The outer layer adopts a stochastic game to find an overall optimal AP selection decision profile for all the WSs, and the inner layer adopts a greedy approximation method to find a locally optimal AP selection decision of WS i for its r_i.

Let $\mathbb{A} = (A_1, A_2, ..., A_{i-1}, A_i, A_{i+1}, ..., A_N)$ denote the AP selection decision profile of all WSs. Also, let $\mathbb{A}_{-i} = (A_1, A_2, ..., A_{i-1}, A_{i+1}, ..., A_N)$ denote the AP selection decision profile of all WSs except WS i. A_i denotes the AP selection decision of WS i. $\mathbb{A}$ can be also denoted as $(A_i, \mathbb{A}_{-i})$. Given the AP selection decision profile of other WSs $\mathbb{A}_{-i}$, WS i would like to choose a locally optimal AP selection decision A_i^* to meet its r_i. WS i adopts A_i^*, which can maximize its Q_i, i.e.,

$$Q_i\left(A_i^*, \mathbb{A}_{-i}\right) \geq Q_i\left(A_i, \mathbb{A}_{-i}\right), \forall\, i \in \mathbb{N}, \tag{15}$$

Furthermore, we can formulate our problem of maximizing the overall QoS AP selection as a strategic game as

$$\Gamma = (\mathbb{N}, \{\overline{\mathbb{A}_i}\}_{i \in \mathbb{N}}, \{Q_i(A_i, \mathbb{A}_{-i})\}_{i \in \mathbb{N}}), \tag{16}$$

where $\mathcal{N}$ is the set of game players, $\overline{A}_i = \{1, \cdots, M\}$ is the set of strategies for game player WS i, and $Q_i(A_i, A_{-i})$ is the revenue function to be maximized by player WS i.

In QASA, each WS first calculates their locally optimal AP selection decision profile in coming time slot and then computes the overall QoS based on this decision profile. If the overall QoS is increased, save this WS to the contention set. Then for any WS in contention set, only one WS whose decision profile has the highest overall QoS wins the update opportunity and other WSs keep unchanged in this time slot. Then update the current decision profile with decision profile of this winner WS.

According to the Nash equilibrium existence theorem, our multi-user AP selection game is a finite player game, in which each WS can choose a pure strategy from a finite set of AP selection strategies, and thus it has a Nash equilibrium [10]. That is, after a finite number of time slots, the algorithm converges when no WS needs to update its AP selection decision profile, and the near-optimal decision profile for all the WSs are obtained [11]. The pseudo code is as follows:

Algorithm QoS-Guaranteed AP selection algorithm(QASA)

1: Initialize: N, M, B, the distance from WS i to AP j $d_{i,j}$, p_i, path loss exponent γ, σ, r_i,

$\overline{A}_i$, the initial A_i is the closest AP to WS i, the contention set $\mathbb{C} = \varnothing$;

2: for each iteration τ

3:    for all WS i do

4:        compute the local optimal AP selection decision profile of WS i (A_i^*, A_{-i}) in next time slot $(\tau + 1)$;

5:        compute $\sum_{i=1}^{N}\sum_{j=1}^{M} Q_{i,j}\lambda_{i,j}(\tau+1)$ based on (A_i^*, A_{-i});

6:        if $\sum_{i=1}^{N}\sum_{j=1}^{M} Q_{i,j}\lambda_{i,j}(\tau+1) > \sum_{i=1}^{N}\sum_{j=1}^{M} Q_{i,j}\lambda_{i,j}(\tau)$ then

7:            save WS i to $\mathbb{C}(\tau)$;

8:        end if

9:    end for

10:   while $\mathbb{C}(\tau) \neq \varnothing$ do

11:       each WS in $\mathbb{C}(\tau)$ contends and only one WS with the biggest $\sum_{i=1}^{N}\sum_{j=1}^{M} Q_{i,j}\lambda_{i,j}$ wins update opportunity

12:       $A_{winner} = A_{winner}^*$

13:       update A_{winner} in A;

14:   end while

15:end for

16: return A and $\sum_{i=1}^{N}\sum_{j=1}^{M} Q_{i,j}\lambda_{i,j}$

---

5   Simulation Results

In this section, the scenario includes M IEEE 802.11b APs randomly distributed in an area of 200 m × 200 m with a minimum distance of 20 meters between them, and N WSs randomly distributed in this area. The bandwidth of the wireless channel B is 20 MHz. The transmit power of the WS i p_i is 10 dBm. The Gaussian thermal noise σ is −55 dBm. We assume that the path loss follows that free-space model, and the path loss exponent γ is set to 3.5. The QoS requirement r_i is randomly generated from 3 to 7 Mbps.

In order to verify our proposed algorithm, we use the exhaustive algorithm as a benchmark for comparison, as shown in Fig. 3. In the exhaustive algorithm, it first checks all possible selection decision profiles of WSs, and then chooses a globally optimal selection decision profile with an overall maximum QoS. Because the exhaustive algorithm has a large amount of computation and is not suitable for the scenario of owing many WSs and APs. Therefore, the number of AP is set to 4, and the number of WSs $N = 1, 2, \ldots, 8$. It can be found that the overall QoS of our proposed QASA is very consistent with exhaustive algorithm results. The numerical results confirm that QASA can provide a near-optimal solution to our problem.

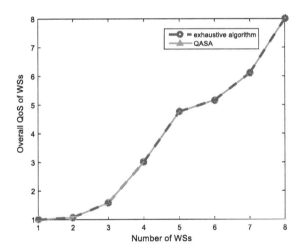

Fig. 3. Comparison between QASA and exhaustive algorithm

In addition, we present the convergence behavior of QASA under different WS numbers, which are $N = 30, 50$ respectively. The number of AP is set to 8. From Fig. 4, one can be seen that the overall QoS of WSs gradually increases as the number of iterations increases. After a finite number of iterations, the QASA converges to a certain value. In addition, when $N = 30$, the number of iteration to reach equilibrium is 22, and when $N = 50$, the number of iteration is 48. It indicates that the more WSs exist, the more iterations are needed.

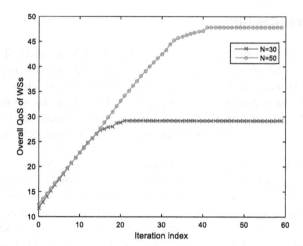

Fig. 4. Iteration comparison of QASA with different number of WSs

In order to verify the performance of our proposed QASA, we compare our QASA with the RSSI algorithm and the maximizing total network throughput algorithm (MTNT) [3]. At this simulation, the number of APs is set to 10. The comparison results are shown in Fig. 5. It can be seen from the simulation results that our QASA can increase the overall QoS by 160% and 10% compared to the RSSI algorithm and MTNT, when the number of WS is 50. In RSSI algorithm, all WSs make decisions alone, and might lead to bad performance and waste many resources. The MTNT does not consider the requirements of WSs and causes unfair distribution. Numerical result of Fig. 5 demonstrates our QASA has better flexibility and adaptability and can effectively improve the overall QoS of WSs.

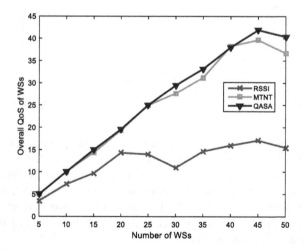

Fig. 5. Comparison results by adopting different AP selection algorithms

6 Conclusions

IEEE 802.11 WLAN is a popular connectivity method. It is important to select a suitable AP for every WS in this network. The actual situation is that every WS has different QoS requirements for the actual acquired throughput. For guaranteeing a fair distribution of the actual throughput according to the QoS requirement of WS, we propose a AP selection algorithm based on maximizing overall QoS criterion for multi-WSs and multi-APs WLAN, and utilize game theory to get the near-optimal AP selection decision. Simulations show that the QASA can effectively improve the overall QoS of WSs for the actual acquired throughput.

In the future work, we will pay attention to predict the movement of WSs and make decisions based on that. Moreover, we will investigate more metrics to define QoS, such as delay and jitter.

References

1. Schwab, D., Bunt, R.: Characterising the use of a campus wireless network. In: IEEE INFOCOM 2004, vol. 2, pp. 862–870. IEEE, Hong Kong (2014)
2. Chen, J., et al.: WLC19-4: effective AP selection and load balancing in IEEE 802.11 Wireless LANs. In: IEEE Global 2006, pp. 1–6. IEEE, San Francisco (2006)
3. Yen, L., Li, J., Lin, C.: Stability and fairness of AP selection games in IEEE 802.11 access networks. IEEE Trans. Vehicular Technol. **60**(3), 1150–1160 (2011)
4. Kim, Y., Kim, M., Lee, S., Griffith, D., Golmie, N.: AP selection algorithm with adaptive CCAT for dense wireless networks. In: 2017 IEEE Wireless Communications and Networking Conference, pp. 1–6. IEEE, San Francisco (2017)
5. Raschellà, A., Bouhafs, F., Mackay, M., Shi, Q., Ortín, J., Gállego, J.R., Canales, M.: AP Selection algorithm based on a potential game for large IEEE 802.11 WLANs. In: IEEE/IFIP Network Operations and Management Symposium (NOMS 2018), pp. 23–27. IEEE, Taipei (2018)
6. Chen, C., Wang, C., Liu, H., et al.: A novel AP selection scheme in software defined networking enabled WLAN. Comput. Electr. Eng. **66**, 288–304 (2017)
7. Ortín, J., Gállego, J.R., Canales, M.: Joint cell selection and resource allocation games with backhaul constraints. Pervasive Mobile Comput. **35**, 125–145 (2017)
8. Heusse, M., Rousseu, F., Berger-Sabbatel, G., Duda, A.: Performance anomaly of 802.11b. In: IEEE INFOCOM 2003 Twenty-second Annual Joint Conference of the IEEE Computer and Communications Societies, vol. 2, pp. 836–843 (2003)
9. Ren, Y., Xie, Z., Ding, Z., Sun, X., Xia, J., Tian, Y.: Computation offloading game in multiple UAV-enabled mobile edge computing networks. IET Commun. (2020). https://doi.org/10.1049/iet-com.2020.0131
10. Nash, J.F.: Equilibrium points in N-person games. Proc. Natl. Acad. Sci. U.S.A **36**(1), 48–49 (1950)
11. Monderer, D., Shapley, L.S.: Potential games. Games Econ. Behav. **14**(1), 124–143 (1996)

Transmission and Load Balancing

Autonomous Positioning Algorithm for UE in Cellular Networks

Yifan Xi$^{(\boxtimes)}$, Hang Long, and Tong Li

Wireless Signal Processing and Network Lab,
Key Laboratory of Universal Wireless Communication, Ministry of Education,
Beijing University of Posts and Telecommunications, Beijing, China
xiyifan@bupt.edu.cn

Abstract. The positioning techniques in wireless cellular networks detect and receive signals transmitted by the user equipment (UE) through base stations (BS), then the network side uses the time difference of arrival (TDOA) values carried in the received signals to calculate UE's position. However, if the network side doesn't report the positioning information to the UE, UE will never know its positioning result. Besides, a BS can accurately locate the UE within its coverage area, the security issue of UE needs to be considered. In order to improve the autonomy and security of UE, in this paper, we propose an autonomous positioning algorithm for the UE in cellular networks. The UE uses TDOA measurements and multiple UE positioning results sent by the network side to inversely calculate the positions of the BS participating in the positioning. After that, the UE can use calculated BS position coordinates and TDOA values measured by UE itself to calculate its position independently. The simulation results show that the method is effective, and the error of the calculated BSs' coordinates is within the acceptable range.

Keywords: Positioning · TDOA · Genetic algorithm

1 Introduction

Since the promulgation of E-911 regulations in 1996 [1], positioning technology has received the attention of companies and institutions in various countries due to its huge commercial potential. With the rapid development of technology of mobile cellular communications, the number of users increased significantly.

According to the characteristics of measured value, the positioning methods of the UE in cellular network can be divided into three categories: field location algorithm, the positioning algorithm based on the incident angle of the signal arrival (AOA) and the positioning algorithm based on the electric wave arrival time (TOA) or time difference of arrival (TDOA). The field location algorithm is based on the proportional relationship between the energy loss of the signal

Supported by National Natural Science Foundation of China under Grant 61628102 and 61931005.

H. Gao et al. (Eds.): ChinaCom 2020, LNICST 352, pp. 447–458, 2021.
https://doi.org/10.1007/978-3-030-67720-6_31

propagating in the channel and the propagation distance [2]. By detecting the field strength values of the transmitted signal and the received signal, the position coordinates of the UE can be roughly estimated. AOA measures the angle of arrival between the UE and the BS. The ray formed from the BS will pass through UE, so that the intersection of two rays is the position of UE [3]. The traveling time of a radio signal between a BS and an UE is the fundamental parameter from which a distance between them is calculated. TOA takes a BS as the center and the distance between the BS and the UE as the radius to make a circle, the intersection of three circles is the position of UE [4]. However, compared with the time difference, the absolute time is more difficult to measure and the accuracy is low. TDOA takes the base stations as the focus and the distance difference between the BS and the UE as the long axis to make a hyperbola, the intersection of two hyperbolas is the signal position [5].

Among these positioning technologies, limited by positioning accuracy requirements and other conditions, the positioning algorithm research mainly focuses on TDOA measurement value estimation. Starting from 3GPP R9, LTE has proposed TDOA positioning technology standards. Later, R14 carried out research on the enhancement of 3G and 4G positioning technology, and enhanced the TDOA positioning technology in the shared physical cell identifier (PCI) scenario. Now 5G uses low-latency and high-precision synchronization technology, which can further improve the accuracy of TDOA [6]. TDOA is still the main technology for 5G positioning [7].

These processes of positioning techniques for wireless cellular networks are basically similar. The UE reports the information required for positioning (such as TDOA measurements) to the network side, and the network side calculates the positioning results. However, these positioning techniques based on the network side still have some problems to be solved. The deficiency is that the UE positioning autonomy is too low, which means if the network side doesn't report the positioning information to the UE, the UE will never know its positioning result. In addition, the network side has the positioning control right and it can accurately locate the positioning information of an UE within its coverage area at any time. The security issue of the UE also needs to be considered. Therefore, an autonomous positioning algorithm for the UE in cellular networks is needed. Without relying on network notifications, the UE can calculate the positioning result by itself and the security of the UE will also be improved.

In order to improve the positioning autonomy and ensure the positioning safety of the UE, we propose an autonomous positioning algorithm for the UE in cellular networks in this paper. UE first uses TDOA measured values and multiple UE positioning results issued by the network side to inversely calculate the position of the BSs participating in the positioning, meanwhile the coordinates of these BSs will be obtained. Also we know that the TDOA measurements are obtained by the UE itself by calculating the time difference between receiving downlink signals of different BSs. Then, the UE can use the calculated BS position coordinates and TDOA measured values to calculate its own position without the BS's help.

The remainder of this paper is organized as follows. On account of the most TDOA positioning technology used currently, a mathematical model of positioning algorithm is given in Sect. 2, and Sect. 3 elaborates the analysis and algorithm, including the application of genetic algorithm in inverse calculation of the BS position. Section 4 shows the simulation results and proves the rationality of analysis. Finally Sect. 5 concludes this paper.

Notations: Uppercase letter X and Y denote the horizontal and vertical coordinates of the BS respectively, and lowercase letter x and y denote the horizontal and vertical coordinates of the UE, respectively. The transpose and inverse of a matrix are expressed as $(\cdot)^{\mathrm{T}}$, $(\cdot)^{-1}$, respectively.

2 System Model

TDOA positioning technique, also called hyperbolic positioning, calculates the position coordinates of target UE by measuring the time difference of the radio wave from the UE to two different base stations [8]. There are two methods for obtaining TDOA measured values. One is based on the difference between TOA measurements of the two base stations. This method requires strict clock synchronization between the UE and the BS. The other is to use related technologies to cross-correlate the signal received by one BS with the signal received by another BS to obtain the TDOA measurements. This method can estimate TDOA measurements when the BS and the UE clocks are not synchronized.

Once a certain TDOA value is obtained, the distance from the UE to the two base stations can be calculated. Several TDOA measured values can form a system of curvature equations about the position of the UE. Then the estimated position of the UE will be captured by solving this system of the curvature equations.

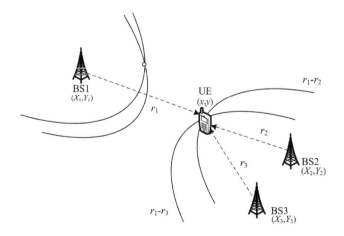

Fig. 1. TDOA-based position method.

According to the geometric principle, the UE must be on the hyperbola with BS$_1$ and BS$_2$ as the focal point and the distance difference $r_{21} = r_2 - r_1$ as the focal length. Similarly, if the distance difference r_{31} between BS$_3$ and BS$_1$ and the UE is obtained, another hyperbola can be constructed. One of the intersection points of these two hyperbolic curves is the position of the UE as shown in Fig. 1. (x, y) is the position coordinate of the UE to be estimated, (X_i, Y_i) is the known position of the i-th BS, and the distance between the UE and the BS$_i$ is

$$r_i^2 = (X_i - x)^2 + (Y_i - y)^2 = K_i - 2X_i x - 2Y_i y + x^2 + y^2 \tag{1}$$

where $K_i = X_i^2 + Y_i^2$.

Let r_{i1} indicates range difference between the distance from the UE to the BS$_i$ and the distance from the UE to the BS$_1$ (service base station).

$$r_{i1} = c\tau_{i1} = r_i - r_1 = \sqrt{(X_i - x)^2 + (Y_i - y)^2} - \sqrt{(X_1 - x)^2 + (Y_1 - y)^2} \tag{2}$$

In the formula, $c = 3 \times 10^8 \, \mathrm{m/s}$ is the propagation speed of the electric wave in the air and τ_{i1} is the measured value of TDOA. These nonlinear system of equations need to be linearized, and the final linear equations can be obtained as follow

$$r_{i1}^2 + 2r_{i1}r_1 = K_i - 2X_{i1}x - 2Y_{i1}y - K_1 \tag{3}$$

where $X_{i1} = X_i - X_1$ and $Y_{i1} = Y_i - Y_1$. Among these equations, only x, y and r_1 are unknown values. Therefore, formula (3) can be regarded as a system of linear equations about the UE position coordinates (x, y) and r_1, and the coordinate to uniquely determine the position of the UE can be obtained by solving the system of equations.

In traditional positioning technologies, most are network-based positioning schemes which detect and receive the signals transmitted by the UE through multiple base stations with known positions and process the TDOA information carried in the received signals. These methods only require UE to send TDOA. However, if the network side doesn't report positioning information to the UE, UE will never know its positioning result. Therefore, an autonomous positioning algorithm for the UE in cellular networks is needed.

3   Application of Genetic Algorithm in Inverse Calculation of Base Station Location

In this paper, we propose an autonomous positioning algorithm for the UE in cellular networks. First the UE uses TDOA measured values and the multiple UE positioning results sent by the network side to inversely calculate the position of the BS participating in the positioning, thereby obtaining the coordinates of the BS. After that, the UE can use the calculated BS position coordinates and TDOA measured values obtained by the UE by calculating the time difference between receiving downlink signals of different BSs to realize its own positioning. In this chapter, we first propose the system model of the inverse base station and then give the solution process of it.

3.1   Inverse Base Station Positioning Model

On the two-dimensional plane, assuming that there are M base stations participate in positioning, the UE obtains N measurement positions. (X_i, Y_i), $i = 1, 2, \cdots, M$ is the position coordinate of the i-th base station BS_i to be solved, and (x_j, y_j), $j = 1, 2, \cdots, N$ is the measurement positioning coordinate of the UE. The distance from the j-th UE positioning (x_j, y_j) to the i-th base station (X_i, Y_i) is R_i^j. Taking the base station with serial number 1 as the serving base station, R_{i1}^j represents the distance difference between the j-th UE positioning to $\mathrm{BS}_i (i \neq 1)$ and BS_1 (serving base station), as

$$R_{i1}^j = R_i^j - R_1^j + c n_{i1}^j \tag{4}$$

and also can be expressed as

$$R_{i1}^j = \sqrt{(X_i - x_j)^2 + (Y_i - y_j)^2} - \sqrt{(X_1 - x_j)^2 + (Y_1 - y_j)^2} + c n_{i1}^j \tag{5}$$

where $i = 1, 2, \cdots, M$, $j = 1, 2, \cdots, N$. In the case of considering error, n_{i1}^j is the error introduced in TDOA measurement.

Taking the inverse calculation of the coordinate of the i-th $(i = 2, 3, \cdots, M)$ BS as an example, inversely calculating the position of the base station (X_i, Y_i) and (X_1, Y_1) is to solve the nonlinear equations composed of N equations as (6). $\mathbf{\Delta R}$, $\mathbf{R}_1$, $\mathbf{R}_i$, $\mathbf{n}$ are N-dimensional vectors formed by N UE measuring and positioning, and can be expressed as

$$\begin{aligned}
\mathbf{\Delta R} &= [R_{i1}^1, R_{i1}^2, \ldots, R_{i1}^N]^{\mathrm{T}}{}_{N \times 1} \\
\mathbf{R}_1 &= [R_1^1, R_1^2, \ldots, R_1^N]^{\mathrm{T}}{}_{N \times 1} \\
\mathbf{R}_i &= [R_i^1, R_i^2, \ldots, R_i^N]^{\mathrm{T}}{}_{N \times 1} \\
\mathbf{n} &= [n_{i1}^1, n_{i1}^2, \ldots, n_{i1}^N]^{\mathrm{T}}{}_{N \times 1}
\end{aligned} \tag{6}$$

substitute into (5) to get

$$\mathbf{\Delta R} = \mathbf{R}_i - \mathbf{R}_1 + c\mathbf{n} = \begin{bmatrix} \sqrt{(X_i - x_1)^2 + (Y_i - y_1)^2} \\ \sqrt{(X_i - x_2)^2 + (Y_i - y_2)^2} \\ \vdots \\ \sqrt{(X_i - x_N)^2 + (Y_i - y_N)^2} \end{bmatrix}$$
$$\begin{bmatrix} -\sqrt{(X_1 - x_1)^2 + (Y_1 - y_1)^2} \\ -\sqrt{(X_1 - x_2)^2 + (Y_1 - y_2)^2} \\ \vdots \\ -\sqrt{(X_1 - x_N)^2 + (Y_1 - y_N)^2} \end{bmatrix} + c\mathbf{n} \tag{7}$$

This paper considers the case of $N \geq 3$, using maximum likelihood to estimate the BS coordinates (X_i, Y_i) and (X_1, Y_1). Since the elements in $\mathbf{R}_i - \mathbf{R}_1$ are known TDOA measurements and n is a normal distribution function with mean 0 and variance σ^2, all elements in $\mathbf{\Delta R}$ follow a normal distribution with mean R_{i1}^j and variance σ^2. Each measurement value is independent, the corresponding likelihood function is

$$\prod_{j=1}^{N}\left[\frac{1}{\sqrt{2\pi}\sigma}\exp\left\{-\frac{(\Delta R_i^j - R_i^j + R_1^j)^2}{2\sigma^2}\right\}\right]$$
$$= \left[\frac{1}{\sqrt{2\pi}\sigma}\right]^N \exp\left(-\frac{(\mathbf{\Delta R} - \mathbf{R}_i + \mathbf{R}_1)^{\mathrm{T}}(\mathbf{\Delta R} - \mathbf{R}_i + \mathbf{R}_1)}{2\sigma^2}\right) \tag{8}$$

Finding the coordinate value that maximizes the maximum likelihood probability is equivalent to solving the following formula

$$(X_i, Y_i, X_1, Y_1) = \arg\min\left[(\mathbf{\Delta R} - \mathbf{R}_i + \mathbf{R}_1)^{\mathrm{T}}(\mathbf{\Delta R} - \mathbf{R}_i + \mathbf{R}_1)\right] \tag{9}$$

Formula (9) is a non-linear function. If the analytical method is used, the process will be very complicated, and bring a huge amount of calculation. In addition, for the inverse calculation of the BS position, since the location difference between the two positions of the mobile terminal and the same base station cannot be obtained, the linear algorithm of the TDOA principle cannot be used to calculate the position of the base station. Therefore for formula (9), the genetic algorithm is used to solve this problem, and the optimal solution is searched in the entire potential solution space to calculate the position of the base station.

3.2 Process of Genetic Algorithm in Inverse Calculation of Base Station Location

Genetic Algorithm (GA) is a kind of random optimization algorithm obtained from the idea of biological inheritance and evolution [9]. GA expresses the problem-solving process as the evolution of chromosomes. Each chromosome represents an individual in the group and is also a solution to the problem [10]. According to the fitness of chromosomes and the corresponding evolutionary rules, through the selection, crossover, and mutation between each generation of chromosomes, it eventually converges to the state of the most suitable environment, and the optimal solution to the problem is obtained [11]. The steps to realize the back-calculation of the location of the base stations using genetic algorithm are as follows.

Initialize Population and Chromosome Coding. First, we need to determine the search space. Obtain the ID of the cell where the UE is located. Due to the uniqueness of the ID, the range of the cell can be seen as the search space. Assume that the upper limit of cell coordinates is x_{max}, y_{max}, the lower limit is x_{min}, y_{min}, Then the BS coordinates (X_i, Y_i), where $i > 1$, and the serving BS coordinate (X_1, Y_1) satisfy the following conditions

$$\begin{cases} x_{\min} \leq X_i \leq x_{\max} \\ y_{\min} \leq Y_i \leq y_{\max} \end{cases} \begin{cases} x_{\min} \leq X_1 \leq x_{\max} \\ y_{\min} \leq Y_1 \leq y_{\max} \end{cases} \tag{10}$$

Then use binary coding for individual coding. This coding method is simple and easy to operate, which is conducive to the realization of crossover and mutation. This algorithm only considers positioning on a two-dimensional plane. Taking (X_i, Y_i) as an example, first encode the base station coordinates (X_i and Y_i) separately, and the encoded bit strings are A and B. Cascade A and B to get the bit string C, then C represents an individual's chromosome. With this coding method, the search space of the genetic algorithm is mapped to the space coordinates within the cell range.

At last, the population and parameters need to be Initialized. Set the population size P and randomly generate a $P \times 4$ dimensional matrix as the initial population within the cell range. Set evolution termination algebra G, crossover rate P_c and mutation rate P_m.

Fitness Function. When Eq. (8) obtains the minimum value, the estimated coordinates of the BS are the best. So the fitness function is taken as

$$f = \frac{1}{(\mathbf{\Delta R} - \mathbf{R}_i + \mathbf{R}_1)^{\mathrm{T}}(\mathbf{\Delta R} - \mathbf{R}_i + \mathbf{R}_1)} \tag{11}$$

The larger the individual fitness value, the closer it is to the optimal solution.

Adaptive Operator. The crossover rate P_c and mutation rate P_m have a great influence on the optimization process of the genetic algorithm, and an appropriate value should be selected to prevent the occurrence of premature convergence.

Genetic Algorithm Process

Select Operation. Use the roulette strategy to select individuals and ensure that highly adaptive individuals can enter the next generation population with a higher probability.

The roulette strategy is to calculate the probability that each individual can be inherited according to the fitness of the individual. According to this probability, individuals in the contemporary population are randomly selected to form the offspring population. The fitness function given in Eq. (10) is judged as a maximization problem, and the steps to solve the maximization problem with this strategy are given below.

Calculate the sum of the fitness values of all individuals in the contemporary population

$$F = \sum_{j=1}^{P} f_j \tag{12}$$

where P is the population size, f_j is the fitness value of the j-th individual. The fitness value of each individual is divided by the total F to obtain the probability that the individual is selected

$$p_j = \frac{f_j}{F} \tag{13}$$

Calculate individual cumulative probability and construct roulette wheel. Then generate a random number in the range of $[0,1]$ when selecting, if the random number is less than or equal to the cumulative probability of the individual (cumulative probability is the individual probability of all individuals in front of the individual in the individual list sum) and greater than the cumulative probability of individual 1, select the individual to enter the progeny population.

Cross Operation. Uniform arithmetic crossover is used in this algorithm. That is, the genes at each locus of two paired individuals are exchanged with the same crossing probability, thereby forming two new individuals.

Mutation Operation. This algorithm uses non-uniform mutation. Randomly disturb the original gene value, and use the disturbed result as the new gene value after mutation. For example, $x = (x_1, x_2, \ldots x_n)$ is the individual to be mutated ($n = 4$ for the algorithm, the mutated individual is X_i, Y_i, X_1, Y_1). For the mutated individual x_k, the new offspring x'_k is

$$x'_k = \begin{cases} x_k + r \cdot (x_{max} - x_k) \cdot \left(1 - \dfrac{g}{G}\right)^b, & a = 0 \\ x_k + r \cdot (x_k - x_{min}) \cdot \left(1 - \dfrac{g}{G}\right)^b, & a = 1 \end{cases} \tag{14}$$

where r is a random number on $[0,1]$, g is the current evolutionary algebra, G is the maximum evolutionary algebra, b is a parameter to determine the non-uniformity, usually 2–5, and a is 0, 1 random number.

According to the above theoretical analysis, the process of genetic algorithm optimization is the change process of objective function and fitness function. During the operation of the genetic algorithm, individuals with low fitness in the population are continuously eliminated and the number of eligible individuals will increase which are closer to the optimal solution of the problem. Then the optimal result of the problem, that is, the position coordinates of the BS can be obtained.

4   Algorithm Simulation and Result Analysis

4.1   Simulation Conditions

This section tests the performance of the previously proposed algorithm in a Gaussian noise environment, and simulates the algorithm with different measurement errors, cell radii, and the mobile terminal acquiring different numbers of position coordinates. The main parameter settings in the paper are as follows: a cellular structure with 3 base stations, the serving base station is BS_1 (0,0), and the remaining BS coordinates are: BS_2 (0,1000) and BS_3 ($500\sqrt{3}$,500). The actual position of the UE takes a random number within the coverage of the three base stations. TDOA measurement error is Gaussian distributed, whose mean value is 0 and the setting standard deviation (converted into distance) are 2 m, 4 m, 6 m, 8 m, and 10 m, respectively.

Genetic algorithm simulation parameters: population size $P = 200$, maximum termination evolution generation $G = 100$, cross rate $P_c = 0.3$, variation rate $P_m = 0.05$, parameter to determine the non-uniform variability $b = 4$.

4.2   Simulation Results and Analysis

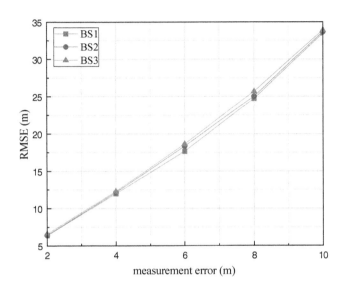

Fig. 2. Inverse calculation results with 3 positioning coordinates

As shown in Fig. 2, under Gaussian mode, the cell radius is 500 m, the measurement error is taken between 2 m to 10 m and the UE uses three positioning positions to inversely calculate the coordinates of the BS. The simulation results show that the positioning error of base station inverse algorithm increases with the increase of measurement error. And since the UE positions are randomly distributed relative to the three base stations, the positioning performance of the three base stations is very close. From the positioning effect, the error values of the three base stations are within the acceptable range and demonstrates excellent positioning performance, so that the feasibility of the inverse algorithm is verified.

As shown in Fig. 3, under Gaussian mode, the cell radius is 500 m, the measurement error is taken between 2 m to 10 m and the UE uses four and five positioning positions to inversely calculate the coordinates of the BS, respectively. And Fig. 4 shows the results using different numbers of UE positions. Compared with Fig. 2, the UE uses more positioning positions to calculate base station coordinates. The simulation results show that the positioning accuracy of the three base stations has been improved. From the perspective of positioning

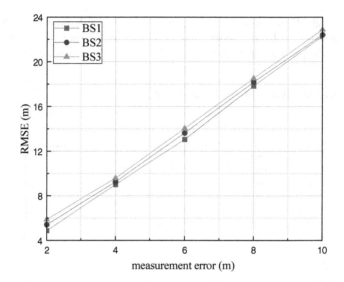

Fig. 3. Inverse calculation results with 4 positioning coordinates

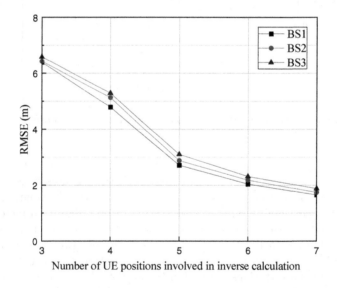

Fig. 4. Inverse calculation results with different number of positioning coordinates

effect, when the UE obtains more position coordinates for inverse calculation, the positioning performance of the algorithm can be improved.

As shown in Fig. 5, under Gaussian mode, the measurement error is taken to be 10 m, the cell radius ranges between 500 m and 2500 m and the UE uses four positioning positions to inversely calculate the coordinates of the BS. The simulation results show that the positioning errors of the three base stations all

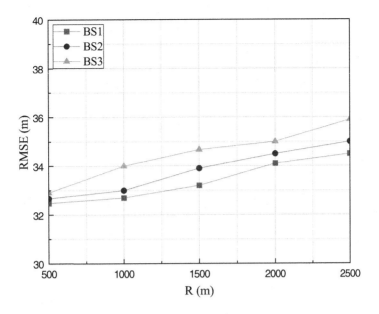

Fig. 5. Relation between standard error and cell radius

decrease as the radius decreases, the errors are all within the acceptable range and the error curve of our algorithm is in a stationary state which demonstrates excellent stability of the algorithm.

5   Conclusion

In order to improve the autonomy and security of the UE, in this paper, we propose an autonomous positioning algorithm for the UE in cellular networks. UE first uses TDOA measured values and multiple UE positioning results issued by the network side to inversely calculate the position of the BSs participating in the positioning, meanwhile the coordinates of these BSs will be obtained. After that, the UE can use the calculated BS position coordinates and TDOA measured values to calculate its own position independently. The simulation results show that: 1. The algorithm is feasible, and the error of the calculated base station position coordinates is within the acceptable range. 2. As the position coordinates of the participating algorithms obtained by UE increase, more accurate base station position coordinates can be obtained. 3. As the cell radius increases, the error curve of our algorithm is in a stationary state which demonstrates excellent stability of the algorithm. It can be seen from the analysis and simulation results that our algorithm is effective and also demonstrates stability and higher positioning accuracy.

References

1. 3GPP TS 36. 355, 3rd Generation Partnership Project. Technical specification group radio access network, E−UTRA; LTE positioning protocol, v10.2.0[S] (2011)
2. Ji, Z.Y., Pi, Y.M., Shu, J.B.: Application and research of mobile location technology based on signal strength in cellular network. GNSS World China **30**(4), 18–22 (2005)
3. Deng, E., Fan, P.Z.: An AOA assisted TOA positioning system. Conf. Commun. Technol. **2**, 1501–1504 (2000)
4. Caffery, J.J.: A new approach to the geometry of TOA location. In: Vehicular Technology Conference, pp. 1943–1949. IEEE (2000)
5. Wang, G., Chen, H.: An importance sampling method for TDOA-based source localization. IEEE Trans. Wirel. Commun. **10**(5), 1560–1568 (2011)
6. Witrisal, K., Meissner, P., Leitinger, E., et al.: High accuracy localization for assisted living: 5G systems will turn multipath channels from foe to friend. IEEE Signal Process. Mag. **33**(2), 59–70 (2016)
7. Ouyang, J., Chen, S.J., Huang, X.M., et al.: High precision localization technology for 5G mobile communication networks. Mobile Commun. **43**(9), 13–17 (2019)
8. Sun, S., Li, H.: Simulation research based on TDOA location technology. Wirel. Commun. Technol. (2002)
9. Holland, J.H.: Adaptation in Natural and Artificial Systems: An Introductory Analysis With Applications to Biology. Univerisity of Michigan Press, Cambridge (1975)
10. Zhou, M., Sun, S.D.: Genetic Algorithms: Theory and Applications. National Defense Industry Press, Beijing (1999)
11. Xuan, G.N., Cheng, R.W.: Genetic Algorithm and Engineering Design. Science Press, Beijing (2000)

Minimize the Cost of Video Transmission Among Cloud Data Center and Edge Cloud CDN Nodes

Pingshan Liu[1,2], Kai Huang[2(✉)], and Guimin Huang[2]

[1] Business School, Guilin University of Electronic Technology, Guilin, China
ps.liu@foxmail.com
[2] Guangxi Key Laboratory of Trusted Software,
Guilin University of Electronic Technology, Guilin, China
1187722529@qq.com

Abstract. With the development of cloud computing, more and more video service providers use services from cloud providers. A video service provider can construct a scalable video streaming platform with high availability by the cloud services. Typically, a video service provider uploads its video data to a cloud data center. Then, the cloud data center distributes the video data to its edge cloud CDN nodes. Usually, the cloud data center links with its edge cloud CDN nodes by high-capacity links, spanning different geographical regions. Video traffic across the cloud data center and the edge cloud CDN nodes of a cloud provider, brings on large operational cost to the cloud provider. How to reduce the video traffic cost is important for a cloud provider. Therefore, to reduce the video traffic cost, we propose a set of algorithms based on network maximum flow and minimum cut, called *Netcut-way*. The proposed *Netcut-way*, charged by the peak-bandwidth billing model, consists of three parts. The first is peak bandwidth calculation. The second is video segment segmentation. The third is video distribution route. Through extensive simulations, we demonstrate that *Netcut-way* can effectively reduce the operational cost of cloud providers in video traffic across data centers.

Keywords: Cloud data center · Edge cloud CDN nodes · Video traffic cost

1 Introduction

In order to provide end users with rich video resources and reduce management costs, video streaming service providers usually combine with cloud service, such as Tencent cloud service, Alibaba CDN cloud service and Amazon Web Services (AWS). For example, Tencent cloud service uses a simple storage service to store all source video data in one cloud data center. Then, Tencent cloud service converts the video data into a certain format, and distributes the formatted video data to its edge cloud CDN nodes in different regions [1]. The edge cloud CDN nodes can provide video streaming services to their nearby end users' terminals. Thus, the video transmission delay can be reduced, and user experience can be improved.

© ICST Institute for Computer Sciences, Social Informatics and Telecommunications Engineering 2021
Published by Springer Nature Switzerland AG 2021. All Rights Reserved
H. Gao et al. (Eds.): ChinaCom 2020, LNICST 352, pp. 459–473, 2021.
https://doi.org/10.1007/978-3-030-67720-6_32

To enable end users to timely obtain reliable video resources and improve the high availability of the network, data resources in cloud data centers should share video resources to the maximum extent [2]. Cloud service providers usually deploy more edge cloud CDN nodes in different regions [3], which are connected by leased high-capacity links. Previous research reveals that traffic costs amount to around 15% of operational costs incurred to a cloud provider [4]. Video traffic among cloud video center and edge cloud CDN nodes usually constitutes a large portion of a cloud service provider's inter-datacenter traffic. Thus, the video traffic of a cloud service provider, across data centers, brings on large operational cost to the cloud service provider. Therefore, how to effectively reduce the video transmission cost for cloud service providers is important.

Because most cloud service providers rely on multiple ISPs to connect to geographically dispersed edge cloud CDN nodes [5], to effectively reduce the video transmission cost across different regions, the costs charged by Internet Service Providers (ISPs) should be reduced or even be minimized. Suppose percentile-based charging models currently used by most ISPs are applied [6], e.g. the 95-th percentile charging model, the previous research has demonstrated that reducing the cost of inter data-datacenter video traffic is feasible [7]. However, the previous research didn't consider the availability of video data [8]. In this paper, we consider the availability of video data, focusing on the case that video data are delivered from one data center to other edge cloud CDN nodes. Therefore, to reduce the video transmission cost among datacenters, we propose a new set of algorithms based on network maximum flow and minimum cut, called *Netcut-way*. *Netcut-way* consists of three parts. The first is peak bandwidth calculation. The second is video segment segmentation. The third is video distribution route. Then, through extensive simulations, we demonstrate that *Netcut-way* can effectively reduce the video transmission cost.

The remainder of this paper is organized as follows. In Sect. 2, we discuss the related work. In Sect. 3, we present the rationale and cloud network model of our work. In Sect. 4, we propose a new set of algorithms based on network maximum flow and minimum cut, called *Netcut-way*. In Sect. 5, the performance of *Netcut-way* is demonstrated. Finally, we conclude the paper in Sect. 6.

2 Related Work

At present, with the development of big data and cloud computing, many cloud service providers distribute multiple edge cloud CDN nodes in different geographical regions, which enables the cloud system to provide a wider range of services with low latency. But some problems have been brought to people's attention. Lei Jiao studied the dynamic OSN on multiple geographically distributed clouds in a continuous cycle [9], and at the same time met the predefined QoS and data availability requirements, thus obtaining the most advanced method to significantly reduce one-time costs on the premise of ensuring QoS and data availability. Yu Wu et al. proposed to carry out effective storage and migration between different cloud sites at an appropriate cost, and adopted the optimal content migration and request distribution based on efficient optimization algorithm to solve the prediction demand [10]. Sem C. Borst proposes to alleviate these huge bandwidth requirements and performance bottlenecks by copying the most popular content closer to the edge of the network, rather than storing it at a central site [11]. Leana Golubchik et al. proposed that under the 95% charging model [12], the first 5% (95% time) of data transmission would not affect the transmission cost. In such an environment, we can greatly reduce costs by allowing some brief delays in data transmission. A. Mahimkar proposed that cloud providers use replication of geographically distributed data centers to improve end-to-end performance and high availability in the case of failure. However, previous studies did not consider the availability of video data. In this paper, the difference of our work is that we consider the availability of video data, and focus on how to transmit video data from one data center to other edge cloud CDN nodes at low cost.

2.1   System Overview

This system adopts the content distribution network model deployed by hierarchical nodes [13]. Figure 1 shows the cloud network model. The innermost layer is cloud video data center, the next layer is edge cloud CDN node, and the outermost layer is user terminal. The cloud data center deploys a stream source server, a central index server, some secondary servers, and so on. Streaming media source servers store many video resources uploaded by video service providers. The central index server maintains the video information of each edge cloud CDN node and responds to client requests [14]. First, the cloud data center distributes video clips to the edge cloud CDN node, making the edge cloud CDN nodes cache part of the video data [15]. Then, when another edge cloud CDN node requests a video clip at a minimum cost on the path, the response node obtains a copy based on the resource replication policy and forwards the copy to the requesting node [16]. In this network system, edge cloud CDN node can request the lost video clip from other nodes and record the updated video clip information of the neighbor node [17].

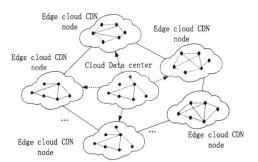

Fig. 1. the cloud network model

2.2 Overlay Structure Construction

Aiming at the video transmission strategy in the cloud, this paper studies how to reduce the cost of video transmission between cloud data center and edge cloud CDN nodes [19]. First, the video service provider uploads the video resources to the cloud data center. When the streaming media source server located in the cloud data center transmits video to the edge cloud CDN node, the cloud service provider rents multiple ISP lines. In this way, ISP service providers will charge cloud service providers based on video traffic. ISP provides peak bandwidth pricing strategy. In this strategy, ISP regularly collect peak bandwidth on leased lines and charge for maximum peak bandwidth. In this way, the paid-for link bandwidth can be utilized as much as possible without adding additional cost, thus reducing the cost of video transmission. This is also the basis for reducing the cost of video transmission between cloud data center and edge cloud CDN node. On this basis, we mainly divided into two stages. In the first phase, the cloud data center sends different video clips to the edge cloud CDN node according to the link's historical peak bandwidth. In the second stage, the minimum cost path between nodes is calculated based on the historical bandwidth peak of the current link. When a certain video fragment is missing from an edge cloud CDN node, it is necessary to select the path with the lowest cost to request the required video fragment from other nodes.

3 Minimum Cost Method for Video Distribution

3.1 Cloud Network Model Analysis

Before the minimum path algorithm for cloud video transmission cost is proposed, we first introduce the cloud network model in this paper regarding the traffic between cloud data center and edge cloud CDN node. The important symbols used in this article are listed in Table 1.

We consider the deployment of cloud data center and edge cloud CDN nodes. Cloud data center and edge cloud CDN nodes are connected through multiple highly available links. We use directed graph $G\ (V,\ E)$ to represent the network deployment model. For each link $\{i, j\} \in E$, cost per unit bandwidth of link $\{i, j\}$ is a_{ij}. We use C_{ij} to represent Maximum available bandwidth capacity of link $\{i, j\}$ and rk to represent the expected video transmission rate. In the case of network congestion, the video transmission rate is lower than the expected transmission rate. (Dk, Sk) represents the set of links from the source node to the target node. According to the historical peak bandwidth of the link, the video to be transmitted is divided into segments of appropriate size, and video segments are transmitted between nodes at the lowest cost. f_{ij}^k represents the bandwidth allocated by video K over the link. $C(x) = C * x$ represents the cost of video transmission on the link. If f_{ij}^k is less than the maximum peak value of transmission, the cost is free.

Table 1. Notations and definitions

Notation	Definition
V	The number of nodes in the network
E	A set of all directed links connecting edge cloud CDN nodes
K	The size of the data stream of the video
(S_k, D_k)	A collection of links for source and destination nodes
rk	Represent the expected video transmission rate
C_{ij}	Maximum available bandwidth capacity of link $\{i, j\}$
a_{ij}	Cost per unit of bandwidth on link $\{i, j\}$
f_{ij}^K	Video stream K allocates bandwidth on link $\{i, j\}$
$maxf_{ij}^K$	Maximum bandwidth peek on link $\{i,j\}$
$d_{ij}(t)$	The bandwidth of other nodes in the network during time t
$cost_{ij}(t)$	The cost of link $\{i, j\}$ over a period

If we use f_{ij} to denote the maximum video transmission bandwidth that a link can allocate, then the total bandwidth of link *{i, j}* can be expressed as follows:

$$f_{ij} = \sum f_{ij}^K(t) + d_{ij}(t) \tag{1}$$

In the above formula, $d_{ij}(t)$ represents the bandwidth of other nodes in the link at a certain time, f_{ij}^K represents video stream K allocates bandwidth on link $\{i, j\}$.

3.2 Algorithm Analysis

Cloud video transmission cost consists of three modules. The first module is the cloud link status monitoring module, which can obtain the historical bandwidth traffic and current link congestion. The second module is the video segmentation module. According to the peak bandwidth, the response node can divide the video into appropriate segment sizes and forward it to the requesting node with the lowest cost path. The third module is the design of the least-cost path algorithm, which calculates the least-cost path between edge CDN nodes in the cloud based on the historical peak bandwidth of each link and the node video stream cache. We use $cost_{ij}(t)$ to represent the cost of the link $\{i, j\}$ transmitted in the network, then the following formula can be obtained:

$$cost_{ij}(t) = a_{ij}maxf_{ij}^K \tag{2}$$

According to Eq. (**5**), the current link bandwidth cost can be rewritten as follows:

$$cost_{ij}(t) = \begin{cases} cost_{ij}(t-1) & f_{ij}(t) \leq max_{t-1}f_{ij} \\ a_{ij}maxf_{ij}^K & otherwise \end{cases} \tag{3}$$

If $K_f(t)$ represents the historical bandwidth peak in the time interval of t-1, then the link {i, j} link bandwidth peak at time t is:

$$f_{ij}^K(t) \leq min\left(\substack{max\\t-1}f_{ij}, c_{ij}(t) - d_{ij}(t)\right), \forall\{i,j\} \tag{4}$$

$$\sum_{j \in V} f_{Skj}^K(t) = \sum_{j \in V} f_{jDk}^K$$

$$f_{ij}^K \geq 0, \forall k \in K_{f(t)}, \forall\{i,j\} \in E$$

Based on the above analysis, in order to reduce the cost of cloud video transmission between edge CDN nodes in cloud, we design a set of *Netcut-way* algorithms. The algorithms are divided into three parts. One is the calculation of peak bandwidth. The second is video segmentation. Third, the minimum cost path.

According to Algorithm 1, peak bandwidth is calculated as follows [20]. First, the system sets the initial capacity value and the initial bandwidth peak. The edge cloud CDN node counts and monitors the link status between the cloud data center and each edge cloud CDN node every 5 min to get the current bandwidth peak of the link. The maximum peak value of the link is updated based on the bandwidth obtained.

Algorithm 1   Determine the maximum link history bandwidth peak algorithm

---

1	Obtain the initial solution $f^1, f^2, \ldots f^i$
2	Set the maximum peak bandwidth is f^K
3	**while $f^i \in (f^1, f^2, \ldots f^{i-1})$ do**
4	**for Each $k \in K_{f(t)}$ do**
5	Find the max flow$(f^1, f^2, \ldots f^{i-1}) = f^K$
6	**if $(f^i > f^K)$ then**
7	update $f^K = f^i$
8	**End if**
9	**End for**
10	**End While**
11	**Return f^k**

If we use $(f_1^K, f_2^K, f_3^K, \ldots, f_i^K) \in K(t)$ means that the node divides the video into segments of appropriate size, and distributes the video to other nodes through the minimum path algorithm. When the node can transmit the video clip with the maximum bandwidth, the following formula can be obtained:

$$\forall \{i,j\} \in E, \sum_k f_i^K \leq c_{ij} \qquad (5)$$

In view of the above analysis, we design Algorithm 2 to solve the problem of node video segmentation. The node can transmit as many free video clips as possible without exceeding the maximum historical peak of the link.

Algorithm 2   Bandwidth allocation based on historical peak bandwidth

---

1   The node divides the video into segments of appropriate size clips $(f_1^K, f_2^K, f_3^K, \dots, f_i^K)$

2   The max capacity bandwidth of the link $C_{ij} (\{i,j\} \in E)$

3   **While** $(f_1^K, f_2^K, f_3^K, \dots, f_i^K)$ **do**

4   **Algorithm 1** obtains the historical bandwidth peak of the link f^K

    if $(f_i^K(t) < f^K)$ **do**

5   The free video transmission clip is $f_i^K(t)$

    else

6   Update $f_i^K(t) = C_{ij} - d_{ij}(t)$

7   **End While**

8   **Return** $(f_1^K, f_2^K, f_3^K, \dots, f_i^K)$

The design of the minimum cost path algorithm is the focus of this paper. Based on maximum flow and minimum cut, we design Algorithm 3 [21]. First, we use Algorithm 2 to solve the video segmentation problem. Cloud data center divides video K into segments $(f_1^K, f_2^K, f_3^K, \dots, f_i^K)$. The edge cloud CDN node saves the video clips distributed by the cloud Data center. Then the edge cloud nodes run the algorithm to find the least cost path to request the lost video clips from other nodes.

Algorithm 3 Maximum network minimum cut minimum cost algorithm

---

1	Set the video resource of cloud center node to K
2	**Algorithm 2** divides video streaming K into segments of appropriate size$(f_1^K, f_2^K, f_3^K, \ldots, f_i^K)$
3	Cloud data center takes video clips $(f_1^K, f_2^K, f_3^K, \ldots, f_i^K)$ on different edge Cloud CDN nodes through different links
4	Edge cloud node saves video clips $(f_1^K, f_2^K, f_3^K, \ldots, f_i^K)$
5	Suppose an edge cloud CDN node requests a video clip f_i^K
6	**Set** a queue **Q** to represent the currently unchecked label point
7	path: Marks the nodes that the current path passes through
8	Repeat:
9	path==null
10	The source point Sk is marked and enters queue Q
11	**While** (Q! = null and Dk is not remarked) **do**
12	Begin:
13	u=Q. remove ()
14	**for** (edge (u, v) that starts from u) **do**
15	**if** (v is the node marked and *(u, v)* which can) **do**
16	Q. add(v), path. add(v)
17	**End if**
18	**End for**
19	**End while**
20	**if** (*Dk* is marked) **do**
21	Modify traffic along the path from Dk
22	**until** (*Dk* marked)
23	**Return** $(f_1^K, f_2^K, f_3^K, \ldots, f_i^K)$

---

4 Performance Evaluation

In this section, we demonstrate the performance of the *Netcut-way* algorithms through simulation. Firstly, the simulation steps and evaluation indexes are introduced. The performance of the algorithms is represented by the figures. Then compared with *Jetway* algorithms, the performance of our algorithms is superior to *Jetway* algorithms

4.1 Simulation Setup

We implemented an event-driven simulator on the PeerSim simulation platform to evaluate the performance of *Netcut-way*. We built a cloud video transmission system and stored all the video clips in the cloud data center. First, different initial transmission delays and initial bandwidth peaks are randomly set in the system link. In the experiment, considering the difficulty of the experiment, we set up 40 edge cloud CDN nodes. At the same time, we also set the video source size to 20 TB, and the expected transmission rate range of video is [3.31,15.72] Mbps. First, the cloud data center splits the appropriate video clips and distributes them to other edge cloud CDN nodes. Edge cloud CDN nodes cache some video clips received. As can be seen from Fig. 2, the edge cloud CDN node contains edge streaming media server and edge index server. The edge index server is responsible for discovering neighbor nodes and finding the minimum cost path to the destination node. The edge streaming server is responsible for storing video stream. Moreover, when other edge cloud CDN nodes request the missing video clips, the edge stream server can respond to the request and generate video clips copies to distribute to the requesting nodes with the least cost path.

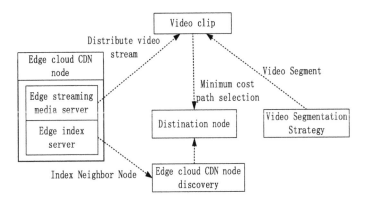

Fig. 2. logic structure of the simulation

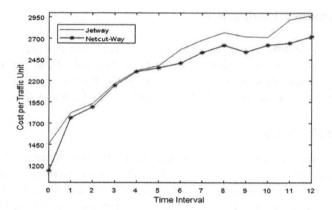

Fig. 3. Based on 100% charge mode per unit cost

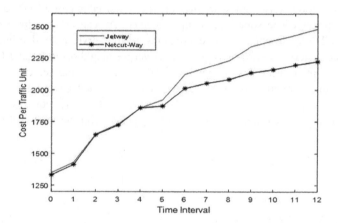

Fig. 4. Based on 95% charge mode per unit cost

4.2   Simulation Result

(1) Unit transportation cost

Unit transport cost is defined as the unit video transmission cost of a system over a period. In the experiment, we recorded the video transmission cost of edge cloud CDN node in different time periods. We consider the maximum capacity of network bandwidth in the case of video segment transmission units on each link, regardless of the importance of the time dimension. Figure 3 and Fig. 4 show the unit transmission costs of *Jetway* and *Netcut-way* in different charging model, respectively. As can be seen from the figure, the unit transmission cost of *Netcut-way* algorithms is lower than that of *Jetway* algorithms

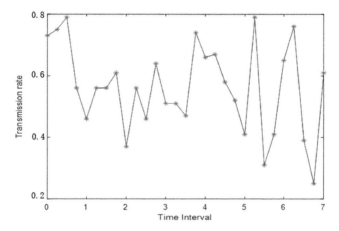

Fig. 5. Video transmission rate of *Netcut-way*

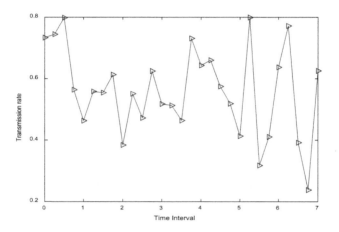

Fig. 6. Video transmission rate of *Jetway*

(2) Video transmission rate

The video transmission rate represents the video transmission rate of edge cloud CDN nodes. In the experiment, we count the transmission rate of the node every 30 s. Figure 5 and Fig. 6 show the node transmission rate in a certain period under the two algorithms. As can be seen from the figure, the video transmission rate of edge cloud CDN node under *Netcut-way* algorithms is basically consistent with that under *Jetway* algorithms. Therefore, our algorithms have little effect on the node transmission rate.

(3) Actual and expected transfer rate ratio

The ratio of the actual transmission rate to the expected transmission rate indicates the smoothness of the video transmission. The cloud data center divides the video into appropriate segments and distributes them to the surrounding edge cloud CDN nodes, so that the edge cloud CDN nodes cache a part of the video clips received from the cloud data center. With the increase of the number of requests and cache video clips, many video clips can be transmitted through the least cost path, which makes edge cloud CDN nodes cache more and more videos. Figure 7 shows the results of *Netcut-way* and Fig. 8 shows the results of *Jetway*. As can be seen from the figure, the transmission efficiency of *Netcut-way* algorithms is higher than that of *Jetway* algorithms.

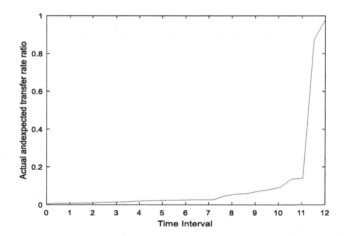

Fig. 7. Actual and expected transfer rate ratio of *Netcut-way*

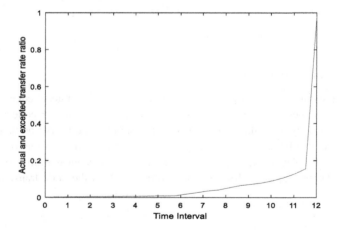

Fig. 8. Actual and expected transfer rate ratio of *Jetway*

(4) Cost reduction percentage

In order to further analyze the cost reduction under *Netcut-way* algorithms and *Jetway* algorithms, we draw the video transmission cost diagram of the two algorithms in a certain period. Figure 9.a shows the implementation of *Jetway* algorithms, and about 75% of the video clips are realized by means of payment. Figure 9.b is implemented under the *Netcut-way* algorithms, and about 71.7% of the video clips are realized through the payment bandwidth. Therefore, it is showed that *Netcut-way* algorithms can further reduce the operating cost of cloud service providers than *Jetway* algorithms.

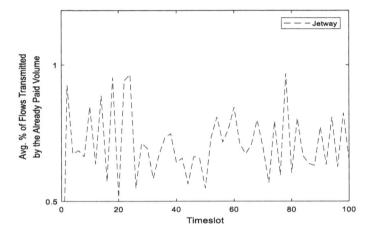

(a) cost reduction percentage of *Jetway* algorithms

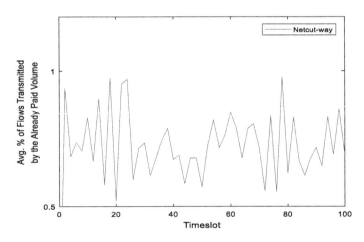

(b) cost reduction percentage of *Netcut-way* algorithms

Fig. 9. Cost reduction percentage

5 Conclusions

Currently, ISP service providers have a variety of pricing methods, such as traffic-based billing, peak bandwidth billing. Cloud service providers typically deploy multiple edge cloud CDN nodes in different geographic locations and rent a highly available ISP link to transmit video data. In order to reduce the cost of video transmission for cloud service providers, this paper proposes the *Netcut-way* algorithms based on peak bandwidth billing model. The algorithms can effectively reduce cloud video transmission cost and cloud video transmission delay between edge cloud CDN nodes. Finally, through the comparison of experimental results of *Jetway* algorithms, it can be demonstrated that *Netcut-way* algorithms can effectively reduce the operating cost of cloud service providers and is more effective than *Jetway* algorithms.

Acknowledgement. The research was supported by the National Natural Science Foundation (No. 61762029, No. U1811264, No. 61662012), Guangxi Natural Science Foundation (No. 2016 GXNSFAA380193), Guangxi Key Laboratory of Trusted Software (No. kx201726).

References

1. Zhao, Z., Sheng, Y., Wang, J., Zhu, M.: An in-network service system based on smart routers and edge devices. In: 2017 20th International Symposium on Wireless Personal Multimedia Communications (WPMC), Bali, pp. 568–574 (2017). https://doi.org/10.1109/wpmc.2017.830187
2. Chen, Y., et al.: Packet cloud: a cloudlet-based open platform for in-network services. IEEE Trans. Parallel Distrib. Syst. **27**(4), 1146–1159 (2016). Article (CrossRef Link)
3. Mahimkar, A., et al.: Bandwidth on demand for inter-data center communication. In: Proceedings 10th ACM Workshop on Hot Topics in Networks (HotNets), pp. 1–6 (2011). Article (CrossRef Link)
4. Tencent. https://www.it610.com/article/2066413.html. Article (CrossRef Link)
5. Greenberg, A., Hamilton, J., Maltz, D.A., Patel, P.: The cost of a cloud: research problems in data center networks. SIGCOMM Comput. Commun. Rev. **39**(1), 68–73 (2009)
6. Zhang, Z., Zhang, M., Greenberg, A., Hu, Y.C., Mahajan, R., Christian, B.: Optimizing cost and performance in online service provider networks. In: Proceedings 7th Conference on Networked Systems Design and Implementation (NSDI), pp. 1–15 (2010)
7. RouteScience Technologies, Route Optimization for Ebusiness Applications. White Paper (2003)
8. Jetway: Minimizing Costs on Inter-Datacenter Video Traffic
9. Jiao, L., Li, J., Xu, T., Du, W., Fu, X.: Optimizing cost for online social networks on geo-distributed clouds. IEEE/ACM Trans. Networking **24**(1), 99–112 (2016). https://doi.org/10.1109/tnet.2014.2359365. Article (CrossRef Link)
10. Zhu, J., et al.: Scaling Service-Oriented Applications into Geo-distributed Clouds (2013). Article (CrossRef Link)
11. Borst, S., Gupta, V., Walid, A.: Distributed caching algorithms for content distribution networks. In: Proceedings-IEEE INFOCOM, pp. 1478–1486 (2010).https://doi.org/10.1109/INFCOM.2010.5461964. Article (CrossRef Link)

12. Golubchik, L., Khuller, S., Mukherjee, K., Yao, Y.: To send or not to send: Reducing the cost of data transmission. In: 2013 Proceedings IEEE INFOCOM, Turin, pp. 2472–2478 (2013). https://doi.org/10.1109/infcom.2013.6567053. Article (CrossRef Link)
13. Sajithabanu, S., Balasundaram, S.R.: Direct push-pull, or assisted push-pull? Toward optimal video content delivery using shared storage-based cloud CDN (SS-CCDN). J. Supercomputing **75**(4), 2193–2220 (2019). Article (CrossRef Link)
14. Hu, H., et al.: Joint content replication and request routing for social video distribution over cloud cdn: a community clustering method. IEEE Trans. Circuits Syst. Video Technol. **26**(7), 1320–1333 (2016). Article (CrossRef Link)
15. Al-Habashna, A., et al.: Distributed cached and segmented video download for video transmission in cellular networks. In: International Symposium on Performance Evaluation of Computer and Telecommunication Systems. IEEE (2016). Article (CrossRef Link)
16. Liu, P., Feng, S., Huang, G.: Bandwidth-availability-based replication strategy for P2P VoD systems. The Comput. J. **57**(8), 1211–1229, August 2013. Article (CrossRef Link)
17. Zhang, Z., Zhang, M., Greenberg, A., Hu, Y.C., Mahajan, R., Christian, B.: Optimizing cost and performance in online service provider networks. In: Proceedings 7th Conference on Networked Systems Design and Implementation (NSDI), pp. 1–15 (2010). Article (CrossRef Link)
18. Laoutaris, N., Sirivianos, M., Yang, X., Rodriguez, P.: Inter-datacenter bulk transfers with NetStitcher. In: Proceedings ACM SIGCOMM (2011). Article (CrossRef Link)
19. Peng, G.: CDN: content distribution network. Research Proficiency Exam Report (2003). Article (CrossRef Link)
20. Ai-Ping, Z., Guang, C.: Analysis of network behavior characteristic and influence factor based on the peak traffic. J. Commun. **33**(10), 117–125 (2012). Article (CrossRef Link)
21. Wang, M., Li, B.: Linear network coding. IEEE Trans. Inf. Theory **49**(2), February 2003. Article (CrossRef Link)

A Link Load Balancing Algorithm Based on Ant Colony Optimization in Data Center Network

Shuqing Ma[1,2(✉)] ⓘ, Hong Tang[1,2], and Xinxin Wang[1,2]

[1] School of Communication and Information Engineering,
Chongqing University of Posts and Telecommunications,
Chongqing 400065, China
masq1018@163.com
[2] Chongqing Key Lab of Mobile Communications Technology,
Chongqing University of Posts and Telecommunications,
Chongqing 400065, China

Abstract. With the continuous increase of types of services and data volume in data center, the traffic loads of some links are excessive, and how to balance the link load and ensure the quality of network service have become research hotpots. However, the traditional link load balancing mechanisms ignore the complexity of network and the Quality of Service (QoS) requirement of the flow when calculating the forwarding paths. Therefore, we propose a link load balancing algorithm based on Ant Colony Optimization (LLBA) in data center network. The algorithm redefines the heuristic function according to the number of elephant flows on the link and the real-time load of links, and updates the pheromones according to the path length. Then, the algorithm customizes the optimal path determination rule according to the path transmission delay and the real-time loads, so as to find a best forwarding path for the current flow under the multiple constraints including the path length, link load, and transmission delay. The experiment results show that, the proposed algorithm improves the link utilization and network throughput effectively, and also reduces the delay and delay jitter to some extent, as compared with the traditional mechanisms.

Keywords: Ant colony optimization · Data center network · Load balancing · Quality of service

1 Introduction

With the rise of cloud computing and big data technology, data center as the information infrastructure, gradually become the core of data computing, transmission, and storage. Data center contains thousands of interconnected servers, and various services such as MapReduce, network games, and search engines and so on. Due to the uneven business distribution and the randomness of data quantity changes, and the loads of different links which resulting in the decline of the performance of data center network [1]. In the meanwhile, with the diversification of Internet business, and users' requirements for service quality are increasing day by day. In order to improve resource utilization,

H. Gao et al. (Eds.): ChinaCom 2020, LNICST 352, pp. 474–489, 2021.
https://doi.org/10.1007/978-3-030-67720-6_33

decrease response time, and ensure service quality while maintaining stable and efficient network operation, thus, a flexible link load balancing strategy is urgently needed [2].

The flow characteristics are the basic basis of making the flow scheduling strategy. Generally, flows in data centers can be categorized into two types: elephant flow and mice flow [3]. Mice flows are often generated by interactive services, such as web page traffic, search queries, and so on, so mice flows take little resource and the number of bytes is numerous. Elephant flows are often generated by bandwidth-demanding services such as massive data synchronization, distributed computing, and video streaming, which live a long time and take advantage of lots of network resource. Since the number of elephant flows is less than 10%, but it takes up more than 90% of the link bandwidth, which is the main factor that cause link load imbalance. Due to elephant flows would take lots of network resources, so elephant flows scheduling is a challenging problem in data centers which may cause congestion. Thus, it is very feasible to design effective link load balancing strategies aim at elephant flows [4].

Equal-Cost Multipath Routing (ECMP) [5] is widely used for traffic scheduling in data center networks. It regards the available paths as equal cost paths and does not distinguish between elephant flows and mice flows. ECMP allocates the corresponding path statically to forward flows by calculating a hash value which typically from transport port numbers, IP addresses of the source and destination, and protocol. But a key limitation of ECMP is that, it is possible to allocate multiple flows to the same path, which causing elephant flows collision and bring the problem of link bottleneck. The appearance of Software Defined Networking (SDN) [6] provides an opportunity for fine-grained traffic scheduling for data center networks to distinguish the size of flows and perceive the global topology, thus making effective routing strategies to balance link load dynamically. There are so many traffic scheduling strategies based on SDN. Some methods which aim at mice flows, such as MiFlO [7]. In MiFlO, the scheduling algorithm which classifies the elephant flows and mice flows in servers and ensures the priority of mice flows. Additionally, there are some link load balancing strategies which aim at elephant flows, such as Hedera [8], Nimble [9], and Mahout [10]. Flow detection is performed in edge switches (Hedera, Nimble) or terminal hosts (Mahout), the basis of judging the size of flow is according to number of bytes transmitted per unit time. If the number of bytes transmitted per unit time that beyond the threshold, we regard this flow as elephant flow. And then calculate a light- load path for elephant flow. These load-balancing strategies calculate the forwarding path are according to the real-time link loads. To some extent, those solutions can extend network's capacity of data transmission, and also improve the utilization rate of the network links, as compared with ECMP.

However, the following problems remain: 1) network flows will occupy as much transmission bandwidth as possible in the links, which resulting in a large transient load fluctuation in the case of multiple elephant flows in a link. Thus, the instantaneous load information acquired by the controller does not completely reflect the load situation of the link. 2) The more links an elephant flow passes, the more link resources it occupies. It is unreasonable to calculate the forwarding path only according to the real-time load of the path, while ignoring the number of links that is unreasonable. 3) The current business requires high quality of network services. Such as multimedia information, which not only requires abundant bandwidth to transmit mass data, but also requires timeliness of data transmission [11]. Therefore, it is necessary to consider multiple factors in traffic scheduling strategy to ensure QoS.

Considering problems of all above, this paper proposes a load balancing algorithm based on ant colony optimization (LLBA) to balance the link load of network effectively and improve the quality of flow transmission. Ant colony optimization algorithm (ACO) with the features of self-organization, positive feedback, and heuristic search. The ACO algorithm is widely used in routing problems [12]. This paper redefines the parameters and operations. Firstly, the top-k shortest path set is selected as the basis of ants routing. Then, according to the number of elephant flow and the real-time residual bandwidth, two heuristic functions are designed to guide ants routing. Additionally, LLBA adopts global pheromone updating method. Finally, the real-time link load and path delay are used as the basis to judge the QoS of the path. After repeated iterations, the best forwarding path of flow is obtained.

The rest of the paper is organized as follows. Section 2 introduces the system architecture of the algorithm. We introduce the detailed design of the algorithm in Sect. 3. The performance evaluation is presented in Sect. 4. Finally, in Sect. 5 we draw our conclusions.

2   System Architecture

The system architecture of the proposed algorithm is showed in Fig. 1, which mainly consists of the following three modules.

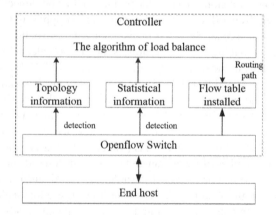

Fig. 1. The system architecture Algorithm.

2.1   Topology Information Detection

The function of this module is to get the network topology, and then obtain the shortest top-k paths between the source address and the destination address of flows. The controller gets the network topology by the link layer discovery protocol (LLDP). The specific steps are as follows:

Step 1: the controller establishes the OpenFlow channel and obtains the physical port information of the switch through the handshake principle.
Step 2: the controller periodically sends LLDP data packets to the switches, and then the switches which receive the LLDP data packets will sent the data packets to the

neighboring switches through the ports. After receiving the LLDP data packets, the neighboring switches will send a Packet-In message to the controller.

Step 3: the controller obtains the connection information of the switch according to the Packet-In message and the saved physical port information, thus getting the whole network topology information.

After obtaining the network topology information, the controller uses the Top-k shortest paths (KSP) algorithm to get the shortest path set according to the source switch address and the destination switch address of the flow.

2.2 Statistical Information Detection

The function of the module is to obtain the flow information and link status information, which mainly includes filtering the flows, counting the number of flows on the path, and calculating the real-time load and transmission delay of the path.

Filter the Flows. The controller polls each OpenFlow switch at a fixed time interval by sending OFPT_STATS_REQUEST message, and then queries and stores all of flow information and port statistics from the switches.

The elephant flows whose number of bytes transmitted per unit time exceeds the threshold that was preset in the algorithm. And we record and store the elephant flows according to the flow information. And we can distinguish the elephant flows and mice flows by formula (1).

$$\varphi = \frac{b_{t_2} - b_{t_1}}{(t_2 - t_1)B_{max}} \tag{1}$$

b_{t_1} and b_{t_2} respectively represent the number of bytes received by the switch at time t_1 and t_2, B_{max} is the maximum available bandwidth of the link. In this paper, we assume that when φ of a flow exceeds 10%, the flow is considered to be an elephant flow. And the flow needs to be dynamically scheduled [8].

Count the Number of Elephant Flows on the Path. The statistical information detection module counts the number of elephant flows on each link in real time. Assuming that the p_{th} path contains n links, and there are Num_i elephant flows on the i_{th} link, the number of elephant flows on the p_{th} path is shown as (2).

$$Num = max(Num_1, Num_2, \cdots, Num_n) \tag{2}$$

Compute the Real-Time Load of Path. The statistical information detection module monitors the traffic information both of each port and link in the network in real time. So the real-time load of path can be shown as (3).

$$Load = max(Load_1, Load_2, \cdots, Load_n) \tag{3}$$

Compute the Transmission Delay of Path. The transmission delay of path can be calculated by the delay of the LLDP packet and Echo packet.

First, as shown by the red arrow in Fig. 2, the controller sends a LLDP packet with timestamp to Switch_A, and instructs Switch_A to forward the packet to Switch_B. After receiving the data packet, switch B does not have a flow entry matching the data packet in the flow table, so the data packet is encapsulated in the Packet-In message and be send to the controller. The controller parses the sending time-stamp of the LLDP Packet from the Packet-In message. The LLDP Packet $delay_1$ can be obtained by subtracting the time stamp from the current system time. Similarly, the reverse $delay_2$ is shown by the blue arrows.

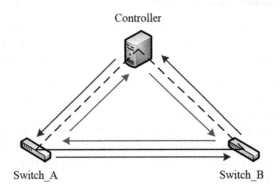

Fig. 2. Delay calculation of topology. (Color figure online)

Then, the controller sends an Echo_request message with the timestamp to Switch_A, and the controller parses the Echo_reply returned by Switch_A. The timestamp is subtracted from the current system time to obtain the round-trip time difference $delay_a$ between the controller and Switch_A. Similarly, the round-trip time difference $delay_b$ from the controller to switch B is obtained.

Then, the i_{th} link transmission delay between Switch_A and Switch_B can be calculated by (4).

$$Delay_i = (delay_1 + delay_2 - delay_a - delay_b)/2 \tag{4}$$

So the total transmission delay of the path is shown as (5).

$$Delay = \sum_{i=1}^{n} Delay_i \tag{5}$$

Flow Table Installed. Based on the topology and statistical information obtained previously, this module calculates the forwarding path for elephant flows, and then sends the new flow tables to the OpenFlow switches through the OFPT_FLOW_MOD message. The switches transmit the elephant flows according to routing information in the flow table.

3 Algorithm Design

Ant colony optimization (ACO) is a bionic algorithm that simulates ant foraging [13]. Each ant releases pheromones along the way of foraging, and the pheromones always evaporate with time. The higher pheromone intensity of path, the closer they get to the food. Other ants can perceive pheromones and tend to choose the path that contain more pheromones. Therefore, these collective behaviors of ants that present a positive feedback mechanism. The more ants pass on a certain path, the more likely the latter ants to choose the path, and finally get a closest path to the food.

The ACO algorithm with the characteristics of distributed computing, information positive feedback, and heuristic search which can solve the combination optimization routing problem [14]. In this paper, we propose a link load- balancing algorithm LLBA based on ACO. The algorithm customs the new best path determination rule. And we find the optimal transmission path for the current flow under the multiple constraints of path length, link-load, and delay, so as to achieve the link load-balancing of each link, and ensure the quality of network service.

3.1 Path Selection Strategy

There are many reachable paths between the two hosts in data center, and it would take many resources to find the optimal path from all of the reachable paths. If the more elephant flows pass through, the more resources are occupied. Therefore, in this paper, firstly, the algorithm obtains the shortest top-k paths to form the path set $SP = \{SP_1, SP_2, \cdots, SP_k\}$ from the information statistics module, and then the ant chooses an optimal path in the path set. The method of all ants select next site is roulette, and the calculation formula is given as (6).

$$p_i^j(t) = \frac{[\tau_i(t)]^{\alpha} * [\eta_i^1(t) * \eta_i^2(t)]^{\beta}}{\sum_{s \in allowed_k} [\tau_s(t)]^{\alpha} * [\eta_s^1(t) * \eta_s^2(t)]^{\beta}}, i \in allowed_k \tag{6}$$

M is the number of ants, $p_i^j(t)$ is the probability that the j_{th} ant chooses the i_{th} path at time t, and $allowed_k$ denotes the optional paths that ants can choose. $\tau_i(t)$ is the pheromone intensity on i_{th} path, $\eta_i^1(t)$ and $\eta_i^2(t)$ are the heuristic functions of path. And α and β are the weight values of pheromone and heuristics function respectively, and representing the relative influence of each factor. The higher the pheromone intensity and the value of the heuristic function, the greater the probability that path can be selected.

3.2 Heuristic Function

In order to solve the load imbalance problem, the path with lighter load should be preferentially selected. However, in actual network, the network flows will preempt the larger transmission bandwidth in the links as much as possible, resulting in large transient load fluctuation under the condition of multiple elephant flows on the link. The instantaneous load information obtained by the controller at a certain time cannot exactly reflect the load situation in the current state of the link. Therefore, the algorithm estimates the path load by the number of elephant flows on the path, and proposes the heuristic function 1 of path as shown in (7), and then proposes the heuristic function 2 according to the real-time link load shown as shown in (8).

$$\eta_i^1(t) = \frac{1}{Num_i + 1} \tag{7}$$

$$\eta_i^2(t) = 1 - Load_i / B_{max} \tag{8}$$

Num_i is the number of elephant flows on the i_{th} path, $Load_i$ is the real-time load on path. And the larger value of $Load_i$ and B_{max}, resulting in the heavier load of the i_{th} path. The smaller the heuristic function of path, the less likely it is that this path be chosen by the ants.

3.3   Pheromone Update Strategy

After completing the path search, ants would release pheromone along the path, and the pheromone also would evaporate automatically with time. This process is called as pheromone updating.

The algorithm adopts global pheromone updating, that is, only update the pheromone increment on the optimal path after all ants complete the path selection and obtain the path solution set $AP = \{AP_1, AP_2, \cdots, AP_m\}$. And pheromone updating is defined as (9) and (10).

$$\tau_i(t+n) = (1 - \rho) * \tau_i(t) + \Delta\tau_i(t) \tag{9}$$

$$\Delta\tau_i(t) = \begin{cases} \frac{Q}{Len_i} & if \quad (i = bestpath) \\ 0 & else \end{cases} \tag{10}$$

The change amount of pheromone intensity on the i_{th} path is calculated by (9), including natural volatility and pheromone increment. ρ is the speed of pheromone volatilization, and $\Delta\tau_i(t)$ is the amount pheromone increment. The pheromone increment is calculated by (10), where Len_i is the length of the best path, and Q is the constant total amount of pheromone released by the ant colony.

The purpose of the LLBA is to balance the network load and guarantee the transfer quality of flows. The delay is an important factor to measure the transfer quality of the flows. In the complex network environment, the path with the least load does not necessarily with the minimum delay, and the delay obtained in real time with randomness. To find the optimal solution, the algorithm takes the real-time load and delay of the path into consideration when choosing the path. So the path state function is proposed as (11). And μ is the weight of delay. According to (11), the smaller the path state value is, the better QoS of the path gets.

$$Cost_i = \mu * delay_i + (1 - \mu) * Load_i \tag{11}$$

Thus, the best path with the minimum cost is shown as (12).

$$bestpath = \underset{i}{argmin}(Cost_1, Cost_2, \cdots, Cost_m) \tag{12}$$

3.4 Algorithm Description

The algorithm description is shown as Table 1.

Table 1. The algorithm description of LLBA.

LLBA: Link load-balancing algorithm based on ACO
Initialize parameters
Calculate the k_shortest_path
While F<Fmax
for each ant m
ant choose path from k_shortest_path
AP.append(path)
Selectbestpath(AP)
Update pheronmone
end while
for each ant m
ant choose path from k_shortest_path
AP.append(path)
selectbestpath(AP)
Print bestpath

Firstly, we initialize the basic parameters of the algorithm, and obtain the path set SP which consists of the shortest top-k paths, according to the source address and the destination address of the flow. Secondly, each ant selects a path from the path set SP according to (6), get the path set AP after all the ants complete the routing. Then, the best path is selected from the path set AP according to (12), and update the pheromone on the path by (9) and (10), the ant colony restarts the next search. Finally, when the number of iterations of the algorithm reaches a certain value, the optimal path is output.

4 Performance Evaluation

4.1 Proof Experiment

It is mentioned in the algorithm that the network flows will preempt the larger transmission bandwidth in the link as much as possible, resulting in a transient load fluctuation under the condition of multiple elephant flows on the link. The instantaneous load information obtained by the controller at a real-time cannot fully reflect the load situation in the current state of the link. To demonstrate this, we used the topology shown in Fig. 3 to simulate the confirmatory experiment. On the 0th second, Host1 sends the first big flow to Host3 by using the Iperf command. On the 20th second, Host2 sends the second big flow to Host4. The monitoring time lasts for 60 s. Note that all links bandwidth of hosts-switches and inter-switches in the topology are set to 1Gbps because of the limited switching ability on a single machine. The experimental results are shown as Fig. 4 and Fig. 5.

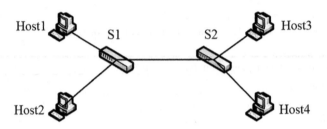

Fig. 3. Confirmatory experiment.

As shown in Fig. 4, between 0 and 20 s, the first flow takes up more than half of the bandwidth of the link. At the 20th second, the second flow is generated, and two flows begin to compete for bandwidth resources, the real-time transmission rate of the first flow gradually decreases. The real-time load of links between S1 and S2 is shown in Fig. 5, we can see that as the number of flows increases, the real-time load of links does not increase significantly, but fluctuates constantly. Therefore, the load of links cannot be determined completely according to the real-time load.

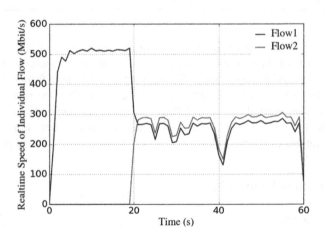

Fig. 4. Real-time speed of individual flow.

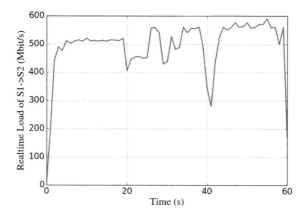

Fig. 5. Real-time load of s1- > s2.

4.2 Comparative Experiment

In order to evaluate the performance of the algorithm, we use the lightweight simulation experiment platform Mininet [15] to build the Fat-Tree network, which is implemented on a Ryu [16] controller. The Fat-Tree topology is shown as Fig. 6. It is highly accepted in data center networks due to its simple structure, ease of deployment, and scalability. In addition, we adopt the test suite of various communication modes of the data center network provided by the working group of Kandula et al. [17] as the traffic generation mode of this experiment.

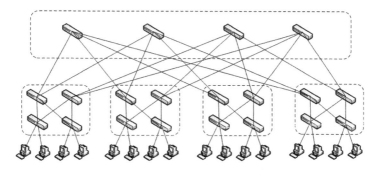

Fig. 6. Fat-tree topology with 4 pods.

Simulation Parameter. In this section, we conduct four series of experiments to show the performance of algorithms in terms of the link utilization, network throughput, round-trip delay, and delay jitter, respectively. We use the Mininet [15] to generate a Fat-tree structure of four core switches and four pods, and all of the link bandwidth between the switches are 1 Gbps. Every host sends flows to other hosts according to the probability of the traffic model, and uses Iperf command to send data to simulate the elephant flows (the default number of send is 1). Each elephant flow lasts for 60 s and occupies the bandwidth between 100 Mbps and 1 Gbps. Mice flows are simulated with a sending interval of 0.1 s by using Ping packets (the default number of packets is 600). Each mice flow has a duration less than 0.5 s and occupies the bandwidth less than 50 Mbps.

We simulate realistic datacenter workloads with 9 types: 0.2, 0.3, 0.4, 0.5, 0.6, 0.7, 0.8, 0.9, 1.0. '0.2' means that 20% of hosts in the network initiate an elephant flow, whereas the remaining hosts do not send elephant flow. Other load conditions are similar. We use the network monitoring software bwm-ng to monitor the real-time network load. The network performance test duration is 60 s. The parameters of the algorithm are shown in Table 2, which are determined after several simulation tests, so as to guarantee the convergence rate of the algorithm and the effect of solving the problem.

Table 2. Parameter setting.

Description	Parameter	Value
Num. of literation	F_{max}	5
Num. of ants	m	10
The initial value of the pheromone	t	1
The weight of pheromone	α	2
The weight value of heuristics function	β	1
The weight value of delay	μ	0.7
Volatile factor	ρ	0.1
Total pheromones	Q	1

Algorithm Effect Test 1: Link Load Balancing. The utilization of links refers to the ratio of the number of links being utilized to the total number of links. And the utilization of links also reflects the utilization of network resources. The higher the link utilization rate, the higher the return on investment of the network and the more balanced the distribution of network traffic in network. The network link is generally bidirectional. In the experiment, a network link is regarded as two opposite directions in the analysis and

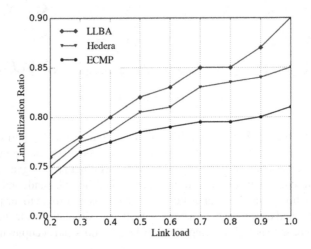

Fig. 7. Network link utilization.

calculation, and the statistical links does not include the directly connected links between the access layer switches and the end hosts. Figure 7 shows that the network link utilization performance of different algorithms as the load increases.

As shown in Fig. 7, due to the ECMP algorithm treats the available paths as the same cost paths and statically schedules all flows, the link utilization is the lowest among these algorithms. Hedera and LLBA both dynamically schedule flows according to the network real-time status and balance the link load to some extent, so the link utilization is significantly improved compared to ECMP. Hedera finds a path randomly based on the real-time load of link, and if one path meets the bandwidth requirement of the flow firstly, and then it will be selected. Although the efficiency of Hedera is high, there are some deficiencies, such as the path with less load may not be selected for a long time. The LLBA algorithm calculates the forwarding path according to the dual constraint of flow number and the real-time load, so the path with lighter load is easier to be selected. Therefore, LLBA gets the higher utilization than Hedera and ECMP. Figure 7 illustrates that the proposed algorithm has the best performance in balancing link load.

Algorithm Effect Test 2: Network Transmission Performance. The average throughput of the network refers to the throughput of per unit time obtained by the network system under the current traffic model. The standardized throughput of the network refers to the ratio of the total throughput obtained by the network system under the current traffic model and the maximum throughput obtained under the ideal traffic model. Both of them are important indicators to evaluate the network transmission performance. Figure 8 shows the average throughput and standardized throughput of different algorithms. We can see from Fig. 8, since the ECMP algorithm statically schedules traffic and it cannot make full use of network resources, the average throughput and standardized throughput are the lowest among these algorithms. Hedera and LLBA adopt flexible flow scheduling strategies, so the average throughput and standardized throughput are improved. The LLBA has a better performance in balancing link load compared with Hedera, so the network throughput and standardized throughput are higher.

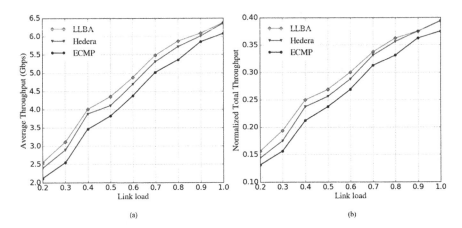

Fig. 8. The network throughput. (a) Average throughput. (b) Normalized total throughput.

Algorithm Effect Test 3: QoS. The first packet round-trip delay refers to the time between the first-packet of the flow is sent from the client to receiving the reply of the first packet from the server. The average round-trip delay refers to the average time between the sending of all data packets from the client to the receiving of the reply from the server. Both of them are important basis for detecting the service quality of the flow. Figure 9 shows that the round-trip delay and average round-trip delay of different algorithms as the load increase. When the switch forwards the non-first packet, which can match the existing flow entries in the flow table, achieving the fast forwarding, and the average round-trip delay is much lower than the first-packet round-trip delay.

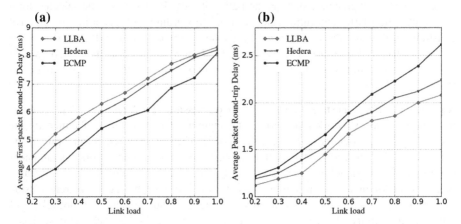

Fig. 9. The round trip delay. (a) Average delay of first-packet. (b) Average delay of packet.

As shown in Fig. 9, since the ECMP algorithm directly forwards the flows without the flow entries being sent by the controller, the round-trip delay of the first packet is the lowest. Both Hedera and LLBA need the controller to calculate the forwarding path of elephant flows and send the flow entries, so the first packet round-trip delay is higher than ECMP. Hedera forwards the to-be-scheduled flow once it finds the path which satisfies the bandwidth requirement, so the efficiency is high. The LLBA algorithm is more accurate in calculating the elephant-flow forwarding path, so the first packet has the higher round-trip delay than other packets. However, as the network load increases, the gap in first packet round-trip delay between LLBA and Hedera is gradually reduced.

LLBA algorithm preferentially selects a path with a smaller transmission delay, thereby effectively reducing the average packet round-trip delay of the flows. ECMP does not consider the link state, equivalently forwards all flows, and it is prone to network congestion due to elephant-flows collisions. Therefore, the average round-trip delay of ECMP is the highest. Hedera reduces network congestion through link load-balancing, but it does not consider the transmission delay of links, so the average round-trip delay is lower than ECMP but higher than LLBA.

The average round-trip delay deviation refers to the average value of the difference between the round-trip delay and the average round-trip delay of each packet over the flows. The smaller the average round-trip delay deviation is, the more concentrated the round-trip delay of the packet is. The average round-trip delay deviation has some similarities with the delay jitter, and services such as audio transmission have higher requirements on these. Figure 10 shows that the average round-trip delay deviation of different algorithms as the load increases.

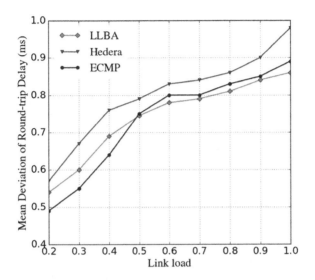

Fig. 10. The round trip delay deviation of flow.

As shown in Fig. 10, when the overall load of the network is light, the delay deviation of the ECMP algorithm is minimal. However, as the load increases, the probability of network congestion is increased due to the flow collision of ECMP, so the delay bias increases sharply. The LLBA and Hedera effectively balance the link load through dynamic scheduling, so the delay deviation increases slowly. Compared with the Hedera, LLBA has better performance in balancing link load, and it can control the transmission delay of elephant flows effectively, so the average round-trip delay deviation of the flows is minimal.

5   Conclusion

In this paper, we analyze the existing traffic scheduling algorithms in data center, and propose a link load balancing algorithm based on ant colony optimization (LLBA). Firstly, the algorithm redefines the heuristic function according to the number of elephant flows and real-time load on the link. Then, updating the pheromone according to the path length. Additionally, the optimal path determination rule is customized according to the path transmission delay and real-time load. Finally, get the optimal transmission path for the current flow after repeated iterations. Simulation experiments show that compared with other algorithms, the proposed algorithm significantly improves link utilization and network throughput, and obtains the lower round-trip delay and the lower average round-trip delay deviation to some extent. However, the upper limit of the link bandwidth in this simulation is 1 Gbps, which cannot meet the bandwidth requirements of the actual data center network. Future work is needed to improve the upper limit bandwidth in our simulations.

Acknowledgment. This work was supported by Program for Changjiang Scholars and Innovative Research Team in university (IRT_16R72).

References

1. Wang, Y., You, S.: An efficient route management framework for load balance and overhead reduction in SDN-based data center networks. IEEE Trans. Netw. Serv. Manage. **15**(4), 1422–1434 (2018)
2. Cong, L., Yong-Hao, W.: Strategy of data manage center network traffic scheduling based on SDN. In: 2016 International Conference on Intelligent Transportation, Big Data & Smart City (ICITBS), Changsha, pp. 29–34 (2016)
3. Wang, J.M., Wang, Y., Dai, X., Bensaou, B.: SDN-based multi-class QoS guarantee in inter-data center communications. IEEE Trans. Cloud Comput. **7**(1), 116–128 (2019)
4. Wang, W., Sun, Y., Zheng, K., Kaafar, M.A., Li, D., Li, Z.: Freeway: adaptively isolating the elephant and mice flows on different transmission paths. In: 2014 IEEE 22nd International Conference on Network Protocols, Raleigh, NC, pp. 362–367 (2014)
5. Zhang, H., Guo, X., Yan, J., Liu, B., Shuai, Q.: SDN-based ECMP algorithm for data center networks. In: 2014 IEEE Computers, Communications and IT Applications Conference, Beijing, pp. 13–18 (2014)
6. Truong, T., Fu, Q., Lorier, C.: FlowMap: Improving network management with SDN. In: NOMS 2016 - 2016 IEEE/IFIP Network Operations and Management Symposium, Istanbul, pp. 821-824 (2016)
7. Qiu, S., Yu, X., Wang, K., Gu, H.: MiFlO: a scheduling algorithm based on mice flows optimization in hybrid data center network. In: 2017 16th International Conference on Optical Communications and Networks (ICOCN), Wuzhen, pp. 1–3 (2017)
8. Al-Fares, M., et al.: Hedera: dynamic flow scheduling for data center networks. In: Proceedings of the 7th USENIX Symposium on Networked Systems Design and Implementation, NSDI 2010, 28–30 April 2010, San Jose, CA, USA DBLP (2010)
9. Curtis, A.R., Kim, W., Yalagandula, P.: Mahout: low-overhead datacenter traffic management using end-host-based elephant detection. In: 2011 Proceedings IEEE INFOCOM, Shanghai, pp. 1629–1637 (2011)
10. Long, L., Binzhang, F., Lixin, C.: Nimble: a fast flow scheduling strategy for OpenFlow networks. Chin. J. Comput. **38**(5), 1056–1068 (2015)
11. Hu, W., Liu, J., Huang, T., Liu, Y.: A completion time-based flow scheduling for inter-data center traffic optimization. IEEE Access **6**, 26181–26193 (2018)
12. Kanthimathi, M., Vijayakumar, D.: An enhanced approach of genetic and ant colony based load balancing in cloud environment. In: 2018 International Conference on Soft-computing and Network Security (ICSNS), Coimbatore, pp. 1–5 (2018)
13. Dorigo, M., Stützle, T.: Ant colony optimization: overview and recent advances. In: Gendreau, M., Potvin, J.-Y. (eds.) Handbook of Metaheuristics. ISORMS, vol. 272, pp. 311–351. Springer, Cham (2019). https://doi.org/10.1007/978-3-319-91086-4_10

14. Wang, C., Zhang, G., Xu, H., Chen, H.: An ACO-based link load-balancing algorithm in SDN. In: 2016 7th International Conference on Cloud Computing and Big Data (CCBD), Macau, pp. 214–218 (2016)
15. Mininet. http://www.mininet.org/
16. Ryu. https://github.com/osrg/ryu
17. Al-Fares, M., Loukissas, A., Vahdat, A.: A scalable, commodity data center network architecture. ACM SIGCOMM Comput. Commun. Rev. **38**(4), 63–74 (2008)

Stereo Matching Based on Improved Matching Cost Calculation and Weighted Guided Filtering

Junxing Xu[✉], Wei He, and Zengshan Tian

School of Communication and Information Engineering, Chongqing University
of Posts and Telecommunications, Chongqing 400065, China
1791821966@qq.com

Abstract. Aiming at the problem that the existing local stereo matching algorithm has low matching accuracy in weak texture, disparity discontinuity and occlusion regions, an improved algorithm based on matching cost calculation and weighted guided filtering is proposed. The algorithm first improves the traditional gradient cost (GRAD) and Census transform, normalizes and fuses these two matching costs to form a new matching cost, then proposes a weighted guided filter based on the Kirsch operator and aggregates the matching cost, finally, the method of the winner-takes-all (WTA) is used to complete the disparity calculation, and we use the method of left and right disparity consistency and the quadratic curve interpolation to complete the disparity optimization and obtain the final disparity map. A large number of experiments prove that the proposed stereo matching algorithm has an average mismatch rate of about 5.45% relative to the standard disparity map on the test platform of Middlebury. Compared with most algorithms, proposed algorithm achieves a good matching effect.

Keywords: Census transform · Adaptive window · Kirsch operator · Weighted guided filtering

1 Introduction

Stereo matching is a process of recovering scene depth information by finding corresponding matching points of two images of the same scene under different viewing angles, and obtaining the disparity pixel by pixel. Stereo matching is widely used in 3D reconstruction [1], digital surface model generation [2], virtual reality and driverless [3].

Scharstei et al. [4] studied and summarized typical stereo matching algorithms and formed a basic stereo matching algorithm framework. The existing stereo matching algorithms are mainly divided into two categories: global stereo matching and local stereo matching algorithms. Global stereo matching algorithm is to solve the optimal solution in the global scope, although the global stereo matching algorithm has a high matching accuracy, the calculation is more complicated, so it has great limitations in practical application. Common global matching algorithms include confidence propagation [5] (BP), graph cut [6] (GC), minimum spanning tree [7] (MST), and split tree [8] (ST). The local stereo matching algorithm uses the neighborhood information in the

H. Gao et al. (Eds.): ChinaCom 2020, LNICST 352, pp. 490–504, 2021.
https://doi.org/10.1007/978-3-030-67720-6_34

window for matching, although the accuracy is poor compared to the global stereo matching, the time complexity is low, and easy to implement, so it is widely used. The steps of local stereo matching algorithm are mainly divided into four steps: matching cost calculation, matching cost aggregation, disparity calculation and disparity optimization. In recent years, the matching accuracy of some excellent local stereo matching algorithms has approached the global stereo matching algorithm, but the local stereo matching algorithm still faces two difficulties: selection of matching cost function and cost aggregation.

One difficulty of the local stereo matching algorithm is the selection of matching cost function, common matching cost functions mainly include pixel-based absolute value of gray difference (AD), gray difference square (SD), gradient-based measurement GRAD, sampling-insensitive BT algorithm [9]; Matching cost based on window mainly includes sum of absolute differences (SAD), sum of squared difference (SSD), normalized cross-correlation (NCC); Matching cost based on non-parametric transform [10] includes Census and Rank transform. GRAD is based on the gradient difference of the image, which can improve the matching accuracy of the weak texture regions to a certain extent, so that the disparity map becomes clearer, but traditional GRAD only considers the gradient value in the x direction, resulting in limited improvement in matching accuracy of weak texture regions. Census transform is an excellent stereo matching method, however, it relies heavily on the gray value of the central pixel, which makes it more sensitive to noise, and the size of support window is fixed, if the size of support window is selected too large, it will contain more texture information, which can improve the matching accuracy, however, the time complexity will also become higher; If the size of support window is too small, the texture information contained is too little, resulting in lower stereo matching accuracy, so selecting an appropriate support window is crucial. At the same time, through a lot of research, some scholars proposed that the improvement of the single measure function cannot further improve the matching accuracy, the fusion of multiple measure functions can significantly improve the matching accuracy, for example, literature [11] proposed a stereo matching algorithm combining SAD and Census transform, which achieves a good matching effect.

Another difficulty of the local stereo matching algorithm is the selection of cost aggregation method. Since the cost calculation only considers the local correlation, it is very sensitive to noise and cannot be directly used to calculate the disparity, therefore, cost aggregation is needed to improve the reliability of stereo matching. Traditional cost aggregation often uses box filtering and gaussian filtering, such filters are less difficult to implement, but the effect of maintaining the edge of the image is poor. In this regard, Yoon et al. [12] proposed to use bilateral filtering for cost aggregation, this method can effectively maintain the edge of the image, but the time complexity is relatively high. He et al. [13] proposed the use of guided filtering in cost aggregation, compared with bilateral filters, guided filtering can better maintain the edge of the image, and the time complexity is also independent of the size of window, but the traditional guided filter does not consider the different textures of different windows, it is simple to use the same regularization parameters for each window, resulting in a poor matching effect of the algorithm on regions with large texture changes.

In view of the above problems, in order to improve the accuracy of stereo matching, this paper proposes a stereo matching algorithm based on improved matching cost and weighted guided filter. The main contribution is: based on the traditional GRAD [14], the gradient values in the y direction and the two diagonal directions are fused to the GRAD, which increases more gradient information and makes the disparity map smoother and clearer. We replace the fixed window of traditional Census transform with adaptive window. The size of the window can be dynamically adjusted according to the standard deviation of the pixels in the window. At the same time, the method of confidence constraint is used to adjust the gray value of the center pixel in the window, and then these improved two matching costs are normalized and fused to achieve complementary advantages, which greatly improves matching accuracy; In the cost aggregation stage, a weighted guided filter based on the Kirsch operator is proposed, proposed algorithm dynamically adjusts the regularization parameter according to the different texture information in each window; Finally, the winner-takes-all (WTA) for disparity calculation, and the disparity refinement and optimization are performed by the method of left and right disparity consistency detection and the quadratic curve interpolation.

2   Matching Cost Calculation

2.1   Improved Census Transform

Census transform belongs to a kind of non-parametric transform algorithm, this algorithm does not use gray-scale information but neighborhood information for matching, so the matching accuracy of this algorithm is higher. The traditional Census transform first selects a pixel, and builds a rectangular window based on the pixel, turning the gray value of other pixels in the window except the central pixel into a bit string, the specific process is: compare the gray value of the pixel in each region with the gray value of the central pixel, if the value is less than the gray value of the central pixel, record it as 1, otherwise as 0. The formula for Census transform is as follows:

$$B_T(p) = \underset{q \in N_p}{\otimes} \xi(I(p), I(q)) \tag{1}$$

$$\xi(I(p), I(q)) = \begin{cases} 1, I(p) \geq I(q) \\ 0, I(p) < I(q) \end{cases} \tag{2}$$

Where p is the central pixel, N_p is the rectangular window with p as the center, q is the neighboring pixel of p in the window, $B_T(p)$ is the converted bit string, and $\otimes$ represents the bit-by-bit connection. The schematic diagram of Census transform is shown in Fig. 1.

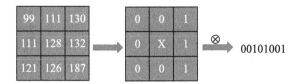

Fig. 1. Schematic diagram of Census transform

Since the Census transform is greatly affected by the central pixel and is susceptible to noise, it is very important to set the center pixel to an appropriate gray value. Some scholars proposed to use the average value of pixels in the window as a reference value, and other scholars proposed to use the median value of all pixels in the window as a reference value, but when the window is greatly affected by noise, the reference values obtained by these two methods are not effective, so an appropriate evaluation mechanism is needed to determine the reference pixel value, this paper proposes a judgment method based on confidence constraints, which improves the matching accuracy of the Census transform. The formula of confidence calculation is shown in (3)(4)(5):

$$R_{ave}(p) = |u(p) - I(p)| \tag{3}$$

$$R_{med}(p) = |m(p) - I(p)| \tag{4}$$

$$R(p) = R_{med}(p) + R_{ave}(p) \tag{5}$$

Where $u(p)$ is the average value in the rectangular window with p as the center pixel, $m(p)$ is the median value of all pixels in the rectangular window, $R_{ave}(p)$ is the mean confidence, $R_{med}(p)$ is the median confidence, and $I(p)$ is the gray value of the center pixel. The smaller the $R(p)$, the higher the confidence level of the gray value of $I(p)$ as the center pixel, which indicates that the gray value of the $I(p)$ as the center pixel of the rectangle is more reliable; When the confidence $R(p)$ is greater than a certain threshold, it is considered unreliable to use $I(p)$ as the central pixel value, and the gray value of the center pixel of the rectangle window needs to be corrected, the correction strategy is shown in (6):

$$I(p) = \begin{cases} \dfrac{u(p) + m(p)}{2} & R(p) > Th \\ I(p) & otherwise \end{cases} \tag{6}$$

When the confidence level $R(p)$ is greater than the confidence threshold Th, the gray value of the center pixel is updated to the average of $u(p)$ and $m(p)$, otherwise the gray value of the center pixel of the rectangle remains unchanged.

Another important factor that affects the matching accuracy of Census transform is the size of the support window. If the size of support window is smaller, the amount of information contained will be less, matching accuracy will be lower; If the size of window is larger, time complexity will become higher. Therefore, this paper proposes an adaptive window method, which achieves a good matching effect, the core idea is:

when the pixel standard deviation in the support window is large enough, the size of the current window is used directly, otherwise the window size is increased to obtain more information, thus achieving the balance of time complexity and matching accuracy. The specific process is shown in Fig. 2:

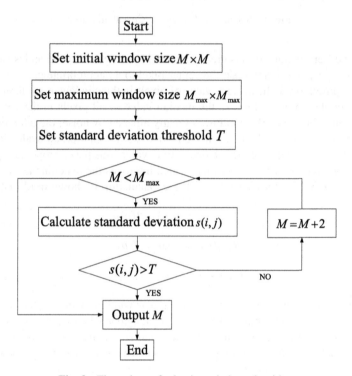

Fig. 2. Flow chart of adaptive window algorithm

The final improved Census transform is as follows:

$$B_T'(p) = \underset{q \in N_p}{\otimes} \xi(I'(p), I(q)) \tag{7}$$

$$\xi(I'(p), I(q)) = \begin{cases} 1, I'(p) \geq I(q) \\ 0, I'(p) < I(q) \end{cases} \tag{8}$$

Where $B_T'(p)$ is the bit string obtained by the improved Census transform, $I'(p)$ is the modified pixel reference value, and the meanings of the remaining symbols are the same as those in (1) and (2). When the disparity is d, the Hamming distance calculation formula of the two bit strings is:

$$H_h(p,d) = \min(Ham \min g(C_{CTL}(p), C_{CTR}(p-d)), \ T_{cen}) \tag{9}$$

Where $H_h(p,d)$ is the matching cost of Census transform, $C_{CTL}(p)$ is the Hamming value of the left image at pixel p, $C_{CTR}(p-d)$ is the Hamming value of the right image at pixel $p-d$, and T_{cen} is the threshold value of the Census transform, and we can come to a conclusion: the smaller the Hamming distance between two pixels, the higher the matching accuracy, otherwise the lower the matching accuracy. In order to further reduce the interference of noise, the matching cost is normalized:

$$C_{cen}(p,d) = 1 - \exp(-H_h(p,d)) \tag{10}$$

2.2 Enhanced Gradient Cost

In order to further obtain a high-precision disparity map, this paper improves the traditional GRAD algorithm, traditional GRAD only considers the gradient value in the x direction, which leads to a limited improvement in the matching accuracy of the weak texture regions, in this regard, this paper proposes an improved GRAD algorithm, the improved algorithm fuses the gradient information in the y and the two diagonal directions, the calculation formula is (11).

$$\begin{cases} grad_x = \partial G/\partial x \\ grad_y = \partial G/\partial y \\ grad_{45°} = \partial G/\partial_{45°} \\ grad_{135°} = \partial G/\partial_{135°} \end{cases} \tag{11}$$

$$grad = \sqrt{(grad_x)^2 + (grad_y)^2 + (grad_{45°})^2 + (grad_{135°})^2} \tag{12}$$

Where G is the gray scale of the image, $grad_x$ is the gradient of the image in the x direction, $grad_y$ is the gradient in the y direction, $grad_{45°}$ and $grad_{135°}$ are the gradients in the two diagonal directions, and $grad$ is the image gradient, then the gradient cost is:

$$grad_g(p,d) = \min(grad_L(p) - grad_R(p-d), \ T_g) \tag{13}$$

Where $grad_g(p,d)$ s the gradient cost when the disparity value of pixel p is d, $grad_L(p)$ is the gradient of the left image at pixel p, $grad_R(p-d)$ is the gradient of the right image at pixel $p-d$, and T_g is the gradient threshold. The normalized similarity measure function is:

$$C_g(p,d) = 1 - \exp(-grad_g(p,d)) \tag{14}$$

2.3 Matching Cost Fusion

In order to achieve the complementary advantages of the two measurement functions, the two measurement functions are weighted and fused, the fused matching costs are as follows:

$$C(p,d) = \lambda_1 C_{cen}(p,d) + \lambda_2 C_g(p,d) \tag{15}$$

Where λ_1 and λ_2 are the weight values of the improved Census transform and the enhanced gradient cost, and $\lambda_1 + \lambda_2 = 1$.

3 Cost Aggregation

After calculating the matching cost of the image, it is necessary to perform cost aggregation on the matching window. The essence of cost aggregation is filtering, which is used to filter out the noise introduced in the matching cost calculation. Among all the methods of cost aggregation, the guided filtering is an excellent one, The process of guiding the filtering algorithm is as follows: suppose I is the guide image, usually the left image is the guide image, the input image is p, U is the filtered image, the filtered linear conversion model is:

$$U_i = a_k I_i + b_k, \quad \forall i \in w_k \tag{16}$$

Where k is the center pixel in the window, w_k is a rectangular window centered on pixel k, U_i and I_i are the corresponding points before and after the guided filtering. a_k and b_k are the linear coefficients, a_k and b_k can be obtained by minimizing (17).

$$E(a_k, b_k) = \sum_{i \in w_k} [(a_k I_i) + b_k - p_i)^2 + \varepsilon a_k^2] \tag{17}$$

Where ε is the regularization parameter, usually $\varepsilon > 0$, and the larger the value, the smoother the filtered image; The smaller the value, the better the edge of filtered image. The solution of (17) can be obtained by the least square method.

$$a_k = \frac{\frac{1}{|w|} \sum_{i \in w_k} I_i p_i - u_k \bar{p}_k}{\sigma_k^2 + \varepsilon} \tag{18}$$

$$b_k = \bar{p}_k - a_k u_k \tag{19}$$

Where u_k and σ_k^2 represent the mean and variance of the guide image I in the window w_k; $|w|$ represents the total number of pixels in the window w_k; $\bar{p}_k$ is the average gray value of the input image p in the window w_k.

However, the traditional guided filtering algorithm does not consider the difference in pixel texture of different windows, and each window uses the same regularization parameters, resulting in limited improvement in matching accuracy. Aiming at the problem, Li et al. [15] improved the guided filtering, and proposed a weighted guide filtering, which defines the pixel variance in the window as a weighted factor, and adaptively adjusts the regularization parameters. However, defining the variance as a weighted factor does not highlight the edges of the image very well, so this paper proposes to define the Kirsch operator [16] as a weighted factor, the Kirsch operator can detect the edges in the image very well, in the edge regions of the image, a larger amplitude response can be obtained, in the smooth region of the image, the amplitude response is close to 0.

The edge weight of Kirsch operator is defined as:

$$\Gamma_I(p) = \frac{1}{N} \sum_{p'=1}^{N} \frac{KIR(p)}{KIR(p')} \tag{20}$$

Where $KIR(p)$ is the edge value of p corresponding to Kirsch operator, I is the guide image, N represents the total number of pixels, (18) can be replaced with (21).

$$a_k = \frac{\frac{1}{|w|} \sum_{i \in w_k} I_i p_i - u_k \bar{p}_k}{\sigma_k^2 + \frac{\varepsilon}{\Gamma_I}} \tag{21}$$

4 Disparity Calculation

The final matching cost obtained after cost aggregation is $C'(p, d)$. We use the WTA strategy to select the minimum matching cost correspondence for each pixel as the initial disparity value.

$$d_p = \arg \min_{d \in D} [C'(p, d)] \tag{22}$$

Where D represents all possible disparity values.

5 Disparity Refinement

There are many mismatched points in the initial disparity map obtained by the WTA method, so post-processing optimization is needed to reduce the number of mismatched points. Post-processing optimization methods mainly include left and right consistency detection, occlusion processing, weighted median filtering and sub-pixel refinement [17, 18]. The way to detect mismatched points through left and right consistency detection is as follows.

$$|d_L(p) - d_R[p - d_L(p)]| < 1 \tag{23}$$

Where $d_L(p)$ is the disparity value of pix p in the left view, and $d_R[p - d_L(p)]$ is the disparity value of pix $p - d_L(p)$ in the right view. If the left and right disparity value are not equal, it is considered as an anomaly point, and at the same time, the anomaly points are divided into occlusion points and mismatched points according to the principle of epipolar geometry [19].

Different interpolation methods are used for the occlusion points and mismatched points obtained by the left and right consistency detection. For the occlusion pixels, since the occlusion regions are usually located in the background of the image, the disparity value from the non-occlusion region pixels is used to interpolate the occlusion point. We search the effective pixel closest to the occlusion pixel in 8 directions, and use its disparity value as the disparity value of the current occlusion point; For mismatched pixels, the disparity value of the effective pixel with the most similar pixel color is used as the disparity value of the mismatched pixel in 8 directions, and the weighted median filter [20] is used to smooth the interpolated disparity map. In order to reduce discontinuity of the disparity map caused by discrete disparity, sub-pixel estimation is performed based on binomial polynomial interpolation [21]. For each pixel, the best sub-pixel interpolation value is:

$$d_{sub} = d - \frac{S(p, d_+) - S(p, d_-)}{2[S(p, d_+) + S(p, d_-) - 2S(p, d)]} \tag{24}$$

Where d is the disparity value after pixel p weighted median filtering; $S(p, d_+)$ and $S(p, d_-)$ are the cost aggregation of pixel p when the disparity is $d + 1$ and $d - 1$, respectively.

6 Experimental Results and Analysis

In order to evaluate the matching effect of the proposed algorithm, this paper uses the data set provided by the Middlebury [22] experimental platform for experiments. The simulation experiments are performed on a computer with a CPU of Intel Core i3-3210M, 2.5 GHz, and 8G of memory, and the programming environment is Matlab R2016a. The parameter settings involved in the experiment are shown in Table 1, the following dates are obtained through many experiments.

Table 1. Parameter settings

Parameter	Value	Parameter	Value	Parameter	Value
Th	6	T_{cen}	0.06	T	5.23
M	0.39	T_g	0.35	λ_2	0.55
M_{max}	11	λ_1	0.45	ε	0.012

6.1 Matching Cost Calculation Experiment

In order to verify the effectiveness of the improved matching cost, this paper selects two cost functions to compare with the proposed cost function, these two cost functions are SAD + GRAD [23] and AD + Census [17]. The experimental images are 4 sets of stereo matching image pairs (Cones, Tsukuba, Teddy, Venus). Figure 3 is the result of different cost functions experiments, the experiments do not undergo any post-processing, Table 2 is error matching rates of different algorithms for different images, Fig. 4 is the histogram of the average mismatch rate of each cost function.

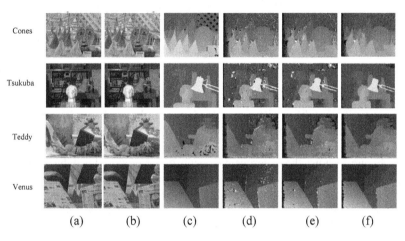

Fig. 3. Disparity map of different cost functions. (a) left image; (b) right image; (c) real disparity map;(d) SAD + GRAD;(e) AD + Census; (f) Proposed

Table 2. Error matching rates of different cost functions for different images %

Algorithm	Cones	Tsukuba	Teddy	Venus
SAD + GRAD	27.165	23.898	24.554	22.441
AD + Census	19.145	17.984	19.689	18.356
Proposed	13.895	11.546	14.262	13.654

As can be seen from Fig. 3, compared with the matching cost of SAD + GRAD and AD + Census, the proposed cost function shows a better matching effect in four different pairs of stereo matching images, and it can be seen from Table 2 and Fig. 4 that the proposed cost function has a lower mismatch rate, which proves the effectiveness of the proposed cost function.

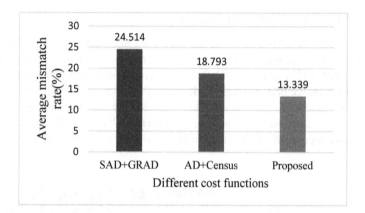

Fig. 4. Average mismatch rate of different cost functions

6.2   Cost Aggregation Experiment

In order to verify that the proposed weighted guided filtering has an excellent effect in maintaining edges, this paper compares literature [15] with the proposed weighted guided filter. The experimental images are 2 sets of stereo matching image (Sawtooh, Barn2). Figure 5 shows the experimental results of the proposed weighted guided filter and literature [15] in 2 different stereo matching image pairs. Figure 5 (a) is the left view of the stereo matching image pair, (b) is the right view of the stereo matching image pair, (c) is the real disparity map, (d) is the disparity map of literature [15], (e) is the disparity map of the proposed weighted guided filter.

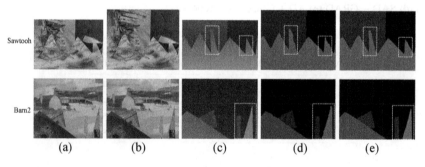

Fig. 5. Disparity map of different cost aggregation algorithms. (a) left view; (b) right view; (c) real disparity map; (d) literature [15]; (e) proposed weighted guided filter

As shown in Fig. 5, in the regions marked by the rectangular frame, the weighted guided filter proposed in this paper has a better effect in maintaining edges than the literature [15], which proves the effectiveness of the improved weighted guided filter proposed in this paper.

6.3   Algorithm Comparison Analysis

In order to further verify the superiority of the proposed algorithm, four common stereo matching algorithms are selected to perform experiments on four sets of stereo matching image pairs (Venus, Cones, Tsukuba, Teddy). These four algorithms include semi-global stereo matching [23] (SGM), cross-scale minimum spanning tree [14] (CS-MST), block matching [24] (BM), BP algorithm. The experimental simulation diagrams are shown in Fig. 6, the error matching rates of different algorithms for different images are shown in Table 3, Table 4, and where *n-occ* represents the ratio of mismatched pixels in non-occluded regions, *all* represent the total pixel mismatch rate, *disc* represents the mismatch rate in the disparity discontinuity regions, Fig. 7 is the histogram of the average mismatch rate of each stereo matching algorithm.

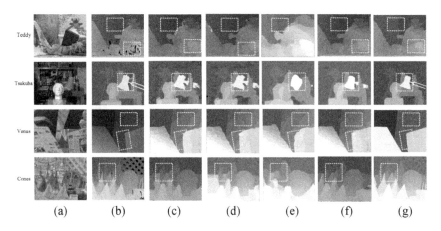

(a)   (b)   (c)   (d)   (e)   (f)   (g)

Fig. 6. Disparity map of different algorithms. (a) Left image (b) Real disparity map (c) CS-MST (d) SGM (e) BM (f) BP (g) Proposed

Table 4. Error matching rates of different algorithms for different images%

Algorithm	Venus			Cones		
	n-occ	all	disc	n-occ	all	disc
CS-MST	8.654	10.562	14.236	10.445	12.886	15.234
SGM	11.565	14.784	17.889	16.223	18.442	23.896
BM	15.447	17.845	21.246	22.445	25.221	30.568
BP	4.245	6.231	9.556	6.315	9.564	11.235
Proposed	3.562	5.556	8.233	4.225	5.253	7.565

Table 3. Error matching rates of different algorithms for different images%

Algorithm	Teddy			Tsukuba		
	n-occ	all	disc	n-occ	all	disc
CS-MST	6.235	9.254	11.556	8.369	10.546	13.456
SGM	7.453	10.545	13.563	10.656	13.895	16.235
BM	20.231	25.235	30.231	25.336	29.256	33.445
BP	5.562	7.236	9.235	5.356	6.745	8.754
Proposed	4.535	5.652	8.123	4.114	5.852	7.565

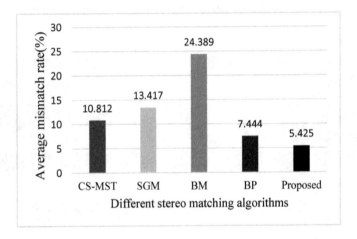

Fig. 7. Average mismatch rate of different stereo matching algorithms

As can be seen from Fig. 6, compared with SGM, CS-MST and other algorithms, the proposed algorithm shows good matching accuracy in weak texture, disparity discontinuity and occlusion regions. As shown in the Cones test chart, the part marked by the rectangular frame is a disparity discontinuity regions, the proposed algorithm achieves a smoother disparity in this regions. The part marked by rectangles in the Teddy test chart are the weak texture and the occlusion regions, the disparity of proposed algorithm in these regions is more clear and smooth, and in Venus test chart, the proposed algorithm is better in the edge regions. Similarly, the algorithm of this paper is better than other algorithms on the test charts of Tsukuba, which proves the effectiveness of proposed algorithm.

As can be seen from Table 3, Table 4 and Fig. 7, compared with other algorithms, proposed algorithm has better matching effect in multiple image regions. The average mismatch rate of CS-MST, SGM, BM, BP and proposed algorithm is 10.812%, 13.417%, 24.389%, 7.444%, 5.425%, respectively, which shows the superiority of proposed algorithm.

7 Conclusion

Based on the existing local stereo matching algorithm, this paper proposes a stereo matching algorithm with higher matching accuracy. In the matching cost calculation stage, two major improvements are made. ① the gradient information in the y direction and the two diagonal directions is fused to the traditional GRAD, which improves matching accuracy of weak texture regions. ② Based on the traditional Census transform, a confidence constraint method is proposed to modify the central pixel value in the support window, at the same time, the size of the support window is dynamically adjusted by the method based on gray standard deviation, then these two cost functions are weighted and fused, which further improves the matching accuracy. In the cost aggregation stage, a weighted guided filtering algorithm based on Kirsch operator is proposed, which maintains the edges of the image. A large number of experimental results show that compared with other algorithms, proposed algorithm can effectively enhance the matching accuracy of weak texture, disparity discontinuity and occlusion regions.

References

1. Shen, S.: Accurate multiple view 3D reconstruction using patch-based stereo for large-scale scenes. IEEE Trans. Image Process. **22**(5), 1901–1914 (2013)
2. Hamzah, R.A., Rahim, R.A., Rosly, H.N.: Depth evaluation in selected region of disparity mapping for navigation of stereo vision mobile robot. In: 2010 IEEE Symposium on Industrial Electronics and Applications (ISIEA), Penang, pp. 551–555 (2010)
3. Irijanti, E., Nayan, M.Y., Yusoff, M.Z.: Fast stereo correspondent using small-color census transforms. In: Proceedings of the 4th International Conference on Intelligent and Advanced Systems, pp. 685–690 (2012)
4. Scharstein, D., Szeliski, R., Zabih, R.: A taxonomy and evaluation of dense two-frame stereo correspondence algorithms. In: Proceedings IEEE Workshop on Stereo and Multi-Baseline Vision (SMBV 2001), Kauai, HI, USA, pp. 131–140 (2001)
5. Sarkis, M., Diepold, K.: Sparse stereo matching using belief propagation. In: 2008 15th IEEE International Conference on Image Processing, San Diego, CA, pp. 1780–1783 (2008)
6. Kolmogorov, V., Zabih, R.: Computing visual correspondence with occlusions using graph cuts. In: Proceedings Eighth IEEE International Conference on Computer Vision. ICCV 2001, Vancouver, BC, Canada, vol. 2, pp. 508–515 (2001)
7. Yang, Q.: A non-local cost aggregation method for stereo matching. In: 2012 IEEE Conference on Computer Vision and Pattern Recognition, Providence, RI, pp. 1402–1409 (2012)
8. Mei, X., Sun, X., Dong, W., Wang, H., Zhang, X.: Segment-tree based cost aggregation for stereo matching. In: 2013 IEEE Conference on Computer Vision and Pattern Recognition, Portland, OR, pp. 313–320 (2013)
9. Birchfield, S., Tomasi, C.: Depth discontinuities by pixel-to-pixel stereo. In: Sixth International Conference on Computer Vision (IEEE Cat. No. 98CH36271), Bombay, India, pp. 1073–1080 (1998)
10. Zabih, R., Woodfill, J.: Non-parametric local transforms for computing visual correspondence. In: 3rd European Conference on Computer Vision, vol. 2, pp. 151–158 (1994)

11. Zhou, J., Ying, W., Meng, L.: A new stereo matching algorithm based on adaptive weight SAD algorithm and Census algorithm. Bull. Survey. Map. **0**(11), 11–15 (2008)
12. Yoon, K.-J., Kweon, I.S.: Adaptive support-weight approach for correspondence search. IEEE Trans. Pattern Anal. Mach. Intell. **28**(4), 650–656 (2006)
13. He, K., Sun, J., Tang, X.: Guided image filtering. IEEE Trans. Pattern Anal. Mach. Intell. **35**(6), 1397–1409 (2013)
14. Zhang, K., et al.: Cross-scale cost aggregation for stereo matching. In: 2014 IEEE Conference on Computer Vision and Pattern Recognition, Columbus, OH, pp. 1590–1597 (2014)
15. Li, Z., Zheng, J., Zhu, Z., Yao, W., Wu, S.: Weighted guided image filtering. IEEE Trans. Image Process. **24**(1), 120–129 (2015)
16. Mani, D.S., Nagaraju, C.: Face recognition based on kirsch compass kernel operator. In: 2017 International Conference on Communication and Signal Processing (ICCSP), Chennai, pp. 1322–1324 (2017)
17. Mei, X., Sun, X., Zhou, M., Jiao, S., Wang, H., Zhang, X.: On building an accurate stereo matching system on graphics hardware. In: 2011 IEEE International Conference on Computer Vision Workshops (ICCV Workshops), Barcelona, pp. 467–474 (2011)
18. Hosni, A., Rhemann, C., Bleyer, M., Rother, C., Gelautz, M.: Fast cost-volume filtering for visual correspondence and beyond. IEEE Trans. Pattern Anal. Mach. Intell. **35**(2), 504–511 (2013)
19. Hirschmuller, H.: Stereo processing by semiglobal matching and mutual information. IEEE Trans. Pattern Anal. Mach. Intell. **30**(2), 328–341 (2008)
20. Ma, Z., He, K., Wei, Y., Sun J., Wu, E.: Constant time weighted median filtering for stereo matching and beyond. In: 2013 IEEE International Conference on Computer Vision, Sydney, NSW, pp. 49–56 (2013)
21. Yang, Q., Yang, R., Davis, J., Nister, D.: Spatial-depth super resolution for range images. In: 2007 IEEE Conference on Computer Vision and Pattern Recognition, Minneapolis, MN, pp. 1–8 (2007)
22. Yang, Q., Wang, L., Yang, R., Stewénius, H., Nistér, D.: Stereo matching with color-weighted correlation, hierarchical belief propagation, and occlusion handling. IEEE Trans. Pattern Anal. Mach. Intell. **31**(3), 492–504 (2009)
23. Hirschmuller, H.: Accurate and efficient stereo processing by semi-global matching and mutual information. In: 2005 IEEE Computer Society Conference on Computer Vision and Pattern Recognition (CVPR'05), San Diego, CA, USA, vol. 2, pp. 807–814 (2005)
24. Hamzah, R.A., Hamid, A.M.A., Salim, S.I.M.: The solution of stereo correspondence problem using block matching algorithm in stereo vision mobile robot. In: 2010 Second International Conference on Computer Research and Development, Kuala Lumpur, pp. 733–737 (2010)

A Computation Offloading Strategy Based on Stackelberg Game for Vehicular Network

Yingdi Dai$^{(\boxtimes)}$, Zhanjun Liu, and Ya Kang

School of Communication and Information Engineering,
Chongqing University of Posts and Telecommunications, Chongqing, China
dongnan163@163.com

Abstract. Edge computing was proposed to offload the computing tasks of the vehicle to the vehicle fog nodes, which can achieve efficient vehicle services and higher utilization of computing resources. However, encouraging vehicles to share resources or execute applications for others remains a sensitive issue due to the selfishness of users. In this paper, we propose an incentive mechanism based on Stackelberg game and the problem of computation offloading of tasks on vehicles is modeled. In order to utilize the idle computing resources of nearby vehicles, for the computation offloading in vehicular networks. Specifically, we introduce multi-hop offloading into vehicle-to-vehicle (V2V) offloading, and model the probability of forwarding packets by assisting vehicles from multiple perspectives in order to ensure the stability and reliability of communication transmission. A distributed iterative algorithm is designed to solve the final Stackelberg equilibrium, so as to minimize the cost of requesting vehicles to execute tasks and maximize the service benefits of assisting vehicles. The simulation results show that the proposed computation offloading strategy has superiorities in improving utilization efficiency, reducing tasks execution latency and enhancing service quality.

Keywords: Vehicular network · Computation offloading · Stackelberg game

1 Introduction

With the advancement of the Internet of Things (IoT) and wireless communication technology, vehicles equipped with advanced sensors and communication devices can build large-scale interactive networks, which has promoted the development of the intelligent transportation systems (ITS) [1]. In the internet of vehicles, vehicles are usually equipped with on-board unit (OBD), which

This work is supported in part by Chongqing University of Posts and Telecommunications Doctoral Fund project, E010A2019017 and the other part by the Science and Technology Research Program of Chongqing Municipal Education Commission (Grant No. KJQN201900645) and PHD Initialted Fund Project (A2009-17).

H. Gao et al. (Eds.): ChinaCom 2020, LNICST 352, pp. 505–518, 2021.
https://doi.org/10.1007/978-3-030-67720-6_35

has a certain computing and storage capacity. However, some emerging vehicle applications require complex task computation and massive data storage. The resource constrained vehicles may be strained by these applications, thus resulting in bottlenecks and making it challenging for vehicles to ensure the required Quality of Service (QoS) [2].

Aiming at the above issue, mobile cloud computing offloading technology, which allows vehicles offload the computing tasks to a powerful cloud server or base station (BS) with mobile edge computing (MEC) servers for execution, and finally return the processing result to the corresponding vehicles [3,4]. However, in the wireless network environment without infrastructure, the deployment of a high performance server for a certain area is faced with problems of site selection and cost problems. Therefore, vehicular cloud computing (VCC) as a new computing paradigm is proposed by integrating the resources of vehicles [5]. A group of vehicles can share their computing resources and storage resources by using VCC technology. Compared with fixed-location mobile edge servers, the offloading scheme in VCC has the advantages of infrastructure independence and economic benefits. Therefore, the work of this paper is carried out on this basis.

Generally, vehicles belong to different private users or different organizations, forwarding data packets or executing computing tasks for others will occupy their computing and communication resources and consume their energy. Every vehicle is rational, they all want to maximize their utility, requesting vehicles will prefer to offload tasks to assisting vehicles with good link quality, low-priced and sufficient resources for execution, while assisting vehicles hope to rent computing resources to provide high-priced requesting vehicles to obtain as much revenue as possible. This interaction between them is a typical Stackelberg game problem. If the selfishness of the vehicle cannot be overcome, the connectivity of the network will become worse, and it will cause serious waste of computing resources. Therefore, the selfishness of vehicles cannot be ignored.

How to efficiently share idle resources is a key issue. In [6], the authors divided the computing task into two subtasks, and jointly optimizes the offloading ratio and computing resource allocation to minimize the latency of task completion. In [7], a federated offloading scheme was proposed that combined vehicle-to-infrastructure (V2I) and V2V communication in MEC-enabled vehicular networks to minimize the total latency. In [8], the authors studied the computation overhead (the weighted sum of the latency and the energy consumption) minimization problem by jointly optimizing communication and computation resources. However, the computation offloading schemes proposed by the above works are all to satisfy the optimization of system performance, but ignore the selfishness of resource providers in the network. Based on this, the authors in [9] established a computation offloading marketplace based on reverse auction mechanism, which provided sellers incentives to lease their idle computation resources and allowed buyers to express diverse preferences for different sellers by unilateral-matching. In [10], the authors proposed a contract-based incentive mechanism to motivate vehicles to share their resources and transform the task assignment problem into a two-sided matching problem between vehicles and user equipment to minimize the network delay. The strategies proposed in [9,10] used incentive mechanism to overcome the selfishness of vehicles, but the matching mode between vehicles is specified, which lead to the insufficient utilization of computing resources in the network. Therefore, in this paper, we analyze the probability of assisting vehicles forwarding request packets from multiple perspectives, and introduce it into multi-master and multi-slave computation offloading game to realize the adaptive transmission of multi-hop link. In this game, the pricing incentive mechanism is used to encourage the assisting vehicles to actively participate in the task execution, and a low-complexity algorithm is used to achieve equilibrium.

This paper is organized as follows. The system model is introduced in Sect. 2. In Sect. 3, we model the computation offloading between requesting vehicles (resource requesters) and the assisting vehicles (resource providers) with a Stackelberg game approach. Numerical results are obtained and analyzed in Sect. 4. Finally, we conclude this paper in Sect. 5.

2 System Model

In this section, the system model include vehicular network model, data transmission model between vehicles, and partial offloading model, then these models will be described.

2.1 Vehicular Network Model

We consider a unidirectional road, where a RSU is located the road that enables vehicles to access the network and sends road information to vehicles, as depicted in Fig. 1. There are N requesting vehicles and M assisting vehicles arriving at the road. The i-th requesting vehicle is defined as $B_i, i \in [1, 2, \ldots, N]$, and the j-th assisting vehicles is defined as $S_j, j \in [1, 2, \ldots, M]$. Each requesting vehicle has a computing task, which can be described as $D_i = \{d_i, h_i, W_i, T_i^{res}, T_i^{max}\}$, where d_i denotes the size of input data (e.g. the program codes and input parameters), h_i is the mapping between data size and CPU cycles, W_i is the required computing resources for computing task. Referring to [11] and [12], we can obtain the number of CPU cycles, W_i for data d_i by $W_i = h_i \cdot d_i$. T_i^{res} denotes the maximum latency of receiving response in a period, T_i^{max} denotes the maximum latency allowed to accomplish the task.

We assume that each vehicle in the network is equipped with a positioning system and a wireless communication module, V2I communication can quickly obtain the required information, V2V communication can transmit information to the vehicles in the distance through multi-hop transmission [13], and can also cover all surrounding vehicles with broadcast technology [14]. The assisting vehicle S_j forwards the request packet to the l-th vehicle of k-th hop, which is expressed as $S_{j^{k_l}}$. If $k = 0$, then $l = 0$ indicates that the request message has not been forwarded by multi-hop, $S_j = S_{j^{k_l=0}}$.

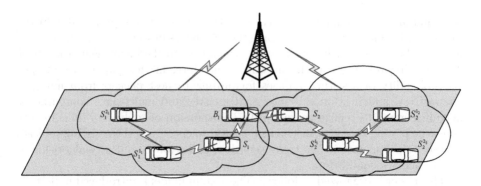

Fig. 1. A computation offloading case in vehicular network.

2.2 Data Transmission Model

The requesting vehicle B_i sends the request packet by broadcasting. If S_j within the communication range of B_i, S_j will compute the computing resources can provide and update the packet, then forward it with a certain probability and feedback the final results to B_i. The data transmission model is shown in Fig. 2.

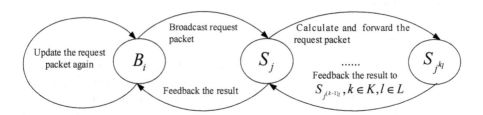

Fig. 2. Data transmission between vehicles.

Since the network environment and the situation of vehicles are relatively complex, it is necessary to determine the probability of assisting vehicles forwarding request packet. We define the forward probability of assisting vehicles includes the following factors:

• *Vehicle Density:* Assume that all vehicles in the network follow a random poisson process with a rate of λ (vehicles per second) to arrive and drive on the road at a speed V, where $V \in [V_{min}, V_{max}]$ is a random variable that independent and uniformly distributed in the distribution interval. The probability density function of moving speed V can be expressed as follows:

$$f_V(x) = \begin{cases} \frac{1}{V_{max} - V_{min}} & , V_{min} \leq x \leq V_{max} \\ 0 & , others \end{cases} \tag{1}$$

Let ρ represents the vehicle density. According to the traffic flow theory [15] $\lambda = \rho v$, the expression of ρ can be obtained by (1):

$$\rho = \lambda E(\frac{1}{V}) = \frac{\lambda ln(V_{max}/V_{min})}{V_{max} - V_{min}} \tag{2}$$

Then the assisting vehicles can estimate the number of vehicles within the communication range of B_i in the current vehicular network environment as: $N_{pre} = \rho R$, where R is the communication radius of the vehicle.

• *Link Expiration Time:* Assuming that the initial position of B_i is (X_i, Y_i), and (X_j, Y_j) is the position of S_j. After moving $\triangle t$ time slot, the position of B_i is changed to $(X_i + V_i \triangle t, Y_i)$, and the position of S_j is changed to $(X_j + V_j \triangle t, Y_j)$. If the distance between vehicles is greater than R, the V2V communication link cannot be established. Therefore, the critical time to ensure the normal communication of the link needs to meet the following:

$$R^2 = [(X_i + V_i T_{LET}) - (X_j + V_j T_{LET})]^2 + (Y_i - Y_j)^2 \tag{3}$$

where T_{LET} is the maximum link expiration time of B_i and S_j. Solve formula (3), T_{LET} can be obtained (select the positive value):

$$T_{LET} = \frac{-(X_i - X_j) \pm \sqrt{R^2 - (Y_i - Yj)^2}}{V_i - V_j} \tag{4}$$

This method is also suitable for two-way driving roads. We can regard the road as a two-dimensional coordinate, and define the driving speed of B_i as the positive speed, and the vehicle with the opposite direction as the negative speed. The link lifetime between vehicles can also be calculated by using (4).

• *Willingness of Assisting Vehicles:* We define the willingness of S_j to participate in computing offloading services as $\theta_j = \alpha_j \frac{Q_j^{max}}{Q_j^{ex}}$, where α_j is the preset constant, Q_j^{max} and Q_j^{ex} respectively represent the largest available computing resources and the already occupied computing resources of S_j.

To sum up, the probability function of assisting vehicles to forward packets can be expressed as follows:

$$F_j = 1 - e^{-\frac{T_{LET} \cdot \widehat{N}}{T_{obs} \cdot N_{pre} \cdot \theta_j}} \tag{5}$$

where $\widehat{N}$ represents the number of request packets received by the assisting vehicle, and T_{obs} is the longest travel time on the observation road. In this paper, the assisting vehicle updates the request packet can reduce the unit resource bid in the original packet according to the forwarding probability. That is the unit resource bid is determined by $p_{i,j}$ and F_j.

2.3   Partial Offloading Model

In this paper, the task of the requesting vehicle can be divided into independent subtasks, which can be executed locally or offloaded to other vehicles.

- *Local Execution:*

$$T_{i,i}^{local} = \frac{W_i - \sum\limits_{j=1}^{M} Q_{j^{k_l}}}{f_i} \tag{6}$$

where f_i is the computing capability of B_i. $T_{i,i}^{local}$ is the local execution time. $Q_{j^{k_l}} = \sum\limits_{k=0}^{K} \sum\limits_{l=0}^{L} q_{j^{k_l}}$ is the total computing resources that can be provided by S_j in combination with other vehicles. $q_{j^{k_l}}$ is the computing resources provided by $S_{j^{k_l}}$, $q_j = q_{j^{k_l}=0}$.

- *V2V Offloading:* Considering the vehicles and vehicles' computational capacity distribute randomly, some vehicles with powerful computation capacity may be located out of 1-hop range. Therefore, we introduce multi-hop offloading into V2V offloading. The total latency for V2V offloading consists of two part: the latency for offloading the data of subtasks and the latency to execute subtasks on assisting vehicles [10]. The latency for offloading the data from B_i to S_j can be expressed as:

$$T_{i,j}^{trans} = max \left\{ \frac{Q_{j^{k_l}}/h_i}{R_i} \mid i \in N, j \in M \right\} \tag{7}$$

where R_i is the transmission rate of B_i, $Q_{j^{k_l}}/h_i$ represents the data size of B_i offloading subtasks to S_j. Since the V2V offloading in this paper is a multi-hop offloading, the latency to execute subtasks on assisting vehicles should be included the maximum latency of subtasks execution and the maximum transmission latency of offloading subtasks on each hop, which can be expressed as:

$$T_{j^{k_l}}^{ex} = max \left\{ \frac{Q_{j^{k_l}}}{f_{j^{k_l}}} \mid k \in K, l \in L \right\} \tag{8}$$

$$T_{j^{k_l}}^{trans} = max \left\{ \frac{q_{j^{(k+1)_l}}/h_i}{R_{j^{k_l}}} \mid k \in K, l \in L \right\} \tag{9}$$

where $R_{j^{k_l}}$ is the transmission rate of the $S_{j^{k_l}}$. Since the amount of computation results data is very little, we are not concerned about it in this paper. Then the latency to execute subtasks on S_j can be expressed as : $T_{i,j}^{comp} = T_{i,j}^{trans} + (T_{j^{k_l}}^{trans} + T_{j^{k_l}}^{ex})$. Finally, the total latency for B_i to execute the whole task in our scheme can be given by :

$$T_i = max\{T_{i,i}^{local}, T_{i,j}^{comp}\} \tag{10}$$

3   Computation Offloading Based on Stackelberg Game

In this section, computation offloading the workload of the requesting vehicle to the assisting vehicle is modeled as a Stackelberg game. The computing resources provided by assisting vehicles are equivalent to the workload that can undertake.

3.1 Requesting Vehicle Side Analysis

The requesting vehicle with computation-intensive application waiting to be executed as the game buyer. Due to unsatisfactory computational capability and limited on-board resources, a buyer is inclined to offload part or entire of the computation workload to an applicable nearby vehicles to reduce the application completion overhead, improving the quality of experience (QoE). We define the cost function for the buyer of computing resources as follows:

$$\min_{p_{i,j}} \; U_i = p_{ci}(W_i - \sum_{j=1}^{M} Q_{j^{k_l}}) + p_{i,j} \sum_{j=1}^{M} Q_{j^{k_l}} + \chi_i \sum_{j=1}^{M} Q_{j^{k_l}}/h_i$$

$$\text{S.t} \qquad p_{i,j} \le p_{ci}, \qquad\qquad \forall i \in N$$

$$0 \le \sum_{j=1}^{M} Q_{j^{k_l}} \le W_i, \quad \forall j \in M, k \in K, l \in L$$

$$T_i \le T_i^{max}, \qquad\qquad \forall i \in N \tag{11}$$

where p_{ci} is the unit workload execution cost of B_i in local, which is a piecewise function:

$$p_{ci} = \begin{cases} p_{base}, & \frac{W_i}{f_i} \le T_i^{max} \\ p_{base} \times \left(\frac{W_i/f_i}{T_i^{max}} \right)^{\eta}, & \frac{W_i}{f_i} > T_i^{max} \end{cases} \tag{12}$$

p_{base} is the basic cost of vehicle to execute unit workload, and η ($\eta \ge 1$) is the price growth coefficient. $p_{i,j}$ is the unit computing resource bid by B_i. χ_i is the unit transmission cost over the V2V link. The unit resource bid of B_i cannot exceed the unit workload execution cost in local. The computing resources purchased by B_i cannot exceed that required to execute the whole task and B_i needs to complete the execution of the task within T_i^{max}.

3.2 Assisting Vehicle Side Analysis

The assisting vehicle has idle computing resources that can lease to buyers, helping to relieve the heavy on-board workload while receiving a reasonable revenue as the game seller. The revenue of S_j can be divided into two parts, the first part is that S_j provids the computing resources of itself under the unit resource bid of B_i:

$$U_j^1 = \theta_j ln\,(1 + p_{i,j}q_j) - c_j q_j \tag{13}$$

And the second part is that S_j reduces the bid price in the request packet, and cooperates with other assisting vehicles to execute the task to obtain the derivative income:

$$U_j^2 = \theta_j ln\left(1 + p_{i,j}(1 - F_j)Q_{j^{(k+1)_l}}\right) - \chi_j \frac{Q_{j^{(k+1)_l}}}{h_i} \tag{14}$$

In summary, the revenue maximization problem of S_j can be expressed as:

$$\min_{q_j} \quad U_j = U_j^1 + U_j^2$$

$$\text{S.t} \quad q_j \leq Q_j^{max} - Q_j^{ex}, \qquad \forall j \in M$$

$$\frac{q_j}{f_j} \leq min(T_{LET}, T_i^{max}), \quad \forall i \in N, j \in M \qquad (15)$$

The computing resources provided by each assisting vehicle cannot exceed its own available resources. The latency for processing computing tasks in the assisting vehicle cannot exceed the link expiration time with the requesting vehicle and the requesting vehicle task processing latency threshold.

3.3 Game Equilibrium Analysis

In this subsection, the inverse derivation method is used to solve the above problems. The optimal workload of the seller under different unit resource bidding is analyzed. Finding the first and second partial derivatives of U_j with respect to q_j can be obtained:

$$\frac{\partial U_j}{\partial q_j} = \frac{\theta_j p_{i,j}}{1 + p_{i,j} q_j} \qquad (16)$$

$$\frac{\partial^2 U_j}{\partial q_j^2} = -\frac{\theta_j p_{i,j}^2}{(1 + p_{i,j} q_j)^2} \qquad (17)$$

The revenue function of U_j is convex and has a maximum value. Using $\frac{\partial U_j}{\partial q_j} = 0$, we can get:

$$q_{j^{k_l}} = \begin{cases} \frac{\theta_j}{c_j} - \frac{1}{p_{i,j}} & p_{i,j} \geq \frac{c_j}{\theta_j} \\ 0 & p_{i,j} < \frac{c_j}{\theta_j} \end{cases} \qquad (18)$$

As the buyer, each requesting vehicle needs to find an optimal quotation to minimize the cost of executing the computing task. When the optimal unit resource bid is determined, the buyer have the minimum cost to execute the task, so the strategy will not be changed. At the same time, each seller have the maximum revenue. At this time, both the buyers and the sellers are in a state of maximizing benefits, and neither side has the motivation and demand to change the strategy again, and the game finally reaches equilibrium.

3.4 Algorithm to Reach Stackelberg Equilibrium

In this subsection, we propose a distributed algorithm for the buyers to select their optimal offloading strategies. For reasonable consideration, we assume that the unit resource bid of each buyer has a range $p_{i,j}^{min} \leq p_{i,j} \leq p_{ci}, i \in N, j \in M$, and the iteration step is $\triangle \delta = 0.01$. The details of the proposed distributed algorithm are illustrated in Algorithm 1:

Algorithm 1. Distributed algorithm for obtaining optimal offloading strategy

---

Input: D_i^{in}, h_i, T_i^{res}, T_i^{max}, f_i, f_j, R_i, R_j, p_{base}, χ_j, η, c_j , R, $\triangle\delta$, $Q_{jk_l}^{max}$, $Q_{jk_l}^{ex}$

Output: $p_{i,j}^*$, $Q_{jk_l}^*$, U_j^*

1: Compute the demand response of the vehicles, set $p_{i,j} = p_{i,j}^{min}$, $\triangle\delta = 0.01$;
2: **for** $i = 1 : M$ **do**
3:     **for** $j = 1 : N$ **do**
4:         Send request packet to S_{jk_l} within the coverage;
5:         compute q_{jk_l} using formula (18) and the transmission time T_{jk_l} of forwarding request message.;
6:         **if** $q_{jk_l} \geq Q_{jk_l}^{max} - Q_{jk_l}^{ex}$ **then**
7:             $q_{jk_l} = Q_{jk_l}^{max} - Q_{jk_l}^{ex}$
8:         **end if**
9:         **if** $T_{jk_l} \geq T_i^{res}$ **then**
10:           $F_{jk_l} = 0$
11:        **else**
12:           $T_{jk_l}^* = T_i^{res} - T_{jk_l}$ back to steps 5 to calculate the workload q_{jk_l} that other vehicles can undertake, until $T_{jk_l}^* \leq 0$
13:        **end if**
14:     **end for**
15: **end for**
16: Calculate $Q_{jk_l}^*$ and U_i^*, update $p_{i,j}^* = p_{i,j}$ then back to steps 5
17: **if** $U_i^* < U_i$ and $p_{i,j}^* < p_{ci}$ **then**
18:     back to steps 16
19: **else**
20:     $P_{i,j}^* = p_{i,j}, q_{jk_l}^* = q_{jk_l}, U_i^* = U_i$
21:     break
22: **end if**

---

4   Simulation Results and Discussions

In this section, we evaluate the performance of the proposed algorithm by simulations. For simulation, a unidirectional 2-lane highway segment is considered. The length of the segment is 1000 m, the vehicle flow rate is 0.25 (vehicles/sec). The idle computing resources of the assisting vehicles are random numbers that meet the normal distribution. The specific simulation parameters are shown in Table 1.

Figure 3 shows the relationship between the maximum transmission hops of V2V link and the input data-size and the load rate of network (the existing workload of all vehicles on this highway segment accounts for the maximum bearable workload). In this simulation, Fig. 3(a) shows the transmission hops of the request link are not consistent although the data-size of each requesting vehicle is the same, what caused this is the randomness distribution of vehicles. Another observation shows that the size of the input data increases leads to an increase in transmission hops, this is because the idle computing resources of assisting vehicles are limited, in order to satisfy the task offloading requirement and obtain more benefits, the assisting vehicles will forward the request packet

Table 1. Simulation parameter settings

Parameter	Value
The size of input data, d_i	$100 \sim 600$ KB
Vehicle communication radius, R	200 m
The mapping between data size and CPU cycles, h_i	$1.8 * 10^4$ cycles/B
Data transmission rate of vehicle, $R_i = R_j$	$5 \sim 6$ MHz
The velocity of B_i and S_j, respectively, V_i, V_j	$80 \sim 100$ Km/h
Price growth coefficient, η	2
Maximum workload of the vehicles, Q_j^{max}	$N(7.0, 1.0)$ Gcycles
Existing workload of vehicles, Q_j^{ex}	$N(4.0, 2.0)$ Gcycles
Cost of executing unit load c_j	0.2
Willingness constant, α	$0.1 \sim 1$
The computational capability of B_i, f_i	$N(2.5, 0.2)$ GHz
The computational capability of S_i, f_j	$N(3.0, 0.2)$ GHz

spontaneously. Figure 3 (b) shows the relationship between the transmission hops and the load rate of network. We assume the input data size of the three requesting vehicles is 500 KB and change the load rate of network. It can be seen that the transmission hop increases with the load rate. This is because the higher the load rate of network, the fewer computing resources available to the assisting vehicles, and the greater the probability that assisting vehicles forward request packet, thus increasing the transmission hops of the request link. Since each vehicle is independent, the intersection in the picture is meaningless.

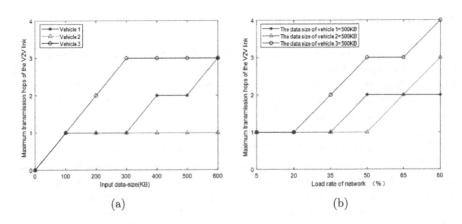

(a)   (b)

Fig. 3. The transmission hops of V2V link

We evaluate our proposed computation offloading with multi-hop execution (MHCO) strategy with two conventional computing offloading strategies. One

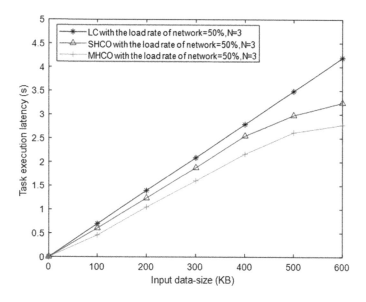

Fig. 4. The task executing latency.

is the task is completely calculated locally (LC), and the other is computation offloading with 1-hop execution (SHCO). Figure 4 presents the task execution latency versus the size of the input data. In this figure, the latency of LC is the largest. Since the computing capability of three requesting vehicle is fixed, they are 2.57 GHz, 2.72 GHz, 2.45 GHZ, the task executing latency using the LC increases linearly. It can be obtained from the simulation data, among the three computing offloading strategies, the task executing latency of SHCO and MHCO is reduced by 15% and 27% respectively compared with LC.

Figure 5 shows the effect of input data-size and the number of requesting vehicles on the cost of requesting vehicles to execute task. We assume that the maximum latency allowed for the three requesting vehicles to accomplish the task is 3.28 s, 2.63 s, 3.76 s. Figure 5(a) shows that when the latency for executing the task is greater than the maximum latency allowed to accomplish the task, the cost increases exponentially. With the increase of data-size, the requesting vehicle needs to increase the bid of unit resources to encourage assisting vehicles to provide computing resources. Compared with LC and SHCO, the MHCO offloading strategy proposed in this paper can forward request packet spontaneously, which can effectively reduce the cost of the requesting vehicles. Set the data-size of the requesting vehicles is 200 KB, 400 KB and 600 KB respectively. In Fig. 5(b), we can observe that the task executing cost is positively correlated with the number of requesting vehicles. This is because when the number of requesting vehicles increases, there is competition between them. The requesting vehicle needs to increase the bid of unit resources to meet the offloading demand.

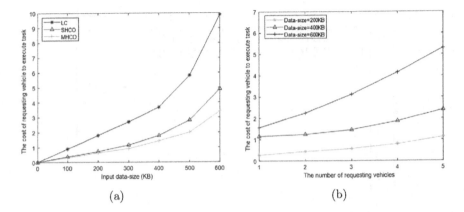

Fig. 5. The cost of requesting vehicles to execute task

In Fig. 6, set the utilization rate of computing resources in the network is 50% and the number of requesting vehicles is three. If the task is completely executed locally, the utilization ratio of computing resources in the network will not change when the computing resources required to execute the task exceed the local remaining resources. Besides, the figure shows that the utilization rate of computing resources in the network using MHCO computing offloading strategy is greater than that of SHCO.

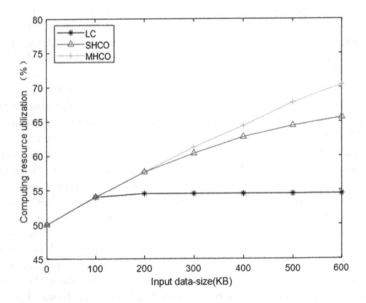

Fig. 6. The computing resource utilization.

5   Conclusion

In this paper, we proposed a computation offloading strategy based on Stackelberg game in vehicular network. The problem of computing task offloading of requesting vehicles was described as a pricing and quantitative problem, and the cost of requesting vehicles to execute task was minimized by game theory. Based on the framework, in order to improve the utilization efficiency of the idle computing resources in the vehicular network, we introduced multi-hop offloading into V2V offloading, and modeled the probability of forwarding request packets by assisting vehicles from multiple perspectives to ensure the stability and reliability of communication transmission. Then, we designed a simple and feasible distributed iterative algorithm to solve the final Stackelberg equilibrium state. Finally, the simulation results showed the effectiveness of our proposed computation offloading strategy can reduce the cost of requesting vehicles to execute tasks, and improve the utilization of computing resources in vehicular network significantly.

References

1. Jia, D., Lu, K., et al.: A survey on platoon-based vehicular cyber-physical systems. IEEE Commun. Surv. **18**(1), 263–284 (2016)
2. Popli, S., Jha, R.K., Jain, S.: A survey on energy efficient narrowband internet of things (NBIoT): architecture, application and challenges. IEEE Access **7**, 16739–16776 (2019)
3. Kaiwartya, O., et al.: Internet of vehicles: motivation, layered architecture, network model, challenges, and future aspects. IEEE Access **4**, 5356–5373 (2016)
4. Mach, P., Becvar, Z.: Mobile edge computing: a survey on architecture and computation offloading. IEEE Commun. Surv. **19**(3), 1628–1656 (2017)
5. Iyer, V., Kumari, R., et al.: Deterministic approach for performance efficiency in vehicular cloud computing. In: 2015 2nd International Conference on Computing for Sustainable Global Development (INDIACom), New Delhi, pp. 78–82 (2015)
6. Zhou, J., Wu, F., Zhang, K., Mao, Y., Leng, S.: Joint optimization of offloading and resource allocation in vehicular networks with mobile edge computing. In: 2018 10th International Conference on Wireless Communications and Signal Processing (WCSP), Hangzhou, pp. 1–6 (2018)
7. Wang, H., Li, X., Ji, H., Zhang, H.: Federated offloading scheme to minimize latency in MEC-enabled vehicular networks. In: IEEE Globecom Workshops (GC Wkshps), Abu Dhabi, United Arab Emirates 2018, pp. 1–6 (2018)
8. Wang, J., Feng, D., Zhang, S.: Computation offloading for mobile edge computing enabled vehicular networks. IEEE Access **7**, 62624–62632 (2019)
9. Liwang, M., Dai, S., Gao, Z., Tang, Y., Dai, H.: A truthful reverse-auction mechanism for computation offloading in cloud-enabled vehicular network. IEEE Internet Things J. **6**(3), 4214–4227 (2019)
10. Zhou, Z., Liu, P., Feng, J., Zhang, Y., et al.: Computation resource allocation and task assignment optimization in vehicular fog computing: a contract-matching approach. IEEE Trans. Veh. Technol. **68**(4), 3113–3125 (2019)
11. Muñoz, O., Pascual-Iserte, A., Vidal, J.: Optimization of radio and computational resources for energy efficiency in latency-constrained application offloading. IEEE Trans. Veh. Technol. **64**(10), 4738–4755 (2015)

12. Wang, Y., et al.: Energy-optimal partial computation offloading using dynamic voltage scaling. In: 2015 IEEE International Conference on Communication Workshop (ICCW), London, pp. 2695–2700 (2015)
13. Sharshembiev, K., Yoo, S., Elmahdi, E.: Broadcast storm mitigation from unintentional misbehavior in vehicular ad hoc networks. In: 2018 IEEE 8th Annual Computing and Communication Workshop and Conference (CCWC), Las Vegas, NV, pp. 925–930 (2018)
14. Cheng, L., Henty, B.E., Stancil, D.D., Bai, F., Mudalige, P.: Mobile vehicle-to-vehicle narrow-band channel measurement and characterization of the 5.9 GHz dedicated short range communication (DSRC) frequency band. IEEE J. Sel. Areas Commun. 25(8), 1501–1516 (2007)
15. Wei-Dong, Y., Ji-Zhao, L., Hong-Song, Z.: A mobility model based on traffic flow theory for vehicular delay tolerant network. In: 2010 IEEE Fifth International Conference on Bio-Inspired Computing: Theories and Applications (BIC-TA), Changsha, pp. 1071–1075 (2010)

Context-Bound Cybersecurity Framework for Resisting Eavesdropping in Vehicle Networks

Longjiang Li[1]([envelope])[ID], Bingchuan Ma[1], Yonggang Li[2], and Yuming Mao[1]

[1] Department of Network Engineering, SICE,
University of Electronic Science and Technology of China,
Chengdu 611731, China
{longjiangli,ymmao}@uestc.edu.cn, 806166563@qq.com
[2] Chongqing University of Posts and Telecommunications,
Chongqing 400065, China
lyg@cqupt.edu.cn

Abstract. Wireless channels that are widely adopted between autonomous vehicles are vulnerable to eavesdropping or interferences, so that attacks on cybersecurity may lead to serious consequences, such as losing control of vehicles. In particular, the cryptographic methods used for information security rely on the strict privacy of keys, which is often difficult to guarantee in a wireless environment. This paper proposes a context-bound cybersecurity framework, which protects communication from eavesdroppers by encrypting critical data with a dynamic context among vehicles. The context is synchronized among vehicles through a progressive encoding method, which makes it difficult for third parties to learn the entire context by eavesdropping through the channels, especially in the case of mobility. The normal vehicles may extract a security key from the context to encrypt and decrypt key data, but it is impossible or overwhelmingly expensive for the third parties to decode the data transmitted due to the lack of the context. Besides, the proposed framework also provides a promising way to resist the upcoming quantum computers, because it will become more and more difficult for third parties to collect the complete context as the context continues to update.

Keywords: Cybersecurity · Context · Vehicle networks · Cryptography

1 Introduction

With the development of 5G and Internet of Things(IoT), the boundaries of communication are increasingly blurred, making data-secure sharing face unprecedented challenges, including message cracking [35], data leakage, data tampering,

---

Funded by the National Natural Science Foundation of China (61273235), and the Defence Advance Research Foundation of China (61400020109).

H. Gao et al. (Eds.): ChinaCom 2020, LNICST 352, pp. 519–536, 2021.
https://doi.org/10.1007/978-3-030-67720-6_36

integrity sabotage, unauthorized access, etc. [16,25,29]. As one of the important application scenarios in IoT, autonomous driving has intensely high requirements for data security, and consequently faces the same dilemma. Especially, autonomous vehicles take wireless signal as the main access carrier, which makes the data transmission process face various threats of attack, such as channel monitoring, information content tampering, counterfeiting, man-in-the-middle forwarding, blocking and replaying [3,11,25,32].

Researches on secure communication can be divided into two categories: physical layer security and network layer cryptography. To approach the secrecy capacity, the former requires a wiretap code whose length is infinite [21], which limits its practical application. In contrast, the latter has no such restriction. Related technologies [32] mainly protect data security by integrating various security cryptography algorithms into data distribution, routing and storage procedures. According to the implementation method, the secure cryptography algorithm can be divided into public key cryptography, symmetric cryptography, anti-quantum cryptography, chaotic cryptography, quantum cryptography, lattice-based cryptography, etc. [8,10,27,30,33]. However, with the development of quantum computers [4], traditional security cryptography methods, such as RSA, are in danger of being cracked. Although post-quantum cryptography [14] has the potential to resist quantum computer attacks, it requires a lot of computational overhead [8,14], which is arduous to apply in a vehicle network with limited computing power [20]. Besides, from a mathematical point of view, most of the existing mainstream security cryptography methods are essentially based on computational complexity problems, and their effectiveness depends only on the privacy of the key [28]. Once the key is cracked or compromised, the communication is no longer secure. Moreover, there are great key exposure risks in the process of key generation, distribution and use due to the emergence of side-channel attacks [7,18,19], cold boot attacks [15], etc. Especially, in the scenario of virtual car platoons, many cars and devices are connected to the network, so that the distribution and management of keys are extremely challenging [4,9,12].

In order to solve those security problems, we propose a context-bound cybersecurity framework (CBCF) for vehicle networking scenarios, which is inspired by the application of blockchain technology that chains records for immutable transactions [17,24]. The main idea to bind the coding process to a cybersecurity context that is a chained structure of private information shared between the source and the sink. The mobility of vehicles brings huge dynamics to the communication process, making it virtually impossible or extremely expensive for a third party to completely grasp the communication context, so that the possibility of key exposure is greatly reduced. On the contrary, through progressive coding, the members of a vehicle network can synchronize the context in time, so as to use the correct key to encrypt or decrypt messages. In essence, the framework utilizes the difficulty of the third party in grasping the whole context to overcome the key exposure problem, and further provide support for cybersecurity in the vehicle network.

Our main contributions are summarized as follows:

– We propose a context-bound cybersecurity framework that can be extended in multiple ways to solve the key exposure problem and protect the data security in vehicle networks by dynamically binding the cybersecurity context.
– Two typical implementations of the proposed framework are given as iterative functions. Besides, we demonstrate how to use the progressive coding method to synchronize the context between source and sink vehicles, and a theoretical analysis for the performance of these two implementations is also provided.
– We conducted experiments to illustrate the ciphering and deciphering process of our framework in vehicle networks. The results show that the ciphertext has good randomness and statistical correlation characteristics with the plaintext.

The rest of the paper is structured as follows. Section 2 presents the system model and introduces the basic idea. In Sect. 3, two typical implementations of context-bound cybersecurity framework are given, whose cybersecurity and performance are also analyzed. Section 4 simulates the encrypted communication process in vehicle networks, and evaluates the statistical characteristics of plaintext and ciphertext. Section 5 discusses the framework's resistance to attacks, scalability, and it's relationship with blockchain. Finally, Sect. 6 concludes the paper.

2 Proposed Context-Bound Cybersecurity Framework

In this section, we first describe the system model and then propose the context-bound cybersecurity framework for vehicle networks.

2.1 Typical Vehicle Networking Scenarios

We consider the problem of encrypted communication in the cross roads scenario, as shown in Fig. 1. This model consists of three major components: Road Side Unit(RSU), vehicles equipped with On Board Unit(OBU) and eavesdroppers.

– RSU: When two of the vehicles cannot communicate directly due to problems such as distance or channel quality, the RSU will serve as a relay node to help them maintain the continuity of the communication process. In order to provide promising services and avoid channel congestion, RSUs are usually deployed at intersections or on roads with heavy traffic [2].
– Vehicles: Vehicles can be regarded as high-speed mobile nodes equipped with OBUs, which makes them capable of communicating with other vehicles [36]. In a vehicle network, every vehicle will establish context-bound security channels with each of the others to maintain the synchronization of information inside the network, and resist third-party eavesdropping attacks at the meanwhile.
– Eavesdroppers: Fig. 1 demonstrates two types of eavesdropper, mobile vehicle eavesdropper E_v, and roadside fixed eavesdropper E_p. The former can eavesdrop on the communication messages of other vehicles during the movement, while the latter can only intercept an exceedingly small part of the messages briefly beside the road.

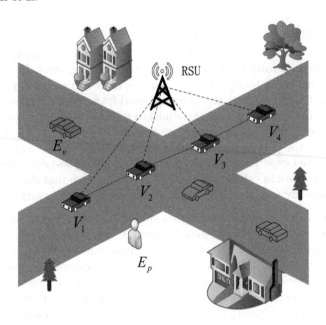

Fig. 1. Encrypted communication in vehicle networking scenarios

2.2   The Main Idea

As shown in Fig. 1, each communication process consists of a source vehicle, a
dynamic wireless channel, and a sink vehicle, in which the source sends a series
of messages to the sink through the channel. The typical solution is to encrypt
the transmitted message by employing a symmetric encryption method, such
as AES, with a series of privacy keys that may be distributed by public key
cryptography, such as RSA, or physical layer key generations. However, keys
may be leaked to eavesdroppers due to various attacks or algorithm limitations.
Once the keys are compromised, any messages encoded with these keys are no
longer secure to eavesdroppers.

In order to overcome these problems, the main idea is to bind the encrypting and decrypting process to a cybersecurity context. As shown in Fig. 2, the
cybersecurity context is constructed by collecting various private random information, such as keys, shared by both parties. For an eavesdropper, it is much
harder to master the dynamic context than to steal a single key, so the security
level of the cryptosystem is greatly improved.

To simplify the presentation, we assume that the channel is error-free, reliable
and orderly. We call the cybersecurity context as Sink Anchor (SA), because the
context represents some kind of information that should only be fully accessible
between the sink and the source. The message sequence is represented as a series
of variables $<x_0, x_1, ..., x_{i-1}, x_i, ...>$, where x_i is called a Sink-Anchoring Coding
unit (abbreviated as a SAC unit), denoted by $x_i \in \Lambda$. Sink Anchor (SA) is
denoted by $\psi \in \Psi$. The encoding process at the source and the decoding process

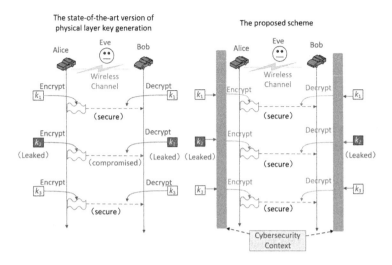

Fig. 2. Comparison between the proposed scheme and the state-of-the-art version of physical layer key generation methods

at the sink can be expressed as a pair of function $<f : \Psi \times \Lambda \to \Lambda, g : \Psi \times \Lambda \to \Lambda>$, where the encoder is

$$f(\psi_i, x_i) \to m_i \tag{1}$$

and decoder is

$$g(\psi_i, m_i) \to x_i \tag{2}$$

in which x_i and m_i are plaintext and ciphertext, respectively.

We declare that an effective CBCF scheme shall possess the following characteristics.

1. The SA, $\psi_i \in \Psi$, is a slice of common information shared by the source and the sink.
2. There is a pair of efficient encoding and decoding functions, $<f(), g()>$, so that $g(\psi_i, f(h_i, x_i)) = x_i$ holds, for any given $x_i \in \Lambda$.
3. Any third party cannot or has to pay an unacceptable price to decode x_i, even if it has the opportunity to intercept quiet a few or all of the ciphertexts.
4. Any third party cannot or has to pay an unacceptable price to know the content of ψ_i.

Note that we assume that $<f(), g()>$ can be publicly known by a third party, but it should be impossible or extremely arduous to know the contents of SA. Feature 1 defines that the essence of the anchor is information. Feature 2 requires the SA be used for efficiently encoding and decoding. Feature 3 emphasizes that the third party should be unable to decipher the message. Feature 4 requires a third party to be unable to crack the SA to support for Feature 3.

In short, the two parties of the communication maintain a context based on some kind of shared information between them to control the coding process, as shown in Fig. 2. Due to the constant movement of vehicles and continuous inter-action between them, the context is always changing, so it's extremely arduous for the eavesdropper to obtain the complete context, consequently, the current message is prevented from being cracked by introducing information integrity problems.

3   Two Typical Implementations of Context-Bound Cybersecurity Framework

Although all kinds of information, such as moving trajectory, topology, recip-rocal channel characteristics and historical communication messages, that are privately shared between the source and the sink can be utilized for construct-ing the SA, we focus on historical communication messages here for brevity.

Here, we propose two typical implementations of context-bound cybersecu-rity framework: AES(Advanced Encryption Standard) enhanced Sink-Anchoring Coding (ASAC) and Pseudo-random generator based Sink-Anchoring Coding (PSAC).

3.1   Iterative Function

For x_i, we define its SA ψ_i as the result of an iterative function $\psi : \Psi \times \Lambda \to \Lambda$ based on some historical messages. Equation (3) and Eq. (4) below represent the iterative formulas of ASAC mode and PSAC mode, respectively:

- ASAC mode:

$$\psi_i = \psi(\psi_{i-1}, x_{i-1})$$
$$= \begin{cases} AES_{key}(\psi_{i-1} \oplus x_{i-1}) & R_{key}(*) \geq \overline{R} \\ \psi_{i-1} & R_{key}(*) < \overline{R} \end{cases} \tag{3}$$

- PSAC mode:

$$\psi_i = \psi(\psi_{i-1}, x_{i-1})$$
$$= \begin{cases} (\psi_{i-1} + x_{i-1} \oplus (R_{seed}(*) \times 2^z)) \ MOD \ 2^z & R_{seed}(*) \geq \overline{R} \\ \psi_{i-1} & R_{seed}(*) < \overline{R} \end{cases} \tag{4}$$

where $R_\alpha(*)$ is a pseudo-random sequence generator function, defined as $R_\alpha : \Lambda \to [0, 1)$, and it is assumed that the generator function implementation can be public, with only the random number seed to be private to the eavesdropper. For the sake of uniformity, we refer to α as the initial parameter of the SA. $\overline{R}$ represents the average sampling rate and can be set to a pseudo-random sequence or constant. As an example, $\overline{R} = 0.1$ indicates that about 90% of the historical messages (historical SAC units) are used in the calculation of ψ_i.

The ASAC mode is implemented based on the one-way function $AES_{key}()$, which is the encoding function of AES. $AES_{key}(x)$ means using AES to encrypt x based on key (Note: only encryption operation is needed, no decryption opera-tion), and $R_{key}(*)$ indicates that the AES key is used as the random number seed of pseudo-random sequence generator function (if the length does not match, fill or truncate it). In addition, the PSAC mode is based on the pseudorandom sequence generator function, while MOD is a modulo operation to ensure the length of ψ_i is equal to that of SAC unit, and the role of $x_{i-1} \oplus (R_{seed}(*) \times 2^z)$ is to randomly flip x_{i-1} by using $R_{seed}(*)$.

3.2   Communicating Process

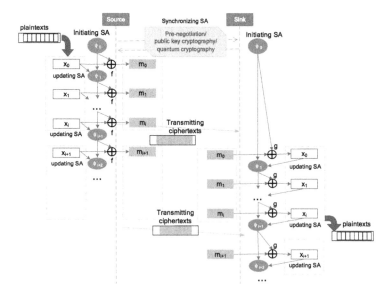

Fig. 3. Sequence diagram of context-bound cybersecurity framework

Figure 3 gives the sequence diagram of both ASAC mode and PSAC mode. At first, a pair of source and sink synchronizes the initial SA and such a synchronization process should be protected by a key exchange mechanism, such as public key cryptography or quantum cryptography, so that the third party cannot know the SA. After starting the communication, the source and sink update the SA according to the same iterative function (i.e., ASAC or PSAC). Since the SA is privately shared between the source and the sink, $f()$ and $g()$ can utilize the SA to construct a key for encryption and decryption of messages, respectively. Specifically, the source encrypt SAC unit one by one with the constantly updated SA through encryption algorithm $f()$, while the sink utilizes the corresponding SA through decryption algorithm $g()$ to decrypt received ciphertexts. For simplicity, we adopt XOR operation for implementing $f()$ and $g()$ () as follows:

$$f(\psi_i, x_i) = \psi_i \oplus x_i \rightarrow m_i \tag{5}$$

$$g(\psi_i, m_i) = \psi_i \oplus m_i \rightarrow x_i \tag{6}$$

where $\oplus$ is a XOR operator, satisfying $0 \oplus 0 = 0;\ 0 \oplus 1 = 1;\ 1 \oplus 0 = 1;\ 1 \oplus 1 = 0$, i.e., same as 0, different as 1.

A simple way to understand cybersecurity is to think of the key cracking process as solving equations. For a third party who is assumed to have eavesdropped k ciphertext units, $<m_{i-k}, m_{m-k+1}, ..., m_{i-1}>$, a brute force deciphering is equivalent to solving the following system of equations.

$$\begin{cases} \psi_{i-k+1} \oplus x_{i-k+1} = m_{i-k+1} \\ \psi(\psi_{i-k+1}, x_{i-k+1}) \oplus x_{i-k+2} = m_{i-k+2} \\ \quad\vdots \\ \psi(\psi...(\psi_{i-k+1}, x_{i-k+1})...), x_{i-1}) \oplus x_i = m_i \end{cases} \tag{7}$$

Note that the equation system (7) consists of k equations, but involves a total of $k + 1$ unknown variables, i.e. $\{\psi_{i-k+1}, x_{i-k+1}, x_{i-k+2}, ..., x_i\}$, so it cannot be solved directly in general, unless there is a special correlation among these equations that can be explored. When k is equal to $i+1$, it is corresponding to the case that the third party has eavesdropped all historical ciphertexts. Under the assumption that the initial SA is unknown, it is also impossible to directly solve the equation system, as the number of unknown variables is still more than the number of equations. For more rigorous theoretical analysis on the cybersecurity, please refer to Sect. 3.4.

3.3   The Necessity of Pseudo-random Generator Function

If a true random number generator is used, the same seed cannot lead to the same sequence, then the source and the sink will not be able to construct feasible encoding and decoding functions, which violates the feature 2 of CBCF scheme.

From Eq. (3) and Eq. (4), it can be seen that the role of pseudo-random generator function in PSAC mode is mainly in three aspects: initializing SA, selecting historical messages, and randomly flipping x_{i-1} by bit. In ASAC mode, there are only two effects of pseudo-random function: initializing SA and selecting historical messages. Using a pseudo-random generator function $R_\alpha(*)$ to select historical messages brings the following advantages:

1. From the perspective of staying security from eavesdroppers. First of all, when the initial parameter α of $R_\alpha(*)$ is unknown to the third party, all historical messages have to be collected since it is impossible to know what historical messages are used for constructing ψ_i, which may trigger difficulties in collection or storage. Secondly, even if the eavesdropper collects all the historical messages, it will not be able to crack the current ciphertext m_i without α, because it cannot know which messages are used to construct ψ_i. Finally, if the eavesdropper does not collect all the messages from the beginning, it will fail to crack the ciphertext because it cannot master all the messages satisfying $R_\alpha(*) \geq \overline{R}$, even if the eavesdropper learns α by cracking $R_\alpha(*)$ after the communication has started for a period of time.

2. From the perspective of the performance of encryption and decryption at the source and the sink. The source and the sink invariably generate the same pseudo-random number sequence, and only need to update the SA for messages satisfying $R_\alpha(*) \geq \overline{R}$. When $\overline{R}$ approaches 0, the sink anchor will be updated for almost every message, thus achieving a similar "one-time pad" effect [13]. When $\overline{R} > 0$, the calculation frequency will be reduced and thus the computing resources are saved.

3.4   Analysis of Cybersecurity

In order to simplify the analysis, we only consider the situation of passive eavesdropping, i.e., eavesdroppers will not interfere with the transmission of messages, so that no message-authentication code (MAC) is needed.

Min-entropy is widely used in modern cryptography for evaluating the guessing probability of a key in the worst case, so it provides another perspective for understanding the cybersecrity. For a random variable X, its min-entropy of is defined as [1]

$$H_\infty(X) \overset{def}{=} -\log_2(\max_{x \leftarrow X} Pr[X = x]) \tag{8}$$

where $x \leftarrow X$ means the operation of sampling a random x according to X. Given two independent random variables, X and Y, it can be proven that

$$H_\infty(X \oplus Y) \geq max(H_\infty(X), H_\infty(Y)) \tag{9}$$

and

$$H_\infty((X + Y) \ MOD \ 2^z) \geq$$
$$max(H_\infty(X \ MOD \ 2^z), H_\infty(Y \ MOD \ 2^z)) \tag{10}$$

where $max()$ is a function that returns the larger value from the two inputs. For simplicity, we assume that the one way function $AES_{key}()$ is entropy-preserving in an ideal case, i.e., $H_\infty(AES_{key}(X)) \equiv H_\infty(X)$. Then, for ASAC mode, we have,

$$H_\infty(\psi_i) \geq max(H_\infty(\psi_{i-1}), H_\infty(x_{i-1}), H_\infty(key)) \tag{11}$$

Thus, during the update process of the SA, the min-entropy is always increasing until its value reaches the absolute maximum limit for z-bit keys. That is to say, the update process of the SA makes it more difficult for eavesdroppers to guess the SA. Therefore, by taking the SA as keys for encryption and decryption, the cybersecurity is indeed enhanced by the ASAC mode.

Likewise, we can draw the same conclusion for the PSAC mode.

3.5   Theoretical Analysis of Performance

In the PSAC mode, each update process of SA includes at most one addition, XOR, multiplication, and modulo operation, so the computational time complexity is O(1). In the ASAC mode, each update process of SA only adds an

XOR operation to the standard AES encryption operation, therefore its computational time complexity can still be considered as O(1). In terms of space complexity, both the source and sink need to maintain a record of SA, and determine whether to update it every time a message is sent or received. The input of computation process is ψ_{i-1} and x_{i-1} while the output is ψ_i. Since there is no need to save earlier historical messages, the storage space overhead is O(1). After encoding, the ciphertext and the plaintext are of equal length, so the additional communication overhead of the protocol is O(0). In contrast, for the eavesdropper, the theoretical computational time complexity of cracking plaintext from the ciphertext is infinite if the initial parameter α is unknown, because the equation system (7) cannot be solved directly unless all historical messages are intercepted and stored.

4   Experiments and Analyses

In order to evaluate the proposed framework, we implemented the scenario, as shown in Fig. 1, and then analyzed the statistical characteristics of plaintext and ciphertext.

4.1   Ciphering and Deciphering in Vehicle Networks

We simulated a vehicle network based on multicast and TCP connections for internal interaction, where each vehicle establishes communication with other vehicles centered on itself. Here, we focus on the communication process of one vehicle V_s in a vehicle network consisting of n vehicles, and we use $V_i (i = 1, ..., s-1, s+1, ..., n)$ to represent the remaining vehicles, as shown in Fig. 4.

The vehicles firstly synchronize the initial SA parameter α by using public key encryption, then establish context-bound secure channels with initialized SA. The whole process can be divided into four steps:

- Step A: V_s generates a pair of RSA keys locally and then sends the public key to the remaining vehicles as a multicast source. After receiving the public key, vehicle V_i will establish a TCP channel with V_s.
- Step B: V_i generates an initial SA parameter α locally, which will be encrypted with the received RSA public key and then synchronized to V_s latter. After V_s decrypting the encrypted parameter α with local RSA private key, both parties will initialize SA according to α.
- Step C: After SA is initialized, V_i will encrypt messages with SA and send them to V_s through the TCP connection, while V_s decrypts using the synchronized SA.
- Step D: Each message transmission will lead to the update of SA according to the iteration function, and both parties will iteratively update SA with the same algorithm (ASAC is used as an example here) in the subsequent communication process.

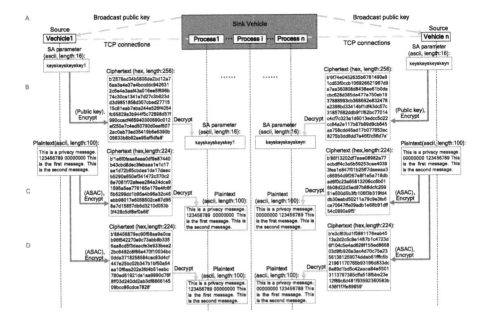

Fig. 4. An example of ciphering and deciphering process (z = 128)

For the iteration and communication with different vehicles, V_s will handle them in a multi-threaded manner. Besides, as it can be seen in Fig. 4, even if the same message is sent continuously, the ciphertext obtained is completely different due to the update of SA, which is benificial for the security of user data.

4.2 Distribution and Randomness of Ciphertext

In order to study the statistical characteristics of the encrypted messages under the proposed cybersecurity framework, we use linear sequence, constant sequence and random sequence as plaintext, which are encrypted in ASAC and PSAC modes, respectively, so as to explore the distribution of the ciphertext. For comparison, we uniformly set the random number seed to 1, the binary length of the coding unit to 128, the average sampling rate $\overline{R}$ to 0.1, and each sequence contains 1000 coding units.

The results are shown in Fig. 5, where Fig. 5(a) represents the distribution of plaintext, Fig. 5(b) represents the distribution of ciphertext in ASAC mode, and Fig. 5(c) represents the distribution of ciphertext in PSAC mode. It can be seen that regardless of the type and serial number of the plaintext, the corresponding ciphertext is evenly distributed in the range of $[0, 2^z - 1]$ whether in ASAC or PSAC mode, reflecting outstanding statistical distribution characteristics. Even at the worst case, assuming that the third party has the opportunity to know the initial parameter α of SA, the only case it can crack the ciphertext is that

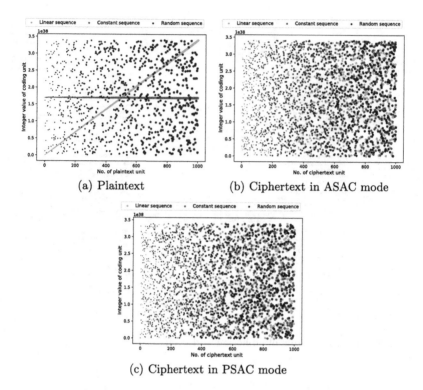

(a) Plaintext   (b) Ciphertext in ASAC mode

(c) Ciphertext in PSAC mode

Fig. 5. Distribution of plaintext and ciphertext

it collects and stores all the ciphertexts before getting α, which is usually not possible for third parties with relatively limited capabilities.

Further, the randomness of ciphertext is also tested. Since the ciphertext is used to ensure the security of information, and any non-random features will reduce the difficulty for third parties to crack the message, the randomness of ciphertext is crucial [22]. For testing, we use the linear sequence generated above as plaintext (1000 coding units), then encrypt it in ASAC mode and PSAC mode, respectively. For the obtained ciphertext, we use a few different algorithms from the NIST public-domain test suite to calculate the $p-value$ of it [23,31], which is an index used to evaluate the randomness of bit sequence, and the sequence is considered to be random when $p-value > 0.01$ (For a detailed representation of the tests and definitions of $p-value$, interested readers can refer to [31]). The results are shown in Table 1, in which we can see that the ciphertext we obtained has a good performance in terms of randomness, which further ensures the effectiveness of our framework.

Table 1. Results from randomness tests on ciphertext

Test	P-value (ASAC)	P-value (PSAC)
DFT test	0.7953	0.3316
Longest run	0.2724	0.7328
Maurer's test	0.3722	0.9793
Non-overlapping	1.0000	1.0000

4.3 Correlation Between Plaintext and Ciphertext

To study the correlation between plaintext and ciphertext, we use random sequence, linearly increasing sequence and linearly decreasing sequence as plaintext respectively to get ciphertext, and then calculate the corresponding Pearson correlation coefficient [5]. The results are shown in Fig. 6.

We use the same seed 1 for pseudo-random generator to produce randomly distributed plaintext sequences, the linear increasing sequence and the linear decreasing sequence. These linear sequences take values uniformly on $[0, 2^z - 1]$, and the range of the coding unit length z is varying in $[32, 512]$, incremented by 32 each time. Figure 6(a) demonstrates the plaintext distribution of three sequences when $z = 512$, where each sequence contains 1000 coding units. Figure 6(b) demonstrates the correlation between plaintext and ciphertext under different coding unit lengths in ASAC mode, and Fig. 6(c) demonstrates the correlation

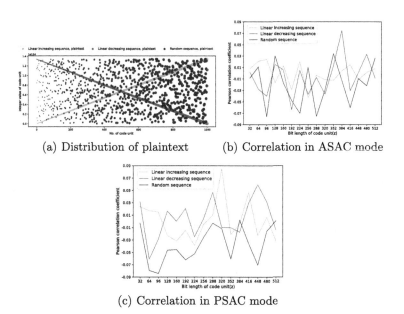

(a) Distribution of plaintext   (b) Correlation in ASAC mode

(c) Correlation in PSAC mode

Fig. 6. Correlation between plaintext and ciphertext

in PSAC mode. The Pearson correlation coefficient in Fig. 6(b) and Fig. 6(c) represent the linear correlation between plaintext and ciphertext, where 1 is indicative of a positive correlation and -1 is indicative of a negative correlation. It can be seen that the correlation coefficients between plaintext and ciphertext of the three different sequences are all within the range of $[-0.09, +0.09]$, which manifests that the encryption result is not sensitive to different values of z, and reflects outstanding statistical correlation characteristics of the proposed framework.

5   Discussions

5.1   Resistance to Attacks

From the perspective of attack games common to cryptographic systems, the quintessential types of attack fall into four categories, including ciphertext-only attacks, known ciphertext attacks, chosen plaintext attacks, and chosen ciphertext attacks [6]. These attacks assume that the ciphertext relies only on the key, consequently, the focus is on attacking the key. However, the encoding and decoding process of our context-bound cybersecurity framework rely on the dynamic context, which means it naturally has the advantage of resisting such attacks. Furthermore, the proposed framework can resist third-party's retrospective attacks on communication fragments. In a retroactive attack, if the third-party cannot crack the message in a timely manner, the interested communication segment can be stored and decrypted when the conditions are met. As an example, the eavesdropper can wait for the future quantum computer to decrypt messages that was captured 10 years ago. Encryption methods such as DES, AES, and RSA all face this type of risk. In contrast, our framework makes the decryption process not only require the initial parameter of SA, but also historical messages, which will force the third-party to have the ability to continuously monitor the channel, collect data and store complete historical messages in addition to powerful computing capabilities, so the cost is extremely great.

5.2   Extensions

As mentioned above, the method of constructing context is not limited to historical messages. As an example, some reciprocal channel characteristics in the vehicle network also provide context information shared between the source and sink, thus can be utilized for our framework. Currently, the fifth generation (5G) telecommunications techniques [34], such as massive multiple-input-multiple-output (MIMO), millimeter wave (mmWave), and non-orthogonal multiple access (NOMA), have been widely explored for physical-layer key generation, but each key is used only once and discarded, resulting in a waste of a large number of precious keys. By weaving these keys into the context, our scheme provides an elegant way to resist key exposure for these methods. One approach is replacing x_{i-1} in Eq. (3) and Eq. (4) with reciprocal channel parameters, then the framework no longer depends on historical messages, but on the channel parameters.

In addition, in Sect. 2.2, we assumed that the channel is error-free, reliable, and not out of order, but the real networking environment may be unreliable or does not need to be absolutely reliable. For instance, audio and video transmission have certain tolerance to channel errors.

To this end, other variations of ASAC and PSAC can be considered, as shown in Fig. 7. We can use Automatic Repeat-reQuest (ARQ) to get an error-free, reliable, and not out of order channel. If ARQ is not available, by replacing x_{i-1} in Eq. (3) and Eq. (4) with fixed values, two weakened modes are obtained: Weak-ASAC mode and Weak-PSAC mode. These weakened modes can be applied to unreliable transmission environment, though some protocol enhancements, e.g., embedding message-authentication codes (MAC), may be needed to handle packet loss and out of order. Therefore, combining with various modes, it is possible to apply our framework to all protocol layers of the vehicle network, including physical layer, link layer, network (IP) layer, transport layer, application layer, etc.

5.3   Relationship with Blockchain

A blockchain is typically an ordered and growing list of blocks that are linked using cryptography. Each block contains a cryptographic hash of the previous block, a timestamp, and transaction data. Essentially, it is a decentralized distributed ledger database where each member independently stores a copy of the blockchain and updates it synchronously [26]. Our framework uses a similar idea of chain, but the context is only shared between legitimate communication peers, i.e., the source and the sink. We take messages and SA as the content of block and carefully designed iterative function as the chain-relationship between blocks to establish a context that is stored in a distributed manner and updated synchronously between the two parties in communication. In natrue, we constructed a dynamic key that changes with the context for message encryption and decryption, so that the risk of key exposure can be overcomed.

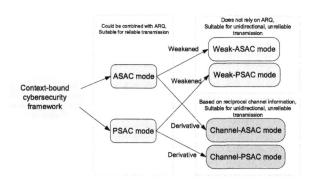

Fig. 7. Six typical modes based on context-bound cybersecurity framework

6 Conclusion

In this paper, we propose a context-bound cybersecurity framework to solve the data security problem in vehicle networking scenarios. By binding the message encryption to the context information in the communication process, it is impossible or extremely hard for third parties to crack the message from the channel or relay. On the basis of the proposed framework, this paper offers two typical implementations in combination with AES and pseudo-random generator: ASAC and PSAC. By experimenting on communication process in a vehicle network consisting of several cars, the simulation results demonstrate that the proposed framework has outstanding statistical correlation and distribution characteristics. The context-bound cybersecurity framework greatly reduces the risk of key exposure by utilizing the mobility of vehicles, thus offers a solution to simplify the technology implementation of mobile user privacy protection and data security sharing. Our approach also provides a promising way to resist the upcoming quantum computers, because it will become more and more difficult for third parties to collect the complete context as the context continues to update.

Besides, we only considered the situation of passive eavesdropping and assumed that the channel is error-free, reliable, and not out of order. The real networking environment may be unreliable or does not need to be absolutely reliable. Moreover, the eavesdropper may actively interfere with the transmission process of the communication by modifying, deleting, inserting, or replaying the message transmission. Thus, some protocol enhancements, e.g., embedding message-authentication codes (MAC), may be needed. These issues need further efforts invested.

References

1. Aggarwal, D., Dodis, Y., Jafargholi, Z., Miles, E., Reyzin, L.: Amplifying privacy in privacy amplification. In: Garay, J.A., Gennaro, R. (eds.) CRYPTO 2014. LNCS, vol. 8617, pp. 183–198. Springer, Heidelberg (2014). https://doi.org/10.1007/978-3-662-44381-1_11
2. Al-Sultan, S., Al-Doori, M.M., Al-Bayatti, A.H., Zedan, H.: A comprehensive survey on vehicular ad hoc network. J. Netw. Comput. Appl. 37, 380–392 (2014)
3. Alaba, F.A., Othman, M., Hashem, I.A.T., Alotaibi, F.: Internet of things security: a survey. J. Netw. Comput. Appl. 88, 10–28 (2017). https://doi.org/10.1016/j.jnca.2017.04.002
4. Arute, F., et al.: Quantum supremacy using a programmable superconducting processor. Nature 574(7779), 505–510 (2019)
5. Benesty, J., Chen, J., Huang, Y., Cohen, I.: Pearson correlation coefficient. In: Cohen, I., Huang, Y., Chen, J., Benesty, J. (eds.) Noise Reduction in Speech Processing, pp. 1–4. Springer, Heidelberg (2009). https://doi.org/10.1007/978-3-642-00296-0_5
6. Biryukov, A., Wagner, D.: Advanced slide attacks. In: Preneel, B. (ed.) EUROCRYPT 2000. LNCS, vol. 1807, pp. 589–606. Springer, Heidelberg (2000). https://doi.org/10.1007/3-540-45539-6_41

7. Brumley, D., Boneh, D.: Remote timing attacks are practical. Comput. Netw. **48**(5), 701–716 (2005). https://doi.org/10.1016/j.comnet.2005.01.010

8. Cao, J., Yu, P., Xiang, X., Ma, M., Li, H.: Anti-quantum fast authentication and data transmission scheme for massive devices in 5G NB-IOT system. IEEE Internet Things J. **6**(6), 9794–9805 (2019)

9. Cao, J., et al.: A survey on security aspects for 3GPP 5G networks. IEEE Commun. Surv. Tutor. **22**(1), 170–195 (2019)

10. Chaudhary, R., Aujla, G.S., Kumar, N., Zeadally, S.: Lattice-based public key cryptosystem for internet of things environment: challenges and solutions. IEEE Internet Things J. **6**(3), 4897–4909 (2019). https://doi.org/10.1109/JIOT.2018. 2878707

11. Conti, M., Dragoni, N., Lesyk, V.: A survey of man in the middle attacks. IEEE Commun. Surv. Tutor. **18**(3), 2027–2051 (2016). https://doi.org/10.1109/COMST. 2016.2548426

12. Cui, C., et al.: Twin-field quantum key distribution without phase postselection. Phys. Rev. Appl. **11**(3), 034053 (2019)

13. Deng, F.G., Long, G.L.: Secure direct communication with a quantum one-time pad. Phys. Rev. A **69**(5), 052319 (2004)

14. Ebrahimi, S., Bayat-Sarmadi, S., Mosanaei-Boorani, H.: Post-quantum cryptoprocessors optimized for edge and resource-constrained devices in IoT. IEEE Internet Things J. **6**(3), 5500–5507 (2019)

15. Guan, L., Lin, J., Ma, Z., Luo, B., Xia, L., Jing, J.: Copker: a cryptographic engine against cold-boot attacks. IEEE Trans. Dependable Secure Comput. **15**(5), 742–754 (2016)

16. Handler, I.: Data sharing defined - really! IEEE Comput. **51**(2), 36–42 (2018). https://doi.org/10.1109/MC.2018.1451659

17. Huh, S., Cho, S., Kim, S.: Managing IoT devices using blockchain platform. In: 2017 19th International Conference on Advanced Communication Technology (ICACT), pp. 464–467. IEEE (2017)

18. Kocher, P., Jaffe, J., Jun, B.: Differential power analysis. In: Wiener, M. (ed.) CRYPTO 1999. LNCS, vol. 1666, pp. 388–397. Springer, Heidelberg (1999). https://doi.org/10.1007/3-540-48405-1_25

19. Kocher, P.C.: Timing attacks on implementations of Diffie-Hellman, RSA, DSS, and other systems. In: Koblitz, N. (ed.) CRYPTO 1996. LNCS, vol. 1109, pp. 104–113. Springer, Heidelberg (1996). https://doi.org/10.1007/3-540-68697-5_9

20. Liu, Y., et al.: Experimental twin-field quantum key distribution through sending or not sending. Phys. Rev. Lett. **123**(10), 100505 (2019)

21. Luzzi, L., Vehkalahti, R., Ling, C.: Almost universal codes for mimo wiretap channels. IEEE Trans. Inf. Theory **64**(11), 7218–7241 (2018). https://doi.org/10.1109/ TIT.2018.2857487

22. Mathur, S., Trappe, W., Mandayam, N., Ye, C., Reznik, A.: Radio-telepathy: extracting a secret key from an unauthenticated wireless channel. In: Proceedings of the 14th ACM International Conference on Mobile Computing and Networking, pp. 128–139 (2008)

23. Maurer, U.M.: A universal statistical test for random bit generators. J. Cryptol. **5**(2), 89–105 (1992). https://doi.org/10.1007/BF00193563

24. Meng, W., Li, W., Yang, L.T., Li, P.: Enhancing challenge-based collaborative intrusion detection networks against insider attacks using blockchain. Int. J. Inf. Secur. **19**(3), 279–290 (2019). https://doi.org/10.1007/s10207-019-00462-x

25. Mollah, M.B., Azad, M.A.K., Vasilakos, A.V.: Secure data sharing and searching at the edge of cloud-assisted internet of things. IEEE Cloud Comput. 4(1), 34–42 (2017). https://doi.org/10.1109/MCC.2017.9

26. Ortega, V., Bouchmal, F., Monserrat, J.F.: Trusted 5G vehicular networks: blockchains and content-centric networking. IEEE Veh. Technol. Mag. 13(2), 121–127 (2018)

27. Pirandola, S., et al.: High-rate measurement-device-independent quantum cryptography. Nat. Photonics 9(6), 397–402 (2015)

28. Preneel, B.: Cryptography and information security in the post-snowden era. In: TELERISE@ ICSE, p. 1 (2015)

29. Renauld, M., Standaert, F.-X.: Algebraic side-channel attacks. In: Bao, F., Yung, M., Lin, D., Jing, J. (eds.) Inscrypt 2009. LNCS, vol. 6151, pp. 393–410. Springer, Heidelberg (2010). https://doi.org/10.1007/978-3-642-16342-5_29

30. Riyadi, M.A., Khafid, M.R.A., Pandapotan, N., Prakoso, T.: A secure voice channel using chaotic cryptography algorithm. In: Proceedings of International Conference on Electrical Engineering and Computer Science (ICECOS), pp. 141–146 (2018)

31. Rukhin, A., Soto, J., Nechvatal, J., Smid, M., Barker, E.: A statistical test suite for random and pseudorandom number generators for cryptographic applications. Technical report, Booz-allen and hamilton inc mclean va (2001)

32. Singh, J., Pasquier, T.F.J., Bacon, J., Ko, H., Eyers, D.M.: Twenty security considerations for cloud-supported internet of things. IEEE Internet Things J. 3(3), 269–284 (2016). https://doi.org/10.1109/JIOT.2015.2460333

33. Wang, Z., Han, Y., Liu, W., Chen, L.: Anti-quantum generalized signcryption scheme based on multivariate and coding. In: Proceedings of Chinese Control and Decision Conference (CCDC), pp. 3587–3594 (2019)

34. Wu, Y., Khisti, A., Xiao, C., Caire, G., Wong, K.K., Gao, X.: A survey of physical layer security techniques for 5G wireless networks and challenges ahead. IEEE J. Sel. Areas Commun. 36(4), 679–695 (2018)

35. Zargar, S.T., Joshi, J., Tipper, D.: A survey of defense mechanisms against distributed denial of service (DDoS) flooding attacks. IEEE Commun. Surv. Tutor. 15(4), 2046–2069 (2013). https://doi.org/10.1109/SURV.2013.031413.00127

36. Zhang, J., Zheng, K., Zhang, D., Yan, B.: AATMS: An anti-attack trust management scheme in VANET. IEEE Access 8, 21077–21090 (2020)

Edge Computing and Distributed Machine Learning

Distributed Unsupervised Learning-Based Task Offloading for Mobile Edge Computing Systems

Jianming Wei, Qiuming Liu$^{(\boxtimes)}$, Shumin Liu, Yiping Zeng, and Xin Xiong

Department of Software Engineering,
Jiangxi University of Science and Technology, Nanchang, China
jianmingweijmw@163.com, liuqiuming@jxust.edu.cn, liushumin_001@163.com,
yipingZzeng@163.com, xiongxinbear@163.com

Abstract. Mobile edge computing has emerged as a new paradigm to enhance computing capabilities by offloading complicated tasks to nearby cloud server. To conserve energy as well as maintain quality of service, algorithms with low time complexity for task offloading is required. In this paper, a multi-user with multiple tasks scenario is considered, taking full account of factors including data size, bandwidth, channel state information, we propose a distributed unsupervised learning-based offloading algorithm for task offloading, where distributed parallel networks are employed to guarantee the robustness of algorithm. Additionally, we exploit a memory pool to store input data and corresponding decisions as key-value pairs. Based on the experience mechanism, the proposed algorithm can omit the step of data calibration compared with supervised learning method. To further reduce the communication cost, we analyze four bandwidth allocation schemes. Results reveal that the channel state-based strategy cost 3% less than data size-based, mixed and user size-based. Besides, the proposed algorithm can save 14, 18% cost than local and edge schemes respectively. Numerous results show that the proposed algorithm can achieve near-optimal decisions timely as well as having high reliability.

Keywords: Mobile edge computing · Distributed unsupervised learning · Energy efficiency · Task offloading

1 Introduction

Since the number of wireless devices has increased dramatically, the wireless network is facing a number of challenges such as mass data, high real-time requirement, limited resources and energy consumption. Meanwhile, traditional cloud computing exposes a lot of problems, such as data block and delay [1]. To solve problems mentioned above, mobile edge computing (MEC) is emerged to alleviate the task computing at cloud server by migrating the task near to the servers [2]. The MEC can effectively solve problems caused by massive tasks offloading

H. Gao et al. (Eds.): ChinaCom 2020, LNICST 352, pp. 539–553, 2021.
https://doi.org/10.1007/978-3-030-67720-6_37

and limited local resources, especially in wireless timely systems such as vehicular communications, augmented reality or virtual reality, so as to maintain the quality of server. However, a good algorithm with low complexity as well as robustness for task offloading in MEC systems is required. Meanwhile, the performance of algorithm for MEC network is influenced by many factors, such as bandwidth, data size, energy harvesting, task priority, etc.

There are mang studies on algorithms for MEC systems. In [3], an algorithm based on distributed deep learning was proposed, it can generate near-optimal offloading solution within short time. In [4], Madej et al. analyzed the resources allocation problem of task scheduling in MEC network, and demonstrated that the performance of task offloading strategy based on priority is better than the first come first serve. In [5], the author considered a vehicle network, where the channel state information feedback is provided with delay, and proposed a low-complexity algorithm for spectrum and power allocation. In [6], the authors considered the case of channel with interference, and obtained that the task offloading strategy is NP hard. Based on game theory, a distributed optimal algorithm was proposed to reduce the delay. Later, the scenario of energy harvesting devices is considered in [7], the authors not only studied the situation of task offloading, but also analyzed the task dropping strategy. To guarantee the quality of experience, a dynamic task offloading algorithm based on Lyapunov was proposed. In [8], by exploiting task buffer, a Lyapunov algorithm was proposed to optimize resource allocation and improve system performance. In [9], the problem of non-convex optimization strategy was simplified to a semi-definite relaxation problem and the sum of energy consumption of multiple users is quantified. In [10], a joint optimization method based on alternating direction multiplier method was proposed, which can obtain the optimal solution through a few iterations. Literature [11] modeled a user with multiple tasks communication scenario, and proposed an optimal strategy by combining semi-definite relaxation and heuristic random mapping method, which can make near-optimal decisions with a few random iterations. In [12], the authors proposed a deep Q-learning algorithm for multi-user with multiple tasks communication, which can generate near-optimal offloading decisions by training extensive samples.

From above studies, we found that the MEC network performance was impacted by all networks parameters, where the works mentioned above only consider partial parameters. To better analyze the role of multiple factors comprehensively, we model a multi-user with multiple tasks scenario, considering data size, bandwidth, and channel state information factors. Taking advantages of deep learning in MEC [13], we propose a distributed unsupervised learning-based offloading (DULO) algorithm. To fix the architecture of network, we set 3 users with 5 tasks as the value of input layers, the number of which is 18. Thus, the number of output layer is 15. Then, we exploit a memory pool to store input data and corresponding decisions as key-value pairs. Particularly, we design parallel networks to ensure the robustness of algorithm. To better evaluate the DULO algorithm, we use 80% of the samples as training samples. With thousands of iterations, parameters of networks are fitted based on experience mechanism. Additionally, we take the rest of samples as testing samples to test the performance of a trained model. Experiments show that bandwidth allocation based on channel state can achieve 3% cost less than other three strategies. Besides, the proposed algorithm can save 14, 18% cost than local and edge schemes respectively, which reveals that DULO can make task offloading decisions with near-optimal results.

The rest of this paper is organized as follows. In Sect. 2, we introduce the network model and formulate problems. In Sect. 3, we propose a distributed unsupervised learning-based task offloading algorithm. Section 4 presents numerical results and evaluates the proposed algorithm.

2 System Model

2.1 Network Model

The one-to-many network cell for edge server and wireless devices (WDs) is shown in Fig. 1. Multiple users offload tasks to the edge server through the access point (AP). After the edge server completes the execution of a task, the edge server transmits results back to the corresponding WD. Since the AP and

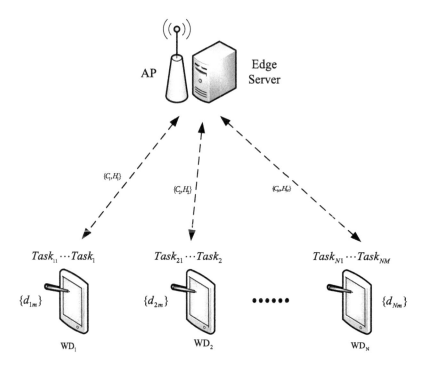

Fig. 1. Network model.

the server communicate with each other by using high speed optical fiber, the communication delay between the AP and the server can be ignored. Under the constraint bandwidth C of the system, C_n and H_n represent the bandwidth and channel gain of user n respectively. We denote $\mathbb{N} = \{1, 2, 3, \cdots N\}$ as a set of communication users, where each WD owns $\mathbb{M} = \{1, 2, 3, \cdots M\}$ tasks, which are mutually independent. We use $Task_{nm}$ to denote m-th task of n-th WD, and denote d_{nm} as data size. Since offloading decision is a binary decision set, we use $x_{nm} = 1$ to represent offloading task to the edge server, otherwise, when $x_{nm} = 0$ represents task being executed locally. Aiming to conserve energy and delay, two sub-models of energy consumption and delay are modeled to simplify the problem.

2.2 Energy Consumption Model

When $Task_{nm}$ is executed locally, we denote E_b as energy consumption for per bit of data processed by local resources. Therefore, the energy consumption E_{nm}^l of $Task_{nm}$ executed locally is given by

$$E_{nm}^l = E_b \times d_{nm}. \tag{1}$$

When task is offloaded to the server, we denote the energy consumption E_{nm}^t for $Task_{nm}$ transmission as

$$E_{nm}^t = P_n \times \frac{d_{nm}}{R_n}. \tag{2}$$

Since the task offloading process is affected by the channel state information, the transmission rate R_n is given by

$$R_n = C_n \times \log(1 + \frac{S_n}{\delta_n}), \tag{3}$$

where $S_n = P_n \times H_n$ and δ_n represents channel noise power of user n. We denote P_n as the transmitting power provided by the n-th WD, and denote S_n as signal power of user n. When the task is completed, results are transmitted to the mobile device. Because the size of results is much smaller than the size of task, the energy consumption for transmitting back is negligible. Therefore, the total energy cost E_{nm}^r for $Task_{nm}$ offloading is given by

$$E_{nm}^r = E_{nm}^t + E_s \times d_{nm}. \tag{4}$$

We use E_s to denote as energy consumption for per bit of data processed by server. When $E_s = 0$, we ignore the energy consumption for sever execution. Similarly, the delay model can also be divided into local execution and cloud execution.

2.3   Delay Model

We denote T_{nm}^l as delay when $Task_{nm}$ is executed locally. We use T_b to denote time cost for per bit of data. Thus, the delay of $Task_{nm}$ for local execution is given by

$$T_{nm}^l = T_b \times d_{nm}. \tag{5}$$

When $Task_{nm}$ is executed by server, time cost T_{nm}^r including transmission delay and execution delay. We denote transmission delay and execution delay as

$$T_{nm}^t = \frac{d_{nm}}{R_n}, \tag{6}$$

and

$$T_{nm}^c = \frac{d_{nm}}{F_c}, \tag{7}$$

respectively, where we denote F_c as the size of data that server can process in per second. Thus, the total delay when n-th WD decides to offload $Task_{nm}$ is given by

$$T_{nm}^r = T_{nm}^t + T_{nm}^c. \tag{8}$$

2.4   Problem Formulation

Through analyzing above two sub-models, the total cost for $Task_{nm}$ is given by

$$Cost_{nm} = (E_{nm}^l + T_{nm}^l)(1 - x_{nm}) + (E_{nm}^r + T_{nm}^r)x_{nm}, \tag{9}$$

where $x_{nm} = \{0, 1\}$, $\forall n, m$. To jointly minimize energy consumption and delay in MEC communication system, we formulate the problem as

$$\min J(\mathbf{d}, \mathbf{h}, \mathbf{c}, \mathbf{x}) = \sum_{n=1}^{N} \Big(\sum_{m=1}^{M} (E_{nm}^l (1 - x_{nm})$$

$$+ E_{nm}^r x_{nm}) + max(\sum_{m=1}^{M} T_{nm}^r, \sum_{m=1}^{M} T_{nm}^l) \Big) \tag{10}$$

$$s.t. \begin{cases} \sum_{n=1}^{N} C_n \leq C & \forall n \in \mathbb{N}, \\ C_n \geq 0 & \forall n \in \mathbb{N}, \\ x_{nm} = \{0, 1\} & \forall n \in \mathbb{N}, m \in \mathbb{M}. \end{cases}$$

To solve the problem J, we have to consider variables including decisions, data size, bandwidth and channel states. The complexity of J mainly depends on the size of N and M, but an algorithm of low time complexity is required. Next, we introduce an algorithm based on unsupervised deep learning (Table 1).

Table 1. Notations

Notation	Definition
$Task_{nm}$	m-th task of n-th user
x_{nm}	The decision of $Task_{nm}$
d_{nm}	Data size of $Task_{nm}$
E_b	Local energy consumption per bit
E_{nm}^l	Energy cost of $Task_{nm}$ for local execution
E_s	Server energy consumption for per bit
E_{nm}^t	Energy cost of $Task_{nm}$ for transmission
E_{nm}^r	Total energy cost of $Task_{nm}$ for edge processing
C_n	Bandwidth allocated to user n
P_n	Transmission power of n-th user
H_n	Channel gain of n-th user
F_c	Edge processing rate
R_n	Transmission rate of n-th user
S_n	Signal power of user n
δ_n	Channel noise power of user n
T_b	Local time consumption per bit
T_{nm}^l	Time cost of $Task_{nm}$ for local execution
T_{nm}^t	Time cost of $Task_{nm}$ for transmission
T_{nm}^c	Time cost of $Task_{nm}$ for server executing
T_{nm}^r	Total time cost of $Task_{nm}$ for edge processing

3   DULO Algorithm

In this section, we employ bandwidth allocation scheme to simplify the complexity of the problem (10) and proposed the DULO algorithm for task offloading, where distributed unsupervised learning are used to generate near-optimal decisions.

Once bandwidth is allocated, the problem (10) will be simplified, which affected by variables including decisions, data size and channel states. There are many proposed schemes on how to efficiently solve the bandwidth allocation problem, in [14] bandwidth allocation is studied. In this paper, we employ data-size-based and channel-state-based allocation schemes respectively, which is given by

$$C_n = \beta \times \frac{d_{nm}}{\sum_{n=1}^N \sum_{m=1}^M d_{nm}} + \gamma \times \frac{H_n}{\sum_{n=1}^N H_n}. \tag{11}$$

Then, the problem (10) is simplified as

$$\min \boldsymbol{J}^*(\mathbf{d}, \mathbf{h}, \mathbf{x}) = \sum_{n=1}^{N} \Big(\sum_{m=1}^{M} (E_{nm}^l (1 - x_{nm})$$
$$+ E_{nm}^r x_{nm}) + max(\sum_{m=1}^{M} T_{nm}^r, \sum_{m=1}^{M} T_{nm}^l) \Big) \tag{12}$$
$$s.t. \begin{cases} \sum_{n=1}^{N} C_n \leq C & \forall n \in \mathbb{N}, \\ C_n \geq 0 & \forall n \in \mathbb{N}, \\ x_{nm} = \{0, 1\} & \forall n \in \mathbb{N}, m \in \mathbb{M}. \end{cases}$$

Since the system has 2^{NM} choices to execute tasks, especially, complexity of many traditional algorithms grows exponentially as users or tasks increase, it is NP hard to get the optimal decision. In order to find the optimal solution in the problem $\boldsymbol{J}^*$, we try to find a function $f(u, v)$, which satisfies the following condition

$$f(d_{ij}, h_{ij}) = x_{ij}, (i \in \mathbb{N}, j \in \mathbb{M}). \tag{13}$$

Deep learning network can solve complicated problems well, therefore, we propose an algorithm based on deep learning. Supervised deep leaning can fit a complex function well, especially, it can learn to find high dimensional features such as image classification. However, it is hard to get massive samples with labels. Therefore, unsupervised learning algorithm is used, as show in Fig. 2.

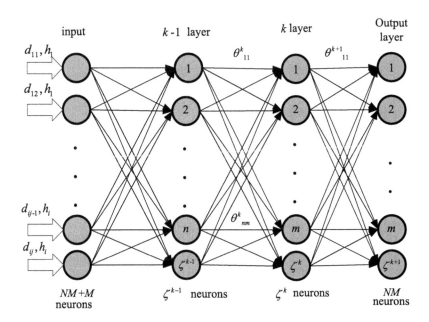

Fig. 2. Unsupervised learning algorithm for offloading network.

We denote data size d_{ij} and channel state information h_i as input values. We use x_{ij} to denote the output value corresponding to the binary offloading decision, if $x_{ij} = 1$, it denotes offloading the task to the server, else, it represents executing the task locally. Concretely, we employ ζ^k neurons as k-th layer and ζ^{k-1} neurons as $k - 1$-th layer, which are fully connected. Then the output of m-th neuron of k-th layer is formulated as

$$z_m^k = \sum_{n=1}^{\zeta^{k-1}} (\theta_{nm}^k \times z_n^{k-1} + b_m^k), \tag{14}$$

where we denote θ_{nm}^k as value of weight of k-th layer, which connects between n-th neuron of $k - 1$-th layer and m-th neuron of k-th layer. To formulate the nonlinear relationship between input and output, we employ relu function to process the output values of hidden layers, which is given by

$$y_m^k = \max(0, z_m^k). \tag{15}$$

Different from relu function, we use the following function to generate binary values of the output

$$x_{ij} = \begin{cases} 1, & y_m^k > 0; \\ 0, & y_m^k < 0. \end{cases} \tag{16}$$

Since it is a network with binary output, we use the cross-entropy function to optimize parameters by thousands of iterations, as

$$L(\boldsymbol{\theta}) = -\mathbf{x}^{\mathrm{T}} \log(f(\mathbf{d}, \mathbf{h})) - (1 - \mathbf{x})^{\mathrm{T}} \log(1 - f(\mathbf{d}, \mathbf{h})). \tag{17}$$

To better fit the problem $\boldsymbol{J}^*$ and get the optimal solution, parameters about the network need to be fitted by backpropagation. The θ_{nm}^k at time t+1 is given by

$$\theta_{nm(t+1)}^k = \theta_{nm(t)}^k - \eta \frac{\partial L}{\partial \theta_{nm(t)}^k}, \tag{18}$$

where η denotes the learning rate.

However, it is not a good substitute for $\boldsymbol{J}^*$ with only one network, and the cost for training samples is time-consuming. Therefore, we employ I networks with same architecture but different parameters to ensure the robustness of the algorithm. Details about distributed unsupervised deep learning algorithm is shown as in Fig. 3.

At the beginning of training model, the parameter values $\boldsymbol{\theta}$ are randomly initialized. Then we have to create a empty memory pool and set the number of iterations. As the input values are operated with formulations (14) and (15) repeatedly, and processed by formulation (16). We get a set of binary offloading decisions. For I pairs offloading decisions, we compute the corresponding cost as candidates. From the number of I candidates, we select the minimum as a best candidate temporarily, as

$$\mathbf{x}_{tmp} = \arg\min_{\tau \in I} J^*(\mathbf{d}, \mathbf{h}, \mathbf{x}_\tau). \tag{19}$$

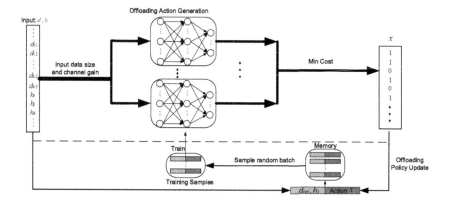

Fig. 3. Architecture of distributed unsupervised learning-based offloading network.

Then, we exploit the memory pool to store $Task_{nm}$ and corresponding decisions $\mathbf{x}_{tmp}$ as key-value pair. Once the temporary offloading decisions $\mathbf{x}_{tmp}$ is obtained, we use it as a temporary label for the task. Along with the iteration, we store better offloading decisions as well as corresponding tasks and discard the old key-value pair in the finite size of memory pool. Based on the experience mechanism, the model can learn to fit optimal parameters after training the network.

As a number of decisions is given, the optimal problem orientated to resources allocation is simplified to a convex problem, the server can better allocate its resources to execute tasks. Algorithm 1 presents the proposed DULO algorithm for task offloading.

Algorithm 1. DULO algorithm for training samples

Require:
　The set of training samples for current batch d, h;
Ensure:
　Offloading decision $\mathbf{x}$;
1: **Initialization:**
2:　　Initialize I networks with random parameters;
3:　　Create an empty memory pool;
4:　　Set the number of iterations K;
5: **while** $j = 1, 2, ..., K$ **do**
6:　　Randomly select values of tasks to load the network;
7:　　Select the minimum value as a good candidate;
8:　　Store tasks and decisions as key-value pairs;
9:　　Backpropagate to fit parameters;
10:　　Update the memory pool;
11: **end while**

---

4 Performance Evaluation

4.1 Performance of Training Model

In this paper, we simulate a small cellular network communication process. To fix the architecture of network, we set 3 users with 5 tasks as input layers, and we split 80% of samples as training samples and the remain as testing samples.

We train the model with different number of networks, taking the optimal solution as the benchmark. As shown in Fig. 4, when we use 2 networks to train model, it finally converges at the gain ratio 0.96 after 28 thousands. When more than 2 networks are used, the performance becomes better. We have to notice that when the number of networks is more than 3, it costs 15 thousands iterations to converge. As the number of networks increases, more data with correct labels can generate. Although, when more than 3 networks are used, the speed of convergence can't get faster, the performance get better when more than 3 networks are used. Generally, when more than 3 networks are used, the DULO performances better in convergence and gain ratio.

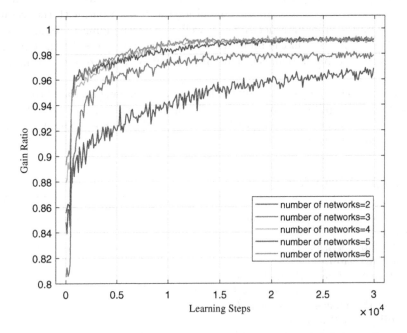

Fig. 4. Performance of training model with different number of networks.

Additionally, we test the performance of the algorithm with different channel states, including large-scale fading, rayleigh fading, rice fading as well as ideal communication system. We can learn from Fig. 5, the performance of training model maintains good with different channel states, it always converges at 0.98.

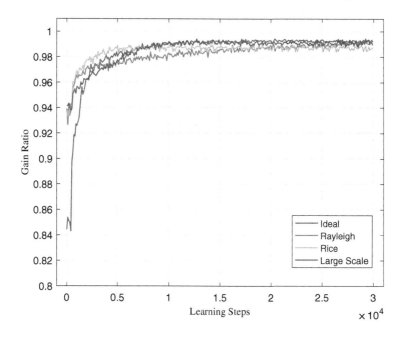

Fig. 5. Performance of training model with different channel states.

But the ideal communication converges faster than other situations. As shown in Fig. 5, when we train the ideal communication, the gain ratio converges after 10 thousands iterations. However, since one more factor considered, the DULO algorithm converges at 0.98 after 20 thousands iterations.

4.2   Performance of Testing Model

In order to evaluate the DULO algorithm better, we use testing samples as input values to analyze the offloading decisions, delay and robustness respectively.

Firstly, we analyze the influence of different weight of β. As shown in Fig. 6, the bandwidth allocation based on data-size strategy cost more than other strategies. When channel-state based strategy is employed, the communication cost 3% less than others. As the value of β increases, the cost goes up. When the β is 0.7, the cost tends to be stable. Generally, the channel state-based strategy costs less, therefore, the channel state factor is valuable.

Then, we compare DULO algorithm with other schemes in Fig. 7. We use enumeration strategy to enumerate all options, and then select the decision corresponding to minimum cost as the benchmark decision. As E_s is the energy consumption for per bit of data processed by server, local cost will not change with its value, edge cost goes up linearly with the value. Results show that the DULO cost less 14, 18% than local and edge schemes respectively, which is the near-optimal solution compared with minimum cost.

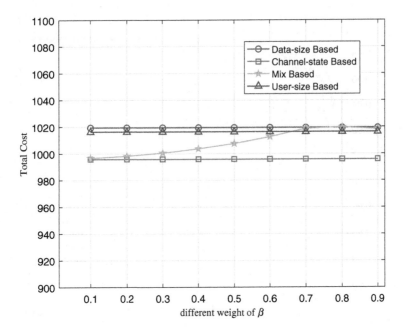

Fig. 6. Total cost under different bandwidth allocation schemes.

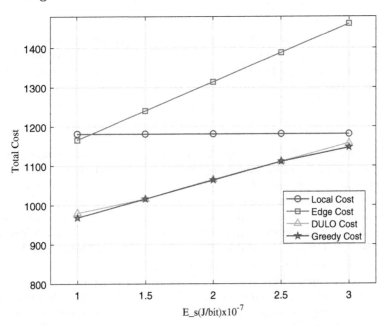

Fig. 7. Total cost with different schemes.

To test whether the algorithm can meet the real-time requirement of the system, time cost with different number of networks is tested. We can learn from Fig. 8, when channel state information is considered, it costs 0.02 s more than the ideal communication, since there are more data to compute. Additionally, time cost goes up with the number of networks. When the number of networks is less than 6, DULO algorithm generates near-optimal solution in 0.2 s. Specially, when the number of networks is 2, we can get a solution within 0.1 s. However, when 2 networks are used in the network, it performances worse than 3 networks. Both in terms of time cost and accuracy, a proper number of networks is required, therefore, more than 2 networks are used in this paper.

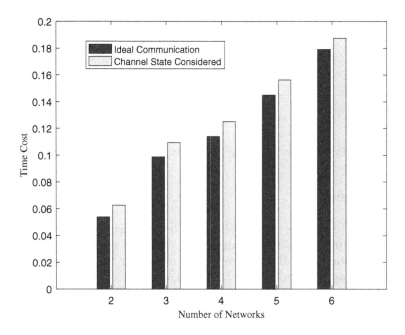

Fig. 8. Time cost for offloading decision with different number of networks.

Besides, we consider different hyper-parameters including number of networks, learning rates and memory sizes to test the robustness of DULO algorithm. As shown in Fig. 9, gain ratio goes up as the number of networks increases, when more than 5 networks are used, the performance of DULO is better than others. As memory pool gets larger, the DULO algorithm performances better, since there are more data with labels to learn for the DULO, which makes the algorithm robust. And when the value of learning rate is 0.03 and memory size is 2048, it has good stability with good performance whatever the number of network is. Taking different hyper-parameters into consideration, we can learn from Fig. 9 that the trained model can generates a near-optimal solution over the gain ratio 0.95.

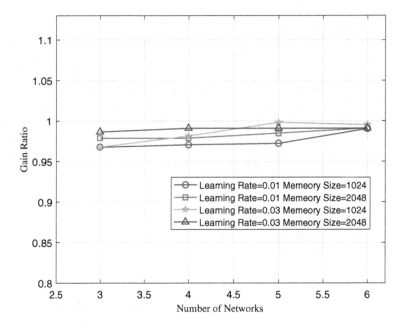

Fig. 9. Performance of testing model with different hyper-parameters.

5   Conclusion

In this paper, to improve quality of service by minimizing energy consumption and delay, we propose a DULO algorithm with known data size and channel states information for MEC networks. By exploiting the advantages of deep learning and parallel architecture, DULO algorithm makes up for the disadvantages of traditional algorithms such as decision delay. Compared with supervised learning, it avoids the complex work of data calibration and has better robustness. Results show that the bandwidth allocation based on channel-state strategy can save 3% cost than other schemes, and the proposed DULO algorithm can achieve a near-optimal decision in a short time less than 0.2 s, which cost 14, 18 less than percent than local and edge strategies respectively.

Acknowledgment. This work was supported in part by National Natural Science Foundation of China (No. 61761021), Natural Science Foundation of Jiangxi Province (Grant No. 20181bab202018, 20202BAB212003), Special Fund for Postgraduate Innovation of Jiangxi Province(No. YC2020-S481), Projects of Humanities and Social Sciences of universities in Jiangxi (JC18224), Jiangxi art planning project (YG2018042) and the Doctoral Research Fund of Jiangxi University of Science and Technology.

References

1. Shamsi, J.A., Khojaye, M.A., Qasmi, M.A.: Data-intensive cloud computing: requirements, expectations, challenges, and solutions. J. Grid Comput. **11**(2), 281–310 (2013)
2. Syamkumar, M., Barford, P., Durairajan, R.: Deployment characteristics of "the edge" in mobile edge computing. In: ACM Special Interest Group on Data Communication, Budapest, pp. 43–49 (2018). https://doi.org/10.1145/3229556.3229557
3. Huang, L., Feng, X., Feng, A., Huang, Y., Qian, L.P.: Distributed deep learning-based offloading for mobile edge computing networks. Mobile Netw. Appl. **23**(6), 1–8 (2018). https://doi.org/10.1007/s11036-018-1177-x
4. Madej, A., Wang, N., Athanasopoulos, N.: Priority-based Fair Scheduling in Edge Computing. arXiv Distributed, Parallel, and Cluster Computing (2018)
5. Liang, L., Kim, J., Jha, S.C.: Spectrum and power allocation for vehicular communications with delayed CSI feedback. IEEE Wireless Commun. Lett. **6**(4), 458–461 (2017)
6. Chen, X., Jiao, L., Li, W.: Efficient multi-user computation offloading for mobile-edge cloud computing. IEEE ACM Trans. Netw. **24**(5), 2795–2808 (2016)
7. Zhao, H., Du, W., Liu, W.: QoE aware and cell capacity enhanced computation offloading for multi-server mobile edge computing systems with energy harvesting devices. In: IEEE SmartWorld, Ubiquitous Intelligence and Computing, Guangdong, pp. 671–678 (2018). https://doi.org/10.29007/mq2s
8. Mao, Y., Zhang, J., Song, S.H.: Power-delay tradeoff in multi-user mobile-edge computing systems. In: IEEE Global Communications Conference, Washington, pp. 1–6 (2016). https://doi.org/10.1109/GLOCOM.2016.7842160
9. Chen, M., Liang, B., Dong, M.: Joint offloading decision and resource allocation for multi-user multi-task mobile cloud. In: International Conference on Communications, Kuala Lumpur, pp. 1–6. Springer, Heidelberg (2016). https://doi.org/10.1109/ICC.2016.7510999
10. Author, B.S.: Computation rate maximization for wireless powered mobile-edge computing with binary computation offloading. IEEE Trans. Wireless Commun. **17**(6), 4177–4190 (2018)
11. Chen, M., Liang, B., Dong, M.: A semidefinite relaxation approach to mobile cloud offloading with computing access point. In: International Workshop on Signal Processing Advances in Wireless Communications, Stockholm, pp. 186–190. IEEE (2015). https://doi.org/10.1109/SPAWC.2015.7227025
12. Huang, L., Feng, X., Qian, L., Wu, Y.: Deep reinforcement learning-based task offloading and resource allocation for mobile edge computing. In: Meng, L., Zhang, Y. (eds.) MLICOM 2018. LNICST, vol. 251, pp. 33–42. Springer, Cham (2018). https://doi.org/10.1007/978-3-030-00557-3_4
13. Li, H., Ota, K., Dong, M.: Learning IoT in edge: deep learning for the internet of things with edge computing. IEEE Network **32**(1), 96–101 (2018)
14. Huang, L., Feng, X., Zhang, C.: Deep reinforcement learning-based joint task offloading and bandwidth allocation for multi-user mobile edge computing. Digital Commun. Netw. **5**(1), 10–17 (2019)

Decoupling Offloading Decision and Resource Allocation via Deep Reinforcement Learning and Sequential Least Squares Programming

Zhihao Xuan[1]($\boxtimes$), Guiyi Wei[1], Zhengwei Ni[1], and Jifa Zhang[2]

[1] Zhejiang Gongshang University, Hangzhou 310018, China
xuanzhihao1997@outlook.com,
{weigy,zhengwei.ni}@zjgsu.edu.cn
[2] Zhejiang University, Hangzhou 310027, China
jfzhang@zju.edu.cn

Abstract. Edge computing is to generate faster network service response and meet the basic needs of the industry in real-time business, application intelligence, security and privacy protection. This paper studies the mobile edge computing network, where the computing power of the edge server (ES) is limited, and multiple user equipment (UE) can offload the thinking to the ES in order to save energy consumption and computing delay. The ES needs to determine which UEs can upload its tasks and need to allocate computing resources for these UEs, so this problem is highly coupled and difficult to calculate. This paper proposes an algorithm based on deep reinforcement learning and Sequential Least Squares Programming (SLSQP), which decouples and solves the problem. Experiments show that the algorithm works well and can be dynamically adjusted according to environmental changes. The comparison with other algorithms also proves that the algorithm has better results and less time-consuming.

Keywords: Edge computing · Offloading · Resource allocation · Reinforcement learning

1 Introduction

1.1 A Subsection Sample

In recent years, with the rapid development of IoT technology and the increasing requirements of software for hardware, IoT devices are facing more and more computer-intensive and delay-sensitive tasks. However, IoT devices are often constrained by their size and power. Factors such as these are not suitable for running these tasks on the device itself.

Edge computing can offload the computing tasks required by the user equipment to the edge computing server with rich computing resources for calculation, in order to reduce the energy consumption of the user equipment and the delay of the computing tasks. Compared with the existing cloud computing, the edge computing server is deployed on the edge of the network, such as a base station or an access point of a

H. Gao et al. (Eds.): ChinaCom 2020, LNICST 352, pp. 554–566, 2021.
https://doi.org/10.1007/978-3-030-67720-6_38

wireless network, and is closer to the user, which can avoid long-distance data transmission from the cloud computing center, thereby reducing the time delay and transmission energy required for computing tasks, and improves the user experience.

However, unlike cloud computing, edge computing servers usually have limited computing power and bandwidth. Therefore, offloading decisions and resource allocation for user equipment has become a hot research issue and a difficulty in edge computing systems. In the case of limited resources, unreasonable judgment and allocation of unloading decisions and resources may lead to increased delay and energy consumption, and may also make the system load unbalanced, affecting the stability of the system.

In order to solve the above problems, many scholars have conducted research in this area. [1] proposes a mobile edge computing scheduling decision algorithm based on multi-round auctions. Its implementation requires multiple rounds of communication between the user equipment and the edge server. There are also some studies to solve the above problems through game theory [2], the method still requires multiple rounds of communication between the device and the server. However, the complexity and communication time are still too high for tasks with delay sensitivity in current edge computing system. There are also many studies that conduct separate studies on computing offloading or resource allocation [3, 4]. However, the two issues should be considered and solved jointly due to embroiled relation between them.

In recent years, machine learning methods have made some breakthroughs in different fields, such as natural language processing [5], data mining [6] and other fields. However, at present, there are few researches on edge computing systems using machine learning methods, and most of the research is still based on Q learning [7], but the table search structure inside Q learning is actually not suitable for multidimensional and highly coupled problems. Some scholars have also used deep learning methods to carry out research on offload decision-making [8], but they require supervised learning of neural network training, and the results cannot adapt to changes in various conditions.

This paper proposes an offloading decision and resource allocation method based on reinforcement learning and convex optimization for the goal of task completion delay and energy consumption optimization in the case of edge computing servers with limited computing power. This method decouples the highly coupled problem of offloading decision and resource allocation into two sub-problems of offloading decision and resource allocation, and uses reinforcement learning and convex optimization methods to solve them. Experiments show that this method has achieved good results, can make decisions in a short time, and has good scalability.

2 System Model and Problem Description

2.1 System Model

The template is used to format your paper and style the text. All margins, column widths, line spaces, and text fonts are prescribed; please do not alter them.

The structure of the edge computing system in this article is shown in Fig. 1, which includes an edge computing server (ES) and N user equipments (UE). These N UEs are the same, denoted by $N = \{1, 2, \ldots, N\}$. The ES and UE are connected through a wireless network, and the transmission delay between them can be ignored. Generally speaking, ES has a stable power supply and fast calculation speed, while UE is the opposite. Therefore, UE can offload the tasks it needs to calculate to ES for calculation, and then receive the calculation result from ES to reduce the energy consumption and time delay required by calculating the task. But for ES, it has limited computing resources F_{max}^{es}, so it needs to allocate computing resources for the UE that decides to perform task offloading. The allocation of computing resources will affect energy consumption and delay. In this article, a binary offloading strategy is used. For a certain UE, all its tasks are either calculated locally or offloaded to ES for calculation. The decision to calculate offloading is represented by a binary variable $x_i \in \{0, 1\}$. Specifically, when $x_i = 0$, it means that the i-th user equipment decides to perform calculations locally, and when $x_i = 1$, it means that the i-th user equipment decides to perform task offloading.

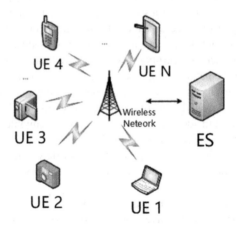

Fig. 1. Edge computing model.

2.2 Task Offloading

The template is used to format your paper and style the text. All margins, column widths, line spaces, and text fonts are prescribed; please do not alter them.

When the UE_i decides to offload all its tasks to the ES for calculation, it first needs to upload data of size d_i to the ES via the wireless network, and suppose the uplink speed of the UE_i and ES direct wireless network in the system model is r_i.

In general, the amount of data in the calculation results is very small and can be ignored compared to the amount of uploaded data. Therefore, this article ignores the energy consumption and time delay caused by downloading the calculation results. The same assumption also exists in [9].

First, model the time required to unload the task. The time it takes for UE_i to upload data of size d_i to ES under the condition of transmission speed r_i is:

$$T_i^t = \frac{d_i}{r_i} \tag{1}$$

The time required for computing in the ES server is:

$$T_i^c = \frac{d_i}{f_i^{es}} \tag{2}$$

Where f_i^{es} represents the computing resources allocated by the ES server to the UE_i, that is, the amount of data that can be processed per unit time. In summary, the total time consumed for task uninstallation is:

$$T_i^{offload} = T_i^t + T_i^c \tag{3}$$

Next, model the energy consumption of task offloading. Assuming that the upload power of each UE is p_{upload}, the energy consumed during the upload task is:

$$E_i^{upload} = p_{upload} * T_i^t \tag{4}$$

From the task is uploaded to the ES to the calculation result is obtained, the UE_i will be in a waiting state. Assuming that the waiting state power of all user equipment during this period is $p_{waiting}$, then the energy consumed by the UE_i during this period is:

$$E_i^{waiting} = p_{waiting} * T_i^c \tag{5}$$

Then the total energy consumed when perform offloading is:

$$E_i^{offload} = E_i^{upload} + E_i^{waiting} \tag{6}$$

2.3   Local Computing

Now to model the delay and energy consumption of local calculations, we use f_{local} to represent the amount of data that the UE_i itself can process per second, so for the UE, the time required for local calculations is:

$$T_i^{local} = \frac{d_i}{f_{local}} \tag{7}$$

Next, model the energy consumption of the local calculation. Assuming that the operating power when computing locally is p_{local}, then the energy consumed is:

$$E_i^{local} = T_i^{local} p_{local} \tag{8}$$

2.4   Problem Description

In order to minimize the delay and energy consumption of all user equipment, this paper uses a linear weighting method to define a weighted function $Cost(\mathbf{d}, \mathbf{x}, \mathbf{r})$ composed of delay and energy consumption to evaluate the performance of the system:

$$Cost(\mathbf{d}, \mathbf{x}, \mathbf{r}) = \sum_{i=1}^{N} \left\{ (1 - x_i)\left(T_i^{local} + \alpha E_i^{local}\right) + x_i \left(T_i^{offload} + \alpha E_i^{offload}\right) \right\}$$

Where $\mathbf{d} = \{d_i | i \in N\}, \mathbf{x} = \{x_i | i \in N\}, \mathbf{r} = \{r_i | i \in N\}$, α represents the weight of energy consumption in the weighting function. Then the question is:

$$Q(\mathbf{d}) = \text{minimize } Q(\mathbf{d}, \mathbf{x}, \mathbf{r}) \tag{9a}$$

Subject to:

$$\sum_{i=1}^{N} r_i \leq R_{max} \tag{9b}$$

$$r_i \geq 0, \forall i \in N \tag{9c}$$

$$x_i \in \{0, 1\} \tag{9d}$$

3   Algorithms Based on Deep Reinforcement Learning and Convex Optimization

For problem (9a), it is a highly coupled problem to find the offloading decision and the allocation of computing resources at the same time. There are multiple parameters that affect each other. Generally speaking, it is difficult to solve such problems, so we decouple the problem, Decomposed into the problem of offloading decision-making and the problem of computing resource allocation.

For the generation of offloading decisions, it is necessary to take the amount of data $\mathbf{d}$ required to be computed for all devices as input, and find the most appropriate offloading strategy x considered by the system. For N devices, there are a total of 2^N types of offloading strategies. If the brute force search method is used in actual situations, the number of selectable offloading strategies will increase exponentially with the increase of the number of devices N. Therefore, the brute force search method is not applicable in actual situations. In order to solve the above problems, this paper uses a method based on deep reinforcement learning to generate unloading decisions.

The structure of deep reinforcement learning in this article is shown in Fig. 2. It roughly contains two modules, namely the offloading decision generation module and the improvement module for offloading decision-making. The offloading decision generation module contains the solution to the problem of computing resource allocation.

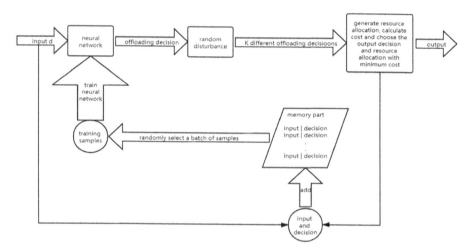

Fig. 2. Overview of our algorithm.

Among them, the offloading decision generation module contains a neural network model, which will generate an offloading decision x according to the input task data $\mathbf{d}$ that each user needs to compute. For this neural network, according to the well-known universal approximation theorem, for a feedforward neural network, even if there is only one hidden layer, it can infinitely approximate any bounded continuous function, so here, we use ReLU function as the hidden layer activation function, and in the output layer, the Sigmoid activation function is used.

After an unloading decision is generated, the system will randomly perturb the unloading decision. Each disturbance will randomly select c users for the unloading decision. If the user's decision is to uninstall, it will be calculated locally, and vice versa. A total of K different unloading decisions are generated.

When a different offloading decision is generated, problem (9a) becomes a computing resource allocation problem:

$$Q(\mathbf{d}, \mathbf{x}) = \text{minimize } Q(\mathbf{d}, \mathbf{x}, \mathbf{r}) \tag{10a}$$

Subject to:

$$\sum_{i=1}^{N} r_i \leq R_{\max} \tag{10b}$$

$$r_i \geq 0, \forall i \in N \tag{10c}$$

$$x_i \in \{0, 1\} \tag{10d}$$

According to these K different offloading decisions, the system calculates the optimal computing resource allocation and the corresponding cost according to the convex optimization scheme and cost function, and outputs the offloading decision with the lowest cost.

In view of the fact that the predecessors have carried out extensive and in-depth research on convex optimization, there are many efficient algorithms in the current research on convex optimization. For the problem of computing resource allocation, this paper uses a convex optimization algorithm named "SLSQP". Method SLSQP uses Sequential Least Squares Programming to minimize a function of several variables with any combination of bounds, equality and inequality constraints. The method wraps the SLSQP Optimization subroutine originally implemented by Dieter Kraft [10].

As for the improved module for offloading decision-making, every time the offloading decision-making module produces the optimal offloading decision, the offloading decision and the calculated task data required by the user will be saved as a sample in the memory. The storage capacity is limited. If the memory is full when the new optimal unloading decision is added, the sample that was added to the memory earliest will be eliminated.

In the improved module for offloading decision-making, every time the offloading decision is generated, if the current number of decisions is exactly a multiple of δ, a batch of samples will be randomly selected from the memory for training, instead of the traditional method Use all data for training. In the training process, this article uses Adam algorithm to update the parameters of the neural network to reduce the average cross entropy loss. Because it uses its own decisions instead of the optimal decisions obtained, the neural network model can continuously improve its own model without supervision and produce better offloading decisions, unlike traditional deep learning that requires optimal offloading decisions for learning.

In particular, limited memory capacity can help improve the efficiency of training, because the newly generated samples are generally better than the old samples. In fact, there are still some other techniques that can help accelerate training, such as distributed importance sampling [11] And priority experience replay [12].

4   Performance Evaluation

In this section, we use simulation experiments to evaluate the performance of the proposed algorithm. In the simulation experiment, we set the number of user equipment N = 10. The speed of computing locally or computing in the server is determined regardless of the task type. The task data volume of each user equipment is uniformly distributed between 5 MB and 25 MB. Edge computing server can process 30 MB of data per second. The communication speed between the user equipment and the edge computing server is 2.5 MB/s. The user's power when sending data is 1 W, and the power when waiting is 0.5 W, while the user's local computing The data processing speed is 3 MB/s, the computing power is 3 W, and the coefficient of energy consumption in the cost function is $\alpha = 1$; we set the capacity of the memory module to 256 and the neural network model to four layers, each time $K = 7$ offloading strategies will be generated, the number of randomly disturbed users is $c = 4$, one learning is performed every 10 predictions, 128 samples are randomly selected from the memory module for each learning, and the learning rate is 0.01. For neural networks, this paper uses a fully connected network consisting of one input layer, two hidden layers, and one output layer. Two hidden layers contain 120 and 80 neurons, respectively. The convex optimization "SLSQP" method is implemented using the corresponding function in the scipy library.

4.1   Convergence Performance

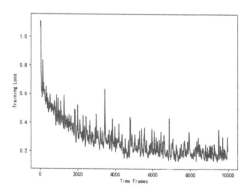

Fig. 3.   Training losses in iterations.

Figure 3 and Fig. 4 are the training loss of the neural network and the ratio of the minimum cost in the iteration to the current cost, respectively. It can be seen from Fig. 3 that the training loss has been continuously reduced from the beginning, and a total of 10,000 iterations have been carried out. When the number of iterations reaches about 4500, the loss basically stabilizes, and the rate of decline slows down significantly. When it reaches about 6500, It basically no longer declines, maintaining between 0.1–0.2. Figure 4 shows the ratio of the minimum cost to the cost of the

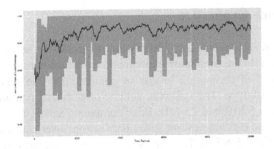

Fig. 4. The ratio of the minimum cost in an iteration to the current policy cost. (Color figure online)

current strategy in the iteration. The minimum cost is obtained by finding all kinds of offloading strategies, and then generating the resource allocation of each strategy according to the "SLSQP" convex optimization algorithm and calculating its cost, and finally taking the smallest cost one; the ratio in Fig. 4 is recorded every 10 iterations, the blue line represents the average ratio of the nearly 20 records, and the blue shading represents the maximum and minimum values of the nearly 20 records; you can see from the figure It can be seen that after 4000 iterations, the ratio basically stabilizes, changing around 0.995.

4.2   Situation When Parameters Change

Figures 5 and 6 are respectively the training loss and the ratio of the minimum cost in the iteration to the current cost when α changes. In particular, when the number of iterations is 5000, α will become 3, and when the number of iterations is 7000, a will Change back to 1.It can be found that the training loss in Fig. 5 has increased significantly at 5000 and 7000 times, and then quickly converged, while the ratio in Fig. 6 is also the same, there has been a significant decline, and both are resumed after about 1000 iterations.

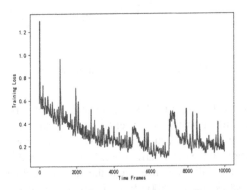

Fig. 5. Training losses in iterations (Change α).

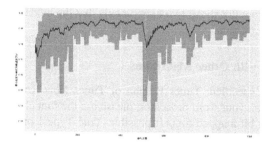

Fig. 6. The ratio of the minimum cost in an iteration to the current policy cost (Change α).

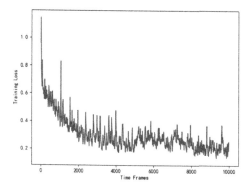

Fig. 7. Training losses in iterations (User equipment switch on and off).

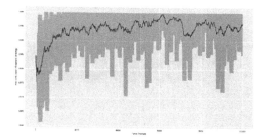

Fig. 8. The ratio of the minimum cost in an iteration to the current policy cost (User equipment switch on and off).

Figures 7 and 8 are respectively the training loss and the ratio of the minimum cost in the iteration to the current cost when shutting down and turning on three user devices when the number of iterations is 5000 and 7000. It can be seen that the impact of switching on and off the user equipment is less than the impact of a change on α. After a small increase in loss and a small decrease in cost ratio, it quickly returns to the previous level.

It can be seen from the above two experiments that the model can quickly adapt to changes in the environment.

4.3   Comparison with Other Algorithms

Wherever Times is specified, Times Roman or Times New Roman may be used. If neither is available on your word processor, please use the font closest in appearance to Times. Avoid using bit-mapped fonts if possible. True-Type 1 or Open Type fonts are preferred. Please embed symbol fonts, as well, for math, etc.

This section compares the proposed algorithm with the following strategies, which are:

ALL_LOCAL: All user devices calculate tasks locally.

ALL_OFFLOAD: All user devices offload tasks to the edge computing server for calculation, and all user devices are equally offloaded.

CD_AVERAGE: The offloading strategy is calculated by the gradient descent method [9], and the offloading resources are equally divided among all user devices.

CD_SLSQP: The offloading strategy is calculated by the gradient descent method, and the computing resources are allocated using the "SLSQP" algorithm.

This paper compares the above algorithms for 10,000 iterations, and the comparison results are shown in Fig. 9.

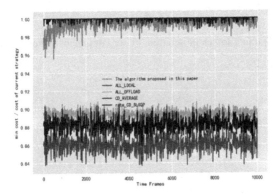

Fig. 9. Results compared with other algorithms

It can be seen from Fig. 9 that ALL_LOCAL, ALL_OFFLOAD, and CD_AVERAGE perform poorly, and the ratio of the minimum cost to the current strategy cost is between 0.9 and 0.84; and the CD_SLSQP algorithm performs best, with the cost ratio closest to 1, which is stable at Around 0.998, and the algorithm proposed in this paper, after a period of convergence, the cost ratio stabilizes at around 0.995.

Table 1. Time comparison between the proposed algorithm and CD_SLSQP

Algorithm	Number of UE		
	10	15	20
Algorithm this paper proposed	0.069 s	0.174 s	0.435 s
CD_SLSQP	0.470 s	2.247 s	9.333 s

It can be seen from Table 1 that the time consumed by the algorithm proposed in this paper to generate an offloading strategy and computing resource allocation plan is much smaller than that of the CD_SLSQP algorithm, but the number of user devices is 10, 15 and 20, respectively. The time-consuming of CD_SLSQP is 6.8, 12.9, and 21.5 times of the algorithm proposed in this paper. It can be seen that the time complexity of CD_SLSQP algorithm is relatively high, and the time consumed will increase greatly when the number of user equipment increases.

5   Conclusion

In this paper, a computational offload decision and resource allocation algorithm based on deep reinforcement learning and convex optimization is proposed to solve a highly coupled problem. The results show that the proposed algorithm has good performance and less time consumption. We also think that this method has good extensibility, and I can easily use this algorithm to solve other similar problems.

References

1. Zhang, H., Guo, F., Ji, H., Zhu, C.: Combinational auction-based service provider selection in mobile edge computing networks. IEEE Access **5**, 13455–13464 (2017). https://doi.org/10.1109/ACCESS.2017.2721957
2. Messous, M., Sedjelmaci, H., Houari, N., Senouci, S.: Computation offloading game for an UAV network in mobile edge computing. In: 2017 IEEE International Conference on Communications (ICC), Paris, pp. 1–6 (2017). https://doi.org/10.1109/icc.2017.7996483
3. Sun, W., Liu, J., Yue, Y., Zhang, H.: Double auction-based resource allocation for mobile edge computing in industrial Internet of Things. IEEE Trans. Industr. Inf. **14**(10), 4692–4701 (2018). https://doi.org/10.1109/TII.2018.2855746.minutes
4. Chen, M., Hao, Y.: Task offloading for mobile edge computing in software defined ultra-dense network. IEEE J. Sel. Areas Commun. **36**(3), 587–597 (2018). https://doi.org/10.1109/JSAC.2018.2815360
5. Otter, D.W., Medina, J.R., Kalita, J.K.: A survey of the usages of deep learning for natural language processing. IEEE Trans. Neural Netw. Learn. Syst., 1–21 (2020). https://doi.org/10.1109/tnnls.2020.2979670
6. Atluri, G., Karpatne, A., Kumar, V.: Spatio-temporal data mining: a survey of problems and methods. ACM Comput. Surv. **51**(4), 41 (2018). Article 83. https://doi.org/10.1145/3161602
7. Wang, X., Wang, C., Li, X., Leung, V.C.M., Taleb, T.: Federated deep reinforcement learning for Internet of Things with decentralized cooperative edge caching. IEEE Internet Things J. **7**, 9441–9455 (2020). https://doi.org/10.1109/jiot.2020.2986803

8. Chen, J., Ran, X.: Deep learning with edge computing: a review. Proc. IEEE **107**(8), 1655–1674 (2019). https://doi.org/10.1109/JPROC.2019.2921977
9. Bi, S., Zhang, Y.J.: Computation rate maximization for wireless powered mobile-edge computing with binary computation offloading. IEEE Trans. Wireless Commun. **17**(6), 4177–4190 (2018). https://doi.org/10.1109/TWC.2018.2821664
10. Kraft, D.: A software package for sequential quadratic programming. 1988. Tech. rep. DFVLR-FB 88-28, DLR German Aerospace Center – Institute for Flight Mechanics, Koln, Germany
11. Loshchilov, I., Hutter, F.: Online batch selection for faster training of neural networks (2015). arXiv:1511.06343
12. Horgan, D., Quan, J., Budden, D., Barth-Maron, G., Hessel, M., van Hasselt, H., Silver, D.: Distributed prioritized experience replay (2018). arXiv:1803.00933

Gradient Based Fast Intra Prediction Algorithm for VVC

Shiyu Wang[(✉)] and Qiang Li

School of Communication and Information Engineering,
Chongqing University of Posts and Telecommunications, Chongqing, China
593948357@qq.com

Abstract. Versatile Video Coding (VVC) achieves better performance with about 40% bitrate reduction compared to H.265/HEVC under the same video quality. The improvement of VVC coding performance is at the cost of increased computational complexity of the VVC encoder. To address this issue, a fast algorithm for VVC intra prediction based on gradient information of coding unit (CU) is proposed in this paper. Experimental results illustrate that the proposed algorithm can save 47.78% coding time on average as compared with VTM7.0 with 2.25% increase in BDBR and 0.09 dB loss in PSNR.

Keywords: VVC · Intra prediction · Gradient · CU partition · Mode decision

1 Introduction

With the rapid development of video technology, people have higher requirements for the visual experience brought by video. High Definition (HD) and Ultra-High Definition (UHD) videos have drawn more attention since they can retain more abundant image details and provide more realistic visual effects. However, the network transmission of HD and UHD videos is faced with a great challenge due to the huge data volume brought by HD and UHD videos with the high resolution and extensive range of luminance. Therefore, there is an urgent need for a coding standard with better performance than the previous generation video standard including High Efficiency Video Coding (HEVC) [1] and other currently coding standards. In order to solve this problem, ISO/IEC Moving Picture Experts Group and ITU-T Video Image Experts Group have jointly established the Joint Video Experts Team (JVET) and have launched an emerging standard, namely Versatile Video Coding (VVC). JVET determined the first draft of the standard [2] and Video Test Model 1 (VTM1) [3] at the San Diego conference in the United States on April 10, 2018. The standardization of H.266/VVC is currently in progress, and it is expected to be completed by 2020.

As shown in Fig. 1, VVC still adopts the hybrid coding structure in HEVC, but new coding tools are introduced into each functional module. For the block partition structure, VVC adopts Quadtree with nested Multi-type tree (QTMT) structure and removes the separation of the CU, Prediction Unit (PU), and Transform Unit (TU) concepts [4]. For the intra prediction, VVC adopts the more refined intra prediction mode with 65 intra prediction directions [5] for square CUs and 85 intra

H. Gao et al. (Eds.): ChinaCom 2020, LNICST 352, pp. 567–581, 2021.
https://doi.org/10.1007/978-3-030-67720-6_39

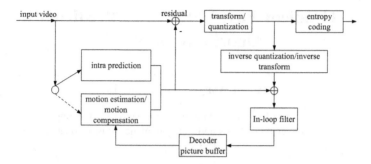

Fig. 1. H.266/VVC coding framework

prediction directions for rectangle CUs [6], to replace the 33 directions in square PUs in HEVC. In order to further improve the compression performance of intra prediction, plenty of advanced coding tools are implemented, e.g. Position-Dependent intra Prediction Combination (PDPC) [7], Multi-Reference Line (MRL) [8] and Multiple Transform Selection (MTS) [9]. PDPC employs a weighted combination of filtered and unfiltered boundary reference samples for intra prediction. The MRL not only uses the nearest reference line as HEVC intra prediction but also uses two additional lines which near the nearest reference line. Although these advanced coding tools enhance the performance of VVC by increasing the compression ratio of VVC, they increase the coding complexity of the encoder. Under All-Intra (AI) test configuration, the intra coding complexity of VVC Test Model (VTM) is 18 times higher than that of HEVC test Model (HM) [10]. Thus, it is meaningful to committed to the research of the optimization of intra prediction in VVC.

There have been a couple of researches to reduce the intra prediction computational complexity with more time saving and negligible loss in BDBR and PSNR. The authors in [11] proposed the fast CU partition algorithm for VVC intra coding based on the correlation between the best partition type of the parent CU and that of its sub-CUs. In [12], a fast QTMT partition decision based on the distribution features of CU size and intra mode was proposed. This paper also introduced a fast intra mode decision algorithm based on gradient descent search. Ting-Lan Lin [13] predicted the depth range of the CU and made a fast decision of binary tree division according to the gradient information. In [14], a fast intra coding algorithm based on the spatial correlation of the CU and its neighboring LCUs was proposed. Although the algorithms mentioned above effectively reduce the encoding time for VCC intra prediction, there are still some defects and room for improvement. First, the samples selected by the fast algorithm based on statistics must have universal applicability. Second, when it comes to videos with complex texture, the performance of fast algorithms based on spatial correlation is significantly degraded.

The rest of the paper is organized as follows: Sect. 2 presents an overview of partition structure and intra prediction in VVC. Section 3 gives a detailed description of the fast intra prediction algorithm for VVC, including the fast CU partition algorithm and fast mode decision algorithm. Experimental results and comparisons with VTM7.0 are shown in Sect. 4. Finally, the conclusions are drawn in Sect. 5.

2 Overview of Partition Structure and Intra Prediction in VVC

2.1 Partition Structure

In order to better adapt to the local texture features of the image, VVC employs QTMT as block partition structure. The partition process of Coding Tree Unit (CTU) is as follows. The CTU is partitioned by Quadtree (QT) at first. Then, the leaf nodes in the previous step can be further partitioned by QT or multi-type tree (MT) until the maximum partition depth is reached.

As illustrated by Fig. 2, QTMT includes QT and MT which consist of Vertical Binary Splitting (SPLIT_BT_VER), Horizontal Binary Splitting (SPLIT_BT_HOR), Vertical Ternary Splitting (SPLIT_TT_VER), and Horizontal Ternary Splitting (SPLIT_TT_HOR). Figure 3 is given as an example of block partition in VVC. The QTMT partition structure of a CTU and the corresponding tree representation are shown in Fig. 3(a) and Fig. 3(b), respectively. The QT partition is indicated by solid black lines, and the MT partition is indicated by blue and red dotted lines.

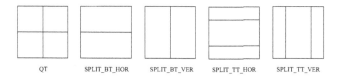

Fig. 2. Five partition types of QTMT structure

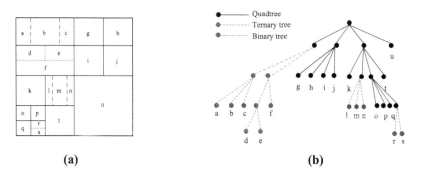

(a)   (b)

Fig. 3. An illustration of QTMT structure. (a) Example of QTMT structure, (b) The corresponding tree representation

There are some restrictions to simplify the CTU partition process. For example, QT is forbidden in the leaf nodes of MT and some redundant partition patterns could potentially result in the same coding block structure are disallowed. Besides, the sizes of the quad-tree and the multi-type tree are limited by several encoding parameters.

2.2   Intra Prediction

VVC adopts 67 intra prediction modes in square CU. Specifically, VVC retains the Planar mode and DC mode of HEVC and adds 32 directional modes which are denoted by red dotted arrows in Fig. 4. As shown in Fig. 4, the 65 directional prediction modes of VVC have the same direction range as the 33 directional modes of HEVC, but the granularity is smaller.

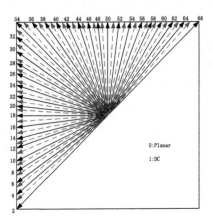

Fig. 4.   67 intra prediction modes

In order to simplify the intra prediction process of VVC, the classical three-stage fast intra mode decision is adopted. Firstly, N modes with least Hadmard cost (*HCost*) is selected through Rough Modes Decision (RMD) to construct the candidate list.

$$HCost = D_{SATD} + \lambda \cdot Bit_{mode} \qquad (1)$$

Where, D_{SATD} is the Sum of Absolute Transform Difference (SATD) of residual signal, which represents the distortion of the current prediction mode, λ is the Lagrangian multiplier, and Bit_{mode} represents the number of bits that is required for encoding in the intra prediction mode. Secondly, Most Probable Modes (MPM) are taken into account to update the candidate list. Thirdly, 2 or 3 modes with less cost are involved in the Rate-Distortion Optimization (RDO) process [15] to select the least RDO cost (*RDCost*) mode which is the optimal mode for CU. The *RDCost* is calculated by (2).

$$RDCost = SSE + \lambda \cdot Bit_{mode} \qquad (2)$$

Where, the *SSE* is the sum of the squared errors between original CU and the reconstructed CU. Although the three-stage fast intra mode decision played a certain role in reducing the complexity of intra prediction, the intra prediction process is still time-consuming. To further reduce the intra prediction coding complexity, VVC adopts two step fast decision in RMD process. First, N modes with least Hadmard cost were selected from 33 prediction modes with the same direction as HEVC. Then, the neighbor modes of the selected N modes are further checked by Hadmard cost.

3   Proposed Fast Intra Prediction Algorithm

3.1   Fast CU Partition Decision Algorithm

In order to reduce the computational burden of encoder through reducing the number of partition patterns that need to be traversed in CU recursive partition process, we explore the new partition types distribution of VTM7.0. We count the QTMT distribution information of six video sequences with different resolutions, motion characteristics and contents from each class. *FoodMarket* from Class A1 has rich scene changes, *CatRobot* from Class A2 with local motion, *BasketballDrive* from Class B with high motion, *PartyScene* from Class C with rich texture, *BQSquare* from Class D includes natural scenery and building structure and *FourPeople* from ClassE with a lot of static background area. Each sequence is coded with QP of 22, 27, 32, 37 and AI configuration. QT1 means QT depth is 1, QT2 means QT depth is 2, QT3 means QT depth is 3, QT4 means QT depth is 4, noMT means without MT.

Table 1. Partition types distribution for different sequences (%).

Sequence	QT1	QT2 +noMT	QT2 +MT	QT3 +noMT	QT3 +MT	QT4 +noMT	QT4 +MT
FoodMarket	41.7	22.9	27.7	4.1	2.2	1.1	0.3
CatRobot	27.5	19.7	43.5	2.6	2.9	1.6	2.2
BasketballDrive	15.3	21.3	40.4	5.6	12.2	3	2.2
PartyScene	0.1	1.6	9.9	6.4	46.8	6.9	28.3
BQSquare	0.1	7.6	19.5	5.1	32.1	7.6	28
FourPeople	41.9	16.8	18.5	7.9	5.7	2.5	6.7
Average	21.1	14.98	26.58	5.28	16.98	3.78	11.28

From Table 1, we can observe that the probability that the MT partition is selected is high. In addition, the regions with relatively complex texture are more likely to be partitioned by MT to achieve better prediction results. Therefore, this algorithm makes a fast decision on MT partition to achieve the purpose of saving coding time. Considering that small CUs do not occupy much encoding time, just the CUs of size 16×16 and above are taken into account.

1) *Extraction of gradient information of CU*

Gradient is often used in edge detection for image processing, which can be applied to CU partition decision [16]. Commonly used gradient operators include Roberts operator, Prewitt operator, and Sobel operator. Roberts operator is the simplest kind of operator, but it will cause the phenomenon of local edge loss, causing the contour edge of the detected object to be discontinuous. Prewitt operator does not have high accuracy for edge positioning, the detected edge is wider, and there are slightly more discontinuities. Sobel operator that performs weighted averaging on the upper, lower, left, and right neighboring pixels of a pixel can provide accurate edge direction information is adopted in this algorithm. Therefore, Sobel operator is adopted in this section to extract the gradient information of CU. The two convolution masks of Sobel operator, G_x and G_y, are described as follows.

$$G_x = \begin{bmatrix} -1 & 0 & 1 \\ -2 & 0 & 2 \\ -1 & 0 & 1 \end{bmatrix} \tag{3}$$

$$G_y = \begin{bmatrix} -1 & -2 & -1 \\ 0 & 0 & 0 \\ 1 & 2 & 1 \end{bmatrix} \tag{4}$$

We add the absolute gradients of the CU in the horizontal and vertical directions to get the absolute total gradient of the CU (G_{total}) as Eq. (5), (6) and (7) are the Calculation formula of the absolute total gradient in horizontal direction (G_{x_total}) and vertical direction (G_{y_total}), respectively.

$$G_{total} = G_{x\_total} + G_{y\_total} \tag{5}$$

$$G_{x\_total} = \sum_{i=0}^{width-1} \sum_{j=0}^{height-1} |G_x(i,j)| \tag{6}$$

$$G_{y\_total} = \sum_{i=0}^{width-1} \sum_{j=0}^{height-1} |G_y(i,j)| \tag{7}$$

Where, $G_x(i, j)$ and $G_y(i, j)$ represent the gradients of each pixel in horizontal and vertical direction respectively. $width$ and $height$ represent the numbers of pixels in each row and each column of CU, respectively. $G_x(i, j)$ and $G_y(i, j)$ are calculated as follows.

$$G_x(i,j) = M_{i,j} \times G_x \tag{8}$$

$$G_y(i,j) = M_{i,j} \times G_y \tag{9}$$

Where, $M_{i,j}$ represents a 3×3 matrix centered on the pixel with coordinates (i, j).

2) *Fast CU partition decision*

In this work, all CUs are classified into three types (horizontal texture CU, vertical texture CU and other CU) according to the gradient information. The partition types of CU are determined as follows.

If the following two conditions are met: ① $G_{total} > TH$; ② $G_{x\_total} > G_{y\_total}$, then classify the CU to horizontal texture CU and skip SPLIT_BT_VER and SPLIT_TT_VER. Otherwise, if the following two conditions are met: ① $G_{total} > TH$; ② $G_{y\_total} > G_{x\_total}$, then classify the CU to vertical texture CU and skip SPLIT_B-T_HOR and SPLIT_TT_HOR. Otherwise, classify the CU to other CU and traverse all partition types.

3) *Derivation of threshold*

The CUs with larger size or richer texture represented by larger G_{total} are highly to be further partitioned. The threshold, namely *TH*, is used to select this type of CU before deciding the partition types of CU. *TH* is defined as follows.

$$TH = \alpha > > (10 - bitdepth) \tag{10}$$

Where, *bitdepth* represents the coding bit depth and α is adjustable parameter. α is derived from three representative test sequences include *FoodMarket*, *BasketballDrive* and *PartyScene*. Figure 5 shows the relationship between α, BDBR and coding time saving (*TS*). Where, BDBR represents the percentage of the code rate that can be saved by a certain encoding method under the same objective quality. The calculation formula of *TS* is as follows,

$$TS = \frac{\left(T_{ref} - T_{pro}\right)}{T_{ref}} \times 100\% \tag{11}$$

Where, T_{ref} and T_{pro} represent the encoding time of VTM7.0 reference software and proposed algorithm, respectively. We can observe from Fig. 5 that the BDBR and *TS* decrease with α and there exists a corner point for curve of BDBR from which the curve starts to be steady. Therefore, this point can be selected as the value of α. α is set to 700 in this work.

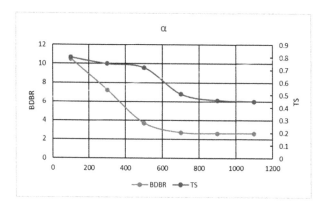

Fig. 5. Relationship between α, BDBR and coding time saving

3.2   Fast Inra Prediction Mode Decision Algorithm

In VVC intra prediction, the most time-consuming part is the calculation of *RDCost*. Although, VVC introduces three-stage fast intra mode decision and two-step fast RMD process, the coding complexity of VVC intra prediction is still very high. A fast intra prediction mode decision algorithm according to gradient information is proposed.

In order to analyse the correlation between the best intra prediction mode and gradient information of CU, six sequences mentioned above are tested (Table 2).

Table 2. Best intra mode distribution for different sequences (%).

Sequence	Sit1	Sit2	Sit3	Sit4	Sit5
FoodMarket	27.55	26.34	3.57	5.25	37.29
CatRobot	28.32	27.15	2.06	1.94	40.53
BasketballDrive	21.54	20.71	3.84	4.61	49.30
PartyScene	26.94	24.60	5.08	5.46	37.92
BQSquare	22.58	26.05	3.95	4.26	43.16
FourPeople	33.57	33.47	4.09	2.46	26.41
Average	26.75	26.39	3.77	3.99	39.10

Sit1 (situation1) means that the CU is classified to horizontal texture CU and the distribution of the best intra prediction mode is between 2 to 33; Sit2 (situation2) means that the CU is classified to vertical texture CU and the distribution of the best intra prediction mode is between 34 to 65; Sit3 (situation3) means that the CU is classified to horizontal texture CU and the distribution of the best intra prediction mode is between 34 to 65; Sit4 (situation4) means that the CU is classified to vertical texture CU and the distribution of the best intra prediction mode is between 2 to 33; Sit5 (situation5) means the best prediction mode is Planar mode or DC mode.

From Table 1,we can find that DC mode, Planar mode and the mode conformed to the major edge of image would probably be the best prediction mode. However, some unexpected situations should not be ignored.

Therefore, a method of reducing the intra prediction modes need to be traversed in RMD according to the gradient information of the CU is proposed. At the first stage of RMD, skip the mode 36, 40, 44, 48, 52, 56, 60, 64, if the CU is horizontal texture CU. Otherwise, skip the mode 4, 8, 12, 16, 20, 24, 28, 32, if the CU is vertical texture CU (Fig. 6).

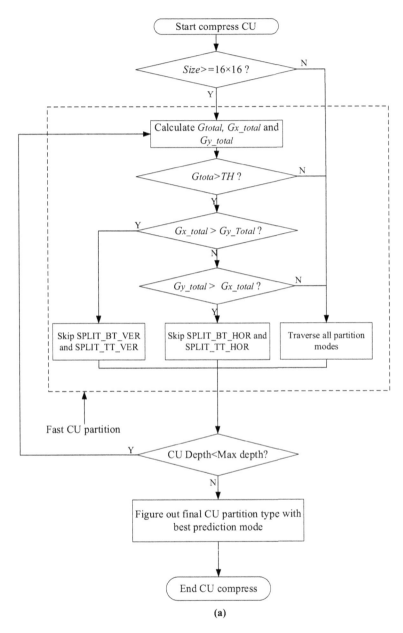

Fig. 6. Flow chart of fast CU partition and mode decision (a) Flow chart of fast CU partition (b) Flow chart of fast mode decision

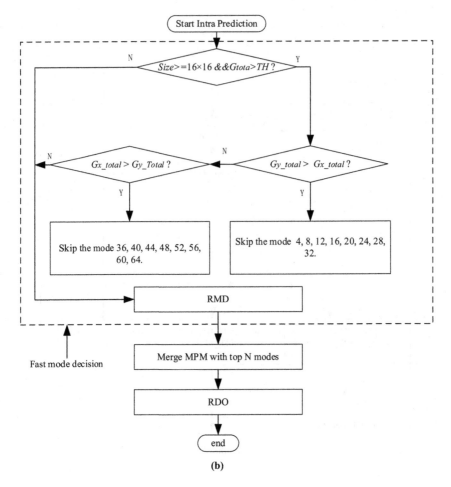

Fig. 6. (*continued*)

4 Overall Algorithm

According to the previous analysis, our fast intra prediction algorithm is described as follows.

1) Start compress CU.
2) Extract gradient information, including absolute total gradient and that in horizontal direction and vertical direction of CU.
3) Fast CU partition decision: if $G_{total} > TH$ and $G_{x\_total} > G_{y\_total}$, then classify the CU to horizontal texture CU and skip SPLIT_BT_VER and SPLIT_TT_VER. Otherwise, if $G_{total} > TH$ and $G_{y\_total} > G_{x\_total}$, then classify the CU to vertical texture CU and skip SPLIT_BT_HOR and SPLIT_TT_HOR. Otherwise, classify the CU to other CU and traverse all partition types.

4) Fast mode decision: skip some modes at the first stage of RMD, according to the gradient information of CU. If the CU is horizontal texture CU, then skip the mode 36, 40, 44, 48, 52, 56, 60, 64. Otherwise, if the CU is vertical texture CU, then skip the mode 4, 8, 12, 16, 20, 24, 28, 32.
5) The CU partition types and best intra prediction mode are decided.

5   Experimental Result

This fast intra prediction algorithm was implemented in VTM7.0 reference software. On the condition of AI, 30 frames of each sequence of all classes are tested with QP value of 22, 27, 32 and 37, afterwards calculate the mean result of them as the final result of each sequence. The test platform is Intel(R) Xeon(R) Gold 6128 CPU @ 3.40 GHz, 64 GB RAM and CentOS7 system.

Coding efficiency is measured with BDBR, BD-PSNR and TS. Where, BD-PSNR represents the difference between the PSNR-Y of the two methods at the given equivalent code rate.

Table 3. Simulation results of the proposed algorithm.

Sequence	BDBR (%)	BD-PSNR (dB)	TS (%)
Tango	1.52	−0.02	45.52
FoodMarket	1.26	−0.02	33.07
Campfire	2.65	−0.08	49.70
CatRobot	1.93	−0.03	34.30
DaylightRoad	1.34	−0.03	40.20
ParkRunning	0.62	−0.04	45.24
MarketPlace	1.78	−0.05	53.16
RitualDance	3.86	−0.19	53.85
Cactus	2.61	−0.09	51.70
BasketballDrive	2.76	−0.06	51.39
BQTerrace	1.90	−0.11	47.94
BasketballDrill	4.14	−0.13	50.51
BQMall	2.47	−0.15	50.47
PartyScene	1.47	−0.12	42.06
RaceHorses	1.67	−0.11	50.17
BasketballPass	3.28	−0.09	50.63
BQSquare	2.43	−0.21	43.62
BlowingBubbles	1.67	−0.10	56.20
RaceHorses	2.04	−0.15	43.27
FourPeople	2.91	−0.16	53.68
Johnny	2.85	−0.11	51.58
KristenAndSara	2.45	−0.12	52.81
Average	2.25	−0.09	47.78

Table 3 presents the results of the proposed algorithm compared with VTM7.0. From this table, we conclude that the proposed algorithm, compared with VTM7.0, can save 47.78% of the encoding time with 2.25% BDBR increase and 0.09 dB BD-PSNR decrease.

(a)

(b)

Fig. 7. CU partition results of different algorithms (a) partition result of VTM7.0 (b) partition result of the proposed algorithm

Figure 7 displays the partition of the first frame from BasketballDrive by using the default algorithm in VTM7.0 and the proposed algorithm separately. For some areas with flat texture, the partition results of the two algorithms are the same. But for areas with more complex textures, the partition results of the two algorithms are different. Although the proposed algorithm skips some partitions in areas with complex textures,

(a)

(b)

Fig. 8. Decoded frames of different algorithms (a) A decoded frame of DaylightRoad by VTM7.0 (b) A decoded frame of DaylightRoad by the proposed algorithm

resulting in a difference from the partition result of the default algorithm in VTM7.0, the performance degradation caused by this difference is almost negligible.

Figure 8 shows the decoded frames from DaylightRoad by using different algorithms. The naked eyes can hardly identify the difference between the two frames in Fig. 8. There is almost no difference between these buildings and vehicles.

6 Conclusions

This paper proposed a fast intra prediction algorithm for VCC includes fast CU partition and fast mode decision by making use of the gradient information of CU to skip some partition modes and intra prediction modes. Experimental results show that the proposed algorithm achieves a significant coding time saving with negligible effect in BDBR and BD-PSNR, compared to VTM7.0. The algorithm proposed in this paper can be applied to scenarios that require high real-time video transmission, such as live broadcast and video communication.

Acknowledgment. This work is supported by the National Natural Science Foundation of China (61571102) and Chongqing Basic and Frontier Research Project (cstc2017jcyjXBX0037).

References

1. Sullivan, G.J., Ohm, J., Han, W., Wiegand, T.: overview of the high efficiency video coding (HEVC) standard. IEEE Trans. Circ. Syst. Video Technol. 22(12), 1649–1668 (2012)
2. Bross, B.: Versatile video coding (Draft 1). Joint Video Experts Team (JVET) of ITU-T SG 16 WP 3 and ISO/IEC JTC 1/SC 29/WG 11 10th Meeting. San Diego, US: Joint Video Exploration Team, Doc. JVET-J1001 (2018)
3. Chen, J., Alshina, E.: Algorithm description for versatile video coding and test model 1 (VTM 1). Joint Video Experts Team (JVET) of ITU-T SG 16 WP 3 and ISO/IEC JTC 1/SC 29/WG 11 10th Meeting. San Diego, US: Joint Video Exploration Team, Doc. JVET-J1002 (2018)
4. Ma, J., et al.: Quadtree plus binary tree with shifting. Joint Video Experts Team (JVET) of ITU-T SG 16 WP 3 and ISO/IEC JTC 1/SC 29/WG 11 10th Meeting. San Diego, US: Joint Video Exploration Team, Doc. JVET-J0035 (2018)
5. Auwera, G.V., Seregin, V., Ramasubramonian, A.K., Karczewicz, M.: CE3: variable number of directional intra modes (Tests 1.1.1 and 1.1.2). Joint Video Experts Team (JVET) of ITU-T SG 16 WP 3 and ISO/IEC JTC 1/SC 29/WG 11 11th Meeting. Ljubljana, SI: Joint Video Exploration Team, Doc. JVET-K0060 (2018)
6. Racapé, F., Rath, G., Urban, F., Zhao, L., Liu, S., Zhao, X., et al.: CE3-related: Wide-angle intra prediction for non-square blocks. Joint Video Experts Team (JVET) of ITU-T SG 16 WP 3 and ISO/IEC JTC 1/SC 29/WG 11 11th Meeting. Ljubljana, SI: Joint Video Exploration Team, Doc. JVET-K0500 (2018)
7. Auwera, G.V., Seregin, V., Said, A., Ramasubramonian, A.K., Karczewicz, M.: CE3: simplified PDPC (Test 2.4.1). Joint Video Experts Team (JVET) of ITU-T SG 16 WP 3 and ISO/IEC JTC 1/SC 29/WG 11 11th Meeting. Ljubljana, SI: Joint Video Exploration Team, Doc. JVET-K0063 (2018)

8. Bross, B., et al.: CE3: multiple reference line intra prediction (Test 1.1.1, 1.1.2, 1.1.3 and 1.1.4). Joint Video Experts Team (JVET) of ITU-T SG 16 WP 3 and ISO/IEC JTC 1/SC 29/WG 11 12th Meeting. Macao, CN: Joint Video Exploration Team, Doc. JVET-L0283 (2018)

9. Zhao, X., Chen, J., Karczewicz, M., Zhang, L., Li, X., Chien, W.: Enhanced multiple transform for video coding. In: Data Compression Conference (DCC). Snowbird, United States, pp. 73–82. IEEE (2016)

10. Li, X., Suehring, K.: AHG report: JEM software development (AHG3). Joint Video Experts Team (JVET) of ITU-T SG 16 WP 3 and ISO/IEC JTC 1/SC 29/WG 11 10th Meeting. San Diego, US: Joint Video Exploration Team, Doc. JVET-J1001 (2018)

11. Fu, T., Zhang, H., Mu, F., Chen, H.: Fast CU partitioning algorithm for H.266/VVC intra-frame coding. In: IEEE International Conference on Multimedia and Expo (ICME). Shanghai, China, pp. 55–60. IEEE (2019)

12. Yang, H., Shen, L., Dong, X., Ding, Q., An, P., Jiang, G.: Low complexity CTU partition structure decision and fast intra mode decision for versatile video coding. IEEE Trans. Circ. Syst. Video Technol. p. 1 (2019)

13. Lin, T., Jiang, H., Huang, J., Chang, P.: Fast binary tree partition decision in H.266/FVC intra coding. In: IEEE International Conference on Consumer Electronics-Taiwan (ICCE-TW). Taichung, China, pp. 1–2. IEEE (2018)

14. Chen, J., Chiu, Y., Lee, C., Tsai, Y.: Utilize neighboring LCU depth information to speedup FVC/H.266 intra coding. In: International Conference on System Science and Engineering (ICSSE). Dong Hoi, Vietnam, pp. 308–312. IEEE (2019)

15. Mir, J., Talagala, D.S., Fernando, A.: Optimization of HEVC λ-domain rate control algorithm for HDR video. In: IEEE International Conference on Consumer Electronics (ICCE). Las Vegas, Vietnam, pp. 1–4. IEEE (2018)

16. Jiang, W., Ma, H., Chen, Y.: Gradient based fast mode decision algorithm for intra prediction in HEVC. In: 2nd International Conference on Consumer Electronics, Communications and Networks (CECNet). Yichang, China, pp. 1836–1840. IEEE (2012)

Performance Analysis of Multipath Mitigation Using Different Anti-multipath Techniques in BPSK and BOC Modulated Signals

Xiyan Sun[1,2], Shaojie Song[1,2(✉)], and Yuanfa Ji[1,2]

[1] Guanxi Key Laboratory of Precision Navigation Technology and Application, Guilin University of Electronic Technology, Guilin 541004, China
ssjl816977054@163.com
[2] National & Local Joint Engineering Research Center of Satellite Navigation and Location Service, Guilin 541004, China

Abstract. Multipath is a major source of positioning error in high precision navigation applications. Narrow correlator method and Gating signal correlator method are two effective methods for BPSK modulation signals multipath mitigation. In this paper, the mathematic model of multipath error are first established, and then multipath mitigation performance of the BPSK signal and BOC modulation signal are analyzed and simulated based on the narrow correlator method in a comparative way. The simulation results show that BOC signal is better than BPSK signal as for multipath mitigation performance. The consistency between narrow correlator and gating signal correlator method is deduced and proven theoretically. And it is concluded that the BOC code tracking loop phase discriminator function has ambiguity using the gating signal multipath mitigation method, so the gating signal multipath mitigation method does not work for BOC signal tracking code loop phase discriminator. Finally, the correctness of theoretical derivation is verified through simulation.

Keywords: Binary offset carrier · Gating signal · Narrow correlator · Multipath mitigation

1 Introduction

Multipath error is one of the most dominant error sources of high precision GPS positioning [1, 2]. Because multipath errors are greatly affected by the environment, and there is no correlation in time and space, so it is difficult to eliminate them by traditional differential methods. Study on multipath has always been a hot spot in the satellite navigation community. The relatively early and effective method for BPSK signal multipath mitigation is NovAtel's narrow correlator technology (NC) [3, 4]. With the implementation of the GPS modernization program and the successive emergence of Galileo and COMPASS systems, a new type of binary offset carrier (BOC) modulation signal [5, 6] has been adopted widely. The BOC signal is generated after the information code is multiplied by the pseudo-code spread spectrum, and then multiplied by a square wave sub-carrier with a higher rate than the pseudo-code rate. In this way, the original spectrum is split into two parts. Due to the splitting of the

H. Gao et al. (Eds.): ChinaCom 2020, LNICST 352, pp. 582–594, 2021.
https://doi.org/10.1007/978-3-030-67720-6_40

spectrum, the overcrowded navigation frequency band can also meet the compatibility of navigation systems to a certain extent, and avoid interference between navigation systems. In addition, the BOC signal has a sharper autocorrelation peak than the BPSK signal (as shown in Fig. 1). Therefore, the BOC signal has better ranging performance and multipath mitigation [7]. The multipath mitigation method used for the BPSK signal is based on the unimodality of the autocorrelation function of the BPSK signal, such as the narrow correlator method and the gating signal correlator method. However, the autocorrelation function of the BOC signal has multiple peaks (as shown in Fig. 1). Whether these multipath mitigation methods can be adapted to the BOC signal remains to be further demonstrated.

Aiming at the above problems, this paper comparatively analyses the multipath mitigation performance of on BPSK and BOC modulation signal using the narrow correlator method. The consistency of the gate signal correlator and narrow correlator method is derived and proved theoretically. The result shows that BOC signal is superior to BPSK signal in multipath mitigation performance by narrow correlation method. The structure of this article is as follows: Sect. 2 introduces the multipath error model of GPS receiver BPSK and BOC signals, and uses the method of formula derivation to visually show the receiver pseudo-range error and influence caused by multipath signals. The performance of multipath suppression for BPSK and BOC signals using two multipath suppression methods such as narrow correlation and gate function be tested in Sect. 3. Section 4 gives a conclusion of this paper.

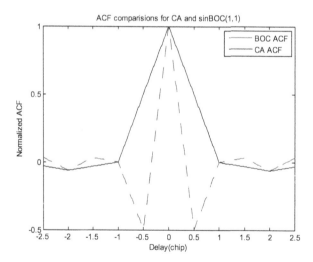

Fig. 1. Comparison of autocorrelation function of the BOC and BPSK signal

2 Multipath Error Model

2.1 Multipath Signal Model

The GPS receiver receives the superimposed signal of the satellite direct signal and the multipath signal reflected by the objects around the receiver, so the BPSK intermediate frequency signal [8] received by the receiver can be expressed as:

$$s(t) = \sum_{i=0}^{M} \alpha_i A d(t) c(t - \tau_i) \cos[w_0 t + \phi_i(t)] \tag{1}$$

Where $i = 0$ is the satellite direct signal, remaining M–1 signals are multipath signals, A is the carrier amplitude, the A value is assigned 1 in order to simplify the analysis process, α_i is the ratio of multipath to direct signal amplitude (attenuation coefficient MDR), α_0 corresponds to direct signal, $d(t)$ is the navigation message, $c(t - \tau_i)$ are the GPS pseudo-random codes with different delays, w_0 is the intermediate frequency of the satellite signal, this paper assumes that the direct signal and the multipath signal have the same frequency, $\phi_i(t)$ is the phase of the i-th signal.

The BOC IF signal received by the receiver can be expressed as:

$$s(t) = \sum_{i=0}^{M} \alpha_i d(t) sc(t) c(t - \tau_i) \cos[w_0 t + \phi_i(t)] \tag{2}$$

Where $sc(t) = sign[\sin(2\pi f_{sc}t)]$ represents sine BOC modulation signal, $sc(t) = sign[\cos(2\pi f_{sc}t)]$ represents cosine BOC modulation signal.

Assuming that the local carrier correctly tracks the frequency of the received signal, the locally generated in-phase signal and quadrature signal can be expressed as:

$$y_i(t) = c(t - \hat{\tau}_0) cos[w_0 t + \hat{\phi}_0] \tag{3}$$

$$y_q(t) = c(t - \hat{\tau}_0) sin[w_0 t + \hat{\phi}_0] \tag{4}$$

Assuming that during the integration process, the navigation message remains unchanged (that is, the meaning of the navigation message can be ignored), that is $d(t) = 1$, the correlation between the received signal and the local signal is

$$IP(\tau) = \frac{1}{T} \int_0^T s(t) y_i(t - \tau) dt$$

$$= \frac{1}{T} \int_0^T \left\{ \sum_{i=0}^{M} \alpha_i c(t - \tau_i) cos[w_0 t + \phi_i(t)] \right\} \left\{ c(t - \hat{\tau}_0) cos[w_0 t + \hat{\phi}_0(t)] \right\} dt \tag{5}$$

$$\approx \sum_{i=0}^{M} \alpha_i R(\hat{\tau}_0 - \tau_i) \frac{1}{2} \left\{ cos[2w_0 t + \phi_i + \hat{\phi}_0] + cos(\phi_i - \hat{\phi}_0) \right\}$$

Filtering out the high frequency part in Eq. 5 can be expressed as

$$IP(\tau) = \sum_{i=0}^{M} 0.5\alpha_i R(\hat{\tau}_0 - \tau_i)cos(\phi_i - \hat{\phi}_0) \tag{6}$$

Where $\hat{\tau}_0$ is the delay estimation of the direct signal, $\hat{\phi}_0$ is the carrier phase estimation of the direct signal.

2.2   Narrow Correlator Method

The traditional GPS receiver uses three correlators (Prompt, Early and Late) and cor-relation techniques to capture and track PRN codes, and thus the time delay estimation of direct signal is $\hat{\tau}_0$. Then, $\hat{\tau}_0$ is tracked by the delay lock loop (DLL) to measure the pseudo-range from the satellite to the receiver. Narrow correlator technology [9] is widely used in navigation receivers as the earlier effective multipath mitigation tech-nology. It improves the anti-multipath capability of code tracking ring by reducing the spacing between early correlator and late correlator.

The correlation function of prompt(P), early(E) and late(L) can be respectively impressed as:

$$IP(\tau) = \sum_{i=0}^{M} \frac{\alpha_i}{2} R(\hat{\tau}_0 - \tau_i)cos(\phi_i - \hat{\phi}_0) \tag{7}$$

$$IE(\tau) = \sum_{i=0}^{M} \frac{\alpha_i}{2} R(\hat{\tau}_0 - \tau_i + \frac{d}{2})cos(\phi_i - \hat{\phi}_0) \tag{8}$$

$$IL(\tau) = \sum_{i=0}^{M} \frac{\alpha_i}{2} R(\hat{\tau}_0 - \tau_i - \frac{d}{2})cos(\phi_i - \hat{\phi}_0) \tag{9}$$

In the traditional receiver, the correlator spacing d is 1Tc generally. However, d is less than 1Tc in the narrow correlator technique. Using the Eq. 8 and Eq. 9, the discrim-inator output expression of narrow correlator technique can be expressed as:

$$D_{narrow} = IE - IL \tag{10}$$

Where the spacing d between the early and late correlator is less than 1Tc.Figure 2 shows the phase discriminator output of BPSK signal and BOC(1,1) signal when the narrow correlator interval d = 0.1Tc.

Since the bandwidth of the actual receiver channel is limited, the peak value of the correlation function will not only become smooth, but also a position shift may occur, under the combined influence of the multipath signal and the limited bandwidth. When d decreases below the inverse of the bandwidth of the receiving channel, the tracking error will tend to be a constant. Therefore, due to the limitation of the receiver channel bandwidth, continuously reducing the correlation spacing d cannot reduce the multi-path error indefinitely.

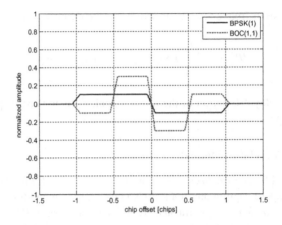

Fig. 2. Phase discriminator output of BPSK and BOC(1,1)

2.3 Gating Signal Correlator Method

In addition to the narrow correlator method, the gating signal correlator method [9, 10] is a signal correlation calculation method similar to the narrow correlator. The locally generated code of the gating signal correlator method is no longer the recurring code generated by the traditional receiver, but one or more gate-like signals, called Gating Signals. There are many ways to construct the gating signal structure, and one of them is shown in and Fig. 10 [9, 10]. The mathematical expression is

$$w(t - \tau) = \sum_{k} c_k g(t - kT_c - \tau) \tag{11}$$

Where c_k is the gating signal amplitude. It can be seen from the above expression that the period of the gating signal is the same as the period of the CA code.

The receiver signal processing process of the gating signal correlator method is as follows. The in-phase(I) branch and quadrature(Q) branch signals of the received BOC and BPSK signal can be expressed as, respectively

$$s_{i\_BOC} = \sum_{i=0}^{M} \alpha_i d(t) sc(t) c(t - \tau_i) \cos[w_0 t + \phi_i(t)] \tag{12}$$

$$s_{q\_BOC} = \sum_{i=0}^{M} \alpha_i d(t) sc(t) c(t - \tau_i) \sin[w_0 t + \phi_i(t)] \tag{13}$$

$$s_{i\_BPSK} = \sum_{i=0}^{M} \alpha_i d(t) c(t - \tau_i) \cos[w_0 t + \phi_i(t)] \tag{14}$$

$$s_{q\_BPSK} = \sum_{i=0}^{M} \alpha_i d(t) c(t - \tau_i) \sin[w_0 t + \phi_i(t)] \tag{15}$$

The symbols in Eq. 12 to Eq. 15 have the same meaning as in Eq. 1. In order to facilitate the derivation of the formula, the multipath signal is not considered. The received signal after the phase changes through rotation can be expressed as:

$$
\begin{bmatrix} y_i(t) \\ y_q(t) \end{bmatrix} = \begin{bmatrix} \cos(\hat{w}_0 t + \hat{\phi}_0) & \sin(\hat{w}_0 t + \hat{\phi}_0) \\ -\sin(\hat{w}_0 t + \hat{\phi}_0) & \cos(\hat{w}_0 t + \hat{\phi}_0) \end{bmatrix} \begin{bmatrix} s_i(t) \\ s_q(t) \end{bmatrix}
$$
$$
= \begin{bmatrix} \cos[(w_0 - \hat{w}_0)t + (\phi_0 - \hat{\phi}_0)] \\ \sin[(w_0 - \hat{w}_0)t + (\phi_0 - \hat{\phi}_0)] \end{bmatrix} \begin{bmatrix} s_i(t) \\ s_q(t) \end{bmatrix} \tag{16}
$$

Where $\hat{w}_0$ is the frequency estimation, $\hat{\phi}_0$ is the phase estimation. Assuming that the carrier frequency has been correctly tracked by the carrier loop, that is $w_0 - \hat{w}_0 = 0$. Let $\phi_0 - \hat{\phi}_0 = \phi_e$, $\tau_0 = 0$, $d(t) = 1$, Eq. 16 can be re-expressed as:

$$\begin{bmatrix} y_i(t) \\ y_q(t) \end{bmatrix} = \alpha_0 sc(t) c(t) \begin{bmatrix} \cos(\phi_e) \\ \sin(\phi_e) \end{bmatrix} \tag{17}$$

$$\begin{bmatrix} y_i(t) \\ y_q(t) \end{bmatrix} = \alpha_0 c(t) \begin{bmatrix} \cos(\phi_e) \\ \sin(\phi_e) \end{bmatrix} \tag{18}$$

The signals after phase rotation are related to the gating signal, BOC signal and CA code, which can be respectively expressed as:

$$I_{yw}(\tau) = \frac{1}{T} \int_0^T y_i(t) w(t - \tau) dt = \frac{1}{T} \alpha_0 R_{xw}(\tau) \cos(\phi_e) \tag{19}$$

$$I_{yx}(\tau) = \frac{1}{T} \int_0^T y_i(t) x(t - \tau) dt = \frac{1}{T} \alpha_0 R_{xx}(\tau) \cos(\phi_e) \tag{20}$$

$$I_{yw}(\tau) = \frac{1}{T} \int_0^T y_i(t) w(t - \tau) dt = \frac{1}{T} \alpha_0 R_{cw}(\tau) \cos(\phi_e) \tag{21}$$

$$I_{yc}(\tau) = \frac{1}{T} \int_0^T y_i(t) c(t - \tau) dt = \frac{1}{T} \alpha_0 R_{cc}(\tau) \cos(\phi_e) \tag{22}$$

If Sine-BOC(1,1) and CA code use narrow correlator method to mitigate multipath effect, their phase discrimination functions can be expressed as:

$$
\begin{aligned}
I_{yx}&\left(t+\frac{d}{2}-\tau\right) - I_{yx}\left(t-\frac{d}{2}-\tau\right) \\
&= \frac{1}{T}\int_0^T y_i(t)x\left(t+\frac{d}{2}-\tau\right)dt - \frac{1}{T}\int_0^T y_i(t)x\left(t-\frac{d}{2}-\tau\right)dt \\
&= \frac{1}{T}\int_0^T y_i(t)\left[x\left(t+\frac{d}{2}-\tau\right) - x\left(t-\frac{d}{2}-\tau\right)\right]dt \\
&= 2I_{yw}(\tau)
\end{aligned}
\tag{23}
$$

$$
\begin{aligned}
I_{yc}&\left(t+\frac{d}{2}-\tau\right) - I_{yc}\left(t-\frac{d}{2}-\tau\right) \\
&= \frac{1}{T}\int_0^T y_i(t)c\left(t+\frac{d}{2}-\tau\right)dt - \frac{1}{T}\int_0^T y_i(t)c\left(t-\frac{d}{2}-\tau\right)dt \\
&= \frac{1}{T}\int_0^T y_i(t)\left[c\left(t+\frac{d}{2}-\tau\right) - c\left(t-\frac{d}{2}-\tau\right)\right]dt \\
&= 2I_{yw}(\tau)
\end{aligned}
\tag{24}
$$

3   Simulation and Performance Analysis

In this section, for the Sine-BOC(1,1) modulation signal, through MATLAB simulation experiments, the multi-path mitigation performance of narrow correlator technique and the gating signal correlation technique on BOC signal is simulated and analyzed, respectively. The simulation conditions are as follows: static environment with only one multipath signal, the amplitude ratio of the multipath signal to the direct signal is - 6 dB and the influence of the receiver RF front-end bandwidth is negligible.

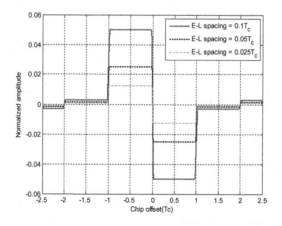

Fig. 3. Narrow correlator discriminator curve of BPSK

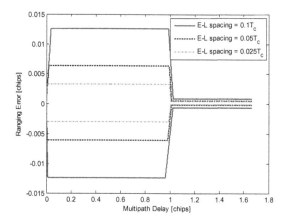

Fig. 4. Multipath error envelope for the BPSK signal with the narrow correlator

Based on the above theoretical derivation and analysis in Sect. 2, the BPSK signal and BOC signal using narrow correlator technique are simulated and analyzed in terms of multipath error. The simulation results are shown in Fig. 3 to Fig. 6.

As can be seen from Fig. 4, the narrow correlator technique has significant effect on BPSK signal multipath mitigation. With the smaller correlator interval, the better the multipath mitigation effect is. What's more, as shown in Fig. 3, the narrow correlation phase discrimination function does not have the phase discrimination ambiguity problem, because there is only one zero crossing point in the range of [−1Tc 1Tc], and there are no other false lock points. However, it is different for the BOC signal. Since the main peak of the autocorrelation function of the BOC signal is sharper than the CA code, its multipath mitigation effect is better relatively.

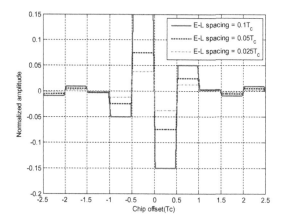

Fig. 5. Narrow correlator discriminator curve of BPSK Sine-BOC(1,1)

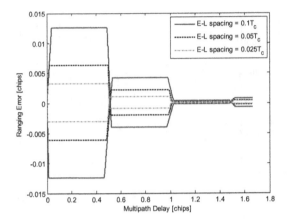

Fig. 6. Multipath error envelope for the Sine-BOC(1,1) signal with the narrow correlator

Figure 6 that Sine-BOC(1,1) has better multipath mitigation effect in the same correlator interval. However, the narrow correlation phase discrimination function of Sine-BOC (1,1) has ambiguity, because there are three zero-crossing points in the range of [−1 1] chips (as shown in Fig. 5), and the coordinates of the three zero-crossing points are −0.5, 0.5 and 0, respectively. This is because there are multiple side peaks in the autocorrelation function of the BOC signal. If the side peaks are eliminated thoroughly and the signal is correctly captured and tracked, BOC has a better multipath mitigation effect than BPSK.

The receiver signal processing block is shown in Fig. 7, the multi-path mitigation effect of the gating signal correlator and the narrow correlator method is shown in Fig. 8–13.

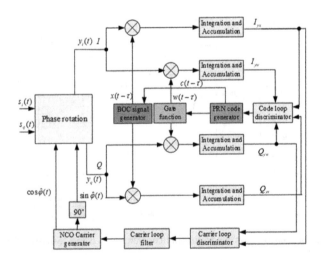

Fig. 7. Gating signal and block program of carrier loop signal processing

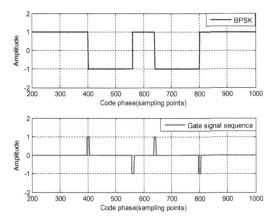

Fig. 8. Gating signal corresponding to BPSK

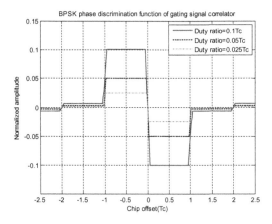

Fig. 9. BPSK phase discrimination function of gating signal correlator

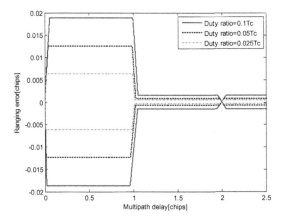

Fig. 10. Multipath error envelope for the BPSK signal with gating signal

Figure 8 is a gating signal corresponding to BPSK. The gating signal has the same period as the CA code, and the influence of the CA code is considered, so the waveform of the gating signal corresponds to the phase inversion position of the CA code. The cross-correlation function of the gating signal and the input signal is directly used as the phase discrimination function of the code tracking loop, phase discrimination curve is shown in Fig. 9 and its multipath error envelope curve is shown in Fig. 10. It can be seen from the error envelope curve that the smaller the duty ratio of the gating signal, the smaller the multipath error.

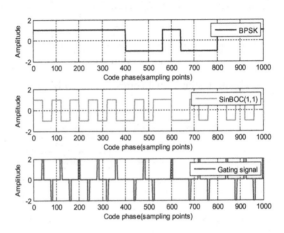

Fig. 11. Gating signal corresponding to Sine-BOC(1,1)

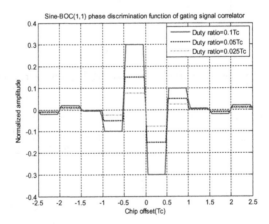

Fig. 12. Sine-BOC(1,1) phase discrimination function of gating signal correlator

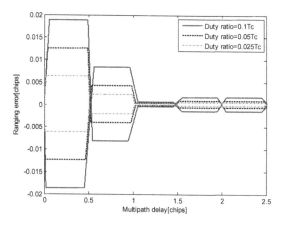

Fig. 13. Multipath error envelope for the Sine-BOC(1,1) signal with gating signal

Figure 11 is the gating signal corresponding to Sine-BOC(1,1). Compared with the BPSK gating signal, it considers the influence of subcarrier. The phase discrimination function is shown in Fig. 12. Like the phase discrimination function of the narrow correlator, the gating signal also has the problem of phase discrimination ambiguity. The multipath error envelope curve is shown in Fig. 13. Compared with the BPSK multipath error envelope curve, the Sine-BOC(1,1) multipath error is smaller,, and the multipath error is half of BPSK in the range of 0.5 to 1 chip, especially.

4 Conclusion

As a new type of navigation signal, BOC has many advantages, such as the sharper main peak and high ranging accuracy. However, multipath error is still one of its main sources of error. In this paper, the multipath mitigation effect of BPSK and BOC signals using narrow correlator method is analyzed and compared. The study shows that the BOC narrow correlator phase discrimination function has phase discrimination ambiguity. The BOC signal must be accurately captured and tracked in order to have a good multipath mitigation effect. According to the theoretical formula derivation and simulation, the gating signal correlator and the narrow correlator method have the same effect on multipath mitigation. When the correlation function of the gating signal can be used as the phase discrimination function of the code tracking loop of the BPSK signal, there is no problem of phase discrimination ambiguity. However, the correlation function of the gating signal as the phase discrimination function of the code tracking loop of the BOC signal has the problem of ambiguity. Therefore, when the traditional narrow correlator and gating signal correlator methods are used for multipath mitigation of BOC signals, measures must be taken to ensure that the signals are used under the premise of correct acquisition and tracking.

References

1. Braasch, M.S., Van Graas, F.: Guidance accuracy considerations for real-time GPS interferometry. In: Institute of Navigation GPS 91 conference, Albuquerque, New Mexico, pp. 373–386, 11–13 Sept. 1991
2. Patrick, C., Fenton, J.J.: The Theory and performance of NovAtel Inc.'s vision correlator. In: ION GNSS 2005 Long Beach, CA (2005)
3. Fenton, P., Falkenberg, B., Ford, T., Ng, K., Van Dierendonck, A.J.: NovAtel's GPS receiver-the high performance oem sensor of the future. In: Proceedings of ION GPS-91, Fourth International Technical Meeting of the Division Satellite of the Institute of Navigation, Albuquerque, NM, pp. 49–58, 11–13 Sept. 1991
4. Van Dierendonck, A.J., et.al.: Theory and performance of narrow correlator spacing in a GPS receiver. J. Instit. Navig. **39**(3), 265–283 (1992)
5. Trautenberg, H.L., Weber, T., Schafer, C.: GALILEO system overview. Acta Astronaut. **55**, 643–647 (2004)
6. Wallner, S., Hein, G., Pany, T., Rodriguez, J.A., Posfay, A.: Interference computation between GPS and Galileo. In: Proceedings of ION GNSS, pp. 861–876, Long Beach, CA (2005)
7. Betz, J.W.: Binary offset carrier modulations for radio navigations. J. Inst. Navig. **48**, 227–246 (2002)
8. Ji, Y.F., Shi, H.L., Sun, X.Y.: A strobe correlator and its multipath suppression performance research. J. Astro. **28**(5), 1094–1099 (2007)
9. Van Dierendonck, A.J., Fenton, P.C., Ford, T.J.: Theory and performance of narrow correlator spacing in a GPS receiver. Navigation **39**(3), 265–283 (1992)
10. Wu, J., Dempster, A.G.: The "BOC-Gated-PRN", A novel unambiguous multipath mitigation technique for BOC (n, n) Modulation Waveform. submitting to IEEE Transactions on Aerospace and Electronic Systems (2008)

Application of Vague Sets and TOPSIS Method in the Evaluation of Integrated Equipment System of Systems

Shangwei Luo, Yonggang Li[(⊠)], and Yanyan Chen

School of Communication and Information Engineering,
Chongqing University of Posts and Telecommunications,
Chongqing 400065, People's Republic of China
1144919304@qq.com, liyg@cqupt.edu.cn

Abstract. There are many uncertain factors in the evaluation process of integrated equipment system of systems (IES), owing to lacking the effective evaluation method. Considering the expert evaluation process is often subjective, so take the combination of entropy weight method, the Gini coefficient weighting method and AHP method are used to calculate the weight of the combat capability index; the expert evaluation information is also vague, and the vague set theory can well describe the support, neutral and opposition information. Therefore, the combination of vague set and TOPSIS method is used to calculate the degree of closeness to measure the importance of IES; Given that the combat process, equipment may be failed. The fault function is introduced to evaluate the contribution of IES dynamically by defining the new fault function and the recurrent fault function. Finally, through the case analysis, it is proved that the proposed algorithm can more accurately evaluate the contribution of IES.

Keywords: Vague set · TOPSIS · IES · Combined weight · Fault function · Combat effectiveness

1 Introduction

IES is a heterogeneous weaponry system coupled upward from different functional nodes and subsystems in accordance with the overall requirements of building an information-based army and winning an information or war. Assessing the operational effectiveness of IES in a reasonable and effective, which can play an indispensable part in optimizing the system structure and accelerating the development of weaponry. It can be able to fight and win the war security. Many experts and scholars have conducted related research and have achieved many results. Literature [1] uses grey theory to analyze the contribution of radar anti-stealth capability. Literature [2] uses data envelopment method for evaluation. Cheng C H [3] appraises according to fuzzy set theory. Gong Y [4] proposed ADC-based assessment method. Shu J S. et al. [5]

This work is supported by the Defence Advance Research Foundation of China under Grant 61400020109.

H. Gao et al. (Eds.): ChinaCom 2020, LNICST 352, pp. 595–612, 2021.
https://doi.org/10.1007/978-3-030-67720-6_41

proposed an evaluation based on Bayesian networks. Liu P. et al. [6] introduced price parameters to appraise the contribution rate of the equipment system. There are also related scholars who carry out simulation-based evaluation methods ah, such as Huang Y Y [7] who proposed a process modeling-based approach. Metin D. et al. [8] proposed an evaluation method based on hierarchical analysis, and Xiao H H. et al. [9] proposed a research method based on the vague set.

In this paper, the contribution of IES is reviewed from a new perspective, using the ordering method TOPSIS [10] (Technique for Order Preference by Similarity to Ideal Solution). A ranking method approximates the ideal solution. The TOPSIS ranks the combat capability of equipped weapon, and the higher the ranking, the greater the contribution of the equipped weapon to IES. However, the expert appraisal information is subjective in the assessment process. From this perspective, this study uses a combination of entropy weighting, Gini coefficient method [11] and hierarchical analysis to determine the indicator weights of equipped weapon. At the same time, the expert evaluation information is ambiguous and the introduction of vague set theory can be a very effective solution to this problem. In addition, the contribution rate should be evaluated taking into account the failure of the weaponry, by constructing nascent and recurring failure functions to produce the failure function. Combined, they determine the contribution of a weapon to the overall system.

2   IES Structure Model

As shown in Fig. 1, IES presents a tree-like hierarchical structure from top to bottom, composed of functionally interconnected and performance complementary equipment weapons. Underlying equipment-level weapon nodes are up-coupled into platform-level equipment, based on their different operational capabilities and characteristics. Platform-level equipment is a subsystem composed of equipment-level weapons for a particular mission, whose function includes sub-team piloting and coordinated attack, such as tank, satellite, and UAV groups. System-level equipment plays a part in leading the entire system in the context of an integrated combat network, coupled from platform-level equipment into a giant complex system.

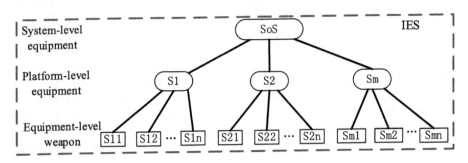

Fig. 1. IES hierarchy

IES is integrated. Locally, we can follow the OODA ring theory and view equipment-level weapons as sensor nodes, decision nodes, and influence nodes. Taken as a whole, IES can be viewed as a black box that performs a particular type of task. It can also be abstracted as a collection of three types of nodes: sensor nodes, decision nodes, and influence nodes. As illustrated in Fig. 2, IES exists to perform a series of dynamic tasks. System-level equipment divides the tasks received and assigns them to platform-level equipment. Platform-level equipment refines the tasks and assigns them to equipment-level weapons. Equipment-level weapons execute missions to detect or attack the target and feed information back to platform-level equipment. The platform level equipment forwards the received message to the system level equipment to finalize the task.

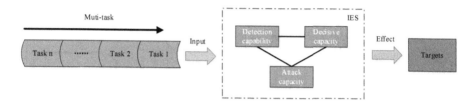

Fig. 2. IES completes tasks dynamically

3 IES Contribution Evaluation

The study of the contribution of weaponry to the system's combat capability should build the system of capability indicators for equipment. As shown in Fig. 3, the capability indicator system in this paper consists of three level 1 capability indicators, namely: detection capability, decisive capability and attack capability.

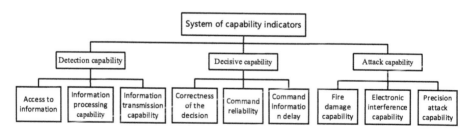

Fig. 3. System of combat capability indicators

Detective capability refers to the ability to acquire, process, and transmit information. It specifically includes the ability to access information, the ability to process information and the ability to transmit information.

Decision capability refers to the ability of aids equipment to make judgments. This includes correctness of decision-making, command reliability and command information delay.

Attack capability is the ability to attack an enemy target and incapacitate it. It specifically includes fire damage capability, electronic interference capability, and precision attack capability.

When evaluating the contribution of equipment in the IES, the characteristics of the IES should be considered. Moreover, the contribution of equipment-level weapons should be calculated through expert evaluation information. As shown in Fig. 4, this paper combines entropy weighting method, the Gini coefficient method and AHP to calculate the weights of combat capability indicators. Considering the uncertainty of expert evaluation information, thus the relative proximity is obtained by the vague set and TOPSIS method. Finally, the joint weaponry failure function collaboratively evaluates the contribution of the weaponry.

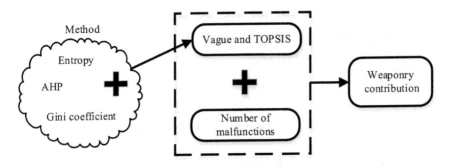

Fig. 4. Evaluation model of contribution degree of IES

3.1   Calculate the Weight of Combat Capability Indicator

The accuracy of the combat capability indicator has a direct impact on the assessment of weaponry. Therefore, it is especially important to discover a reasonable way to get the weight of the combat indicator. Considering that, IES is a complex system with many uncertainties in the battlefield; the expert assesses the combat indicators based on empirical judgment, which is also subjective. Information entropy is used to measure the amount of information contained in the indicator. It takes objective data as a landing point and talking in terms of data; The Gini coefficient reflects the accuracy of the objective data and adequately conveys the information of the objective data; to sum up, this paper uses combination of entropy weighting, Gini coefficient method and AHP to get weaponry weights. This approach takes into consideration subjectivity as well as expressing objectivity.

Due to in the process of obtaining the weapon combat capability indicator, the expert evaluation information is vague, so the evaluation indicator should be unified. Expert evaluation information is vague and needs to be measured by unified and standardized indicators. Based on previous research, this paper uses the method $1 \sim 9$ scale. It expresses the importance of current operational capability indicators relative to the higher level. As shown in Table 1.

Table 1. Relative importance judgment of the method 1–9-scale table

Relative importance	Extremely important	Very important	Important	Slightly important	Equally important
Quantified value	9	7	5	3	1

When evaluating the contribution of weaponry in the IES, there are a total of m combat indicators. It can be denoted by the set $V = (v_1, v_2, \cdots, v_n)$. During the evaluation process, n military experts evaluate m combat indicators. Since the metrics for each combat indicator are different, the evaluation indicators need to be standardized, which will result in an evaluation matrix B. where the set of experts $U = (u_1, u_2, \cdots, u_n)$. The evaluation matrix B is:

$$B = \begin{bmatrix} a_{11} & a_{12} & \cdots & a_{1m} \\ a_{21} & a_{22} & \cdots & a_{2m} \\ \vdots & \vdots & \ddots & \vdots \\ a_{n1} & a_{n2} & \cdots & a_{nm} \end{bmatrix} \tag{1}$$

Through expert analysis, the judgment matrix C is given:

$$C = \begin{bmatrix} \beta_{11} & \beta_{12} & \cdots & \beta_{1m} \\ \beta_{21} & \beta_{22} & \cdots & \beta_{2m} \\ \vdots & \vdots & \ddots & \vdots \\ \beta_{n1} & \beta_{n2} & \cdots & \beta_{nm} \end{bmatrix} \tag{2}$$

The Gini Coefficient Weighting Method

The Gini coefficient is a quantitative measure of the degree of income distribution disparity. It is widely used in the analysis of income distribution differences within the population. The formula is shown in the Eq. (3) [13]:

$$\Delta = \frac{1}{n(n-1)} \sum_{i=1}^{n} \sum_{j=1}^{n} |Y_i - Y_j|, 0 < \Delta < 2\lambda \tag{3}$$

where Δ is the value of the Gini coefficient, n is the sample size, Y_i is the income level of group i, and λ is the income expectation.

Each column of evaluation matrix B represents a different combat capability indicator, and each row represents the expert's evaluation of the indicator in that row. When using the Gini coefficient method to obtain the weights of combat indicators, this study will treat the n rows of evaluation information corresponding to a column of the evaluation matrix as different income situations in order to calculate the weights of combat capability indicators. The steps of the solution are as follows.

Step1: calculation the Gini coefficient of the IES indicator.

$$
G_k = \begin{cases} \dfrac{\sum\limits_{i=1}^{n}\sum\limits_{j=1}^{n}\left|Y_{ki}-Y_{kj}\right|}{2n^2\lambda_k}, & \lambda_k \neq 0 \\[4ex] \dfrac{\sum\limits_{i=1}^{n}\sum\limits_{j=1}^{n}\left|Y_{ki}-Y_{kj}\right|}{n^2-n}, & \lambda_k = 0 \end{cases} \tag{4}
$$

where G_k is the Gini coefficient for indicator k, n is the total number of data for the indicator, Y_{ki} is the i-th data for indicator k, and λ_k is the expectation for indicator k.

Step2: normalization of G_k obtains the weights of the k-th indicator ω_k''.

$$
\omega_k'' = \frac{G_k}{\sum\limits_{i=1}^{m} G_i} \tag{5}
$$

Linear Weighting to obtain the Combat Indicator Weights
In this study, linear weighting will be used to obtain the weights of the combat capability indicators. The use of linear weighting can easily express the subjectivity and objectivity of expert evaluation information. Moreover, this can ensure the accuracy of the combat capability indicators. According to literature [12], we can obtain the entropy weight $W' = (\omega_1', \omega_2', \cdots, \omega_n')$. By Eq. (5), the Gini coefficient weight are obtained $W'' = (\omega_1'', \omega_2'', \cdots, \omega_n'')$. According to literature [14], we can apply the AHP to find the weight $W''' = (\omega_1''', \omega_2''', \cdots, \omega_n''')$. Finally, we obtain the weight of the weaponry operational capability indicator $W = 1/3(W' + W'' + W''')$.

3.2 Vague and TOPSIS Evaluate the Contribution of the Integrated Equipment System

Introduction to Vague
Vague sets [15] is an extension of fuzzy sets. Cao and Buehree proposed vague sets based on fuzzy sets theory [16]. They argue that the membership of each element can be divided into supporting and opposing sides, i.e., a truth-membership and a false-membership. From an objective point of view, a vague set provides evidence for and against. Through supporting and against arguments, neutral evidence can be derived. It can be seen that vague sets are more realistic and more graphic than fuzzy sets in describing the objectivity, and better describes the uncertainty of the data source. Using vague sets to represent the subjectivity, uncertainty and ambiguity of the expert assessment information is more precise than fuzzy sets and more flexible in treatment as they are widely used in solution selection [17].

A vague set A in $X = (x_1, x_2, \cdots, x_n)$ is characterized by a truth-membership function $t_A(x_i)$ and a false-membership function $f_A(x_i)$. $t_A(x_i)$ is a lower bound on the grade of membership of x_i derived from the evidence for x_i, and $f_A(x_i)$ is a lower bound on the negation of x_i derived from the evidence against x_i. $t_A(x_i)$ and $f_A(x_i)$ are interrelated and have some relationship, where $t_A(x_i) + f_A(x_i) \leq 1$, $x_i \in [0, 1]$. In other words, that is $t_A : X \rightarrow [0, 1]$ and $f_A : X \rightarrow [0, 1]$. This approach bounds the grade of membership of any variable $x_i \in X$ to a subinterval $[t_A(x), 1 - f_A(x)]$ of $[0, 1]$.

Evaluate the Contribution of Weaponry using Vague Set

There are many uncertainties in the sources of information used to assess the contribution of weaponry, and the experts' evaluation has both qualitative and quantitative indicators. At the same time, different experts will have different evaluations, and some will choose to abstain. Therefore, in order to ensure the accuracy of the assessment, the chosen method should be able to portray the three aspects of support, opposition and neutrality. This paper combines vague sets with the TOPSISS method, which aptly expresses these three aspects. The steps of the algorithm for assessing the contribution of weaponry using vague sets and the TOPSIS method are divided into the following main steps.

Step1: converting evaluation information into the vague value.

Differences in the outline and physical meaning of evaluation indicators should be taken into account in the process of converting evaluation information into the vague values. Based on previous studies, the evaluation indicators are usually divided into benefit, cost and fixed target indicators. As can be seen from Fig. 3, this paper deals only with benefits-based indicators.

S_{ij} is the degree to which the weaponry q_j, under the combat indicator v_i, fulfils a task. It is a benefit interval data indicator and can be expressed as $[x_{ij}, y_{ij}]$. It is converted into a vague value by Eqs. (14) and (15) [18].

$$t_{ij} = \frac{x_{ij}^p}{x_{jmax}^p}\left(1 + \frac{y_{ij}^p - x_{ij}^p}{x_{jmax}^p}\right) \tag{6}$$

$$f_{ij} = \left(1 - \frac{y_{ij}^p}{x_{jmax}^p}\right)\left(1 + \frac{y_{ij}^p - x_{ij}^p}{x_{jmax}^p}\right) \tag{7}$$

where $x_{jmax} = \max\left(x_{1j}, y_{1j}, x_{2j}, y_{2j}, \cdots, x_{mj}, y_{mj}\right), p \in N^+$. Moreover, in this paper, p equals two.

Step2: constructing the vague decision matrix M.

The vague set is formed based on expert evaluation scores and is represented by matrix $Q = \{Q_1, Q_2, \cdots, Q_m\}$, where Q_i denotes the valuation of the weaponry q_i under the operational capability indicator v_i, expressed as a vague value:

$$Q_i = \left\{(v_1, T_{i1}), (v_2, T_{i2}), \cdots, (v_n, T_{ij})\right\} \tag{8}$$

where $T_{ij} = [t_{ij}, 1 - f_{ij}]$, t_{ij} is the grade of support of the operational indicator S_j for the weaponry. In addition, f_{ij} is the degree of negative reaction of the combat indicator S_j for the weaponry q_i.

From the formula $t_A(x_i) + f_A(x_i) \leq 1$, $x_i \in [0, 1]$, let $\lambda_{ij} = 1 - f_{ij}$. So, the formula (6) can be expressed by the following matrix M.

$$M = \begin{bmatrix} [t_{11}, \lambda_{11}] & [t_{12}, \lambda_{12}] & \cdots & [t_{1n}, \lambda_{1n}] \\ [t_{21}, \lambda_{21}] & [t_{22}, \lambda_{22}] & \cdots & [t_{2n}, \lambda_{2n}] \\ \vdots & \vdots & \ddots & \vdots \\ [t_{m1}, \lambda_{m1}] & [t_{m2}, \lambda_{m2}] & \cdots & [t_{mn}, \lambda_{mn}] \end{bmatrix} \tag{9}$$

Step 3: Identification of the ideal weaponries.

The TOPSIS method [19] is an ordering that approach the ideal solution. The TOPSIS value for each weaponry is obtained by determining the positive ideal solution (PIS) and negative ideal solution (NIS).

The ideal weaponry is selected based on the vague matrix M. The greater the similarity between the weaponry to be evaluated and the combat capability of the ideal weaponry, the greater the contribution of the weaponry to IES.

Vague sets contain three aspects of information, support, oppose and abstain. The matrix M is converted into a suitable matrix V for the combat indicator through Eq. (10). Use V_{ij} to represent the suitability of weaponry q_i for the combat indicators [20].

$$V_{ij} = (t_{ij} - f_{ij}) + (\partial_{ij} - \beta_{ij})\pi_{ij} \tag{10}$$

where ∂_{ij} and β_{ij} are the sort parameter, $\partial_{ij} \in [0, 1]$, $\beta_{ij} \in [0, 1]$. When ∂_{ij} is not equal to β_{ij}, let ∂_{ij} equal to t_{ij} and β_{ij} equal to f_{ij}. π_{ij} is the part that abstained.

$$\pi_{ij} = 1 - t_{ij} - f_{ij} \tag{11}$$

Positive ideal and negative ideal solutions are obtained from the fit matrix V, where $V_j^+ = \max\limits_{1 \leq i \leq m} V_{ij}, V_j^- = \min\limits_{1 \leq i \leq m} V_{ij}, 1 \leq j \leq n$.

$$VPIS = (V_1^+, V_2^+, \cdots, V_n^+) \tag{12}$$

$$VNIS = (V_1^-, V_2^-, \cdots, V_n^-) \tag{13}$$

VPIS is the vague solution for the positive ideal weaponry corresponding to the decision matrix. VNIS is the vague solution for the negative ideal weaponry corresponding to the decision matrix.

Step 4: Combined vague and TOPSIS calculations to assess the distance to best-case solution D_i^+ and worst-case solution D_i^- of the weaponry q_i.

$$D_i^+ = 1 - \frac{1}{n} \sum_{j=1}^{n} \omega_j M([t_{ij}, 1 - f_{ij}], VPIS), i = 1, 2, \cdots, m \tag{14}$$

$$D_i^- = 1 - \frac{1}{n}\sum_{j=1}^{n} \omega_j M([t_{ij}, 1 - f_{ij}], VNIS), i = 1, 2, \cdots, m \qquad (15)$$

where $M(x, y)$ can be expressed by the following equation.

$$M(x, y) = 1 - \frac{|t_x - t_y - f_x + f_y|}{8} - \frac{|t_x - t_y + f_x - f_y|}{4} - \frac{|t_x - t_y| + |f_x - f_y|}{8} \qquad (16)$$

Step 5: Calculating the close degree of proximity of weaponry $S(q_i)$.

$$S(q_i) = \frac{(D_i^+)^2}{(D_i^-)^2 + (D_i^+)^2} \qquad (17)$$

Step6: Rank $S(q_i)$. The higher the ranking, the greater the contribution to IES.

3.3   The Least-Squares Method to Derive the Number of Faults

In this paper, IES faults are divided into new and recurrent faults, and the probability density functions of new faults and recurrent faults are derived qualitatively by formulas.

$$g(t) = g_1(t) + g_2(t) \qquad (18)$$

where t represents the cumulative number of hours worked on the weapon. $g(t)$, $g_1(t)$, $g_2(t)$ denote equipment failure rate, new failure rate, recurrence failure rate at time t, respectively.

The probability of exposure to a new fault per unit of time is called the probability of new failure and it is displayed using a multi-exponential function, as in Eq. (19)

$$g_1(t) = A_1 e^{-B_1 t} + A_2 e^{-B_2 t} + \cdots + A_n e^{-B_n t} = \sum_{n=1}^{\infty} A_n e^{-B_n t}, \; n \in N^+ \qquad (19)$$

where A_n is the parameter strength, B_n is the shape parameter, and both parameters are greater than zero. $g_2(t)$ represents the probability of a failure occurring again within a unit of time after a failure.

$$g_2(t) = \int_0^t g_1(x)\varphi(t - x)dx \qquad (20)$$

where $\varphi(t)$ can be represented by Eq. (21).

$$\varphi(t) = \sum_{j=1}^{\infty} \partial_j e^{-b_j t} \qquad (21)$$

where j is a positive integer, ∂_j is a weight parameter, b_j is a shape parameter, and both parameters are greater than zero.

Bringing Eqs. (25), (26) and (27) into Eq. (24) derives the total number of malfunctions that occur with the weaponry working for t hours.

$$G(t) = \int_0^t g(x)dx = \int_0^t g_1(x)dx + \int_0^t g_2(x)dx \tag{22}$$

Assuming the weaponry works at time t_k, a total of q failures occurs, and the corresponding time for each failure is $\tau_1, \tau_2, \cdots, \tau_q$. A total of p faults is exposed, and the time of occurrence of each fault is $t_1, t_2, \cdots, t_p$.

The value of the highest fit will be obtained by least squares. As shown in Eq. (23), we first take the discrepancy sum of the total number of the weaponry new failures. Then, make the partial derivatives of the parameter strength A_n and shape parameter B_n in p equal to 0.

$$\varphi = \sum_{k=1}^p (G_1(t_k - k))^2 \tag{23}$$

$$\begin{cases} \dfrac{\partial \varphi}{\partial A_i} = 0 \\ \dfrac{\partial \varphi}{\partial B_i} = 0 \end{cases} \tag{24}$$

A set of solutions to φ can be derived by Eq. (24), denoted by the vector as $\alpha = (A_1, A_2, \cdots, A_n)$. Similarly, the probability density function of the failure attenuation function is obtained by the least square's method, and finally the probability density function of the failure of the weaponry is obtained. The number of malfunctions of the weaponry working in the time interval $[t_a, t_b]$ is derived from Eq. (25).

$$Con_v = \int_{t_a}^{t_b} g(t)dt \tag{25}$$

3.4 Assessment Model of the Contribution Rate of the Weapon Equipment System

The close degree of proximity of weaponry $S(q_i)$ and the number of failures Con_v are derived from Sects. 3.2 and 3.3, respectively. Assuming that the cost of breakdown repair λ_i is proportional to the cost of weaponry δ_i (RMB).

$$\chi_i = \lambda_i \times \delta_i \times con_v \tag{26}$$

where χ_i is the cost of losses due to possible malfunction of weaponry during the mission. Since the costs are economic, the smaller the better, the contribution of weaponry is:

$$C_i = \beta \frac{S(q_i)}{\sum\limits_{j=1}^{n} S(q_i)} + (1-\beta)\left(1 - \frac{\chi_i}{\sum\limits_{j=1}^{n} \chi_i}\right) \tag{27}$$

where C_i represents the degree of contribution of weapon q_i to the completion of a mission in the IES. Since $S(q_i)$ and χ_i are different in the outline of indicators, β is used to adjust the corrections to ensure accuracy. Finally, it is normalized.

4 Case Analysis

4.1 Obtain the Weight of Combat Indicators

The system of the combat capability indicators for weaponry, shown in Fig. 3, contains three first-level combat capability indicators, namely, detection capability, decision capability and attack capability. The experts use the 1–9 scale method to evaluate the three first-level combat capability indicators. In addition, we can get the judgment matrix B by them.

$$B = \begin{bmatrix} 1 & 3 & 1/5 \\ 1/3 & 1 & 1/7 \\ 5 & 7 & 1 \end{bmatrix}$$

Eight experts in the relevant fields were invited to rate the relative importance of the three first-level combat capability indicators to obtain the evaluation matrix C.

$$C = \begin{bmatrix} \text{expert1} & 9.0 & 9.3 & 7.4 \\ \text{expert2} & 8.2 & 8.6 & 8.4 \\ \text{expert3} & 9.8 & 9.0 & 6.6 \\ \text{expert4} & 8.4 & 7.5 & 5.5 \\ \text{expert5} & 8.6 & 8.1 & 9.4 \\ \text{expert6} & 7.8 & 9.2 & 6.6 \\ \text{expert7} & 9.2 & 7.5 & 7.5 \\ \text{expert8} & 8.3 & 8.5 & 9.0 \end{bmatrix}$$

The Gini Coefficient Solution Weight
The Gini coefficients of the first-level combat capability are calculated from Eqs. (4) and (5) and, which are shown in Table 2.

Table 2. Gini coefficient weight assignment

First level combat capability indicator	Detective capability	Decision capability	Attack capability
Expectation value λ_k	8.6625	8.4625	7.5500
Gini coefficient G_k	0.0768	0.0812	0.1854
Weight W_k''	0.236	0.2365	0.5399

Linear Weighting to obtain the Combat Indicator Weights

Firstly, using the evaluation matrix B can derive the entropy weight [12] of combat capability indicators. Next, the hierarchical analysis weights [14] is derived by determining the matrix C. Finally, the linear weighting method is used to derive the weight of the first-level combat capability indicator, as shown in Table 3.

Table 3. Weight distribution table

Methods	Detective capability	Decision capability	Attack capability
The entropy weight	0.1209	0.1648	0.7143
The Gini weight	0.236	0.2365	0.5399
AHP	0.1884	0.0810	0.7306
Linear weighting	0.1818	0.1608	0.6574

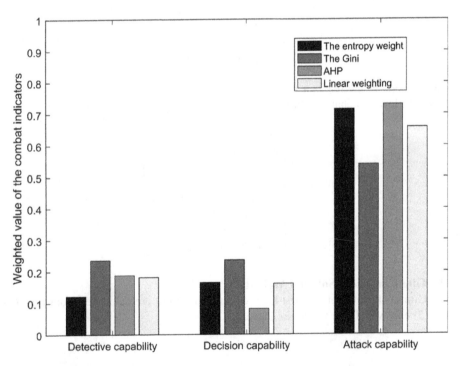

Fig. 5. Combat capability index weight

As shown in Fig. 5, it is unreasonable to obtain the weight of the combat capability indicator by AHP, ignoring the objectivity of the evaluation information. Linear weighting finds a balance between subjectivity and objectivity. The simulation graph shows that the linear weightings are always between subjectivity (AHP) and objectivity (the entropy weight and the Gini coefficient). This can prove linear weighting consider both subjectivity and objectivity. Moreover, it has high reliability. Given the weights of the secondary combat indicators, the combined weights are derived by multiplying them with the first-level combat capability indicators, as Table 4 shows.

Table 4. Table of the combat capability indicators

The first-level combat capability indicator	Weight	The second-level combat capability indicator	Weight	Combined weight
Detective capability	0.1608	Access to information	0.3	0.04824
		Information processing capability	0.5	0.0804
		Information transmission capability	0.2	0.03216
Decisive capability	0.6574	Correctness of the decision	0.35	0.23009
		Command reliability	0.25	0.16435
		Command information delay	0.4	0.26296
Attack capability	0.1818	Fire damage capability	0.55	0.09999
		Electronic interference capability	0.2	0.03636
		Precision attack capability	0.25	0.04545

4.2   Finding the Closeness of Vague Sets

Due to space limitations, this paper calculates the level of combat capability satisfaction in terms of the combat capability, as shown in Fig. 6.

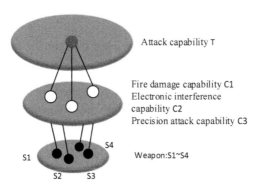

Fig. 6. Topological relationship of attack capability

In Fig. 6, T represents attack capability, C1, C2, C2 respectively represent different combat capability indicators, and S1, S2, S3, and S4 represent different weaponry.

The expert evaluates the satisfaction of the equipment in completing its mission based on experience. Looking at the basic information of the attack weaponry's status, we can get the cumulative working time (100 min) of the weapon and the amount of money spent on purchasing the weapon (millions of dollars). Table 5 shows the evaluation information and basic status of the attack weaponry.

Table 5. Table of evaluative information and basic status of weaponries

Weaponry	Attack capability			Time	Cost
	C1	C2	C3		
S1	70–75	81–90	84	5–6	2
S2	85	60–75	88	15–16	3
S3	84–92	82–87	82–90	25–16	4
S4	85–90	83–85	86–94	50–51	5

The expert evaluation is ambiguous and therefore supports interval scoring to better reflect and express this ambiguity. In this paper, using vague sets and TOPSIS method, it is possible to deal with evaluation values as single and interval values. The value of the fuzzy decision matrix M is obtained by Eqs. (6), (7), (9) and the result is shown in Table 6.

Table 6. Vague value of combat equipment

Weaponry	Attack capability		
	C1	C2	C3
S1	70–75	81–90	84
S2	85	60–75	88
S3	84–92	82–87	82–90
S4	85–90	83–85	86–94

From Table 4, it can be seen that the weight vector for the secondary combat capability indicator is $W = (0.35, 0.25, 0.4)$. The VPIS and VNIS are derived by Eqs. (12), (13). The results are VPIS = ([0.9723, 1.0], [0.9639, 1.0], [0.9503, 0.9757]) and VNIS = ([0.6285, 0.6358], [0.5556, 0.6181], [0.7986, 0.7986]).

Calculation of the distance between the striking equipment and the positive and negative ideals by Eqs. (18), (19), (20). Then according to the Eq. (21) to calculate the close degree of proximity of the weaponry $S(q_i)$, the results are shown in Table 7.

Table 7. Similarity and closeness of combat equipment

Weaponry	D_i^-	D_i^+	$S(q_i)$
S1	0.9490	0.9008	0.4740
S2	0.9446	0.9040	0.4780
S3	0.8701	0.9766	0.5575
S4	0.8679	0.9776	0.5592

Table 7 shows that S4 > S3 > S2 > S1 when damage to equipment is not taken into account. From the table you can get the best results for weaponry S4. Nevertheless, that is clearly, not how we measure the combat contribution of IES. Under resource-constrained conditions, we should take a comprehensive view of the problem, consider equipment failures, and choose the optimal strategy. In a war, the cost of equipment malfunction should be a secondary, but necessary. Number of malfunctions of the weaponry can be calculated by formula (25). Then Eq. (26) takes the cost of weaponry and finally the combined contribution is obtained through (27). When $\beta = 0.9$, the data results are shown in Table 8. The number of failures is Con, and the maintenance cost is Cos (10,000 Yuan).

Table 8. Consider the comprehensive contribution of the malfunction

Weaponry	Cos	Con	Ranking of $S(q_i)$	The comprehensive contribution
S1	0.9490	0.9008	0.4740	0.2255
S2	0.9446	0.9040	0.4780	0.2369
S3	0.8701	0.9766	0.5575	0.2661
S4	0.8679	0.9776	0.5592	0.2715

From the results in Table 8, it can be seen that the number of failures decreases over time as reliability grows, taking into account changes in the state of the technology. When $\beta = 0.90$, the overall contribution rate is ranked as S4 > S3 > S2 > S1. It is consistent with the ranking of the contribution of the weaponry to IES when failure is not considered.

Considering the effect of different values of β, assuming that the four strike equipment have equal cumulative working hours, and work in the same time period, and make $Con = 1$, so that the weaponry S1, S2, S3, and S4 have the battle losses of 20,000, 30,000, 40,000, and 50,000, respectively.

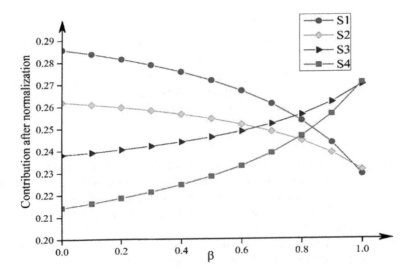

Fig. 7. Topological relationship of strike capability

As shown in Fig. 7, when the number of failures is constant, the contribution of the weaponry to the overall changes with the change of the β value. When $\beta = 0$, S1 has the best effect. As the β value rises, the impact of cost becomes smaller and smaller, and more attention is paid to the degree of weaponry completing the mission. When $\beta = 1$, it is equivalent not to considering the influence caused by the number of failures. It is only related to the closeness of the weaponry, which is consistent with the $S(q_i)$ normalized results in Table 7. As the closeness coefficients of weaponry S1 and S2 are relatively low, with the increase of β value, the impact of cost becomes weaker and weaker, its contribution to the entire system is also lower and lower, and the curve shows a downward trend. On the contrary, S3 and S4 show an upward trend.

5 Conclusion

As a complex system, the integrated equipment system has many uncertainties, and the expert assessment of IES is somewhat subjective. This article makes full use of the evaluation information and uses the combination of objective and subjective methods to comprehensively obtain the weight of combat indicators. The literature does not consider the impact of new and recurring failures on the equipment system when considering the contribution of equipment system effectiveness. Aiming at the problem of equipment body failure, this paper introduces new failures and recurring failures to measure the number of failures. Viewing that the expert evaluation score is vague, it has three forms: support, opposition, and abstention. The TOPSIS method and the vague set express these three forms. In view of this, this paper considers the contribution of IES using a combination of vague and TOPSIS when considering equipment failures. It provides new ideas for evaluating IES. Moreover, Simulation shows that the algorithm in this paper can be used to find the balance point between loss cost and combat capability, and provide a theoretical basis for decision makers to select suitable weapons and equipment and accelerate equipment development.

References

1. Shi, J.P., Hu, G.P., Li, T.: Evaluation of anti-stealth ability of radar on improved grey correlation algorithm. J. Harbin Inst. Technol. **47**(3), 116–121 (2015)
2. Meng, Y.L., Chen, G.M., Han, R.F.: Contribution rate assessment of equipment system capability in early warning counter attack combat. Fire Control & Command Control **44**(7), 27–32 (2019)
3. Cheng, C.H.: Evaluating weapon systems using ran-king fuzzy numbers. Fuzzy Sets & Syst. **107**(1), 25–35 (1999)
4. Gong, Y., Liu, Y.Q., Zhu, R.G.: Effectiveness valuation of data link countermeasure reconnaissance based on ADC and improved cloud model. Fire Control Command Control **44**(8), 111–115 (2019)
5. Shu, J.S., Yao, Q., Wu, J., et al.: Evaluation of conventional missile anti-ship combat effectiveness based on bayesian network. Fire Control & Command Control **44**(1), 114–118 (2019)
6. Liu, P., Zhao, D., Tan, Y.J., et al.: Multi-task oriented contribution evaluation method of weapon equipment system of systems. Syst. Eng. Electron. **41**(8), 1763–1770 (2019)
7. Huang, Y.Y.: A methodology of simulation and evaluation on the operational effectiveness of weapon equipment. In: 2009 Chinese Control and Decision Conference, pp. 131–136. IEEE, Guilin (2009)
8. Metin, D., Yavuz, S., Nevzat, K.: Weapon selection using the ahp and topsis methods under fuzzy environment. Expert Syst. Appl. **36**(4), 8143–8151 (2009)
9. Xiao, H.H., Ji, J.Y., Xu, B.: Evaluation of submarine combat effectiveness based on vague set. J. Naval Aero. Astro. Univ. **30**(1), 78–82 (2015)
10. Kuo, T.: A modified TOPSIS with a different ranking index. Euro. J. Oper. Res. **260**(1), 152–160 (2017)
11. Peng, X., Lin, L., et al.: Approach for multi-attribute decision making based on gini aggregation operator and its application to carbon supplier selection. IEEE Access **7**, 164152–164163 (2019)
12. Xiao, Q., He, R., Ma, C., et al.: Evaluation of urban taxi-carpooling matching schemes based on entropy weight fuzzy matter-element. Appl. Soft Comput. **81**, 105493 (2019)
13. Bowles, S., Carlin, W.: Inequality as experienced difference: a reformulation of the Gini coefficient. Econ. Lett. **186**, 108789 (2020)
14. Eddie, W.L., Cheng.: Analytic Hierarchy Process. Encyclopedia of Biostatistics. John Wiley & Sons (2016)
15. Zhang, Q.H., Xie, Q., Wang, G.Y.: A survey on rough set theory and its application. CAAI Trans. Intell. Technol. **1**(4), 323–333 (2016)
16. Kaur, A., Kacprzyk, J., Kumar, A.: A brief introduction to fuzzy sets and fuzzy systems. Springer, Fuzzy Transportation and Transshipment Problems (2020)
17. Wang, W.P.: Research on the linguistic information multi-criteria decision making based on vague sets. Economic Science Press (2013)

18. Li, G.X., Wang, H.X., et al.: Study on transforming formulas from interval valued data to vague valued data. Comput. Eng. Appl. **46**(23), 56–58 (2010)
19. Zuo, W., Cao, Y., Li, Y., An, B.: An evaluation method based on TOPSIS for urban rail transit power supply system. In: 2019 IEEE PES Asia-Pacific Power and Energy Engineering Conference (APPEEC), pp. 1–5. IEEE, Macao (2019)
20. Xiang, Y., Sheng, J.B., Yuan, H., et al.: Research on degrading and decommissioning assessment of reservoir in China. Sci. Sin. Tech. **45**(12), 1304–1310 (2015)

A Transmission Design via Reinforcement Learning for Delay-Aware V2V Communications

Siyuan Yu, Nong Qu, Yizhong Zhang, Chao Wang$^{(\boxtimes)}$, and Fuqiang Liu

Department of Information and Communication Engineering,
Tongji University, Shanghai 201804, China
{1832914,1910624,1930707,chaowang,liufuqiang}@tongji.edu.cn

Abstract. We investigate machine-learning-based cross-layer energy-efficient transmission design for vehicular communication systems. A typical vehicle-to-vehicle (V2V) communication scenario is considered, in which the source intends to deliver two types of messages to the destination to support different safety-related applications. The first are periodically-generated heartbeat messages, and should be transmitted immediately with sufficient reliability. The second type are randomly-appeared sensing messages, and are expected to be transmitted with limited latency. Due to node mobility, accurate instantaneous channel knowledge at the transmitter side is hard to attain in practice. The transmit channel state information (CSIT) often exhibits certain delay. We propose a transmission strategy based on the deep reinforcement learning technique such that the unknown channel variation dynamics can be learned and transmission power and rate can be adaptive chosen according to the message delay status to achieve high energy efficiency. The advantages of our method over several conventional and heuristic approaches are demonstrated through computer simulations.

Keywords: Cross-layer transmission design · Vehicular communication · Deep reinforcement learning

1 Introduction

Traffic congestion, road safety, and energy shortage have become severe issues in the modern transportation system. Supported by the great progress made in the Internet of things (IoT), high-performance cloud/edge computing, and LTE/5G radio access technologies [10], intelligent transportation system (ITS) has been widely accepted as the promising solution and has attracted tremendous attentions in both academia and industry. As one of the key ITS technologies, vehicular networking enables traffic information, sensing data, and control demands to be shared, so that various ITS services can be developed [7]. However, such applications are in general safety-related and have diverse characteristics. Message transmissions have to be conducted among highly mobile terminals,

© ICST Institute for Computer Sciences, Social Informatics and Telecommunications Engineering 2021
Published by Springer Nature Switzerland AG 2021. All Rights Reserved
H. Gao et al. (Eds.): ChinaCom 2020, LNICST 352, pp. 613–627, 2021.
https://doi.org/10.1007/978-3-030-67720-6_42

subject to stringent delay and reliability requirements [6]. Realizing high-quality communication, especially vehicle-to-vehicle (V2V) transmission, is challenging.

Efficient transmission design in wireless systems has been extensively investigated for years. The conventional designing procedure is mainly based on channel state information (CSI) in the PHY layer. Although such an approach is sufficiently good for most modern wireless communication scenarios, it may not be able to satisfy the demands in vehicular communication systems due to a number of reasons. First, CSI does not reflect the time that source messages have already waited before satisfactory transmission opportunities are available. This may cause unbounded transmission delay. Involving the queue state information (QSI) in the MAC layer also into the decision-making process would enable *delay-aware transmission design*. Second, a sufficient amount of instantaneous CSI at the transmitter side is often assumed. Attaining such channel knowledge can be managed in relatively static wireless environments [14], but is challenging in dynamic ITS. Third, existing transmission design normally focuses on only one single type of application. In vehicular networks, however, multiple types of applications with diverse quality of service (QoS) requirements always co-exist. Therefore, new models and solutions are needed.

Our earlier work [4] investigates a multi-user V2V communication network in which each information source desires to send two types of delay-limited messages with different characteristics to its destination. We propose a cross-layer delay-aware transmission design using both CSI and QSI through the Lyapunov optimization theorem. It is shown that the performance can be much better than conventional CSI-based approaches. Nevertheless, the solution is established based on a knowledge of instantaneous transmit CSI (CSIT). In practice, a common approach to attaining CSIT is to exploit feedback from the receiver. In highly mobile networks, channel knowledge conveyed in the feedback normally can only reflect a delayed version of the true channel condition. In this case, transmission design based on instantaneous CSIT would not be applicable.

The past few years have witnessed explosive advances in artificial intelligence (AI) technologies. Utilizing machine learning techniques to facilitate wireless systems design has already attracted attentions in the communications research society [2,3]. As a main branch of machine learning, reinforcement learning, especially when combined with deep neural networks (DNNs), has been proven to be capable of solving a wide range of challenging sequential optimization problems (see, e.g., [5,8,11,12,15]). In this paper, we investigate the potential of applying reinforcement learning for enabling V2V transmission design.

Specifically, we consider a typical V2V transmission system, in which two types of messages with different QoS requirements are delivered to support safety-related applications. The first type are heartbeat messages which should be transmitted immediately with high reliability. The second type are environment sensing messages that should be sent with limited delay. The small-scale channel fading coefficients change across time-division slots following a fixed but unknown distribution. Only a delayed version of the CSIT is available. It is expected that the transmission can be realized with

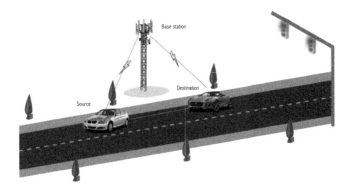

Fig. 1. System model

the maximal energy efficiency, i.e., one joule energy enables the maximum information delivery. We propose formulating the design optimization problem as a finite Markov decision process (MDP), and apply a deep Q-network (DQN) algorithm to solve it. Simulation results show that, our method is able to achieve close performance as using the method presented in [4] with perfect channel knowledge, and outperform three heuristic solutions in the imperfect CSI environment. The advantages of machine-learning-based cross-layer delay-aware transmission design are thus demonstrated.

2   System Model

We consider a typical V2V communication scenario with source S and destination D, as shown in Fig. 1. S desires to send two types of delay-limited messages to D. They have very different characteristics and QoS requirements, to support different safety-related ITS applications. The operations of the whole system are conducted in multiple time-division *transmission blocks*, each of which consists of T unit time slots. At the beginning of each time slot, S executes transmission. At the end of the slot, D provides certain feedback to update the knowledge of S regarding channel and message delivery status.

The first type of messages, termed *type-1 messages*, arrive in S periodically with a fixed data rate r bit/slot. An example of such messages is the heartbeat messages that provide D with the real-time status of S. They should be transmitted immediately. Otherwise the contained information would become stale. For each transmission block, a sufficient proportion of the messages (e.g., 70%) are expected to successfully reach D, as a transmission *reliability requirement*. Let $\phi[t] = 1$ denote that, at time slot t, S successfully delivers a type-1 message to D. Otherwise, $\phi[t] = 0$. The reliability requirement can hence be written as

$$\frac{1}{T}\sum_{t=1}^{T}\phi[t] \geq \phi_0, \tag{1}$$

where ϕ_0 is a constant specified according to the application.

The second type of messages, termed *type-2 messages*, arrive in S randomly. The arrival data volume $a[t]$ (in bits) at time slot t follows a stationary random process (e.g., Poisson with parameter λ bit/slot). An example of these messages is the environment sensing data collected from S's on-board sensors. Sharing them with D can help extend the environment perception capability of D. These messages can be temporarily stored in the source queue $\mathcal{Q}$. But the queuing delay must be limited. Such a *latency requirement* is posed by a finite maximum queue length Q_0 (queue length exceeding Q_0 results in overflow and loss of data). Let $Q[t]$ denote the instantaneous queue length at S and $b[t]$ denote the data volume that the type-2 messages leave the queue, at time slot t. Within each transmission block, at the end of time slot t, the queuing dynamics of $\mathcal{Q}$ under the latency requirement is

$$Q[t] = \max\{Q[t-1] - b[t], 0\} + a[t] \leq Q_0, \tag{2}$$

for all $t \in \{1, 2, \cdots, T\}$ and some $Q[0] \leq Q_0$.

The message transmissions are conducted in a narrow-band block fading environment. The fading coefficient between S and D at time slot t is denoted by $h[t]$, which remains fixed in each time slot, but changes across different slots. To simplify the problem, we assume that the channel gain $\|h[t]\|$ can be discretized into L different levels, the set of which is denoted by $\mathcal{H} = \{g_1, g_2, \cdots, g_L\}$. At time slot $t - 1$, if $\|h[t-1]\| = g_i$, then at the next time slot t, the channel gain changes to $\|h[t]\| = g_j$ $(i, j \in \{1, 2, \cdots, L\})$ with transition probability $\Pr\{\|h[t]\| = g_j \mid \|h[t-1]\| = g_i\} = p_{i,j}$. We consider a stationary environment such that the channel transition probabilities remain fixed, but are unknown at both the source and destination. Due to the mobility of the vehicles, the source has only a delayed channel knowledge. Specifically, at time slot $t-1$, the destination estimates the channel coefficient using the training sequence sent by the source. Assume the estimation is sufficiently accurate. At the end of the slot, D feeds its estimation result (and also D's decoding status) to S. Hence at the beginning of time slot t, S knows only $h[t-1]$.

Based on available knowledge regarding channel and queue conditions, and the past transmission status of the two types of messages, at the beginning of any time slot t, S chooses to use one of M codebooks to encode its message to a unit-power signal $x[t]$. Let the set of data rates of the M codebooks be $\mathcal{R} = \{R_1, R_2, \cdots, R_M\}$. In addition, S selects a power level from set $\mathcal{P} = \{P_1, P_2, \cdots, P_N\}$ to send the signal. As a result, at time slot t, the received signal at D can be expressed as

$$y[t] = \sqrt{P[t]} h[t] x[t] + n[t], \tag{3}$$

where $P[t]$ is the transmit power, and $n[t]$ denotes additive white Gaussian noise (AWGN) with power N_0. The mutual information between S and D is thus (with bandwidth B)

$$I[t] = B \log_2 \left(1 + \frac{P[t]\|h[t]\|^2}{N_0}\right). \tag{4}$$

Further, the source S can choose its *encoding action*, i.e., whether encoding only one type of messages or both types. Specifically, if $R[t] \geq r$, then S has two choices to form $x[t]$. First, both types of messages are transmitted. In this case, $x[t]$ represents r bits of a type-1 message and $R[t] - r$ bits of type-2 message (i.e., reducing the queue length by $b[t] = R[t] - r$ bits). Second, only the type-2 message is encoded so that $x[t]$ represents $R[t]$ bits of type-2 message (i.e., reducing the queue length by $b[t] = R[t]$ bits). Certainly, if the chosen data rate $R[t] < r$, there is only one encoding action: All the $R[t]$ bits in $x[t]$ are from the type-2 message, so that the queue length is reduced by $b[t] = R[t]$ bits. We use a binary indicator $\sigma[t] = 1$ to denote the encoding action that S transmits both messages (including the case $b[t] = 0$, i.e., only a type-1 message is encoded), and use $\sigma[t] = 0$ to denote that S transmits only the type-2 message.

Due to the imperfect CSIT, the transmission of $x[t]$ may not be successful. Assume that the M channel codes adopted by S are sufficiently strong. Then if $R[t] \leq I[t]$, the destination D can correctly recover the transmitted source message. Otherwise, if $R[t] > I[t]$, correct decoding is not possible. Use binary indicator $\psi[t]$ to represent these events. At the end of each time slot t, the destination D feeds $\psi[t] = 1$ (decoding success) or $\psi[t] = 0$ (decoding failure) to S. (By this means, $\phi[t]$ in (1) can be found by $\phi[t] = \psi[t]\sigma[t]$.)

We aim to find an energy-efficient transmission strategy, such that the V2V link can choose its encoding action, transmission power and rate based on its queue state and delayed channel knowledge, to satisfy the desired message delivery requirements in each transmission block with maximized energy efficiency η, defined as the ratio of sum effective data rate to sum power consumption. Mathematically, we aim to solve the following optimization problem.

$$\text{maximize} : \eta = \frac{\bar{R}}{\bar{P}} = \frac{\sum_{t=1}^{T} \psi[t]R[t]}{\sum_{t=1}^{T} P[t]} \tag{5}$$

$$\text{s.t.} : (1) \text{ and } (2) \tag{6}$$

$$P[t] \in \mathcal{P}, R[t] \in \mathcal{R}, \sigma[t] \in \{0, 1\}. \tag{7}$$

This stochastic optimization problem is hard to solve, especially due to unavailability of perfect knowledge regarding CSIT and environment dynamics. We propose to apply the deep reinforcement learning technique to fulfil the task.

3   MDP Formulation

To solve the energy-efficient transmission design problem for the considered delay-ware V2V communications, we first define a finite episodic MDP that reflects the optimization problem (5). Afterwards, a DQN algorithm is adopted to solve the MDP. Following the aforementioned transmission process, each episode represents one transmission block with T time slots. At the beginning of time slot t ($t \in \{1, 2, \cdots, T\}$), the agent (the decision-maker, i.e., in our case, the source S) takes an action $\boldsymbol{a}[t]$, based on the environment state $\boldsymbol{s}[t-1]$ observed at the end of the time slot $t - 1$. At the end of time slot t, the agent receives a reward $r[t]$ from the environment as a response to its action, and also observes the updated state $s[t]$. The procedure continues until the completion of a block at time slot T. An MDP is specified by five elements: state space, action space, rewards, transition probabilities, and discount factor. We elaborate them as follows.

3.1   State Space

The state space of our MDP is defined as:

$$\mathcal{S} = \left\{ s | s = \left[s^{m1}, s^{m2}, s^{ch}, s^{bk} \right] \right\},\tag{8}$$

where s^{m1} represents the reliability state of type-1 messages, s^{m2} represents the queue state of type-2 messages, s^{ch} represents the channel state, and s^{bk} is the block state.

In particular, at the end of any time slot $t \in \{1, 2, \cdots, T\}$ (i.e., after taking action $a[t]$), the value s^{m1} reflects how much the reliability requirement (1) has been satisfied and is defined as the total number of type-1 messages that have been successfully delivered so far:

$$s^{m1}[t] = \sum_{i=1}^{t} \phi[i] = s^{m1}[t-1] + \phi[t],\tag{9}$$

with initial value $s^{m1}[0] = 0$.

The value s^{m2} reflects how much the latency requirement (2) has been satisfied and is set as the current queue length:

$$s^{m2}[t] = Q[t] = \max\{s^{m2}[t-1] - b[t], 0\} + a[t].\tag{10}$$

$s^{m2}[0]$ can be any value in set $\{0, 1, \cdots, Q_0\}$, since there may be data waiting in the queue before a new block starts.

The value s^{ch} reflects the channel quality:

$$s^{ch}[t] = |h[t-1]|.\tag{11}$$

The initial state $s^{ch}[0]$ can be chosen arbitrarily from $\mathcal{H}$. Clearly, for transmission decision-making at time slot t, $s^{ch}[t]$ represents delayed CSIT.

Finally, the value s^{bk} is set as the number of remaining time slots in the block:

$$s^{bk}[t] = T - t = s^{bk}[t-1] - 1,\tag{12}$$

with initial value $s^{bk}[0] = T$. The block state $s^{bk}[t]$ represents the urgency level of taking actions to guarantee the QoS requirements of the two types of messages while achieving the maximum energy efficiency upon completion of the block.

The size of the state space can be extremely large, since s^{m2} is unbounded. We notice that if any $s^{m2}[t] > Q_0$, the latency constraint (2) is violated and overflow occurs. Carrying on transmissions in the block does not help and thus such a situation should be avoided. This is also the case when $\frac{1}{T}\left(s^{\mathrm{m1}}[t] + s^{\mathrm{bk}}[t]\right) < \phi_0$, since the reliability constraint (1) cannot be satisfied regardless of the remaining actions. Therefore, to reduce the size of state space and enable efficient training, we define an extra "abnormal terminal state" s^+. At the end of any time slot t, if either $s^{\mathrm{m2}}[t] > Q_0$ or $\frac{1}{T}\left(s^{\mathrm{m1}}[t] + s^{\mathrm{bk}}[t]\right) < \phi_0$ occurs, the state enters s^+ and the episode ends. By this means, the size of the state space is limited to $\frac{1}{2}(T - T \cdot \phi_0 + 1) \times (T + T \cdot \phi_0 + 1) \times |H| \times (Q_0 + 1)$. In fact, it is possible to further decrease the state space by (possibly non-linearly) quantizing the ranges of s^{m1} (i.e., $[0, T]$), s^{m2} (i.e., $[0, Q_0]$), s^{bk} (i.e., $[0, T]$) into K_1, K_2, and K_3 sub-regions, respectively. Tuning the parameters K_1, K_2, K_3 leads to trade-off between performance and training complexity.

3.2   Action Space

The action space of our MDP is defined as

$$\mathcal{A} = \left\{ \boldsymbol{a} \middle| \boldsymbol{a} = \left[a^{\mathrm{pw}}, a^{\mathrm{rt}}, a^{\mathrm{m1}}, a^{\mathrm{m2}} \right] \right\}. \tag{13}$$

The values a^{pw} and a^{rt} represent the power level and data rate the agent chooses from $\mathcal{P}$ and $\mathcal{R}$ respectively to send signal, i.e., at the beginning of time slot t the source selects $a^{\mathrm{pw}}[t] = P[t] \in \mathcal{P}$, and $a^{\mathrm{rt}}[t] = R[t] \in \mathcal{R}$.

a^{m1} and a^{m2} are binary indicators that reflect the encoding action. Specifically, when a^{rt} is chosen to be $a^{\mathrm{rt}}[t] > r$, setting $a^{\mathrm{m1}}[t] = 1$ and $a^{\mathrm{m2}}[t] = 1$ means that both type-1 (with rate r) and type-2 (with rate $a^{\mathrm{rt}}[t] - r$) messages are sent (equivalent to $\sigma[t] = 1$). If only type-2 messages are encoded (with rate $a^{\mathrm{rt}}[t]$), one has $a^{\mathrm{m1}}[t] = 0$ and $a^{\mathrm{m2}}[t] = 1$ (equivalent to $\sigma[t] = 0$). In addition, if $a^{\mathrm{rt}}[t] = r$, the source either transmits the type-1 message (indicated by $a^{\mathrm{m1}}[t] = 1$ and $a^{\mathrm{m2}}[t] = 0$, equivalent to $\sigma[t] = 1$) or the type-2 message (indicated by $a^{\mathrm{m1}}[t] = 0$ and $a^{\mathrm{m2}}[t] = 1$, equivalent to $\sigma[t] = 0$). Finally, if $a^{\mathrm{rt}}[t] < r$, only the type-2 message can be encoded and one has $a^{\mathrm{m1}}[t] = 0$ and $a^{\mathrm{m2}}[t] = 1$ (equivalent to $\sigma[t] = 0$). Therefore, the size of action space is less than $2MN$.

3.3   Rewards

The rewards reflect how good the chosen actions are in terms of achieving the optimization objective (5) while guaranteeing the constraints in (6). It is the basis for the agent to learn a good policy. To this end, we form the reward $\mathfrak{R}$ by four parts: energy efficiency reward $\mathfrak{R}^{\mathrm{ee}}$, reliability reward of type-1 messages $\mathfrak{R}^{\mathrm{m1}}$, latency reward of type-2 messages $\mathfrak{R}^{\mathrm{m2}}$, and penalty for entering the abnormal termal state $\mathfrak{R}^{\mathrm{ab}}$.

Specifically, at the end of time slot t ($t \in \{1, 2, \cdots, T\}$), the energy efficiency reward of taking action $\boldsymbol{a}[t]$ in state $\boldsymbol{s}[t-1]$ is defined as the incremental energy efficiency, i.e.,

$$\mathfrak{R}^{\mathrm{ee}}[t] = \frac{\sum_{i=1}^{t} \psi[i] a^{\mathrm{rt}}[i]}{\sum_{i=1}^{t} a^{\mathrm{pw}}[i]} - \mathfrak{R}^{\mathrm{ee}}[t-1], \tag{14}$$

where the initial value $\mathfrak{R}^{ee}[0] = 0$ and if for any t, $\sum_{i=1}^{t} a^{pw}[i] = 0$, we set $\mathfrak{R}^{ee}[t] = 0$. By this means, at the end of the transmission block, the accumulative reward is $\sum_{t=1}^{T} \mathfrak{R}^{ee}[t] = \frac{\sum_{i=1}^{T} \psi[i] a^{rt}[i]}{\sum_{i=1}^{T} a^{pw}[i]} = \frac{\sum_{i=1}^{T} \psi[i] R[i]}{\sum_{i=1}^{T} P[i]}$, which is the achievable energy efficiency. Maximizing $\sum_{t=1}^{T} \mathfrak{R}^{ee}[t]$ is the same as maximizing the original objective function (5).

The reliability reward of type-1 messages is defined as the incremental successful transmission ratio of the messages, i.e.,

$$\mathfrak{R}^{m1}[t] = \frac{\sum_{i=1}^{t} \psi[i] a^{m1}[i]}{T} - \mathfrak{R}^{m1}[t-1], \tag{15}$$

in which the initial value $\mathfrak{R}^{m1}[0] = 0$. The accumulative reward after T time slots is $\sum_{t=1}^{T} \mathfrak{R}^{m1}[t] = \frac{\sum_{i=1}^{T} \psi[i] a^{m1}[i]}{T} = \frac{\sum_{i=1}^{T} \phi[i]}{T}$. Maximizing this value is the same as maximizing the actual ratio of delivering type-1 messages defined in (1).

The latency reward of type-2 messages is defined as the reduction of the queue length (as a fraction of Q_0) resulted from taking action $\boldsymbol{a}[t]$:

$$\mathfrak{R}^{m2}[t] = -\left(\frac{s^{m2}[t]}{Q_0} - \frac{s^{m2}[t-1]}{Q_0} \right). \tag{16}$$

Maximizing the accumulative reward at the end of the block $\sum_{t=1}^{T} \mathfrak{R}^{m2}[t] = -\frac{Q[T]}{Q_0}$ minimizes the final queue length.

Finally, a penalty is imposed if at the end of any time slot t the system enters the abnormal terminal state, since the transmission requirements are violated. This penalty is set to

$$\mathfrak{R}^{ab}[t] = \begin{cases} -\Gamma & \text{for } \boldsymbol{s}^{+} \\ 0 & \text{otherwise} \end{cases}, \tag{17}$$

where Γ is a large positive number.

The total reward of our MDP at time slot t is set to

$$\mathfrak{R}[t] = \gamma_1 \mathfrak{R}^{ee}[t] + \gamma_2 \mathfrak{R}^{m1}[t] + \gamma_3 \mathfrak{R}^{m2}[t] + \mathfrak{R}^{ab}[t], \tag{18}$$

where γ_1, γ_2, and γ_3 are weighting parameters with $\gamma_1 + \gamma_2 + \gamma_3 = 1$. We can tune these parameters to specify our preference in maximizing the achievable energy efficiency, maximizing successful transmission ratio of the type-1 messages, or minimizing the queue length at the end of each transmission block. The special case of setting $\gamma_2 = \gamma_3 = 0$ leads to the original optimization problem (5).

3.4   Transition Probabilities and Discount Factor

The probability that state $\boldsymbol{s}$ enters state $\boldsymbol{s}'$, providing reward $\mathfrak{R}$, after the agent takes action $\boldsymbol{a}$, is:

$$p_{\boldsymbol{s},\boldsymbol{s}',\mathfrak{R}}^{\boldsymbol{a}} = \Pr\left\{ \boldsymbol{s}', \mathfrak{R} | \boldsymbol{s}, \boldsymbol{a} \right\}, \quad \forall \boldsymbol{s}, \boldsymbol{s}' \in \mathcal{S}, \boldsymbol{a} \in \mathcal{A}. \tag{19}$$

Certainly such transition probabilities are unknown. since the channel transmission probabilities are not available. The MDP has to be solved by a model-free algorithm.

Since our MDP is episodic and we consider the rewards attained in different time slots to be equally important, the discount factor is set to be 1. The total return is thus

$$G = \sum_{t=1}^{T} \mathfrak{R}[t]. \tag{20}$$

Solving the MDP is to find the strategy that maximizes G.

4   Deep Reinforcement Learning Algorithm

We apply DQN, which is a model-free reinforcement learning method that combines the Q-learning algorithm with DNNs, to solve the finite MDP presented in the above section.

Q-learning is a tabular off-policy reinforcement learning algorithm that establishes a Q-table to infer the optimal policy, i.e., the optimal mapping from states to the probability distribution over actions. Each element in the Q-table $\tilde{Q}(s_t, a_t)$ represents the value of an action-value function $q(s_t, a_t)$, the expected return of taking action a_t in state s_t, and then following policy π:

$$q(s_t, a_t) = E_\pi \left[G_t = \sum_{i=t+1}^{T} \mathfrak{R}_i \middle| s_t, a_t \right]. \tag{21}$$

During the training process, for each sampled training episode, Q-learning keeps updating the Q-table according to

$$\tilde{Q}(s_t, a_t) \leftarrow \tilde{Q}(s_t, a_t) + \alpha \big(\mathfrak{R}_t + \gamma \max_{a \in \mathcal{A}} \tilde{Q}(s_{t+1}, a) - \tilde{Q}(s_t, a_t) \big), \tag{22}$$

in which action a_t is sampled from an ϵ-greedy policy according to the Q-table, $\mathfrak{R}_t$ and s_{t+1} are respectively the observed reward and new state, γ is discount factor, and $\alpha \in [0, 1]$ represents learning rate. Upon convergence, the optimal policy can then be derived from the Q-table.

The issue with conventional Q-learning is that when the state space and/or action space are large, the storage and update of Q-table demand large memory space and long convergence time in the training process [9]. DQN targets this problem by applying DNNs to approximate the action-value functions. However, this may further cause convergence and stability problems [12]. First, small updates to the action-value function $q(s_t, a_t)$ may change the on-going policy significantly, which may mislead the agent to prefer a set of actions with correlated data. To handle this issue, reference [13] proposes the target network technique, which implements two networks with the same architecture but different network weight updating frequencies.

In addition, the input data of the DNNs are often highly correlated due to the ϵ-greedy sampling policy. The environment transfers from each state to a

Table 1. Simulation parameters

Parameter	Value
Type-1 message reliability requirement ϕ_0	$\{0.6, 0.7, 0.8\}$
Type-2 message queue length limit Q_0	$\{10, 8\}$
Transmit power levels $\mathcal{P}$	$\{0, 1, 2, 4, 8, 16, 32\}$
Transmit rates $\mathcal{R}$	$\{0, 1, 2, 3\}$
Abnormal terminal state penalty $-\Gamma$	-10
Initial states of the two messages $(s^{m1}[0], s^{m2}[0])$	$(0, 0)$

certain set of next states with high probability. Hence gradient decent may be conducted frequently based on similar and correlated inputs. Such a fact can cause strictly sub-optimal or unstable training results. Experience replay [12] stores the agent's experiences at each time slot in a data-set. In each episode, a batch containing fixed pieces of experience is selected randomly. The correlation between experience is broken.

The complete algorithm that we apply to solve our MDP is constructed following that in [12]. In the next section, we implement it, using Tensorflow [?], on an example problem. Our experiments apply DNNs with 5 layers. The input and output layers consist of 4 and 35 neurons respectively, taking the state at each time slot and the corresponding action-value of each available action as inputs and outputs. Each of the three hidden layers has 20 neurons. The learning rate is chosen to be $\alpha = 0.005$ and the mini-batch size is 300. The optimization method is Adam [?].

5   Performance Evaluation

We use computer simulation experiments to demonstrate the effectiveness of our method (termed DQN scheme). In the simulations, the block size is set to $T = 10$ slots. Both bandwidth and noise power are normalized to be $B = 1$ and $N_0 = 1$. The two types of messages respectively have fixed and Poisson distributed rates with $r = 1$ and $E[a[t]] = 0.6$ bit/slot. The channel gains are assumed to be $\mathcal{H} = \{0.359, 0.644, 0.866, 1.073, 1.286, 1.525, 1.830, 2.355\}$, approximately corresponding to the $\frac{1}{16}, \frac{3}{16}, \cdots, \frac{15}{16}$ percentiles of a standard Rayleigh distribution respectively. The channel transition probabilities are set to $p_{i,i} = 0.5$ $\forall i \in \{1, \cdots, 8\}$, $p_{i,i\pm 1} = 0.2$ $\forall i \in \{2, \cdots, 7\}$, $p_{i,i\pm 2} = 0.05$ $\forall i \in \{3, \cdots, 6\}$, $p_{1,2} = p_{8,7} = 0.4$, $p_{1,3} = p_{8,6} = p_{2,4} = p_{7,5} = 0.1$. The remaining parameters described in Sects. 3 and 4 are displayed in Table 1. The size of the state space of our MDP is 3960 and for each state there are maximally 35 possible actions to choose. Enumerating all combinations of actions to find the optimal policy is very computationally expensive.

We compare our DQN scheme with four baseline methods. The first follows the scheme proposed in [4] and is termed Lyapunov optimization (LO) scheme.

Accurate instantaneous CSI at the source is assumed to be available. The Lyapunov optimization theorem is applied to transform the reliability and latency requirements (1) and (2) into a penalty term to the objective function. This allows the sequential optimization problem to be solved greedily in each time slot. The method is essentially for continuous power and rate allocation problems with large block length. To apply it in our problem, the parameters are first carefully selected to make sure that the penalty term is sufficiently large. The demanded reliability level of type-1 messages and queue length of type-2 messages within the limited block length are hence guaranteed. The power and rate pair in $\mathcal{P}$ and $\mathcal{R}$ that is closest to its solution is chosen to conduct the transmission.

The other three approaches do not assume accurate CSIT and carry out their transmissions in heuristic fashions. First, an FR (fixed-rate) scheme always transmits with rate $R[t] = 2$ using the smallest power level according to the delayed channel knowledge. If the reliability requirement (1) is not yet satisfied, both type-1 and type-2 messages are transmitted (each with rate 1 bit/slot). Otherwise, only type-2 messages are sent with rate 2 bit/slot. Although the transmission rate is chosen to be larger than the expected sum delivery rate of the two messages, 1.6 bit/slot, due to the lack of accurate CSIT, transmission errors may occur. This limits the transmission rate of the type-2 message to be small and thus may cause increasing queue length and then potential overflow.

The EQ (empty-queue) scheme targets addressing the above issue by always trying to empty the source queue. Specifically, if the reliability requirement (1) is not satisfied, both type-1 and type-2 messages are transmitted, with rates 1 bit/slot and $\min\{Q[t-1], 2\}$ bit/slot respectively, using the smallest power level derived according to the available channel knowledge. Otherwise, after sufficient type-1 messages are delivered successfully, only type-2 messages are sent with rate $\min\{Q[t-1], 3\}$ bit/slot. Compared with the FR scheme, the EQ scheme intends to reduce the possibility of overflow. But it may send messages with a high data rate and easily lead to unsuccessful transmission (due to channel outage). Further, the method does not naturally guarantee the requirements (1) and (2) to be satisfied, either.

To ensure the performance requirements of the two messages without accurate CSIT, a WS (worst-case scenario) scheme works similarly to the EQ scheme, but chooses its power level according to the worst-case scenario (the smallest channel gain that has non-zero transfer probability from the observed value). For example, if the delayed channel gain is 1.286, the power is chosen by assuming that the true value of the unknown channel gain is the worst case 0.866. Clearly, the scheme always satisfactorily delivers the two types of messages, at the cost of potentially a low achievable energy efficiency. For fair comparison, we generate 10^5 test blocks of channel gains and arrival rates of the type-2 messages. The average achievable energy efficiency and the number of successful transmission blocks of the five schemes are compared.

We choose the weighting parameters in (18) as $\gamma_1 = 1$ and $\gamma_2 = \gamma_3 = 0$, i.e., our DQN scheme aims to solve the original energy efficiency maximization

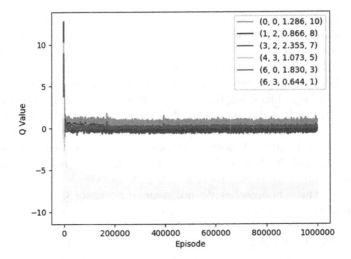

Fig. 2. Learning curve for six example states.

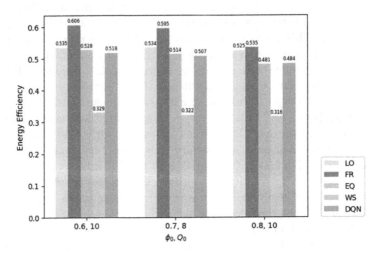

Fig. 3. Energy efficiency comparisons.

problem (5). The training of our algorithm is conducted using 10^6 episodes (blocks). We randomly select 6 out of the 3960 possible states. For each update of the DNN weights, these 6 states are input to the network and each generates 35 outputs (i.e., approximated action-value functions). The average, over all 35 actions, of the outputs associated with these states are displayed in Fig. 2. From the figure it is seen that our algorithm successfully converges. In what follows, we discuss the results obtained using the 10^5 test blocks.

Figure 3 illustrates the average achievable energy efficiency of the five schemes with three different sets of QoS requirements: $(\phi_0, Q_0) =$

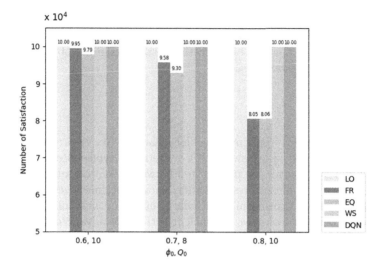

Fig. 4. Satisfaction of transmission constraints.

$(0.6, 10), (0.7, 8), (0.8, 10)$. If for any test block, a transmission scheme results in violation of the requirements (1) and (2), the energy efficiency attained by this scheme in such a failed block is treated as zero (and hence reduces the overall average test energy efficiency). The numbers of successful transmission blocks (i.e., the constraints (1) and (2) are both satisfied) are displayed in Fig. 4. It can be seen from Fig. 3 that the proposed DQN scheme obtains much better performance than the WS scheme, due to its ability of learning the environment changing dynamics and then finding a good solution for the sequential decision-making problem (5). There is no need to always prepare for the worst-case scenario and thus the energy efficiency can be significantly improved. The performance of the DQN scheme is comparable to the LO scheme, even though only delayed channel knowledge is available for making transmission decisions.

In addition, Fig. 3 shows that the FR scheme and EQ scheme may attain even higher energy efficiency. But one can see from Fig. 4 that both approaches lead to violation of the transmission requirements of the two types of messages. Therefore, their high achievable energy efficiency (derived using only the successful transmission blocks) does not serve as the evidence of their usefulness. Clearly, our DQN scheme provides satisfactory performance for all test blocks. Under different QoS requirements, its achievable energy efficiency also changes. This implies that policies are learned to be adaptive to different environments. This is different from the LO and WS schemes that always apply the same transmission strategy. The above observations clearly show the advantages of the proposed transmission design. Note that in our paper, we consider the energy efficiency defined in (5) as the objective for a delay-aware V2V communication system. One can follow the designing process and extend the system model to other scenarios with different performance metric, e.g., minimized power consumption. As mentioned earlier, changing the choices of γ_1, γ_2, and γ_3 in (18) also provides tradeoff between the overall designing objective and the performance of individual messages.

6   Conclusion

We have investigated applying machine learning to assist in cross-layer transmission design for delay-aware V2V systems. A typical scenario, in which multiple types of messages are transmitted between a vehicle source-destination pair with imperfect transmitter-side channel knowledge, has been considered. We have shown that by transforming an energy-efficiency maximization problem to a finite MDP, we can solve it efficiently through advanced reinforcement learning techniques, with achievable performance much better than several heuristic solutions. The advantages of combining machine learning with wireless communication design has been demonstrated. In this paper, we have focused on discretized state and action spaces. Systems with multiple source-destination pairs and continuous state/action spaces are currently under investigation.

Acknowledgement. This work was supported in part by the National Natural Science Foundation of China under Grant 61771343, and the Intelligent Connected Vehicle Pilot Demonstration Project under grant 2019B090912002.

References

1. Cao, B., Zhang, L., Li, Y., Feng, D., Cao, W.: Intelligent offloading in multi-access edge computing: a state-of-the-art review and framework. IEEE Commun. Mag. **57**(3), 56–62 (2019)
2. Jiang, C., Zhang, H., Ren, Y., Han, Z., Chen, K., Hanzo, L.: Machine learning paradigms for next-generation wireless networks. IEEE Wirel. Commun. **24**(2), 98–105 (2016)
3. Zhang, C., Patras, P., Haddadi, H.: Deep learning in mobile and wireless networking: a survey. IEEE Commun. Surv. Tutor. **21**(3), 2224–2287 (2019)
4. Lan, D., Wang, C., Wang, P., Liu, F., Min, G.: Transmission design for energy-efficient vehicular networks with multiple delay-limited applications. In: 2019 IEEE Global Communications Conference (GLOBECOM), Waikoloa, USA, December 2019
5. Hasselt, H., Guez, A., Silver, D.: Deep reinforcement learning with double Q-learning. In: Thirtieth AAAI Conference on Artificial Intelligence, Phoenix, Arizona, USA, February 2016
6. Seo, H., Lee, K., Yasukawa, S., Peng, Y., Sartori, P.: LTE evolution for vehicle-to-everything services. IEEE Commun. Mag. **54**(6), 22–28 (2016)
7. Dar, K., Bakhouya, M., Gaber, J., Wack, M., Pascal, L.: Wireless communication technologies for ITS applications. IEEE Commun. Mag. **48**(5), 156–162 (2010)
8. Crites, R., Barto, A.: Improving elevator performance using reinforcement learning. In: Advances in Neural Information Processing Systems, pp. 1017–1023 (1996)

9. Sutton, R., Barto, A.: Introduction to Reinforcement Learning. MIT Press, Cambridge (2018)
10. Li, S., Xu, L., Zhao, S.: 5G internet of things: a survey. J. Ind. Inf. Integr. **10**, 1–9 (2018)
11. Lillicrap, T., et al.: Continuous control with deep reinforcement learning. CoRR abs/1509.02971 (2015)
12. Mnih, V., et al.: Playing Atari with deep reinforcement learning. NeurIIPS (2013)
13. Mnih, V., et al.: Human-level control through deep reinforcement learning. Nature **518**(7540), 529–533 (2015)
14. Sun, W., Ström, E., Brännström, F., Sou, K., Sui, Y.: Radio resource management for D2D-based V2V communication. IEEE Trans. Veh. Technol. **65**(8), 6636–6650 (2015)
15. Zhan, W., Luo, C., Wang, J., Wang, C., Min, G., Duan, H., Zhu, Q.: Deep reinforcement learning-based offloading scheduling for vehicular edge computing. IEEE Internet Things J. **7**(6), 5449–5465 (2020)

Deep Learning

Efficient Architecture for Convolution and Softmax Function in Deep Learning Accelerator

Zhenyu Jiang[1]([⊠]), Zhifeng Zhang[1], Haoqi Ren[1], and Jun Wu[2]

[1] College of Electronic and Information Engineering, Tongji University,
Shanghai, China
{1832925, zhangzf, renhaoqi}@tongji.edu.cn
[2] School of Computer Science, Fudan University, Shanghai, China
wujun@fudan.edu.cn

Abstract. Convolutional neural network (CNN) has been widely used in deep learning. However, the **hardware** consumption **of the convolutional neural network** is very large. Traditional Central Processing Units (CPUs) and Graphic Processing Units (GPUs) are inefficient and expensive for neural network, so an efficient hardware design is required. The proposed design based on Digital Signal Processor (DSP) has rapid operating speed and strong computation ability for training and inference of CNN. In this paper, the hardware architecture of convolution and softmax function is specially optimized. Winograd algorithm can reduce multiplications of convolution, thus decreases hardware complexity, since multiplication is much more complex in hardware implementation than addition. The softmax function is also simplified by replacing divider by subtractor and logarithmic function which cost fewer resources. The proposed hardware architecture dramatically decreases the complexity and hardware resources.

Keywords: Convolutional neural network · Hardware architecture · Convolution · Winograd algorithm · Softmax function

1 Introduction

With the great success achieved by deep learning techniques in many areas, the convolutional neural network (CNN) is widely used in many fields [1]. CNN is one of the most popular algorithms for deep learning and generally consists of an input layer, multiple convolutional layers, multiple pooling layers, a fully connected layer, and an output layer [2]. The convolutional layer is used to extract the feature information from the input image [3], it mainly consists of convolutional kernels, and mapping the input map to the output feature map [4], which is the most computationally intensive module of the convolutional neural network.

Common convolutional neural network models are very large in terms of computation and parameters, so both training and inference usually require high-performance hardware to provide support for compute and storage requirements [5]. Traditional CPUs and GPUs can be used to execute convolutional neural network

H. Gao et al. (Eds.): ChinaCom 2020, LNICST 352, pp. 631–643, 2021.
https://doi.org/10.1007/978-3-030-67720-6_43

algorithms, but they are slow and have low performance. FPGA, because of the flexibility of programming and the parallelism of computing, is often more favored, but the speed and energy consumption ratio are worse compared to ASICs [6]. To reduce the training and inference time of neural networks, the hardware needs to be specifically optimized.

Google proposed a tensor processing unit (TPU) that can run multiple machine learning algorithms [7]. Performance is 15 to 30 times better than the general-purpose CPUs and GPUs, and 30 to 80 times less power consumption. Compared with GPUs, TPUs use low-precision (8-bit) computing, which significantly reduces power consumption and speeds up operation while maintaining accuracy. Also, the TPU uses pulsed arrays, which are optimized to reduce I/O operations.

Digital Signal Processor (DSP) has strong parallel computing capabilities with SIMD (single instruction multiple data) and VLIW (very long instruction word) functions [8], so it is suitable for massively parallel computation of deep learning neural networks [9]. This paper proposes a deep learning accelerator (DLA) based on DSP, which acts as a DSP co-processor and specializes in neural network operations. The Winograd algorithm is used to accelerate the convolution operations and reduces the MAC resources required for the convolution computation [10]. Totally 256 MAC cells, each contains four 16-bit multipliers. The proposed accelerator supports the direct convolution of 2×2 convolution kernels as well as the 3×3 Winograd convolution. The softmax function is also specially optimized to ensure accuracy while reducing hardware resources.

2   The Overview of the Accelerator

2.1   Overall Architecture

The accelerator can be divided into the following modules based on the function. Configuration registers are for storing the weight of kernels, size of the feature map, channels, and other main parameters. The control unit controls the behavior of the accelerator according to the configuration parameters, mainly including operation requests, data read requests, and write responses. The input interface is for receiving the input feature map and parameters. The pre-processing module is for Winograd matrix pre-processing, and the MAC array performs matrix multiplication operations. The POOL and RELU perform pooling and activation respectively. The FC performs the fully connected function to get the final results.

Figure 1 shows the architecture of DLA. First, the configuration registers need to be configured. The input and output interfaces communicate with the processor via the AXI bus by DMA. At the pre-processing layer, Winograd pre-processing of the matrix is performed, as well as the floating-point to fixed-point conversion to reduce the network data storage and data transmission bandwidth. After padding and rearranging, data will be sent to the MAC array to complete the convolution operation. After pooling and convolution, data will be sent to the full connection layer for feature extraction. The final results will be transferred through the output interface to SRAM.

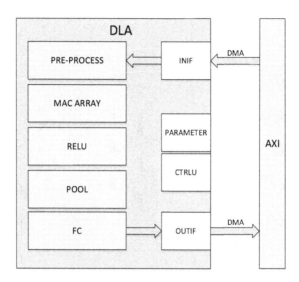

Fig. 1. The architecture of the DLA

2.2   Pre-processing Layer

In the pre-processing layer, the floating-point data will be transferred to fixed-point format first, which supports up to 16-bit fixed-point operation. Then the padding of the image will be done according to the length and width in the configuration register. The image data will then be pre-processed for the Winograd algorithm.

The convolution operation of an image can be represented by Fig. 2 as an example of a 3×3 convolution, the kernel is smoothed over the padding image with a step of strip. Multiply the elements of the image by ones of the convolution kernel at the corresponding positions and then add them together to get the result at this position. Convolution results can be got by iterating over the image following the above steps.

The process of convolution can be written in the form of matrix multiplication [11].

$$r_1 = K \cdot G \tag{1}$$

Where $K = [k_{11}, k_{12}, k_{13}, k_{21}, k_{22}, k_{23}, k_{31}, k_{32}, k_{33}]$,
$G = [g_1, g_2, g_3, g_4, g_5, g_6, g_7, g_8, g_9]^T$.

As we can see, each convolution requires at least 9 multiplications and 8 additions. Multiplication is much more complex to implement in hardware than addition. Therefore, to reduce the number of multiplications, the Winograd algorithm is deliberately introduced.

First, take the 1D convolution as an example. Input $K = [k_0, k_1, k_2, k_3]$ and the convolution kernel is $G = [g_1, g_2, g_3]^T$. Then the convolution can be written as the following matrix Multiplication form:

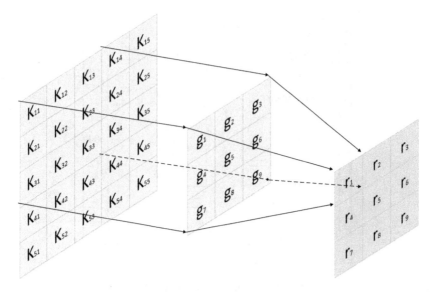

Fig. 2. Matrix convolution diagram

$$F(2,3) = \begin{bmatrix} k_0 & k_1 & k_2 \\ k_1 & k_2 & k_3 \end{bmatrix} \begin{bmatrix} g_1 \\ g_2 \\ g_3 \end{bmatrix} = \begin{bmatrix} r_0 \\ r_1 \end{bmatrix} \tag{2}$$

Since there are many repeating elements regularly distributed in the matrix, the result can be written in the following form:

$$F(2,3) = \begin{bmatrix} k_0 & k_1 & k_2 \\ k_1 & k_2 & k_3 \end{bmatrix} \begin{bmatrix} g_1 \\ g_2 \\ g_3 \end{bmatrix} = \begin{bmatrix} a_1 + a_2 + a_3 \\ a_2 - a_3 - a_4 \end{bmatrix} \tag{3}$$

Where $a_1 = (k_0 - k_2)g_0$, $a_2 = (k_1 + k_2)\frac{g_0 + g_1 + g_2}{2}$, $a_4 = (k_1 - k_3)g_2$, $a_3 = (k_2 - k_1)\frac{g_0 - g_1 + g_2}{2}$. A total of 4 multiplications and 12 additions are required. Since the elements of the convolution kernel are fixed in advance, these additions only need to be calculated once. The hardware implementation of multiplication is more complex than addition. The proposed convolution calculation is more resource-efficient than the direct convolution calculation.

Two-dimensional convolution can be written in the following form. Take for example a 4 × 4 image and a 3 × 3 convolution kernel. The result of their convolution

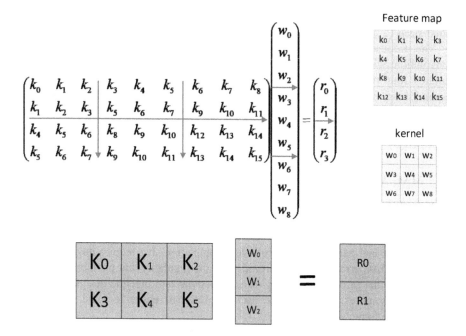

Fig. 3. Matrix convolution partition

can be expressed as $F = KW$. The convolution matrix can be represented by dividing K and W into the following submatrices. Then R_0 and R_1 can be express by:

$$R_0 = (A_0 + A_1 + A_2), \ R_1 = (A_1 - A_2 - A_3) \qquad (4)$$

Where,

$$A_0 = (K_0 - K_2) \cdot W_0,$$

$$A_1 = (K_1 + K_2) \cdot \frac{W_0 + W_1 + W_2}{2},$$

$$A_2 = (K_2 - K_1) \cdot \frac{W_0 - W_1 + W_2}{2},$$

$$A_0 = (K_1 - K_3) \cdot W_2 \qquad (5)$$

Since the pattern of the data repetition distribution of each sub-matrix is the same, the number of multiplication can also be reduced by the Winograd algorithm. Each sub-matrix multiplication requires four multiplications, so the total multiplication times are 16, which greatly reduces the multiplication complexity compared with the direct convolution requiring 36 multiplications.

The convolutional layer of the convolutional neural network performs a 3D convolution, which is equivalent to summing the 2d convolutional results on each channel

[12]. For feature maps of different sizes, they can be divided into overlapping tiles of equal-sized (2 × 3), and perform Winograd convolution on each tile.

The proposed DLA supports both 3 × 3 and 2 × 2 convolution kernels. 3 × 3 convolution will be pre-processed by the Winograd algorithm. Each sub-matrix requires four multiplications to get the result. The 2 × 2 direct convolution also requires four multiplications for each sub-matrix. In order to maintain the hardware regularity of the MAC array and reduce the complexity of the algorithm, 2 × 2 convolution adopts the direct convolution method while 3 × 3 convolution is pre-processed by Winograd algorithm and is rearranged as input into the MAC array.

2.3   MAC Array

MAC array is used to perform convolution operations. Each MAC cell in the array has the same size and function. The rearranged data is transmitted and calculated in adjacent processing units (PE). Convolution computation contains a large number of highly regular and parallelizable repetitive computations that require huge computing power. The MAC array takes full advantage of the regularity and parallelism, data are passed in parallel to the PE array by using a modular and pipeline design, so that the architecture is more regular and the backend design is easier to route, thus greatly increasing the frequency.

Taking 2 × 2 convolution operation as an example, each MAC cell contains 4 PEs which can perform a 16-bit fixed-point multiplication operation. The convolution kernel parameters are pre-stored in the configuration register and only need to be read once at the beginning. The 2 × 2 convolution takes the direct convolution method, the results of

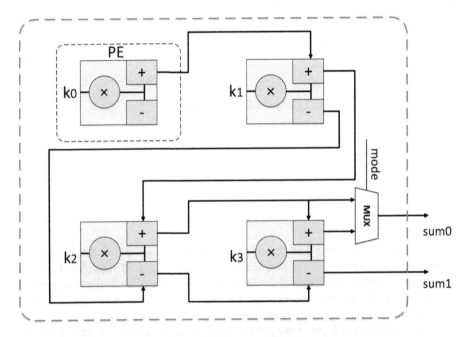

Fig. 4. MAC cell architecture

multiplying the input and weight in each PE can be added directly to get the final result. The 3 × 3 Winograd convolution also requires 4 multiplication operations and multiplexes the same MAC cell with 2 × 2 convolution. However, the 3 × 3 Winograd convolution operation has two results that cannot be added directly while 2 × 2 convolution has only one result. The basic architecture of the MAC cell which combines the 3 × 3 convolution with the 2 × 2 convolution operation is shown in Fig. 4.

The mode bit in the configuration register indicates whether a 3 × 3 or a 2 × 2 convolution operation is used. If mode indicates a 2 × 2 direct convolution, sum0 is output as the only result. If 3 × 3 mode is indicated, the results are sum0 and sum1.

2.4   Softmax Function Implementation

After convolution, pooling and activation, the final layer of the deep neural network is usually a fully connected layer and softmax function (classification network). Softmax function maps the output of the fully connected layer to probabilities, ensuring that the sum of the output is 1 [13].

The softmax function can be expressed by:

$$f(x_i) = \frac{e^{x_i}}{\sum_{j=1}^{N} e^{x_j}}, (i = 1, 2, \ldots, N) \tag{6}$$

where $x_1, x_2, \ldots, x_N$ are the output vectors of the fully connected layer. The hardware

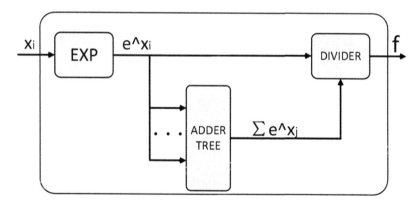

Fig. 5. The architecture of softmax function direct implementation

architecture that directly implements the softmax function can be expressed in the following form.

The division operation is very complex in hardware implementation. In order to simplify the implementation of the softmax function, the above formula can be expressed in the following form [14].

$$f(x_i) = \frac{e^{x_i - x_{max}}}{\sum_{j=1}^{N} \exp(x_j - x_{max})}$$

$$= \exp\left(x_i - x_{max} - \ln\left(\sum_{j=1}^{N} e^{x_j - x_{max}}\right)\right) \qquad (7)$$

The above equation converts division to addition (subtraction) as well as logarithmic operation and prevents overflow from the calculation $(x_i - x_{max} \leq 0)$. The sorting block aims to find out the maximum value of x_i. An example of a 4-input sorting block consisting of multilevel compare and select units is shown in Fig. 6. Compared with the original direct computation architecture, divisors are replaced by subtractors and logarithmic units which have much lower hardware complexity. The hardware architecture of softmax function is shown in Fig. 7. Since the path length of logarithmic units and subtractors are shorter than the data paths of divisors, the critical path length of the proposed architecture can be significantly shortened.

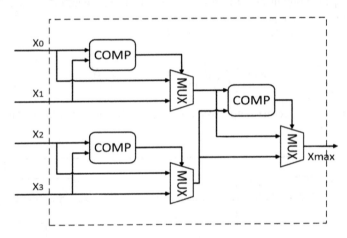

Fig. 6. The architecture of sorting block

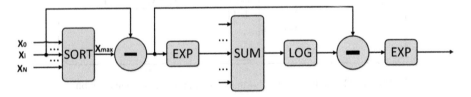

Fig. 7. The architecture of softmax function

Exponential calculation of e can be usually implemented by using the lookup table. This method is the fastest with the least error, but the largest in area and resources. Especially when multiple exponential units are computed in parallel, hardware resource consumption is extremely significant. In order to reduce the hardware resources of the exponential function, the following method is used to simplify the exponential computation.

Set $F = \exp(x_i - x_{max})$, since $x_i - x_{max} \leq 0$, $F \in (0, 1]$. Then

$$F = 2^{(x_i - x_{max}) \log_2 e} = 2^{-(x_{max} - x_i) \log_2 e} \tag{8}$$

Since $\log_2 e$ is a constant, it is not necessary to use conventional multipliers to calculate $(x_{max} - x_i) \log_2 e$, but instead a modified constant multiplier can be used. The approximate value of $\log_2 e$ in binary can be expressed as 1.0111. If a certain bit of the multiplier is 1, simply shift the multiplicand to the left by corresponding bits, and then add up all the shifting results.

Set the integer and decimal parts of $(x_{max} - x_i) \log_2 e$ to be u and v respectively.

$$F = 2^{-(u + v)} = 2^{-u} \times 2^{-v} \tag{9}$$

The above equation converts the exponential operation of e to an exponential operation of 2 and a shift operation.

Since $v \in (0, 1]$, it is possible to fit the function $f(v) = 2^{-v}$ with a linear function. We use the function $f(v) = kv + b$ to approximate the function [15], where k and b are different when v is in different ranges. By varying the segmentation range, a design with adjustable precision can be achieved. With this fitting method, the implementation of the softmax function can be highly efficient and low-complexity. In the proposed design, two decimal bits of v are chosen as the selection bits, and there are $2^2 = 4$ kinds of corresponding fitting functions, the number of both parameters k and b are also 4. Through this method, the exponential implementation function can be greatly simplified. The basic architecture of the EXP implementation unit is shown in Fig. 8.

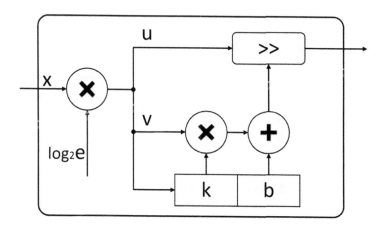

Fig. 8. Exponential unit diagram

Set $S = \sum_{j=1}^{N} e^{x_i - x_{max}}$. Obviously $S > 1$. Set $m \in (1, 2]$, n is a non-negative integer. Then S can be expressed by the following equation:

$$S = m \cdot 2^n \tag{10}$$

So,

$$\ln S = \ln 2^n + \ln m = \ln 2 \cdot (n + \log_2 m) \tag{11}$$

Powers of 2 can be calculated simply by shifting. The values of m and n then need to be determined. According to the characteristic of binary numbers, it is only necessary to know the first non-zero bit of the binary number S, y, and the position of the decimal point, z, then $n = y - z$. Shift S to the right by n bits to get the value of m.

Also using the approximation method to calculate $\log_2 m$, since $m \in (1, 2]$, we can take the approximate function as $f(m) = m-1$. Then

$$\ln S = \ln 2 \cdot (n + m - 1) \tag{12}$$

Again since ln2 is a constant, a modified constant multiplier can be used. In combination with the above modules, the final result of $\ln S$ can be calculated. Logarithm calculation module is shown in Fig. 9.

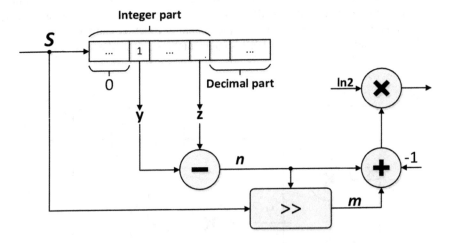

Fig. 9. Logarithm unit diagram

3 Results

Table 1 shows the hardware resource consumption of the direct convolution and Winograd convolution respectively. Both are synthesized in TSMC 28 nm technology by the tool Design Compiler of synopsys and each performs 256 3 × 3 convolutions. As can be seen, the Winograd algorithm leads to significant reduction in gate cell count and area.

Table 1. Hardware resource of convolution

Algorithm	Area (um2)	Cell count
Direct convolution	181639	200152
Winograd convolution	112546	144474

Table 2 compares the resource consumption and performance of the direct calculation architecture with the improved architecture of softmax function, also synthesized in TSMC 28 nm technology by Design Compiler. And Fig. 10 shows the precision comparison of lookup table and the proposed design for exponential function. The proposed design reduces the hardware complexity while maintaining accuracy.

Table 2. Comparison of different softmax function architectures

	Path length	Area (um2)	Cell count
Propose design	0.59	15926	10467
Direct design	1.26	34102	20374

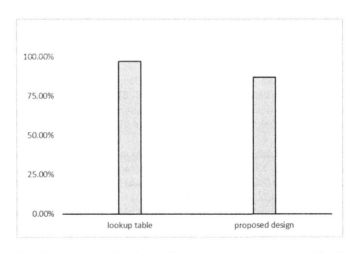

Fig. 10. Accuracy of lookup table and the proposed exponential unit

4   Conclusion

This paper proposed a deep learning accelerator based on DSP. The overall architecture of the accelerator is introduced. Data is pre-processed for the Winograd algorithm in the proposed design, also the MAC array is specially optimized. Softmax function is simplified to reduce the hardware complexity. The design's performance is evaluated. The results show a significant resources reduction and energy efficiency improvement of the proposed design.

Acknowledgement. The authors would like to thank the editors and the reviewers for providing comments and suggestions for this paper. This work was supported in part by the National Natural Science Foundation of China under Grant Nos. 61831018, 61901199, and 61631017, and Guangdong Province Key Research and Development Program Major Science and Technology Projects under Grant 2018B010115002.

References

1. Graves, A., Jaitly, N.: Towards end-to-end speech recognition with recurrent neural networks. In: International Conference on Machine Learning, pp. 1764–1772 (2014)
2. Jogin, M., et al.: Feature extraction using convolution neural networks (CNN) and deep learning. In: 2018 3rd IEEE International Conference on Recent Trends in Electronics, Information and Communication Technology (RTEICT). IEEE (2020)
3. Wang, Z., Chen, J., Wang, X.: Convolutional neural network for image feature extraction based on concurrent nested inception modules. In: 2019 15th International Conference on Computational Intelligence and Security (CIS). IEEE (2020)
4. Li, R., et al.: A convolutional neural network with mapping layers for hyperspectral image classification. IEEE Trans. Geoence Remote Sens. **58**(5), 3136–3147 (2020)
5. Long, P.M., Sedghi, H.: Size-free generalization bounds for convolutional neural networks (2019)
6. Shao, R., Zhong, S., Yan, L.: ASIC-based architecture for the real-time computation of 2D convolution with large kernel size. In: International Symposium on Multispectral Image Processing and Pattern Recognition International Society for Optics and Photonics (2015)
7. Jouppi, N.P., et al.: In-datacenter performance analysis of a tensor processing unit. In: 2017 ACM/IEEE 44th Annual International Symposium on Computer Architecture (ISCA), pp. 1–12. IEEE (2017)
8. Ren, H., Zhang, Z., Wu, J.: SWIFT: a computationally-intensive DSP architecture for communication applications. Mobile Networks Appl. **21**(6), 974 (2016)
9. Spagnolo, F., et al.: Designing fast convolutional engines for deep learning applications. In: 2018 25th IEEE International Conference on Electronics, Circuits and Systems (ICECS). IEEE (2019)
10. Meng, L., Brothers, J.: Efficient Winograd Convolution via Integer Arithmetic (2019)
11. Asgari, B., Hadidi, R., Kim, H.: Proposing a fast and scalable systolic array for matrix multiplication. In: 2020 IEEE 28th Annual International Symposium on Field-Programmable Custom Computing Machines (FCCM). IEEE (2020)
12. Wang, Z., Lan, Q., He, H., Zhang, C.: Winograd algorithm for 3D convolution neural networks. In: Lintas, A., Rovetta, S., Verschure, P.F.M.J., Villa, A.E.P. (eds.) ICANN 2017. LNCS, vol. 10614, pp. 609–616. Springer, Cham (2017). https://doi.org/10.1007/978-3-319-68612-7_69

13. Yuan, B.: Efficient hardware architecture of softmax layer in deep neural network. In: System-on-chip Conference. IEEE (2017)
14. Zhu, D., et al.: Efficient Precision-adjustable architecture for softmax function in deep learning. IEEE Trans. Circuits Syst. II Express Briefs, **99**, 1 (2020)
15. Farabet, C., et al.: Hardware accelerated convolutional neural networks for synthetic vision systems. In: IEEE International Symposium on Circuits and Systems. IEEE (2010)

Two-Stage Task Planning Based on Resource Interchange in Space Information Networks

Runzi Liu[1,3] , Jing Li[2] , Xiang Ji[1] , Weihua Wu[2(✉)] , Di Zhou[2] , and Yan Zhang[2,3]

[1] School of Information and Control Engineering,
Xi'an University of Architecture and Technology, Xi'an 710055, China
rzliu@xauat.edu.cn, xxiang_ji@163.com
[2] School of Telecommunications Engineering, Xidian University, Xi'an 710071, China
{lijing,whwu,zhoudi,yanzhang}@xidian.edu.cn
[3] Science and Technology on Communication Networks Laboratory,
CETC the 54th Research Institute, Shijiazhuang 050081, China

Abstract. The high computational complexity and the mismatch between the space-time distribution of tasks and resources raise great challenges on task planning in space information networks (SINs). This paper studies the task planning problem in SINs by exploiting resource interchange to handle the bottleneck resources. First of all, we use time-varying resource graph to capture the dynamic coordination relationship among resources in SIN. Then, we explore resource interchange and derive its quantitative condition. On this basis, an optimization model for task planning based on resource interchange is formulated. Furthermore, we decompose the task planning problem into two stages for global optimization and local adjustment, and develop the algorithms respectively. Finally, simulation results show that compared with existing works the proposed algorithm strikes a better balance between the number of completed tasks and computational complexity.

Keywords: Space information networks · Resource interchange · Task planning · Resource representation

1 Introduction

Space information network is a network system which could acquire, deliver and process spatial information in real time [1]. Due to the prominent features such as wide coverage and flexible networking free from geographical restrictions [2,3], SINs have played important roles in both military and civilian fields in recent years. There are many types of resources (e.g., communication, computation, storage, observation resources, etc.) in SINs, and complex tasks always require coordination of multiple types of resources [4]. Moreover, due to the high-speed movement of space nodes, the distribution of resources in the network is dynamic

© ICST Institute for Computer Sciences, Social Informatics and Telecommunications Engineering 2021
Published by Springer Nature Switzerland AG 2021. All Rights Reserved
H. Gao et al. (Eds.): ChinaCom 2020, LNICST 352, pp. 644–659, 2021.
https://doi.org/10.1007/978-3-030-67720-6_44

and non-uniform in both space and time dimension. Similarly, the space-time distribution of task arrivals is also of great randomness [5]. With the increasing task demands, the temporal shortage of some kinds of resources in certain region becomes more and more obvious, which brings great challenges to the timely completion of tasks.

Faced up with this problem, some intrinsic properties of SIN such as resource mobility [6] and emerging technologies such as software defined payloads [7] are utilized to compensate the bottleneck resources timely. Literature [6] and [8] study the mode of utilizing resource mobility such as resource interchange and resource aggregation, and show that the utilization of resource mobility can effectively improve the service capability of SINs. On this basis, [9] proposes a task planning strategy that considers the interchange of storage and communication resources of SINs. By designing a time-space dimensional resource representation model of software defined satellite networks, literature [10] characterizes the transformation among network resource functions, and thus proposes a resource management method.

Although the above methods [9,10] have good performance in terms of compensating the bottleneck resources, there still exists a lot room for improvement. Firstly, these methods only exploit resource mobility qualitatively, while the quantitative study of conditions for its utilizations such as resource interchange is still lacking. Secondly, in spite of enhancing the flexibility of the matching between resource combinations and tasks, the utilization of resource interchange also increases the space of feasible solution. As task planning of SIN is a complex combinatorial optimization problem, this significantly increases the computational complexity of solving problem. Most of the existing works either consider a network with limited size or design heuristic algorithm which decomposes the large scale task planning problem into multiple sub-problems according to time or scheduled resources, but ignores the correlation between resources [10,11].

To this end, this paper proposes a two-stage task planning strategy based on resource interchange. Specifically, we first use time-varying resource graph model to capture the dynamic coordination relationship among resources in SIN. Then, we explore the mechanism of resource interchange and propose its quantitative condition. On this basis, an optimization model for task planning based on resource interchange is formulated. Furthermore, we decompose the task planning problem into two stages, named coarse grained global optimization and fine-grained local adjustment, and design algorithms for them, respectively. Finally, simulation results show the effectiveness of the proposed algorithms in terms of task completion rate and computational complexity.

2   System Model

2.1   Network Scenario

We consider a space information network (as shown in Fig. 1), which consists:

- A set of earth observation satellites distributed on low earth orbits, denoted by $OS = \{os_1, os_2, ..., os_n, ...\}$. Each earth observation satellite is equipped with an imager, a solid state mass storage and two transceivers (to communicate with ground stations and data relay satellites respectively).
- A set of data relay satellites on the geostationary earth orbit, denoted by $RS = \{rs_1, rs_2, ..., rs_n, ...\}$.
- A set of ground stations $GS = \{gs_1, gs_2, ..., gs_n, ...\}$.
- A data processing center, denoted by dc.
- A set of observation targets distributed randomly on earth, denoted by $OB = \{ob_1, ob_2, ..., ob_n, ...\}$.

The set of nodes in SIN is denoted as $Nd = OS \cup RS \cup GS \cup OB \cup \{dc\}$.

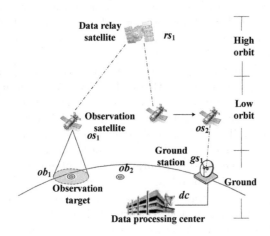

Fig. 1. Network scenario.

2.2   Resource Model

Owing to the high-speed movement of satellite nodes, the topology and coordination relationship among different resources in SINs is of great dynamic. With the advantage of characterizing the resources of dynamic networks in time-space dimensional, the time-extended graph model proposed by Fulkerson [12] and its varieties have been widely used in SINs in recent years [13]. In order to utilize resource interchange, we consider the time-varying resource graph (TVRG) model, which can not only jointly represent multiple kinds of resources of SINs, but also take resource mobility into account.

To construct the TVRG of SIN, we first divide the planning horizon into K equal-length time slots, each with duration τ. As shown in Fig. 2, time-varying resource graph $G_K(V, A)$ is a K-layered directed graph, wherein each layer represent to the topology of SIN of corresponding time slot. V and A represent the set of vertices and arcs in TVRG, respectively.

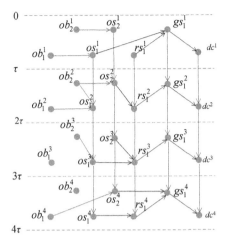

Fig. 2. The time-varying resource graph.

The vertex set V in TVRG is composed of the temporal replicas of the nodes in SIN for every time slots, which can be expressed as $V = V_{OB} \cup V_{OS} \cup V_{RS} \cup V_{GS} \cup V_{DC}$, wherein $V_{OB}, V_{OS}, V_{RS}, V_{GS}$ and V_{DC} are the sets of the temporal replicas of observation targets, observation satellites, relay satellites, ground stations and data processing center, respectively. Specifically, the set of the temporal replicas of observation targets is expressed as

$$V_{OB} = \{ob_i^k | 1 \le k \le K, 1 \le i \le |OB|\}, \tag{1}$$

wherein ob_i^k represents the observation target ob_i in the kth time slot. Similarly, we have $V_{OS} = \{os_i^k | 1 \le i \le |OS|, 1 \le k \le K\}$, $V_{RS} = \{rs_i^k | 1 \le i \le |RS|, 1 \le k \le K\}$, $V_{GS} = \{gs_i^k | 1 \le i \le |GS|, 1 \le k \le K\}$ and $V_{DC} = \{dc^k | 1 \le k \le K\}$, wherein os_i^k, rs_i^k, gs_i^k and dc^k represent observation satellite os_i, data relay satellite rs_i, ground station gs_i and data processing center dc in the kth time slot, respectively.

The arcs in TVRG represent different types of resources in the network, of which the set is defined as $A = A_O \cup A_L \cup A_S$, wherein A_O, A_L and A_S are the set of observation, transmission and storage arcs, respectively. The observation arcs model the opportunities for earth observation satellites to acquire data from observation targets, i.e.,

$$A_O = \{(ob_i^k, os_j^k) | lc(ob_i) \in R(os_j, k), 1 \le i \le |OB|, 1 \le j \le |OS|, 1 \le k \le K\}, \tag{2}$$

where $lc(ob_i)$ represents the location of observation target ob_i, and $R(os_j, k)$ represents the visible range of observation satellite os_j in the kth time slot. The capacity of observation arc $(ob_i^k, os_j^k) \in A_O$ is the maximum amount of data that observation satellite os_j can acquire from observation target ob_i in the kth slot, i.e., $C(ob_i^k, os_j^k) = ro_j \cdot \tau$, where ro_j represents the data acquisition rate of the imager in os_j. The set of transmission arcs is expressed as $A_L = A_{ol} \cup A_{fl}$, wherein A_{ol} represents the set of opportunities for earth observation satellites communicates with data relay satellites or ground stations, and A_{fl} represents the set of fixed links from data relay satellites to their specific ground stations and from ground stations to data processing center, i.e.,

$$A_{ol} = \{(os_i^k, rs_j^k) | lc(os_i) \in R(rs_j, k), os_i^k \in V_{OS}, rs_j^k \in V_{RS}\}$$
$$\cup \{(os_i^k, gs_j^k) | lc(os_i) \in R(gs_j, k), os_i^k \in V_{OS}, gs_j^k \in V_{GS}\} \tag{3}$$

$$A_{fl} = \{(rs_i^k, gs_{rg(i)}^k) | rs_i^k \in V_{RS}, gs_{rg(i)}^k \in V_{RS}\}$$
$$\cup \{(gs_i^k, dc^k) | 1 \le i \le |GS|, 1 \le k \le K\}, \tag{4}$$

wherein $rg(i)$ denotes the index of the specific ground station for data relay satellite rs_i. The capacity of a transmission arc is the maximum of data that the corresponding link can transmit in a time slot, which is given by $c(v_i^k, v_j^k) = rd(v_i, v_j)$, where $rd(v_i, v_j)$ is the transmission rate of the link (v_i, v_j). The storage arc represents the ability of storage resources to store data, the set of storage arcs is expressed as

$$A_S = \{(v_i^k, v_i^{k+1}) | v_i^k \in V_{OS} \cup V_{RS} \cup V_{GS} \cup V_{DC}, 1 \le k \le K - 1\} \tag{5}$$

The storage arc capacity is the volume of the storage of corresponding nodes.

The capacity of the arcs in TVRG represents the size of the ability of corresponding resources, which are measured by uniform unit bit. Moreover, the flows in TVRG represent the task execution process in SIN. Specifically, the set of arcs passed by a flow represents the combination of resources required to complete the corresponding task, and the value of the flow represents the volume of the data acquired, stored, and transmitted by the task.

Definition 1 (Resource Combination). *A set of consecutively connected arcs from vertex v_s^k to v_d^l in TVRG, denoted by $rc(v_s^k, v_d^l) = \{a_1, a_2, ..., a_m\}$, which satisfies*

$$start(a_1) = v_s^k, end(a_m) = v_d^l,$$
$$start(a_{i+1}) = end(a_i), \forall 1 \le i \le m - 1, \tag{6}$$

is referred to as a resources combination from node v_s to node v_d in SIN, where $start(a_i)$ and $end(a_i)$ represent the start and end point of arc a_i, respectively.

2.3   Quantitative Condition for Resource Interchange in SINs

In SINs the bottleneck resources can be compensated by the transformation or replacement with non-scarce resources, which is referred to as resource interchange [6,8]. More specifically, the resource mobility improves the freedom of coordination relationship between different resources in SINs. It provides additional opportunities for different resources to replace each other to complete of tasks, so that the shortage of certain types of resources can be supplemented by other types of resources that are relatively surplus at the same time. For example, on account of the mobility of satellites, data can be carried long distance through network in their storage, therefore some communication links can be replaced by mobile storage resources to a certain extent.

Note that the same service capability is the essential condition for different resources or resource combinations to interchange with each other to complete tasks. In order to uniformly quantify the service capability of resource combination in SINs, we define two metrics: 1) spatial displacement, 2) transferred data volume. More specifically, the spatial displacement of resource is a vector representing the position difference of resources, which may be caused by resources mobility or the difference of the locations of different resources. Let $llv_i^k = (lov_i^k, lav_i^k)$ denote the location of node v_i in the kth slot of SIN, where lov_i^k and lav_i^k represent the longitude and latitude of v_i in the kth slot, respectively. The spatial displacement from vertex v_i^k to v_j^m, denoted by llv_{ij}^{km}, is defined as

$$llv_{ij}^{km} = (lov_j^m - lov_i^k, lav_j^m - lav_i^k). \tag{7}$$

The transferred data volume of a resource is the amount of data that the resource can handle in a unit time, which equals to the capacity of the arc corresponding to the resource. For example, the transferred data volume of communication resource (v_i^k, v_j^k) is number of bits transmitted by the link (v_i, v_j) in the kth time slot (i.e., $c(v_i, v_j)$). Similarly, the transferred data volume of storage resource (v_i^k, v_j^{k+1}) is $c(v_i^k, v_j^{k+1})$.

Based on the definition of spatial displacement and transferred data volume resources, the service capability measurement of a resource can be defined. The service capability of resource (v_i^k, v_j^m) is defined as the product of its spatial displacement and transferred data volume resources, i.e.,

$$sc_{ij}^{km} = c(v_i^k, v_j^m) \cdot llv_{ij}^{km}. \tag{8}$$

As the amount of data can be processed by resource combination $rc(v_s^p, v_d^q)$ is the minimum capacity of the arcs therein, i.e., $min\{c(v_i^k, v_j^m) | (v_i^k, v_j^m) \in rc(v_s^p, v_d^q)\}$, the service capability of resource combination $rc(v_s^p, v_d^q)$ is defined as

$$min\{c(v_i^k, v_j^m) | (v_i^k, v_j^m) \in rc(v_s^p, v_d^q)\} \cdot llv_{sd}^{pq}. \tag{9}$$

With the uniform "transferred data volume spatial displacement" measurement for the service capability of resource combination, we can decide whether different resource combinations can be interchange or not, thereby providing a premise for effective utilization of resource interchange in SINs.

3 Problem Formulation and Decomposition

3.1 Problem Description

There are a set of observation tasks $OM = \{om_1, om_2, ...\}$ to be planned within the plan horizon Th. The requirement for each task can be represented by a four-dimensional tuple, i.e., $om_i = [ob_i, da_i, ts_i, te_i]$, where ob_i is the observation target of task om_i, da_i represents the amount of data required to be acquired by task om_i, and $[ts_i, te_i]$ represents the effective execution window of task om_i, i.e., ts_i is the earliest start time of the task, and te_i is the latest end time.

SIN completes the tasks according to the plans that are made off-line by the operation control center in advance. A task plan indicates whether a task requirement is successfully planned or rejected, and specifies the detailed data acquisition, storage and transmission process of tasks (e.g., the observation satellites, observation the ground station or data relay satellite receive data, and transmission windows allocated to each task). The goal of task planning is to maximize the number of successful planed tasks, of which all the requirements are satisfied.

3.2 Problem Formulation

As we have discussed in Sect. 2.2, the flows in TVRG represent the task execution processes in SIN. Therefore, the task planning problem in SIN can be modeled into the multi-community flow problem in TVRG. To represent the SIN task requirements into flow constraints in TVRG, task om_i is modeled as flow $ob_i^k \to dc^l$, which is originated from vertex ob_i^k and destined to vertex dc^l in TVRG, where l and k satisfy $\lceil ts_i/\tau \rceil \le k \le l \le \lfloor te_i/\tau \rfloor$. The set of flows corresponding to the feasible execution processes of task om_i is given by

$$F_i = \{ob_i^k \to dc^l | \lceil ts_i/\tau \rceil \le k \le l \le \lfloor te_i/\tau \rfloor\}. \tag{10}$$

The set of flows for all the tasks in SIN is expressed as $\mathcal{F} = \bigcup_{1 \le i \le |OM|} F_i$. Let $w(f)$ denote the value of flow f in TVRG, and $w(v_i^k, v_j^l, f)$ represent the value of f on the arc (v_i^k, v_j^l). Furthermore, boolean variable δ_i denote whether task om_i is successfully planned. The task planning problem maximizing the number of successfully planned tasks can be modeled as follows.

$$\textbf{P1}: \qquad \max \sum\nolimits_{1 \le i \le |OM|} \delta_i$$

subject to

$$C1: \sum_{f \in F_i} w(f) = \delta_i \cdot da_i, \ \forall 1 \le i \le |OM|$$

$$C2: \sum_{(v_i^k, v_j^l) \in A} w(v_i^k, v_j^l, f) - \sum_{(v_j^l, v_i^k) \in A} w(v_j^l, v_i^k, f)$$

$$= \begin{cases} -w(f) \ v_i^k = s(f), f \in F \\ 0 \ v_i^k \in V - \{s(f), d(f)\}, f \in F \\ w(f) \ v_i^k = d(f), f \in F \end{cases}$$

$$C3: \sum_{f \in F} w(v_i^k, v_j^l, f) \le C(v_i^k, v_j^l), \ \forall (v_i^k, v_j^l) \in A_S \cup A_{fl}$$

$$C4: \sum_{f \in F} w(v_i^k, v_j^l, f) \le C(v_i^k, v_j^l) \cdot y(v_i^k, v_j^l), \forall (v_i^k, v_j^l) \in A_O \cup A_{ol}$$

$$C5 : \sum_{(os_i^k, rs_j^k) \in A_{ol}} y(os_i^k, rs_j^k) \leq 1, \forall os_i^k \in V_{OS}$$

$$C6 : \sum_{(os_i^k, gs_j^k) \in A_{ol}} y(os_i^k, gs_j^k) \leq 1, \forall os_i^k \in V_{OS}$$

$$C7 : \sum_{(os_i^k, v_j^k) \in A_{ol}} y(os_i^k, v_j^k) \leq 1, \forall v_j^k \in V_{RS} \cup V_{GS}$$

$$C8 : \sum_{(ob_i^k, os_j^k) \in A_O} y(ob_i^k, os_j^k) \leq 1, \forall os_j^k \in V_{OS}$$

In P1, constraint C1 specifies the amount of data should be delivered for each successfully planned task. C2 is the flow conservation constraint, which ensures that in each time slot, the amount of data sent out plus the amount of data stored in a node is equal to the amount of remaining data in the storage at the end of the previous time slot plus the amount of newly arrived data, and $s(f)$ and $d(f)$ respectively represent the source and destination of flow f in TVRG. C3 is the capacity constraint for the storage arcs and fixed transmission arcs, which models the impact of the limited amount of storage and communication resources on the task execution process in SIN. C4 is the capacity constraint for the observation arcs and opportunity transmission arcs, wherein boolean variable $y(v_i^k, v_j^k)$ $(\forall (v_i^k, v_j^k) \in A_O \cup A_{ol})$ represents whether resource (v_i^k, v_j^k) is scheduled or not. C5–C8 are resource scheduling conflict constraints, which are imposed by the limited service capability of onboard antenna/imagers. For example, because the onboard single access antenna can point to only one satellite/ground station at the same time, the schedules of a antenna communicating with different nodes in the same slot conflict with each other, even if all these nodes are in its coverage range. Similarly, the schedules of an imager observing different targets in one slot conflict with each other, too. Specifically, C5 and C6 restrict that each observation satellite only communicates with one data relay satellite and ground station in one slot, respectively. C7 ensures that each data relay satellite or ground station can only receive the data from one observation satellite in one slot. C8 imposes that each imager can only observe one targets in one slot.

By solving problem P1, the optimization task plan can be obtained through the optimized value of variables. More specifically, the value of δ_i determines whether the task is planned successfully, and $w(v_i^k, v_j^l, f)$ indicates which resource is occupied by the task corresponding to flow f in each time slot.

3.3   Problem Decomposition

As we can see from problem P1, variables $w(f)$ and $w(v_i^k, v_j^l, f)$ are continuous variables, and the variables δ_i and $y_{ov}(v_i^k, v_j^l)$ are integer variables, and the optimization subjective and constraints are linear. Therefore, P1 is a mixed integer linear programming problem [14], which is a NP-hard problem in general [15]. With the growth of the size of SIN, the number of feasible resource combinations for each task request increases exponentially, which leads to an explosive increase of the feasible solution space and thus raises great difficulties to find the global optimization of the problem. Moreover, due to the conflicting relationship among resources, it is inefficient to solve P1 by simply decomposing it into small-scale sub-problems according to time or type of resources.

With the consideration that the complexity of solving problem P1 mainly comes from the combinational relationship of feasible resource combinations, this paper aims to reduce the computational complexity by exploring the relationship among resource combinations. It can be observed that the conflict relationship among resources has correlation in time dimension. For example, if communication resources (os_1^k, gs_1^k) and (os_2^k, gs_1^k) conflict with each other in the kth slot, they would will also conflict with each other in $k + 1$th slot with a high probability. In order to quantify the correlation between resource combinations, we employ the concept of independent degree of resource combination proposed in [8].

Definition 2 (Resource Combination Independence). *The independence between resource combinations p_1 and p_2 is defined as the minimum time distance $d(a_1, a_2)$ between the arcs a_1 and a_2 which respectively belong to p_1 and p_2 and correspond to the same imager or link, i.e.,*

$$Z(p_1, p_2) = \min\{d(a_1, a_2) | \forall a_1 \in p_1, \forall a_2 \in p_2\}, \tag{11}$$

wherein the time distance of resources $a_1 = (v_{i1}^k, v_{j1}^k)$ and $a_1 = (v_{i2}^t, v_{j2}^t)$ is expressed as follows:

$$d(a_1, a_2) = \begin{cases} \infty, & v_{i1} \neq v_{i2} \text{ or } v_{j1} \neq v_{j2} \\ |k - t|, & v_{i1} = v_{i2} \text{ and } v_{j1} = v_{j2} \end{cases}. \tag{12}$$

The value of $Z(p_1, p_2)$ reflects the degree of independence between resource combination p_1 and p_2. The larger $Z(p_1, p_2)$ is, the less temporal correlation between the two resource combinations. That is to say, for the feasible resource combinations of a task with weak independence, if one resource combination conflict with a resource combination of other tasks, the other resource combinations would conflict with the same resource combination with a high probability. Therefore, it is inefficient to directly search the optimization task plan from all the feasible resource combinations with the same weight. To reduce the complexity of solving problem P1, we explore the correlation among resource combinations, and divide the solving process into two stages:

1. **Coarse grained global planning:** For each task, sample the feasible resource combinations to obtain a candidate resource combination set of which the independence between any two resource combinations is no less than n. Then, search for the optimal task plan from the candidate resource combination set, thereby reducing the complexity of global optimization.
2. **Fine-grained local adjustment:** Adjust the global planning results locally by utilizing resource interchange, so that more tasks can be successfully planned.

4 Two-Stage Task Planning Based on Resource Interchange

4.1 Coarse Grained Global Planning Algorithm

In the coarse grained global planning stage, we first sample the feasible resource combinations for each task and obtain a candidate resource combination set of which the independence of any two resource combination is no less than n. Let P_{se} denotes the set of candidate resource combinations of SIN, which is expressed as

$$P_{se} = \bigcup_{1 \leq i \leq |OM|} P_i, \tag{13}$$

where P_i is the candidate resource combination set of task om_i. Then, problem P1 is transferred into a problem P2, which aims at matching the task requests in OM with a set of conflict-free resource combinations in P_{se} to maximize the number of successful planned tasks.

$$\textbf{P2} : \max \sum_{1 \leq i \leq |OM|} \delta_i$$

$$s.t. \ \ C1 : \delta_i = \sum_{1 \leq k \leq |P_i|} \phi_{i,k}, \ \forall 1 \leq i \leq |OM|$$

$$C2 : \sum_{1 \leq k \leq |P_i|} \phi_{i,k} \leq 1, \ \forall 1 \leq i \leq |OM|$$

$$C3 : \phi_{i,k} + \sum_{p_{j,l} \in O(p_{i,k})} \phi_{j,l} \leq 1, \ \forall p_{i,k} \in P_{se}$$

$p_{i,k}$ represents the kth candidate resource combination of task om_i, and boolean variable $\phi_{i,k}$ represents whether $p_{i,k}$ is scheduled. Constraint C1 and C2 impose that there is at most one resource combination is allocated to each task. Constraint C3 ensures that the scheduled resource combinations are not conflict with each other. $O(p_{i,k})$ represents the set of all resource combinations that conflict with the resource combination $p_{i,k}$, and the conflict relationship between resource combinations is defined as follows.

Definition 3 (Conflict of Resource Combinations). *Two resource combination $p_{i,k}$ and $p_{j,l}$ are referred to as conflict with each other, if one of the following conditions is satisfied*

1. *there exist $(v_{m1}^r, v_{n1}^r) \in p_{i,k}$ and $(v_{m2}^t, v_{n2}^t) \in p_{j,l}$, which satisfy $v_{m1}^r = v_{m2}^t$ and $v_{m1}^r \in V_{OS}$;*
2. *there exist $(v_{m1}^r, v_{n1}^r) \in p_{i,k}$ and $(v_{m2}^t, v_{n2}^t) \in p_{j,l}$, which satisfy $v_{n1}^r = v_{n2}^t$ and $v_{n1}^r \in V_{OS} \cup V_{RS} \cup V_{GS}$.*

In order to solve P2, we construct a resource combination conflict graph $RCG(V_C, E_C)$, as shown in Fig. 3. Each of the vertices in RCG represents a candidate resource combination, i.e., $V_C = P_{se}$. The edges of RCG represent the

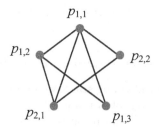

Fig. 3. The resource combination conflict graph.

conflicting relationship among the resource combinations. Specifically, for two resource combinations $p_{i,k}$ and $p_{j,l}$, if one of the following conditions is satisfied, there is an edge between the vertices they corresponding to in RCG:

1. Resource combinations $p_{i,k}$ and $p_{j,l}$ conflict with each other;
2. $i = j$ and $k \neq l$.

With the resource combination conflict graph, problem P2 can be modeled into an maximum independent set problem. Based on the largest degree first rule for the maximum independent set problem [16], a global optimization algorithm based on candidate resource combinations is designed, of which the details is shown in Algorithm 1.

Algorithm 1. Global optimization algorithm

Input: $RCG(V_C, E_C)$;
Output: P_{su};
 1: Initialize $P_{su} = \emptyset$
 2: **while** $V_C \neq \emptyset$ **do**
 3:    Select the vertex with largest degree in V_C denoted as p_0;
 4:    $P_{su} \leftarrow P_{su} \bigcup \{p_0\}$;
 5:    Remove vertex p_0 and all adjacent vertices from V_C;
 6:    Remove all the edges associated with the deleted vertices from set E_C;
 7: **end while**

4.2   Local Adjustment Based on Resource Interchange

After the global planning stage, we adjust the resource allocation result locally through resource interchange to make more tasks successfully planned. The main idea is as follows. Firstly, the resource shortage degree of the unplanned tasks are quantified. Then, we adjust the local resource allocation to release the occupied resources for the unplanned tasks by resource interchange in the descending order of their resource shortage degree.

Algorithm 2. Local Adjustment algorithm

Input: $G_K(V, A), OM_w, G'_K(V', A')$;
Output: OM_y;
1: Initialize $n \leftarrow 1, OM_y \leftarrow OM - OM_w$
2: Calculate the source shortage degree of the tasks in OM_w and sort them in descending order;
3: **while** $n \leq |OM_w|$ **do**
4:     **for** each $a \in P_{Zn}$ **do**
5:         $om_0 \leftarrow fom(a)$;
6:         **if** Exist the path p_z from $ob_0^{\lceil ts_0/\tau \rceil}$ to $dc^{\lfloor te_0/\tau \rfloor}$ in $G'_K(V', A')$ **then**
7:             Replace the resource combination allocated to task om_0 with p_z;
8:             Update $G'_K(V', A')$;
9:             $P_{Zn} \leftarrow P_{Zn} - \{a\}$;
10:        **end if**
11:     **end for**
12:     **if** $P_{Zn} = \emptyset$ **then**
13:         Find the shortest path from $ob_n^{\lceil ts_n/\tau \rceil}$ to $dc^{\lfloor te_n/\tau \rfloor}$ in $G'_K(V', A')$ as the resource combination allocated to task om_n;
14:         $OM_y \leftarrow OM_y \bigcup \{om_n\}$;
15:         Update $G'_K(V', A')$;
16:     **end if**
17:     $n \leftarrow n + 1$
18: **end while**

The resource shortage degree of an unplanned task om_i is defined as the minimum shortage degree of its feasible resource combinations, i.e.,

$$\eta_i = \min_{p_{i,k} \in P_i} \eta_{i,k}, \tag{14}$$

where $\eta_{i,k}$ denotes the number of resources occupied by the planned task in the kth feasible resource combination of task om_i, which can be calculated by following steps:

1. Assign weights to the arcs in TVRG: If the resources corresponding to the arc have been occupied by planned tasks, the weight is assigned 1; otherwise, the weight is assigned 0.
2. For unplanned task om_i, find the shortest path between $ob_i^{\lceil ts_i/\tau \rceil}$ and $dc^{\lfloor te_i/\tau \rfloor}$ on the weighted TVRG, and the resource shortage degree η_i is defined as the sum weight of shortest path.

The local adjustment algorithm based on resource interchange is shown in Algorithm 2. OM_y and OM_w denote the successfully planned tasks and unplanned tasks, respectively. $G'_K(V', A')$ is the sub-graph of $G_K(V, A)$ which excludes the arcs representing the resources occupied by planned tasks. P_{Zn} represents the set of occupied resources in the feasible resource combination with the minimum shortage degree of unplanned tasks om_n. $fom(a)$ denotes the task that occupies resource a.

5   Simulation Results

We conduct a simulation scenario of SIN based on STK (Satellite Tool Kit), which includes 10 low-orbit earth observation satellites distributed in sun-synchronous orbits with altitude from 505 km to 645 km and inclination from 97.4° to 98.5°, 3 data relay satellites distributed in geosynchronous orbit, with longitudes of 16.65°, 76.95° and 176.76° respectively, 100 observation targets randomly distributed on the earth's surface, and three ground stations located in Beijing, Sanya, Kashi and one ground processing center. The planning horizon is 24 h, and 60–300 observation tasks are randomly generated.

In order to verify the performance of the proposed resource interchange based task planning algorithm (RITPA) in terms of task completion, resource utilization, and computational complexity, we employ the following two algorithms for comparison:

1. Time-oriented decomposition based task planning algorithm (TDTPA), which decomposes the task planning problem by dividing the plan horizon into multiple time intervals.
2. Resource-oriented decomposition based task planning algorithm (RDTPA), which decomposes the task planning problem into two stages for observation and data transmission to plan separately.

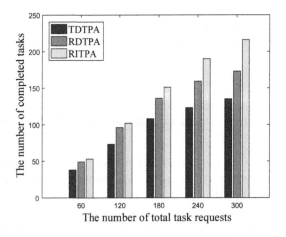

Fig. 4. The number of completed tasks v.s. task number.

Figure 4 depicts the number of completed tasks for the three algorithms with varying task requests. It can be observed that RITPA performs the best, while TDTPA has the worst performance. The reason is that under the temporal decomposition of TDTPA, each sub-planning only focus on the completion of the task requests in current interval and ignores their impact to the tasks in

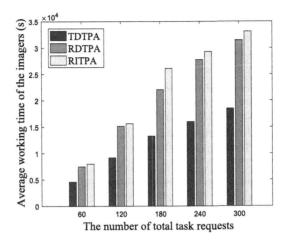

Fig. 5. Average working time of the imagers v.s. task number.

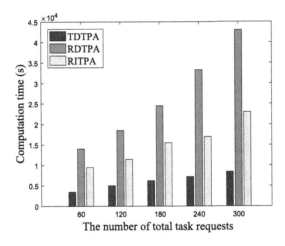

Fig. 6. Computation time v.s. task number.

subsequent intervals. Since the both planning stages of RDTPA are based on the whole planning horizon, it has better performance than TDTPA. However, because the coordination relationship between the observation and communication resources are omitted in RDTPA, some tasks which have been successful planed in the observation stage may be failed in the second stage due to lacking feasible communication resources. RITPA plans globally in both time and the resource dimension in the first stage, and adjust the plans locally in the second stage, which is the main reason why it is superior to TDTPA and RDTPA.

In order to compare the utilization of resources for the three algorithms, Fig. 5 illustrates the average working time of the imagers with varying number of task requests. It can be observed that RITPA has the highest utilization of the observation resource, while that of TDTPA is the lowest. This is because that TDTPA plan least tasks successfully. Moreover, due to the separate planning observation and data transmission in RDTPA, some tasks cannot be delivered to DPC in time after successful observation. This is the main reason that the gap between the imaing time of RDTPA and RITPA is smaller than the gap between the completed task number of those algorithms.

Figure 6 compares the computation time of the three algorithms with varying task requests. It can be observed that TDTPA requires the least computational time, followed by RITPA, and RDTPA requires the most computational time. This is because that TDTPA divides the task requests into time intervals to be planned chronologically, thereby lead to small computational complexity of each sub-planning. Although RDTPA decomposes the planning into two stages, the optimization of both stages is based on the entire planning horizon, which leads to the highest computational complexity of the three algorithms. RITPA proposed in this paper samples feasible resource combinations in the global planning stage, thereby striking a good balance between the performance and computational complexity.

6 Conclusion

This paper proposes a two-stage task planning based on resource interchange in SINs. Specifically, we first we explore the mechanism of resource interchange through TVRG model, and then propose the quantitative condition of it. On this basis, an optimization model for task planning based on resource interchange is formulated. Furthermore, we decompose the task planning problem into two stages: coarse grained global optimization and fine-grained local adjustment, and develop the solving algorithms respectively. Simulation results show that compared with the existing works, the proposed algorithm strikes a better balance between task completion performance and computational complexity.

Acknowledgement. This work is supported by the National Natural Science Foundation of China (61701365, 61801365, 61971327 and 62001347), China Postdoctoral Science Foundation (2018M643581, 2019TQ0241 and 2020M673344), Natural Science Foundation of Shaanxi Province (2019JQ-152, 2020JQ-686), Young Talent fund of University Association for Science and Technology in Shaanxi, China (20200112), Postdoctoral Foundation in Shaanxi Province of China, and the Fundamental Research Funds for the Central Universities.

References

1. Yu, Q., Meng, W., Yang, M.: Virtual multi-beamforming for distributed satellite clusters in space information networks. IEEE Wirel. Commun. **23**(1), 95–101 (2016)
2. Jiang, C., Wang, X., Wang, J.: Security in space information networks. IEEE Commun. Mag. **53**(8), 82–88 (2015)

3. Zhang, Z., Jiang, C., Guo, S.: Temporal centrality-balanced traffic management for space satellite networks. IEEE Trans. Veh. Technol. **67**(5), 4427–4439 (2018)
4. Wang, Y., Sheng, M., Zhuang, W.: Multi-resource coordinate scheduling for earth observation in space information networks. IEEE J. Sel. Areas Commun. **36**(2), 268–279 (2018)
5. Deng, B., Jiang, C., Kuang, L.: Two-phase task scheduling in data relay satellite systems. IEEE Trans. Veh. Technol. **67**(2), 1782–1793 (2017)
6. Sheng, M., Zhou, D., Liu, R.: Resource mobility in space information networks: opportunities, challenges, and approaches. IEEE Network **33**(1), 128–135 (2019)
7. Mast, A.W.: Reconfigurable software defined payload architecture that reduces cost and risk for various missions. In: IEEE Aerospace Conference 2011, pp. 1–5 (2011)
8. Zhu, Y., Liu, R., Sheng, M.: Utilization and analysis of resource mobility in space information networks. J. Commun. Inf. Netw. **4**(1), 67–77 (2019)
9. Liu, R., Zhu, Y., Zhang, Y.: Resource mobility aware hybrid task planning in space information networks. J. Commun. Inf. Netw. **4**(4), 107–116 (2019)
10. Jia, Z., Sheng, M., Li, J.: Joint optimization of VNF deployment and routing in software defined satellite networks. In: IEEE 88th Vehicular Technology Conference (VTC-Fall) 2018, pp. 1–5 (2018)
11. Zhu, W., Hu, X., Xia, W.: A two-phase genetic annealing method for integrated earth observation satellite scheduling problems. Soft. Comput. **23**(1), 181–196 (2019)
12. Ford, L.R., Fulkerson, D.R.: Flows in Networks. Princeton University Press, Princeton (1962)
13. Wang, P., Zhang, X., Zhang, S.: Time-expanded graph-based resource allocation over the satellite networks. IEEE Wirel. Commun. Lett. **8**(2), 360–363 (2019)
14. Bertsekas, D.P.: Nonlinear programming. J. Oper. Res. Soc. **48**(3), 334–334 (1997)
15. Garey, M.R., Johoson, D.S.: Computers and Intractability. W. H. Freeman, New York (2002)
16. Gibbons, A.: Algorithmic Graph Theory. Cambridge University Press, Cambridge (1985)

Low Latency Wireless Communication System Implemented on a Software-Defined Radio Platform

Yujie Liu[1], Jun Yu[1], Fusheng Zhu[2], Wenru Zhang[2], and Jun Wu[3,4(✉)]

[1] College of Electronic and Information Engineering,
Tongji University, Shanghai, China
1833020@tongji.edu.cn, yhbbs7808@163.com
[2] Guangdong Communications and Network Institute, Guangzhou, China
{zhufusheng, zhangwenru}@gdcni.cn
[3] School of Computer Science, Fudan University, Shanghai, China
wujun@fudan.edu.cn
[4] Pengcheng Laboratory, Shenzhen, China

Abstract. Low latency communication has attracted much interest in recent years due to the emergence of new types of delay-sensitive applications. Ultra-Reliable and Low-Latency Communication (URLLC) is also considered as one of the important use-cases in 5G cellular system. Traditional wireless communication system is usually optimized for high data throughput, but cannot satisfy the requirement of strict latency threshold. Semi-persistent scheduling and TTI shortening are two methods to address this problem. We implemented these methods by making modification based on OpenAirInterface framework, and build up a realistic cellular network system using software-defined radio technology. Experimental results show a dramatic latency reduction comparing with the baseline LTE scheme. By integrating these effective methods, we implemented a practical wireless communication system that can provide low latency transmission service.

Keywords: Low latency · Software-defined radio · URLLC · OpenAirInterface

1 Introduction

The 5G wireless communication system has been developed rapidly in recent years, and has been gradually used commercially worldwide. One big difference between 5G and 4G is that 5G is expected to support multiple new applications and use-cases rather than traditional communication scenarios [1]. To meet the requirement of new applications, three different service categories have been introduced: enhanced mobile broadband (eMBB), ultrareliable and low-latency communications (URLLC), and massive machine-type communications (mMTC) [2, 3].

The initial phase of 5G deployment is eMBB, which provides higher throughput and data rate with moderate latency. High resolution video streaming is the main application of eMBB, such as AR/VR/MR media, 4 K/8 K video, panoramic video, etc.

H. Gao et al. (Eds.): ChinaCom 2020, LNICST 352, pp. 660–669, 2021.
https://doi.org/10.1007/978-3-030-67720-6_45

With the development of factory automation, autopilot vehicles, smart grid, Intelligent Traffic System (ITS), etc., traditional wireless system has an urgent motivation to evolve and adapt to new applications. These applications usually don't need high data rate transmission, but has a stringent requirement of latency and reliability. URLLC is a typical scenario with such requirements [4, 5]. Figure 1 shows these new applications and their requirements of latency and reliability [6].

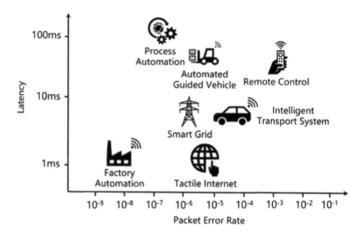

Fig. 1. URLLC use-cases and their requirements of packet error rate and latency

There are many enabling technologies of URLLC, such as short TTI, caching, densification, joint scheduling, grant-free uplink, NOMA, MEC/FOG/MIST and network slicing [7–10]. We would like to survey the related work as follow.

Ashraf et al. proposed a URLLC scheme for factory automation scenario [11]. They used multiple technologies such as shorter TTI, shorter processing time, Semi-Persistent Scheduling (SPS), Instant Uplink Access (IUA) and low complexity channel coding. Simulation results showed good improvement of latency and reliability in a factory deployment model.

Wang et al. gave a comprehensive evaluation of grant-free uplink transmission [12]. Data collision is an unavoidable problem in grant-free transmission, and would thus affect reliability. There are many factors that may affect the performance of grant-free transmission: number of active UEs, HARQ retransmission times, UE detection, SNR, packet arrival rate, etc. Evaluation results showed that a contention-based grant-free transmission scheme could well meet the reliability requirement within the latency bound.

Anand et al. proposed a scheme of joint scheduling of URLLC and eMBB traffic [13]. Superposition, puncturing, overlapping and mini-slot approaches are used to multiplex eMBB and URLLC traffic. Simulation results showed that this joint problem had structural properties, then these properties enabled clean decompositions and corresponding algorithms with theoretical guarantees.

According to our survey and investigation, most of these schemes have been evaluated only by simulations, so we come up with the idea to implement a low latency scheme on a software-defined radio (SDR) system. we implement a realistic wireless system and focus on optimizing the latency to reduce the overall network delay. By doing this, we can evaluate the real-time performance of our scheme, and can use this system as a platform to deploy delay-sensitive applications.

The rest of this paper is organized as follows. In Sect. 2, we analyze the composition of overall radio latency and propose two ways to reduce latency: semi-persistent scheduling and short TTI. In Sect. 3, we described the design and deployment of our system. Section 4 shows the experimental results of our scheme. In Sect. 5, we conclude our work and discuss the direction of future work.

2 Analysis of Latency Composition

A wireless system protocol stack consists of several layers (PHY, MAC, RLC, RRC, PDCP, etc.), and there are many procedures in each layer. In this paper, we focus on MAC and PHY layer, and mainly analyze two factors which contribute to overall transmission latency: scheduling scheme and transmission time interval (TTI).

2.1 Scheduling Scheme

Scheduling is a key feature of MAC layer. In traditional cellular system, the base station acts as the central control entity to allocate time and frequency domain resources, and coordinate all the UEs attaching to this base station by signaling on control channels.

Different scheduling schemes may have different impact on transmission latency. Traditional LTE system uses dynamic scheduling scheme as default, which is fit for mobile broadband traffic with no strict requirement on latency.

In dynamic scheduling scheme, downlink signaling and corresponding downlink data can be sent on the same frame of transmission, so downlink latency is not a critical factor in overall latency. However, in uplink transmission, a "request and grant" procedure is performed when UE has uplink data to send, resulting in a longer latency. Optimization of uplink scheduling scheme is a critical point to reduce the overall latency.

From Fig. 2(a), we can find that the traditional dynamic scheduling has following steps [14]:

1. When an uplink packet arrives at UE's MAC layer at subframe t, a Schedule Request (SR) will be sent at subframe t + 4 on PUCCH.
2. The base station receives the SR at subframe t + 4, and scheduling algorithm is used to make decisions for radio resource allocation. If UE's uplink request is granted, base station will send a specific type of Downlink Control Information (DCI) at subframe t + 8 on PDCCH, allocating radio resource for UE's uplink traffic.
3. UE receives DCI at subframe t + 8, and process the data queueing in buffer. The uplink packet will actually be sent at subframe t + 12.
4. If the uplink packet is correctly received by base station, an ACK would be sent at subframe t + 16, informing UE of a successful uplink transmission.

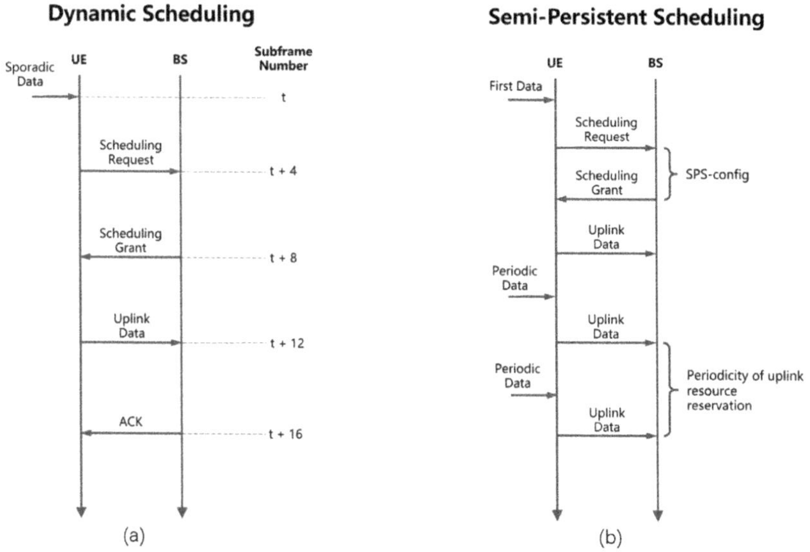

Dynamic Scheduling   **Semi-Persistent Scheduling**

Fig. 2. Procedures of dynamic scheduling and semi-persistent scheduling

Such signaling procedure increases the latency. The overhead would be even bigger when transmitting large amount of small size packet data.

Semi-persistent scheduling (SPS) is a commonly-used scheduling scheme to reduce uplink latency. Figure 2(b) shows that when SPS is applied, the "request and grant" signaling procedure is performed only once at the first attempt of transmission. Afterwards, the uplink data can be sent on pre-allocated radio resource without performing scheduling procedure again. The radio resource is allocated in a periodic manner, and the period of SPS is configurable by the base station. User equipment would wait for its turn to transmit uplink data.

SPS is intended to transmit periodic data, such as VoIP traffic and sensor network data, and the period of SPS could be adjusted to match the production rate of uplink data. For sporadic uplink data, SPS can also be applied to reduce latency, but at the cost of spectral efficiency, because some resource may be over-allocated to user equipment when there is no uplink data to transmit.

2.2 Transmission Time Interval

Transmission Time Interval (TTI) is a very important time parameter in cellular network. TTI is the minimum time unit for many procedures such as channel coding, interleaving and scheduling, so TTI duration would impact the radio latency.

TTI is mainly decided by the design of frame structure. In 4G LTE, TTI is fixed to 1 ms due to the static frame structure, which is equal to the duration of 1 subframe (14 OFDM symbols with normal cyclic prefix). In 5G NR, numerology is introduced to adjust frame structure. Numerologies represent different configurations of Subcarrier Spacing (SCS). According to the theory of OFDM, OFDM symbol duration need to be inversely proportional to SCS to maintain the orthogonality of adjacent subcarriers, which means symbol duration could be shortened with SCS expansion. This new frame structure gives a possibility to shorten the duration of TTI. The baseline configuration of SCS is 15 kHz, which is same with the LTE SCS, and SCS can be scaled with a factor of 2^μ, where $\mu = [0, 1, 2, 3, 4, 5]$. Table 1 shows some combinations of numerology, SCS, OFDM symbol duration and TTI.

Table 1. Numerologies and corresponding frame parameters

Numerology (μ)	Subcarrier Spacing (SCS)	OFDM symbol duration	TTI length	TTI duration
0	15 kHz	66.7 µs	14 symbols with normal CP	1 ms
1	30 kHz	33.3 µs	14 symbols with normal CP	0.5 ms
2	60 kHz	16.7 µs	14 symbols with normal CP	0.25 ms
3	120 kHz	8.33 µs	14 symbols with normal CP	0.125 ms
...	...	...	...	...

Apparently, shorter TTI would have better latency performance. TTI consists of several OFDM symbols, so there are two ways to reduce the duration of TTI: reduce the number of OFMD symbols within a TTI, or reduce the duration of an OFDM symbol. For the former case, mini-slot is introduced in 5G NR as a smaller transmission unit. UE can be scheduled either on slot level (duration of 14 OFDM symbols) or on mini-slot level (duration of several OFDM symbols that can be configured). Mini-slot is a good way to shorten the duration of TTI, but requires lots of modifications of frame structure and signal processing procedures. For the latter case, by increasing the numerology, we can shorten the duration of OFDM symbol, and thus shorten the TTI. When numerology increases, the frame structure is expanded in frequency domain, and is compressed in time domain. In our experiment, we choose to increase numerology and keep the number of symbols within TTI unchanged. Figure 3 shows the frame structures of different numerologies.

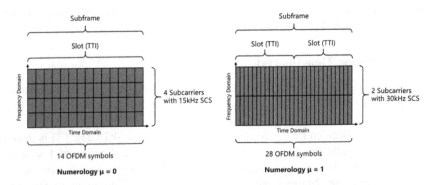

Fig. 3. 5G NR Frame structure of numerology 0 and 1 on time and frequency domain

Shorter TTI have better performance on latency, but TTI can't be reduced infinitely. Shorter TTI would result in higher inter-symbol interference and lower efficiency of channel coding, interleaving, spectrum utilization. A tradeoff should be made to choose a proper TTI to meet the requirement of particular applications.

3 System Design and Deployment

Software-defined radio (SDR) is a kind of radio communication system which implements most of radio signal processing by means of software running on a CPU or DSP, instead of running on specific hardware in traditional wireless system. SDR technology gives more flexibility and convenience on developing radio systems. Researchers can modify the procedures and functions by simply modifying software code, and it's very suitable for implementing prototype system or verifying new schemes. The hardware and software of our SDR system are:

- Host Machine: DELL PowerEdge 8700
- OS: Ubuntu 16.04
- Radio Device: Xilinx Virtex-6 FPGA + Analog Devices AD9361
- SDR Software: Eurocom OpenAirInterface

Our radio frontend device is built up with a Xilinx Virtex-6 FPGA evaluation board and Analog Devices AD9361 radio transceivers, and use PCI-E interface to stream samples to the host machine.

Three host machines are needed by our SDR system. Host Machine 1 runs the components of LTE EPC: HSS, MME and SPGW; Host Machine 2 and 3 are plugged with radio devices and run the softmodem program of eNode and UE respectively. Host Machine 1 and 2 are connected to the same router to provide a backhaul link.

Channel condition is a critical factor in realistic wireless transmission. If channel condition fluctuates a lot, some packets could not transmit successfully at first attempt, and retransmissions would be triggered frequently, causing large fluctuation of latency. To create a relatively stable testing environment, we keep the antennas of UE and eNodeB at a close distance with direct sight to provide a good channel condition, so that most packets can be sent successfully in one transmission.

The SDR software we used in our experiment is OpenAirInterface (OAI). OAI is an open source software radio framework initiate by Eurocom [15], which implements full protocol stack of LTE UE, eNodeB and EPC, forming a complete LTE SDR system. Because the 5G NR version of OAI is under development, we still use the 4G LTE version of OAI as framework and make our modification based on 4G LTE.

Dynamic scheduling is the default option in OAI, and uplink scheduling is triggered by Scheduling Request (SR) or Buffer Status Report (BSR). Following items show our modifications of the scheduling algorithm:

- Disable the uplink scheduling request and grant procedure.
- Use subframe number as index and allocate uplink resource to UE periodically to implement the behavior of SPS.

- Set different period of SPS. For example, if period is set to 1, data can be transmitted on every TTI; if period is set to 3, there is only 1 transmitting TTI in every 3 TTIs.

Next, we try to adjust the transmission time interval to minimize latency. Because we are still using LTE version of OAI as framework, and LTE doesn't support adjustable subcarrier spacing, so we need to set it manually. We set the subcarrier spacing to 15 kHz or 30 kHz by reconfiguring the radio device on both UE and eNodeB side. By doing this, we can make the LTE system running at 1 ms and 0.5 ms TTI respectively.

Figure 4 shows the full deployment of our SDR system.

Fig. 4. Full deployment of our SDR system.

4   Experimental Results

First, we give a standard LTE configuration as baseline:

Subcarrier Spacing: 15 kHz
OFDM symbol duration: 66.7 μs
Multiplex Mode: FDD
TTI duration: 1 ms
Scheduling scheme: dynamic scheduling (DS)

In realistic wireless system, it's not easy to measure the latency between UE and base station directly, so we use round-trip time (RTT) of ping as the indicator of latency. Ping is initiate by UE, and the latency includes transmissions of a Scheduling Request, a Scheduling Grant, a ping request and a ping reply. Ping is based on IP packet, and it can better represent the latency that applications may experience in mobile system. The total round-trip time of ping can be calculated by the following equation:

$t_{total} = t_{SR} + t_{SG} + t_{REQ} + t_{REP}$

t_{total}: the total round-trip time of ping;

t_{SR}: time of uplink Scheduling Request transmission;

t_{SG}: time of downlink Scheduling Grant transmission;

t_{REQ}: time of uplink ping request transmission;

t_{REP}: time of downlink ping reply transmission.

When a UE is attached to the base station, a virtual network interface would be created on UE's host machine, and an IP address would be assigned to this UE by the EPC, and this UE is now able to exchange IP traffic with the network. We ping the core network's packet gateway (PGW) from UE for at least 200 times at one second interval, and take the average ping RTT as the value of latency. All the latency in the following parts of this article is calculated in this way.

To verify the impact of different packet size to the network latency, we disabled the function of dynamic radio resource allocation, and manually set some parameters to let each transmission has the same transport block size (TBS). In our experiment, we set the number of uplink resource blocks to 6 RB, and set uplink Modulation and Coding Scheme (MCS) to 8, so that the uplink TBS is fixed to 101 Bytes. When the size of an uplink packet exceeds TBS, multiple transmissions would be needed for one packet (64 Bytes: 1 transmission; 256 Bytes: 3 transmissions; 1024 Bytes: 11 transmissions), and this would have certain impact on latency when using large period SPS scheme.

First of all, we validated the performance of shorter TTI. From Fig. 5, we can find that there is a proportional relation between TTI and ping latency. When TTI is shortened from 1 ms to 0.5 ms, the ping latency also cut in half. We also tried to shorten the TTI to 0.25 ms, but UE failed to complete the random access procedure and couldn't attach to base station. This may be caused by the increasing Inter Symbol Interference (ISI) and the limitation of hardware's capability.

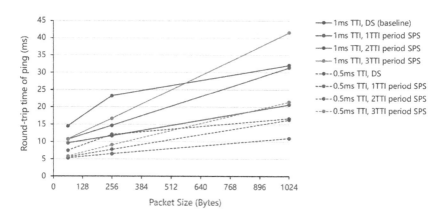

Fig. 5. Latency performance of different TTI duration and scheduling scheme combinations.

Secondly, we validated the performance of optimized uplink scheduling scheme. We applied the modifications mentioned in Sect. 3 to replace the original dynamic scheduling (DS). From Fig. 5, we can find that 1 ms period SPS can reduce the latency by about 5 ms comparing with the baseline when transmitting small packet, and this is contributed by removing the scheduling request and grant procedure, so there is no t_{SR} and t_{SG} components in overall latency.

With Fig. 5,if the uplink packet size is small enough to be sent in one transmission, SPS could have a better latency performance, but if packet size is as large as several times of TBS and the period of SPS is large, SPS performance may be worse than DS. This experiment proved that SPS fitted for periodic small packet traffic. Setting the uplink SPS period to 1 TTI will have the best latency performance because all the data packets could be transmitted immediately upon arrival of MAC layer, but at the cost of reduced spectral efficiency and resource utilization.

By combining shortened-TTI and SPS together, we achieved a minimum 5.3 ms RTT of 64 Bytes ping packet. With this latency level, we have reduced about 65% of network latency comparing with the LTE baseline profile.

5   Conclusion and Discussion

In this paper, we analyzed two major factors contributing to network latency in wireless system: scheduling scheme and transmission time interval. We implemented semi-persistent scheduling and short TTI schemes on a software-defined radio platform and verified the performance. Experimental results show a notable improvement on network latency. The overall delay can be reduced significantly comparing with the baseline LTE system, making it feasible to deploy delay-sensitive applications.

As part of our future work, we also investigate other methods to reduce wireless network latency, such as mini-slot frame structure and HARQ RTT shortening. By exploiting the potential of all these methods, we expect to reduce the latency furthermore, and build up a practical system that can meet the strict latency requirement of URLLC.

Acknowledgement. The authors would like to thank the editors, reviewers and people who provided comments and suggestions for this paper. This work was supported in part by the National Natural Science Foundation of China under Grant Nos. 61831018, and 61631017, and Guangdong Province Key Research and Development Program Major Science and Technology Projects under Grant 2018B010115002.

References

1. Li, Z., Shariatmadari, H., Singh, B., et al.: 5G URLLC: design challenges and system concepts. In: 15th International Symposium on Wireless Communication Systems (ISWCS), Lisbon (2018)
2. GPP TR 38.913, Study on Scenarios and Requirements for Next Generation Access Technologies. 3GPP (2017)
3. GPP TR 36.881, Study on latency reduction techniques for LTE. 3GPP (2016)
4. GPP TR 38.802, Study on New Radio (NR) Access Technology, Physical Layer Aspects. 3GPP (2017)
5. International Telecommunication Union (ITU), IMT Vision – Framework and overall objectives of the future development of IMT for 2020 and beyond. Telecommunication Standardization Sector of ITU (2015)

6. Peisa, J.: 5G techniques for ultra-reliable low latency communication. In: IEEE Conference on Standards for Communications & Networking (CSCN). IEEE, Helsinki (2017)
7. Sachs, J., Wikström, G., Dudda, T., et al.: 5G radio network design for ultra-reliable low-latency communication. IEEE Network **32**(2), 24–31 (2018)
8. Fehrenbach, T., Datta, R., Göktepe, B., et al.: URLLC services in 5G low latency enhancements for LTE. In: IEEE 88th Vehicular Technology Conference (VTC-Fall), pp. 1–6. IEEE, Chicago (2018)
9. Shariatmadari, H., Duan, R., Iraji, S., Li, Z., Uusitalo, M.A., Jäntti, R.: Resource allocations for ultra-reliable low-latency communications. Int. J. Wireless Inf. Netwk. **24**(3), 317–327 (2017). https://doi.org/10.1007/s10776-017-0360-5
10. Pocovi, G., Soret, B., Pedersen, K.I., et al.: MAC layer enhancements for ultra-reliable low-latency communications in cellular networks. In: IEEE International Conference on Communications (ICC) Workshop. IEEE, Paris (2017)
11. Ashraf, A., Aktas, I., Eriksson, E., et al.: Ultra-reliable and low-latency communication for wireless factory automation: From LTE to 5G. In: IEEE 21st International Conference on Emerging Technologies and Factory Automation (ETFA), pp. 1–8. IEEE, Berlin (2016)
12. Wang, C., Chen, Y., Wu, Y., et al.: Performance evaluation of grant-free transmission for uplink URLLC services. In: 85th Vehicular Technology Conference. IEEE, Sydney (2017)
13. Anand, A., De Veciana, G., Shakkottai, S.: Joint scheduling of URLLC and eMBB traffic in 5G wireless networks. In: IEEE International Conference on Computer Communications, pp. 1970–1978. IEEE, Honolulu (2018)
14. GPP TS 36.213, Evolved Universal Terrestrial Radio Access (E-UTRA), Physical layer procedures. 3GPP (2017)
15. OpenAirInterface, https://www.openairinterface.org/

A Deep Learning Compiler
for Vector Processor

Pingping Pan[1], Jun Wu[2(✉)], Songyuan Zhao[1], Haoqi Ren[1],
and Zhifeng Zhang[1]

[1] Department of Computer Science, Tongji University, Shanghai, China
pppgary@163.com,
{1830835, renhaoqi, zhangzf}@tongji.edu.cn
[2] School of Computer Science, Fudan University, Shanghai, China
wujun@fudan.edu.cn

Abstract. The technical route of machine learning compiler generally refers to the application of automatic or semi-automatic code generation in the optimization process instead of hand-optimization. This paper presents a deep learning compiler (DLCS) for target vector processor based on LLVM framework, which lowers deep learning (DL) models to an intermediate representation (IR) of two levels. The high-level IR realizes target-independent optimizations including kernel fusion, data replacement and data simplification, while the low-level IR allows the compiler to perform target-dependent optimizations, such as Eight-Slots VLIW and special intrinsic function. The proposed compiler customizes the architecture description of target vector processor to achieve a high-quality automatic code generation. We evaluate the performance comparison between DLCS and hand-optimization when deploying ResNet-18 model and MobileNet model to the target vector processor. Experimental results show that DLCS offers Multi-slot parallel performance for target vector processor and achieves speedups ranging from $1.5\times$ to $3.0\times$ over existing frameworks backed by hand-optimized libraries.

Keywords: Deep learning compiler · Target optimization · Code generation · Vector processor

1 Introduction

Artificial Intelligence (AI) [1] is undoubtedly a hot topic at present, most of which focus on algorithm. However, engineering is crucial when we put AI into practical application.

A DL algorithm needs to go through two steps from theoretical model to practical application: training [2] and inference [3]. Training refers to the process of guiding DL models through the algorithm to train the data and making it available. At present, training technology has been quite sophisticated, where current mainstream DL

H. Gao et al. (Eds.): ChinaCom 2020, LNICST 352, pp. 670–687, 2021.
https://doi.org/10.1007/978-3-030-67720-6_46

frameworks include TensorFlow [4, 5], PyTorch [6], MXNet [7], Caffe [8], etc. Inference is the process of compiling and deploying DL models to a target hardware. As shown in Fig. 1, inference framework or DL compiler [9] works as a bridge to connect top frameworks and target hardware, which convert the flow graph of DL

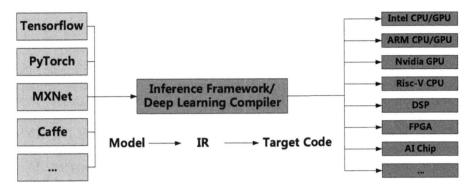

Fig. 1. Inference framework

models into a graph IR [10] and performs optimization operations, such as cycle scheduling [11] and operator fusion [12], and generates machine codes of each target device as required.

On the one hand, the market situation of DL compiler is diverse but not uniform. DL models are usually deployed to a variety of target devices including embedded CPUs, GPUs, DSPs, FPGAs, and ASICs. It is difficult to maintain an efficient inference performance in all target devices. Major hardware vendors have launched their own inference framework, such as Intel OpenVINO [13], ARM NN [14], NV TensorRT [15], but the major equipment manufacturers do not have a framework of generality. For example, the operator functions generated by various DL frameworks are not fully supported, especially the TensorFlow operators, where a DL model can run in one target device but not another.

On the other hand, most DL algorithms are computationally and memory-intensive, which require tens of megabytes of parameter storage and hundreds of millions of operations. It is difficult to compile and deploy DL models to embedded or battery-limited systems such as smartphones and smart glasses. However, DSP has strong computing power, which is more suitable for processing computation-intensive tasks comparing with other embedded processors. Target vector processor [16, 17] is an embedded high performance digital signal processor, which not only has the special convolution instruction based on large-scale matrix multiplication, but also realizes the eight-slots parallel technology of Vary Long Instruction Word (VLIW) [18] and the 2560 bits operations of Single Instruction and Multiple Data (SIMD) [19].

This paper presented the design of DLCS, a deep learning compiler for the independently designed target vector processor, which perform target optimizations through lowering DL models into a high-level IR and a low-level IR. DLCS takes advantage of model pruning to get lightweight model and customize the architecture description of target vector processor to achieve a high-quality automatic code generation.

2 Relate Works

TVM [9, 20] is an automated end-to-end optimizing compiler for deep learning that exposes both graphic-level and operation-level optimizations to provide performance portability to deep learning workloads across different hardware back-ends. TVM addresses optimization challenges for deep learning, such as advanced operator fusion, mapping to arbitrary hardware primitives, and memory latency hiding [21]. TVM automates low-level program optimization into hardware features by using a new, learning-based cost modeling approach to rapidly explore code optimization. TVM serves primarily as a bridge between the top framework and the target device. A trained model can generate optimized operators and inference code for different target platforms through TVM, thus realizing the deployment of multiple target platforms for a single model. The automatically generated operators can be further optimized by hand. In addition, one of the attractions of TVM is that its automatically generated code is comparable to those of hand-optimized acceleration libraries.

Accelerated Linear Algebra (XLA) [22] is a compiler that optimizes the computation of TensorFlow. The input language for XLA is called High Level Optimizer (HLO) [23] which is equivalent to the compiler's IR. The layering criteria for TVM IR are computational graphs and operators, while the layering criteria for HLO IR is device independent and device dependent. XLA improves server and mobile platform speed, memory usage, and portability. XLA performs target hardware-independent optimization and analysis [24], such as the common subexpression elimination (CSE) [25], process fusion of operators independent of the hardware back-end, and buffer analysis for memory allocation at run time for computation. XLA performs target hardware-dependent optimization and analysis, which will be targeted at the specific information and requirements of the hardware target, such as operator fusion on the XLA GPU back-end, and determining how to divide the calculation into streams. In addition, the backend can pattern match certain operators or combinations of them to the optimized library call. Finally, XLA does code generation for specific target hardware. The CPU, including the GPU used by XLA, uses LLVM [26] to generate low-level IR, optimization, and code generation. The hardware back-end emits the LLVM IR [27] required to represent the XLA HLO calculation in an efficient manner, and then calls LLVM to emit native code from the LLVM IR. The GPU back-end currently supports NVIDIA GPU through the LLVM NVPTX back-end, while the CPU back-end supports multiple CPU instruction set architecture (ISA).

Graph lowering compiler (Glow) [28] is a heterogeneous hardware-oriented machine learning compiler. It provides a practical compilation method that generates highly optimized code for multiple targets. Glow reduces the traditional neural network data flow diagram to an intermediate representation of a two-phase strongly-type [29]. The advanced intermediate representation allows the optimizer to perform domain-specific optimizations. Lower level instructions-based addressing only intermediate representations allow the optimizer to perform memory-related optimizations such as instruction scheduling [30], static memory allocation [31], and copy elimination [32]. The intermediate representation at the lowest level allows the optimizer to perform machine-specific code generation to take advantage of specialized hardware features. Glow features a lowering phase in which the compiler can support a large number of input operators and a large number of hardware targets by eliminating the need to implement all operators on all targets. The purpose of the lowering phase is to reduce the input space and allow the new hardware back end to focus on a small number of linear algebra primitives.

Deep Learning Virtual Machine (DLVM) [33] is a compiler infrastructure designed for modern deep learning systems. DLVM applies a multi-stage compiler optimization strategy to high-level linear algebra and low-level parallelism, performs domain-specific transformations, reduces the overhead of front-end languages, and acts as a host for the research and development of neural network domain-specific languages (DSLs). DLVM IR is the core language of the system, which uses static single assignment (SSA) forms, control flow diagrams, advanced types (including first-order tensor types), and a set of linear algebraic operators combined with a generic instruction set. The system supports a variety of domain-specific analyses and transformations, such as inverse pattern algorithmic differentiation [34], algorithmic differential checkpoints, algebraic simplification [35], and linear algebraic fusion. The complete DLVM software stack, including sample front-end neural network DSLs, and DLVM Core contains essential components for an optimizing compiler: intermediate representation and passes.

3   DLCS Core

As shown in Fig. 2, DLCS IR adopts a two-layer optimized structure, including high level IR and low level IR. Low level IR refers to the LLVM IR, which is mainly used for memory-related optimization such as instruction scheduling, static memory allocation, and copy elimination. High level IR refers to Graph IR, which is mainly used for the optimization of computational graphs, such as kernel fusion, data displacement compilation optimization, data layout transformation.

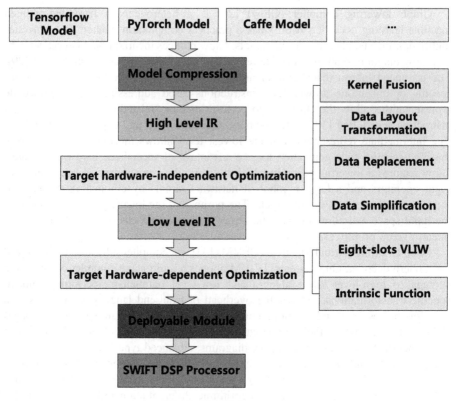

Fig. 2. DLCS framework

3.1   Target Optimization

Target Hardware-Independent Optimization

Data Simplification and Data Layout Transformation. Data simplification mainly includes node pruning and constant folding. Node pruning refers to the removal of identity nodes related to the training phase in the deployment phase, which simplify the computational graph and saving the operation cost. Constant folding optimization is to replace the nodes in the computational graph whose output values can be determined with constants in advance. On the one hand, the output of shape, rank and size in the computational graph is only related to the shape of the output tensor, which has nothing to do with the specific data input and can be calculated in advance. On the other hand, the output of a node whose input is all constant can also be calculated in advance.

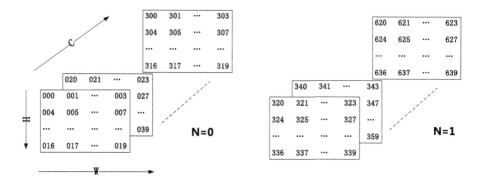

NCHW: 000 001 002 003 004 ... 018 019 020 ... 318 319 320 ... ...
NHWC: 000 020 ... 300 001 021 ... 283 303 004 ... 319 320 340 ...
CHWN: 000 030 001 321 ... 003 323 004 324 ... 019 339 020 ... ...

Fig. 3. Data layout: NCHW, NHWC, CHWN

The data layout transformation optimization mainly refers to the storage mode of transforming tensors to adapt to the target hardware architecture. The target hardware of DLCS is the target vector processor, and the optimal storage format is NCHW [23], where N represents the number of images in a batch, H represents the number of pixels in the direction of vertical height, W represents the number of pixels in the direction of horizontal width, and C represents the number of channels, as shown in the Fig. 3. The data layout optimization done by DLCS is to convert various data layout (NHWC [23] and CHWN [23]) to NCHW storage format. The data layout order of NCHW is [W-H-C-N], where the first element is 000 and the second element is along the W direction from 001 to 003. The second direction is along the H direction from 004 to 019. Next direction is along the C direction from 020 to 319. The last direction is along the N direction from 320 to 639.

Kernel Fusion. Kernel fusion, also known as operator fusion, is to fuse multiple operators in the computational graph into a single kernel. For example, (1) and (2) can be fused into (3), therefor, the entire data flow diagram can be completed with only one kernel. Moreover, by properly designing the placement of input and output data of different kernel functions, such as using shared memory or registers on GPU, the data transmission efficiency and the overall computing performance can be greatly improved. General operations can be divided into three types: one-to-one mapping, complex fusion and infusion. Multiple injective operators can be merged into another injective operator. We can fuse element-level operators into the output of operators which are complex and fusible, like conv2d operator.

$$Z = op1 \ (X, \ Y) \tag{1}$$

$$T = op2 \ (Z) \tag{2}$$

$$T = op2 \ (op1 \ (X, \ Y)) = op3 \ (X, \ Y) \tag{3}$$

Data Replacement. The deployable objects generated by DLCS mainly include parameters and computational graphs, which are serialized by binary to a final output file. As shown in Fig. 4, it's common to simply put the data into an unsigned char array and compile the corresponding file. However, if a large amount of model data is trained, the deployment and compilation phase will take a lot of time. The optimization proposed in this section is to replace the serialized data by placing the flag information in the unsigned char array. This flag information is found in the resulting file and replaced with the original binary serialized data. Finally, we can directly deploy the resulting file to the target vector processor to achieve the deployment of DL models.

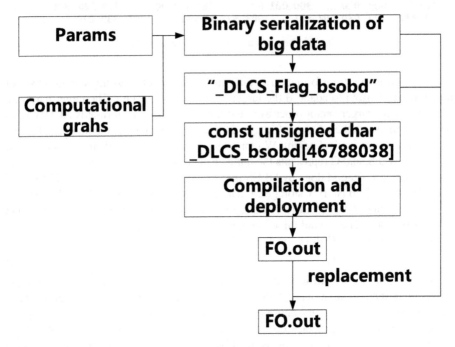

Fig. 4. Data replacement

Target Hardware-dependent Optimization
Eight-slots VLIW. Eight-slots VLIW mainly includes VLIW detection and VLIW coding. VLIW detection refers to the determination of which instructions can form a VLIW package. The judgment of detection is based on whether the empty slots can be allocated and whether the instructions have a dependency with each other. VLIW encoding refers to assigning a value to the slot encoding bit of an instruction in the same VLIW package.

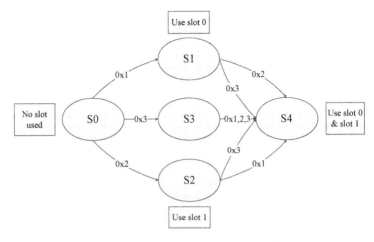

Fig. 5. DFA state transition diagram of two-slots VLIW

We use Deterministic Finite Automaton (DFA) to describe the first detection condition. Given an input instruction sequence in a two-slot state transition diagram, as shown in the Fig. 5, the automatic machine performs a state-to-state transition according to the state transition equation. If an instruction in the input instruction sequence causes the automaton to stop in an unacceptable state, the automaton rejects the instruction. The DFA is reset to start a new round of judgment after the previously accepted instructions forming a VLIW package. As shown in the Table 1, the slot allocation of each instruction would be judged with the VLIW constraint.

Table 1. Instruction slot constraint of target vector processor

Instruction type	Slot 0	Slot 1	Slot 2	Slot 3	Slot 4	Slot 5	Slot 6	Slot 7
Control operations	√	√						
Scalar operation	√						√	
Data transfer operations	√		√		√	√	√	√
Vector operation			√	√				
Loading operations					√	√		
Storage operations							√	√

After completing the first condition of the VLIW detection, we enable a separate compile pass to complete the dependency determination of the detection. The data dependence that we need to judge mainly includes true dependence and output dependence, as shown in the Fig. 6. If one of the two detection conditions is not satisfied, the detection will be exited, and then all the instructions that the previous detection satisfies will be arranged in the same VLIW package.

The input of VLIW encoding is the VLIW package sequence obtained after VLIW detection. We iterate over each VLIW package and assign values to the slot encoding

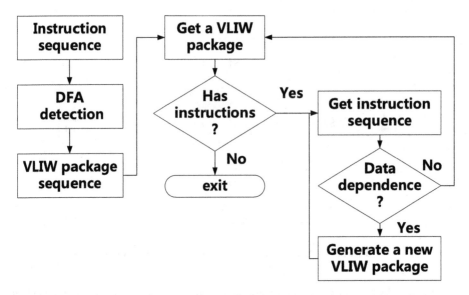

Fig. 6. VLIW detection

bits of all instructions inside. The instruction set of target vector processor is 32-bits encoding mode, where InstBit{31:29} is the slot encoding bits. InstBit{31} indicates whether the instruction is the last instruction in the VLIW packet, while InstBit{30:29} means instruction slot allocation, where InstBit{31:29} = {00} means that slot 0, which has priority, and slot 1 can be allocated to this instruction; InstBit {31:29} = {01} means that slot 2, which has priority, and slot 3 can be allocated to this instruction; InstBit{31:29} = {10} means that slot 4, which has priority, and slot 5 can be allocated to this instruction; InstBit{31:29} = {11} means that slot 6, which has priority, and slot 7 can be allocated to this instruction. Table 2 is an example of VLIW encoding where a detected eight-slot VLIW packet contains five instructions of ABCDE, and their slot encoding bits are 000,001,011,010,111 respectively.

Table 2. An example of VLIW encoding

	Slot 0	Slot 1	Slot 2	Slot 3	Slot 4	Slot 5	Slot 6	Slot 7	Slot allocated	Slot encoding bit
A	√	√							Slot 0	000
B	√	√							Slot 1	001
C			√	√					Slot 3	011
D			√						Slot 2	010
E	√	√					√	√	Slot 6	111

Intrinsic Function and SIMD. Intrinsic function is a special source-language-related function that is not implemented by a call instruction with an offset address, but the compiler. We need to support the call to intrinsic function in DLCS Lower IR. For some special instructions of the target vector processor, as shown in the Fig. 7, we can directly use intrinsic function to map functions from source code to DLCS Lower IR.

```
//#include "vector_support.h"
typedef int    v4i    __attribute__((vector_size(16)));
typedef short v8s    __attribute__((vector_size(16)));
typedef char  v16c   __attribute__((vector_size(16)));
v4i v4i_a = {1, 2, 3, 4},   v4i_b = {3, 3, 3, 3};
v8s v8s_a = { 1, 2, 3, 4, 5, 6, 7, 8 },   v8s_b = { 1, 2, 3, 4, 5, 6, 7, 8};
v16c v16c_a = {  1,  2,  3,  4, 5, 6, 7, 8, 9, 10, 11, 12, 13, 14, 15, 16 };
v16c v16c_b = { 13, 14, 15, 16, 5, 6, 7, 8, 9, 10, 11, 12, 1, 2, 3, 4 };
int main(int argc, char const *argv[]){
        ...
        v4i v4i_dst = __builtin_swift_dsp_vmax_i(v4i_a, v4i_b);
        v8s v8s_dst = __builtin_swift_dsp_vmax_s(v8s_a, v8s_b);
        v16c v16c_dst = __builtin_Swift_dsp_vmax_c(v16c_a, v16c_b);
        ...
}
```

Fig. 7. The example of the intrinsic function of target vector processor in the C language

As shown in Fig. 8, on the front-end of DLCS, we mainly define three types of vectors, the properties of the intrinsic function and the corresponding function nodes. The total bit width of the three vector types is 128 bits, including v4i (i128 = 4xi32), v8s (i128 = 8xi16) and v16c (i128 = 16xi8). The intrinsic function node mainly defines the types of parameters and return values.

On the backend of DLCS, we mainly implement the intrinsic function of the instruction template definition, vector register definition, the mapping and legalization of intrinsic instruction from front-end to back-end. The instruction template definition process mainly makes use of the architecture description language (ADL) to describe the various attributes of the instruction, such as instruction encoding, slot assignment, operand type, matching pattern, node mapping and assembly string. The detailed code of how to define the instruction template will be introduced in the code generation section. Each property declaration of vector register in back-end matches the three vector types of the front-end.

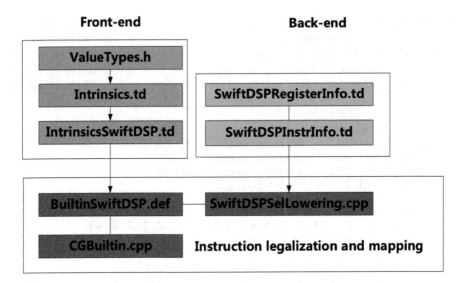

Fig. 8. The files involved in implementing the intrinsic function on front-end and back-end

3.2 Code Generation

Code generation for embedded processors mainly includes operation legalization, instruction selection, instruction scheduling, register allocation, optimization, target code generation. DLCS is based on the LLVM framework, which has already realized the generic parts of code generation. Therefor this section mainly explains the code generation for three special cases that DLCS Lower IR and instruction set of target vector processor are not compatible.

Architecture Description of Target Vector Processor
Description of Register. Before we discuss operational incompatibilities, we use the TableGen tool provided by LLVM to realize architecture description of target vector processor, including the description of instructions and registers. We use the register base class provided by LLVM to describe register properties. As shown in Table 3, the register of target vector processor mainly contains 32 scalar registers and 16 vector registers.

Table 3. The description of register heap for target vector processor

Tag	Register	Bit wide	Counts	Lowest limit of addressing	Highest limit of addressing
GR	General	32	32	000000	011111
VR	Vector	160	16	100000	101111
OFF	Offset	32	4	110100	110111
BAR	Base address	32	4	111000	111011
MR	Modular	32	4	111100	111111

Figure 9 depicts three parts: the definition of the general register, the allocation of the general register, and the calling convention for the register. The general register definition needs to specify information such as the namespace and encoding bit width. Since each register is an instance of a register class, we instantiate each register. We use multiple inheritance to inherit the base class of the scalar register or vector register to generate the definition register. We define registers (V0, V1, A0–A7, VR0–VR15) and classify them into general register class, vector register class and special register class. In the calling convention of register, we stipulate i32 as scalar standard type, and extend i8 and i16 types to i32 for unified processing. We specify that i32 are stored in the general register class, while v4i, v8s, and v16c are stored in the vector register class (VR0–VR15). We specify that the alignment of the function call stack is 4 bytes (i32) and 16 bytes (v4i, v8s and v16c).

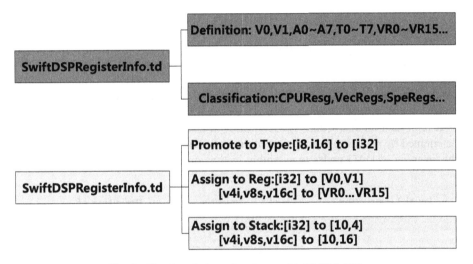

Fig. 9. The description of registers with LLVM ADL

Description of Instructions. We extend the instruction template based on the instruction base class template provided by LLVM. According to the complexity of the instruction set of SWFIT DSP, the instruction template can be divided into four levels of the instruction template, as shown in Fig. 10, where the definition of each level of the instruction template is inherited from the previous level of the template. The level 1 instruction base class template mainly declares parameters common to all instructions and some flag bit settings related to the target hardware. The second level instruction base class template is divided into four common types: single register class, double registers class, three registers class and special registers class. The third level instruction base class template is divided according to the instruction operation code bit width. The level 4 instruction base class template gives a template for one-to-one relationships with each instruction, as shown in Fig. 10.

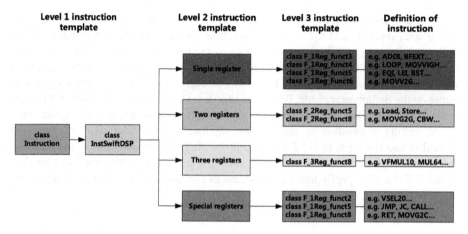

Fig. 10. Inheritance diagram of instruction class

Figure 11 shows the template inheritance diagram of the "VMAX40" instruction, and the corresponding detailed description code is shown in Fig. 12. The level 3 instruction base class template defines the instruction encoding format, assembly strings, matching patterns, and other parameters. We need to define an additional DAG node corresponding to the instruction in the process of defining the instruction. A DAG node represents an instruction. The realization of DAG node and instruction mapping is determined by matching pattern. One of the important steps in target code generation is the mapping of DAG nodes to assembly instructions for the target hardware.

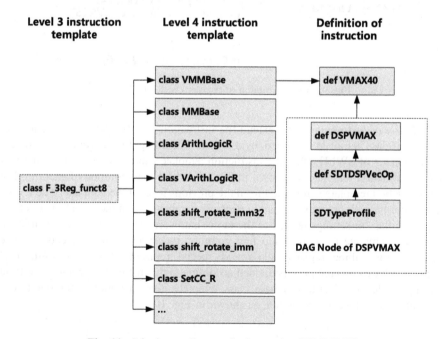

Fig. 11. Inheritance diagram for instruction "VMAX40"

```
//SwiftDSPInstrFormats.td
...
class F_3Reg_funct8<bits<3> type,bits<3> op, bits<8> funct,        dag outs, dag ins, string asmstr,
      list<dag> pattern, InstrItinClass itin>:InstDSP<outs, ins, asmstr, pattern, itin>{
            bits<6> ra;
            bits<6> rb;
            bits<6> rc;
            let Inst{31-29} = type;
            let Inst{28-26} = op;
            let Inst{25-20} = ra;
            let Inst{19-14} = rb;
            let Inst{13-8} = rc;
            let Inst{7-0}  = funct;
}
...
//SwiftDSPInstrInfo.td
...
def SDTDSPVecOp : SDTypeProfile<1,2,[SDTCisSameAs<0,1>,SDTCisSameAs<0,2>,SDTCisVec<0>]>;
def DSPVMAX :       SDNode<"DSPISD::VMAX",SDTDSPVecOp,[SDNPCommutative]>;

class VMMBase<bits<3> typeop,bits<3> op, bits<8> inner_op,        string instr_asm, SDNode OpNode,
            InstrItinClass        itin, RegisterClass RC>:F_3Reg_funct8<typeop,op,inner_op,
            (outs RC:$ra),        (ins RC:$rb, RC:$rt),lstrconcat(instr_asm, "\t$ra, $rb, $rt"),
            [(set RC:$ra, (OpNode RC:$rb, RC:$rt))],itin>{
            let isCommutable = 1;
}
def VMAX40 :VMMBase<5,1,0b01001001,"vmax40",DSPVMAX,ALU32_V_SLOT23,VecRegs>;
...
```

Fig. 12. Description definition code for instruction "VMAX40"

Incompatible Instruction. As shown in Table 4, the first incompatibility is that DLCS Lower IR supports operation X but target vector processor does not. We generally divide X into several combinations of operations supported by target vector processor, such as (4) where target vector processor supports operations of Y, X1 and X2. The second incompatibility is that DLCS Lower IR does not support operation X but target vector processor does. We solved this problem by intrinsic functions in target hardware-dependent optimization. The third incompatibility refers to the case where the bit width of the same operation X, supported by DLCS Lower IR and target vector processor, is different, which is called operation legalization problem. DLCS uses a compile pass to legalize operations. On the basis of DLCS Lower IR and target vector processor instruction set, we analyze which instruction is illegal operation, and then describe the legalization process through a set of interfaces as (5).

$$X = Y \ (X1, \ X2) \qquad (4)$$

$$\text{SetOperationAction (Operation, Type, LegalizaAction)} \qquad (5)$$

"Operation" is an instruction of DLCS Lower IR, and "Type" is the operand Type of the instruction, and "LegalizeAction" is the processing when legalizing the instruction. Corresponding to the previous situation that the instruction bit width does not match, we designed three handling methods: "Promote" represents this operation should be put in a larger type; "Expand" is to try to expand this to other operations, otherwise use a libcall; "Custom" is to use legitimate functions to implement custom lowering.

Table 4. Incompatibility cases related to operation X

	Case 1	Case 2	Case 3
DLCS Lower IR	Support	No	Support
Instruction set of Target Vector Processor	No	Support	Support
The bit width of the instruction			Different

4 Evaluation

We compare the effects of various target optimization operations proposed in this paper and hand-optimization operations on the compilation speed of the DL model, and we take the performance value of hand-optimization as the benchmark. As shown in Fig. 13, the compiled object of the comparison experiment are conventional convolution and pooling with input size (16, 28, 28) and convolution kernel (32, 16, 3, 3), and deep convolution with input size (128, 14, 14) and convolution kernel (256, 128, 3, 3). Experimental results show that all target optimization operations can improve the performance of execution speed, in which SIMD optimization brings the smallest acceleration ratio about 1.1×. Data simplification and data layout transformation can lead to significant improvements in execution speeds up to 2.7×.

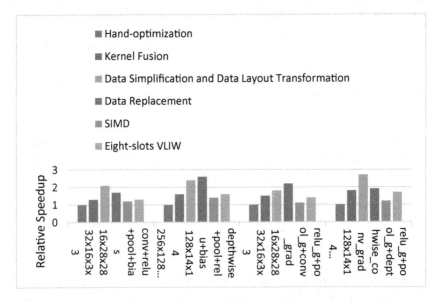

Fig. 13. Performance comparison between hand-optimization and target optimizations

Finally, we evaluate the whole compilation system, including target optimization and code generation. ResNet-18 and MobileNet are used as DL models in the comparison experiment in R1–R5 phase and M1–M5 phase. The performance index is the relative speedup of the two models on the target vector processor. Figure 14 shows a comparison of the performance of the two models deployed on the target vector processor with DLCS and hand-optimization. Experimental results show that DLCS offers Multi-slot parallel performance for target vector processor and achieves speedups ranging from 1.5× in R5 phase to 3.0× in M4 phase over existing frameworks backed by hand-optimized libraries, where the compilation performance improvement of DLCS is generally over 2.0× during most phases of the two DL model compilation.

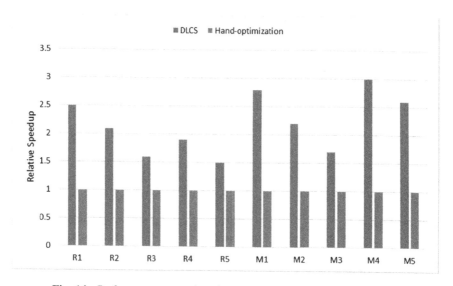

Fig. 14. Performance comparison between DLCS and hand-optimization

5   Conclusion

This paper presented the design of a DL compiler (DLCS) for the independently designed target vector processor. DLCS lowers the deep learning model to an intermediate representation of two levels and realizes the target hardware-independent optimization and target hardware-dependent optimization, so that the performance of the model finally deployed on the target vector processor is far superior to that of hand-optimization. We hope that this work will encourage more research on DL compilation methods for self-developed embedded chips and open-up new opportunities for self-developed embedded system hardware/software co-design technologies.

Acknowledgment. The authors thank the editors and the anonymous reviewers for their invaluable comments to help to improve the quality of this paper. This work was supported in part by the National Natural Science Foundation of China under Grant Nos. 61831018 and 61631017, and Guangdong Province Key Research and Development Program Major Science and Technology Projects under Grant 2018B010115002.

References

1. Suchman, L.A., Trigg, R.H.: Artificial intelligence as craftwork (1993)
2. Moore, R.C., Lewis, W.: Intelligent selection of language model training data. In: Proceedings of the ACL 2010 Conference Short Papers. Association for Computational Linguistics, pp. 220–224 (2010)
3. Yao, L., Mimno, D., McCallum, A.: Efficient methods for topic model inference on streaming document collections. In: Proceedings of the 15th ACM SIGKDD International Conference on Knowledge Discovery and Data Mining, pp. 937–946 (2009)
4. Abadi, M., Barham, P., Chen, J., et al.: TensorFlow: a system for large-scale machine learning. In: 12th USENIX Symposium on Operating Systems Design and Implementation (OSDI 2016), pp. 265–283 (2016)
5. Abadi, M., Agarwal, A., Barham, P., et al.: TensorFlow: large-scale machine learning on heterogeneous distributed systems. arXiv preprint arXiv:1603.04467 (2016)
6. Paszke, A., Gross, S., Massa, F., et al.: PyTorch: an imperative style, high-performance deep learning library. In: Advances in Neural Information Processing Systems, pp. 8024–8035 (2019)
7. Chen, T., Li, M., Li, Y., et al.: MXNet: a flexible and efficient machine learning library for heterogeneous distributed systems. arXiv preprint arXiv:1512.01274 (2015)
8. Jia, Y., Shelhamer, E., Donahue, J., et al.: Caffe: convolutional architecture for fast feature embedding. In: Proceedings of the 22nd ACM International Conference on Multimedia, pp. 675–678 (2014)
9. Chen, T., Moreau, T., Jiang, Z., et al.: TVM: an automated end-to-end optimizing compiler for deep learning. In: 13th USENIX Symposium on Operating Systems Design and Implementation (OSDI 2018), pp. 578–594 (2018)
10. Adachi, Y., Kumano, T., Ogino, K.: Intermediate representation for stiff virtual objects. In: Proceedings Virtual Reality Annual International Symposium 1995. IEEE, pp. 203–210 (1995)
11. Yoo, H., Shim, M., Kim, D.: Dynamic duty-cycle scheduling schemes for energy-harvesting wireless sensor networks. IEEE Commun. Lett. 16(2), 202–204 (2011)
12. Chan, C.H., Tahir, M.A., Kittler, J., et al.: Multiscale local phase quantization for robust component-based face recognition using kernel fusion of multiple descriptors. IEEE Trans. Pattern Anal. Mach. Intell. 35(5), 1164–1177 (2012)
13. Kustikova, V., et al.: Intel distribution of OpenVINO toolkit: a case study of semantic segmentation. In: van der Aalst, W.M.P., et al. (eds.) AIST 2019. LNCS, vol. 11832, pp. 11–23. Springer, Cham (2019). https://doi.org/10.1007/978-3-030-37334-4_2
14. Arm, N.N.: SDK. https://developer.arm.com/products/processors/machine-learning/arm-nn
15. Vanholder, H.: Efficient Inference with TensorRT (2016)
16. Ren, H., Zhang, Z., Wu, J.: SWIFT: a computationally-intensive DSP architecture for communication applications. Mobile Netw. Appl. 21(6), 974–982 (2016)
17. Wang, W.: Functional verification for SWIFT DSP based on software-based simulation and FPGA. In: International Conference on Wireless Communications, Networking and Applications
18. Ellis, J.R.: Bulldog: a compiler for VLIW architectures. Yale Univ., New Haven, CT (USA) (1985)
19. Nassimi, D., Sahni, S.: Data broadcasting in SIMD computers. IEEE Trans. Comput. 100(2), 101–107 (1981)
20. Chen, T., Moreau, T., Jiang, Z., et al.: TVM: end-to-end optimization stack for deep learning. arXiv preprint arXiv:1802.04799 (2018)

21. Maiyuran, S., Garg, V., Abdallah, M.A., et al.: Memory access latency hiding with hint buffer: U.S. Patent 6,718,440, 6 April 2004
22. Leary, C., Wang, T.: XLA: TensorFlow, compiled. TensorFlow Dev Summit (2017)
23. Larsen, R.M., Shpeisman, T.: TensorFlow Graph Optimizations (2019)
24. Chadha, P., Siddagangaiah, T.: Performance analysis of accelerated linear algebra compiler for TensorFlow
25. Yao, C.Y., Chen, H.H., Lin, T.F., et al.: A novel common-subexpression-elimination method for synthesizing fixed-point FIR filters. IEEE Trans. Circuits Syst. I Regul. Pap. **51**(11), 2215–2221 (2004)
26. Lattner, C., Adve, V.: LLVM: a compilation framework for lifelong program analysis & transformation. In: International Symposium on Code Generation and Optimization, CGO 2004. IEEE, pp. 75–86 (2004)
27. Lattner, C.A.: LLVM: an infrastructure for multi-stage optimization. University of Illinois at Urbana-Champaign (2002)
28. Rotem, N., Fix, J., Abdulrasool, S., et al.: Glow: graph lowering compiler techniques for neural networks. arXiv preprint arXiv:1805.00907 (2018)
29. Sivalingam, K., Mujkanovic, N.: Graph compilers for AI training and inference
30. Gibbons, P.B., Muchnick, S.S.: Efficient instruction scheduling for a pipelined architecture. In: Proceedings of the 1986 SIGPLAN symposium on Compiler construction, pp. 11–16 (1986)
31. Kogure, M.: Static memory allocation system: U.S. Patent 5,247,674, 21 September 1993
32. Gopinath, K., Hennessy, J.L.: Copy elimination in functional languages. In: Proceedings of the 16th ACM SIGPLAN-SIGACT Symposium on Principles of Programming Languages, pp. 303–314 (1989)
33. Wei, R., Schwartz, L., Adve, V.: DLVM: a modern compiler infrastructure for deep learning systems. arXiv preprint arXiv:1711.03016 (2017)
34. Griewank, A., Walther, A.: Evaluating derivatives: principles and techniques of algorithmic differentiation. SIAM (2008)
35. Buchberger, B., Loos, R.: Algebraic Simplification. Computer Algebra, pp. 11–43. Springer, Vienna (1982). https://doi.org/10.1007/978-3-7091-7551-4_2

An Accurate Frequency Estimation Algorithm by Using DFT and Cosine Windows

Jinyu Liu[ID], Lei Fan[(⊠)][ID], Renqing Li, Wenbo He, Nian Liu, and ZhanHong Liu

School of Information Science and Engineering, Dalian Polytechnic University, Dalian, China
fanlei@dlpu.edu.cn

Abstract. Sinusoidal signal frequency estimation is one of the fundamental problems in signal processing, and it is widely used in wireless communication, signal processing, navigation, radar and so on. In this paper, an interpolation frequency estimation algorithm based on Discrete Fourier Transform (DFT) and cosine windows is proposed. Firstly, the sampling sequence of the signal is multiplied by a cosine window. Then, N-point DFT is used to search the position of the maximum spectral line and get the coarse estimation of frequency. Finally, the accurate frequency estimation is obtained by DFT interpolation of the maximum spectral line and the two Discrete-Time Fourier Transform (DTFT) samples on the left and right of the maximum spectral line. According to the simulation results, the performance of the proposed algorithm is better than that of MV-IpDTFT(3) algorithm, MV-IpDTFT(2) algorithm and Candan algorithm. The effect of harmonic interference on the frequency estimation results can be effectively suppressed.

Keywords: Frequency estimation · Interpolation · DFT · Cosine windows · Signal processing

1 Introduction

Frequency estimation is a fundamental subject that has been widely studied and applied in wireless communication, signal processing, navigation, radar and so on. For example, as the oscillation frequency produced by crystal oscillator may deviate from the nominal frequency, and the relative motion between the transmitter and receiver generally leads to Doppler frequency shift, the carrier frequency offset generally exists in communication system. The carrier frequency offset will destroy the orthogonality between subcarriers, resulting in inter subcarrier interference and system performance degradation. Therefore, it is necessary to estimate the carrier frequency offset

This work is supported by the Provincial Natural Science Foundation Guidance Plan of China under Grant 2019-ZD-0294.

H. Gao et al. (Eds.): ChinaCom 2020, LNICST 352, pp. 688–697, 2021.
https://doi.org/10.1007/978-3-030-67720-6_47

accurately, which is very important for multi carrier system such as Orthogonal Frequency Division Multiplexing (OFDM) system. For another example, for a low orbit earth satellite, there is obvious Doppler frequency shift between the satellite and the ground station. The Doppler frequency shift must be accurately estimated by the satellite ground station, because its estimation accuracy determines the positioning and orbit determination accuracy of the satellite monitoring system. In satellite navigation receiver, carrier frequency estimation is a key problem in the process of satellite signal acquisition and tracking. These applications are based on single tone frequency sinusoidal signal model. It is necessary to estimate the frequency of sinusoidal signal. And the frequency estimation performance directly affects the performance of these applications.

Specifically, in the background of additive white Gaussian noise, well known frequency estimation algorithms can be divided into two categories, the time domain estimation algorithm and the frequency domain estimation algorithm. The time domain estimation algorithm includes the algorithm based on autocorrelation, the algorithm based on least squares and the algorithm based on maximum likelihood. The maximum likelihood algorithm has high accuracy and reaches the Cramer-Rao lower bound (CRLB), but it is computationally expensive and difficult for real-time realization [1]. The frequency domain estimation algorithms are mainly based on Discrete Fourier Transform (DFT) [2–7, 13]. They have relatively small amount of computation, and significant Signal-to-Noise Ratio (SNR) gain for sinusoidal signals. And they can be applied for real-time applications. For the frequency estimation algorithms based on DFT, the coarse estimation is carried out by searching the maximum spectral line position, and then the accurate estimation value is obtained according to different algorithms. Candan algorithm [2] used the maximum spectral line and its left and right spectral lines to achieve accurate estimation. The algorithm of Aboutanios and Mulgrew (A&M) used two spectrum lines for accurate estimation [3]. When $N = 1024$ and $SNR = 0$ dB, the estimation variance is only 1.0147 times of CRLB. Lei Fan algorithm [4] obtained the accurate frequency estimation by the interpolation of the primary spectral line and two auxiliary spectral lines located at arbitrary position within the main lobe or the first sidelobe of the frequency spectrum. The influence of the interval between the primary spectral line and the auxiliary spectral lines on the Mean Square Error (MSE) is analyzed through theoretical analysis and simulation experiments.

In the references mentioned above, the frequencies of sinusoidal signal are estimated in the additive white Gaussian noise background. However, in practical application, there are both additive white Gaussian noise and interference signals. For this situation, the appropriate window function can be selected for multiply the sampling sequence, and then the signal frequency is estimated based on DFT interpolation algorithm, which can effectively reduce the interference caused by useless spectrum components [8–12]. In [10], Candan derived the estimation expression of the algorithm in [2] when arbitrary window functions were used, and only needed to select different deviation correction coefficients according to the specific window function. When there is a single strong interference signal and additive Gaussian white noise, the algorithm can effectively reduce the impact of interference signal. In [11], Belega derived the frequency estimation expressions of Candan algorithm in [2] and AM algorithm in [3] when windows were used, obtaining MV-IpDTFT(2) algorithm and MV-IpDTFT(3) algorithm respectively, and derived the calculation formula of frequency estimation variance. Through the simulation experiment, the performance of different windowing algorithms in the background of harmonic interference and additive white Gaussian noise is analyzed.

In this paper, an accurate interpolation frequency estimation algorithm based on DFT and cosine windows is proposed. Firstly, the sampling sequence of the signal is multiplied by a cosine window. Then, N-point DFT is used to search the position of the maximum spectral line, and get the coarse estimation of frequency. Finally, the accurate frequency estimation is obtained by DFT interpolation of the maximum spectral line and the two Discrete-Time Fourier Transform (DTFT) samples on the left and right of the maximum spectral line. It can be seen from the simulation results, the performance of the proposed algorithm is better than the other methods, and the influence of harmonic interference on frequency estimation results is effectively suppressed.

2 Proposed Algorithm

This part mainly describes an accurate interpolation frequency estimation method based on DFT and cosine windows. The model of frequency sinusoidal signal in the background of additive white Gaussian noise is

$$s(t) = Ae^{j(2\pi f_0 t + \theta_0)} + z(t) \tag{1}$$

where A is the amplitude of the complex sinusoid, f_0 is the frequency, θ_0 is the primary phase, $z(t)$ is the noise, N is the number of sampling and f_s is the sampling frequency. After sampling, we have

$$s(n) = Ae^{j(2\pi \frac{f_0}{f_s} n + \theta_0)} + z(n), n = 0, 1, 2, \ldots, N - 1 \tag{2}$$

We perform DFT on $s(n)$, and get

$$S[k] = Ae^{j\theta_0} \sum_{n=0}^{N-1} e^{j2\pi \frac{f_0}{f_s} n} e^{-j\frac{2\pi}{N} nk} + Z[k] \tag{3}$$

Let $S[l]$ denote the maximum spectrum line in the frequency spectrum indexed by the DFT. l is the index value of the maximum spectrum line, $\Delta f = f_s/N$ is frequency resolution, and $s_0(n)$ is signal without noise. For ease of expression, the DFT sample value at the position $f = (l+p)\Delta f$ can be denoted as:

$$S_p = \sum_{n=0}^{N-1} s_0(n) e^{-j2\pi fn} \Big|_{f=(l+p)\Delta f} \tag{4}$$

When there are both additive white Gaussian noise and narrowband interference signals, the time-domain windowing algorithm can be used to reduce the interference caused by useless spectrum components and improve the anti-interference performance of the estimation algorithm. Cosine windows are widely used in many references [10, 11], and their expression are

$$w(n) = \sum_{h=0}^{H-1} (-1)^h a_h \cos(2\pi \frac{h}{N} n), n = 0, 1, \ldots, N-1 \tag{5}$$

where $H \geq 1$ is the number of window coefficients a_h. Maximum sidelobe decay window (MSD window) belongs to cosine windows. When H is constant, the sidelobe decay rate of MSD window is as high as $6(2H-1)$ dB/octave. The coefficients of H-term MSD window can be expressed as:

$$a_0 = \frac{C_{2H-2}^{H-1}}{2^{2H-2}}, a_h = \frac{C_{2H-2}^{H-h-1}}{2^{2H-3}}, h = 1, 2, \ldots, H-1 \tag{6}$$

Multiply $s(n)$ by cosine window, we have

$$s_w(n) = s(n) \cdot w(n), n = 0, 1, \ldots, N-1 \tag{7}$$

Then we perform DTFT on $s_w(n)$, and get

$$S_w(\lambda) = AW(\lambda - v)e^{j\varphi} \tag{8}$$

where $v = l + \delta$ is quantized value of the signal frequency, l is the index value of the maximum spectrum line. δ denotes the normalized frequency offset, and its value range is $[-0.5, 0.5]$. If it is assumed that the number of sampling points obtained is large enough ($N \gg 1$), then we have

$$W(\lambda) = \frac{N \sin(\pi\lambda)}{\pi} \sum_{h=0}^{H-1} (-1)^h a_h \frac{\lambda}{\lambda^2 - h^2} e^{-j\pi\lambda} = \tilde{W}(\lambda)e^{-j\pi\lambda} \tag{9}$$

in which

$$\tilde{W}(\lambda) = \frac{N \sin(\pi\lambda)}{\pi} \sum_{h=0}^{H-1} (-1)^h \alpha_h \frac{\lambda}{\lambda^2 - h^2} \geq 0 \tag{10}$$

Calculating the derivation of (10), we have

$$\tilde{W}'(\lambda) = \frac{N}{\pi} [\sin(\pi\lambda) + \pi\lambda \cos(\pi\lambda)] \sum_{h=0}^{H-1} \frac{(-1)^h a_h}{\lambda^2 - h^2} - \frac{2N}{\pi} \lambda^2 \sin(\pi\lambda) \sum_{h=0}^{H-1} \frac{(-1)^h a_h}{(\lambda^2 - h^2)^2} \tag{11}$$

In [4], the estimation formula for the normalized frequency offset δ is expressed as follows:

$$\hat{\delta} = \frac{N}{\pi}\tan^{-1}\left\{\mathrm{Re}\left[\frac{(|S_i| - |S_{-i}|) \cdot \sin(\frac{\pi i}{N})}{(|S_i| + |S_{-i}|) \cdot \cos(\frac{\pi i}{N}) - 2\cos(\pi i) \cdot |S_0|}\right]\right\} \tag{12}$$

When $i = 0.1$, denoting the curly bracket as U and replacing the corresponding spectral lines in (12) by $S_w(l+0.1)$, $S_w(l-0.1)$ and $S_w(l)$, we have

$$U = \frac{[|S_w(l+0.1)| - |S_w(l-0.1)|] \cdot \sin(\frac{\pi}{10N})}{[|S_w(l+0.1)| + |S_w(l-0.1)|] \cdot \cos(\frac{\pi}{10N}) - 2\cos(\frac{\pi}{10}) \cdot |S_w(l)|} \tag{13}$$

In (8), let $\lambda = l+0.1$, $\lambda = l - 0.1$ and $\lambda = l$, we have

$$|S_w(l+0.1)| = |A|\,\tilde{W}(0.1 - \delta) \tag{14}$$

$$|S_w(l-0.1)| = |A|\,\tilde{W}(-0.1 - \delta) \tag{15}$$

$$|S_w(l)| = |A|\,\tilde{W}(-\delta) \tag{16}$$

Substituting (14), (15) and (16) into (13) to simplify the formula, we have

$$U = \frac{[\tilde{W}(0.1 - \delta) - \tilde{W}(-0.1 - \delta)] \cdot \sin(\frac{\pi}{10N})}{[\tilde{W}(0.1 - \delta) + \tilde{W}(-0.1 - \delta)] \cdot \cos(\frac{\pi}{10N}) - 2\cos(\frac{\pi}{10}) \cdot \tilde{W}(-\delta)} \tag{17}$$

Carrying out the first order Taylor series expansion for $\tilde{W}(0.1 - \delta)$, $\tilde{W}(-0.1 - \delta)$ and $\tilde{W}(-\delta)$ near 0.1, -0.1 and 0 respectively, and ignoring the higher order term, we have

$$\tilde{W}(0.1 - \delta) = \tilde{W}(0.1) + \tilde{W}'(0.1)(-\delta) = \tilde{W}(0.1) - \tilde{W}'(0.1)\delta \tag{18}$$

$$\tilde{W}(-0.1 - \delta) = \tilde{W}(-0.1) + \tilde{W}'(-0.1)(-\delta) = \tilde{W}(0.1) + \tilde{W}'(0.1)\delta \tag{19}$$

$$\tilde{W}(-\delta) = \tilde{W}(0) + \tilde{W}'(0)(-\delta) \tag{20}$$

Substituting (18), (19) and (20) into (17), we have

$$U = \frac{\tilde{W}'(0.1) \cdot \sin(\frac{\pi}{10N})}{\cos(\frac{\pi}{10}) \cdot \tilde{W}(0) - \tilde{W}(0.1) \cdot \cos(\frac{\pi}{10N})} \cdot \delta \tag{21}$$

The estimation expression of δ can be obtained as

$$\hat{\delta} = \frac{\cos(\frac{\pi}{10}) \cdot \tilde{W}(0) - \tilde{W}(0.1) \cdot \cos(\frac{\pi}{10N})}{\tilde{W}'(0.1) \cdot \sin(\frac{\pi}{10N})} \cdot U = r \cdot U \tag{22}$$

$\tilde{W}(0)$ and $\tilde{W}(0.1)$ can be calculated according to (10), and $\tilde{W}'(0.1)$ can be calculated according to (11). Substituting $\tilde{W}(0)$, $\tilde{W}(0.1)$ and $\tilde{W}'(0.1)$ into (22), we have

$$r = \frac{\cos(\frac{\pi}{10}) \cdot \alpha_0 - \cos(\frac{\pi}{10N}) \cdot \sin(\frac{\pi}{10}) \cdot \frac{1}{\pi} \cdot \sum_{h=0}^{H-1} (-1)^h \frac{\alpha_h \cdot 0.1}{0.01 - h^2}}{\left\{ [\frac{1}{\pi} \sin(\frac{\pi}{10}) + \frac{1}{10} \cos(\frac{\pi}{10})] \sum_{h=0}^{H-1} \frac{(-1)^h \alpha_h}{0.01 - h^2} - \frac{1}{50\pi} \sin(\frac{\pi}{10}) \sum_{h=0}^{H-1} \frac{(-1)^h \alpha_h}{(0.01 - h^2)^2} \right\} \cdot \sin(\frac{\pi}{10N})} \tag{23}$$

For two term MSD windows, we have $r_{2MSD} = 134.97$. For three term MSD windows, we have $r_{3MSD} = 232.91$. We take the iterative procedures as shown in Table 1.

Table 1. The proposed algorithm steps.

Algorithm: Proposed an interpolation frequency estimation algorithm
1   Get $s_w(n) = s(n) \cdot w(n)$, $n = 0,1,...,N-1$
2   Perform N-point DFT of $s_w(n)$
3   Find l and let $l\Delta f$ be the rough estimation of signal frequency
4   Calculate $S_w(l+0.1)$ and $S_w(l-0.1)$, via $S_w(l+p) = \sum_{n=0}^{N-1} s_w(n) e^{-j2\pi n \frac{l+p}{N}}, p = \pm 0.1$
5   Calculate $\hat{\delta}_1$ with $S_w(l)$, $S_w(l+0.1)$ and $S_w(l-0.1)$, via (22)
6   Calculate $S_w(l+\hat{\delta}_1)$, $S_w(l+\hat{\delta}_1+0.1)$ and $S_w(l+\hat{\delta}_1-0.1)$, via $S_w(l+p) = \sum_{n=0}^{N-1} s_w(n) e^{-j2\pi n \frac{l+p}{N}}, p = \hat{\delta}_1, \hat{\delta}_1 \pm 0.1$
7   Calculate $\hat{\delta}_2$ with $S_w(l+\hat{\delta}_1)$, $S_w(l+\hat{\delta}_1+0.1)$ and $S_w(l+\hat{\delta}_1-0.1)$, via (22)
8   The frequency estimate is $\hat{f} = (l + \hat{\delta}_1 + \hat{\delta}_2)\Delta f$

3 Simulation Results

In order to verify the performance of the proposed algorithm, we carry out simulation analysis in the presence of harmonic interference and additive white Gaussian noise. Meanwhile, for the sake of finding the difference between the performance of the proposed algorithm and that of the competitive algorithms, this part are conducted to compare the performance of the proposed algorithm with that of Candan algorithm [10], MV-IpDTFT(2) algorithm [11] and MV-IpDTFT(3) algorithm [11].

When there are harmonic interference signals, the Total Harmonic Distortion (THD) is used to describe the intensity of the interference signal, and it is defined as follows

$$THD = \sqrt{\frac{\frac{1}{2}\sum\limits_{M=2}^{M_{max}} A_M^2}{\frac{1}{2}A_1^2 + \frac{1}{2}\sum\limits_{M=2}^{M_{max}} A_M^2}} \tag{24}$$

where A_1 is the amplitude of fundamental tone wave and A_M ($M = 2, 3, \cdots, M_{max}$) is the amplitude of each harmonic. In the following simulation experiment, it is assumed that there are second, third and fourth harmonics, and the ratio of amplitudes is 4:2:1. The phase of fundamental tone and each harmonic is uniformly distributed on interval $[0, 2\pi]$.

When $N = 128$, $SNR = 50dB$, $THD = 5\%$, Fig. 1 show the MSE of δ with respect to the signal frequency quantization value v of the proposed algorithm and MV-IpDTFT(3) algorithm [11]. Two iterations are carried out for both algorithms. The quantized value v is in the interval $[2.51, 12.1]$. We calculate a point with a step of

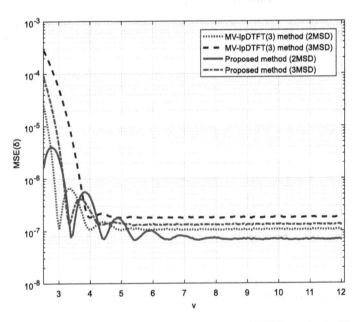

Fig. 1. Simulated MSE of the proposed method and MV-IpDTFT(3) method with respect to v ($N = 128$, $SNR = 50dB$, $THD = 5\%$)

1/16. It can be seen from Fig. 1, when $v > 5.5$, the MSE(δ) of the proposed algorithm with two term MSD window is the lowest. When $2.51 \leq v \leq 5.5$ and $|\delta|$ is close to 0.5, the MSE(δ) of the proposed algorithm with two term MSD window is the lowest. When $2.51 \leq v \leq 5.5$ and $|\delta|$ is close to 0, the MSE(δ) of MV-IpDTFT(3) algorithm with two term MSD windows is the lowest.

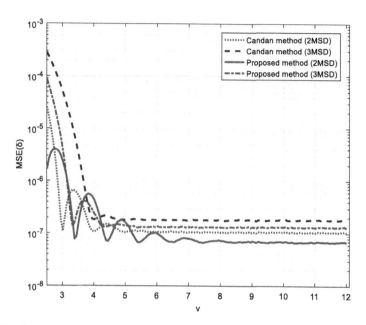

Fig. 2. Simulated MSE of the proposed method and Candan method with respect to v ($N = 128$, $SNR = 50dB$, $THD = 5\%$)

When $N = 128$, $SNR = 50dB$, $THD = 5\%$, Fig. 2 show the MSE of δ with respect to the signal frequency quantization value v of the proposed algorithm and Candan algorithm [10]. Two iterations are carried out for both algorithms. The quantized value v is in the interval [2.51, 12.1]. We calculate a point with a step of 1/16. It can be seen from Fig. 2, the MSE(δ) of Candan algorithm is similar to that of MV-IpDTFT(3) algorithm. When $v > 5.5$, MSE(δ) of the proposed algorithm with two term MSD window is the lowest. When $2.51 \leq v \leq 5.5$ and $|\delta|$ is close to 0.5, the MSE (δ) of the proposed algorithm with two term MSD window is the lowest. When $2.51 \leq v \leq 5.5$ and $|\delta|$ is close to 0, the MSE(δ) of Candan algorithm with two term MSD windows is the lowest. It can also be seen from Fig. 1 and Fig. 2 that for the same estimation algorithm, when H is smaller, the MSE(δ) is lower, and the estimation performance is better.

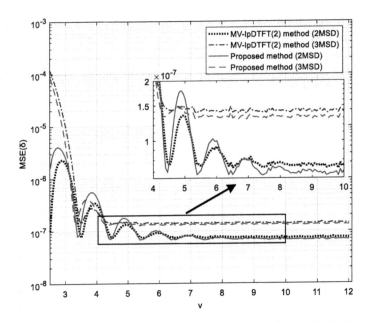

Fig. 3. Simulated MSE of the proposed method and MV-IpDTFT(2) method with respect to v ($N = 128$, $SNR = 50dB$, $THD = 5\%$)

When $N = 128$, $SNR = 50dB$, $THD = 5\%$, Fig. 3 show the MSE of δ with respect to the signal frequency quantization value v of the proposed algorithm and MV-IpDTFT(2) algorithm [11]. Two iterations are carried out for both algorithms. The quantized value v is in the interval [2.51, 12.1]. We calculate a point with a step of 1/16. It can be seen from Fig. 3, when $v > 6$, MSE(δ) of the proposed algorithm with two term MSD window is the lowest. When $v \leq 6$, in most cases, the MSE(δ) of MV-IpDTFT(3) algorithm and the proposed algorithm with two term MSD windows are lower than the other algorithms. It can also be seen from Fig. 3 that for the same estimation algorithm, when H is smaller, the MSE(δ) of frequency estimation is lower, and the estimation performance is better.

4   Conclusion

In this paper, an interpolation frequency estimation algorithm based on DFT and cosine windows is proposed. Firstly, the sampling sequence of the signal is multiplied by a cosine window. Then, N-point DFT is used to get the coarse estimation of frequency. Finally, the accurate frequency estimation is obtained by DFT interpolation of the maximum spectral line and the two DTFT samples on the left and right of the maximum spectral line. It can be seen from the simulation results, the performance of the proposed algorithm is better than that of MV-IpDTFT(3) algorithm, MV-IpDTFT(2) algorithm and Candan algorithm. When $2.51 \leq v \leq 5.5$ and $|\delta|$ is close to 0.5, the MSE (δ) of the proposed algorithm with two-term MSD window is lower than MV-IpDTFT

(3) algorithm and Candan algorithm. When $v > 6$, MSE(δ) of the proposed algorithm with two-term MSD window is lower than the competing algorithms. The proposed algorithm can actively suppress the effects of the harmonic interference signal. It can be used in complex applications with both additive white Gaussian noise and harmonic interference.

References

1. Rife, D.C., Boorstyn, R.R.: Single-tone parameter estimation from discrete-time observations. IEEE Trans. Inform. Theory **55**(9), 591–598 (1974)
2. Candan, C.: Analysis and further improvement of fine resolution frequency estimation method from three DFT samples. IEEE Sig. Process. Lett. **20**(9), 913–916 (2013)
3. Aboutanios, E., Mulgrew, B.: Iterative frequency estimation by interpolation on fourier coefficients. IEEE Trans. Sig. Process. **53**(4), 1237–1242 (2005)
4. Fan, L., Qi, G.Q.: Frequency estimator of sinusoid based on interpolation of three DFT spectral lines. Sig. Process. **144**, 52–60 (2018)
5. Yang, C., Wei, G.: A noniterative frequency estimator with rational combination of three spectrum lines. IEEE Trans. Sig. Process. **59**(10), 5065–5070 (2011)
6. Djukanović, S., Popović-Bugarin, V.: Efficient and accurate detection and frequency estimation of multiple sinusoids. IEEE Access **7**, 1118–1125 (2019)
7. Serbes, A.: Fast and efficient sinusoidal frequency estimation by using the DFT coefficients. IEEE Trans. Commun. **67**(3), 2333–2342 (2019)
8. Andria, G., Savino, M., Trotta, A.: Windows and interpolation algorithms to improve electrical measurement accuracy. IEEE Trans. Instrum. Meas. **38**(4), 856–863 (1989)
9. Belega, D., Dallet, D., Petri, D.: Accuracy of sine wave frequency estimation by multipoint interpolated DFT approach. IEEE Trans. Instrum. Meas. **59**(11), 2808–2815 (2010)
10. Candan, C.: Fine resolution frequency estimation from three DFT samples: case of windowed data. Sig. Process. **114**, 245–250 (2015)
11. Belega, D., Petri, D.: Frequency estimation by two- or three-point interpolated Fourier algorithms based on cosine windows. Sig. Process. **117**, 115–125 (2015)
12. Shin, D., Kwak, C., Kim, G.: An efficient algorithm for frequency estimation from cosine-sum windowed DFT coefficients. Sig. Process. **166**, 245–250 (2020)
13. Fan, L., Qi, G.Q., Xing, J., Jin, J.Y., Liu, J.Y., Wang, Z.S.: Accurate frequency estimator of sinusoid based on interpolation of FFT and DTFT. IEEE Access **8**, 44373–44380 (2020)

Workshop on Data Intensive Services based Application

Research on Construction of Measurement Matrix Based on Welch Bound

Han Zhang, Song Xiao$^{(\boxtimes)}$, and Hongping Gan

State Key Laboratory of Integrated Services Networks(ISN),
School of Telecommunications Engineering, Xidian University, Xi'an, China
xiaosong@mail.xidian.edu.cn

Abstract. Compressive sensing (CS) is a new theory of data acquisition and reconstruction. It permits the data of interest being sampled at a sub-Nyquist rate, meanwhile still allowing perfect reconstruction of data from highly incomplete measurements. During this process, the construction of measurement matrix is undoubtedly the key point. However, the traditional random measurement matrices, though having good performance, are difficult to implement in hardware and lack the ability of dealing with large signals. In this paper, we construct a series of novel measurement matrices (HWKM and HWCM) based on Welch bound, by sifting the basis matrix based on Hadamard matrix. Therefore, the proposed matrices are deterministic measurement, which can be easily designed in hardware. Specially, it is proved to have low coherence, which can even approach to Welch bound. Experimental results show that the proposed matrices, compared with traditional measurement matrices, not only have considerable reconstruction performance in terms of reconstruction error and the signal-to-noise ratio, but also accelerate recovery time.

Keywords: Compressive sensing · Measurement matrix · Welch bound

1 Introduction

Compressive sensing (CS) is an innovative framework for data acquisition and reconstruction, which uses the sparse structure of most signals typically in the time or transform domain to break through the limitations of the traditional Nyquist sampling band width. Since CS was first proposed by Candès [1], Tao [2] and Donoho [3] in 2004, it has aroused widespread concern in the academic community, which is widely used in various fields, such as information theory, wireless communication and data processing. All the time, CS mainly has three research directions [4]: sparse representation of signals, the design of measurement matrices and recovery (such as orthogonal matching pursuit, OMP) algorithm.

The sparse representation of the signal is the premise of the CS framework. Moreover, measurement matrix plays a dual role in the process of CS, which not only ensures that we can capture and preserve valuable original information

H. Gao et al. (Eds.): ChinaCom 2020, LNICST 352, pp. 701–712, 2021.
https://doi.org/10.1007/978-3-030-67720-6_48

of the signal with fewer measurements possibly, but also influences the performance of the recovery. Finally, it is also important to design an effective and fast algorithm to recover the signal. Although these three issues are related, it is generally recognized that designing a stable measurement matrix is helpful for several other aspects, so we need pay attention to the construction of a stable measurement matrix. This paper mainly describes the construction of the measurement matrix, which mainly combines the advantages and disadvantages of existing matrices and the properties of measurement matrix itself. Specifically, the restricted isometry property(RIP) is an essential criterion for measurement matrix properties.

Nowadays, measurement matrices are mainly divided into two categories: the first type is random measurement matrices, such as random Gaussian matrix, random Bernoulli matrix [5], partial Hadamard binary matrix, and partial Fourier matrix, which can satisfy RIP with great probability. However, random matrices have the disadvantages of uncertainty and waste of storage resources, which limit their practical applications. For example, the storage space required for random Gaussian matrix elements is large, and its unstructured nature leads to complicated calculations. The second category is deterministic measurement matrices which can overcome these shortcomings. Commonly, there are Toeplitz matrix [6], circulant matrix [7], and polynomial deterministic matrix [8]. However, the construction of the deterministic measurement matrices is hard and is difficult to handle large signals. Many researchers use some techniques to construct deterministic measurement matrices. For example, reference [9] points out that a good low density parity check (LDPC) code for primitive graphs is proposed, LDPC) check matrix can be used as measurement matrix; reference [10] uses m sequence to construct measurement matrix with better performance; reference [11] uses Berlekamp-Justesen code to construct measurement matrix. Therefore, how to combine the advantages and disadvantages of the traditional matrices to construct a novel measurement matrix is the key point of this paper.

The RIP simplifies the investigation of CS reconstruction and provides the best guarantee currently known. However, it is very difficult to judge whether a measurement matrix satisfies RIP [12]. In order to reduce the complexity of the problem, it is another key point to find an alternative method that can easily implement the RIP condition. R Baraniuk [12] pointed out that if the measurement matrix and sparse basis are guaranteed to be irrelevant, the measurement matrix satisfies the RIP condition with a high probability. In this paper, it is found that starting from the lower bound of correlation, some theories have proved that the performance of measurement matrix is better when the correlation bound is close to Welch bound [13]. In order to reduce the randomness of random matrix and simplify the construction of measurement matrix, this paper proposes a set of measurement matrices satisfying the optimal Welch bound. Moreover, theoretical verification and experimental analysis show that the measurement matrices proposed in this paper have good performance.

2   CS Theoretical Basis

2.1   CS Model

For an original finite-length signal $x \in R^N$, assuming $\{\psi_i\}_{i=1}^N$ is a set of orthogonal basis vectors of R^N, the finite-length signal x can be linearly expressed as:

$$x = \sum_{i=1}^{N} s_i \psi_i \tag{1}$$

where $s_i = \langle x, \psi_i \rangle = \psi^T x$, Eq. 1 can also be written as the following matrix form:

$$x = \Psi s \tag{2}$$

where $\Psi = [\psi_1, \psi_2, \cdots, \psi_N] \in R^{N \times N}$, and $\Psi \Psi^T = \Psi^T \Psi = I$, $s = [s_1, s_2, \cdots, s_N]^T$, assuming that the column vector s has only $K(K \leq N)$ nonzero coefficients, the signal x is said to be K sparse under the base matrix Ψ. At this time, the measurement matrix $\Phi \in R^{M \times N}(M \leq N)$ that is not related to the orthogonal basis matrix Ψ can be used for projection compression:

$$y = \Phi x \tag{3}$$

It can be obtained that there are M linear observations $y \in R^M (M \leq N)$. From the signal sparsity, it can be known that the information contained in the small linear projection is sufficient to reconstruct the signal x. Substituting Eq. 2 into Eq. 3 gives

$$y = \Phi \Psi s = As \tag{4}$$

where $\Phi \in R^{M \times N}$ is called measurement matrix or observation matrix, $\Psi \in R^{N \times N}$ is called sparse matrix or transform matrix, $A = \Phi \Psi$, $A \in R^{M \times N}$ is called sensing matrix or information operator.

For formula 3 , since $M \leq N$, recovering x from y is obviously a ill-conditioned problem. After being converted to formula 4 , although it is still ill-conditioned, it becomes a reality to recover the signal x from the sensing matrix A, because s is K sparse and $K \leq N$. The focus of this paper is to construct measurement matrix. Since we assumed that the original signal x itself is sparse and the sparse matrix Ψ is a identity matrix, then we only need to ensure that the measurement matrix Φ satisfies the conditions. On top of this, we can reconstruct x accurately by using the l_0 optimization:

$$\begin{cases} \min_x \|x\|_0 \\ s.t. y = \Phi x \end{cases} \tag{5}$$

2.2   Restricted Isometry Property and Correlation Property

For the case where the original signal x is a sparse signal, the measurement matrix Φ needs to meet the Restricted Isometry Property (RIP).

Definition 1 [14]. *The RIP parameter δ_k of the measurement matrix Φ is defined as the minimum value δ satisfying the following formula:*

$$(1 - \delta) \left\| x \right\|_2^2 \leq \left\| \Phi x \right\|_2^2 \leq (1 + \delta)\|x\|_2^2 \tag{6}$$

where x is a K sparse signal. If $\delta_k < 1$, the measurement matrix Φ is said to satisfy the K order RIP.

However, judging whether a given measurement matrix has RIP properties is a combination complexity problem. In order to reduce the complexity of the problem, the judgment of the RIP property of the matrix can be converted into correlation discrimination [15], Spark discrimination, etc. [16]. This article mainly analyzes and constructs the measurement matrix based on the correlation. From [16], it can be known that the measurement matrix with low correlation satisfies the RIP property, so the following description will be made on the correlation.

Definition 2. *For a matrix $\Phi = (\phi_1, \phi_2, \ldots, \phi_N) \in R^{M \times N}$, its correlation $\mu(\Phi)$ is defined as follows:*

$$\mu(\Phi) = \max_{1 \leq i \neq j \leq N} \frac{|\langle \phi_i, \phi_j \rangle|}{\|\phi_i\|_2 \bullet \|\phi_j\|_2} \tag{7}$$

where the inner product of the vector is represented $\langle \phi_i, \phi_j \rangle = \phi_i^T \phi_j$.

Obviously, when the column vectors of the matrix are unitized, $\mu(\Phi) = \max_{1 \leq i \neq j \leq N} |\langle \phi_i, \phi_j \rangle|$, the correlation of the matrix Φ is the maximum value of the inner product of any two columns. Since the low correlation measurement matrix satisfies the RIP property, our task comes to find a inner product with a suitable lower bound. At this point, Welch bound provides us with a good solution.

Theorem 1. *For a matrix $\Phi \in R^{M \times N}(M \leq N)$, after its column vectors being unitized, and its correlation satisfies $\mu(\Phi) \geq \sqrt{\frac{N-M}{M(N-1)}}$, that is, the inner product of any two columns satisfies this lower bound. This theorem gives a lower bound of correlation function, also called Welch bound.*

In order to build a measurement matrix, the premise, from the RIP to the correlation and eventually to the lower bound, is simplified step by step. Next, we will focus on the construction of low-randomness measurement matrices that satisfie Welch bound.

3   Construction of Low-Randomness Measurement Matrices Satisfying Welch Bound

3.1   Generation of Measurement Matrices

Based on Hadamard matrix, and screened by Welch bound, we can get a measurement matrix of size $M \times N(N = 2^n)$, which is the Hadamard Welch Matrix

(HWKM) we need. However, when the number of matrix rows or columns is too large, the correlation of this matrix may not satisfy Welch bound. In this case, we can solve a small matrix satisfying the condition by Kronecker integral, then construct the measurement matrix HWKM of size $\Phi \in R^{M \times N}(M = m^p, N = n^p, p > 1)$. The concrete steps to realize are as follows:

Step 1: first, we select a size of $N \times N$ Hadamard matrix, and randomly select M rows from it to generate a partial Hadamard matrix. Second, replace the -1 value in the matrix with 0 to generate a 0–1 matrix, and unit the column vector to obtain the matrix $\Phi = (\phi_1, \phi_2, \ldots, \phi_N) \in R^{M \times N}$, then determine whether its correlation satisfies Welch bound for screen. If the correlation is satisfied, the matrix is required to be output, otherwise, step 2 is performed.

Step 2: If the value of M or N is too large, we may not find the matrix that satisfies the Welch bound condition. In this case, we can get the basis matrix by Kronecker decomposition for the matrix $\Phi \in R^{M \times N}$ of known size [17]. The basis matrix $S \in R^{m \times n}(M = m^p, N = n^p)$ that satisfies Welch bound is smaller and easier to find.

Step 3: If the matrix $S = (s_1, s_2, \ldots, s_n) \in R^{m \times n}$ satisfies the Welch bound condition and its column vectors are unitized, $\|s_i\| = 1(i = 1, 2, \ldots, n)$, then the correlation of any two columns $s_i, s_j(i, j = 1, 2, \ldots, n)$ of the matrix S $\mu = \max |\langle s_i, s_j \rangle| \geq \sqrt{\frac{n-m}{m(n-1)}}$. Through mathematical verification, the matrix Φ expanded by Kronecker product of matrix S also satisfies the Welch bound, which is the measurement matrix HWKM we seek.

From the above construction process, it can be seen that the matrix filtered by Welch bound meets the requirements of the measurement matrix, reduces randomness and improves efficiency. At the same time, due to the characteristics of the family of low correlation sequences, the matrix HWKM $\Phi \in R^{M \times N}(M \leq N)$ that meets the Welch bound has been verified to meet the low correlation requirement of the measurement matrix. The matrix Hadamard Welch Circulant Matrix (HWCM) $\Phi \in R^{M \times N}(M \leq N)$ constructed by cyclic shift has also been verified to meet the low correlation requirement of the measurement matrix. The two kinds of measurement matrices constructed above are filtered by the Welch bound to reduce the randomness, and the matrices are verified to meet the low correlation characteristic of the measurement matrix, which guarantee the reconstruction probability, reduce the calculation time, and also have some improvement in hardware.

3.2 Theoretical Analysis

In this section, we mainly carry out theoretical analysis. First, the 0–1 matrix $\Phi = (\phi_1, \phi_2, \ldots, \phi_N) \in R^{M \times N}$ of size $M \times N$, after column vectors ϕ_i being unitized, it can be known from Theorem 1 that its correlation must satisfy the Welch bound:

$$\mu(\varPhi) \geq \sqrt{\frac{N - M}{M(N - 1)}} \qquad (8)$$

Judging the Welch bound of the constructed basis matrix, if Eq. 8 is satisfied, the measurement matrix that meets our need can be obtained. Of course, if the value of M or N is too large, the basis matrix that satisfies the condition may not be found. At this time, through the Kronecker product decomposition of the matrix $\varPhi \in R^{M \times N}$ of known size [16], a smaller basis matrix $S \in R^{m \times n}(M = m^p, N = n^p)$ that satisfies Welch bound is easier to find. If the matrix S satisfies Welch bound, the matrix $\varPhi$ expanded by its Kronecker product also satisfies Welch bound, which is the measurement matrix we seek, and it is named HWKM matrix.

It is assumed that the basis matrix $S = (s_1, s_2, \ldots, s_n) \in R^{m \times n}$ satisfies the Welch bound and its column vectors $s_i = (x_1, x_2, \ldots, x_m)(i = 1, 2, \ldots, n)$ have been unitized, $\|s_i\|_2^2 = \sum_{j=1}^{m} x_j^2 = 1 \quad (i = 1, 2, \cdots, n)$, then the correlation function of any two column vectors $s_i, s_j(i, j = 1, 2, \ldots, n)$ of the matrix S satisfies $\mu_{i,j} = \max |\langle s_i, s_j \rangle| \geq \sqrt{\frac{n-m}{m(n-1)}}$. The matrix $\varPhi \in R^{M \times N}(M = m^p, N = n^p, p \geq 1)$ is extended by the Kronecker product, and the correlation function between any two columns $i + qn, j + qn(q = 0, 1, \ldots, p - 1)$ is:

$$\mu_{i+gn,j+gn} = \left(\sum_{l=1}^{m} x_l^2\right)^p \cdot \mu_{i,j} = \mu_{i,j} \geq \sqrt{\frac{n - m}{m(n - 1)}} \geq \sqrt{\frac{n^p - m^p}{m^p(n^p - 1)}} \qquad (9)$$

It can be proved that the matrix $\varPhi \in R^{M \times N}(M = m^p, N = n^p, p > 1)$ also meets Welch bound.

$$\sqrt{\frac{n - m}{m(n - 1)}} \geq \sqrt{\frac{n^p - m^p}{m^p(n^p - 1)}} \qquad (10)$$

Using Mathematical Induction to prove as follows:

1. When $p = 1$, inequality 10 was equal, which obviously holds.
2. Assuming when $p = k$, inequality 10 holds, then $\sqrt{\frac{n-m}{m(n-1)}} \geq \sqrt{\frac{n^k - m^k}{m^k(n^k - 1)}}$.
3. Well, when $p = k + 1$,

$$\sqrt{\frac{n^{k+1} - m^{k+1}}{m^{k+1}(n^{k+1} - 1)}} = \sqrt{\frac{(n-m)(n^k + mn^{k-1} + \cdots m^k)}{m^{k+1}(n-1)(n^k + n^{k-1} + \cdots 1)}}$$
$$= \sqrt{\frac{n-m}{m(n-1)}} \cdot \sqrt{\frac{n^k + mn^{k-1} + \cdots m^k}{m^k(n^k + n^{k-1} + \cdots 1)}} = \sqrt{\frac{n-m}{m(n-1)}} \cdot \sqrt{\frac{n^k + mn^{k-1} + \cdots m^k}{m^k n^k + m^k n^{k-1} + \cdots m^k}}$$

Because $n^k + mn^{k-1} + \cdots + m^k < m^k n^k + m^k n^{k-1} + \cdots + m^k$, $\sqrt{\frac{n^k + mn^{k-1} + \cdots m^k}{m^k n^k + m^k n^{k-1} + \cdots m^k}} < 1$, which is $\sqrt{\frac{n^{k+1} - m^{k+1}}{m^{k+1}(n^{k+1} - 1)}} < \sqrt{\frac{n-m}{m(n-1)}}$.

Therefore, when $p = k + 1$, inequality 10 also holds, and the proposition is proved. Therefore, formula 9 holds, that is, the matrix $\varPhi \in R^{M \times N}(M = m^p, N = n^p, p > 1)$ obtained by the Kronecker product expansion also meets the

Welch bound. So, the measurement matrix Φ is what we seek, which is named the HWKM matrix.

The matrix HWKM $\Phi \in R^{M \times N}(M \leq N)$ meets the Welch bound condition and has the characteristics of a low correlation sequence family. The matrix HWCM $A \in R^{N \times MN}(M \leq N)$ constructed by cyclic shift has been verified to still meet the low correlation requirement of the measurement matrix.

The matrix HWKM can be regarded as a sequence set of $(N, M, \mu_{\max})$, where $\mu_{\max}$ is the maximum correlation value of the sequence set. Along with Welch bound [13], the maximum value of the correlation function for a sequence set S consisting of M sequences with period N satisfies $\mu_{\max} \geq \sqrt{\frac{N-M}{M(N-1)}}$.

Through cyclic shift, we can construct a measurement matrix from a sequence set. Specifically, for the above mentioned sequence set $S = \left\{ s_i = \{s_i(t)\}_{t=1}^{t=N} : 1 \leq i \leq M \right\}$, with $s_1^1 = (s_1(1), s_1(2), \ldots, s_1(N))^T$, where T represents the transpose of the vector, after one cyclic shift, with $s_1^2 = (s_1(N), s_1(1), \ldots, s_1(N-1))^T$, and so on, after k cyclic shifts $(2 \leq k \leq N)$, we can get:

$$s_1^k = (s_1(N-k), s_1(N-k+1), \cdots, s_1(N), s_1(1), s_1(2), \cdots, s_1(N-k-1))^T \tag{11}$$

From that we can see that by cyclic shift, we can construct a matrix A of size $N \times NM$:

$$A = \left[s_1^T, s_2^T, \cdots, s_M^T, L\left(s_1^T\right), L\left(s_2^T\right), \cdots, L\left(s_M^T\right), \cdots, L^N\left(s_1^T\right), L^N\left(s_2^T\right), \cdots, L^N\left(s_M^T\right) \right] \tag{12}$$

At this time, the correlation function of the matrix A is:

$$\mu(A) = \mu_{\max} \geq \sqrt{\frac{N-M}{M(N-1)}} > \sqrt{\frac{NM-N}{N(MN-1)}} \tag{13}$$

which meets Welch bound condition. So the matrix $A \in R^{N \times MN}(M \leq N)$ is the measurement matrix HWCM we seek. Among them,

$$\sqrt{\frac{N-M}{M(N-1)}} > \sqrt{\frac{NM-N}{N(MN-1)}} \tag{14}$$

Inequality (14) is proved as follows:

Because $M \leq N$, $N - M > M - 1$, that is $\sqrt{\frac{N-M}{M(N-1)}} > \sqrt{\frac{M-1}{N(N-1)}}$. Also because $MN - M < MN - 1$, $\sqrt{\frac{N-M}{M(N-1)}} > \sqrt{\frac{M-1}{N(N-1)}}$. So $\sqrt{\frac{N-M}{M(N-1)}} > \sqrt{\frac{M-1}{MN-1}} = \sqrt{\frac{MN-N}{N(MN-1)}}$, that is, inequality (14) is proved, and formula (13) holds.

It is verified that the measurement matrices that constructed above all meet the requirement of low correlation. The measurement matrix HWKM constructed by Welch bound screening effectively reduce the randomness, and the measurement matrix HWCM are easier to implement in hardware.

4   Simulation Results

In this section, firstly, we analyze the RIP of random Gaussian matrix, partial Hadamard matrix and measurement matrices HWKM and HWCM constructed in this paper.

As seen in Fig. 1, we can clearly find that the measurement matrices HWKM and HWCM have similar properties. From the last two subgraphs of each row, we can find that the diagonal entries of the Gram matrix converge to 1, and the nondiagonal entries float around 0, which means that the measurement matrices HWKM and HWCM satisfy the RIP [18].

Then we compare traditional random Gaussian matrix, partial Hadamard matrix, low-randomness measurement matrices HWKM and HWCM constructed in this paper, and analyze the results both in noiseless and noisy environments.

Assuming that the original signal $x \in R^{256}$, the measurement matrix Φ with size of 64×256 and sparsity k, where $k \in \{5, 10, 15, 20, 25, 30, 35, 40, 45, 50, 55, 60, 65\}$, use the above four matrices to perform signal reconstruction separately, and use Matlab 2018b to generate 2000 averages for each sparsity k. This section aims to analysis and contrast the results from three aspects: recovery error, signal-to-noise ratio (SNR) and recovery time.

This experiment uses the basic tracking algorithm (BP) to obtain the recovery signal x^* and other results. The recovery error is defined by $||x - x^*||^2$, and the SNR is defined by:

$$SNR = 10 \log_{10} \left(\frac{||x||^2}{||x - x^*||^2} \right) dB \qquad (15)$$

For a signal x, we say x^* is a perfect recovery if $SNR(x) \geq 100$.

4.1   Signal Reconstruction in Noiseless Environment

For random Gaussian matrix, partial Hadamard matrix, low-randomness measurement matrices HWKM and HWCM which constructed in this paper, assuming that all with a size of 64×256 and under noiseless condition, we mainly compare the recovery error, the SNR and the recovery time.

As seen in Fig. 2, when the sparsity $k \leq 40$, the measurement matrices HWKM and HWCM constructed in this paper have smaller recovery error, higher SNR, shorter recovery time than random Gaussian matrix and partial Hadamard matrix.

4.2   Signal Reconstruction in Noisy Environment

For random Gaussian matrix, partial Hadamard matrix, low-randomness measurement matrix HWKM and HWCM which constructed in this paper, assuming that all with a size of 64×256 and under noisy condition, we mainly compare the recovery error, SNR and the recovery time.

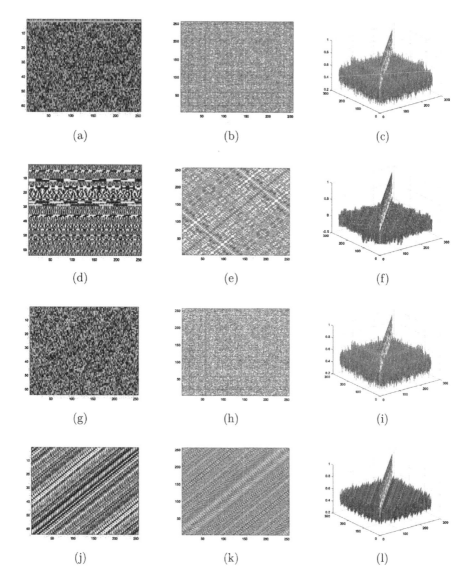

Fig. 1. First column is the imagec functions of random Gaussian matrix, partial Hadamard matrix, HWKM and HWCM, as well as the last two subgraphs of each row are the contour functions and mesh functions of their Gram matrix.

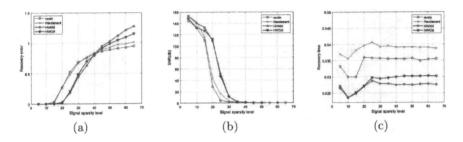

Fig. 2. Performance comparisons for recovering real singal using random Gaussian matrix, partial Hadamard matrix, HWKM and HWCM in noiseless environment: (a) Recovery errors with different sparsity, (b) SNR with different sparsity, (c) Recovery time with different sparsity.

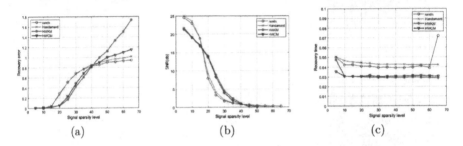

Fig. 3. Performance comparisons for recovering real singal using random Gaussian matrix, partial Hadamard matrix, HWKM and HWCM in noisy environment: (a) Recovery errors with different sparsity, (b) SNR with different sparsity, (c) Recovery time with different sparsity.

As seen in Fig. 3, in the case of sparsity $15 \leq k \leq 40$, which is just within a relatively more suitable sparsity range, the measurement matrices HWKM and HWCM constructed in this paper have smaller recovery error, higher SNR, and significant reduction in time than random Gaussian matrix and partial Hadamard matrix.

Through the above simulation, it is found that the measurement matrices HWKM and HWCM constructed in this paper have better reconstruction performance than the random Gaussian matrix and partial Hadamard matrix in the appropriate sparsity range under the same conditions, no matter whether it is noisy or noiseless.

5   Conclusion

In this paper, we propose a series of novel measurement matrices(HWKM and HWCM) based on Welch bound, which use the Hadamard matrix as the basis matrix and perform a series of screening and construction. It is easy to come to the conclusion that, under the same experimental situation, the measurement matrices HWKM and HWCM have preferable performance than random Gaussian matrix and partial Hadamard matrix, whether the background is noisy or not. In addition, within the appropriate sparsity range, compared to typical constructions, the proposed measurement matrices in this paper can facilitate the reestablishment of signals and could perform more practically and effectively. Although CS has been implemented in many areas, many algorithms still have room for improvement. Next step, test will be carried out in its application in Wireless Sensor Networks to optimize the process of data collection.

Acknowledgements. This paper is supported by NSFC (No. 61372069).

References

1. Candès, E.J.: Compressive sampling. In: Proceedings of the International Congress of Mathematicians, vol. 3, pp. 1433–1452. Madrid, Spain (2006)
2. Candès, E.J., Tao, T.: Near-optimal signal recovery from random projections: universal encoding strategies? IEEE Trans. Inf. Theory **52**(12), 5406–5425 (2006)
3. Donoho, D.L.: Compressed sensing. IEEE Trans. Inf. Theory **52**(4), 1289–1306 (2006)
4. Jiao, L.C., Yang, S.Y., Liu, F.: Review and prospect of compressed sensing. Chin. J. Electron. **39**(7), 1651–1662 (2011)
5. Zhang, G.S., Jiao, S.H., Xu, X.L.: Compressed sensing and reconstruction with Bernoulli matrices. In: Proceedings of the 2010 IEEE International Conference on Information and Automation. Harbin, China (2010)
6. Haupt, J., Bajwa, W.U., Raz, G.: Toeplitz compressed sensing matrices with applications to sparse channel estimation. IEEE Trans. Inf. Theory **56**(11), 5862–5875 (2010)
7. Yin, W., Morgan, S., Yang, J.F.: Practical compressive sensing with Toeplitz and circulant matrices. In: Proceedings of the 2010 Visual Communications and Image Processing(VCIP). Huangshan, China (2010)
8. Gilbert, A., Indyk, P.: Sparse recovery using sparse matrices. Proc. IEEE **98**(6), 937–947 (2010)
9. Zhang, J., Han, G., Fang, Y.: Deterministic construction of compressed sensing matrices from protograph LDPC codes. IEEE Sig. Process. Lett. **22**(11), 1960–1964 (2015)
10. Dang, K., Ma, L.H., Tian, Y.: Construction of the compressive sensing measurement matrix based on m sequences. J. Xidian Universit **42**(2), 186–192 (2015)
11. Xia, S., Liu, L., Liu, X.J.: Deterministic constructions of compressive sensing matrices based on berlekamp-justesen codes. J. Electron. Inf. Technol. **37**(4), 763–769 (2015)
12. Baraniuk, R.: A lecture on compressive sensing. IEEE Sig. Process. Mag. **24**(4), 118–121 (2007)
13. Welch, L.: Lower bounds on the maximum cross correlation of signals (Corresp.). IEEE Trans. Inf. Theory **20**(3), 397–399 (1974)
14. Wang, Q., Li, J., Shen, Y.: Overview of deterministic measurement matrix construction algorithms in compressed sensing. Chin. J. Electron. **41**(10), 2041–2050 (2013)
15. Donoho, D.L., Huo, X.: Uncertainty principles and ideal atomic decomposition. IEEE Trans. Inf. Theory **47**(7), 2845–2862 (1999)

16. Elad, M., Bruckstein, A.M.: A generalized uncertainty principle and sparse representation in pairs of bases. IEEE Trans. Inf. Theory **48**(9), 2558–2567 (2002)
17. Lin, D.H.: Method of finding matrix Kronecker integral solution. J. Minjiang Univ. **28**(5), 7–9 (2007)
18. Gan, H.P., Xiao, S.: Chaotic binary sensing matrices. Int. J. Bifurcation Chaos **29**(9), 1–20 (2019)

An Improved HMFCW Algorithm for Ranging in RFID System

Zengshan Tian, Shuwen Wu$^{(\boxtimes)}$, Liangbo Xie, and Xixi Liu

School of Communication and Information Engineering,
Chongqing University of Posts and Telecommunications, Chongqing, China
`1014295949@qq.com`

Abstract. Since the measured phase is usually a wrapped phase during ranging, the phase-based ranging method needs to solve the ambiguity so as to obtain the true phase. HMFCW (heuristic multi-frequency continuous wave) algorithm provides the same tolerance of error for the observed phase of different frequencies. When the error of phase is within the tolerance and the range of ranging is less than the period, the phase error tolerance method can get the correct cycle number, and achieve a ranging accuracy of centimeter. However, when phase errors of some frequencies are large, HMFCW algorithm may have difficulty in solving the integer ambiguity, which leads to the decrease of ranging accuracy. In this paper, an improved HMFCW algorithm based on HMFCW algorithm is proposed. The improved HMFCW algorithm calculates the average value of the clustering phase results to eliminate the phases with large errors, and performs the cycle calculation to obtain the ranging value. Simulation results show that improved HMFCW algorithm can solve the problem of error in the cycle ambiguity solution caused by the point with large phase error effectively, and improve the ranging accuracy.

Keywords: Phase cycle ambiguity · Improved HMFCW algorithm · The point with large phase error · Phase error tolerance · Ranging

1 Introduction

With the development of radio frequency identification (RFID) technology, the energy required to activate the tag has become lower and lower, and the reading range of the reader has become larger. Researchers have gradually realized the importance of ultra-high frequency radio frequency identification (UHF RFID) ranging technology and have carried out related researches. UHF RFID ranging methods based on time of arrival (TOA), received signal strength (RSS), and phase of arrival (POA) have been developed recently [1]. The TOA-based ranging method requires strict synchronization of clock between devices, and the cost is relatively high. RSS-based ranging methods can be divided into model ranging method [2] and fingerprint ranging method [3]. However, RSS is greatly affected by the indoor environment and has poor stability, the accuracy of it is low usually. As for the ranging method based on the phase of arrival, the phase has the ambiguity of the cycle in the RFID system and cannot be used for ranging directly [4]. The current phase-based UHF RFID ranging system can be

H. Gao et al. (Eds.): ChinaCom 2020, LNICST 352, pp. 713–728, 2021.
https://doi.org/10.1007/978-3-030-67720-6_49

divided into two categories according to the scene: tag or antenna of reader in the ranging system move or the tag and reader antenna are stationary.

Many ranging systems with phase-based UHF RFID rely on the movement of the tag or antenna of reader to locate the tag. E. Di Giampaolo et al. installed the antenna of reader on the mobile robot, and locate the tag by fusing the information of phase in the tag and the mileage information of the robot [5]. A. Buffi et al. built a synthetic aperture array by moving objects on the conveyor belt, combined with the information of phase in the tags to locate the object [6]. Hong Chao of Huazhong University of Science and Technology installed the antenna of reader on the robot, and obtained data of phase in different positions with the help of the robot's movement, and obtained the true phase by using the ambiguity resolution algorithm based on shooting method. The final ranging accuracy is decimeter [7]. The ranging system that relies on the movement of the tag or antenna of reader can restore the true phase more accurately, and its ranging accuracy can achieve a ranging accuracy of decimeter. However, this type of system has certain requirements for the movement of the tag or antenna and cannot locate static target.

When the tag and antennas of reader are stationary, the phase difference of arrival (PDOA) method is used to solve the ambiguity commonly. The PDOA method is divided into two types mainly: Frequency Domain (FD) and Spatial Domain (SD) [8]. The FD-PDOA method refers to measuring the phase of tags under carriers with different frequencies, and using PDOA to calculate the distance between the tag and the reader. Li proposed a ranging algorithm based on multi-frequency carrier phase [9], which combined the Chinese remainder theorem on the basis of the difference of phase in dual-frequency to select a combination of frequency that is more conducive to solving the ambiguity and obtains a larger Non-ambiguous distance. The SD-PDOA method is to measure the phase of tag in multiple antennas at the same frequency, and solve the position of tag relative to the antenna through PDOA. Liu uses multiple antennas placed side by side to measure the phase of the tag, constructs a hyperbolic equation set through PDOA, and then obtains the target position by finding the intersection point of the hyperbola [10]. Compared with the RSS, the information of phase is more accurate and stable, and the ranging method based on PDOA has higher accuracy, which can achieve a ranging accuracy of meter or even the decimeter. In recent years, Yunfei Ma in Cornell proposed a method to expand the bandwidth of UHF RFID system [11], which created favorable conditions for solving ambiguity, and he proposed HMFCW (heuristic multi-frequency continuous-wave) on the basis of expanding bandwidth. The ambiguity solving algorithm achieves a ranging accuracy of centimeter, which improves the accuracy of UHF RFID ranging technology greatly. Nevertheless, when the error of phase at some frequencies is large, HMFCW algorithm may have the problem of solving the ambiguity, which causes the ranging accuracy to decrease.

In order to solve this problem, this paper conducts a more in-depth study on the basis of HMFCW algorithm, and proposes improved HMFCW algorithm, when the phases have excessive error, improved HMFCW algorithm can eliminates the influence of it and improves the accuracy of ranging. Simulation results show that when the phases contain points with large phase error, the cumulative probability of HMFCW algorithm within the range of 0–0.3 cm is 52.3%, while improved HMFCW algorithm is 71.3%. Experiment results also show the same conclusion.

The rest of this paper is organized as follows: Sect. 2 introduces the overall architecture of the RFID system; Sect. 3 introduces the theory of HMFCW algorithm; Sect. 4 gives a detailed description and analysis of the improved HMFCW algorithm, and simulation results are provided to demonstrate the effectiveness of the proposed algorithm; Finally, Sect. 5 summarizes the work of this paper.

2 Brief Review

2.1 RFID System Architecture

The RFID system proposed by this paper is shown in Fig. 1. The main functions of the reader are to communicate with tag and to be responsible for reading or writing information of tag [12]. The reader usually obtain the signal phase by method of I/Q signal complex demodulation [13]. The computer is the control center, which is used to issue relevant instructions to reader and set relevant parameters, store and process information of tag. The information of tag read by the reader can also be displayed in the computer [14]. The system consists of two kinds of transmitters and receiver. Transmitter 1 is an Impinj R420 reader to activate tag, and transmitter 2 and receiver are USRP devices to locate tag. Impinj transmits a continuous high-power radio frequency signal with frequency of f_1(902−928 MHz) to activate the tag. USRP transmits a low-power signal with frequency of f_2(750−930 MHz) by frequency hopping, and the step of frequency hopping is 10 MHz. Frequency hopping can expand the bandwidth of system. Other USRP receives the data whose frequency is f_2 returned from tag, and the data is used for ranging.

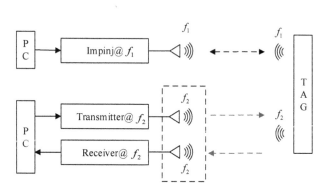

Fig. 1. RFID system architecture

After the signal sent by the USRP arrives at the tag, the tag modulates the stored ID information onto the signal and the signal reaches the receiver through backscattering. The receiver performs coherent demodulation on the received signal and demodulates the signal. Computer obtains the phase after relevant processing. The phase is determined by the sum of the distances from the USRP to the tag and the tag to the receiver.

Finally, the obtained phases at different frequencies are substituted into the ranging algorithm in the Matlab software to calculate the position of tag.

2.2 RFID Phase Ranging Theory and HMFCW Algorithm

When electromagnetic waves propagate in space, the energy gradually attenuates, and its phase also changes periodically. The distance can be estimated through the energy attenuation or change of phase, but the distance estimated by the energy attenuation will cause huge errors of ranging, so this section introduces the changing law of the phase in the electromagnetic wave propagation process, which provides the basis for the algorithm for solving the integer ambiguity.

The propagation speed of electromagnetic waves in the air is the speed of light c. Electromagnetic waves have two important properties: frequency f and wavelength λ. The relationship between them is $\lambda = c/f$. Electromagnetic waves propagate in all directions in a periodic oscillation manner. The phase will change 2π when the electromagnetic wave propagates for a distance of one wavelength in space. It can be seen that the true change of phase θ caused by the distance d is

$$\theta = 2\pi\frac{d}{\lambda} = 2\pi ft \tag{1}$$

The signal generated by the USRP is an electromagnetic wave. So other shifts of phase, except the change of phase caused by the distance, will be also added in during the propagation. The true change of phase from the USRP to the receiver through the tag is

$$\varphi = -w_c\tau + \varphi_S - \varphi_L + \varphi_T \tag{2}$$

where $w_c\tau$ is the change of phase caused by the distance of propagation, but it becomes a negative value during the demodulation process, $\varphi_S - \varphi_L$ is the initial difference of phase between the transmitted signal and the received signal, and φ_T is the offset of phase introduced by tag. Due to the periodicity of the sinusoidal signal, the phase of the tag measured by the receiver is not the true phase but a wrapped phase ϕ. The relationship between it and the true phase φ is

$$\phi = \varphi + 2n\pi \tag{3}$$

where n is the integer of cycles. The phase $\phi \in (-\pi, \pi]$ ϕ is called the observed phase, the difference between ϕ and the real phase φ is an integer multiple of 2π.

In the real change of phase φ, the distance-related part is only $w_c\tau$, so the influence of interference phase needs to be eliminated before ranging, $\varphi_S - \varphi_L$ and φ_T are fixed value under the same system, and can be eliminated by calibration. For the convenience of discussion, "observed phase" and "true phase" in the following both refer to the value after eliminating the fixed phase offset.

The frequency allocated to UHF RFID in China is very limited (920−925 MHz). So the phase difference of arrival in frequency domain (FD-PDOA) method will cause a large error when ranging. Yunfei Ma proposed a frequency hopping scheme to expand bandwidth [11], in which one USRP transmits a high-power signal to provide energy for the tag, meanwhile, other USRPs transmit low-power signals to achieve channel estimation by frequency hopping method. This program does not interfere with other electromagnetic communications and expand system bandwidth. The bandwidth of signals using frequency hopping can reach 220 MHz, creating a condition for solving the integer ambiguity.

The USRP obtains the phases at different frequencies f through frequency hopping, and the wavelengths of these frequencies are different, so the distance can be estimated by combining the phases of different frequencies. According to Eqs. (2) and (3), the relationship between RFID phase and distance is derived, and it can be expressed as:

$$
\begin{cases}
(\dfrac{\phi_1}{2\pi} + n_1) \times \lambda_1 = d \\[2mm]
(\dfrac{\phi_2}{2\pi} + n_2) \times \lambda_2 = d \\[2mm]
(\dfrac{\phi_i}{2\pi} + n_i) \times \lambda_i = d \\[2mm]
\quad \cdots \\[2mm]
(\dfrac{\phi_n}{2\pi} + n_n) \times \lambda_n = d
\end{cases}
\tag{4}
$$

where d represents the propagation distance of the signal, ϕ_i, n_i, λ_i represent the observed phase, the integer of cycles, and the wavelength of the frequency of f_i, and n_i is a non-negative integer. In the Eq. (4), n_i and d are unknown, ϕ_i and λ_i are known, $i = (1, 2, \ldots, n)$, and the Eq. (4) is an underdetermined equation, but only non-negative integers can be used in n_i, and the range of d is also limited, so that n_i and d can be solved.

In Fig. 2. Assuming that the reader obtains the observed phases at 4 different frequencies, the distance of the observed phase is obtained from $\phi_i/(2\pi) \times \lambda_i$, and it adds $0, 1, 2, \ldots, N_{max}$ times wavelength to get multiple distance. There is a value very close to the actual distance, and the N_{max} is determined by the ranging range. Since the true distance is fixed, the distances each frequency closed to each other are clustered into one category, and the category with the smallest variance among the clustering results is selected to be the result of ranging. Literature [15] proposed HMFCW algorithm to solve the integer ambiguity. The algorithm provides the same error tolerance for the observed phase at different frequencies. The theory and specific steps of HMFCW integer ambiguity solution algorithm will be introduced in detail below.

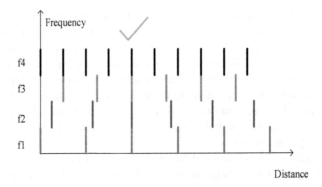

Fig. 2. Theory of solving the integer ambiguity

The numbers of cycle at each frequency constitutes an array of cycle. Among all the clustering results, the type closest to the true distance is considered as the correct cycle number. Other numbers of cycle are called wrong cycle number. In particular, the two numbers closest to the correct cycle number (one on each side) are called adjacent cycle number. HMFCW algorithm obtains the correct cycle number through correlation operations, thereby realizing ranging.

In order to introduce the parameter phase error, the estimated distance is expressed as

$$d^{(i)} = d + \frac{\Delta\phi_i}{2\pi} \times \lambda_i \tag{5}$$

where d represents the true distance and $\Delta\phi_i$ represents the observed phase error at frequency f_i. Assuming that the absolute value of the error of observed phase at all frequencies is less than $\Delta\phi$, the distances of any two frequencies are subtracted and substituted into Eqs. (4) and (5) to obtain

$$
\begin{aligned}
\left| d^{(i)} - d^{(k)} \right| &= \left| n'_i \times \lambda_i + \frac{\phi_i}{2\pi} \times \lambda_i - n'_k \times \lambda_k - \frac{\phi_k}{2\pi} \times \lambda_k \right| \\
&= \left| \frac{\Delta\phi_i}{2\pi} \times \lambda_i - \frac{\Delta\phi_k}{2\pi} \times \lambda_k \right| < \frac{\Delta\phi}{2\pi} \times (\lambda_i + \lambda_k) \forall i, k
\end{aligned}
\tag{6}
$$

where n'_i represents the actual cycle number. It is known that the correct cycle number satisfies the Eq. (6), but when the $\Delta\phi$ setting is too large, the wrong cycle number will also satisfy the Eq. (6). In order to distinguish the correct cycle number from multiple numbers of cycle and make $\Delta\phi$ as large as possible, $\Delta\phi$ are derived following.

Supposing there is another cycle number satisfying Eq. (6), then

$$-\Delta\phi < \frac{2\pi \times \left((n_i' + b_i) \times \lambda_i + \frac{\phi_i}{2\pi} \times \lambda_i - (n_k' + b_k) \times \lambda_k - \frac{\phi_k}{2\pi} \times \lambda_k\right)}{\lambda_i + \lambda_k} < \Delta\phi \tag{7}$$

$$\forall i, k, \exists b_i, b_k, \sum_i b_i^2 > 0$$

By changing the form of Eqs. (6) and (7), we can get:

$$\frac{\pi \times |b_i \times \lambda_i - b_k \times \lambda_k|}{\lambda_i + \lambda_k} < \Delta\phi \; \forall i, k, \exists b_i, b_k, \sum_i b_i^2 > 0 \tag{8}$$

If Eq. (8) deduces a contradiction, and according to proof by contradiction, Eq. (7) does not exist, then there will only be a correct cycle number. Since the gap between the distances of the adjacent cycle number is smallest than other number except the correct number, the unique solution can be guaranteed when the adjacent cycle number fails to pass the judgement (provided that there is enough frequency), so $b_i = 1$, $b_k = 1$, get

$$\Delta\phi_{\max} = \frac{\pi \times (f_{\max} - f_{\min})}{f_{\max} + f_{\min}} \tag{9}$$

where f_{max} represents the maximum frequency of frequency hopping signals, and f_{min} represents the minimum frequency. The equation above is the most reasonable value of $\Delta\phi$. It obtains the maximum value under the condition that the cycle number to be calculated is correct, and the value $\Delta\phi_{\max}$ is called the error tolerance.

Limiting the cycle number through the range of ranging, and judging the distance by one frequency with another, and getting the expression:

$$\begin{cases} 2\pi \times \dfrac{\left| n_i \times \lambda_i + \frac{\phi_i}{2\pi} \times \lambda_i - n_k \times \lambda_k - \frac{\phi_k}{2\pi} \times \lambda_k \right|}{(\lambda_i + \lambda_k)} < \Delta\phi \; \forall i, k \\ n_i \leq \dfrac{R_{\max}}{\lambda_i} \; n_i \in N \end{cases} \tag{10}$$

where n_i represents the cycle number to be found, $\Delta\phi$ represents the judgement threshold, and its initial value is taken as $\Delta\phi_{\max}$, $R_{\max}$ represents the maximum distance of ranging, N represents a natural number.

Equation (9) expresses the relationship between error tolerance and frequency. Substituting the observed phase and its frequency into Eq. (10), the correct cycle numbers can be obtained when the error of observed phase at all frequencies is less than $\Delta\phi_{\max}$, then the ranging result can be acquired. However, the error of measured phase may exceed the error tolerance. In that case, the measure taken is to enlarge $\Delta\phi$ gradually until a cycle number passes. The accuracy of ranging can achieve a ranging accuracy of centimeter when getting the correct cycle number. When the phase error is bigger than the error tolerance, the result is mostly adjacent numbers of cycle, so the ranging accuracy is about at the decimeter.

To summarize HMFCW algorithm, the steps are as follows:

1. Add the distance of the observed phases to an integer multiple of the wavelength to obtain multiple distances, classify the distances that are close at each frequency into one category, and record the number of the cycle of each category.
2. Substitute various corresponding phase and cycle number into Eq. (10) for judgement, where the value of $\Delta\phi$ at the first judgement is $\Delta\phi_{max}$.
3. If step 2) can get a cycle number, then go to step 4). If no cycle number passes, then the decision threshold $\Delta\phi$ will be enlarged by 1.02 times and return to step 2).
4. Obtain the distance at each frequency from the observed phase and the cycle number through the judgement, and use these distances to obtain the final ranging result.

The ranging effect of HMFCW algorithm is related to the frequency of frequency hopping and bandwidth. After actual measurement, the RFID system built in this paper has good quality of signal in 750–930 MHz and can decode successfully. Therefore, the simulation and experiment are all within this range. At the same time, this paper uses equal interval frequency hopping.

3   Improved HMFCW Algorithm

The system changes the frequency by frequency hopping, collects the phase of the tag at the same position and different frequencies, and then substitutes the phase and its frequency into the error tolerance method to obtain the result of ranging. However, in the actual process of acquiring phase, the error of phase at certain frequencies may be too large, which is called the point with large phase error. If the point with large phase error is substituted into HMFCW algorithm, it may cause large error of ranging in the solving integer ambiguity.

In order to explain the impact of the point with large phase error on solving the integer ambiguity better, the situation of solving the integer ambiguity before and after adding the point with large phase error is shown by Fig. 3. The parameters used for drawing are as follows: the range of frequency is from 750 MHz to 930 MHz, 7 frequencies at equal intervals in this range, the distance is 4.5 m, and the phase error is taking 0.8 radians and −1.2 radians on 900 MHz, 930 MHz respectively.

Substituting the true phase and its frequency to get Fig. 3(a), and substituting the phase of points with large error phase to get the Fig. 3(b). A column of points with the same color are a class of clustering result in the figure. The result is shown by Matlab. For example, the third clustering result in Fig. 3(a) is marked with "3" at 810 MHz and 930 MHz, which means that these two points cannot pass the judgement at the same time. The mark is "3" because the smaller frequency (810 MHz) is the third frequency in array of frequencies when all frequencies are arranged from small to large. All the judgements are indicated by marks. In Fig. 3(a), only the fourth class of clustering result has no mark, indicating that the fourth class of clustering result is passed the decision of Eq. (10). This result of ranging is consistent with the true distance. In Fig. 3 (b), the phase errors at 900 MHz and 930 MHz are relatively large. The clustering results of the true distance cannot pass the judgement at these two frequencies, while the adjacent clustering results (the third class) passed. So the final result of ranging is one wavelength smaller than the true distance. The results in figure show that the point with large phase error may cause the correct cycle number to fail to pass the judgement, and the adjacent cycle number can pass the judgement because its difference of distance is smaller, which causes the ranging error to become larger. Therefore, it is necessary to process the points with large phase error.

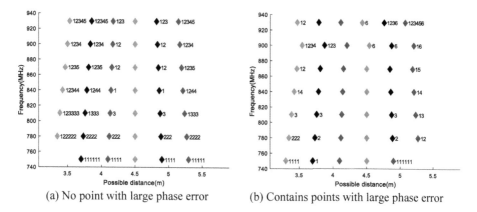

(a) No point with large phase error   (b) Contains points with large phase error

Fig. 3. The influence of the point with large phase error on solving the integer ambiguity

In the correct cycle number, the point with larger phase error will cause the distance to differ greatly from the mean value of the clustering result, so the point with large phase error can be filtered out by the method of averaging. However, if too many frequencies are filtered out, the error margin will be reduced, resulting in two or more ranging results appearing at the same time. If too few frequencies are filtered out, the point with large phase error cannot be filtered out completely, and the error of ranging is still large. To solve these problems, this paper proposes an improved HMFCW algorithm based on the phase error tolerance method: setting an initial value for the critical value of magnification of error tolerance at the beginning, if the magnification of error tolerance exceeds the certain critical value, the phase point with the largest phase error is removed and the remaining distances are judged. This process iterates until the clustering result passes the threshold. The algorithm calculates the ratio of the points with large phase error in the measured phase first generally, and then adjusts the critical value to ensure that the ratio of the filtered phase in the algorithm is close to this value. The pseudocode of improved HMFCW algorithm is shown in the Table 1 shown.

Table 1. The pseudocode of improved HMFCW algorithm

Input: $(f_1, f_2, \ldots f_m)$, $(\phi_1, \phi_2, \ldots \phi_m)$, R_{max}, *bound* ;

// $(f_1, f_2, \ldots f_m)$: Frequency

// $(\phi_1, \phi_2, \ldots \phi_m)$: Measurement phase

// R_{max} : Maximum ranging range

// *bound* : Critical value of error tolerance

Output: Distance;

1: $\quad \phi_{max} = \pi \dfrac{f_{max} - f_{min}}{f_{max} + f_{min}}$, $\phi_{max}' = \phi_{max}$;

   // Calculate the error tolerance, and the amplified error tolerance is ϕ_{max}'

2: $\quad \lambda = c / f$;

   // Calculate wavelength of each frequency

3: $\quad N_{max} = \dfrac{R_{max}}{\lambda_m} + 1$;

   // Calculate the maximum integer of cycle

4: $\quad$ for $n_m = 0 : N_{max}$

5: $\quad d_{temp} = n_m \times \lambda_m + \dfrac{\phi_m}{2\pi} \times \lambda_m$

6: $\quad\quad$ for $i = 1 : m - 1$

7: $\quad\quad\quad n_i = round \left(\dfrac{d_{temp} - \dfrac{\phi_i}{2\pi} \times \lambda_i}{\lambda_i} \right)$;

   // Round is a rounding function, clustering the distances under different frequencies

8: $\quad\quad\quad$ end for

9: $\quad$ end for

10: $\quad$ while (temp) do

11: $\quad\quad$ if $\phi_{max}' - \phi_{max} > bound$

12: $\quad\quad\quad [f, \phi] = update(f, \phi)$;

   // Remove the farthest value from the clustering result

13: $\quad\quad\quad \phi_{max} = \phi_{max}'$;

   // Update the amplified value of the error tolerance

14: $\quad\quad$ end if

15: $\quad\quad$ if $2\pi \times \dfrac{\left| n_i \times \lambda_i + \dfrac{\phi_i}{2\pi} \times \lambda_i - n_k \times \lambda_k - \dfrac{\phi_k}{2\pi} \times \lambda_k \right|}{(\lambda_i + \lambda_k)} < \phi_{max}' \quad \forall i, k$

16: $\quad\quad d = \dfrac{1}{m} \sum_{i=1}^{m} n_i \times \lambda_i + \dfrac{\phi_i}{2\pi} \times \lambda_i$;

   //Calculate the mean value and use it as the final result

(continued)

Table 1. (*continued*)

17:	$temp = 0$; //The end of cycle flag
18:	else
19:	$\phi_{max}' = \phi_{max} \times ampli$ //Enlargement decision threshold
20:	end if
21:	end while
22:	$ratio = \dfrac{length(f)}{m}$ //Record frequency utilization

4 Simulation and Experiment Results

4.1 Simulation Results

In the phase-based UHF RFID tag ranging system, the PDOA-based method is usually used to solving ambiguity of phase. Considering that this paper uses frequency hopping technology to expand the bandwidth, it is better to replace the frequency in the FD-PDOA method with the maximum and minimum frequencies of the frequency hopping to increase the bandwidth, improved method is named WB-PDOA. This paper will use WB-PDOA method and improved HMFCW algorithm for comparison.

In order to test the ranging accuracy of these algorithm, a multipath channel model is constructed using Matlab. The channel impulse response can be expressed as:

$$h(t) = \sum_{i=1}^{N} a_i \delta(t - t_i) \tag{11}$$

where a_i and t_i are the amplitude and delay of the i path respectively, and N represents the number of paths. The statistical characteristics of these parameters obey a certain probability distribution. Generating the amplitude of each path from the Rice distribution, setting the amplitude of direct path is 3 times larger the mean value of amplitude of multipath, and generating the arrival time interval of multipath from the exponential distribution to get the delay of each path. The average arrival time interval of the path is 10 ns [16].

Obtaining the phase of all frequencies through this model. The phase at each frequency has a 15% probability of being replaced with a point with large phase error to simulate the actual receiving phase. The error of the point with large phase error is within (0.5, 2) radians that obey uniform distribution. Performing 1000 times of simulation with or without points with large phase error, calculating the phase error at a frequency of 900 MHz, and obtaining the cumulative distribution function of the phase error as shown in Fig. 4. It can be seen from the figure that the maximum phase error of the phase generated by the multipath model is about 1 rad. In the case of points with large phase error, the maximum phase error reaches 2 rad, and 10% of the phase error is distributed in the range of (1, 2).

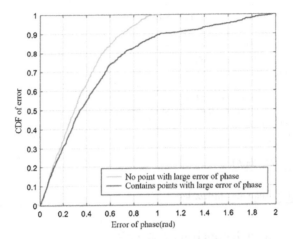

Fig. 4. Cumulative probability distribution of phase error

In the two cases above, calculating the ranging error of improved HMFCW algorithm. The juxtaposed algorithms are WB-PDOA algorithm and HMFCW algorithm. It can be seen from the Fig. 5 that when there is no point with large phase error, the improved HMFCW algorithm and the HMFCW algorithm have similar ranging errors, and the maximum error of both algorithms does not exceed 38.5 cm. The maximum error of the WB-PDOA algorithm is 47.2 cm, and its ranging error is larger than the improved HMFCW algorithm and the HMFCW algorithm. When there are points with large phase error, the improved HMFCW algorithm has a higher accuracy than HMFCW algorithm, and the maximum error of both algorithms does not exceed 73.3 cm. The cumulative probability of improved HMFCW algorithm is 18.7% higher than HMFCW algorithm when the error of ranging is within the range of (0, 3) cm, which shows that improved HMFCW algorithm can indeed eliminate the influence of points with large phase error, while the ranging error of the WB-PDOA algorithm is still larger than the improved HMFCW algorithm and the HMFCW algorithm.

Comparing Fig. 5(a) with Fig. 5(b), we can see that the cumulative probability of HMFCW algorithm within 0–0.3 cm is 52.3% with points with large phase error, and the improved HMFCW algorithm is 71.3%, which shows that the points with large phase error reduce the correctness of HMFCW algorithm significantly, and improved HMFCW algorithm can be more effective to eliminate the influence of point with large phase error. In addition, the cumulative probability of improved HMFCW algorithm in Fig. 5(b) increases significantly when the ranging error is 16–20 cm. This is because the algorithm may lead to the consequences that the correct number and adjacent cycle number passing through judgement simultaneously after eliminating points with large phase error. And the error will be about half a wavelength after taking the average of the two clustering results.

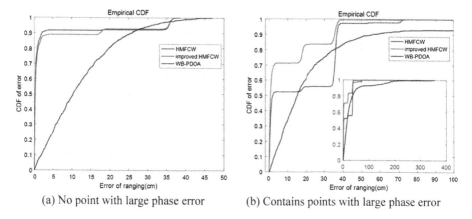

(a) No point with large phase error   (b) Contains points with large phase error

Fig. 5. Comparison of ranging error in the two cases

4.2   Measurement Results

The algorithm is verified through experiments, and the place of experiment is shown in Fig. 6.

Fig. 6. Place of experiment

The antennas of USRP and Impinj are fixed on a wall of 3.1 m high, 3 receivers are placed side by side, and the height of it is 1.5 m. All USRPs are close to wall. The heights of tag and receiver are same. The yellow marks in the Fig. 6 are test points. The test points are marked with different numbers, and they are arranged at equal interval. The interval between two adjacent points is 0.5 m. 25 positions are tested. The height of the tag remains unchanged when the position of tag is changed.

HMFCW algorithm and improved HMFCW algorithm are used to complete this experiment. The performance of the two algorithms is evaluated by the error of ranging. In the experiment, the measured phase at most frequencies is close to the theoretical phase, and the difference of phase is within 0.3 radians. While at a few frequencies, measured phases are far from the theoretical phase, and the difference of phase exceeds 0.7 radians. That's say there are points with large phase error in the measured phase.

By substituting the measured phase into HMFCW algorithm and improved HMFCW algorithm, the ranging can be obtained. The results of ranging in the first 10 positions of antenna 1 are shown in Table 2. And ranging errors of each antenna at the 25 positions are shown in Fig. 7.

Table 2. Ranging results of the first 10 positions of antenna 1

Number of position	True distance (cm)	Result of HMFCW (cm)	Error of HMFCW (cm)	Result of Improved HMFCW (cm)	Error of Improved HMFCW (cm)
1	338.6	338.5	0.1	338.8	0.2
2	340.3	379.5	39.2	324.1	16.2
3	364.8	360.4	4.4	360.4	4.4
4	419.8	420.3	0.5	420.3	0.5
5	483.3	521.5	38.2	521.5	38.2
6	417.2	415.3	1.9	415.3	1.9
7	418.0	456.1	38.1	455.8	37.8
8	442.3	443.3	1	443.6	1.3
9	486.9	450.9	36	468.7	18.2
10	543.9	544.0	0.1	543.6	0.3

By observing the ranging results in Fig. 7, it can be found that the ranging error of improved HMFCW algorithm is concentrated in half of one wavelength. This is because the ranging error is mainly determined by the cycle number. Improved HMFCW algorithm removes the frequencies with the points of large phase error, which may lead to only the correct cycle numbers and the adjacent cycle number can pass the judgement. So the errors are concentrated in the half of one wavelength. When the ranging error of HMFCW algorithm is about one wavelength, improved HMFCW algorithm improves about half of the situation; when the consequence of HMFCW algorithm is correct, improved HMFCW algorithm can also be correct, two algorithms can remain consistent at this situation.

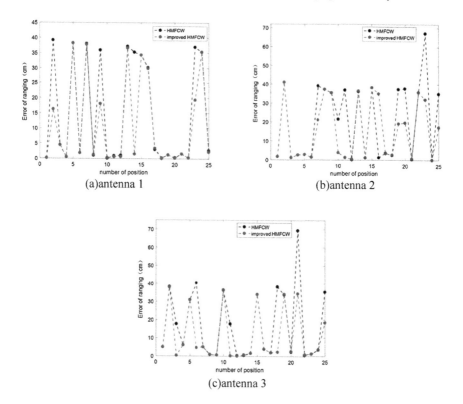

(a)antenna 1   (b)antenna 2

(c)antenna 3

Fig. 7. Ranging error of each antenna

5 Conclusion

This paper proposes improved HMFCW algorithm to solve the problem of points with large phase error. Improved HMFCW algorithm calculates the mean value of the clustering results, then eliminate the phase with large error, and solve the integer ambiguity to obtain the consequence of ranging. Simulation results show that when the phases contain points with large phase error, the cumulative probability of HMFCW algorithm within the range of 0–0.3 cm is 52.3%, while improved HMFCW algorithm is 71.3%. Experiment results also show the same conclusion. It means that improved HMFCW algorithm can eliminate the influence of points with large phase error effectively and improve the ranging accuracy.

Acknowledgements.   This work was supported partly by the National Natural Science Foundation of China (No. 61704015), and General program of Chongqing Natural Science Foundation (special program for the fundamental and frontier research) (No. cstc2019jcyj-msxmX0108).

References

1. Bingbing, Y., Wenbo, R., Bolin, Y., Yang, L.: An indoor positioning algorithm and its experiment research based on RFID. Int. J. Smart Sens. Intell. Syst. **7**(2) (2017)
2. Meng, Q.B., Ju, L., Jin, J., Li, W.X.: Active RFID indoor location system based on RSSI ranging correction. In: Advanced Materials Research, pp. 449–454 (2013)
3. Buffi, A., et al.: RSSI measurements for RFID tag classification in smart storage systems. IEEE Trans. Instrum. Meas. **67**(4), 894–904 (2018)
4. Yan, S.L., Dong, W.: UHF RFID localization based on phase map. pp. 1463–1466 (2014)
5. Giampaolo, E.D., Martinelli, F.: Mobile robot localization using the phase of passive UHF RFID signals. IEEE Trans. Ind. Electron. **61**(1), 365–376 (2014)
6. Buffi, A., Nepa, P., Lombardini, F.: A phase-based technique for localization of UHF-RFID tags moving on a conveyor belt: performance analysis and test-case measurements. IEEE Sens. J. **15**(1), 387–396 (2014)
7. Hong, C.: Research of positioning algorithm for RFID tags based on unwrapping phase. Huazhong University of Science and Technology, Wuhan (2018)
8. Ma, Y., Wang, B., Pei, S., Zhang, Y., Zhang, S., Yu, J.: An indoor localization method based on AOA and PDOA using virtual stations in multipath and NLOS environments for passive UHF RFID. IEEE Access **6**, 31772–31782 (2018)
9. Li, X., Zhang, Y., Amin, M.G.: Multifrequency-based range estimation of RFID Tags. In: IEEE International Conference on RFID. Nanjing, China, pp. 458–475. IEEE (2009)
10. Liu, T., et al.: Anchor-free backscatter positioning for RFID tags with high accuracy. In: IEEE INFOCOM 2014-IEEE Conference on Computer Communications. Toronto, pp. 978–993. IEEE (2014)
11. Ma, Y., Selby, N., Adib, F.: Minding the billions: ultra-wideband localization for deployed RFID tags. In: Proceedings of the 23rd Annual International Conference, pp. 248–260 (2017)
12. Ying, C., Fu-hong, Z.: A system design for UHF RFID reader. In: 11th IEEE International Conference on Communication Technology, Hangzhou, pp. 301–304(2008)
13. Scherhäufl, M., Pichler, M., Müller, D., Ziroff, A., Stelzer, A.: Phase-of-arrival-based localization of passive UHF RFID tags. In: IEEE MTT-S International Microwave Symposium Digest (MTT), Seattle, WA, pp. 1–3 (2013)
14. Li, L.-L., Wang, F.J., Zhang, H.G.: Realization of serial communication between computer and RFID. Mach. Build. Autom. pp. 28–129 (2009)
15. Ma, Y., Hui, X., Kan, E.C.: 3D real-time indoor localization via broadband nonlinear backscatter in passive devices with centimeter precision. In: Proceedings of the 22nd Annual International Conference on Mobile Computing & Networking. California: ACM, pp. 216–229 (2016)
16. Arulogun, O.T., Falohun, A.S., Akande, N.O.: Radio frequency identification and internet of things: a fruitful synergy. Curr. J. Appl. Sci. Technol. pp. 1–16 (2017)

Efficient Unmanned Aerial Vehicles Assisted D2D Communication Networks

Wenson Chang[1](✉) [ID], Kuang-Chieh Liu[2], Zhao-Ting Meng[3], and Li-Chun Wang[4]

[1] Department of Electrical Engineering, National Cheng-Kung University, Tainan, Taiwan R.O.C.
`wenson@ee.ncku.edu.tw`
[2] Realtek Semiconductor Co. Limited, Hsinchu, Taiwan R.O.C.
`x3334226x@gmail.com`
[3] SiliTai Electronics Co. Limited, Zhubei, Taiwan R.O.C.
`fateric885522@gmail.com`
[4] Department of Electrical Engineering, National Chiao-Tung Unversity, Hsinchu, Taiwan R.O.C.
`lichun@faculty.nctu.edu.tw`

Abstract. Nowadays, mounting a base-station (BS) onto an unmanned aerial vehicle (UAV) has become a new dimension for constructing the new generation of wireless communication networks. For example, dynamically constructing an UAV-BS network can provide some instantaneous and emergent communication services when the infrastructure of the cellular network is destroyed owing to some devastating disasters. In this paper, we aim to deploy a scalable and self-organized UAV-assisted device-to-device (D2D) communication network using a minimum number of UAV-BSs (UBSs) under some link quality constraints. Specifically, a sequential UBS deploying algorithm is designed to guarantee the signal quality for the links between the UBSs and ground terminals, and those between UBSs and central controller. Via the simulation results, it is interesting to find that how to deploy a proper number of UBSs rather than constructing a highly connected UBS network is the key to guarantee higher spectrum efficiency for the UBS-assisted D2D networks.

Keywords: UAV-assisted network · Fiedler value · D2D communications · 3D network · Laplacian matrix

This paper was jointly supported by Ministry of Science and Technology (under the Grant Numbers 108-2634-F-009-006, 109-2634-F-009-018 and 109-2221-E-006-176-MY2 through Pervasive Artificial Intelligence Research (PAIR) Lab and wireless communications and network Lab (WCNLAB), Taiwan) and QualComm Tech. INC. (under the Contract Numbers NAT-408931 and NAT-435536.).

H. Gao et al. (Eds.): ChinaCom 2020, LNICST 352, pp. 729–741, 2021.
https://doi.org/10.1007/978-3-030-67720-6_50

1 Introduction

Providing ubiquitous and high-quality communication services has always been the ultimate goal for the new generation of wireless communication systems. As the technology of unmanned aerial vehicle (UAV) becomes more and more mature, the UAV-assisted communications has emerged to be a new dimension of wireless communication networks. Generally speaking, mounting a base-station (UBS) on to a UAV (i.e., the namely UBS) can efficiently reconstruct the communication network when the infrastructure is devastated by large-scale disaster [1,2]. Also, it can be utilized to extend the coverage areas and collect data in some hazardous environments [3]. In some special events, the UBS can provide additional capacity and on-demand communication services [4,5]. More recently, incorporating UBSs into the framework of mobile edge computing further extends the frontier of high-speed and low-latency services [6,7].

In principle, the performance of an UAV-assisted network highly depends on two pivotal factors, i.e., the connectivity between UAVs [8–11] and its deployment scheme, including the flying trajectory arrangement [12–19]. In the aspect of interconnectivity between UAVs, the authors in [8] used the second eigenvalue of the Laplacian matrix (i.e., so-called Fiedler value [20]) to evaluate the connectivity of the UBS network. And thus, the main principle of deploying an UBS is to guarantee the signal-to-noise power ratio (SNR) requirements for the mobile terminals (MTs) on the ground, while achieving the highest Fiedler value for the whole network. In [9], a centralized algorithm was proposed to deploy a proper number of UBSs to serve a given number of MTs under the constraint of the inter-connectivity of UBSs. However, the methods of overall deployment and then one-by-one eliminating the unnecessary connections for MTs and idle UBSs based on the so-called degrees (i.e., the connection opportunities for each MT and UBS, respectively) cannot guarantee the minimum number of UBSs and the maximum connectivity between UBSs. In [10], the UAVs were deployed as the relays to reconstruct the post-disaster communication network. Specifically, three phases were designed to select the optimal positions, including the network traversing, coverage hole detection and deployment phases, so that the damaged network can be recovered autonomously. In [11], the interconnected UAVs were used to facilitate the data delivery for the intelligent transportation systems. Generally, the UAVs can develop the connections for the sparsely connected network on the ground. Then, with aid of UAVs, the routing path which avoids obstructions at the cost the limited overheads can be constructed and maintained.

In the aspect of UAV deployment, the reverse neural network model was applied in [17] to properly deploy UAVs such that the network coverage and overall throughput can be improved. In [12], the UAV-assisted three-hop transmission scheme was designed to deliver packets between two fixed sites on the ground. Therein, the optimal UAV trajectory was designed by taking several constraints into account (including the constraints of the transmit power, UAV mobility, collision avoidance and information causality) so that the end-to-end throughput can be maximized. To solve the non-convex problem, the so-called alternating maximization and successive convex optimization techniques were applied to develop an iterative trajectory algorithm. In [13], multiple solar-powered UBSs were deployed to improve the energy-efficiency of the heterogeneous network on the ground. To achieve this, an linear programming problem was formulated by taking the constraints of quality-of-service (QoS) and cells' capacity into consideration. With aid of UBSs, the small-cells on the ground can be switched into the sleep mode so as to minimize the energy consumption. In [14], to maximize the QoS for users while guaranteeing the backhaul connections, the probability density function of users' locations was taken into account when deciding the locations of UBSs. In [16], the multiple UBSs were sequentially deployed according to a incycloduction-spiral path until all the ground users can access more than one UBSs. In this way, the unnecessary UBS deployments can be eliminated. In [18], the coexistence of UAV-assisted and device-to-device (D2D) networks were considered for two cases: (1) analytically deciding the optimal altitude of the fixed UAVs according to the density of D2D users such that maximal coverage of UAV-assisted network can be achieved; and (2) analytically deciding the minimum stop points of mobile UAVs to reach a better tradeoff between the delay and coverage of UAV-assisted network, and the outage probability of D2D network. In [19], the backhaul connections between the UAVs and gateway were developed by deploying the multi-hop UAV-assisted networks. Under this scenario, the non-convex optimization problem was formulated to maximize the downlink throughput subject to the constraints of flow conservation. Then, using the mentioned alternating maximization and successive convex optimization techniques were applied to decide the locations of UAVs, and the allocations of bandwidth and transmission power, respectively.

According to the above discussions, we found that jointly consider the interconnectivity between UBSs, and that between UBSs and backhual links under the QoS constraints for all the links (including links between the UBSs and the ground users) was neglected from the literature. Thus, in this paper, we aim to develop a scalable and self-organized UBSs network to facilitate the D2D communications under the QoS constraints for all the links. To accomplish this goal, the minimal number of UBSs are sequentially deployed to serve the ground users under the mentioned constraints for all the links; and at the meantime, the interconnectivity between UBSs can be guaranteed. According to the simulation results, we find that how to deploy a proper number of UBSs rather than constructing a highly connected UBS network is the key to guarantee higher spectrum efficiency for the UBS-assisted D2D networks.

The rest of this paper is organized as follows. In Sect. 2, the system model, including the signal model and definition of the interconnectivity between UBSs are introduced. In Sect. 3, the optimization problem of UBS deployment will be formulated and solved by proposing a sequential UBS deployment algorithm. At last, Sects. 4 and 5 render the simulation results and concluding remarks, respectively.

2 System Model

Consider that a set of UAV-BS (denoted by $\boldsymbol{\Omega} = \{\Omega_i\}$ $\forall i = 1, \cdots, N$) are deployed to serve M groups of mobile terminal (denoted by $\mathbf{M} = \{\mathbf{m}_j\}$ $\forall j = 1, \cdots, M$); and each group consists of $|\mathbf{m}_j|$ MTs (denoted by $\mathbf{m}_j = \{m_{jk}\}$ $\forall k = 1, \cdots, |\mathbf{m}_j|$), where $|\mathbf{m}_j|$ calculates the number of elements in the set $\mathbf{m}_j$. Moreover, a UBS controller (denoted by Ω_o) is deployed to govern all the activities of UBSs, including the collections and exchanges for all the data and control information. Figure 1 demonstrates a sketch for the considered system scenario, where $N = 3$ UBSs are deployed to serve $M = 5$ groups of MTs; and within each group there are $|\mathbf{m}_j| = 5$ $\forall j = 1, \cdots, M$ MTs.

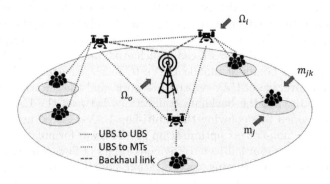

Fig. 1. System model

2.1 Signal Model

To begin with, the orthogonal channels are allocated to all the links between (1) two arbitrary neighboring UBSs, (2) UBS and MT and (3) UBS and controller Ω_o [8,13]. Also, according to [8,9,13,18], the impact of small-scale fading is generally neglected for the air-to-air and air-to-ground channel environments; and that, solely the effect of the path-loss is considered to characterize these two environments. Therefore, with LoS channel, the signal-to-noise power ratio (SNR) for the link between Ω_i and Ω_j (denoted by $\mathcal{L}_{ij}$) can be defined as

$$\gamma_{ij} = \frac{P_\Omega d_{ij}^{-\alpha}}{\sigma^2}, \tag{1}$$

where d_{ij} denotes the length of the link $\mathcal{L}_{ij}$ (i.e., the geographical distance between Ω_i and Ω_j); α and P_Ω stand for the path-loss exponent and transmission power of UAV, respectively; σ^2 represent the variance of the zero-meaned additive white Gaussian noise (AWGN). However, in addition to the LoS channel environment, it is highly possible to have non-LoS (NLoS) channel between UBSs and ground terminals, e.g., the MT or the controller Ω_o, respectively.

Thus, referring to [13], the path-loss for the air-to-ground link between Ω_i and m_{jk} (denoted by $\mathcal{L}_{ijk}$) can be defined as

$$g_{ijk}^{\varphi} = 20 \log_{10} \left(\frac{4\pi \hat{d}_{ijk}}{\lambda_0} \right) + \xi_{\varphi} \text{ (dB)}, \tag{2}$$

where $\hat{d}_{ijk}$ denotes the length of $\mathcal{L}_{ijk}$; λ_o represent the wave-length corresponding to the carrier frequency; $\varphi \in \{\text{LoS, NLoS}\}$ indicates the LoS and NLoS cases, respectively. By analogy, the path-loss g_{io}^{φ} between Ω_i and Ω_o (i.e., the link $\mathcal{L}_{io}$) can also be defined by substituting the length of $\mathcal{L}_{io}$ (say d_{io}) into (2).

To characterize the random effect of LoS and NLoS channel for the air-to-ground link, the LoS probability for the link $\mathcal{L}_{ijk}$ (denoted by $P_{LoS}(\mathcal{L}_{ijk})$) can be defined as

$$P_{LoS}(\mathcal{L}_{ijk}) = \left(\frac{1}{1 + \nu_1 \exp\left(-\nu_2 \left[\theta_{ijk} - \nu_1\right]\right)} \right), \tag{3}$$

where ν_1 and ν_2 reflect the environmental effects; θ_{ijk} denotes the elevation angle between Ω_i and m_{jk}. By definition, the NLoS probability can be defined as $P_{NLoS} = 1 - P_{LoS}$. Note that the LoS probability for the link $\mathcal{L}_{io}$ can also be defined by replacing θ_{ijk} of (3) with θ_{io} (i.e., the elevation angle between Ω_i and Ω_o). Then, the average path-loss for the link $\mathcal{L}_{ijk}$ can be calculated as

$$\bar{g}_{ijk} = \sum_{\varphi \in \{\text{LoS, NLoS}\}} P_{\varphi}(\mathcal{L}_{ijk}) g_{ijk}^{\varphi} . \tag{4}$$

Similarly, the average path-loss $\bar{g}_{io}$ for the link $\mathcal{L}_{io}$ can also be defined by replacing $\mathcal{L}_{ijk}$ of (4) with $\mathcal{L}_{io}$. At last, the SNR for the links $\mathcal{L}_{ijk}$ and $\mathcal{L}_{io}$ (denoted by γ_{ijk} and γ_{io}) can be obtained by substituting $\bar{g}_{ijk}$ and $\bar{g}_{io}$ into $d_{ij}^{-\alpha}$ of (1), respectively.

2.2 Connectivity of UBS Network

To well describe connectivity of the UBS network, a graph $\mathcal{G}(\mathcal{V}, \mathcal{E})$ can be constructed by using each UBS as each vertex; and an edge can then be constructed if the corresponding two UBSs (i.e., two vertices in other words) are reachable to each other (i.e., the distance in-between should be less than the serving distance d_{Ω}), where $\mathcal{V}$ and $\mathcal{E}$ denote the set of vertices and connected edges, respectively. To facilitate the presentation, let $V_i \in \{\mathcal{V}\}$ denote the vertex corresponding to Ω_i; and similarly, the edge $E_{ij}^{(\ell)} \in \{\mathcal{E}\}$ stands for the ℓ-th edge in $\mathcal{E}$ connecting V_i and V_j. Moreover, let's define edge vector $\mathbf{a}_\ell = [a_{\ell,i}]^T$ corresponding to $E_{ij}^{(\ell)}$ $\forall a_{\ell,i} \in \{0,1\}$, $i = 1, \cdots, |\mathcal{V}|$, $\ell = 1, \cdots, |\mathcal{E}|$, where $|\mathcal{V}|$ and $|\mathcal{E}|$ calculates the number of elements in the set $\mathcal{V}$ and $\mathcal{E}$, respectively. Then, with $E_{ij}^{(\ell)}$, it results in $a_{\ell,i} = a_{\ell,j} = 1$, which means V_i is in the coverage area $\mathcal{A}_j$ of V_j (denoted by $V_i \in \mathcal{A}_j$) and vice versa (denoted $V_j \in \mathcal{A}_i$). Note that each edge vector can only have two non-zero elements for describing its corresponding edge.

Then, the Laplacian matrix $\mathbf{L}$ can be defined as

$$\mathbf{L} = \mathbf{A}\mathrm{diag}(\mathbf{w})\mathbf{A}^T = \sum_{\ell=1}^{|\mathcal{E}|} w_\ell \mathbf{a}_\ell \mathbf{a}_\ell^T , \tag{5}$$

where the matrix $\mathbf{A}$ is constructed by using the edge vectors; $\mathrm{diag}(\mathbf{w})$ is the diagonal matrix constructed by using the weight vector $\mathbf{w} = \{w_\ell\}$ $\forall \ell = 1, \cdots, |\mathcal{E}|$ [8,20]. Note that according to (5), one can know that $\mathbf{L}$ is semi-positive definite, which means all the corresponding eigenvalues are non-negative. In fact, its first smallest eigenvalue is zero; and the second smallest one (denoted by $\lambda_2(\mathbf{L})$) is defined as the so-called Fielder value. According to the graph theory, the Fielder value can be used to characterize the connectivity of the network. By definition, the larger the Fielder value, the better the connectivity of the network. As $\lambda_2(\mathbf{L}) = 0$, it is said that the network is completely disconnected. It should be noticed that as the connectivity increases, more edges can be constructed. That means there will be more edge vectors and consequently the size of the Laplacian matrix $\mathbf{A}$ increases (e.g., more UBSs are hovering in the air).

3   Problem Formulation and UBS Deployment Scheme

In this paper, we aim to develop an efficient UBS-assisted D2D network by deploying the minimum number of UBSs. To this end, the positions of the hovering UBSs (denoted by $\mathbf{U} = \{(x_i, y_i, z_i)\} \forall \Omega_i \in \mathbf{\Omega}$) should be properly decided such that the considered groups of MTs on the ground can be well-severed, where (x_i, y_i, z_i) means the coordination of the Ω_i within the three-dimensional space. To clarity, the term "well-served" means two conditions are satisfied: (1) the SNR constraints for all the MTs belonging to the considered MT groups, and (2) the SNR constraints for the link between Ω_i and Ω_o. Moreover, the connectivity of the UBS network should be maintained as well. In this fashion, the self-organized D2D network can be developed. Then, the optimization problem of interest can be formulated as

$$\max_{\mathbf{U}} \min_{\Omega} \ \lambda_2(\mathbf{L}(\mathbf{U})) \tag{6}$$
$$\text{s. t. } \gamma_{ijk} \geq \gamma_{th} \ \forall \ \Omega_i \in \mathbf{\Omega} \ , \ m_{jk} \in \mathbf{m}_j \in \mathbf{M}$$
$$\gamma_{io} \geq \gamma_{th} \ \ \forall \ \Omega_i \in \mathbf{\Omega} \ .$$

where γ_{th} is the SNR threshold. It should be noticed that to well maintain the connectivity between UBSs, the weight w_ℓ for $E_{ij}^{(\ell)}$ is designed to be $d_{ij}^{-\alpha}$.

Prior to developing the UBS deployment algorithm, several terminologies are defined as follows.

1. $\mathcal{A}_{oj}$: the square area formed by using Ω_o and the center of $\mathbf{m}_j$ as the two vertices on the diagonal.
2. $\mathcal{A}_{ij}$: the areas in which the UBS Ω_i can well-serve the MT groups $\mathbf{m}_j$.
3. $\mathbf{\Phi}_i$: the set of MT groups which can be well-served by the UBS Ω_i.

4. $\Pi_{\mathbf{m}_j \in \Phi_i} \mathcal{A}_{ij}$: the intersected area in which the UBS Ω_i can well-serve the MT group $\mathbf{m}_j \in \Phi_i$.
5. d_{jn}: the distance between the centers of the MT groups $\mathbf{m}_j$ and $\mathbf{m}_n$.
6. d_{jo}: the distance between the centers of the MT groups $\mathbf{m}_j$ and Ω_o.

Moreover, two conditions are make: (1) all the UBSs are at the same height h_Ω m (i.e., $z_i = h_\Omega \ \forall \Omega_i \in \Omega$); and (2) the coordinate plane at $z_i = h_\Omega$ is discretized with space granularity μ point/m^2.

Now, to solve (6), we develop the sequential UBS deployment scheme as listed in Algorithm 1. Observing Algorithm 1, one can find that Lines 6 and 10 dominate the computational complexity. Specifically, the two while-loops on Lines 4 and 9 requires computations of order $\mathcal{O}(M^2)$; whereas searching for the "well-served areas" on Lines 6 and 10 maximally needs to scanning $\mathcal{A}_{oj} \times \mu$ points over the mentioned coordinate plane. To sum up, Algorithm 1 needs computations of order $\mathcal{O}(M^2 \times \mathcal{A}_{oj} \times \mu)$, which mainly depends on the granularity (i.e., μ) and distribution of the MT groups (i.e., $\mathcal{A}_{oi}$ in other words), respectively.

It should be noticed that Line 3 guarantees the maximal area of $\mathcal{A}_{oi}$ for the searching processes on Lines 6 and 10. Accordingly, it gives the most opportunities for Lines 10–12 to serve the most MT groups; and for Line 15 to maximize the Fiedler value $\lambda_2(\mathbf{L}(\mathbf{U}))$. This somehow verifies the optimality of the proposed UBS deploy algorithm with the prescribed order of complexity. Moreover, the proposed algorithm can be scalable by interatively treating each UBS as a new controller to the further deploy some additional UBSs for extending the overall coverage area. That means these additional UBSs can keep connected to the original central controller via some multi-hop (rather than one-hop) transmission path.

4 Simulation Result

In this section, we verify the effectiveness of the proposed sequential UBS deployment algorithm by comparing with the counterpart in [8] in terms of the number of hops, SE and Fiedler value. Note that in [8], each UBS is deployed for serving each MT group under the condition of maximizing the Fiedler value; whereas, in the proposed scheme, a UBS can possibly serve multiple MT groups. For fair performance comparison, the simulation environment is built by referring to that in [8]. In addition, to make it more practical, the scale of the 3D network is expanded to cover a square area of 1000×1000 m^2. Also, the back haul connection to the UBS controller is newly included (i.e., the constraint of γ_{io} in (6)). Moreover, the effects of the LoS and NLoS channel environment in [13] is considered as well. Table 1 summarizes all the simulation parameters; and all the results are obtained by averaging over 1,000 randomly generated network topologies.

4.1 Number of Hops

Figure 2 shows the impact of the SNR requirement γ_{th} on the number of hops for the UAV-assisted D2D communication networks, where (a) $\gamma_{th} = 40$ dB, (b)

Algorithm 1: Sequential UBS Deployment Algorithm

Input: M;
Output: Ω, **U**;

1 Let $\Omega = \mathbf{U} = \hat{\mathbf{M}} = \emptyset$;
2 $i = 1$;
3 Sort all the MT groups $\mathbf{m}_\ell$ $\forall \ell = 1, \cdots, M$ according to $\check{d}_{\ell o}$; and restore the results back to **M** in the descending order;
4 **while** $\mathbf{M} \neq \emptyset$ **do**
5   Consider the first MT group from the top of **M** and let it be $\mathbf{m}_j$;
6   Find $\mathcal{A}_{ij}$ within $\mathcal{A}_{oj}$;
7   Sort all the MT groups $\mathbf{m}_n \in \mathbf{M}$ $\forall n \neq j$ according to the distance $\check{d}_{jn}$ to $\mathbf{m}_j$ in the ascending order and store the results into $\hat{\mathbf{M}}$;
8     $\boldsymbol{\Phi}_i = \mathbf{m}_j$; $k = 1$;
9   **while** $\hat{\mathbf{M}} \neq \emptyset$ **do**
10   Find $\mathcal{A}_{ik}$ within $\mathcal{A}_{ij}$ for $\mathbf{m}_k \in \hat{\mathbf{M}}$;
11   **if** $\Pi_{\mathbf{m}_n \in (\boldsymbol{\Phi}_i \cup \mathbf{m}_k)} \mathcal{A}_{in} \neq \emptyset$ **then**
12           $\boldsymbol{\Phi}_i = \boldsymbol{\Phi}_i \cup \mathbf{m}_k$;
13        $\hat{\mathbf{M}} = \hat{\mathbf{M}} \setminus \mathbf{m}_k$;
14        $k = k + 1$;
15   Search for a point (x_i, y_i, z_i) nearest to Ω_o within $\Pi_{\mathbf{m}_n \in \boldsymbol{\Phi}_i} \mathcal{A}_{in}$ so that $\lambda_2(\mathbf{L}(\mathbf{U}))$ can be maximized;
16     $\Omega = \Omega \cup \Omega_i$;
17     $\mathbf{U} = \mathbf{U} \cup (x_i, y_i, z_i)$;
18     $\mathbf{M} = \mathbf{M} \setminus \boldsymbol{\Phi}_i$;
19     $i = i + 1$;
20 Return $\{\Omega, \mathbf{U}\}$;

$\gamma_{th} = 45\,\text{dB}$ and (c) $\gamma_{th} = 50\,\text{dB}$, respectively. In the simulations, the transmitting and receiving ends of a D2D pair are randomly deployed into two different MT groups, respectively. Observing these figures, one can apparently find that the number of hops increases as γ_{th} grows. There are two reasons for this phenomenons. Firstly, with higher requirements of link quality (i.e., γ_{th} in other words), the UBSs tend to hovering in the air near their served MT groups; and consequently, the distance between two arbitrary UBSs increases. Therefore, the connectivity of the UBS network reduces; and more hops are needed to develop a routing path for a D2D pair. Secondly, for the proposed scheme, more UBSs should be deployed to provide links with higher quality (as demonstrated in the following Fig. 3), which results in more required hops as well. Furthermore, in either cases of γ_{th}, less hops are required to develop the routing paths for the D2D network using the proposed scheme. For the example with $\gamma_{th} = 40\,\text{dB}$, approximately one hop can be saved for 80% of the D2D transmissions.

Table 1. Simulation parameters

Parameter	Value
Operational mode of UBS	Half-duplex transmissions
Average additional loss ξ_{LoS}	1 dB [13]
Average additional loss ξ_{NLoS}	12 dB [13]
Environment effects ν_1 and ν_2	9.6 and 0.29 [13]
Wavelength λ_0	0.125 m [13]
Path loss exponent α	2 [8]
The transmission power of UAV P_Ω	0.1 W [8]
Variance of AWGN σ^2	−174 dBm/HZ
System bandwidth	1 MHz
UAV height h_Ω	40 m
Area of network topology	$1000 \times 1000\,\text{m}^2$
Number of MT groups M	5
Number of MT in each group $|\mathbf{m}_j|$	$5\ \forall j = 1, \cdots, 5$
Radius of MT group	20 m
Serving distance of UBS d_Ω	100 m
Space granularity μ	$1\ 1/\text{m}^2$

4.2   Spectrum Efficiency and Fiedler Value

Figure 3 shows the impact of the SNR requirement γ_{th} on the (a) 10-percentile SE (denoted by $SE_{10\%}$) and (b) Fiedler value for the UAV-assisted D2D communication networks, respectively. Herein, the so-called 10-percentile SE is obtained by properly taking an threshold for the probability cumulative distribution function of SE (i.e., the *cdf* curve of SE) so that 90% of the SE can be higher than γ_{th}. As explained in Fig. 2, a higher γ_{th} can lead to longer distance between any two arbitrary UBSs, which lowering the connectivity of the UBS network. Moreover, as the γ_{th} keeps increasing, more UBSs are required when the proposed scheme is applied; and consequently the Fielder value can rise. However, the higher Fiedler values don't always means the higher SE. De facto, it is interesting to find that the UBS network with lowest Fiedler value (i.e., the loosely connected UBS network in other words) can achieve the highest $SE_{10\%}$. This phenomenon is somehow contradicted to the general intuition and the goal of [8,20]. Therefore, one can say that how to deploy a proper number of UBSs rather than constructing a UBS network with higher Fielder value is the key to guarantee a higher SE for the UBS-assisted D2D networks.

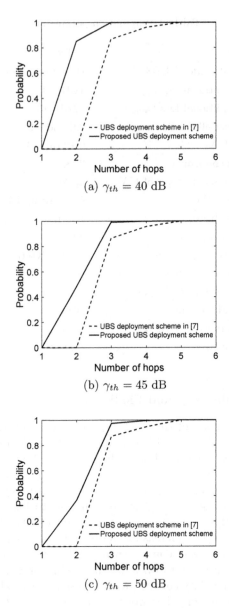

(a) $\gamma_{th} = 40$ dB

(b) $\gamma_{th} = 45$ dB

(c) $\gamma_{th} = 50$ dB

Fig. 2. Impact of the SNR requirement γ_{th} on the number of hops for the UAV-assisted D2D communication networks, where (a) $\gamma_{th} = 40$ dB, (b) $\gamma_{th} = 45$ dB and (c) $\gamma_{th} = 50$ dB, respectively.

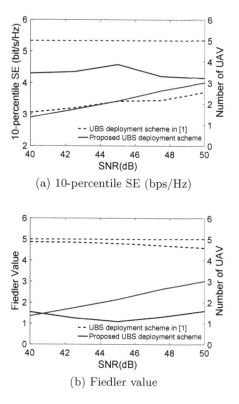

(a) 10-percentile SE (bps/Hz)

(b) Fiedler value

Fig. 3. Impact of the SNR requirement γ_{th} on the (a) 10-percentile SE (denoted by $SE_{10\%}$) and (b) Fiedler value for the UAV-assisted D2D communication networks, respectively.

5 Conclusions and Future Works

In this paper, we have developed a scalable UBS deployment algorithm to construct an UAV-assisted D2D communication network. Using the proposed algorithm, a self-organized 3D network can be constructed to provide some instantaneous and emergent communication services, especially for the cases when the infrastructure of cellular network become malfunctioned. Specifically, to guarantee the reliability and efficiency, a minimum number of UBSs are properly deployed under the SNR constraints for links between UBSs and ground terminals. Moreover, the links between the UBSs and center controller are included, as well. The efficiency of the self-organized 3D network has been verified by simulations in terms of the number of hops and spectrum efficiency, respectively. Most importantly, an counterintuitive phenomenon has been observed that a highly connected UBS network cannot always guarantee a higher spectrum efficiency. Instead, how to deploy a proper number of UBSs is key for constructing a high efficient UBS-assisted D2D network. It is well-known that steadily hovering in the air for a long period of time is fatal to the success of the UBS-assisted communication network. One important factor behind is the energy-efficiency (EE) of the UBS, including its energy-harvest (ER) capability. Therefore, incorporating the EE as well as the ER method into the proposed scheme can be an important topic for further study.

References

1. Zhao, N., et al.: UAV-assisted emergency networks in disasters. IEEE Wirel. Commun. **26**(1), 45–51 (2019)
2. Erdelj, M., Natalizio, E., Chowdhury, K.R., Akyildiz, I.F.: Help from the sky: leveraging UAVs for disaster management. IEEE Perv. Comput. **16**(1), 24–32 (2017)
3. Tazibt, C.Y., Achir, N., Muhlethaler, P., Djamah, T.: UAV-based data gathering using an artificial potential fields approach. In: IEEE Vehicular Technology Conference (VTC-Fall), pp. 27–30, August 2018
4. Li, Y., Cai, L.: UAV-assisted dynamic coverage in a heterogeneous cellular system. IEEE Netw. **31**(4), 56–61 (2017)
5. Wang, X., Feng, W., Chen, Y., Ge, N.: Sum rate maximization for mobile UAV-aided internet of things communications system. In: IEEE 88th Vehicular Technology Conference (VTC-Fall), pp. 27–30, August 2018
6. Zhou, F., Wu, Y., Hu, R.Q., Qian, Y.: Computation rate maximization in UAV-enabled wireless-powered mobile-edge computing systems. IEEE J. Sel. Areas Commun. **36**(9), 1927–1941 (2018)
7. Zhang, X., Zhong, Y., Liu, P., Zhou, F., Wang, Y.: Resource allocation for a UAV-enabled mobile-edge computing system computation efficiency maximization. IEEE Access **7**, 113 345–113 354 (2019)
8. Abdel-Malek, M.A., Ibrahim, A.S., Di, X., Mokhtar, M.: Optimum UAV positioning for better coverage-connectivity tradeoff. In: IEEE Annual International Symposium on Personal, Indoor, and Mobile Radio Communications (PIMRC), pp. 1–5, September 2017
9. Zhao, H., Wang, H., Wu, W., Wei, J.: Deployment algorithms for UAV airborne networks toward on-demand coverage. IEEE J. Sel. Areas Commun. **36**(9), 2015–2031 (2018)
10. Park, S.-Y., Shin, C.S., Jeong, D., Lee, H.: DronenetX: network reconstruction through connectivity probing and relay deployment by multiple UAVs in ad hoc networks. IEEE Trans. Veh. Technol. **67**(11), 11 192–11 207 (2018)
11. Oubbati, O.S., Chaib, N., Lakas, A., Lorenz, P., Rachedi, A.: UAV-assisted supporting services connectivity in urban VANETs. IEEE Tran. Veh. Technol. **68**(4), 3944–3951 (2019)
12. Zhang, G., Yan, H., Zenga, Y., Cui, M., Liu, Y.: Trajectory optimization and power allocation for multi-hop UAV relaying communications. IEEE Access **6**, 48 566–48 576 (2018)
13. Alsharoa, A., Ghazzai, H., Kadri, A., Kamal, A.E.: Energy management in cellular hetnets assisted by solar powered drone small cells. In: IEEE Wireless Communications and Networking Conference (WCNC), pp. 1–6, March 2017
14. Savkin, A.V., Huang, H.: Deployment of unmanned aerial vehicle base stations for optimal quality of coverage. IEEE Wirel. Commun. Lett. **8**(1), 321–324 (2019)
15. Lagum, F., Bor-Yaliniz, I., Yanikomeroglu, H.: Strategic densification with UAV-BSs in cellular networks. IEEE Wirel. Commun. Lett. **7**(3), 384–387 (2018)

16. Lyu, J., Zeng, Y., Zhang, R., Lim, T.J.: Placement optimization of UAV-mounted mobile base stations. IEEE Commun. Lett. **21**(3), 604–607 (2017)
17. Sharma, V., Bennis, M., Kumar, Y.: UAV-assisted heterogeneous networks for capacity enhancement, pp. 1207–1210, June 2016
18. Mozaffari, M., Saad, W., Bennis, M., Debbah, M.: Unmanned aerial vehicle with underlaid device-to-device communications: performance and tradeoffs. IEEE Trans. Wireless Commun. **15**(6), 3949–3963 (2016)
19. Li, P., Xu, J.: UAV-enabled cellular networks with multi-hop backhauls: placement optimization and wireless resource allocation. In: IEEE International Conference on Communication Systems (ICCS), pp. 110–114, December 2018
20. Ibrahim, A.S., Seddik, K.G., Liu, K.J.R.: Connectivity-aware network maintenance and repair via relays deployment. IEEE Trans. Wireless Commun. **8**(1), 356–366 (2009)

Early Stopping for Noisy Gradient Descent Bit Flipping Decoding of LDPC Codes

Li Zhang, Nan Liu[✉][iD], Zhiwen Pan, and Xiaohu You

National Mobile Communications Research Laboratory, Southeast University,
Nanjing 210096, China
{220170901,nanliu,pzw,xhyu}@seu.edu.cn

Abstract. A new early stopping criterion is proposed for the Noisy Gradient Descent Bit Flipping (NGDBF) Decoding of Low-Density Parity-Check (LDPC) codes to reduce the number of decoding iterations. The new criterion is based on the number of flipped bits at certain iterations, and has extremely low complexity. It is shown in the simulation results that the proposed early stopping criterion can significantly reduce the number of decoding iterations at low signal-to-noise ratios (SNRs), and only a slight bit error rate (BER) performance decrease is experienced at high SNRs.

Keywords: GDBF · Bit flipping · LDPC · Early stopping · Iterative decoding

1 Introduction

Since its proposal in 1963 by Gallager [3], Low-Density Parity-Check (LDPC) codes has gained considerable research attention over the past years. In [3], two types of decoding algorithms were proposed. The first type consists of soft decision algorithms, such as Belief Propagation (BP) algorithm [3], Min-Sum (MS) algorithm [2] and Offset Min-Sum (OMS) algorithm [1]. The second type consists of hard decision algorithms, known as Bit-Flipping (BF) algorithms.

The soft decision algorithms have an excellent bit error rate (BER) performance, but this comes at the cost of high computation complexity. The iterative soft decision decoder will not stop until a valid codeword is found by parity checking or the preset maximum number of iterations is reached. In many cases, the maximum number of iterations exceed one hundred for better BER performance. However, what often happens at low signal-to-noise ratios (SNRs) is, the valid codeword can not be found. Under these scenarios, the soft decision algorithms will continue to run until the maximum number of iterations is exceeded, and therefore, a lot of energy and time are wasted especially when the maximum number of iterations is large. Therefore, how to detect and stop the decoding of undecodable blocks at an early stage is worth studying.

H. Gao et al. (Eds.): ChinaCom 2020, LNICST 352, pp. 742–752, 2021.
https://doi.org/10.1007/978-3-030-67720-6_51

Various kinds of early stopping criterion for soft decision algorithms have been proposed. In [6], the new criterion is based on the convergence of the mean magnitude (CMM) of the log-likelihood ratio messages at the output of each decoding iteration. The CMM early stopping criterion can detect and stop the undecodable blocks very well, but at the cost of high computation complexity. In [8], the proposed stopping criterion is based on the variations of the number of satisfied parity-check constraints in the BP decoder. This criterion has a lower complexity compared with the CMM, but it may cause a performance loss at high SNRs. Similarly, an efficient criterion based on the evolution of the number of reliable variable nodes is proposed in [11]. In [5], a new method based on the intelligent combination of check equations and temporary hard decisions has been proposed, and it can be pipelined with the updating of the check nodes.

Since soft decision algorithms have high complexity, early stopping criterions for these algorithms were popular. In comparison, the hard decision algorithms such as the BF algorithm [3] were very simple, and therefore, the energy and time saved by using early stopping does not seem to be worth the cost. However, in recent years, many variants of BF algorithms which employ soft information have been investigated to improve the BER performance, such as Gradient Descent Bit Flipping (GDBF) algorithms [10] and Noisy Gradient Descent Bit Flipping (NGDBF) algorithms [9]. These algorithms have a better BER performance than previous BF algorithms at the cost of complexity increase. Hence, early stopping for these BF algorithms, which has higher complexity, is worth studying. In [4], an early stopping criterion for the Adaptive Threshold Bit Flipping (ATBF) algorithm, which is a variant of GDBF algorithm, has been proposed. It is based on the threshold of the ATBF algorithm, and thus termed the Early Stopping Adaptive Threshold Bit Flipping (ES-ATBF) algorithm.

In this paper, we propose a new early stopping criterion for the multi-bit NGDBF (M-NGDBF) algorithm [9] with a fixed threshold. M-NGDBF algorithm assumes the knowledge of the SNR from an external SNR estimator, but such knowledge is not always available. In absence of an external SNR estimator, we propose a new early stopping criterion that decides whether to stop the decoder at a certain iteration based on the number of flipped bits in the corresponding iteration. As a result, our proposed early stopping criterion is very simple to implement and has low complexity. We believe it is important that the early stopping criterions for the BF decoding algorithms should be much simpler than that for the BP decoding algorithms, given the lower complexity of BF decoding algorithms compared to the BP decoding algorithms. Simulation results show that the proposed early stopping criterion can significantly reduce the number of decoding iterations at low SNRs with an extremely small amount of complexity increase, and at high SNRs, only a slight BER performance degradation is incured.

2   System Model

Let $\mathbf{H}$ denote a binary parity check matrix with $m \times n$ dimensions and $n > m \geq 1$. We consider the set of LDPC codes $\mathcal{C}$ which is represented by

$$\mathcal{C} \triangleq \{\mathbf{c} \in F_2^n : \mathbf{H}\mathbf{c} = 0\}.$$

In the present paper, we define the set of bipolar codes

$$\hat{\mathcal{C}} \triangleq \{(1 - 2c_1), (1 - 2c_2), ..., (1 - 2c_n)\}$$

corresponding to $\mathcal{C}$. Evidently, $\mathcal{C}$ is mapped to $\hat{\mathcal{C}}$ from binary $(0, 1)$ to bipolar $(+1, -1)$. Assume that the codes are transmitted over a binary input AWGN channel and the transmission is defined by

$$\mathbf{y} = \hat{\mathbf{c}} + \mathbf{z},$$

where $\hat{\mathbf{c}} \in \hat{\mathcal{C}}$ and $\mathbf{z}$ is a white Gaussian noise vector. Each element of $\mathbf{z}$ is an independent and identically distributed Gaussian random variable with zero-mean and variance $N_0/2$, where N_0 is the noise spectral density.

The hard decision vector $\mathbf{x} \in \{+1, -1\}^n$ is obtained by

$$x_k = \text{sign}(y_k), \quad k = 1, 2, ..., n.$$

Let T be the maximum iteration number specified by the algorithm. In each iteration, one or more bits may be flipped according to different algorithms. In this paper, we only consider the M-NGDBF algorithm.

The parity check matrix $\mathbf{H}$ can also be defined by a Tanner graph with m Check Nodes (CNs) and n Variable Nodes (VNs). Let h_{ij} be the (i, j)-th element of the parity check matrix $\mathbf{H}$, $i \in [1, m]$, $j \in [1, n]$. If $h_{ij} = 1$, the i-th CN is said to be linked with the j-th VN. Let

$$N(i) \triangleq \{j \in [1, n] : h_{ij} = 1\}, \quad i = 1, 2, ..., m$$

be the VNs linked with the i-th CN. The CNs linked with the j-th VN is represented similarly as

$$M(j) \triangleq \{i \in [1, m] : h_{ij} = 1\}, \quad j = 1, 2, ..., n.$$

The parity check conditions of the codeword can be written as

$$s_i \triangleq \prod_{j \in N(i)} x_j, \quad i = 1, 2, ..., m,$$

which is also called the bipolar syndrome. A codeword is said to be a legitimate codeword if and only if every syndrome component is equal to $+1$.

3 Early Stopping Criterion for Multi-bit Noisy Gradient Descent Bit Flipping Decoding

3.1 Preliminary

In this subsection, we will briefly review the Noisy Gradient Descent Bit Flipping Algorithm.

In [10], the GDBF algorithm is proposed by regarding the maximum likelihood (ML) problem as the gradient descent optimization problem of an objective function. Inspired by the ML decoding problem, the objective function of the GDBF algorithm, which employs the syndrome component as a penalty term, is proposed. The objective function is defined by

$$f(\mathbf{x}) = \sum_{k=1}^{n} x_k y_k + \sum_{i=1}^{m} s_i. \tag{1}$$

Thus, the codeword that solves the ML problem is also the solution that maximizes the objective function. The local inversion function is obtained by taking the partial derivative of a particular symbol x_k, and it is defined by

$$E_k = x_k \frac{\partial f(\mathbf{x})}{\partial x_k} = x_k y_k + \sum_{i \in M(k)} s_i. \tag{2}$$

To increase the objective function (1), every x_k whose E_k is under the inversion threshold is flipped. To improve the performance of the GDBF algorithm, a noisy type of GDBF called NGDBF was proposed in [9]. The inversion function of the NGDBF algorithm is defined by

$$E_k = x_k y_k + w \sum_{i \in M(k)} s_i + q_k, \tag{3}$$

where w is the parameter of the syndrome weight and q_k, $k = 1, 2, ..., n$ are independent and identically distributed Gaussian random variables from the distribution $\mathcal{N}(0, \eta^2 N_0/2)$, and η is the noise scale parameter. The details of the M-NGDBF algorithm can be described in Algorithm 1.

Algorithm 1. M-NGDBF Algorithm [9]

1: Initialization: Set $t = 0$.
2: Compute bipolar syndrome $s_i = \prod_{j \in N(i)} x_j$ for $i = 1, 2, ..., m$. If $s_i = +1$ for all $i \in [1, m]$, output $\mathbf{x}$ and stop; Otherwise, $t = t + 1$.
3: Calculate the inversion function at each VN using (3).
4: Flip every bit whose $E_k < \theta$, where θ is the inversion threshold.
5: If $t < T$, repeat steps 2 to 4 till a valid codeword is detected; otherwise output $\mathbf{x}$ and exit.

Note that Algorithm 1 is the non-adaptive version of the M-NGDBF algorithm [9]. The adaptive version is similar, except in Step 4, the inversion threshold is a function of the SNR. Only the non-adaptive version is written out above, as we are focusing on the scenario where there is no external SNR estimator, and as a result, the decoder does not know the SNR of the channel. When the SNR of the channel is not available, devising an early stopping criterion is meaningful for the M-NGDBF algorithm as we may prevent the time and energy wasted in running the maximum number of iterations for an undecodable block at low SNRs.

3.2 Proposed Early Stopping Criterion

In absence of an external SNR estimator, the variance of the q_k, $k = 1, 2, ..., n$ in Algorithm 1 can no longer depend on N_0. Thus, we assume that the i.i.d. Gaussian random variables q_k, $k = 1, 2, \cdots, n$ follow the distribution $\mathcal{N}(0, \eta^2 \sigma^2)$, where η is the noise scale parameter in [9] and σ is the scale parameter used in early stopping criterion.

To design an early stopping criterion that is meaningful to the M-NGDBF algorithm, the simplicity and easiness of implementation is key. Figure 1 illustrates the average number of flipped bits for the success and failure of M-NGDBF decoding at different iterations and SNRs. As can be seen, the average number of flipped bits in the failure cases of M-NGDBF decoding falls more slowly as the number of iterations increases and remains at a higher value than that of the successful decoding cases, especially during the first twenty iterations. For example, the average number of flipped bits in the decoding failure cases at 2 dB hardly drops and almost all of them exceed 140 throughout the iterations.

Based on this observation, we use the number of flipped bits at certain iterations to predict and decide whether to stop the decoder. Algorithm 2 presents the details of the early stopping criterion for the M-NGDBF algorithm we propose. The algorithm starts out the same way as the M-NGDBF algorithm, and in Step 5, the early stopping criterion is being checked. Let S denote the set of iterations in which we will check the number of flipped bits. If the current iteration is in S, we record the number of flipped bits in this iteration and check this number against a predefined threshold λ. If the number of flipped bits exceeds λ, we deem that this block will not be decoded successfully and we stops the algorithm. Otherwise, the algorithm proceeds normally as in the M-NGDBF algorithm. Hence, S and λ are two parameters of the proposed algorithm, and their values shall be obtained empirically through simulations.

We comment here that the proposed early stopping criterion has a very low complexity and only runs for a few times, i.e., only when the iteration number is in S. Hence, the cost of the proposed early stopping criterion is extremely small.

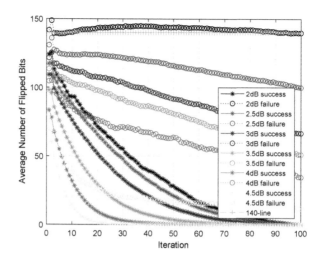

Fig. 1. Average number of flipped bits for success and failure of M-NGDBF decoding on the PEGReg 504×1008, rate-$\frac{1}{2}$, regular LDPC code at different iterations and SNRs, with parameters $\omega = 0.75$, $\eta = 1$, $\sigma = 0.8$, $\theta = 0.3$ and $T = 100$.

Algorithm 2. M-NGDBF Algorithm with Early Stopping

1: Initialization: Set $t = 0$ and $l = 0$.

2: Compute bipolar syndrome $s_i = \prod_{j \in N(i)} x_j$ for $i = 1, 2, ..., m$. If $s_i = +1$ for all $i \in [1, m]$, output $\mathbf{x}$ and stop; Otherwise, $t = t + 1$.

3: Calculate inversion function at each VN using (3).

4: Flip every bit whose $E_k < \theta$, where θ is the inversion threshold.

5: If $t \in \mathcal{S}$, denote l as the number of bits flipped in Step 4 for the current iteration; if $l \geq \lambda$, output $\mathbf{x}$ and exit.

6: If $t < T$, repeat steps 2 to 5 till a valid codeword is detected; otherwise output $\mathbf{x}$ and exit.

---

4   Simulation Results

Simulation results are based on PEGReg 504×1008 regular $(3,6)$, $\frac{1}{2}$-rate LDPC code from MacKay's online encyclopedia [7] on the AWGN channel with the Binary Phase Shift Keying (BPSK) modulation. The Bit Error Rate (BER), Block Error Rate (BLER) and average number of iterations (ANI) for the M-NGDBF algorithms with and without early stopping are compared. As mentioned above, due to the absence of an external SNR estimator, all the parameters of the M-NGDBF algorithms are set to constant. Therefore, the parameters are determined as $\omega = 0.75$, $\eta = 1$, $\sigma = 0.8$ and $\theta = 0.3$. The maximum number of decoding iterations T is set to 100. The M-NGDBF decoder with early stopping criterion is indicated by ES-M-NGDBF.

Figure 2 and Fig. 3 present the BER/BLER and ANI for the M-NGDBF and ES-M-NGDBF algorithms. The ES-M-NGDBF algorithm with $\mathcal{S} = \{1, 10, 20\}$ and $\lambda = 140$ has almost the same BER and BLER performance as the

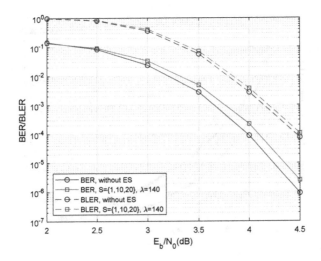

Fig. 2. BER and BLER performance for the M-NGDBF and ES-M-NGDBF with $\mathcal{S} = \{1, 10, 20\}$ and $\lambda = 140$.

M-NGDBF algorithm at low to middle SNRs, and a little performance degradation at high SNRs. Since the complexity of the proposed early stopping criterion is extremely low, and it is designed for the BF algorithm, a certain BER and BLER performance degradation at high SNRs is acceptable. Note that the ANI of ES-M-NGDBF with $\mathcal{S} = \{1, 10, 20\}$ and $\lambda = 140$ drops to about 30%, 50% and 80% of the ANI of M-NGDBF at 2 dB, 2.5 dB and 3 dB, respectively, in Fig. 3. The number of iterations hardly decreases at high SNRs, because the original number of iterations is relatively small, and the probability of decoding failure is quite low.

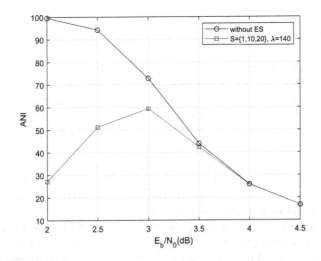

Fig. 3. ANI for the M-NGDBF and ES-M-NGDBF with $\mathcal{S} = \{1, 10, 20\}$ and $\lambda = 140$.

The ES-M-NGDBF algorithm is sensitive to the precise variance of the threshold λ. The value of the threshold λ is found empirically through a numerical search. This search may produce different values for different codes and algorithms. Some examples are shown in Figs. 4 and 5. According to Fig. 4, with the same set $\mathcal{S}$, the ES-M-NGDBF with $\lambda = 140$ has less BER and BLER performance loss than that with $\lambda = 130$, but the decline in ANI is also smaller as shown in Fig. 5. Since the BER performance degradation should not be too much, these results indicate that the optimal λ is typically around the average number of flipped bits for failure decoding at the specific SNR, such as the line with a value of 140 in Fig. 1.

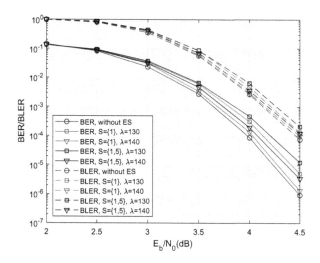

Fig. 4. Sensitivity of BER and BLER performance for the ES-M-NGDBF relative to the parameter λ.

As with the value of threshold λ, the set $\mathcal{S}$ is found through an empirical search. Figures 6 and 7 represents the result of an example with parameter $\lambda = 140$. When the number of elements in the set $\mathcal{S}$ increases, the value of ANI at low SNRs become smaller, and the BER performance loss at high SNRs become bigger. It is shown in the Fig. 7 that the decreases in ANI at low SNRs is relatively large when the number of elements in the set $\mathcal{S}$ increases from 1 to 2, and from 2 to 3. When it is further increased, the ANI at low SNRs hardly decreases. Obviously, there is a tradeoff between the number of elements in the set $\mathcal{S}$ and the ANI at low SNRs. Since the BER and BLER performance degradation should not be too large, it is best to set the number of elements in set $\mathcal{S}$ to 3 in the ES-M-NGDBF. The values of the three elements in the set $\mathcal{S}$ are obtained through simulation.

Table 1 shows the performance analysis of ES-M-NGDBF with $\mathcal{S} = \{1, 10, 20\}$ and $\lambda = 140$. The maximum number of error blocks for the simulation is 5000 at 2 dB, 3000 at 2.5 dB, 1000 at 3 dB, 500 at 3.5 dB, 200 at

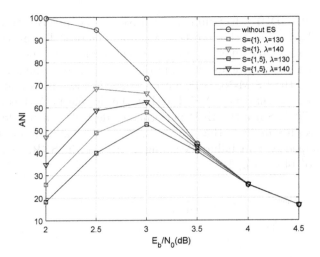

Fig. 5. Sensitivity of ANI for the ES-M-NGDBF relative to the parameter λ.

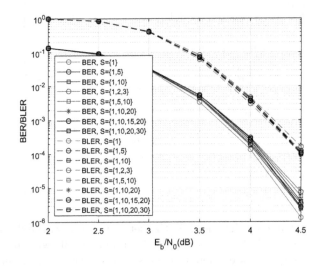

Fig. 6. Sensitivity of BER and BLER performance for the ES-M-NGDBF relative to the parameter $\mathcal{S}$, with parameter $\lambda = 140$.

4 dB and 100 at 4.5 dB. ES_isRight_counter and ES_counter represent the number of correct and total early stopping respectively, and ES_isRight_P and ES_P are their probability. ES_miss_counter represent the number of failure decoding which is not stopped by the proposed criterion. The ES_isRight_P is close to 1 at 2 dB, which means that the proposed criterion is an efficient filter at low SNRs. The ES_P and ES_isRight_P decrease while the ES_miss_counter increases from low to high SNRs, which means that the probability of failure decoding becomes smaller and the detection of early stopping becomes more difficult with the rise

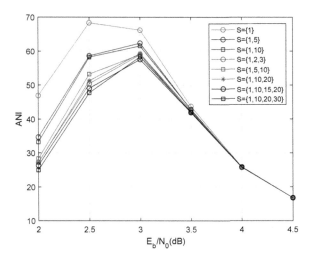

Fig. 7. Sensitivity of ANI for the ES-M-NGDBF relative to the parameter $\mathcal{S}$, with parameter $\lambda = 140$.

of SNR. Therefore, there is a certain BER and BLER performance loss at high SNRs. With high-complexity early stopping criterions, the performance loss at high SNRs may be avoided, but given our proposed early stopping criterion is extremely low complexity, we feel that the slight performance degradation at high SNRs is tolerable.

Table 1. Performance analysis of proposed early stopping

	Signal-to-Noise Ratio (SNR)					
	2 dB	2.5 dB	3 dB	3.5 dB	4 dB	4.5 dB
ES_isRight_counter	3836	1614	345	107	26	2
ES_counter	3851	1679	411	188	75	15
ES_isRight_P	0.9961	0.9613	0.8394	0.5691	0.3467	0.1333
ES_P	0.7585	0.4672	0.1655	0.0265	0.0014	0.0000
ES_miss_counter	1149	1321	589	312	125	85

5   Conclusion

In this paper, an early stopping criterion, which is based on the number of flipped bits for certain iterations, is proposed for the M-NGDBF decoding algorithm. With an extremely small amount of complexity increase, the proposed early stopping criterion can significantly reduce the number of decoding iterations at low SNRs, while the BER and BLER performance at high SNRs is slightly decreased.

Acknowledgment. This work is partially supported by the National Key Research and Development Project under Grants 2019YFE0123600 and 2020YFB1806800, the National Natural Science Foundation of China under Grants 62071115, and the European Union's Horizon 2020 research and innovation programme under the Marie Skłodowska-Curie grant agreement No 872172 (TESTBED2 project: www.testbed2.org).

References

1. Chen, J., Dholakia, A., Eleftheriou, E., Fossorier, M.P.C., Hu, X.: Reduced-complexity decoding of LDPC codes. IEEE Trans. Commun. **53**(8), 1288–1299 (2005). https://doi.org/10.1109/TCOMM.2005.852852
2. Chen, J., Fossorier, M.P.C.: Near optimum universal belief propagation based decoding of low-density parity check codes. IEEE Trans. Commun. **50**(3), 406–414 (2002). https://doi.org/10.1109/26.990903
3. Gallager, R.G.: Low-density parity-check codes. IRE Trans. Inf. Theory **8**(1), 21–28 (1962). https://doi.org/10.1109/TIT.1962.1057683
4. Ismail, M., Ahmed, I., Coon, J., Armour, S., Kocak, T., McGeehan, J.: Low latency low power bit flipping algorithms for LDPC decoding. In: 21st Annual IEEE International Symposium on Personal, Indoor and Mobile Radio Communications, pp. 278–282, September 2010. https://doi.org/10.1109/PIMRC.2010.5671820
5. Jiang, M., Dou, J.: An efficient stopping criterion of LDPC decoding for optical transmission. In: 9th International Conference on Optical Communications and Networks (ICOCN 2010), pp. 275–278, October 2010. https://doi.org/10.1049/cp.2010.1204
6. Li, J., You, X., Li, J.: Early stopping for LDPC decoding: convergence of mean magnitude (CMM). IEEE Commun. Lett. **10**(9), 667–669 (2006). https://doi.org/10.1109/LCOMM.2006.1714539
7. Mackay, D.J.C.: Encyclopedia of sparse graph codes. http://www.inference.org.uk/mackay/codes/data.html. Accessed 20 2014
8. Shin, D., Ha, J., Heo, K., Lee, H.: A stopping criterion for low-density parity-check codes. IEICE Trans. Commun. Lett. **E91-B**(4), 1145–1148 (2008)
9. Sundararajan, G., Winstead, C., Boutillon, E.: Noisy gradient descent bit-flip decoding for LDPC codes. IEEE Trans. Commun. **62**(10), 3385–3400 (2014). https://doi.org/10.1109/TCOMM.2014.2356458
10. Wadayama, T., Nakamura, K., Yagita, M., Funahashi, Y., Usami, S., Takumi, I.: Gradient descent bit flipping algorithms for decoding LDPC codes. IEEE Trans. Commun. **58**(6), 1610–1614 (2010). https://doi.org/10.1109/TCOMM.2010.06.090046
11. Wang, H., Fan, G., Kuang, J.: A hybrid low complexity decoding of LDPC codes. In: IET 3rd International Conference on Wireless, Mobile and Multimedia Networks (ICWMNN 2010), pp. 108–112, September 2010. https://doi.org/10.1049/cp.2010.0630

Pilot Allocation Scheme Based on Machine Learning Algorithm and Users' Angle of Arrival in Massive MIMO System

Min Yu$^{(\boxtimes)}$, Si Yuan Li, and Dong Feng Chen

School of Communication and Information Engineering,
Chongqing University of Posts and Telecommunications, Chong Qing, China
120588660@qq.com

Abstract. Massive MIMO system has attracted attention due to it's significant improvement in system capacity and spectrum utilization. Pilot pollution greatly limited the performance of Massive MIMO system. To optimize the pilot pollution in Massive MIMO system. In this paper, a pilot allocation scheme based on machine learning algorithm and users' angle of arrival is proposed. The scheme firstly classified all users according to whether the users' angle of arrival overlaps with each other. It randomly assigned pilot sequences to users whose angle of arrival do not overlap with each other. Secondly, it used machine learning algorithm to classify users whose angle of arrival overlap with each other into interfering group and non-interfering groups based on users' location information. We assign orthogonal pilots to users in the interfering group and randomly assign pilot sequences to users in the non-interfering group. Simulation results show that when the number of antenna reached 300, the pilot efficiency can be increased by about 11.67%. The pilot allocation scheme proposed in this paper can effectively suppress the impact of pilot pollution on the performance of Massive MIMO system, improve pilot efficiency and reduce pilot overhead.

Keywords: Pilot allocation · K-means clustering · Users' angle of arrival · Massive MIMO

1 Introduction

Massive MIMO system is one of the key technologies of Fifth Generation (5G) mobile communication systems. It is configured a large number of antennas on the base station side. The system capacity can be greatly increased. It brings high energy efficiency and high spectrum efficiency [1]. Massive MIMO technology has attracted much attention because of these advantages. However, the above-mentioned advantages of massive MIMO technology depend on being able to correctly analyze the channel state information. To accurately estimate the uplink channel, terminals should use mutually orthogonal pilot sequences. However, due to the surge in the number of antennas, the number of orthogonal pilots is limited by the channel coherence time. It leads to a limited number of orthogonal pilots. The pilot sequence is inevitably reused between neighboring cells. The interference caused by non-orthogonal pilot sequences is called

© ICST Institute for Computer Sciences, Social Informatics and Telecommunications Engineering 2021
Published by Springer Nature Switzerland AG 2021. All Rights Reserved
H. Gao et al. (Eds.): ChinaCom 2020, LNICST 352, pp. 753–765, 2021.
https://doi.org/10.1007/978-3-030-67720-6_52

pilot pollution. Pilot pollution greatly limits the system performance. Therefore, it is particularly important to study how to reduce the impact of pilot pollution. In order to solve the problem, many scholars have done a lot of research. N. Akbar et al. proposed to use line-of-sight interference between users to classify users, and assign non-orthogonal pilots to users with low line-of-sight interference. It can suppress pilot pollution to a certain extent [2]. Zhu, X. et al. proposed that according to certain standards, each cell is divided into a central area and an edge area, users in the edge area are assigned orthogonal pilot sequences, users in the central area reuse the pilot sequences [3]. Although pilot pollution can be reduced in this way, the gain of pilot multiplexing will become smaller. X. Nie et al. proposed a location-aware pilot assignment algorithm, which used the line-of-sight interference between users to classify users, assigned non-orthogonal pilot sequences to users with low line-of-sight interference, assigned orthogonal pilot sequences to users with high line-of-sight interference. It can suppress pilot contamination to a certain extent [4]. Kim, K. et al. proposed to distinguish users by defining users' similarity, thereby establishing the basis of distribution. They designed the angle of arrival positioning method to obtain users' location information, monitored and estimated the distance between users in different cells based on their location information to represent the potential interference of the transmission power between users. According to the similarity and distance between users, the pilot allocation strategy is implemented [5]. It can effectively reduce the problem of pilot pollution and improve pilot efficiency. X. Zhao et al. proposed to use machine learning algorithm in the pilot allocation strategy, used the exhaustive method to obtain the optimal pilot allocation scheme as the training sequence, and used the training sequence to obtain the pilot allocation model [6]. However, the complexity of this method is relatively high. Zhu Xu dong et al. proposed a low-complexity pilot allocation scheme based on matching algorithm. This scheme divided the users of the target cell into a strong user group and a weak user group and used a minimum-maximum matching method for pilot allocation. In the weak user group, the Hungarian algorithm is used for pilot allocation. The pilot allocation method is designed to ensure the fairness of strong user groups [7]. S. Ma et al. proposed a pilot allocation scheme that used the Signal to Interference Noise Ratio (SINR) of harmonic to quantify the fairness of all users in the network [8]. On the other hand, T. Van Chien et al. proposed a pilot allocation scheme, which did not model the pilot design as a combined allocation problem, but used basic pilot and treated the correlation power coefficient as a continuous optimization variable [9]. In order to reduce the pilot overhead in the massive MIMO system, X. Xiong et al. suggested that pilot is reused by using the characteristic that the energy of the channel is concentrated in a small number of specific areas of the angle domain and the delay domain. When there is pilot pollution in the system, the solution analyzed the asymptotic behavior and designed an inter-ference cancellation (IC) precoder to ensure the user's Qos to the greatest extent [10]. C. Hu et al. combined pilot allocation and semi-blind channel estimation and proposed a sector-based pilot allocation method, which includes cross-sector pilot allocation and intra-sector pilot optimization. When there are a large number of users in each cell, the method can reduce the complexity of searching for optimal pilot allocation [11].

In massive MIMO system, pilot pollution greatly limited the performance of the massive MIMO system. The methods to mitigate pilot pollution are mainly from the following three aspects: pilot allocation, precoding and channel estimation. Pilot allocation is one of the mainstream methods.

The main contributions in this paper are listed as follows:

a. Different from the previous way, the solution proposed in this article grouped users multiple times. According to whether the users' angles of arrival overlapped with each other, users were divided into n + 1 categories. The first n categories represented users whose angles of arrival did not overlap, and the n + 1-th category represented users whose angles of arrival partially overlapped or completely overlapped.
b. Because of the large number of users in massive MIMO systems, it used an unsupervised machine learning algorithm, which suitable for massive data processing—K-means clustering algorithm to group the n + 1th users into two groups according to the interference cost. The function measured the amount of interference between users. It divided users with high interference into interfering group and users with low interference into non-interfering group. Finally performed different pilot allocation schemes for users in different groups.
c. Grouping users multiple times to reduce the number of users processed in each step. Compared with the previous pilot allocation scheme, it reduced pilot overhead to a certain extent.

The remainder of this paper is organized as follows. Section 2 describes the system model. Section 3 introduces pilot allocation scheme. Section 4 analyzes experimental results. Finally, conclusions are given in Sect. 5.

2 System Model

It is configured a large number of antennas on the base station side. We consider the UL of a multi-cell Massive MIMO system with L cells. Each cell consists of a BS equipped with M antennas that serves K single-antenna users. $\mathbf{h}_{(j,k)l} \in C^{M \times 1}$ is the channel vector from the k-th user in the j-th cell to the BS in the l-th cell. The channel is modeled using shadow fading in a uniform linear massive MIMO antenna array as:

$$\mathbf{h}_{(j,k)l} = \sqrt{\frac{\beta_{(j,k)l}}{P}} \sum_{p=1}^{P} a(\theta^p_{(j,k)l}) \partial^p_{(j,k)l} \tag{1}$$

Where P is the number of multipath bars $a(\theta) = [1, e^{-2\pi jd/\lambda_c \sin \theta}, \ldots,$ $e^{-j2\pi d(M-1)/\lambda_c \sin \theta}]^T$ is the antenna array vector in the direction θ. $\partial^p_{(j,k)l} \sim CN(0,1)$ is the complex Gaussian gain on the P-th path of the signal. d is the spacing between antennas λ_c is signal wavelength. $\beta_{(j,k)l}$ is a large-scale fading factor, which includes path loss and shadow fading. It can be expressed as: $\beta_{(j,k)l} = z_{(j,k)l}/r^\partial_{(j,k)l}.r_{(j,k)l}$ is the distance between the k-th user in the j-th cell and the base station that in the l-th cell. ∂ is path loss fading coefficient. $z_{(j,k)l}$ is shadow fading factor.

2.1 Uplink Pilot Transmission Process

During the uplink pilot transmission process, the signal $\mathbf{y}_l \in \mathbb{C}^{M \times \tau}$ received by the base station in the l-th cell can be expressed as:

$$\mathbf{y}_l = \sqrt{\rho_r \tau} \sum_{l=1}^{L} \sum_{k=1}^{K} \varphi_k \mathbf{h}_{(i,k)l}^T + \mathbf{n}_l \tag{2}$$

Where ρ_r is the average power sent by the user. $\mathbf{h}_{(i,k)l}$ is the channel vector from the kth user in the i-th cell to the base station in the l-th cell. $l = 1, 2, \ldots, L$. $n_l \in \mathbb{C}^{M \times \tau} \sim CN(0, \delta^2)$ is additive white gaussian noise. $\varphi_k \in \mathbb{C}^{\tau \times 1}$ is pilot sequence, which satisfy $\varphi_k^H \varphi_{k'} = \sigma_{kk'}$ and $\sigma_{kk'} = \{^{1,\, k=k'}_{0,\, k \neq k'}$ When $\sqrt{\rho_r \tau} = 1$, the base station received signal in the l-th cell can be expressed as:

$$\mathbf{y}_l = \mathbf{P}_l \mathbf{h}_{l,l} + \mathbf{n}_l \tag{3}$$

Where $\mathbf{P}_l = [\varphi_1, \varphi_2, \ldots, \varphi_K]^T \in \mathbb{C}^{\tau \times K}$ is the pilot sequence sent by the users in the l-th cell $\mathbf{h}_{i,l} \in C^{K \times M}$ represents the channel state information matrix. In the massive MIMO system, the signal received by the base station in the l-th cell can be expressed as:

$$\mathbf{y}_l = \mathbf{P}_l \mathbf{h}_{l,l} + \sum_{i=1, i \neq l}^{L} \mathbf{P}_l \mathbf{h}_{i,l} + \mathbf{n}_l \tag{4}$$

Using the Least squares (LS) channel estimation algorithm to obtain the channel estimation matrix of the base station which in the l-th cell for the users of this cell:

$$\hat{\mathbf{h}}_{l,l} = \mathbf{P}_l^{-1} \mathbf{y}_l = \sum_{k=1}^{K} \mathbf{h}_{(l,k)l} + \sum_{k=1}^{K} \sum_{i=1, i \neq l}^{L} \mathbf{h}_{(i,k)l} + \mathbf{n}'_l \tag{5}$$

It can be seen from formula (5) that the base station in l-th cell obtained the CSI of the users in the l-th cell and the CSI of the users in other cell who sent the same pilot. Pilot pollution is caused by users in other cells which sent the same pilot.

3 Pilot Allocation Scheme Based on Machine Learning and User Arrival Angle

We proposed a massive MIMO pilot allocation scheme based on machine learning algorithm and users' angle of arrival. Firstly, it according to whether the users' angle of arrival overlaps to classify users. The users whose angles of arrival are non-overlapping were classified into the first n sets. The remaining users were classified into the $n + 1$-th set. For the $n + 1$-th set, the unsupervised machine learning algorithm– K-means clustering algorithm was used to cluster users according to their location information.

It further divided users into interfering group and non-interfering group. Finally, different pilot allocation schemes were performed for users in different groups. For the first n sets, the pilot sequences were randomly allocated. Pilot sequences were multiplexed in each user set. Orthogonal pilot sequences were allocated for users in the interfering group and non-orthogonal pilot sequences were randomly allocated for users in the non-interfering group.

3.1   Classify Users Based on Their Angles of Arrival

Firstly we classified users according to whether their angles of arrival overlapped with each other. The model of users' angle of arrival was shown in the Fig. 1:

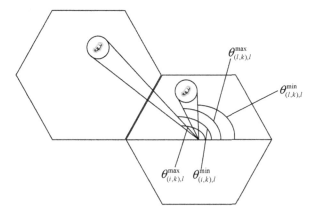

Fig. 1. User angle of arrival

$$\theta_{(i,k),l}^{\mu} = \arctan(\frac{[x_{(i,k)}]_2 - [x_l]_2}{[x_{(i,k)}]_1 - [x_l]_1}) \tag{6}$$

$$\theta_{(i,k),l}^{\delta} = \arcsin(\frac{r}{\|x_{(i,k)} - x_l\|}) \tag{7}$$

Where $\theta_{(i,k),l}^{\min} = \theta_{(i,k),l}^{\mu} - \theta_{(i,k),l}^{\delta}$, $\theta_{(i,k),l}^{\max} = \theta_{(i,k),l}^{\mu} + \theta_{(i,k),l}^{\delta}$. And r is the user's scattering radius. From theorem 1 of [6], we can see that when the angle of arrival of the interfering user $[\theta_{(i,k),l}^{\min}, \theta_{(i,k),l}^{\max}]$ does not overlap with the angle of arrival of the expected user $[\theta_{(l,k),l}^{\min}, \theta_{(l,k),l}^{\max}]$ at all. Even if the used pilot sequence is non-orthogonal, there will be no interference. Users are classified according to whether the angles of arrival overlap with each other. There are n + 1 categories, in which $C_1, C_2, \ldots, C_n$ is the users whose angles of arrival do not overlap each other and C_{n+1} is the users whose angles of arrival overlap partially or completely overlap.

3.2   User Classification Based on Machine Learning Algorithm

For users in C_{n+1}, the K-means clustering algorithm was further used to classify users based on users' location information. The users in C_{n+1} users were further divided into interference groups and non-interference groups.

- Set the cost function

We assumed that the angle of arrival of the target user is θ_a, the angle of arrival of the interfering user is θ_b, the signal wavelength is λ, the number of antennas at the base station is M and the antenna spacing is D. The target user's guidance vector of the *Pth* path is:

$$\partial(\theta_a^p) = \sum_{m=1}^{M} \exp(2\pi i(m-1)\frac{D}{\lambda}\cos(\theta_a^p)) \tag{8}$$

The interfering user's guidance vector of the *Pth* path is:

$$\partial(\theta_b^p) = \sum_{m=1}^{M} \exp(2\pi i(m-1)\frac{D}{\lambda}\cos(\theta_b^p)) \tag{9}$$

The distance between the target user and the interfering user is:

$$d_{ab} = \|x_a - x_b\| \tag{10}$$

We defined the interfering cost function as:

$$
\begin{aligned}
J_{ab} &= \frac{1}{d_{ab}M}\partial(\theta_a^p)\partial(\theta_b^p) \\
&= \frac{1}{d_{ab}M}\sum_{m=1}^{M}\exp(2\pi i(m-1)\frac{D}{\lambda}(\cos(\theta_a^p)-\cos(\theta_b^p)))
\end{aligned}
\tag{11}
$$

When the target user's angle of arrival and the interfering user's angle of arrival partially overlapped or completely overlapped. There was $\cos(\theta_a^p) = \cos(\theta_b^p)$. When the number of antenna port on the base station side was large enough: there was $\lim_{M\to\infty} J_{ij} = \frac{1}{d_{ij}}$. Therefore, when the interfering user's angle of arrival and the target user's angle of arrival partially overlapped or completely overlapped, the amount of interference from the interfering user to the target user was inversely proportional to its relative distance. We assumed that: $\cos(\theta_i^p) = \cos(\theta_j^p) = 1$, $d \in [10, 2000]$, $M = 64$. The simulation results were as shown in the Fig. 2:

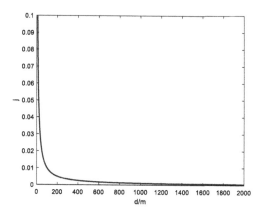

Fig. 2. Interference cost function

It can be seen from the figure that when $d \approx 1000$ and $J \approx 0.002$, the amount of interference between users is almost negligible. Based on the above results, for set C_{n+1} users, it further used the unsupervised machine learning algorithm–K-means clustering algorithm to classify users according to users' location information. The users were further divided into K clusters. These clusters with a distance between cluster centers greater than 1000 m were divided into non-interfering group, and these clusters with a distance between cluster centers less than 1000 m were divided into interfering groups.

- K-means clustering algorithm

The K-means clustering algorithm included: a user space position acquisition unit, a cluster center calculation unit and a convergence decision unit. Firstly, the central control unit obtained the user's location information. Secondly, the cluster center calculation unit calculated the cluster center based on the position information. Finally, the convergence decision unit judges whether to output the clustering result. The K-means clustering algorithm is shown in Table 1:

The steps of K-means clustering algorithm are as follows:

1. Initial input: K is the number of clusters. ε is a minimum value used to determine when the clustering ends. C_{n+1} is the user's location information who need to cluster.
2. First step, randomly selected the location information of K users as the initial clustering center. And the preferred value of K was set to 6, because the number of neighboring cells of each cell is 6. (x_1, y_1, z_1), (x_2, y_2, z_2), (x_3, y_3, z_3), (x_4, y_4, z_4), (x_5, y_5, z_5), (x_6, y_6, z_6), as the initial clustering center.
3. According to the distance between the set C_{n+1} users and the cluster center, $d = \sqrt{(x_i - x_j)^2 + (y_i - y_j)^2 + (z_i - z_j)^2}$, the users were clustered into the cluster with the closest cluster center. And update the cluster center: $C_j = \overline{D_j}$.
4. Third step calculated the objective function to determine whether the two clustering centers have converged. If they have converged, then end to the clustering;

Table 1. K-means clustering algorithm

K-means clustering algorithm
1:   **input:** K, ε, C_{n+1}
2:   **Parameter initialization:** Clustering center $C_1 = (X_1, Y_1, Z_1), ..., C_K = (X_K, Y_K, Z_K)$
3:   **Calculate the distance of each user relative to the cluster center:** i is the cell index, j is the user index $$d[i][j] = \sqrt{(X_i - X_j)^2 + (Y_i - Y_j)^2 + (Z_i - Z_j)^2}$$
4:   **Attributing the user with the smallest distance to a cluster:** $D_j = \min(d[i]) \cup (X_j, Y_j, Z_j)$
5:   **Update cluster center:** $C_j = \overline{D_j}$
6:   **Calculate decision parameters** E_i: $$E_i = \sum_{i=1}^{6} \sum_{x \in C_i} \lVert x - u_i \rVert^2 \quad u_i = \frac{1}{\lvert C_i \rvert} \sum_{x \in C_i} x$$
7:   **Stop iteration condition:** $\lvert E_{i+1} - E_i \rvert < \varepsilon$
output: C_j

otherwise, update the clustering center and return to step 2 to re-cluster according to the current clustering center. The objective function is:

$$E = \sum_{i=1}^{6} \sum_{x \in C_i} \lVert x - u_i \rVert^2, u_i = \frac{1}{\lvert C_i \rvert} \sum_{x \in C_i} x.$$ Where u_i is the mean vector of cluster C_i and

the distance from each sample point to the mean point. Determine whether the last two adjacent objective functions $\lvert E_2 - E_1 \rvert < \varepsilon$ hold. Where ε is a minimum value. And the difference between the criterion functions of two adjacent iterations is less than a minimum value, which indicated that the sum of squared errors within the cluster has converged and the clustering end.

- Pilot allocation

Pilot sequences were randomly allocated to the first n types users and each group of pilot sequences was multiplexed in each user set. The users in the interfering group were assigned strictly orthogonal pilot sequences and the users in the non-interfering group were randomly assigned pilot sequences.

4 Performance Analysis

It can be seen from formula (5) that the base station used zero-forcing detection algorithm to receive the signal can be expressed as:

$$
\begin{aligned}
x_{l,k} &= (h_{(l,k)l})^H \left(\sum_{i=1}^{L} \sum_{k=1}^{K} h_{(i,k)l} x_{i,k} + N_l \right) \\
&= (h_{(l,k)l})^H h_{(l,k)l} x_{l,k} + (h_{(l,k)l})^H \sum_{i=1,i\neq l}^{L} \sum_{k=1}^{K} h_{(i,k)l} x_{i,k} \\
&\quad + (h_{(l,k)l})^H N_l
\end{aligned}
\tag{12}
$$

Where $(h_{(l,k)l})^H h_{(l,k)l} x_{l,k}$ is the desired signal $(h_{(l,k)l})^H \sum_{i=1,i\neq l}^{L} \sum_{k=1}^{K} h_{(i,k)l} x_{i,k}$ is interfering signal. $(h_{(l,k)l})^H N_l$ is noise. The uplink SINR can be expressed as:

$$
SINR_{l,k} = \frac{|E\{(h_{(l,k)l})^H h_{(l,k)l}\}|^2}{\sum_{i=1,i\neq l}^{L} \sum_{k=1}^{K} |E\{(h_{(l,k)l})^H h_{(i,k)l}\}|^2 + \delta^2 |E\{(h_{(l,k)l})^H\}|^2}
\tag{13}
$$

Users' uplink reachable rate is:

$$
R_{<l,k>} = E\{\log_2(1 + SINR_{(l,k)})\}
\tag{14}
$$

The spectrum efficiency of the lth cell can be expressed as:

$$
SE_l = (1 - \mu) \sum_{k=1}^{K} R_{<l,k>}
\tag{15}
$$

Where $\mu = \frac{\tau}{T}$ represents the loss of frequency efficiency due to uplink pilot transmission. τ is the pilot length and T is the channel coherence time. The pilot efficiency is defined as: $\gamma = \frac{R}{O}$ $R = \sum_{l=1}^{L} \sum_{k=1}^{K} R_{<l,k>}$ is the uplink reach sum rate. O represents the number of orthogonal pilot sequences used by the algorithm in the multi-cell massive MIMO system.

5 Simulation Analysis

5.1 Simulation Parameter Setting

In order to better analyze the performance of the pilot allocation scheme proposed in this paper, we conducted simulation analysis based on the MATLAB simulation platform. The simulation used a massive MIMO system composed of L cells in TDD mode. Each cell center had a base station configured with M antennas and containing K randomly distributed users. The simulation parameters were set as follows (Table 2):

Table 2. Simulation parameters

Simulation parameters	Set value
Number of cells L	7
Number of antennas M	100 ~ 500
Number of users K	32
Cell radius $/m$	500
SNR $/dB$	20
Shadow fading coefficient	8
Pilot length τ	5
Path loss index	3

5.2 Analysis of Simulation Results

We assumed that the positions of users in a cell follow a random distribution. Random distribution of user positions can better simulate the actual situation and verify the universality of the algorithm. The location of users in a cell in the system is shown in Fig. 3:

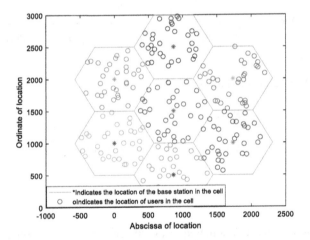

Fig. 3. User Distribution

Figure 4 shows the pilot efficiency of different pilot allocation algorithms with different numbers of antennas. It can be seen that the pilot allocation algorithm proposed in this paper further performed K-means clustering algorithm to classify C_{n+1} users from the pilot allocation scheme that based on users' arrival of angle. And different pilot allocation schemes were performed for different groups of users. Compared with the pilot allocation scheme based on users' angle of arrival, the scheme proposed in this paper has higher pilot efficiency and less pilot overhead. Compared with the pilot allocation scheme based on large-scale fading factors, the scheme proposed in this paper is more refined in user grouping. Not only does the user's location information be used for grouping, but also the user's arrival angle is used for grouping, which greatly improved the pilot efficiency.

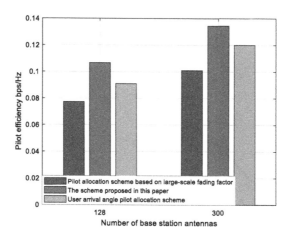

Fig. 4. Pilot efficiency of different pilot allocation algorithms with different numbers of antennas

Figure 5 shows that the users' minimum uplink reachable rate varies with the numbers of antennas on the base station side. The simulation experiments in this paper compared the simulation results of the pilot allocation scheme with the random pilot allocation scheme and the pilot allocation scheme based on user arrival angle. When the number of antennas is about 100 to 200, the users' minimum uplink reachable rate increases almost linearly with the number of antennas. When the number of antennas gradually increases, the growth rate of the users' minimum uplink reachable rate slows down.

The performance of the massive MIMO system will not be infinitely improved with the increase of the number of antennas at the base station side and the performance of the massive MIMO system will approach a fixed value. Under the condition of the same number of antennas, the minimum uplink achievable rate of the randomly allocated pilot scheme is the smallest, followed by the pilot allocation scheme proposed in this paper, and the minimum uplink achievable rate of the pilot allocation scheme based on user arrival angle is the largest. Because of the scheme proposed in this paper further clustered C_{n+1} users based on the users' grouping result of the pilot assignment algorithm based on the user arrival angle. And it further divided the C_{n+1} into interfering group and non-interfering group. The users in the interfering group multiplexed the pilot sequence. On the one hand, compared with the pilot allocation scheme based on user arrival angle, the scheme proposed in this paper further clustered users and further performed pilot multiplexing to reduce pilot overhead. On the other hand, compared with the random pilot allocation scheme, the pilot allocation scheme proposed in this paper greatly increased the minimum uplink reachable rate of users, reduced the impact of pilot pollution on system performance and improved system performance. The pilot allocation scheme proposed in this paper makes a compromise between improving system performance and reducing pilot overhead. Under the condition of acceptable system performance loss, the pilot efficiency is effectively improved and the pilot overhead is reduced.

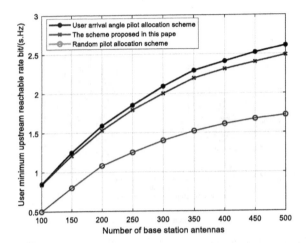

Fig. 5. Users' minimum uplink reachable rate varies with the numbers of antennas

6 Conclusion

This paper proposed a pilot allocation scheme based on machine learning algorithm and users' angle of arrival in a massive MIMO system. The scheme includes: user arrival angle grouping, K-means clustering grouping and pilot allocation. The scheme firstly classifies users according to whether their angles of arrival overlap with each other. It classifies non-overlapping users into the first n sets and classifies the remaining users into n + 1th sets. For users in the n + 1 set, the unsupervised machine learning algorithm is used to cluster users according to their location information. The users in the n + 1 set are further divided into interfering group and non-interfering group. Finally, different pilot allocation schemes are performed for users in different groups. For the first n sets, pilot sequences are randomly allocated and pilot sequences are multiplexed in each set. Orthogonal pilots are allocated to users in the interfering group. Non-orthogonal pilot sequences are randomly allocated to users in the non-interfering group. Simulation results show that, on the one hand, the pilot allocation scheme based on machine learning algorithm and users' angle of arrival proposed in this paper can effectively suppress the impact of pilot pollution on the performance of massive MIMO systems and effectively improve system performance compared with random pilot allocation schemes. On the other hand, compared with the pilot allocation scheme based on user arrival angle, the pilot allocation scheme based on machine learning algorithm and users' angle of arrival proposed in this paper can effectively improve the pilot efficiency and reduce the pilot overhead. It is suitable for massive MIMO systems with extremely tight pilot resources.

References

1. Rajmane, R.S., Sudha, V.: Spectral efficiency improvement in massive MIMO systems. In: 2019 TEQIP III Sponsored International Conference on Microwave Integrated Circuits, Photonics and Wireless Networks (IMICPW), Tiruchirappalli, India, pp. 357–360 (2019)
2. Akbar, N., Yan, S., Yang, N., Yuan, J.: Location-aware pilot allocation in multicell multiuser massive MIMO networks. IEEE Trans. Veh. Technol. 67, 7774–7778 (2018)
3. Zhu, X., Wang, Z., Qian, C., et al.: Soft pilot Reuse and multi-cell block diagonalization precoding for massive MIMO systems. IEEE Trans. Veh. Technol. 65(5), 3285–3298 (2015)
4. Nie, X., Zhao, F.: Joint pilot allocation and pilot sequence optimization in massive MIMO systems. IEEE Access 8, 60637–60644 (2020)
5. Kim, K., Lee, J., Choi, J.: Deep learning based pilot allocation scheme (DL-PAS) for 5G massive MIMO systems. IEEE Commun. Lett. 22(4), 828–831 (2018)
6. Zhao, X., Yao, Y., Xu, L., et al.: A Low-complexity and high-efficient pilot allocation scheme based on user grouping in massive MIMO system. In: 2019 IEEE 19th International Conference on Communication Technology (ICCT), Xi'an, China, pp. 658–663 (2019)
7. Xudong, Z., Lingdong, D., Zhaocheng, W., et al.: Weighted-graph-coloring-based pilot decontamination formulticell massive MIMO systems. IEEE Trans. Veh. Technol. 66(3), 2829–2834 (2017)
8. Ma, S., Xu, E.L., Salimi, A., Cui, S.: A novel pilot assignment scheme in massive MIMO networks. IEEE Wirel. Commun. Lett. 7(2), 262–265 (2018)
9. Van Chien, T., Björnson, E., Larsson, E.G.: Joint pilot design and uplink power allocation in multi-cell massive MIMO systems. IEEE Trans. Wirel. Commun. 17(3), 2000–2015 (2018)
10. Xiong, X., Jiang, B., Gao, X., You, X.: QoS-guaranteed user scheduling and pilot assignment for large-scale MIMO-OFDM systems. IEEE Trans. Veh. Technol. 65(8), 6275–6289 (2016)
11. Zhu, X., Dai, L., Wang, Z.: Graph coloring based pilot allocation to mitigate pilot contamination for multi-cell massive MIMO systems. IEEE Commun. Lett. 19(10), 1842–1845 (2015)
12. Raharya, N., Hardjawana, W., Al-Khatib, O., et al.: Pursuit learning-based joint pilot allocation and multi-base station association in a distributed massive MIMO network. IEEE Access 8, 58898–58911 (2020)
13. He, X., Song, R., Zhu, W.: Pilot allocation for distributed-compressed-sensing-based sparse channel estimation in MIMO-OFDM systems. IEEE Trans. Veh. Technol. 65(5), 2990–3004 (2016)
14. Ruoyu, Z., Honglin, Z., Jiayan, Z., et al.: Hybrid orthogonal and non-orthogonal pilot distribution based channel estimation in massive MIMO system. J. Syst. Eng. Electron. 29(5), 881–898 (2018)
15. Larsson, E.G., Edforsn, O., Tufvesson, F., et al.: Massive MIMO for next generation wireless systems. IEEE Commun. Mag. 52(2), 186–195 (2014)

A Method of Node Importance Measurement Base on Community Structure in Heterogeneous Combat Networks

Zhaofeng Yang[✉], Yonggang Li, and Jinyu Liu

School of Communication and Electronic, Chongqing University of Posts and Telecommunications, Chongqing, People's Republic of China
yangfengzmc2012@gmail.com

Abstract. The measurement of the importance for the nodes is of great significance to the test and simulation for Heterogeneous Combat Networks (HCN), combat situation assessment and other topics. Due to the complexity of equipment types and styles in such system, traditional algorithms (degrees, betweenness, closeness, eigenvectors) are difficult to achieve both speed and accuracy in identifying the important nodes of Heterogeneous Combat Networks. This paper fully considers the heterogeneity of combat system nodes, and proposes an evaluation model based on community structure, IEBC (importance evaluation based on community), which can measure the importance of each node. We form functional modules (FM) by distinguishing the function of nodes. Then divide the network into communities according to the concentration of FM. Finally, we compare IEBC with traditional ranking models (e.g., degree centrality). After simulation calculation, compared with other algorithms, IEBC takes into account the balance of efficiency and accuracy at the same time.

Keywords: Identification of key nodes · SDI military operation chain · Community detection

1 Introduction

Heterogeneous Combat Networks (HCN) [1] addresses all key aspects of joint operations and military action, which integrates the combat platforms, weapon systems, intelligence reconnaissance, command control, and logistics support systems into an integrative combat system. Because the diversity of missions in network combat and the complexity of combat environment have placed increasing demands on the reliability of HCN. The demand for the reliability of HCN is getting higher and higher, because the diversity of missions in network combat and the complexity of combat environment. Since the nodes and links of such network are vulnerable to targeted attacks by the enemy, it is particularly important to analyze the importance of nodes in HCN [2].

This work is supported by the Defence Advance Research Foundation of China under Grant 61400020109.

H. Gao et al. (Eds.): ChinaCom 2020, LNICST 352, pp. 766–775, 2021.
https://doi.org/10.1007/978-3-030-67720-6_53

At present, the research on the importance of complex network nodes in the military field is still in the preliminary stage. Opsahl *et al.* analyzed the key node identification method based on node centrality, which is measured according to the topological position of the node [3]. But the method is strongly dependent on the network topology. Martin *et al.* proposed that the node betweenness can reflect the dynamic characteristic of the network [4], but it is restricted by the complexity of the algorithm complexity. Freeman [5] improved the important node identification method based on betweenness, and the computational efficiency was reduced, but the versatility was poor. A method that take the approximate flow betweenness of nodes as an indicator for centrality measure is proposed by Liu *et al.* [6].

Most of above-mentioned improvements are designed for calculating betweenness centrality, but HCN is a typical open complex giant system which is composed of many community structures with different sizes. This system will continue to evolve as the battlefield environment changes. The current key node identification method can't be fully applicable to the combat network [7]. Therefore, we investigate the function of different nodes and the community of the nodes. Moreover, the role of nodes in different combat tasks may be different, so the importance of nodes will be constantly changing.

We divide the nodes of HCN into Sensor entities (S), Decider entities (D), Influential entities (I) according to the requirements of combat missions. The completion of specific combat mission benefits from the combination of S, D, and I nodes. Because these nodes form a complete chain of reconnaissance, decision-making and attack. Such S, D, and I nodes will form an observe, orient, decide, and act operational cycle (OODA loop) [8, 9], which forms a functional module (FM) that completes the corresponding tasks. FM are considered building blocks for combat networks. We use a framework that identifies clusters of FM and develop a method, Importance evaluation based on the community (IEBC), that can evaluate node's importance base on community structure in HCN.

The paper is organized as follows: in second part, we introduce the composition of HCN, community structure and functional module. A community identification algorithm and the IEBC framework based on the result of community dividing is proposed in third part, while we performed simulation and result analysis in fourth section. In "Conclusion" section, we summarize our final remarks.

2   Network Structure and Motif

The traditional combat system network is an abstraction of the command and control relationship. It constructs a tree diagram based on such relations in military operations, where leaf nodes are combat entities with different capabilities, and other nodes are centers of command and control at different layers [10] (see Fig. 1).

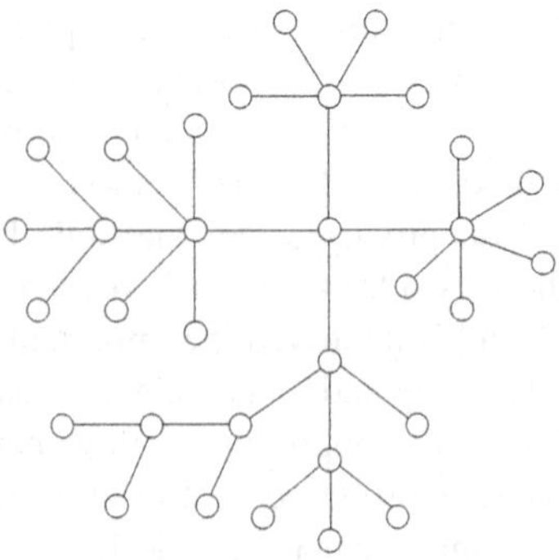

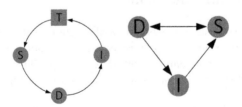

Fig. 1. The traditional combat system network.

With the advent of the information age, the form of combat has changed from traditional platform-centric warfare to network-centric warfare. Information systems have greatly improved combat capabilities. All combat equipment makes full use of information systems to achieve the integration of interconnection and interoperability. All entities share battlefield intelligence, jointly perceive the battlefield situation, and adjust the operation plan in time according to changes in the battlefield situation. Thus, we suggest that HCN is a distributed network.

According to the heterogeneity of nodes in combat network, we divide the nodes into three part [10]. 1) Sensor entities (S). 2) Decider entities (D). 3) Influential entities (I).

Once sensor node (S) finds target, it will transmit the target information to the decider node (such as operational center). The operational center (D) comprehensively analyzes the received information and then issues instructions to the influencer (I) to execute military operations on the target. Finally, the offensive and defensive engagement nodes (I) attack the target. Such a loop is shown in Fig. 2. Therefore, A functional chain of military operations: $S \rightarrow D \rightarrow I$ is formed on the target node, where T is the node in the enemy network. In addition, we consider that the decision maker D will adjust the sensor S according to the changing combat environment. Moreover, the sensor S will also re-detect the combat situation, after the node I conducts military operations on the target T. Thus, four edges should be considered in the function module (Fig. 2 right): $S \rightarrow D, D \rightarrow S, D \rightarrow I, I \rightarrow S$.

Fig. 2. (left) An operation loop of SDI. (right) The Function Module used in this paper.

The combat area can be continuously expanded based on the FM. The combat space is divided into different communities, then various communities coordinated operations (see in Fig. 3). More precisely, node 24, 22 and 23 compose a FM, and the yellow community consists of four FM.

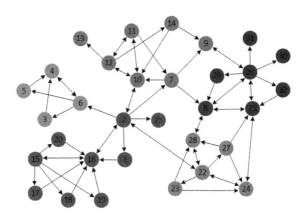

Fig. 3. Combat network detected by Function Modules. (Color figure online)

3 Importance Measurement

3.1 Community Detection

To evaluate the importance of nodes based on the community structure, we must first accurately identify the community structure of the network. The current community detection algorithms [11–13] were mostly for homogeneous networks, and didn't take into account the heterogeneous characteristics of the nodes in Heterogeneous Combat Networks. HCN is depended on FM to complete a military operation. Thus, the community detection of the combat network should take FM as a combined unit. After the detection, the number of functional modules in each community is roughly the same, and the functional modules are destroyed as little as possible. Here we benefit from the high-order clustering of complex networks to define a module conduction rate (MCR). Our algorithm has two goals for finding communities in the network.

- There should be as many nodes as possible in the FM of each community.
- The number of broken FM should be as few as possible.

The above two requirements can be reflected through MCR. When the conduction rate is the smallest, the result of community division is the best [14]. MCR define as follow:

$$RatioCut_M(S_1, S_2, \ldots, S_k) = \frac{1}{2} \sum_{i=1}^{k} \frac{W_M(S_i, \overline{S_i})}{vol_M(S_i)} \tag{1}$$

where $\overline{S}_i$ is the remaining part after removing S_i (see Fig. 4), $W_M(S_i, \overline{S}_i)$ is the number of broken FM, which at least one node in a FM is in S_i and another node is in $\overline{S}_i$. Also, $vol_M(S_i)$ is number of nodes belonging to function module M in S_i.

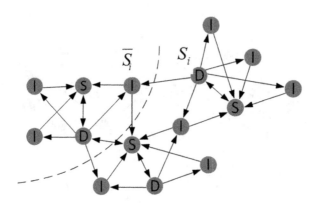

Fig. 4. Community detection process based on functional modules.

As the number of nodes increases, the types of community divisions will increase exponentially. It is an NP problem to accurately calculate the minimum module conduction rate. Therefore, we propose an approximate optimization algorithm to obtain an approximate optimal solution. A spectral graph clustering algorithm based on eigenvector is put forward. This method has high calculation efficiency, is easy to implement, and has approximately optimal mathematics guarantees for the obtained clustering. The approach consists of modular high-order clustering algorithm (algorithm Ψ) and improved K-means clustering sub-algorithm (algorithm Φ). The approach detect community as follow:

- Step 1: Given a network and functional module M, form the module adjacency matrix W_M. Element in matrix: $(W_M)_{ij}$ is the number of modules that contain both node i and j.
- Step 2: Calculate the diagonal matrix of the module adjacency matrix:

$$(D_M)_{ii} = \sum_j (W_M)_{ij} \qquad (2)$$

- Step 3: Calculate the Laplacian transformation matrix of the module adjacency matrix:

$$\zeta_M = D_M^{1/2}(D_M - W_M)D_M^{1/2} \qquad (3)$$

- Step 4: Calculate the k eigenvectors corresponding to the smallest non-zero eigenvalue of the ζ_M: $Z_1, Z_2, \ldots, Z_k$. Then, arrange k eigenvectors to form an $N \times K$ matrix Z. Take

$$Y_{ij} = \frac{Z_{ij}}{\sqrt{\sum_{j=1}^{k} Z_{ij}^2}} \qquad (4)$$

as N k-dimensional row vectors.

- Step 5: The ith row vector of matrix Y represents the node i. Then use the sub-algorithm Φ to group each node.

The sub-algorithm Φ is an improvement of the k-mean algorithm, including the following steps:

- Step 1: Select the initial community center node according to the two indicators of high centrality and discrete distribution. Centrality of node is defined as:

$$\rho_i = (D_M)_{ii} \qquad (5)$$

where $(D_M)_{ii}$ is degree of FM, showing in eq. (3). In order to measure the dispersion of nodes in the network, we calculate the minimum value of the distance between a node and other nodes with higher centrality by using the minimum distance $\delta_i (i = 1, 2, \ldots, n)$:

$$\delta_i = \min_{j: \rho_j > \rho_i} (d_{ij}) \qquad (6)$$

where d_{ij} is the Euclidean distance between the representative vector (V_k) of node i and node j in matrix Y. If the node has the largest centrality, it is more likely to be selected as the initial central node than other adjacent nodes. Therefore, this article specifies the minimum distance of the node as:

$$\delta_k = \max(\delta_i), i \neq k \qquad (7)$$

Considering the two characteristics comprehensively, we select the top k nodes with the largest $\rho_i \delta_i$ as the initial central nodes.

- Step 2: Divide the node into the community to which the nearest central node belongs, by calculating the Euclidean distance between each node and each center node. Then calculate the MCR at this time.
- Step 3: Calculate the average distance between each node and other nodes in each community, and update the node with the smallest average distance as the center node of the community.
- Step 4: Repeat steps 2 and 3, if the center node no longer changes or the number of iterations reaches the upper limit, stop the algorithm. The division result with the smallest MCR is the approximate optimal solution.

3.2   Importance Evaluation Based on Community

The IEBC sorting algorithm implementation process is shown below, the algorithm calculates the IEBC value of each node through a matrix.

- Step 1: Given the network model adjacency matrix and the detect communities by approach proposed in chapter 3.1.
- Step 2: Compute the number of communities connected to nodes i (V_{ic}).
- Step 3: Generate list of IEBCs, and IEBC of each node is count by

$$IEBC_i = \sum_{w=1}^{c} \frac{S_w}{N} V_{ic} \tag{8}$$

where c is the aggregate number of communities in HCN, S_w is the number of nodes in community w.

- Step 4: Sort the IEBC value of each node.

4   Simulation Results and Analysis

There are six kind of links in HCN, as shown in Fig. 5. Combined with the actual military operation, different connection probabilities are given to different links. The connection probability is defined as the ratio of the actual number of links to the number of possible links between two nodes. For example, the connection probability of $D \rightarrow S$ is:

$$P_{D \rightarrow S} = \frac{N_{D \rightarrow S}}{N_D N_S} \tag{9}$$

where $N_{D \rightarrow S}$ is the actual number of links for $D \rightarrow S$, N_D, N_S is the number of nodes S and D.

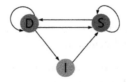

Fig. 5. A simple FM of heterogeneous combat networks.

In this paper, associate with corresponding military applications, 300 S nodes, 150 D nodes, and 550 I nodes are randomly generated. And the given connection probability is: $P_{S \rightarrow S} = 0.025$, $P_{S \rightarrow D} = 0.04$, $P_{D \rightarrow S} = 0.03$, $P_{D \rightarrow D} = 0.185$, $P_{D \rightarrow I} = 0.1$, $P_{I \rightarrow S} = 0.02$.

The ER topology model of HCN constructed according to the above connection probability is shown in Fig. 6.

Communities are divided according to the FM shown in Fig. 2. Then apply several importance ranking methods (Degree centrality, Betweenness centrality, Closeness centrality, Eigenvector centrality [15]) and IEBC to rank the node importance. According to the importance of ranking results, attack the nodes of HCN. As a node is damaged, some nodes connect with it cannot communicate each other, causing the entire HCN to split into many independent connected giant components. We use the number of nodes in maximum connected component (H) and the number of surviving functional modules (N_M) to evaluate the functional robustness of the damaged HCN [10]. The simulation results are as follow figures.

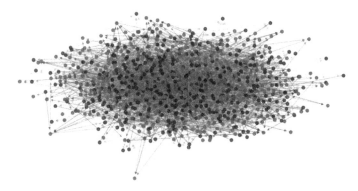

Fig. 6. The ER topology model of HCN. Red represents node D, blue represents node S, and gray represents node I. (Color figure online)

As we can seen in Fig. 7, the trend of H is sharply reduced, and then gradually flattened, which fully illustrates the effectiveness of the algorithm for nodes sorting. In the process of increasing f and attacking nodes with the IEBC measurement result, the decreasing speed of H is faster than other traditional algorithms in the figure. After the important nodes identified by IEBC are damaged, H is already less than 100, when the fraction of f increases to 0.3.

Therefore, it means that the important nodes detected by IEBC are the intermediary nodes with the most links. These intermediate nodes connect communities to each other and are distributed discretely. The existence of these nodes maximizes the spread of information in the community. When such nodes are attacked, the communities will separate, and the size of maximum connected component changes the fastest.

Figure 8 reports that as f increases, attacks on important nodes identified by IEBC N_M decrease the fastest. When the attack intensity is greater than 0.77, it decreases to 0. Therefore, it shows that the nodes detected by IEBC are more important than the nodes identified by degree centrality, betweenness centrality, closeness centrality, and eigenvector centrality. The nodes detected by IEBC are more likely a D node, because N_M decline the fastest as the D node under attack. When all D nodes are damaged, the number of FM reduces to 0.

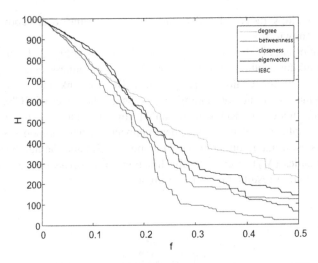

Fig. 7. The number of nodes in surviving maximum connected component (*H*) varies with the attack intensity (*f*).

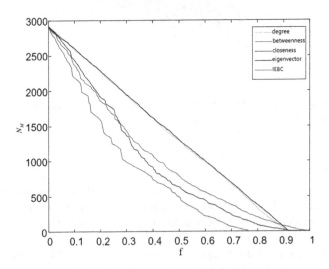

Fig. 8. The number of surviving FM (N_M) varies with attack intensity (*f*).

5 Conclusion

According to the size of the largest connected component and the number of remaining effective functional modules with the attack intensity, the simulation analysis results show that the effectiveness of IEBC is better than the other methods when measuring the importance of nodes in HCN. IEBC can accurately and quickly determine the key nodes in HCN. In military operations against the enemy, precision strikes can be carried out on key nodes, so that the enemy's system operational effectiveness will rapidly reduced, and the enemy's military communications network will paralyze. Further more, it can also focus on protecting important nodes in our HCN to ensure that the system can continue to maintain its original functions when attacked by the enemy.

References

1. Li, J., Jiang, J., Yang, K., et al.: Research on functional robustness of heterogeneous combat networks. IEEE Syst. J. **13**(2), 1487–1495 (2018)
2. Lu, X.Q., Yang, X.D.: A invulnerability assessment based on importance-degree of node for weighted network under attack with partial information. In: 11th International Conference on Wireless Communications, Networking and Mobile Computing, pp. 1–5. IET, Shanghai (2015)
3. Opsahl, T., Agneessens, F., Skvoretz, J.: Node centrality in weighted networks: Generalizing degree and shortest paths. Soc. Netw. **32**(3), 245–251 (2010)
4. Martin, T., Zhang, X., Newman, M.E.: Localization and centrality in networks. Phys. Rev. E **90**(5–1), 052808 (2014)
5. Freeman, L.C.: Centrality in social networks conceptual clarification. Soc. Netw. **1**(3), 215–239 (1978)
6. Liu, R.J., Guo, S.Z., et al: An approximate flow betweenness based centrality measure for complex network. IEICE Trans. Inf. Syst. **E96-D**, 727–730 (2013)
7. He, H, Li, Z.F., Wang, W.P., et al.: Research on critical node analysis method of new combat SoS. In: IEEE International Systems Engineering Symposium 2018, pp. 1–7. IEEE, Rome (2018)
8. Daniel, F.: A Vision So Noble: John Boyd, the OODA Loop, and America's War on Terror. CreateSpace Independent Publishing Platform (2010)
9. Li, E.Y., Gong, J.X., et al.: Node importance analysis of complex networks for combat systems based on function chain. J. Command Control **4**(01), 42–49 (2018)
10. Yu, J., Wang, W., Zhang, G., Guo, N.: Research on joint operations command system based on complex network. Fire Control Command Control **36**(2), 5–10 (2011)
11. Boccaletti, S., Latora, V., Moreno, Y., et al.: Complex networks: Structure and dynamics. Phys. Rep. **424**(4–5), 175–308 (2006)
12. Fortunato, S.: Community detection in graphs. Phys. Rep. **486**(3–5), 75–174 (2010)
13. Duch, J., Arenas, A.: Community detection in complex networks using extremal optimization. Phys. Rev. E **72**(2), 027104 (2005)
14. Benson, A.R., Gleich, D.F., Leskovec, J.: Higher-order organization of complex networks. Science **353**(6295), 163–166 (2016)
15. Lü, L.Y., Chen, D.B., Ren, X.L., et al.: Vital nodes identification in complex networks. Phys. Rep. **650**, 1–63 (2016)

Research on Timing Synchronization Algorithm of Cell Search in 5G NR System

Hang Jiang[1(✉)], Longhan Cao[2], Zhizhong Zhang[1],
and Bingguang Deng[1]

[1] Communication Networks Testing Technology Engineering Research Center,
Chongqing Universting of Posts and Telecommunications, Chongqing, China
s180131053@stu.cqupt.edu.cn
[2] Key Laboratory of Control Engineering, Chongqing Communication
University, Chongqing, China

Abstract. For the 5G NR systems in the Third Generation Partnership Project (3GPP), the computation of synchronous signal detection algorithms is very complex. Based on the characteristics of the domain synchronous signal, this paper proposed a differential and superimposed interdependent cross correlation detection algorithm, which only performed a differential interdependent processing of the received signal with the sum of the local PSS sequence to obtain the position of the relevant peak and detects the position of the coarse synchronous point, and further performed local correlation of the coarse synchronous point, and detected the precise synchronous point according to the maximum peak, which reduced the computational complexity. A theoretical derivation and a comparative complexity analysis between the traditional cross correlation and the improved algorithms shows that the differential and superposition cross correlation joint detection algorithm has high efficiency, low complexity and strong resistance to frequency deviation, which meets the synchronization requirements of 5G NR systems.

Keywords: 5G NR system · Timing synchronization · Superposition · Differential

1 Introduction

5G New Radio (NR) is the fifth generation mobile communication technology standard developed by the Third Generation Partnership Project (3GPP) Organization. Since 2019, many countries and regions around the world have launched their own 5G services and 5G has become the mainstream mobile communication standard. The first step of cell search is to realize the detection and reception of Primary Synchronization Signal (PSS) [1]. The accuracy of detection directly affects the establishment of communication links, and 5G NR systems have more stringent requirements for frequency offset. Therefore, the study of timing synchronization algorithms with high precision and low complexity is of utmost importance.

The typical downlink timing synchronization algorithms in Long Term Evolution (LTE) systems includes autocorrelation algorithm based on PSS signal [2],

© ICST Institute for Computer Sciences, Social Informatics and Telecommunications Engineering 2021
Published by Springer Nature Switzerland AG 2021. All Rights Reserved
H. Gao et al. (Eds.): ChinaCom 2020, LNICST 352, pp. 776–785, 2021.
https://doi.org/10.1007/978-3-030-67720-6_54

interconnection algorithm based on PSS signal [3] and maximum release algorithm based on cyclic prefix (CP) [4–6]. 5G NR systems have more differences than LTE systems, although the algorithms in LTE systems can be used directly. The poor performance is due to the characteristics of 5G NR systems. In this paper, the existing timing synchronization algorithm was improved to meet the requirements of resisting frequency deviation and complexity of 5G NR systems in accordance with the characteristics of 5G NR systems.

In order to solve the problems in 5G NR system, this paper proposed an improved algorithm based on the traditional cross correlation algorithm, which first superimposed local signals into a set of local signals, then differential cross correlation was operated with the received signals to detect the peak, after obtaining the coarse synchronization point and sub-carrier configuration parameter μ, finally the coarse synchronization point before and after down sampling was formed into a sequence that is cross correlated with the local signals to obtain the precision synchronization point and cell group ID. This algorithm can improve the performance of timing synchronization detection against frequency deviation, and effectively reduce the complexity of implementation.

2 PSS

The number of physical cell IDs in 5G NR systems increased from 504 to 1008 and was divided into 336 groups of 3 each is different from LTE.

$$N_{ID}^{cell} = 3N_{ID}^{(1)} + N_{ID}^{(2)} \tag{1}$$

In the formula: $N_{ID}^{(1)} \in \{0, 1, 2, ..., 335\}$ is called the cell group ID and the formula $N_{ID}^{(2)} \in \{0, 1, 2\}$ is called the intracell group ID.

In the 5G NR system, the domain synchronous signal does not to use the Zadoff-Chu sequence in the LTE system, although the Zadoff-Chu sequence has good autocorrelation characteristics, its ability to resist frequency deviation is poor, the length of 127 Binary Phase Shift Keying (BPSK) modulation of the M sequence are adopted, which has good autocorrelation characteristics and good cross correlation characteristics.

The master synchronous signal is generated as follows.

$$PSS(n) = 1 - 2x(m)$$
$$m = \left(n + 43N_{ID}^{(2)}\right) \bmod 127, 0 \leq n < 127 \tag{2}$$
$$x(i + 7) = (x(i + 4) + x(i)) \bmod 2$$

In the formula: x(6) = 1, x(5) = 1, x(4) = 1, x(3) = 0, x(2) = 1, x(1) = 1, x(0) = 0.

3 PSS Synchronization Analysis and Improvement

3.1 Traditional Intercorrelation Algorithm

PSS has had a good cross correlation propety that can achieve timing synchronization by receiving signals directly correlated with local PSS signals [7, 8]. After detecting the relevant peaks, the metric function is.

$$C_\mu(n) = \left| \sum_{k=1}^{K} r(k+n) \cdot pss^*_{(q,\mu)}(k) \right| \quad q = 0, 1, 2 \tag{3}$$

In the formula: $PSS_{(q,\mu)}(k)$ is the group q PSS signal when the locally generated subcarrier control parameter is μ; $r(k)$ is the received signal.

The PSS coarse synchronization position and intra-cell group ID $N_{ID}^{(2)}$ are obtained by taking the maximum value of the set of three related values:

$$\left\{ \hat{\theta}, N_{ID}^{(2)} \right\} = \arg\max_{\{n,\mu\}} \left\{ |S_\mu(n)|, n = 0, 1, 2... \right\} \tag{4}$$

3.2 Piecewise Correlation Synchronization Algorithm

In practical applications, there are inevitable problems such as carrier deviation. The traditional cross correlation algorithm can not effectively solve the frequency deviation, the literature [9] proposed a synchronization algorithm related to the PSS signal segmentation, which had good frequency deviation resistance ability in the case of large frequency deviation, the metric function is.

$$C_\mu(n) = \sum_{m=0}^{M-1} \left| \sum_{k=0}^{L-1} pss^*_{(q,\mu)}(k+mL) \cdot r(k+n+mL) \right|^2 \tag{5}$$

In the formula: $q = \{0, 1, 2\}$, $\mu = \{0, 1, 2, 3, 4\}$; $r(n)$ is the N-point reception data obtained after narrowband filtering and downsampling, M is the number of segments and L is the length of each segment.

3.3 Improvement and Analysis of Timing Synchronization Algorithm

In order to reduce the complexity of the local interconnection algorithm and improve the frequency deviation resistance performance of the algorithm, this paper improved the traditional cross correlation algorithm and proposed a differential and superimposed interconnection joint synchronization algorithm, which effectively reduced the algorithm calculation volume and improved the cross correlation resistance performance. The overall design flow of the timed synchronous improvement algorithm is shown in Fig. 1.

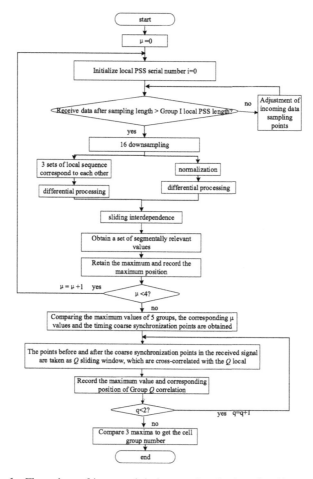

Fig. 1. Flow chart of improved timing synchronization algorithm design

First, the received signal is downsampled from the local PSS signal at 16 times the sampling rate to reduce the computation. As shown in Fig. 2, the coarse synchronization points detected are exactly the same as those obtained by the traditional cross correlation algorithm, and only one cross correlation operation is needed. The three local PSS signals after downsampling were expressed as $PSS_{(0,\mu)}(k)$, $PSS_{(1,\mu)}(k)$ and $PSS_{(2,\mu)}(k)$, respectively. Three sets of local PSS signals are processed by corresponding supersition. The following are the formula:

$$PSS_\mu(k) = PSS_{(0,\mu)}(k) + PSS_{(1,\mu)}(k) + PSS_{(2,\mu)}(k) \tag{6}$$

Assuming that there is just frequency deviation in channel and other conditions are perfect, the formula of received signal is.

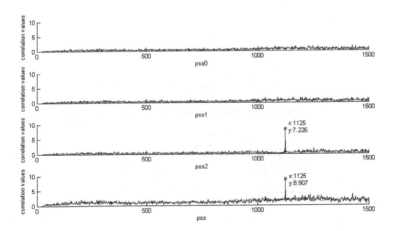

Fig. 2. Correlation between received signal and local pss signal

$$r(n) = s(n)e^{j2\pi\varepsilon n/N} \tag{7}$$

In the formula, $s(n)$ is the transmit signal, $\varepsilon = \Delta f / f_{sc}$ is the normalized frequency deviation, Δf is the value of the frequency deviation between transmit signal and received signal, and f_{sc} is the sub-carrier interval.

The differential cross correlation of the received signal and the superimposed local PSS signal is operated with the following expression.

$$
\begin{aligned}
C_\mu(n) &= \left| \sum_{k=0}^{K-1} r(k+n)r^*(k+n-1)(PSS_\mu(k)PSS_\mu^*(k-1))^* \right| \\
&= \left| \sum_{k=0}^{K-1} e^{j2\pi\varepsilon/N} s(k+n)s^*(k+n-1)(PSS_\mu(k)PSS_\mu^*(k-1))^* \right| \\
&= \left| \sum_{k=0}^{K-1} s(k+n)s^*(k+n-1)(PSS_\mu(k)PSS_\mu^*(k-1))^* \right|
\end{aligned} \tag{8}
$$

Compared with the traditional cross-correlation algorithm, the differential cross-correlation algorithm can eliminate $e^{j2\pi\varepsilon/N}$ items and reduce the influence of frequency offset as well as improve the detection performance of timing synchronization.

In the 5G NR system, $\mu = \{0, 1, 2, 3, 4\}$, therefore, five cross correlation is necessary to obtain a set of five maximum peaks and determine the corresponding μ value and coarse synchronization point of the maximum. Here is the formula.

$$\{d, \mu\} = \arg\max\{|C_\mu(n)|\} \tag{9}$$

Although the improved coarse synchronization algorithm can obtain the coarse synchronization position of the PSS in the OFDM symbol, it can not accurately determine the precision synchronization position and ID $N_{ID}^{(2)}$ within the cell group. In order to solve this problem, this paper took the coarse synchronization point before and after Q points in the received signal before sampling as a sliding window, and the three groups of local signals before sampling to do cross correlation. The formula is.

$$R_q^{'}(h) = \sum_{t=1}^{T} \tilde{r}(t+h)\tilde{S}_q^*(h) \quad q = 0,1,2 \tag{10}$$

In the formula, $\tilde{r}(h)$ is the Q-point sliding sequence before and after the coarse synchronization point; $\tilde{S}_q(h)$ is the local sequence before the three sets of downsampling.

By comparing the three maximum values in the set of related values, the sliding window maximum value position $\hat{\theta}$ can be obtained, and then combined with the coarse synchronization point, the precision synchronization position and the cell group ID $N_{ID}^{(2)}$ can be obtained, the formula is:

$$\{\hat{\theta}, N_{ID}^{(2)}\} = \arg\max\{R_q(h), h = 0, 1, 2, \ldots, 63\} \tag{11}$$

4 Simulation Results

In order to verify the performance of the algorithm, the algorithm proposed in this paper was compared with the traditional cross correlation algorithm, and the simulation is carried out using MATALB software under different conditions, the simulation parameters are shown in Table 1.

Table 1. Simulation parameter table

Parameter	Values
μ	1
Channel bandwidth/MHz	100
Frequency of sampling/MHz	122.88
FFT points	4096
ε	0.2/1.2
Channel model	AWGN channel

Figure 3 and 4 show the correlation peak plots of the conventional cross correlation algorithm at $\varepsilon = 0.2$ and $\varepsilon = 1.2$ respectively, and Fig. 5 shows the correlation peak plots of the improved algorithm of this paper at $\varepsilon = 1.2$. It can be seen in Fig. 3 and 4 that the traditional algorithm has better detection performance at small frequency range, with obvious peak, but at large frequency range, the peak is not obvious and performance is poor, but a comparison between Fig. 4 and 5 show that the improved algorithm still has obvious peak at large frequency range.

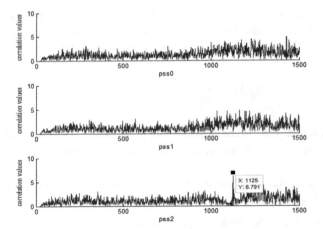

Fig. 3. Peak value of traditional algorithm when $\varepsilon = 0.2$

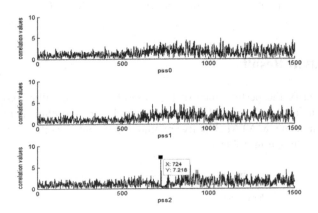

Fig. 4. Peak value of traditional algorithm when $\varepsilon = 1.2$

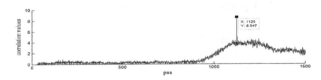

Fig. 5. Peak value of the improved algorithm when $\varepsilon = 1.2$

The calculation above obtained the coarse synchronization point position is 1125 and the corresponding position before the downsampling is 17985. The length of Q point before and after the interception can be shown in Fig. 6, but when $Q = 64$, the detection accuracy rate can reach 1. Figure 7 shows the cross correlation between Q point length of the received signal and three groups of local signals to detect the maximum value of the three groups of related sets and the position is 36. The cell group ID $N_{ID}^{(2)} = 2$ and precision synchronization point position is $17985 - (64/2 - 36) = 17989$.

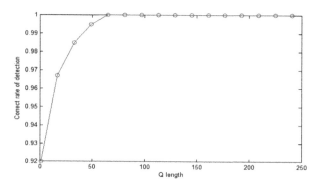

Fig. 6. The accuracy of precision synchronization point detection of main synchronization signals with different Q values

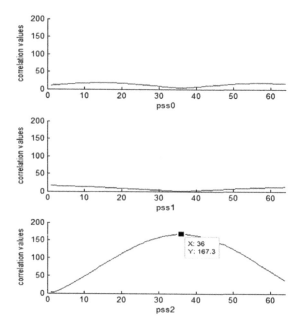

Fig. 7. Precision synchronization point timing position

Figure 8 represents the probability of detection between the improved and inter-related algorithms when $\varepsilon = 0$ and $\varepsilon = 1.2$. It can be seen from the figure that the performance of the improved algorithm in this paper is better than that of the cross correlation algorithm regardless of $\varepsilon = 0$ or $\varepsilon = 1.2$. When the signal-to-noise ratio is constant, the detection probability decreases with the increase of frequency deviation, and when the frequency deviation is constant, the detection probability increases with the increase of signal-to-noise ratio.

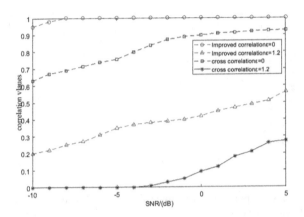

Fig. 8. Detection probability of different ε values for different algorithms

In 5G systems, there are five different cases of μ values, and in this paper, μ values, intra-cell group IDs, and timing synchronization points are determined by comparing with the peak sizes of 15 sets of related sequences. In the synchronization process, the computational complexity of the different algorithms is analyzed when $\mu = 1$, subcarrier interval 30 kHz, sampling frequency 122.88 MHz. Data of 38400 point lengths in the post-downsampling half frame were selected for comparative analysis. The complex multiplication and complex addition calculations of the traditional cross correlation algorithm are 29753344 and 29634048 respectively. In this paper, the computational quantities of complex multiplication and complex addition of the improved algorithm are 10092544 and 10050048 respectively. The detailed computational complexity data are shown in Table 2. As can be seen from the table, the number of operations of the improved algorithm complex multiplication and complex addition is reduced by 66.47% and 66.09% compared to the traditional cross correlation algorithm. From the overall amount of computation, it can be concluded that the algorithm in this paper has a 66.28% reduction in computation compared to the traditional correlation algorithm.

Table 2. Calculation complexity analysis and comparison

Method	Multiplication	Addition	Sum
Traditional correlation algorithm	29753344	29634048	59387392
Improved algorithm	10092544	10050048	20142592

5   Conclusion

In this paper, we improved the PSS timed synchronization algorithm through the analysis of the characteristics of PSS sequences and the shortcomings of the traditional synchronization algorithm. We propose the differential and superposition correlation joint detection algorithm, and design a detailed timed synchronization detection algorithm based on this algorithm. Among them, the computational complexity of the algorithm is reduced about 66% compared with the traditional cross correlation algorithm and fast correlation detection can be achieved in case of large frequency deviation conditions. The simulation results show that in the AWGN channel, the probability of detection of the improved algorithm is significantly improved compared to the traditional cross correlation algorithm with a large frequency shift, and it needs less computation, besides, it is much easier to implement. The simulation is performed only under the AWGN channel. Readers can evaluate it under different channel environments and simulation parameters.

Acknowledgment. It is supported by the Science and Technology Major Project in Chongqing (R&D and application of 5G road test instruments: No. cstc2019jscx-zdztzxX0002).

References

1. Pushpalata, T., Chaudhari, S.Y.: Need of physical layer security in LTE: analysis of vulnerabilities in LTE physical layer. In: 2017 International Conference on Wireless Communications, Signal Processing and Networking (WiSPNET). IEEE (2017)
2. Qian, X., Deng, H., He, H.: Joint synchronization and channel estimation of ACO-OFDM systems with simplified transceiver. IEEE Photonics Technol. Lett. 30(4), 383–386 (2018)
3. Wang, J., Zhang, Q., Zhang, Z.: Antioxidant activity of sulfated polysaccharide fractions extracted from Laminaria japonica. Int. J. Biol. Macromol. 42(2), 0–132 (2008)
4. Das, A., Mohanty, B., Sahu, B.: A NOVEL CAZAC sequence based timing synchronization scheme for OFDM system. In: 2018 International Conference on Wireless Communications, Signal Processing and Networking (WiSPNET), Chennai, pp. 1–4 (2018)
5. Jian, D., Wu, H., Gao, W., Jiang, R.: A novel timing synchronization method based on CAZAC sequence for OFDM systems. In: 2018 IEEE International Conference on Signal Processing, Communications and Computing (ICSPCC), Qingdao, pp. 1–5 (2018)
6. Jeon, Y., Park, H., Choi, E.: Synchronization and cell search procedure in 3GPP 5G NR systems. In: 2019 21st International Conference on Advanced Communication Technology (ICACT). Pyeong Chang Kwangwoon_Do:, pp. 475–478. IEEE (2019)
7. Aymen, O., Mohammed, S., Abdelmohsen, A., et al.: Synchronization procedure in 5G NR systems. IEEE Access 7, 41286–41295 (2019)
8. Zhao, Y., Cao, J., Li, Y.: An improved timing synchronization method for eliminating large doppler shift in LEO satellite system. In: 2018 IEEE 18th International Conference on Communication Technology (ICCT), Chongqing, pp. 762–766 (2018)
9. Zhang, Z., Liu, J., Long, K.: Low-complexity cell search with fast PSS identification in LTE. IEEE Trans. Veh. Technol. 61(4), 1719–1729 (2012)

Correction to: Land Cover Classification and Accuracy Evaluation Based on Object-Oriented Spatial Features of GF-2

Xiaomao Chen, Jiakun Li, and Yuanfa Ji

Correction to:
Chapter "Land Cover Classification and Accuracy Evaluation Based on Object-Oriented Spatial Features of GF-2" in:
H. Gao et al. (Eds.): *Communications and Networking*, LNICST 352,
https://doi.org/10.1007/978-3-030-67720-6_21

The original version of this chapter was revised: The name of Yuanfa Li has been corrected to Yuanfa Ji.

The updated version of this chapter can be found at
https://doi.org/10.1007/978-3-030-67720-6_21

© ICST Institute for Computer Sciences, Social Informatics and Telecommunications Engineering 2021
Published by Springer Nature Switzerland AG 2021. All Rights Reserved
H. Gao et al. (Eds.): ChinaCom 2020, LNICST 352, p. C1, 2021.
https://doi.org/10.1007/978-3-030-67720-6_55

Author Index

Printed in the United States
by Baker & Taylor Publisher Services